Principles of Corrosion
and Corrosion Control

Published in the Butterworth-Heinemann/IChemE Series

For further information about the Series: books.elsevier.com/icheme

IChemE is the hub for chemical, biochemical and process engineering professionals world-wide. With offices in the UK, Australia and Malaysia, the Institution works at the heart of the process community, promoting competence and a commitment to sustainable development, advancing the discipline for the benefit of society and supporting the professional development of members. For further information visit www.icheme.org.

PRINCIPLES OF CORROSION ENGINEERING AND CORROSION CONTROL

ZAKI AHMAD

IChem^E

AMSTERDAM · BOSTON · HEIDELBERG · LONDON · NEW YORK · OXFORD
PARIS · SAN DIEGO · SAN FRANCISCO · SINGAPORE · SYDNEY · TOKYO
Butterworth-Heinemann is an imprint of Elsevier

Butterworth-Heinemann is an imprint of Elsevier
Linacre House, Jordan Hill, Oxford OX2 8DP, UK
30 Corporate Drive, Suite 400, Burlington, MA 01803, USA

First edition 2006
Reprinted 2007

Notice
No responsibility is assumed by the publisher for any injury and/or damage to persons
or property as a matter of products liability, negligence or otherwise, or from any use
or operation of any methods, products, instructions or ideas contained in the material
herein. Because of rapid advances in the medical sciences, in particular, independent
verification of diagnoses and drug dosages should be made

British Library Cataloguing in Publication Data
A catalogue record for this book is available from the British Library

Library of Congress Cataloging-in-Publication Data
A catalog record for this book is available from the Library of Congress

ISBN: 978-0-7506-5924-6

For information on all Butterworth-Heinemann publications
visit our website at books.elsevier.com

Transferred to Digital Printing in 2010.

Working together to grow
libraries in developing countries
www.elsevier.com | www.bookaid.org | www.sabre.org

ELSEVIER   BOOK AID
International   Sabre Foundation

I dedicate this book to the memory of my most beloved son Intekhab Ahmad who breathed his last on April 20, 2004.

"We are from Almighty God and unto him we return"

Intekhab was the greatest driving force behind this book and may Almighty God rest his soul in peace and grant him the highest level in paradise.

Contents

5 CATHODIC PROTECTION 271

6 CORROSION CONTROL BY INHIBITION 352

7   COATINGS   382

Supporting material accompanying this book

A full set of accompanying multiple choice questions and worked solutions for this book are available for teaching purposes.

Please visit http://www.textbooks.elsevier.com and follow the registration instructions to access this material, which is intended for use by lecturers and tutors.

PREFACE

The phenomenon of corrosion is as old as the history of metals and it has been looked on as a menace which destroys metals and structures and turns beauty into a beast. Our human civilization cannot exist without metals and yet corrosion is their Achilles heel. Although familiarity with corrosion is ancient, it has been taken very passively by scientists and engineers in the past. Surprisingly, it is only during the last six decades that corrosion science has gradually evolved to a well-defined discipline. Corrosion Science and Engineering is now an integral part of engineering curriculum in leading universities throughout the world. With the rapid advances in materials in the new millennium, the demand for corrosion engineering courses has dramatically increased. This has necessitated the need for the publication of new books. Professor U. R. Evans, Prof. H. H. Uhlig and Prof. M. Fontana wrote a classical generation of basic text books covering the fundamentals of corrosion science and engineering. These books served as texts for decades and some of them are still being used. Several new books in corrosion have been published in recent years to cater to the needs of corrosion science and engineering students. As a teacher of corrosion engineering for the last twenty-five years, I found the material to be deficient in corrosion engineering content. However, sufficient coverage was given to the understanding of corrosion science. In this book, chapters on cathodic protection, materials selection, concrete corrosion and coatings have been written to cater to the needs of corrosion engineering students as well as corrosion engineers. These chapters contain simple and sufficient information to enable students to design corrosion preventive measures. A large number of illustrative problems are given in the chapter on cathodic protection to show how simple cathodic protective systems may be designed. The chapter on material selection is devoted to an understanding of the art of selection of materials for corrosive environment and applying the knowledge of corrosion prevention – the objective of corrosion engineering students. Concrete corrosion is a global problem and of particular interest to civil, chemical and mechanical engineering students. The chapter on boiler corrosion would be of specific interest to corrosion engineering students and corrosion engineers who desire to refresh their knowledge of the fundamentals of boiler corrosion and water treatment. The chapter on concrete corrosion has been added in view of the global interest in concrete corrosion. It presents the mechanism of rebar corrosion, preventive measures and evaluation methods in a simplified form with eye-catching illustrations. And the unique feature of the book is the follow-up of each chapter by keywords, definitions, multiple-choice questions, conceptual questions and review questions. A solution manual will soon be available to students containing solutions of problems and answers to multiple-choice questions. These are intended to test the readers comprehension of the principles covered in the text. I have put all my lifetime teaching experience into writing this book for corrosion engineering students in the sophomore or junior year. Graduate students lacking background in corrosion will also benefit from the book. It is expected that the students would be able to understand the principles of corrosion science and engineering in a simple and logical manner and apply them for solutions to corrosion engineering problems. This book is written with a new approach and new philosophy and it is hoped that it will fulfill their aspirations. While writing this book, I passed through the most turbulent period of my life with the loss of my most beloved son Intekhab Ahmad who passed away suddenly on April 20, 2004 leaving a sea of unending tears and sadness in my life. It was followed by my own sickness, operation and desertions of some of my closest ones. I am grateful to Almighty Allah that I passed through this traumatic period and am able to complete the book. The success of my efforts will depend on how well this book is received by the students and the corrosion community. This book will not only be found very useful by corrosion engineering students but also by corrosion scientists and engineers in their problems in their professional capacity and those interested in corrosion.

ACKNOWLEDGMENTS

The author appreciates the valuable support provided by King Fahd University of Petroleum & Minerals (KFUPM), Dhahran, Kingdom of Saudi Arabia in writing this book. I thank Dr Faleh Al-Sulaiman, Chairman ME Department for his encouragement. The author wishes to thank Prof. G. Hocking, Professor, Imperial College of Science and Technology for his dedicated support in technical editing of the script. The author thanks Mr Irfan Hussaini (now a graduate student at Pennsylvania State University, USA) for making all the figures of the textbook which was a gigantic task and very well accomplished by him. I was indeed fortunate to have Mr Syed Ishrat Jameel, Secretary, ME Department, KFUPM to type the various drafts of the book. I sincerely thank him. I thank Mr Abdul Aleem B. J., Lecturer, ME Department, KFUPM for proof reading of chapters, figures and coordinating with the publisher. I wish to thank Mr Mohammed Abbas, my graduate student, for his dedicated support in correction of scripts, figures, checking equations, problems and solutions. I thank my teacher and mentor Dr J. C. Scully, Ex-Reader, Leeds University, who inspired me to adopt corrosion as my career. I have been inspired by the works of Prof. Mars G. Fontana, R. N. Perkins, R. W. Staehle, P. A. Schweitzer, Prof. Ronald. M. Latanision, Prof. A. Ashby and do not claim any originality of ideas. The author wishes to thank the numerous publishers, professional societies and corrosion consultants for granting permission for copyrighted photographs and figures. Special thanks to American Society of Metals (ASM), American Society of Testing and Materials (ASTM), National Association of Corrosion Engineers (NACE), British Corrosion Journal, British Standard Institute, John Wiley and Sons Inc., McGraw Hill, Chapman & Hall and others. I appreciate the cooperation of Corrintec, USA and Texaco, Houston Research Center and Sigma Paints Company for permission to use figures. I appreciate the moral support given by my sons Manzar Ahmad, Assistant Professor, SSUET, Karachi and Intessar Ahmad, University of Adelaide, Australia. Finally, I thank my creator, God the Almighty, who gave me the strength and patience to complete this book.

INTRODUCTION TO CORROSION

1.1 HISTORICAL BACKGROUND

'Lay not up for yourselves treasures upon earth where moth and rust doth corrupt and where thieves break-through and steal.'

(Mathew 6:14)

The word corrosion is as old as the earth, but it has been known by different names. Corrosion is known commonly as rust, an undesirable phenomena which destroys the luster and beauty of objects and shortens their life. A Roman philosopher, Pliny (AD 23–79) wrote about the destruction of iron in his essay 'Ferrum Corrumpitar.' Corrosion since ancient times has affected not only the quality of daily lives of people, but also their technical progress. There is a historical record of observation of corrosion by several writers, philosophers and scientists, but there was little curiosity regarding the causes and mechanism of corrosion until Robert Boyle wrote his 'Mechanical Origin of Corrosiveness.'

Philosophers, writers and scientists observed corrosion and mentioned it in their writings:

- Pliny the elder (AD 23–79) wrote about spoiled iron.
- Herodotus (fifth century BC) suggested the use of tin for protection of iron.
- Lomonosov (1743–1756).
- Austin (1788) noticed that neutral water becomes alkaline when it acts on iron.

- Thenard (1819) suggested that corrosion is an electrochemical phenomenon.
- Hall (1829) established that iron does not rust in the absence of oxygen.
- Davy (1824) proposed a method for sacrificial protection of iron by zinc.
- De la Rive (1830) suggested the existence of microcells on the surface of zinc.

The most important contributions were later made by Faraday (1791–1867) [1] who established a quantitative relationship between chemical action and electric current. Faraday's first and second laws are the basis for calculation of corrosion rates of metals. Ideas on corrosion control started to be generated at the beginning of nineteenth century. Whitney (1903) provided a scientific basis for corrosion control based on electrochemical observation. As early as in eighteenth century it was observed that iron corrodes rapidly in dilute nitric acid but remains unattacked in concentrated nitric acid. Schönbein in 1836 showed that iron could be made passive [2]. It was left to U. R. Evans to provide a modern understanding of the causes and control of corrosion based on his classical electrochemical theory in 1923. Considerable progress towards the modern understanding of corrosion was made by the contributions of Evans [3], Uhlig [4] and Fontana [5]. The above pioneers of modern corrosion have been identified with their well known books in the references given at the end of the chapter. Corrosion laboratories established in M.I.T., USA and University of Cambridge, UK, contributed significantly to the growth and development of corrosion science and technology as a multi disciplinary subject. In recent years, corrosion science and engineering has become an integral part of engineering education globally.

1.2 DEFINITIONS

Corrosion is a natural and costly process of destruction like earthquakes, tornados, floods and volcanic eruptions, with one major difference. Whereas we can be only a silent spectator to the above processes of destruction, corrosion can be prevented or at least controlled. Several definitions of corrosion have been given and some of them are reproduced below:

(A) Corrosion is the surface wastage that occurs when metals are exposed to reactive environments.
(B) Corrosion is the result of interaction between a metal and environments which results in its gradual destruction.
(C) Corrosion is an aspect of the decay of materials by chemical or biological agents.
(D) Corrosion is an extractive metallurgy in reverse. For instance, iron is made from hematite by heating with carbon. Iron corrodes and reverts to rust, thus completing its life cycle. The hematite and rust have the same composition (Fig. 1.1).
(E) Corrosion is the deterioration of materials as a result of reaction with its environment (Fontana).
(F) Corrosion is the destructive attack of a metal by chemical or electrochemical reaction with the environment (Uhlig).

Despite different definitions, it can be observed that corrosion is basically the result of interaction between materials and their environment. Up to the 1960s, the term corrosion was restricted only to metals and their alloys and it did not incorporate ceramics, polymers, composites and semiconductors in its regime. The term corrosion now encompasses all types of natural and man-made materials including biomaterials and nanomaterials, and it is not confined to metals and alloys alone. The scope of corrosion is consistent with the revolutionary changes in materials development witnessed in recent years.

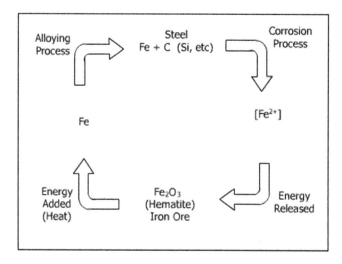

Figure 1.1 Refining-corrosion cycle

1.3 CORROSIVE ENVIRONMENT

Corrosion cannot be defined without a reference to environment. All environments are corrosive to some degree. Following is the list of typical corrosive environments:

(1) Air and humidity.
(2) Fresh, distilled, salt and marine water.
(3) Natural, urban, marine and industrial atmospheres.
(4) Steam and gases, like chlorine.
(5) Ammonia.
(6) Hydrogen sulfide.
(7) Sulfur dioxide and oxides of nitrogen.
(8) Fuel gases.
(9) Acids.
(10) Alkalies.
(11) Soils.

It may, therefore, be observed that corrosion is a potent force which destroys economy, depletes resources and causes costly and untimely failures of plants, equipment and components.

1.4 CONSEQUENCES OF CORROSION

Some important consequences of corrosion are summarized below:

■ Plant shutdowns. Shutdown of nuclear plants, process plants, power plants and refineries may cause severe problems to industry and consumers.
■ Loss of products, leaking containers, storage tanks, water and oil transportation lines and fuel tanks cause significant loss of product and may generate severe accidents and hazards. It is well-known that at least 25% of water is lost by leakage.
■ Loss of efficiency. Insulation of heat exchanger tubings and pipelines by corrosion products reduces heat transfer and piping capacity.
■ Contamination. Corrosion products may contaminate chemicals, pharmaceuticals,

dyes, packaged goods, etc. with dire consequences to the consumers.
■ Nuclear hazards. The Chernobyl disaster is a continuing example of transport of radioactive corrosion products in water, fatal to human, animal and biological life.

The magnitude of corrosion would depend upon the sensitivity of a particular metal or alloy to a specific environment. For instance, copper corrodes rapidly in the presence of ammonia and it is a serious problem in agricultural areas. Many historical statues made from bronze have been destroyed by ammonia released from fertilizers. Environmental conditioning offers one method of controlling corrosion, such as the use of inhibitors and oil transmission pipelines.

1.5 COST OF CORROSION

In a study of corrosion cost conducted jointly by C. C. Technologies Inc., USA [6], Federal Highway Agencies (FHWA), USA [7] and National Association of Corrosion Engineers [8], the direct corrosion cost was estimated to be around 276 billion US dollars, approximately 3.1% of the national gross domestic product. Based on an extensive survey conducted by Battelle Columbus Laboratories, Columbus, Ohio, USA and National Institute of Standards and Technology (NIST), in 1975, the cost was estimated to be 82 billion US dollars, which would have exceeded 350 billion US dollars in view of price inflation over the last twenty-five years. Because of the long time involved in conducting cost structure, it is not possible to update the information every year. However, both studies show that corrosion costs are staggering and a figure of about 350 billion US dollars appears to be a reasonable estimate for another two to three years. At least 35% of the above amount could have been saved by taking appropriate corrosion control measures. In UK, the corrosion cost is estimated to be 4–5% of the GNP [4]. In Japan, the cost of corrosion is estimated to be 5258 trillion Yen per year. For most industrialized nations, the average corrosion cost is 3.5–4.5%

of the GNP. Below are some startling figures of corrosion losses:

- The corrosion cost of gas and liquid transmission pipelines in USA exceeds seven billion US dollars. The figure for the major oil producing countries in the Gulf region are not known, however the cost expected to be very high because of highly corrosive environment in the region [8].
- The corrosion-free life of automobiles in the coastal regions of Arabian Gulf is about six months only [9].
- Nearly 95% of concrete damage in the Arabian Gulf coastal region is caused by reinforcement corrosion and consequent spalling of concrete [10].
- It is estimated that 10% of all aircraft maintenance in USA is spent on corrosion remediation [11].
- Major annual corrosion losses to the tune of £350 million in transport, £280 million in marine, £250 million in buildings and construction and £180 million in oil and chemical industries, have been reported in UK [12]. These are uncorrected 1971 figures.
- About $120 billion is spent on maintenance of aging and deteriorating infrastructures in USA [13].
- Automotive corrosion costs 23.4 billion US dollars annually in USA [8].
- Every new born baby in the world now has an annual corrosion debt of $40.

1.6 BREAKDOWN OF SPENDING ON CORROSION

The petroleum, chemical, petrochemical, construction, manufacturing, pulp and paper and transportation (railroad, automotive and aerospace) industries are the largest contributors to corrosion expenditure. The cost of corrosion differs from country to country. For instance in USA, the transportation sector is the largest sector contributing to corrosion after public utilities, whereas in the oil producing countries, such as the Arabian Gulf countries, petroleum and petrochemical industries are the largest contributor to corrosion expenditure. The highway sector in USA alone includes 4 000 000 miles of highways, 583 000 bridges, which need corrosion remediation maintenance [8]. The annual direct corrosion cost estimated to be 8.3 billion US dollars. The direct corrosion of transportation sector is estimated to be 29.7 billion US dollars. It includes the corrosion cost of aircraft, hazardous materials transport, motor vehicles, railroad car and ships [8]. In the oil sector, drilling poses severe hazards to equipment in the form of stress corrosion cracking, hydrogen induced cracking and hydrogen sulfide cracking [6]. In USA alone this sector costs more than 1.2 billion US dollars. The cost is very staggering in major oil producing countries, like Saudi Arabia, Iran, Iraq and Kuwait. The direct cost of corrosion to aircraft industry exceeds 2.2 billion US dollars [8].

Corrosion has a serious impact on defense equipment. In the Gulf War, a serious problem of rotor blade damage in helicopter was caused by the desert sand. The thickness of the blade was reduced to 2–3 mm in some instances. The desert erosion–corrosion offered a new challenge to corrosion scientists and engineers. The storage of defense equipment is a serious matter for countries with corrosive environments, such as Saudi Arabia, Malaysia and Southeast Asia. Humidity is the biggest killer of defense hardware. Storage of defense equipment demands minimum humidity, scanty rainfall, alkaline soil, no dust storms, no marine environment and minimal dust particles.

From the above summary, it is observed that corrosion exists everywhere and there is no industry or house where it does not penetrate and it demands a state of readiness for engineers and scientists to combat this problem.

1.7 CORROSION SCIENCE AND CORROSION ENGINEERING

The term science covers theories, laws and explanation of phenomena confirmed by inter-subjective observation or experiments. For instance, the explanation of different forms of corrosion, rates of corrosion and mechanism of corrosion is provided by corrosion science. Corrosion science is a 'knowing why' of corrosion. The term engineering, contrary to science, is directed towards an action for a particular purpose under a set of directions and rules for action and in a well-known phrase it is 'knowing how.' Corrosion engineering is the application of the principles evolved from corrosion science to minimize or prevent corrosion. Corrosion engineering involves designing of corrosion prevention schemes and implementation of specific codes and practices. Corrosion prevention measures, like cathodic protection, designing to prevent corrosion and coating of structures fall within the regime of corrosion engineering. However, corrosion science and engineering go hand-in-hand and they cannot be separated: it is a permanent marriage to produce new and better methods of protection from time to time.

1.8 INTER-DISCIPLINARY NATURE OF CORROSION

The subject of corrosion is inter-disciplinary and it involves all basic sciences, such as physics, chemistry, biology and all disciplines of engineering, such as civil, mechanical, electrical and metallurgical engineering.

1.9 CORROSION EDUCATION

The subject of corrosion has undergone an irreversible transformation from a state of isolated and obscurity to a recognized discipline of engineering. From the three universities in USA which offered courses in corrosion in 1946, corrosion courses are now offered by almost all major technical universities and institutions in USA, UK, Europe, Southeast Asia, Africa and Japan. Corrosion is now considered as an essential component of design. Learned societies like National Association of Corrosion Engineers,

European Federation of Corrosion, Japan Society of Corrosion Engineers and others are playing leading role in the development of corrosion engineering education. Detailed information on corrosion education, training centers, opportunities in corrosion can be found in various handbooks and websites. Some sources of information are listed in the bibliography. As a consequence of cumulative efforts of corrosion scientists and engineers, corrosion engineering has made quantum leaps and it is actively contributing to technological advancement ranging from building structures to aerospace vehicles.

1.10 FUNCTIONAL ASPECTS OF CORROSION

Corrosion may severely affect the following functions of metals, plant and equipment:

(1) **Impermeability**: Environmental constituents must not be allowed to enter pipes, process equipment, food containers, tanks, etc. to minimize the possibility of corrosion.
(2) **Mechanical strength**: Corrosion should not affect the capability to withstand specified loads, and its strength should not be undermined by corrosion.
(3) **Dimensional integrity**: Maintaining dimensions is critical to engineering designs and they should not be affected by corrosion.
(4) **Physical properties:** For efficient operation, the physical properties of plants, equipment and materials, such as thermal conductivity and electrical properties, should not be allowed to be adversely affected by corrosion.
(5) **Contamination:** Corrosion, if allowed to build up, can contaminate processing equipment, food products, drugs and pharmaceutical products and endanger health and environmental safety.
(6) **Damage to equipment:** Equipment adjacent to one which has suffered corrosion failure, may be damaged.

Realizing that corrosion effectively blocks or impairs the functions of metals, plants and equipment, appropriate measures must be adopted to minimize loss or efficiency of function.

1.10.1 HEALTH, SAFETY, ENVIRONMENTAL AND PRODUCT LIFE

These can involve the following:

(1) **Safety**: Sudden failure can cause explosions and fire, release of toxic products and collapse of structures. Several incidents of fire have been reported due to corrosion causing leakage of gas and oil pipelines. Corrosion adversely affects the structural integrity of components and makes them susceptible to failure and accident. More deaths are caused by accidents in old cars because of the weakening of components by corrosion damage. Corrosion has also been a significant factor in several accidents involving civil and military aircraft and other transportation vehicles. Corrosion failure involving bridges, ships, airports, stadiums are too numerous to be mentioned in detail in this chapter and recorded in the catalog of engineering disasters [11].

(2) **Health**: Adverse effects on health may be caused by corroding structures, such as a plumbing system affecting the quality of water and escaping of products into the environment from the corroded structures.

(3) **Depletion of resources**: Corrosion puts a heavy constraint on natural resources of a country because of their wastage by corrosion. The process of depletion outweighs the discovery of new resources which may lead to a future metal crisis similar to the past oil shortage.

(4) **Appearance and cleanliness**: Whereas anesthetics numb the senses, aesthetics arouse interest, stimulate and appeal to the senses, particularly the sense of beauty. A product designed to function properly must have an aesthetic appeal. Corrosion behaves like a beast to a beauty. It destroys the aesthetic appeal of the product and damages the product image which is a valuable asset to a corporation. Surface finishing processes, such as electroplating, anodizing, mechanical polishing, electro polishing, painting, coating, etching and texturing all lead to the dual purpose of enhancement of aesthetic value and surface integrity of the product.

(5) **Product life**: Corrosion seriously shortens the predicted design life, a time span after which replacement is anticipated. Cars have, in general, a design life of twelve years, but several brands survive much longer. A DC-3 aircraft has a design life of twenty years but after sixty years they are still flying. The Eiffel Tower had a design life of two years only, and even after one hundred years it is still a grand symbol of Paris. The reason for their survival is that the engineers made use of imaginative designs, environmental resistant materials and induction of corrosion-free maintenance measures. Distinguished and evocative designs always survive whereas designs of a transitory nature deteriorate to extinction with time. Design life is a process of imagination, material selection and corrosion-free maintenance.

(6) **Restoration of corroded objects**: Objects of outstanding significance to natural history need to be preserved. Many historical structures have been lost through the ravages of corrosion. One recent example is the call for help to restore the revolutionary iron-hulled steamships Great Britain built in 1843. It has been described as mother of all modern ships, measuring 3000 feet in length and weighing 1930 tons. A plea for £100 000 has been made for its restoration.

1.11 FIVE GOOD REASONS TO STUDY CORROSION

(1) Materials are precious resources of a country. Our material resources of iron, aluminum, copper, chromium, manganese, titanium, etc. are dwindling fast. Some day there will be an acute shortage of these materials. An impending metal crisis does not seem

anywhere to be a remote possibility but a reality. There is bound to be a *metal crisis* and we are getting the signals. To preserve these valuable resources, we need to understand how these resources are destroyed by corrosion and how they must be preserved by applying corrosion protection technology.

(2) Engineering knowledge is incomplete without an understanding of corrosion. Aeroplanes, ships, automobiles and other transport carriers cannot be designed without any recourse to the corrosion behavior of materials used in these structures.

(3) Several engineering disasters, such as crashing of civil and military aircraft, naval and passenger ships, explosion of oil pipelines and oil storage tanks, collapse of bridges and decks and failure of drilling platforms and tanker trucks have been witnessed in recent years. Corrosion has been a very important factor in these disasters. Applying the knowledge of corrosion protection can minimize such disasters. In USA, two million miles of pipe need to be corrosion-protected for safety.

(4) The designing of artificial implants for the human body requires a complete understanding of the corrosion science and engineering. Surgical implants must be very corrosion-resistant because of corrosive nature of human blood.

(5) Corrosion is a threat to the environment. For instance, water can become contaminated by corrosion products and unsuitable for consumption. Corrosion prevention is integral to stop contamination of air, water and soil. The American Water Works Association needs US\$ 325 billion in the next twenty years to upgrade the water distribution system.

QUESTIONS

CONCEPTUAL QUESTIONS

1. Explain how corrosion can be considered as extractive metallurgy in reverse.
2. List five important consequences of corrosion.
3. Which is the most common cause of corrosion damage, corrosion fatigue, stress corrosion cracking or pitting corrosion?
4. Describe with an example how corroded structures can lead to environmental pollution.
5. Does corrosion affect humans? If so, explain how.
6. Describe two engineering disasters in which corrosion played a leading role.
7. State two important corrosion websites.
8. How can corroded structures be injurious to human health?
9. Name three cities in Southeast Asia and the Middle East which have the most corrosive environment.
10. What is the best way to minimize the corrosion of defense equipment during storage?
11. What is the relationship between depletion of natural resources and corrosion?

REFERENCES

[1] Walsh, F. (1991). Faraday and his laws of electrolysis. *Bulletin of Electrochem*, **7**, 11, 481–489.
[2] Schönbein, C. (1936). *Pogg. Ann.*, **37**, 390.
[3] Evans, U.R. (1972). An Introduction to Metallic Corrosion, 2nd ed. London: Arnold.
[4] Uhlig, H.H. (1985). Corrosion and Corrosion Control, 3rd ed. New York: John Wiley and Sons.
[5] Fontana, M.G. (1986). Corrosion Engineering, 3rd ed. New York: McGraw-Hill Book Company.
[6] C. C. Technologies Laboratories, Inc. (2001). Cost of corrosion and prevention strategies in the United States, Ohio: Dublin, USA.
[7] Federal Highway Administration (FHWA), Office of the Infrastructure and Development (2001). Report FHWA-RD-01-156.
[8] National Association of Corrosion Engineers (NACE) (2002). *Materials Performance*, Special Issue, Houston, Texas, USA, July. Jointly with C. C. Technologies and FHWA.
[9] Ahmad, Z. (1996). Corrosion phenomena in coastal area of Arabian Gulf. *British Corrosion Journal*, **31**, (2), 191–197.
[10] Rashid-uz-Zafar, S., Al-Sulaiman, G.J. and Al-Gahtani, A.S. (1992). *Symp. Corrosion and the Control*, Riyadh, Saudi Arabia, May, 110.
[11] Tullmin, M.A.A., Roberge, P.R., Grenier, L. and Little, M.A. (1990). *Canadian Aeronautics and Space Journal*, **42**, (2), 272–275.

[12] Hoar, T.P. (1971). Report of the Committee on Corrosion and Protection, London: HMSO.

[13] Latanisian, R.M., Leslie, G.G., McBrine, N.J., Eselman, T. *et al.* (1999). Application of practical ageing management concepts to corrosion engineering, Keynote Address. *14th ICC*, Capetown, South Africa, 26 Sep–10 Oct.

WEBSITES

[20] www.intercorrosion.com
[21] www.learncorrosion.com
[22] www.nace.org
[23] www.iom3.org

GENERAL REFERENCES

[14] Hackerman, N. (1993). A view of the history of corrosion and its control. In: Gundry, R.D. ed. *Corrosion 93 Planery and Keynote Lectures*, Texas: NACE, Houston, 1–5.

[15] Pliny (1938). *Natural History of the World*. London: Heinemann.

[16] Hoare, T.P. (1971). Report of the Committee on Corrosion and Protection. London: HMSO.

[17] Uhlig, H.H. (1949). *Chemical and Engineering News*, **97**, 2764.

[18] N. B. S. (1978). *Corrosion in United States*, Standard Special Publication.

[19] Bennett, L.H. (1978). *Economic Effects of Metallic Corrosion in USA*, Special Publication 511-1, Washington, DC.

SOFTWARE

[24] NACE: Basic Corrosion (National Association of Corrosion Engineers, Houston, Texas), Course on CD-Rom, 2002.

[25] Corrosion Survey Data Base (COR.SUR), access via NACE website, NACE, Houston, Texas, 2003.

[26] Rover Electronic Data Books™, William Andrew, Inc., New York, USA, 2002.

[27] Peabody, A.W. *Control of Pipeline Corrosion*, 2nd Edition, Ed. R. Bianchette, NACE, Houston, Texas, 2003.

[28] Corrosion Damage: A Practical Approach, NACE, Houston, Texas, 2003.

CHAPTER 2

BASIC CONCEPTS IN CORROSION

For corrosion to take place, the formation of a *corrosion cell* is essential. A corrosion cell is essentially comprised of the following four components (Fig. 2.1).

- Anode
- Cathode
- Electrolyte
- Metallic path.

Anode: One of the two dissimilar metal electrodes in an electrolytic cell, represented as the negative terminal of the cell. Electrons are released at the anode, which is the more reactive metal. Electrons are insoluble in aqueous solutions and they only move, through the wire connection into the cathode. For example, in a battery, zinc casing acts as the anode. Also in a Daniel cell, zinc is the anode as oxidation occurs on it and electrons are released (Fig. 2.2). Corrosion nomenclature is the opposite of electroplating nomenclature, where an anode is positive and the cathode is negative.

Cathode: One of the two electrodes in an electrolytic cell represented as a positive terminal of a cell. Reduction takes place at the cathode and electrons are consumed. Example, carbon electrode in a battery, copper electrode in a Daniel cell. Figure 2.3 shows the reduction of hydrogen ion. The electron is always a reducing agent.

Electrolyte: It is the electrically conductive solution (e.g. salt solution) that must be present for corrosion to occur. Note that pure water is a bad conductor of electricity. Positive electricity passes from anode to cathode through the electrolyte as cations, e.g. Zn^{++} ions dissolve from a zinc anode and thus carry positive current away from it, through the aqueous electrolyte.

Metallic Path: The two electrodes are connected externally by a metallic conductor. In the metallic conductor, '*conventional*' current flows from $(+)$ to $(-)$ which is really electrons flowing from $(-)$ to $(+)$. Metals provide a path for the flow of conventional current which is actually passage of electrons in the opposite direction.

Current Flow: Conventional current flows from anode $(-)$ to cathode $(+)$ as Zn^{++} ions through the solution. The current is carried by these positive charged ions. The circuit is completed by passage of electrons from the anode $(-)$ to the cathode $(+)$ through the external metallic wire circuit (outer current).

Electron Flow:

$$H^+ + e \rightarrow H, \quad 2H \rightarrow H_2 \uparrow$$

Although the anode (e.g. Fe or Zn) is the most negative of the two metals in the cell, this reaction does not occur there because its surface is emanating Fe^{++} ions which repel H^+ ions from discharging there. The circuit is completed by negative ions $(-)$ which migrate from the cathode $(+)$, through the electrolyte, towards the anode $(-)$. They form

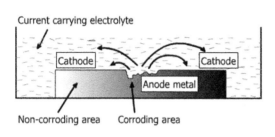

Figure 2.1   Corrosion cell in action

$Fe(OH)_2$ when they enter the cloud of Fe^{++} ions coming from the anode.

Anions: Migrate towards the anode (OH^-) but precipitate as $Fe(OH)_2$ before reaching it.

Cations: Migrate towards the cathode (Fe^{2+}).

Current flow in an electrochemical cell is shown in Fig. 2.4.

2.1 ANODIC AND CATHODIC REACTIONS

The anode is the area where metal is lost. At the anode, the reactions which take place are oxidation reactions. It represents the entry of metal ion into the solution, by dissolution, hydration or by complex formation. It also includes precipitation of metal ions at the metal surface. For example $Fe^{2+} + 2OH^- \rightarrow Fe(OH)_2$. Ferrous hydroxide or rust formation on steel surface is a common example. Some more examples are:

(a) $2Al + 6HCl \rightarrow 2AlCl_3 + 3H_2 \uparrow$
(b) $Fe + 2HCl \rightarrow FeCl_2 + H_2 \uparrow$

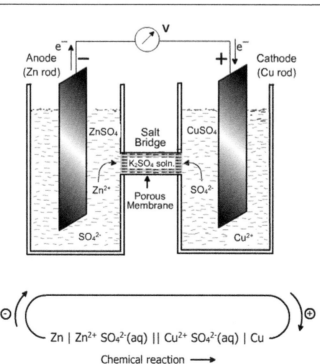

Figure 2.2   A galvanic cell (Daniel cell)

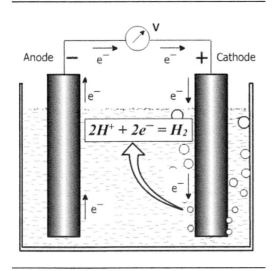

Figure 2.3 Figure showing the reduction of hydrogen in an acid electrolyte

(c) $Zn + H_2SO_4 \rightarrow ZnSO_4 + H_2 \uparrow$
(d) $Zn + HCl \rightarrow ZnCl_2 + H_2 \uparrow$

The reactions (a–d) involve the release of hydrogen gas. All the reactions shown above involve oxidation to a higher valence state. Reactions (a–d) can be written in terms of electron transfer as below:

(a) $Al \rightarrow Al^{3+} + 3e$
(b) $Fe \rightarrow Fe^{2+} + 2e$
(c) $Zn \rightarrow Zn^{2+} + 2e$
(d) $Zn \rightarrow Zn^{2+} + 2e$

Anodic reaction in terms of electron transfer is written as

$$M \rightarrow M^{n+} + ne$$

2.2 ANODIC REACTIONS CHARACTERISTICS

(1) Oxidation of metal to an ion with a charge.
(2) Release of electrons.
(3) Shift to a higher valence state.

The process of oxidation in most metals and alloys represents corrosion. Hence, if oxidation is stopped, corrosion is stopped.

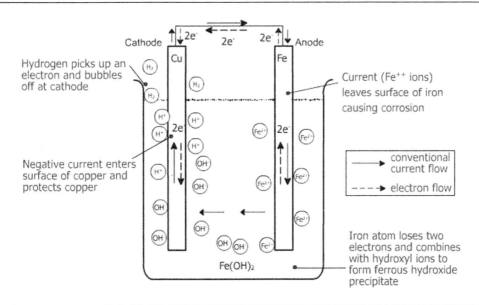

Figure 2.4 Current flow in an electrochemical cell

2.3 Cathodic Reactions Characteristics

Cathodic reactions are *reduction* reactions which occur at the cathode. Electrons released by the anodic reactions are consumed at the cathode surface. Unlike an anodic reaction, there is a decrease in the valence state. The most common cathodic reactions in terms of electrons transfer are given below:

(a) $2H^+ + 2e \rightarrow H_2 \uparrow$ (in acid solution)
(b) $O_2 + 4H^+ 4e \rightarrow 2H_2O$ (in acid solution)
(c) $2H_2O + O_2 + 4e \rightarrow 4OH^-$ (in neutral and alkaline solutions)
(d) $Fe^{3+} + e \rightarrow Fe^{+2}$ (metal ion reduction in ferric salt solutions)
(e) Metal deposition: $M^{2+} + 2e \rightarrow M$
$$Ni^{++} + 2e \rightarrow Ni$$
$$Cu^{2+} + 2e \rightarrow Cu$$
(f) Bacterial reduction of sulfate: $SO_4^{2-} + 8H^+ + 8e \rightarrow S^- + 4H_2O$

2.4 Types of Corrosion Cells

There are several types of corrosion cells:

(1) Galvanic cells
(2) Concentration cells
(3) Electrolytic cell
(4) Differential temperature cells.

2.4.1 Galvanic Cells

The galvanic cell may have an anode or cathode of dissimilar metals in an electrolyte or the same metal in dissimilar conditions in a common electrolyte. For example, steel and copper electrodes immersed in an electrolyte (Fig. 2.5), represents a galvanic cell. The more noble metal copper acts as the cathode and the more active iron acts as an anode. Current flows from iron anode to copper cathode in the electrolyte.

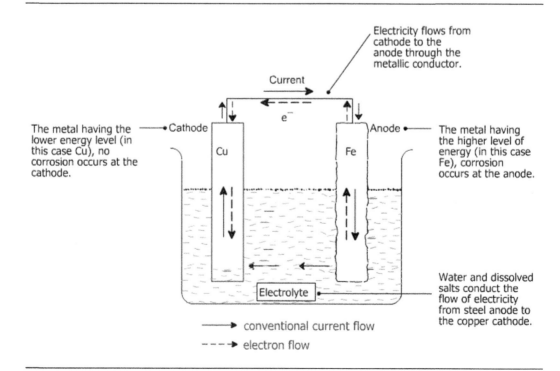

Figure 2.5 Typical galvanic cell

2.4.2 CONCENTRATION CELLS

This is similar to galvanic cells except with an anode and cathode of the same metal in a heterogeneous electrolyte. Consider the corrosion of a pipe in the soil. Concentration cells may be set up by:

(a) Variation in the amount of oxygen in soils.
(b) Differences in moisture content of soils.
(c) Differences in compositions of the soil.

Concentration cells are commonly observed in underground corroding structures, such as buried pipes or tanks (Fig. 2.6). The inequality of dissolved chemicals causes a potential difference which establishes anode in the more concentrated region and cathode in the less concentrated region.

2.4.3 ELECTROLYTIC CELL

This type of cell is formed when an external current is introduced into the system. It may consist of all the basic components of galvanic cells and concentration cells plus an external source of electrical energy.

Notice that anode has a (+) polarity and cathode has (−) polarity in an electrolytic cell, where external current is applied. This is the type of cell set up for electrically protecting the structures by cathodic protection. The polarity of an electrolytic cell is opposite to that in a galvanic (corrosion) cell (Fig. 2.7).

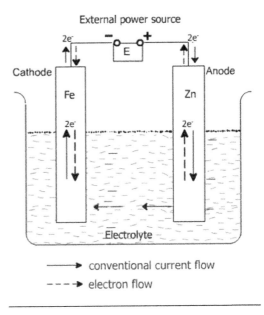

Figure 2.7   Electrolytic cell. The cathode and anode can be any metal

2.5 MECHANISM OF CORROSION OF IRON

Consider a piece of iron exposed to humid air which acts as an electrolyte. Fe^{2+} ions are released from the anode by oxidation and OH^- ions from the cathode by reduction on the metal surface.

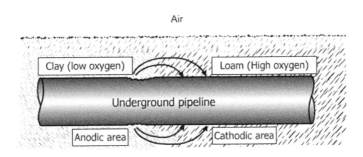

Figure 2.6   Concentration cell formation in an underground pipeline

The negative and positive ions combine

$$Fe^{++} + 2OH^- \rightarrow Fe(OH)_2$$
$$\text{(white green precipitate)}$$

$Fe(OH)_2$ is insoluble in water and separates from the electrolyte. A more familiar name of $Fe(OH)_2$ is *rust*.

Details of reactions involved in the corrosion of iron-based materials is given below:

(1) $Fe + H_2O \leftrightarrows FeO + 2H^+ + 2e^-$ Forms monolayer of FeO islands
(2) $Fe + 2H_2O \rightarrow Fe(OH)_2 + 2H^+ + 2e$
(3) $3FeO + H_2O \rightarrow Fe_3O_4 + 2H^+ + 2e$
    (Black)   (Magnetite)
(4) $2Fe_3O_4 + H_2O \rightarrow 3(\gamma - Fe_2O_3) + 2H^+ + 2e^-$
          (Brown)
(5) $2[\gamma - Fe_2O_3] + 3H_2O \rightarrow 6(\gamma - FeOOH)$
    (Brown)   (Yellow hydrated oxide)

The formation of rust, $Fe(OH)_2$, is shown in Fig. 2.8.

2.6 CONCEPT OF FREE ENERGY

In this section the relationship between free energy and equilibrium constant is shown. The contribution made by one mole of any constituent, A, to the total free energy, G, of a mixture is G_A, which may be represented by

$$G_A = G_A^\circ + RT \ln a_A \qquad (2.1)$$

where
$$a_A = \text{activity of the substance,}$$
$$T = \text{absolute temperature,}$$
and $$R = \text{gas constant.}$$

From the Gibbs–Helmholtz equation:

$$G = H - TS \qquad (2.2)$$

Substituting $H = U + PV$ from the First Law of Thermodynamics

$$F = PV - TS \qquad (2.3)$$

Differentiating:

$$dG = dU + PdV + VdP - TdS - SdT \qquad (2.4)$$

But for a reversible process,

$$dU = q_{rev} - PdV \text{ at constant pressure and}$$

$$q_{rev} = TdS \text{ at constant temperature} \qquad (2.5)$$

So,

$$dG = TdS - PdV + PdV + VdP$$
$$- TdS - SdT$$

i.e. $dG = VdP - SdT$, for a reversible process at constant pressure and temperature. (2.6)

At constant pressure, dP vanishes and this becomes:

$$\frac{dG}{dT} = -S$$

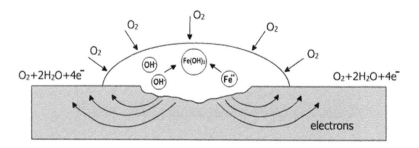

Figure 2.8 Formation of rust in seawater

and at constant temperature dT vanishes and:

$$dG = VdP$$

However, the ideal gas equation for 1 mole is

$$PV = RT$$

$$V = \frac{RT}{P} \rightarrow dG = \left(\frac{RT}{P}\right)dP, \text{ since } dG = VdP$$
$$(2.7)$$

Integration between the limits G'_A, G_A and P'_A, P_A gives:

$$\int_{G'_A}^{G_A} dG = \int_{P'_A}^{P_A} \left(\frac{RT}{P}\right)dP$$
$$(2.8)$$

$$G_A - G'_A = RT \ln\left(\frac{P_A}{P'_A}\right)$$

If G_A is taken to refer for standard conditions,

$$G_A - G'_A = RT \ln P_A$$
$$(2.9)$$
$$\text{or} \quad G_A = G_A^* = RT \ln P_A$$

If activities are used instead of pressures, then as in equation (2.1)

$$G_A = G_A^\circ + RT \ln a_A$$

Applying ΔG for the reaction

$$aA + bB \leftrightarrow cC + dD$$

gives

$$\Delta G = (cG_c + dG_d) - (aG_A + bG_B) \quad (2.10)$$

Using expression (2.1) for G_A, G_B, G_C and G_D:

$$\Delta G = c\left[G_C^\circ + RT \ln a_C\right] + d\left[G_D^\circ + RT \ln a_D\right]$$
$$- a\left[G_A^\circ + RT \ln a_A\right] - b\left[G_B^\circ + RT \ln a_B\right]$$

So, for 1 mole of forward reaction

$$\Delta G = \Delta G^\circ + RT \ln\left(\frac{a_C^c a_D^d}{a_A^a a_B^b}\right)$$

(Van't Hoff reaction isotherm) (2.11)

At equilibrium

$$\Delta G = 0$$

and

$$\left(\frac{a_C^c a_D^d}{a_A^a a_B^b}\right) = K, \text{ where } K \text{ is the equilibrium constant.}$$

$$0 = \Delta G^\circ + RT \ln K$$

$$\Delta G^\circ = -RT \ln K \quad (2.12)$$

$$(\because \Delta G^\circ \text{ is always } = -RT \ln K)$$

2.6.1 FREE ENERGY (THE DRIVING FORCE OF A CHEMICAL REACTION)

A chemical reaction at constant temperature and pressure will only occur if there is an overall decrease in the free energy of the system during the reaction. Consider the following two reactions for example:

(a) $ZnO(s) + C(s) \rightarrow Zn(s) + CO(g)$ at 1373 K
$\Delta S^\circ = +285 \, JK^{-1} \, mol^{-1}$
$\Delta H^\circ = +349.9 \, kJ \cdot mol^{-1}$

(b) $Fe(s) + \frac{1}{2}O_2(g) \rightarrow FeO(s)$
$\Delta S^\circ = -71 \, JK^{-1} \, mol^{-1} \, FeO$
$\Delta H^\circ = +265.5 \, kJ \cdot mol^{-1} \, FeO$

In example (b), both ΔH° and ΔS° work in opposite signs to each other in terms of direction of energy change, however, both of them proceed spontaneously as indicated. It is important to know which one would decide the direction of the reaction (ΔH° or ΔS°). We must, therefore, introduce a single function expressing the combined effect of change of both ΔH and ΔS. Of the total enthalpy of the system only a part is converted to useful work, which is called free energy (ΔG). It can be defined in terms of ΔH, ΔS and T. The relationship between the three is deduced as

$$\Delta H^\circ - \Delta G^\circ = T\Delta S^\circ \qquad (2.13)$$

which is a statement of the Second Law of Thermodynamics and may be rearranged to:

$$\Delta G = \Delta H - T\Delta S \qquad (2.14)$$

ΔG is thus a driving force for a reaction to occur. Reactions at constant temperature and pressure proceed in a direction which tends to cause a decrease in free energy. ΔG can be used to predict the feasibility of a reaction as shown below.

For a chemical reaction where the reactants ($A+B$) react to yield the product ($C+D$) according to

$$A+B \rightarrow C+D \qquad (2.15)$$

the free energy change of the reaction can be expressed as

$$\Delta G = G_P - G_R \qquad (2.16)$$

$$\Delta G = (H_P - H_R) - T(S_P - S_R) \qquad (2.17)$$

$$\Delta G = \Delta H - T\Delta S \qquad (2.18)$$

where

P = products
R = reactants.

(a) If the value of free energy change is negative, the reactions will be spontaneous and take place from left to right ($A+B \rightarrow C+D$).
(b) If the value of the free energy change is zero, the reaction is at equilibrium ($A+B \rightleftarrows C+D$).
(c) If the value of the free energy change is positive, the reaction will proceed in the reverse direction ($A+B \leftarrow C+D$).

It is not possible to obtain the absolute value of free energy for a reaction. However, the change in free energy can be measured.

Illustrative Example 2.1

Calculate ΔG° for the following reaction at 500 K.

$$CuO(s) + H_2(g) \rightarrow Cu(s) + H_2O$$

Solution:

$$\Delta H^\circ_{500} = -87\,kJ \cdot mol^{-1}$$

$$\Delta S^\circ_{500} = +47\,JK^{-1}mol^{-1}$$

$$\Delta G = \Delta H^\circ - T\Delta S^\circ$$

$$= -87\,000 - (500 \times 47)\,J \cdot mol^{-1}$$

$$= 110.5\,kJ \cdot mol^{-1}$$

ΔG° is negative ($-ve$), hence the reaction is feasible.

Ever if ΔG° were positive, a reaction producing gas pressures of products of less than unit activity (=1 bar, for gases) is possible. The sign of ΔG° only predicts spontaneity if all activities are unity.

2.6.2 CELL POTENTIALS AND EMF

Electrochemical cells generate electrical energy due to electrochemical reactions. The electrical energy available is

$$\text{Electrical energy} = \text{volts} \times \text{current} \times \text{time}$$

$$= \text{volts} \times \text{coulombs}$$

$$= EQ$$

where

$Q = nF$
n is the number of electrons involved in the chemical reaction
F is Faraday's constant = 96 500 C (g equiv.)$^{-1}$
    The gram equivalent is the number of moles divided by the number of electrons involved in the reaction
E is electromotive force (emf) of the cell (volts).

Any work performed can only be done through a decrease in free energy of the cell reaction, hence,

$$\Delta G = -nFE$$

where

ΔG = free energy change (joules)
E = standard potential of the reaction (volts).

Illustrative Example 2.2

(1) Calculate the standard enthalpy change for the following reaction at 298 K:

$$2PbS(s) + 3O_2(g) \rightarrow 2PbO(s) + 2SO_2(g)$$

Solution:
Given that the standard enthalpies of formation are:

$$\Delta H^{\circ}_{298} PbS = -94.5 \, kJ \cdot mol^{-1}$$

$$\Delta H^{\circ}_{298} PbO(s) = -220.5 \, kJ \cdot mol^{-1}$$

$$\Delta H^{\circ}_{298} SO_2(g) = -298 \, kJ \cdot mol^{-1}$$

Rewrite the equation:

$$2PbS(s) + 3O_2(g) \rightarrow 2PbO(s) + 2SO_2(g)$$

$$\Downarrow \qquad \Downarrow \qquad \Downarrow \qquad \Downarrow$$

$$2(-94.5) \quad 3(0) \quad 2(-220.5) \; 2(-298.0)$$

The standard enthalpy of formation of an element is taken as zero at 298 K.
Using equation:

$\Delta H = H_2 - H_1$, where H_2 = total enthalpy of formation of products and H_1 is for the reactants

$$\Delta H = [2(-220.5) + 2(-298)] - [2(-94.5) + 0]$$

$$= -848.0 \, kJ$$

(2) For the equilibrium $ZnO(s) + C(s) \Leftrightarrow Zn(s) + CO(g)$

(a) Deduce the direction in which the reaction would be feasible at 300 and 1200 K, when all reactants and products are in their standard states (i.e. unit activities).

(b) Deduct Kp for the reaction using the relationship $\Delta G^{\circ} = \Delta H^{\circ} - T\Delta S^{\circ}$ for the reaction.

Given enthalpies and entropies of reaction:

$$\Delta H_{300} = +65 \, kJmol^{-1}$$

$$\Delta S_{300} = +13.7 \, JK^{-1}mol^{-1}$$

$$\Delta H_{1200} = +180.9 \, kJ \cdot mol^{-1}$$

$$\Delta S_{1200} = +288.6 \, JK^{-1}mol^{-1}$$

Solution:
(a) For reaction at 300 K:

$$(1) \quad \Delta G^{\circ} = 65\,000 - (300 \times 13.7)$$
$$= 60\,890 \, J \cdot mol^{-1}$$

For reaction at 1200 K:

$$(2) \quad \Delta G^{\circ} = 180\,900 - (1200 \times 288.6)$$
$$= -165\,420 \, J \cdot mol^{-1}$$

As ΔG° is negative for the reaction at 1200 K, the reaction proceeds from left to right in the forward direction at 1200 K. The forward reaction is not feasible at 300 K to produce 1 bar of CO_2 from 1 bar of CO: but it would form much lower CO_2 pressures spontaneously. Such low CO_2 pressures would not be sufficient to '*push back the atmosphere*' and so effectively no Zn would be produced.

(b) Calculation of K_p.
The equilibrium constant (K) is given by

$$\Delta G = \Delta G^{\circ} + RT \ln \left(\frac{a_C^c a_D^d}{a_A^a a_B^b} \right) \quad (2.19)$$

but the term in parenthesis is not K, but is any arbitrary non-equilibrium activity for which ΔG is the free energy change if such a reaction was carried out.

At equilibrium $\Delta G = 0$, and then $a_C^c a_D^d / a_A^a a_B^b = K$ equilibrium constant $= \Delta G° + RT \ln K$.

$$So, \Delta G° = -RT \ln K \qquad (2.20)$$

Equation (2.19) can be rewritten as:

$$\Delta G = \Delta G° + RT \ln \left(\frac{P_C P_D}{P_A P_B} \right)$$

where P is partial pressure (actual) assuming A, B, C and D are gases. The above equation connecting ΔG and K is a form of van't Hoff isotherm.

Now calculate K_p

$$\ln K_p = -\frac{\Delta G°}{RT}$$

At 1200 K,

$$\log_{10} K_p = \frac{-165420}{2.303 \times 8.314 \times 1200}$$

$$= 7.197$$

So, $K_p = 1.574 \times 10^7$.

Similarly at 300 K, $K_p = 2.512 \times 10^{-11}$. The value of K_p at 1200 shows that the reaction proceeds fast (left to right) compared to the reaction at 300 K, which proceeds at a negligible rate. At 1200 K, bright red hot, reaction kinetics are fast and it is safe to interpret a large K value as a meaning that reaction is fast. But the reaction $2H_2 + O_2 = 2H_2O$ has a very large K value at room temperature, yet is negligibly slow (unless sparked), due to the very slow kinetics at room temperature. K is a thermodynamic factor, but kinetics may allow only a negligible rate.

2.7 REVERSIBLE ELECTRODE POTENTIAL

When a metal is immersed in an electrolyte, a dynamic equilibrium is established across the interface with a potential difference between the metal and electrolyte. The atoms of the metal, M, ionize producing aquo-ions, $M^{n+}(aq)$, and electrons, ne, according to

$$M = M^{n+}(aq) + ne \qquad (2.21)$$

where M represents metal atoms. The metal is left with a negative charge and its positively charged metal ions, $M^{n+}(aq)$, in the electrolyte are attracted back towards the metal surface. Thus, a potential difference and dynamic equilibrium between the metal and the solution is established. The atoms of the metal continue to ionize until the displacement of electrical charges produced balances the tendency of the metallic atoms to ionize into the electrolyte.

Consider, for example, a piece of zinc metal in water. Zinc dissolves producing positively charged zinc ions (cations):

$$Zn \rightarrow Zn^{2+} + 2e \qquad (2.22)$$

The zinc ions in the solution remain very close to the metal surface. The zinc metal becomes negatively charged as the positive ions leave its surface. The excess electrons on the zinc surface orient themselves opposite to a layer of zinc ions of equal and opposite charge on the water side of the zinc/water interface. Such a process leads to the formation of an electrical double layer of about 1 nm (10^{-7} cm) thickness along the metal/solution interface (Fig. 2.9).

Figure 2.9A shows the metal/solution interface at the moment of immersion and the formation of double layer is shown in Figs 2.9B and C. The double layer shown in Figs 2.9B and C is formed as a result of attraction between the negative ions (anions) and positive ions (cations) on one hand, and repulsion between similarly charged ion, anions or cations on the other hand. As a result of the above interactions, cations diffuse amongst the anions until an equilibrium is established between the metal and solution, and between the bulk of the solution and the layer adjacent to the metal surface. The plane passing through the ions absorbed on the metal surface is called the *Inner Helmholtz*

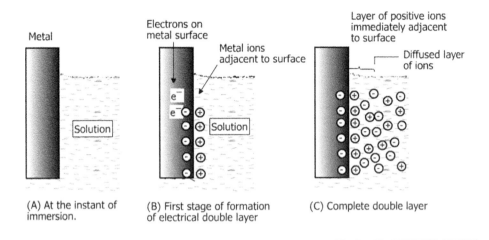

Metal

Electrons on
metal surface

Metal ions
adjacent to surface

Layer of positive ions
immediately adjacent
to surface

Diffused layer
of ions

Solution

Solution

(A) At the instant of
immersion.

(B) First stage of formation
of electrical double layer

(C) Complete double layer

Figure 2.9   Representation of electrical double layer at a metal/solution interface

plane, whereas the plane passing through the center of solvated cation is called the *Outer Helmholtz* plane, which also marks the beginning of a diffuse layer when an excess of charges is neutralized (Fig. 2.9C). At this stage, it is sufficient to understand the meaning of double layer to know the electrode potential. Returning to equation (2.21), it should be noted that under the equilibrium conditions in the system, the rate of ionization on the metal surface:

$$M \rightarrow M^{n+} + ne \qquad (2.23)$$

becomes equal to the rate of discharge across the double layer:

$$M^{n+} + ne \rightarrow M \qquad (2.24)$$

In other words, the rate of forward reaction (equation (2.23)) becomes equal to the rate of backward reaction (equation (2.24)) on the metal electrode. This is described as an exchange current.

The double layer described above is responsible for the establishment of potential difference between the metal and solution, and this potential difference is referred to as '*Absolute Electrode Potential*.' More details will be given in Chapter 3.

2.7.1 EXAMPLES OF REVERSIBLE CELLS

Figure 2.2 shows a Daniel cell which is a good example of a reversible cell. A Daniel cell can be presented as follows:

$$Zn_{metal}/M_{Zn^{2+}}//M_{Cu^{2+}}/Cu_{metal} \qquad (2.25)$$

where:

$$M_{Zn^{2+}} = 1.0 \text{ mol of zinc salt, as for example}$$
$$ZnSO_4 \text{ per liter}$$
$$M_{Cu^{2+}} = 1.0 \text{ mol of } CuSO_4 \text{ per liter.}$$

When writing down a cell, it is common practice to put the anode on the left-hand side and the cathode on the right-hand side. The anode is the electrode where oxidation reaction takes place, and the cathode is the electrode where reduction reaction takes place. Accordingly, the reactions in a Daniel cell can be written as:

$$Zn \rightarrow Zn^{2+} + 2e \text{ (Oxidation at anode)} \quad (2.26)$$

$$Cu^{2+} + 2e \rightarrow Cu \text{ (Reduction at cathode)} (2.27)$$

An emf of 1.10 V is produced in a Daniel cell. This potential is equal to the difference in potentials between the anode and the cathode. It is called the '*Reversible Cell Potential*' (E_{rev}) or '*Equilibrium Potential*' (E_{eq}).

2.7.2 DIFFERENCE BETWEEN REVERSIBLE POTENTIAL AND STANDARD POTENTIAL

If a metal is immersed in a solution of its own ions, such as zinc in $ZnSO_4$ solution, or copper in $CuSO_4$ solution, the potential obtained is called the *reversible potential* (E_{rev}).

If, on the other hand, the substances taking part in the process are in their standard states, such that the activities of the metallic ions are equal to unity or gases are at 1 bar pressure, the potentials obtained are called 'Standard Electrode Potentials' (Table 2.1). A standard potential refers to the potential of pure metal measured with reference to a hydrogen reference electrode. The details of reference electrodes are provided later in this chapter.

Table 2.1 Standard reduction potentials, 25°C (Modified from Uhlig)

Electrode reaction	$E°$ (V)
$Au^{+3}+3e=Au$	1.50
$Pt^{++}+2e=Pt$	ca 1.2
$Hg^{++}+2e=Hg$	0.854
$Pd^{++}+2e=Pd$	0.987
$Ag^{+}+e=Ag$	0.800
$Hg_2^{++}+2e=2Hg$	0.789
$Cu^{+}+e=Cu$	0.521
$Cu^{++}+2e=Cu$	0.337
$2H^{+}+2e=H_2$	0.000
$Pb^{++}+2e=Pb$	−0.126
$Sn^{++}+2e=Sn$	−0.136
$Mo^{+3}+3e=Mo$	ca −0.2
$Ni^{++}+2e=Ni$	−0.250
$Co^{++}+2e=Co$	−0.277
$Tl^{+}+e=Tl$	−0.336
$In^{+3}+3e=In$	−0.342
$Cd^{++}+2e=Cd$	−0.403
$Fe^{++}+2e=Fe$	−0.440
$Ga^{++}+3e=Ga$	−0.53
$Cr^{+3}+3e=Cr$	−0.74
$Zn^{++}+2e=Zn$	−0.763
$Nb^{+3}+3e=Nb$	ca −1.1
$Mn^{++}+2e=Mn$	−1.18

Electrode reaction	$E°$ (V)
$Zr^{+4}+4e=Zr$	−1.53
$Ti^{++}+2e=Ti$	−1.63
$Al^{+3}+3e=Al$	−1.66
$Hf^{+4}+4e=Hf$	−1.70
$U^{+3}+3e=U$	−1.80
$Be^{++}+2e=Be$	−1.85
$Mg^{++}+2e=Mg$	−2.37
$Na^{+}+e=Na$	−2.71
$Ca^{++}+2e=Ca$	−2.87
$K^{+}+e=K$	−2.93
$Li^{+}+e=Li$	−3.05
$FeO_4^{--}+8H^{+}+3e=Fe^{+3}+4H_2O$	1.9
$Co^{+3}+e=Co^{++}$	1.82
$PbO_2+SO_4^{--}+4H^{+}+2e=$ $PbSO_4+2H_2O$	1.685
$NiO_2+4H^{+}+2e=Ni^{++}+2H_2O$	1.68
$Mn^{+3}+e=Mn^{++}$	1.51
$PbO_2+4H^{+}+2e=Pb^{++}+2H_2O$	1.455
$Cl_2+2e=2Cl^{-}$	1.3595
$Cr_2O_7^{--}+14H^{+}$ $+6e=2Cr^{+3}+7H_2O$	1.33
$O_2+4H^{+}+4e=2H_2O$	1.229
$Br_2(I)+2e=2Br^{-}$	1.0652
$Fe^{+3}+e=Fe^{++}$	0.771
$O_2+2H^{+}+2e=H_2O_2$	0.682
$I_2+2e=2I^{-}$	0.5355
$O_2+2H_2O+4e=4OH^{-}$	0.401
$Hg_2Cl_2+2e=2Hg+2Cl^{-}$	0.2676
$AgCl+e=Ag+Cl^{-}$	0.222
$SO_4^{--}+4H^{+}+2e=H_2SO_3+H_2O$	0.17
$Cu^{++}+e \rightarrow Cu^{+}$	0.153
$Sn^{+4}+2e=Sn^{++}$	0.15
$AgBr+e=Ag+Br^{-}$	0.095
$Cu(NH_3)_2^{+}+e=Cu+2NH_3$	−0.12
$Ag(CN)_2^{-}+e=Ag+2CN$	−0.31
$PbSO_4+2e=Pb+SO_4$	−0.356
$HPbO_2^{-}+H_2O+e=Pb+3OH^{-}$	−0.54
$2H_2O+2e=H_2+2OH^{-}$	−0.828
$Zn(NH_3)_4^{++}+2e=Zn+4NH_3$	−1.03

2.7.3 HALF CELLS

The term half cell is frequently used in electrochemistry. To illustrate the formation of half cells, consider a Daniel cell, shown in Fig. 2.2. In the cell shown, zinc rod is immersed in one molar solution of zinc sulfate, and copper rod in one molar solution of copper sulfate. The two solutions, zinc sulfate and copper sulfate, are separated by a porous pot, which prevents them from mixing but allows electrical contact. As shown in the figure, Zn in $ZnSO_4$ (1.0 M) forms one half cell and Cu in $CuSO_4$ (1.0 M) forms another half cell. Each half cell has its own potential.

2.7.4 REACTIONS IN THE CELL

In Daniel cell, zinc dissolves and a potential is set up between the zinc ions and zinc metal in the zinc half cell. In the copper half cell, copper ions are deposited on the copper metal and a potential is set up. Once the half cells are in equilibrium, no further deposition or dissolution would occur. If, now, the two half cells are joined by a conductor, as shown in Fig. 2.2, electrons would flow from zinc (anode) to copper (cathode). As a result of electron flow, the equilibrium in the half cells is distributed, and, therefore zinc will dissolve further according to the anodic reaction (oxidation),

$$Zn \rightarrow Zn^{2+} + 2e \qquad (2.28)$$

and copper ions will discharge from copper sulfate solution in order to remove the excess electrons according to the cathodic reaction (reduction of Cu^{++} ions to Cu metal):

$$Cu^{2+} + 2e \rightarrow Cu \qquad (2.29)$$

The overall reaction of the cell is the sum of the above two half cell reactions:

$$Zn + Cu^{2+} \rightarrow Zn^{2+} + Cu \qquad (2.30)$$

The emf of the cell is the algebraic sum of the potentials of the half cells or '*Standard Electrode Potential*:'

$$E_{cell} = E_{Zn/Zn^{2+}} + E_{Cu^{2+}/Cu} \qquad (2.31)$$

The above cell (Daniel cell) can also be represented as:

$$Zn(s)/ZnSO_4(1.0\,M)//CuSO_4(1.0\,M)/Cu(s) \qquad (2.32)$$

2.7.5 STANDARD HYDROGEN POTENTIAL (SHE) AND MEASUREMENT OF SINGLE ELECTRODE POTENTIALS

Absolute single electrode potential is a characteristic property of a metal. All metals have characteristic electrode potentials. Absolute single electrode potentials cannot be measured directly. They can be measured with respect to a standard electrode such as standard hydrogen electrode (SHE). The standard hydrogen electrode is a widely used standard electrode. It is arbitrarily assumed to have zero potential at all temperatures by definition. Thus, it provides a zero reference point for electrode potentials. The construction of the standard hydrogen electrode is shown in Fig. 2.10. As shown in the figure,

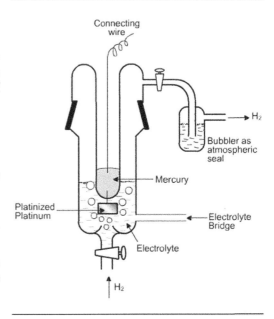

Figure 2.10 Hydrogen electrode

a platinum electrode is dipped in a solution containing hydrogen ions at unit activity. To obtain hydrogen ions at unit activity, a solution of one mol per liter HCl can be used. To obtain unit activity of hydrogen gas, hydrogen is passed at one atmospheric pressure over platinum plate in the solution. The SHE is expressed as

$$H_2 \rightarrow 2H^+ + 2e \qquad (2.33)$$

The most acceptable method of obtaining standard electrode potentials is by comparing the electrode potential of metals with the standard hydrogen electrode. Since the SHE has zero electrode potential at all temperatures by definition, the electrode potential of a metal is numerically equal to the emf of the cell formed by SHE and the metal electrode. In other words, the emf of the cell represents the electrode potential of the half cell formed by the metal with respect to the standard hydrogen electrode. In such a cell, reaction on the hydrogen electrode is oxidation and reaction on the other electrode is reduction. Such a cell can be expressed as:

$$Pt(s)/H_2(1\,atm)//H^+, (a_{H^+} = 1)//M/M^{n+} \qquad (2.34)$$

The activity or concentration of the M^{n+} ion solution needs to be specified, as cell potential will vary with it.

The electrode reaction for such a cell can be written as (considering reduction reactions as convention):

$$M^{n+} + ne = M \qquad (2.35)$$

for metal electrodes, and

$$nH^+ + ne = \frac{n}{2}H_2 \qquad (2.36)$$

for the standard hydrogen electrode (SHE).

The cell reaction is obtained by taking the difference of equations (2.35) and (2.36) above:

$$M^{n+} + \frac{n}{2}H_2 = M + nH^+ \qquad (2.37)$$

Reaction (2.37) is the complete reaction for the cell expressed in equation (2.34) above, where

one of the electrodes is SHE and the other is the metal electrode of which we want to measure the single electrode potential. The emf of the cell expressed in equation (2.34) is given by:

$$E_{cell} = E_{M^{n+}/M} - E^{\circ}_{H^+/H_2} \qquad (2.38)$$

where $E^{\circ}_{H^+/H_2}$ (SHE), the electrode potential of standard hydrogen electrode and is assumed equal to zero at all temperatures by convention. Therefore, equation (2.38) becomes:

$$E_{cell} = E_{M^{n+}/M} \qquad (2.39)$$

2.7.6 THE POTENTIAL (EMF) OF A CELL

The emf of any cell is, therefore given by

$$E_{cell} = \left[E_{Right} - E^{\circ}(SHE) \right]_{Right}$$
$$- \left[E_{Left} - E^{\circ}(SHE) \right]_{Left} \qquad (2.40)$$

and

$$E_{cell} = E_{Right} - E_{Left} \qquad (2.41)$$

As the potential of SHE cancels, no matter what value it is, the emf of the cell, E_{cell}, is not affected. The cell potential (emf) is positive if the left-hand electrode is negative and the right-hand electrode, positive. Consequently, oxidation would occur on the left-hand electrode, and reduction on the right-hand electrode.

2.8 CONCENTRATION CELL

In a concentration cell, the emf results from transportation of an electrolyte (anion, cation or both) from the more concentrated solution to the more dilute solution. Here the metal is in contact with two half cells having the same electrolyte at two different concentrations. Such a cell is represented, for example, by two zinc electrodes immersed in zinc sulfate solutions at different concentrations:

$$Zn(s)/Zn^{2+}(0.1\,M)//Zn^{2+}(1.0\,M)/Zn(s) \quad (2.42)$$

Another example of a concentration cell is two standard hydrogen electrodes immersed into two HCl solutions of different concentrations:

$$Pt(s)H_2(g,\,1\,atm)/HCl(a_2)//$$

$$HCl(a_1)/H_2(g,\,1\,atm)/Pt(s) \quad (2.43)$$

In the above cell, HCl is in two different concentrations. The activity (molality $\times$ activity coefficient) a_1 is greater than activity a_2; $a_1 > a_2$. Several types of concentration cells are encountered in corrosion. For example, a concentration cell is formed if one end of a pipe is exposed to soil and the other end to air. The end of the pipe in air is exposed to a high concentration of oxygen than the end of the pipe in the soil. The formation of a concentration cell leads to differential aeration corrosion in buried structures in the soil.

2.9 LIQUID JUNCTION POTENTIAL

For accurate measurement of emf, two conditions must be satisfied:

(a) The cell reaction must be reversible and
(b) no current must be drawn from the cell.

The boundary between two electrolytic solutions with different concentrations is a source of *irreversibility* in measuring the emf of the cells. The potential difference developed at the boundary is called 'Liquid Junction Potential.'

The potential difference is caused by the migration of ions from one electrolyte to another electrolyte. Let us take the case of concentrated HCl forming a junction with dilute HCl. Both H^+ and Cl^- ions diffuse from concentrated HCl to dilute HCl. The hydrogen ion moves faster and, therefore, the dilute solution becomes positively charged due to ingress of H^+ ions from the stronger acid solution and the more concentrated

solution thus acquires a relative negative charge. The consequent potential difference across the double layer, which is a positively charged layer in the dilute solution side and a negatively charged layer in the concentrated solution, may effect the emf of the cell seriously in certain cases. It is generally believed that the difference in potential is caused by the difference in the rate of diffusion of oppositely charged ions. In the above example, the charge acquired by a dilute solution is that of the faster moving ion (H^+); chloride ions (Cl^-) diffuse much slower. The magnitude of the liquid junction potential may affect the reversible potential positively or negatively depending on the mobilities of the ions. The most common method of removing liquid junction potential is to insert a KCl salt bridge between the electrolytes. Porous barriers are also used. The introduction of salt bridge minimizes the liquid junction potential. The KCl contained in the bridge, allows the migration of K^+ and Cl^- ions which carry most of the current and migrate with the same mobility.

The potential of zinc is more negative, hence it is considered as an anode ($Zn \rightarrow Zn^{2+} + 2e$) and the potential of nickel is less negative than zinc, hence it is considered as a cathode ($Ni^{2+} + 2e \rightarrow Ni$). The sign of the potential is changed for zinc as it oxidizes. From the potential obtained ($+0.51\,V$), it is clear that the reaction will proceed spontaneously.

2.10 APPLICATION OF FREE ENERGY TO CORROSION CELLS

It is known that corrosion reactions produce electrical energy. The amount of work done by a cell is equal to the quantity of electrical energy which it generates under constant pressure, temperature and concentration of the reaction. In an electrochemical reaction, electrical energy available is equal to the product of potential of the cell and the quantity of electricity involved (volts $\times$ amperes $\times$ time). That is, Electrical energy $= E \times Q$. This is equal to the net work done by the cell. From Faraday's law, Q is one Faraday (F) for each gram equivalent of the reactants. For n gram equivalent of the reactants, Q is equal to nF. Work can only be performed if the free energy of the cell is decreased, as shown in Section 2.6.2:

$$\Delta G = -nFE \qquad (2.44)$$

where

ΔG = change in Gibbs free energy of a cell in joules per mole
n = number of electrons involved in the reaction
E = emf of the cell in volts.

If all substances are in their standard states (i.e. at unit activity) the expression (2.44) becomes:

$$\Delta G^\circ = -nFE^\circ \qquad (2.45)$$

Illustrative Example 2.3
For the cell $[Cu/Cu^{2+}(1.0\,M)]//[Zn^{2+}(1.0\,M)/Zn]$, and the cell reaction $Cu + Zn^{2+} \rightarrow Cu^{2+} + Zn$:

$$\text{If} \quad \Delta G^\circ_{298^\circ K} = -147.5 \frac{kJ}{mol^{-1}} \text{ for } Zn^{2+}$$

$$\text{and} \quad \Delta G^\circ_{298^\circ K} = 63.35 \frac{kJ}{mol^{-1}} \text{ for } Cu^{2+}$$

Solution of the Problem:

$$\Delta G^\circ_{Reaction} = \Sigma n_i \Delta G(i) - \Sigma n_j \Delta G(j)$$

where
i = products
j = reactants
n = number of moles

$$\therefore \Delta G^\circ_{Reaction} = 2[63.35 + 0 - (-147.5 + 0)]$$

So,

$$\Delta G^\circ_{Reaction} = 425.1 \frac{kJ}{mol}$$

Since,

$$\Delta G^\circ_{Reaction} = -nFE$$

$$= -2 \times 96\,500 \times E$$

Hence,

$$E^\circ_{Reaction} = -0.002\,V$$

Reaction does not take place simultaneously.

2.11 NERNST EQUATION

The Nernst equation expresses the emf of a cell in terms of activities of products and reactants taking place in the cell reaction. Consider a general cell reaction:

$$M_1 + M_2^{n+} \rightleftharpoons M_2 + M_1^{n+} \qquad (2.46)$$

M_1 and M_2 represent metal electrodes, such as Cu and Zn, in a cell, and the above reaction (2.46) can be written as:

$$Cu + Zn^{2+} \rightleftharpoons Zn + Cu^{2+} \qquad (2.47)$$

The free energy change (ΔG) of a reaction is given by the difference in the molar free energy of products and reactants as shown in Section 2.6.

$$\Delta G = \Sigma n_i \Delta G(i) - \Sigma n_j \Delta G(j) \qquad (2.48)$$

Therefore, the free energy change for the reaction (2.46) can be expressed as:

$$\Delta G^\circ = (G^\circ_{M_2} + G^\circ_{M_1^{n+}}) - (G^\circ_{M_1} + G^\circ_{M_2^{n+}}) \qquad (2.49)$$

If the substances are in their standard states the above expression becomes:

$$\Delta G^\circ = (G^\circ_{M_2} + G^\circ_{M_1^{n+}}) - (G^\circ_{M_1} + G^\circ_{M_2^{n+}}) \qquad (2.50)$$

Free energies of any metal, such as M_1 at its standard state and any arbitrary state are related through the following equation:

$$G_{M_1} - G^\circ_{M_1} = RT \ln a_{M_1} \qquad (2.51)$$

where a_{M_1} is the activity of metal M_1, R is the universal gas constant, and equal to 8.314 joules/degree/mole, and T is absolute temperature in degrees Kelvin, °K, which is $(273.16 + t°C)$.

Subtracting equation (2.50) from equation (2.49), we get:

$$\Delta G - \Delta G° = \left(G_{M_2} - G°_{M_2}\right) + \left(G_{M_1^{n+}} - G°_{M_1^{n+}}\right)$$
$$- \left(G_{M_1} - G°_{M_1}\right) - \left(G_{M_2^{n+}} - G°_{M_2^{n+}}\right)$$

Substituting from equation (2.51), we obtain:

$$\Delta G - \Delta G° = RT \ln a_{M_2} + RT \ln a_{M_1}$$
$$- RT \ln a_{M_1} - RT \ln a_{M_2^{n+}},$$

$$\Delta G - \Delta G° = RT \ln \left[\frac{a_{M_2} a_{M_1^{n+}}}{a_{M_1} a_{M_2^{n+}}} \right] \qquad (2.52)$$

where the 'a' values are any arbitrary activity values, for which ΔG is the corresponding free energy of that reaction. $\Delta G°$ is the free energy change for the reaction involving standard states (unit activities).

As $\Delta G = -nFE$, and also $\Delta G° = -nFE°$, from equations (2.44) and (2.45), respectively, the above equation (2.52) becomes:

$$E - E° = \frac{RT}{nF} \ln \left[\frac{a_{M_1} a_{M_2^{n+}}}{a_{M_2} a_{M_1^{n+}}} \right] \qquad (2.53)$$

whence

$$E - E° - \frac{RT}{nF} \ln \left[\frac{a_{product}}{a_{reactants}} \right]$$

Suppose

$$M \rightarrow M^{z+} + ze$$

$$E_M = E°_M - \frac{RT}{nF} \ln \left[a_M^{z+} \times z a_e^z \right]$$

$$= E°_M - \frac{RT}{nF} \ln \left[a_M^{z+} \right]$$

because a_M and $a_e = 1$.

At 25°C

$$E_M = E°_M - \frac{0.059}{2} \log \left[a_M^{z+} \right]$$

Equation (2.53) is a general form of the Nernst equation. It expresses the emf involved in an electrochemical reaction, such as equation (2.46) in terms of R, T, F, n activities of products and activities of reactants. However, the IUPAC sign convention requires the above minus sign to be replaced by a plus sign. The Nernst equation is re-written to comply with the IUPAC sign convention, as:

$$E = E° + \frac{RT}{nF} \ln \frac{\text{oxidized state activities product}}{\text{reduced state activities product}}$$

which is written *regardless* of which way round the redox reaction is written, Hence, for $Zn = Zn^{++} + 2e^-$, and for $Zn^{++} + 2e^- = Zn$,

$$E = E° + \frac{RT}{nF} \ln a_{Zn^{++}}.$$

2.11.1 APPLICATION OF NERNST EQUATION TO A CORROSION REACTION

A corrosion reaction can be considered as composed of two half cell reactions. One of the half cell reactions corresponds to '*oxidation reaction*' taking place on the '*anode*,' and the other half cell reaction corresponds to '*reduction reaction*' taking place on the '*cathode*' of the cell. The contribution of each half cell reaction to the Nernst expression can be derived as follows:

$$M_1 \rightarrow M_1^{n+} + ne \quad (I) \qquad (2.54)$$

oxidation reaction at anode, and

$$M_2^{n+} + ne \rightarrow M_2 \quad (II) \qquad (2.55)$$

reduction reaction at cathode.

Equation (2.53) can also be re-arranged into expressions to correspond to reactions (I) and (II):

$$E = E_1 + E_2 \tag{2.56}$$

$$E = \left[E_1^\circ + \frac{RT}{nF} \ln\left(\frac{a_{M_1}}{a_{M_1}^{n+}} \right) \right]$$
$$\tag{2.57}$$
$$+ \left[E_2^\circ + \frac{RT}{nF} \ln\left(\frac{a_{M_2}^{n+}}{a_{M_2}} \right) \right]$$

where

$E^\circ =$ standard potential, and $E^\circ = E_1^\circ + E_2^\circ$

$$E_1 = E_1^\circ + \frac{RT}{nF} \ln \left[\frac{a_{M_1}}{a_{M_1}^{n+}} \right] \tag{2.58}$$

$$E_2 = E_2^\circ + \frac{RT}{nF} \ln \left[\frac{a_{M_2}^{n+}}{a_{M_2}} \right] \tag{2.59}$$

and

$$E = E^\circ + \frac{RT}{nF} \ln \left[\frac{a_{M_1} a_{M_2}^{n+}}{a_{M_1}^{n+} a_{M_2}} \right] \tag{2.60}$$

Equations (2.58) and (2.59) are Nernst equations for the half cell reactions and equation (2.60) is the Nernst equation for the complete cell reaction. Equation (2.60) can be put into a mathematical form:

$$E = E^\circ + \frac{RT}{nF} \ln \frac{\text{activities of reactants}}{\text{activities of products}} \tag{2.61}$$

But, in addition to this statement (2.61), a sign convention is necessary as already mentioned and this will be discussed in the next section. Since

$$\frac{RT}{nF} \ln = \frac{2.303 \times 8.314 \times (273 + 25)}{96\,500}$$
$$\tag{2.62}$$
$$= 0.05915 \text{ at } 25^\circ C$$

Equation (2.60) is more commonly used in the form of

$$E = E^\circ + \frac{0.05915}{n}$$
$$\times \log \frac{\text{product of activities of reactants}}{\text{product of activities of products}} \tag{2.63}$$

The expression (2.61) can also be written with the activities inverted, as

$$E = E^\circ - \frac{RT}{nF} \ln \frac{\text{product of activities of products}}{\text{product of activities of reactants}} \tag{2.64}$$

But this equation is incomplete without a sign convention.

Quite separate from the above considerations, standard Redox potentials are tabulated with reactive metals like Zn negative instead of positive, which accords with the experimental fact that Zn will *actually* become negative because when Zn^{++} ions leave it, then two electrons will appear in the Zn metal, to maintain a charge balance.

Such reduction potentials are given in Table 2.1. By convention the cell reaction is written as if the oxidation process is occurring at the electrode on the left and reduction on the right. This practice is in accordance with the International Union of Pure and Applied Chemistry (IUPAC) Convention held in 1953 and modified in 1960.

2.12 SIGN CONVENTION

The older American Convention has been to use oxidation potentials rather than reduction potentials. The present practice is to use reduction potentials instead of oxidation potentials in accordance with the recommendation of IUPAC (International Union of Pure and Applied Chemistry) and existing European practice – the general convention is that the cell emf is given by:

$$E_{cell}^\circ = E_{Right}^\circ - E_{Left}^\circ$$

The standard hydrogen electrode (SHE) is the reference point. The construction of hydrogen electrode is shown in Fig. 2.10. By definition, SHE has zero potential at 25°C. It is assumed that oxidation occurs at the hydrogen electrode so that it is placed at the left-hand side of a cell diagram. If a cell is composed of SHE and some other electrode, then the measured voltage of the cell is the electrode potential of the second electrode:

$$E_{cell} = E_x - 0$$

$$E_{cell} = E_x$$

For example, in a galvanic cell made of a hydrogen electrode and a zinc electrode $[Zn/Zn^{2+} (1.0\,M)]$, the measured voltage is $E = -0.763$ volts.

$$E_{cell}^\circ = E_{Zn}^\circ - E_{H_2}^\circ$$

$$E_{cell}^\circ = -0.763\,V$$

In the older textbooks, *hydrogen electrode was placed on the right, and the cell potential becomes:*

$$E_{cell} = E_{H_2} - E_x$$

The reaction was made oxidation and the sign was changed. Thus the electrode potential $E_{Zn/Zn^{++}}^\circ$ is $+0.763$ volts according to old convention.

$$E_{cell}^\circ = E_{H_2}^\circ - \left(E_{Zn}^\circ\right), \quad E_{Zn}^\circ = +0.763\,V$$

$$E_{cell}^\circ = E_{Zn}^\circ$$

$$E_{cell}^\circ = 0.763\,volts$$

which is opposite to that shown above. This should clarify the confusion caused in student's minds by different conventions. The simplest procedure avoids specifying '*right*' and '*left*' by just writing:

$$E = E^\circ + \frac{RT}{nF} \ln \frac{\text{oxidized state activities product}}{\text{reduced state activities product}}$$

which is written regardless of which way round the chemical equation is written.

2.12.1 CELL EMF AND THE DIRECTION OF SPONTANEOUS REACTION

The following is a step-by-step procedure to apply the above sign convention to a cell reaction in order to determine cell emf and direction of spontaneous reaction correctly.

(1) Write the right-hand electrode reaction. The right-hand electrode reaction is always *reduction* and the electron is always a reducing agent:

$$M_R^+ + e^- \rightarrow M_R$$

And reduction electrode potential is E_R°.

(2) Write the left-hand electrode reaction. The left-hand electrode reaction is *oxidation*. However, we write it as reduction reaction because of the convention we are using:

$$M_L^+ + e^- \rightarrow M_L$$

And the reduction electrode potential is E_L°.

(3) Subtract the left-hand electrode reaction and potential from those of right-hand electrode to obtain cell reaction and cell potential:

$$M_R^+ + M_L = M_R + M_L^+$$

and $E_{cell}^\circ = E_R^\circ - E_L^\circ$

For example, calculate E° for the Daniel cell. From the table of reduction potentials, write the corresponding reactions:

$$Cu^{2+} + 2e \rightarrow Cu \quad E^\circ = 0.337\,(R)$$

$$Zn^{2+} + 2e \rightarrow Zn \quad E^\circ = -0.763\,V(L)$$

$$E_{cell} = E_R^\circ - E_L^\circ$$

$$= 0.337 - (-0.763\,V)$$

$$= 1.100\,V$$

Another example: Using a cell of Pb^{2+}/Pb and Sn^{2+}/Sn, calculate E°_{cell}:

$$E^\circ_{cell} = E^\circ_R - E^\circ_L$$

$$L \rightarrow R$$

$$R \rightarrow L$$

$$= -0.126 - (-0.136)$$

$$= 0.10 \, V$$

$$M^+ + e \rightarrow M$$

$$Zn \rightarrow Zn^{++} + 2e$$

$$Cu^{++} + 2e \rightarrow Cu$$

2.12.2 POSITIVE AND NEGATIVE EMF

A positive emf $(E > 0)$ signifies a deficiency of electrons on the right-hand electrode because of the extraction of electrons from that electrode.

Therefore, the tendency of the right-hand electrode is reduction (acceptance of electron; $M_R E + e = M_R$, if $E > 0$). At the same time, positive emf signifies an excess of electrons on the left-hand electrode. Therefore, the electrode reaction is oxidation ($M_L = M_L^+ + e$) on that electrode. Thus, if $E > 0$, the cell reaction has a tendency to proceed from left to right: that is, electrons are produced by the left-hand electrode and consumed by the right-hand electrode. It also indicates that the reaction is spontaneous.

Conversely, if $E < 0$, the reaction at the right-hand electrode is oxidation and the reaction at the left-hand electrode reduction. The tendency of the reaction would, therefore, be to proceed from right to left and the reaction would not be spontaneous. Figure 2.11 explains the flow of electrons in the two situations explained above. The same convention can also be understood in terms of flow of positive current (positive charged ions). It is to be noted that according to the convention, if a cell is short circuited, and the positive current flows through the electrolyte from left to right, the emf is positive. In this case the left-hand electrode is the anode and the right-hand electrode

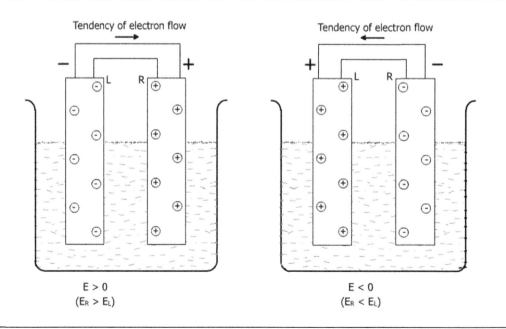

| E > 0 | E < 0 |
| ($E_R > E_L$) | ($E_R < E_L$) |

Figure 2.11 Implications of the sign convention

Table 2.2 Current flow conventions

emf	Current flow	Anode	Cathode
$+$	$L \to R$	L	R
$-$	$R \to L$	R	L

the cathode. If the current flows from right to left, the emf is negative. Table 2.2 summarizes the current flow convention in a cell.

From knowledge of cell potential it is thus possible to predict whether a reaction would proceed spontaneously or not. From knowledge of potential of half cells, and the cell potentials, it is possible to predict whether a metal is an anode or cathode or whether a corrosion reaction is likely to take place or not. An understanding of electrode potential is of fundamental importance to the understanding of corrosion mechanism. This is illustrated in typical problems given below.

Illustrative Problem 2.4
Two half cell reactions are given below:

$$Cu^{2+} + 2e \rightleftarrows Cu, \quad E° = 0.34 \, (V)$$

$$Zn^{2+} + 2e \rightleftarrows Zn, \quad E° = -0.763 \, (V)$$

Their reduction potentials are given opposite to each reaction. Calculate:

(a) the emf of the cell and
(b) show the spontaneous cell reaction.

Solution:
In Fig. 2.2, the left-hand electrode (Zn) is the anode and the right-hand electrode (Cu) is the cathode according to the convention we studied earlier on. Therefore, zinc will undergo oxidation and copper reduction as shown below. Hence:

(a) $E°_{cell} = E°_R - E°_L$ (according to IUPAC convention)
$E°_{cell} = E°_{Cu} - E°_{Zn}$
$\quad\quad = 0.337 - (-0.763)$

$E°_{cell} = +1.100 \, volts$
(b) Spontaneous cell reaction is given by:
cell reaction = cathodic reaction = anodic reaction

$$Cu^{2+} + Zn = Zn^{2+} + Cu$$

From the table of standard reduction potentials (Table 2.1), the reduction potential of Cu is $+0.34$ volts and that of Zn is -0.76 volts. The emf of the cell is positive. Hence the reduction is spontaneous and it should proceed from left to right.

If the position of the electrodes are now interchanged such that the copper electrode is placed on the left and the zinc electrode on the right, the sign of the cell emf will change. Here, zinc electrode being on the right is treated as cathode and the copper on the left as anode according to the convention discussed earlier in the chapter. Therefore:

$$Zn^{2+} + 2e \rightleftarrows Zn, \quad E°_{Zn} = -0.763 \, V$$

$$Cu^{2+} + 2e \rightleftarrows Cu, \quad E°_{Cu} = +0.34 \, V$$

Subtracting anodic reaction from cathodic reaction to obtain the cell reaction

$$Zn^{2+} + Cu \rightleftarrows Zn + Cu^{2+} \text{ and } E°_{cell} = -1.10 \, V$$

The emf obtained is now negative which indicates that the reaction is not spontaneous and the current flows from right to left. The left-hand electrode where the current originates is, therefore, the *anode*, and the right-hand electrode, the *cathode*. The polarity of the cell is, therefore, clearly established.

Illustrative Problem 2.5
Calculate the reversible potential for a zinc electrode in contact with $ZnCl_2$ when the activity of zinc is $a_{Zn^{2+}} = 10^{-3}$, Use IUPAC convention.

Solution:
We write the half cell reaction

$$Zn^{2+} + 2e \leftrightarrows Zn, \quad E°_{Zn} = -0.76 \, V$$

Apply Nernst equation:

$$E_{Zn} = E°_{Zn} + \frac{RT}{nF} \ln \left[a_{Zn^{2+}} \right]$$

$$E_{Zn} = -0.76 + \frac{0.059}{2} \log \left[a_{Zn^{2+}} \right]$$

$$E_{Zn} = -0.76 + 0.03 \log \left[10^{-3} \right]$$

$$E_{Zn} = -0.85 \, V$$

Illustrative Problem 2.6
Show that for the reduction reaction given below:

$$2H^+ + 2e \rightleftharpoons H_2, \quad E_H = -0.0591\,pH$$

Solution:
The reaction is $2H^+ + 2e \rightleftharpoons H_2$. Using Nernst equation we obtain for the above reaction:

$$E_{(H^+/H_2)} = E^\circ_{(H^+/H_2)} + \frac{RT}{nF} \ln \frac{[\text{reactants}]}{[\text{products}]}$$

or $\quad E_{(H^+/H_2)} = E^\circ_{(H^+/H_2)} + \frac{RT}{nF} \ln \frac{[\text{oxidized state}]}{[\text{reduced state}]}$

(IUPAC − Nernst)

Changing to 2.303 log and substituting for

$$R = 8.314\,J/(mol \cdot K), \quad F = 96\,490\,C,$$

$$T = 298^\circ K, \quad 2.303RT/F = 0.0591\,V.$$

The Nernst expression above can, therefore, be written as:

$$E_{(H^+/H_2)} = E^\circ_{(H^+/H_2)} + 0.059 \log[H^+]$$

Now substituting $pH = -\log(H^+)$ in the above expression we get:

$$E_{(H^+/H_2)} = -0.0591\,pH$$

Illustrative Problem 2.7
Show that for the reaction: $O_2 + 2H_2O + 4e \rightleftharpoons 4(OH^-)$,

$$E_{(O_2/OH^-)} = E^\circ_{(O_2/OH^-)} - 0.0591 \log a_{OH^-}$$

Solution:
The reaction is $O_2 + 2H_2O + 4e \rightleftharpoons 4(OH^-)$ (in a basic solution), and

$$E_{(O_2/OH^-)} = E^\circ_{(O_2/OH^-)} + \frac{RT}{nF} \ln \frac{a_{O_2}}{\left(a_{OH^-}\right)^4}$$

Note that O_2 is the oxidized state and OH^- is the reduced state, for the above reaction (from IUPAC–Nernst expression).

The activity (a_i) of an ion is defined by $a_i = M_1 \gamma_i$, where M_1 = molality (mol/kg) of the ions in the solution, and a_i is the activity coefficient. The activity coefficient a_i for a given concentration can be found from a table of activity coefficients given at the end of the chapter. In dilute solutions, the activity coefficient is taken to be unity.

If we take $a_{O_2} = P_{O_2}/P^\circ = 1$, when $P_{O_2} = P^\circ = 1$ bar, standard state, and $R = 8.314\,J/mol \cdot K$, $F = 96\,485\,C(g\ equiv.)^{-1}$, $T = 298^\circ K$, and the term $2.303RT/F = 0.0591$, and substitute in the above expression, the oxygen electrode potential becomes:

$$E_{(O_2/OH^-)} = E^\circ_{(O_2/OH^-)} - 0.0591 \log a_{(OH^-)}$$

It can similarly be shown that in *acid solution* for a reaction, $O_2 + 2H_2O + 4e \rightarrow 2H_2O$, the oxygen electrode potential becomes:

$$E_{(O_2)} = E^\circ_{(O_2)} + \frac{RT}{4nF} \ln \frac{a_{(O_2)}\,(a_{H^+})^4}{a_{H_2O}}$$

Substituting the values for the constants as above, we get:

$$E_{(O_2)} = E^\circ_{(O_2)} + 0.0592 \log a_{H^+}$$

or since $pH = -\log a_{H^+}$,

$$E_{(O_2)} = E^\circ_{(O_2)} - 0.0592\,pH$$

Illustrative Problem 2.8
Calculate the potential of oxygen electrode at $pH = 14.0$.

Solution:
Oxygen electrode reaction: in basic and neutral environment is expressed as:

$$O_2 + 2H_2O + 4e \rightarrow 2H_2O$$

Electrode potential equation for the electrode at 25°C is given by Illustrative Problem 2.7:

$$E_{O_2} = E^\circ_{O_2} - 0.0592 \log a_{(OH^-)}$$

where

$$\log a_{(OH^-)} = -14 - \log a_{(H^+)}$$

$$= -14 + pH$$

(from the knowledge of pH)

Therefore,

$$E_{O_2} = E_{O_2}^{\circ} - 0.0592(-14 + pH)$$

Substitute $E_{O_2}^{\circ} = 0.401$ (from Table 2.1).
Therefore,

$$E_{O_2} = 0.401 + 14 \times 0.0592 - 0.0592\,pH$$

$$E_{O_2} = 0.401 + 0.828 - 0.0592\,pH$$

and so at pH $= 14$,

$$E_{O_2} = 0.401 + 0.828 - 0.828$$

and $E_{O_2} = 0.401$

Illustrative Problem 2.9

In the cell reaction given below, what is the ratio of the activities of ionic species required to make the polarity reverse?

$$Fe^{2+} + Sn \rightarrow Sn^{2+} + Fe$$

Solution:
Cell reaction:

$$Fe^{2+} + Sn \rightarrow Sn^{2+} + Fe$$

Ratio of activities to make the polarity reverse at $25°C$.
Standard cell potential:

$$E_{cell}^{\circ} = E_{Fe^{2+}/Fe}^{\circ} - E_{Sn^{2+}/Sn}^{\circ}$$

$$= -0.441 - (-0.140)$$

$$= -0.441 + 0.140$$

$$= -0.301\,volts$$

The reaction in this case is non-spontaneous. That is, Fe^{2+}/Fe electrode will act as '*anode*' and Sn^{2+}/Sn electrode will act as '*cathode*,' the reverse case of what was assumed in the problem.
To reverse the polarity:
Overall cell potential is given by:

$$E_{cell} = E_{cell}^{\circ} + \frac{RT}{nf} \ln \frac{[a_{Reactant}]}{[a_{Product}]}$$

$$= -0.301 + \frac{2.303RT}{nF} \log \frac{a_{Fe^{2+}}\, a_{Sn}}{a_{Sn^{2+}}\, a_{Fe}}$$

where: $a_{Sn} = 1$ and $a_{Fe} = 1$.

$$E_{cell} = -0.301 + \frac{2.303RT}{nF} \log \frac{a_{Fe^{2+}}}{a_{Sn^{2+}}}$$

$$E_{cell} = 0, \quad \text{At the turn of polarity.}$$

Therefore,

$$0.301 = \frac{2.303RT}{nF} \log \frac{a_{Fe^{2+}}}{a_{Sn^{2+}}}$$

$$\log \left[\frac{a_{Fe^{2+}}}{a_{Sn^{2+}}} \right] = \frac{0.301 \times nF}{2.303RT}$$

Assuming the reaction is taking place at $t = 25°C$, $n = 2$, $T = 298°K$, $R = 8.315$ J/mol and $F = 96\,500$ C (g equiv.)$^{-1}$
Then

$$\log \left[\frac{a_{Fe^{2+}}}{a_{Sn^{2+}}} \right] = \frac{0.301 \times 2 \times 96\,500}{2.303 \times 8.314 \times 298}$$

$$= 6.4 \times 10^{-11} \text{ at } 25°C$$

At $20°C \cong 5 \times 10^{-11}$

Illustrative Problem 2.10

The emf of a cell made of Zn (anode) and H_2 electrode (cathode) immersed in 0.7 M $ZnCl_2$ is $+0.690$ volts. What is the pH of the solution?

Solution:
Given:

$$[Zn^{2+}] = 0.7\,M, \text{ For } Zn^{2+},$$

$$\text{activity coefficient, } \gamma = 0.6133$$

$$\therefore \quad a_{Zn^{2+}} = 0.7 \times 0.6133 = 0.424$$

$$(Right) \rightarrow 2H^+ + 2e^- \rightleftharpoons H_2$$

$$(Left) \rightarrow Zn^{2+} + 2e^- \rightleftharpoons Zn$$

$$E_{Zn}^{\circ} = -0.762\,V$$

$$E_{cell} = E_R - E_L$$

$$= E_{H_2}^{\circ} - \frac{0.0592}{2} \left[\log(pH_2) - \log \left(a_{H^+}^2 \right) \right]$$

$$- \left\{ E_{Zn}^{\circ} \frac{0.0592}{2} \left[\log[a_{Zn}] - \log[a_{Zn^{2+}}] \right] \right\}$$

$$\Rightarrow E_{cell} = -\frac{0.0592}{2}[2\,pH] + 0.762$$

$$+\frac{0.0592}{2}[-\log(0.42)]$$

$$0.690 = -0.0296[2\,pH] + 0.762$$

$$+0.0296(0.376)$$

$$\Rightarrow 0.0296[2\,pH] = 0.762 - 0.690 + 0.011 = 0.083$$

$$\Rightarrow 2\,pH = \frac{0.083}{0.0296} \Rightarrow 2\,pH = 2.8$$

$$\Rightarrow pH = 1.4$$

Illustrative Problem 2.11

Calculate the theoretical tendency of nickel to corrode in deaerated water of pH=8. Assume the corrosion products are H_2 and $Ni(OH)_2$ and the solubility product is 1.6×10^{-16}.

Solution:
Given: $K_{sp} = 1.6 \times 10^{-16}$, pH=8, $E^\circ_{Ni} = -0.25\,V$
Since, pH+pOH=14 $\Rightarrow$ pOH=14−8=6

$$[Ni^{2+}] = \frac{K_{sp}}{[OH^-]^2} = \frac{1.6 \times 10^{-16}}{10^{-12}} = 1.6 \times 10^{-4}$$

$$(R) \rightarrow 2H^+ + 2e^- \rightarrow H_2$$

$$(L) \rightarrow Ni^{2+} + 2e^- \rightarrow Ni$$

$$\Rightarrow E_{cell} = E^\circ_{H_2} - \frac{0.0592}{2}\left[\log(pH_2) - 2\log(a_{H^+})\right]$$

$$= -\left\{\left[E^\circ_{Ni} - \frac{0.0592}{2}\left[\log(a_{Ni})\right.\right.\right.$$

$$\left.\left.\left. - \log(a_{Ni^{2+}})\right]\right]\right\}$$

$$\Rightarrow E_{cell} = -0.0296[2\,pH] + 0.25$$

$$+0.0296\left[-\log(1.6 \times 10^{-4})\right]$$

$$\Rightarrow E_{cell} = -0.0296[2 \times 8] + 0.25$$

$$+0.0296[4 - \log(1.6)]$$

$$= -0.111\,volts$$

$$E_{cell} = -0.111\,volts$$

Illustrative Problem 2.12

Calculate the pressure of H_2 required to stop corrosion of iron immersed in 0.1 M $FeCl_2$, pH=4.

Solution:

$$(R) \rightarrow 2H^+ + 2e^- \rightarrow H_2$$

$$(L) \rightarrow Fe^{2+} + 2e^- \rightarrow Fe$$

$$E^\circ_{Fe} = +0.44$$

$$[a_{Fe^{2+}}] = \gamma \times [Fe^{2+}]$$

$$= 0.1 \times 0.75$$

$$E_{cell} = E_R - E_L$$

$$E_{cell} = E^\circ_{H_2} - 0.0296\left[\log(pH_2) - \log(a_{H^+})^2\right]$$

$$-\left[E^\circ_{Fe} - 0.0296\left(\log(a_{Fe}) - \log(a_{Fe^{2+}})\right)\right]$$

Given: pH=4 $\Rightarrow [a_{H_2}] = 10^{-4}$

$$\Rightarrow E = E^\circ - 0.0296\left[\frac{\log(pH_2)\,[a_{Fe^{2+}}]}{[H^+]^2}\right]$$

$$0 = 0.44 - 0.0296\left[\log(7.5 \times 10^6)\,pH_2\right]$$

$$= \frac{0.44}{0.0296} = \log[7.5 \times 10^6]\,pH_2$$

$$= (7.5 \times 10^6)\,pH_2 = 7.32 \times 10^{14}$$

$$pH_2 = 9.7 \times 10^7\,atm$$

Illustrative Problem 2.13

Calculate if silver would corrode when immersed in 0.5 M $CuCl_2$ to form solid AgCl. What is the corrosion tendency?

Solution:

$$E^\circ_{Cu} = 0.337\,V, \quad \gamma_{CuCl_2} = 0.7$$

$$E^\circ_{AgCl} = 0.220$$

$$\Rightarrow a_{Cu^{2+}} = 0.47 \times 0.5 = 0.233$$

$$= 0.47 \times 2 \times 0.5 = 0.47$$

$$(R) \rightarrow Cu^{2+} + 2e^- \rightarrow Cu$$

$$(L) \rightarrow 2AgCl + 2e^- \rightarrow 2Ag^+ + 2Cl^-$$

$$E_{cell} = E^{\circ}_{Cu} - E^{\circ}_{AgCl} - \frac{0.0592}{2}\big[[\log(a_{Cu})$$

$$- \log(a_{Cu^{2+}})] - [\log(Cl^-)^2$$

$$- \log(AgCl)]\big]$$

$$\Rightarrow E_{cell} = 0.337 - 0.220 - 0.0296\big[-\log(a_{Cu^{2+}})$$

$$- 2\log(a_{Cl^{2+}})\big]$$

$$= 0.117 - 0.0296[0.632 - 2(-0.327)]$$

$$= 0.079\,\text{volts}$$

$$E_{cell} = 0.079\,\text{volts}$$

2.13 REFERENCE ELECTRODES

2.13.1 HYDROGEN ELECTRODE

The hydrogen electrode is used as a reference for electrode potential measurements. Theoretically, it is the most important electrode for use in aqueous solutions. The reversible hydrogen electrode in a solution of hydrogen ions at unit activity exhibits a potential, which is assumed to be zero at all temperatures.

The electrode consists of a platinum wire immersed in a solution (Fig. 2.10) containing hydrogen ions and saturated with hydrogen gas. Platinum is immersed completely in aqueous-arsenic free hydrochloric acid, and hydrogen gas free from oxygen and carbon monoxide is bubbled to the platinum surface. Slowly, the air is displaced by hydrogen and the reversible potential is achieved. Unfortunately, this electrode has some drawbacks. First, the reversibility of hydrogen electrode cannot be maintained in oxidizing media. Second, if a current is withdrawn from the electrode, the electrode acts as an anode because of the ionization of gas molecules. Also, the electrode is fragile and delicate to handle.

The electrode potential for hydrogen E_{H_2} can be determined as follows:

$$2H^+ + 2e \rightleftarrows H_2 \tag{2.65}$$

$$E_{H^+/H_2} = E^{\circ}_{H^+/H_2} + \frac{RT}{nF} \log \frac{(a_{H^+})^2}{(p_{H_2})} \tag{2.66}$$

where a_{H^+} is activity of hydrogen ions, and p_{H_2} is hydrogen partial pressure.

At one atmosphere pressure (p_{H_2}), $a_{H_2} = 1$, and $E^{\circ}_{H^+/H_2} = 0$ by definition. Therefore,

$$E_{H^+/H_2} = 0.059 \log (a_{H^+}) \tag{2.67}$$

or in terms of pH,

$$E_{H^+/H_2} = -0.059\,\text{pH} \tag{2.68}$$

2.13.2 SILVER–SILVER CHLORIDE ELECTRODE

This electrode is composed of a silver wire coated with silver chloride and immersed in a solution of chloride ions (Fig. 2.12).

The chloride equilibrium is given by:

$$AgCl \rightleftarrows Ag^+ + Cl^- \tag{2.69}$$

Two other reactions involve a dynamic equilibrium between deposition and dissolution of silver together with solubility equilibrium between silver chloride and its ions.

The metallic silver reaches equilibrium with silver ions according to the following reaction:

$$Ag^+ + e \rightleftarrows Ag \tag{2.70}$$

The overall electrode reaction is, therefore, given by:

$$AgCl + e \rightleftarrows Ag + Cl^- \quad (E^{\circ}_{AgCl/Ag} = 0.024\,\text{V}) \tag{2.71}$$

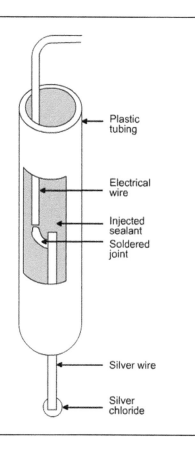

Figure 2.12 Silver–Silver chloride reference electrode

The electrode potential, $E_{Ag/AgCl}$, is given by:

$$E_{Ag/Cl} = E^\circ_{Ag/AgCl} + \frac{RT}{nF} \ln \frac{a_{AgCl}}{a_{Ag} \times a_{Cl^-}} \quad (2.72)$$

$$a_{Ag} = 1$$
$$a_{AgCl} = 1$$

Therefore,

$$E_{Ag/Cl} = E^\circ_{Ag/AgCl} - \left(\frac{2.303RT}{F}\right) \log a_{Cl^-} \quad (2.73)$$

or

$$E_{Ag/Cl} = E^\circ_{Ag/AgCl} - 0.0592 \log a_{Cl^-} \quad (2.74)$$

The equation holds at 25°C. It can also be written in the following form:

$$E_{Ag/AgCl} = 0.224 - 0.0592 \log a_{Cl^-} \quad (2.75)$$

At low concentration $\log a_{Cl^-}$ can be replaced by pH as Cl^- is provided by HCl acid, $[Cl^-] = [H^+]$ and hence $-\log a_{Cl^-}$ can be replaced by $-\log[H^+]$. Therefore,

$$-\log[H^+] = pH$$

Hence,

$$E_{Ag/AgCl} = 0.224 - 0.0592 \, pH \quad (2.76)$$

The following are the values of $E_{Ag/AgCl}$ for different HCl concentrations:

Concentration (M)	Electrode potential (volts)
0.1	0.28
0.01	0.34
0.001	0.40

2.13.3 THE CALOMEL ELECTRODE

It is the most commonly used reference electrode. It has a constant and reproducible potential. The electrode basically consists of a platinum wire dipped into pure mercury which rests in a paste of mercurous chloride and mercury. The paste is in contact with a solution of potassium chloride which acts as a salt bridge to the other half of the cell (Fig. 2.13).

The most commonly used concentrations of KCl are 0.1 N, 1.0 N and 3.5 N and saturated KCl. The saturated calomel is used when the liquid junction potential is to be kept low. The potential of electrode at 25°C is 0.241 V in saturated KCl solution.

Mercurous chloride is slightly soluble, and it is in equilibrium with mercurous ions according to:

$$Hg^+ + e = Hg$$

The overall equilibrium is expressed by:

$$Hg_2Cl_2 + 2e \rightleftharpoons 2Hg + 2Cl^-$$

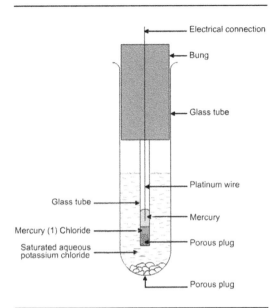

Electrical connection

Bung

Glass tube

Platinum wire

Glass tube

Mercury

Mercury (1) Chloride

Porous plug

Saturated aqueous
potassium chloride

Porous plug

Figure 2.13 A saturated calomel reference electrode

The mercurous chloride and mercury are at unit activity. Therefore, the electrode potential can be written as:

$$E_{Calomel} = E° + \frac{RT}{2F}\ln\left(\frac{1}{a_{Cl^-}}\right)$$

or $\qquad E_{Calomel} = E° - \dfrac{0.0592}{2}\log a_{Cl^-}$

The value of $E°$ for the half cell reaction of calomel electrode is 0.267 V. Thus, the electrode potential becomes:

$$E_{Calomel} = 0.267 - \frac{0.0592}{2}\log\left(a_{Cl^-}\right)$$

Electrode Potentials of Calomel Electrode

(Standard hydrogen electrode taken as reference)

Electrode	Potential (volts)
Hg/Hg$_2$Cl$_2$/KCl (Sat.)	0.2444
Hg/Hg$_2$Cl$_2$/KCl (1.0 N)	0.289
Hg/Hg$_2$Cl$_2$/KCl (0.1 N)/Salt Bridge	0.3356

2.13.4 COPPER–COPPER SULFATE ELECTRODE

This is a reference electrode which is easy, robust and stable. It is used mainly in cathodic protection measurements, such as the measurement of pipe-to-soil potential. It has a lower accuracy than other electrodes used for laboratory work. It consists of copper metal placed in a solution containing copper sulfate and copper sulfate crystals placed in a non-conducting holder with a porous plug (Fig. 2.14). The copper sulfate crystals maintain the solution at a fixed ion concentration. Necessary contact with the earth is made through the porous plug. It is easily recharged when it becomes contaminated.

The equation for the copper–copper sulfate electrode potential is given by:

$$E°_{Cu-CuSO_4} = 0.316 + 0.0009(25°C)\,volts \quad (2.77)$$

The reaction of the Cu–CuSO$_4$ half cell is

$$Cu^{2+} + 2e \rightleftarrows Cu \quad (2.78)$$

and the electrode potential:

$$E_{Cu^{2+}/Cu} = E_{Cu^{2+}/Cu} + \frac{0.0592}{2}\log a_{(Cu^{2+})} \quad (2.79)$$

where

$a_{(Cu^{2+})}$ = activity of copper which is unity,
$E°_{Cu}$ = 0.34 V at 25°C.

A saturated solution of 1.47 M CuSO$_4$ at 25°C is used.

$a_{Cu^{2+}}$ = [Molarity (M) × Activity coefficient γ]
$a_{Cu^{2+}}$ = 1.47 × 0.037 (γ is found from the table of activity coefficients – Table 2.3)

So, $a_{Cu^{2+}}$ = 0.051

Substituting in the above equation we obtain

$$E_{Cu^{2+}/Cu} = 0.34 + \frac{0.0592}{2}\log(0.051) \quad (2.80)$$

$$E_{Cu^{2+}/Cu} = 0.30 \text{ volts} \quad (2.81)$$

$E_{Cu^{2+}/Cu} = 0.316 + 0.009$ (T°C) volts, as seen earlier in this section (equation (2.77)).

Table 2.3 Activity coefficients of strong electrolytes (M = molality)

M	0.001	0.002	0.005	0.01	0.02	0.05	0.1	0.2	0.5	1.0	2.0	3.0	4.0
HCl	0.966	0.952	0.928	0.904	0.875	0.830	0.796	0.767	0.758	0.809	1.01	1.32	1.76
HNO$_3$	0.965	0.951	0.927	0.902	0.871	0.823	0.785	0.748	0.715	0.720	1.17	0.876	0.982
H$_2$SO$_4$	0.830	0.757	0.639	0.544	0.453	0.340	0.265	0.209	0.154	0.130	0.124	0.141	0.171
NaOH	–	–	–	–	–	0.82	–	0.73	0.69	0.68	0.70	0.77	0.89
KOH	–	–	0.92	0.90	0.86	0.82	0.80	–	0.73	0.76	0.89	1.08	1.35
AgNO$_3$	–	–	0.92	0.90	0.86	0.79	0.72	0.64	0.51	0.40	0.28	–	–
Al(NO$_3$)$_3$	–	–	–	–	–	–	0.20	0.16	0.14	0.19	0.45	1.0	1.2
BaCl$_2$	0.88	–	0.77	0.72	–	0.56	0.49	0.44	0.39	0.39	0.44	–	–
CaCl$_2$	0.89	0.85	0.785	0.725	0.66	0.57	0.515	0.48	0.52	0.71	–	–	–
Ca(NO$_3$)$_3$	0.88	0.84	0.77	0.71	0.64	0.54	0.48	0.42	0.38	0.35	0.35	0.37	0.42
CdCl$_2$	0.76	0.68	0.57	0.47	0.38	0.28	0.21	0.15	0.09	0.06	–	–	–
CdSO$_4$	0.73	0.64	0.50	0.40	0.31	0.21	0.17	0.11	0.067	0.045	0.035	0.036	–
CsCl	–	–	0.92	0.90	0.86	0.79	0.75	0.69	0.60	0.54	0.49	0.48	0.47
CuCl$_2$	0.89	0.85	0.78	0.72	0.66	0.58	0.52	0.47	0.42	0.43	0.51	0.59	–
CuSO$_4$	0.74	–	0.53	0.41	0.31	0.21	0.16	0.11	0.068	0.47	–	–	–
FeCl$_2$	0.89	0.86	0.80	0.75	0.70	0.62	0.58	0.55	0.59	0.67	–	–	–
In$_2$(SO$_4$)$_3$	–	–	–	0.142	0.092	0.054	0.035	0.022	–	–	–	–	–
KF	–	0.96	0.95	0.93	0.92	0.88	0.85	0.81	0.74	0.71	0.70	–	–
KCl	0.965	0.952	0.927	0.901	–	0.815	0.769	0.719	0.651	0.606	0.576	0.571	0.579
KBr	0.965	0.952	0.927	0.903	0.872	0.822	0.777	0.728	0.665	0.625	0.602	0.603	0.622
KI	0.965	0.951	0.927	0.905	0.88	0.84	0.80	0.76	0.71	0.68	0.69	0.72	0.75
K$_4$Fe(CN)$_6$	–	–	–	–	–	0.19	0.14	0.11	0.067	–	–	–	–
K$_2$SO$_4$	0.89	–	0.78	0.71	0.64	0.52	0.43	0.36	–	–	–	–	–
LiCl	0.963	0.948	0.921	0.89	0.86	0.82	0.78	0.75	0.73	0.76	0.91	1.18	1.46
LiBr	0.966	0.954	0.932	0.909	0.882	0.842	0.810	0.784	0.783	0.848	1.06	1.35	–
LiI	–	–	–	–	–	–	0.81	0.80	0.81	0.89	1.19	1.70	–

LiNO$_3$	0.966	0.953	0.930	0.904	0.878	0.834	0.798	0.765	0.743	0.76	0.84	0.9	—
MgCl$_2$	—	—	—	—	—	—	0.56	0.53	0.52	0.62	1.05	2.1	—
MgSO$_4$	—	—	—	0.40	0.32	0.22	0.18	0.13	0.088	0.064	0.055	0.064	—
NiSO$_4$	—	—	—	—	—	—	0.18	0.13	0.075	0.051	0.041	—	—
NH$_4$Cl	0.961	0.944	0.911	0.88	0.84	0.79	0.74	0.69	0.62	0.57	—	—	—
NH$_4$Br	0.964	0.949	0.901	0.87	0.83	0.78	0.73	0.68	0.62	0.57	—	—	—
NH$_4$NO$_3$	0.959	0.942	0.912	0.88	0.84	0.78	0.73	0.66	0.56	0.47	—	—	—
(NH$_4$)$_2$SO$_4$	0.874	0.821	0.726	0.67	0.59	0.48	0.40	0.32	0.22	0.16	—	—	—
NaF	—	—	0.93	0.90	0.87	0.81	0.75	0.69	0.62	—	—	—	—
NaCl	0.966	0.953	0.929	0.904	0.875	0.823	0.780	0.730	0.68	0.66	0.67	0.71	0.78
NaBr	0.966	0.955	0.934	0.914	0.887	0.844	0.800	0.740	0.695	0.686	0.734	0.826	0.934
NaI	0.97	0.96	0.94	0.91	0.89	0.86	0.83	0.81	0.78	0.80	0.95	—	—
NaNO$_3$	0.966	0.953	0.93	0.90	0.87	0.82	0.77	0.70	0.62	0.55	0.48	0.44	0.41
Na$_2$SO$_4$	0.887	0.847	0.778	0.714	0.641	0.53	0.45	0.36	0.27	0.20	—	—	—
NaClO$_4$	0.97	0.95	0.93	0.90	0.87	0.82	0.77	0.72	0.64	0.58	—	—	—
PbCl$_2$	0.86	0.80	0.70	0.61	0.50	—	—	—	—	—	—	—	—
Pb(NO$_3$)$_2$	0.88	0.84	0.76	0.69	0.60	0.46	0.37	0.27	0.17	0.11	—	—	—
RbCl	—	—	0.93	0.90	—	—	0.76	0.71	0.63	0.58	0.54	0.54	0.54
RbAc	—	—	—	—	—	—	0.73	0.65	0.52	0.42	—	—	—
TlNO$_3$	—	—	—	—	—	0.77	0.70	0.60	—	—	—	—	—
TlClO$_4$	—	—	—	—	—	0.79	0.73	0.65	0.53	—	—	—	—
TlAc	—	—	—	—	—	0.80	0.74	0.68	0.59	0.51	0.44	0.40	0.38
ZnCl$_2$	0.88	0.84	0.77	0.71	0.64	0.56	0.50	0.45	0.38	0.33	0.44	0.40	—
ZnSO$_4$	0.70	0.61	0.48	0.39	—	—	0.15	0.11	0.065	0.045	0.036	0.04	—

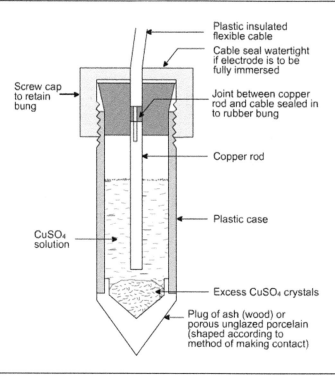

Figure 2.14 Reference copper–copper sulfate electrode

The following is the conversion table for important reference electrodes (Table 2.4):

Table 2.4 Conversion table for important reference electrodes

Potential *vs*		
Hydrogen (mV)	Copper–copper sulfate (mV)	Silver chloride (mV)
−400	−84	−170
−425	−109	−203
−450	−134	−228
−470	−159	−253
−500	−184	−278
−525	−209	−303
−550	−234	−320
−575	−259	−353
−600	−284	−370
−625	−309	−403
−650	−334	−428

Conversion of $Cu/CuSO_4$ electrode potential into $Ag/AgCl$ and hydrogen electrode potentials is expressed by the following equations:

(a) emf (*vs* $Cu/CuSO_4$) = emf (*vs* $Ag/AgCl$) − 50 mV
(b) emf (*vs* $Cu/CuSO_4$) = emf (*vs* H_2/H^+) − 316 mV

The conversion table for reference electrodes is shown in Table 2.4.

Illustrative Problem 2.14
Convert −0.900 V (on SCE, Sat.) to the SHE scale.

Solution:

$$-0.900\,V \text{ (on SCE, Sat.)} = -0.900 + 0.242\,V$$
$$\text{(on SHE)}$$

$$= -0.658\,V \text{ (on SHE)}$$

Illustrative Problem 2.15
Convert −0.916 V (on SCE. 1.0 N) to the SHE scale.

Solution:

$$-0.916\,V \text{ (on SCE, 1.0 N)} = -0.916 + 0.280\,V$$

$$\text{(on SHE)}$$

$$= -0.636\,V \text{ (on SHE)}$$

Illustrative Problem 2.16
What is the potential on SHE scale, for an electrode which is at a potential of $-0.920\,V$ relative to a Ag/AgCl reference in 0.01 N KCl at 25°C.

Solution:
Cell reaction: $AgCl + e \rightarrow Ag + Cl^-$ from equation (2.63)

Cell potential: $E_{Ag^+/Ag} = 0.22 + 0.059 \log\left[\dfrac{1}{a_{Cl^-}}\right]$

or

$$E_{Ag^+/Ag} = -0.222 - 0.059 \log\left[a_{Cl^-}\right]$$

$$a_{Cl^-} = [\gamma_\pm]\left[Cl^-\right] = (0.901)(0.01)$$

where

$\gamma_\pm$	=	activity coefficient for chloride ions
$\gamma_\pm$	=	0.901
Cl^-	=	concentration of chloride ions
Cl^-	=	0.01 N
a_{Cl^-}	=	9×10^{-3}

$$E_{Ag^+/Ag} = -0.222 - 0.059 \log\left[9 \times 10^{-3}\right]$$

$$E_{Ag^+/Ag} = 0.343\,V \text{ with respect to SHE.}$$

Therefore,

$$-0.920\,V \text{ (on Ag/AgCl)} = (-0.920 + 0.343)\,V$$

$$= -0.577\,V$$

$$\text{(with respect to SHE)}$$

Illustrative Problem 2.17
The following reduction reactions are given:

(a) $O_2 + 2H_2O + 4e = 4OH^-$
(b) $O_2 + 4H^+ + 4e = 2H_2O$

Show that the single electrode potential for each reaction at 25°C has the same potential dependence.

Solution:
For reaction (a):

$$E_{O_2/OH^-} = E^\circ_{O_2/OH^-} + \frac{0.0592}{4}\left[\log P_{O_2}\right.$$

$$\left. - \log a^4_{OH^-}\right]$$

$$E_{O_2/OH^-} = 0.401 + \frac{0.0592}{4}\log P_{O_2}$$

$$- 0.0592 \log a_{OH^-}$$

Replacing $\log a_{OH^-} = pH - 14$ in the above equation:

$$E_{O_2/OH^-} = 0.401 - 0.0592[pH - 14]$$

$$+ 0.0592 \log P_{O_2}$$

$$E_{O_2/OH^-} = 1.23 - 0.0592\,pH + \frac{0.059}{4}\log P_{O_2}$$

For reaction (b):

$$E_{O_2/H_2O} = E^\circ_{O_2/H_2O} + \frac{0.0592}{4}\left[\log P_{O_2} + \log a^4_{H^+}\right]$$

$$E_{O_2/H_2O} = 1.23 + 0.0592 \log a_{H^+} + \frac{0.0592}{4}\log P_{O_2}$$

$$E_{O_2/H_2O} = 1.23 - 0.0592\,pH + \frac{0.0592}{4}\log P_{O_2}$$

2.13.5 EMF AND GALVANIC SERIES

Table 2.1 gives the standard electrode potentials of metals with reference to standard hydrogen electrode (SHE) which is arbitrarily defined as zero. Potentials between metals are determined by taking the absolute difference between their standard potentials. The determination of standard electrode potential is shown Fig. 2.15.

As shown above, the electrode potential of two different metals in an electrode can be compared. Each metal in contact with an electrolyte of its ion forms a half cell. The most practical method of obtaining reliable and consistent value of relative electrode potential is to compare the value of each half cell with a common reference electrode.

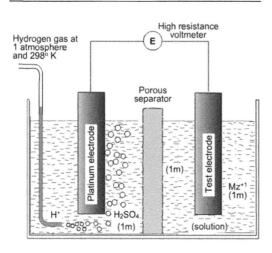

Figure 2.15   Method of determination of standard electrode potential

This common electrode is the standard hydrogen electrode which consists of a platinum wire platinized by electrolysis surrounded by a solution having a H^+ ion activity of $(a_{H^+} = 1)$ of one and enveloped in a hydrogen gas at 1 atmosphere pressure. The potential of the hydrogen electrode is given by:

$$Pt(s)|H_2(1\,atm)|H^+(a_{H^+} = 1), E = 0\,V$$

The electrode potential of all metals is compared with the standard hydrogen electrode and it is called the standard electrode potential ($E°$). Between two metals, such as zinc and aluminum, aluminum is more active than zinc $[E°_{Al} = -1.66\,V, E°_{Zn} = -0.763\,V]$. A metal with a more negative potential has a higher tendency to corrode (dissolve) than a metal with a less negative potential, although kinetic factors may intervene.

If the potential of a metal is less than hydrogen potential, reduction rather than oxidation takes place (electrons are gained), $M^+ + e \rightarrow M$. Metals which correspond to relatively lower standard potentials $E°$ are called *active metals* and metals which corresponds to relatively higher standard potential or less negative potentials are called *noble metals*.

Cu, Ag, Au are examples of noble metals, whereas K, Li and Mg are between iron

$(E°_{Fe} = -0.443\,V)$ and zinc $(E°_{Zn} = -0.763\,V)$; zinc is more active than iron. Between copper $(E°_{Cu} = +0.334\,V)$ and silver $(+0.80\,V)$, Ag is more noble than copper. If the electrode reactions occurring on different metals at room temperature are arranged in accordance with the value of standard potentials, an electrochemical series at $25°C$ is obtained (Table 2.1).

2.13.6 SUMMARY OF CHARACTERISTICS OF EMF SERIES (TABLE 2.1)

(1) Metals with large positive potentials are called '*noble*' metals because they do not dissolve easily. Examples are copper, silver, gold, etc. The potential of a noble metal is preceded by a positive (+) sign.

(2) The electrode potentials are thermodynamic quantities and have little relevance to potential of metals in solution encountered in service.

(3) The emf series lists only the electrode potentials of metals and not alloys. Alloys are not considered in the emf series.

(4) The emf series is based on pure metals. The more active metals, such as Na, Mg, Al, Zn are called '*active metals*.'

(5) Alloys are not considered in the emf series.

(6) From the reversible electrode potential in the standard emf series of metals, it is possible to predict whether a particular metal will spontaneously dissolve.

(7) It gives an indication of how active the metal is but does not necessarily predict corrosion accurately for reason to be explained later in the next chapter.

2.13.7 APPLICATION OF EMF SERIES

Following are useful applications of emf series:

(1) A less electropositive metal would displace a more electropositive metal from one of its salts in aqueous solution. For instance, if a rod of zinc is placed in a solution of copper

sulfate, zinc would dissolve in the solution and copper would be discharged:

$$Zn + CuSO_4 \rightarrow ZnSO_4 + Cu$$
or $\quad Zn \rightarrow Zn^{++} + 2e$ (oxidation)
$\quad\quad Cu^{++} + 2e \rightarrow Cu$ (reduction)

Consequently, copper will deposit on the zinc rod.

(2) Electrode potentials indicate also the tendency of cations in aqueous solutions to be reduced at a cathode. For example, silver ions are reduced more readily than cupric ions, because silver is more electropositive.

(3) Metal ions above hydrogen are more readily reduced than the hydrogen ions with 100% efficiency.

(4) The metals in the series with high positive potentials are recognized as metals with good corrosion resistance. They show a little tendency to pass from a metallic state to an ionic state. Conversely, metals with high negative potentials show a tendency to corrode, but whether they corrode or not would depend on other factors also. For instance, iron has a potential of -0.440 V and indicates tendency to corrode, but if it develops a film of oxide it would not remain active, and hence, it would not corrode. This effect is used in formulating stainless steels, which are covered by an invisible oxide film.

(5) The metal with a more negative potential is generally the anode, and the metal with a less negative potential, the cathode. If zinc and aluminum are coupled, aluminum would become the anode ($E_{Al}^{\circ} = -1.67V$) and zinc ($E_{Zn}^{\circ} = -0.763V$) the cathode. It may be expected that in the presence of metals which are more negative than hydrogen in the emf series, hydrogen reduction would be the preferred process. That is, however, not always the case. Metals as negative as zinc ($E_{Zn}^{\circ} = -0.763$ V) can be plated from an acid solution without liberation of hydrogen. In the case of aluminum, however, hydrogen evolution would be the preferred process, and, therefore, aluminum is deposited by electrolysis of a non-aqueous melt in order not to give any chance for the liberation of hydrogen which may enter the metal and cause its embrittlement. The emf series is also useful for the electrolytic refining of the metals.

2.13.8 LIMITATIONS OF EMF SERIES

The following are the obvious limitations of the EMF series:

(1) The emf series lists only pure metals which have only a limited use in engineering applications. Alloys are of major interest to engineers rather than pure metals.

(2) The electrode potential has little relevance to potentials of metals in solutions, in which the potential of interest is the corrosion potential and not the electrode potential of the metal.

(3) The position of metals in the emf series is determined by the equilibrium potential of the metal with the concentration of ions at unit activity. Prediction about galvanic coupling can only be made when the two metals forming the galvanic couple have their ionic activities at unity. If the above conditions are not met, accurate prediction of galvanic conversion by emf series would not be possible. The activity of ions in equilibrium with a given metal vary with environment and, therefore, accurate predictions of galvanic corrosion are not generally possible.

(4) The emf series predicts the tendency to corrode but it cannot predict whether corrosion would actually take place. For instance, on the basis of some negative potential, iron shows tendency to corrode, however, if it develops a passive film in some environment it would not corrode.

(5) The emf series cannot always predict the effect of environment. For instance, in food cans, the polarity of tin may be critical. The potential of iron ($E_{Fe}^{\circ} = -0.44$) is more negative than the potential of tin ($E_{Sn}^{\circ} = -0.1369$). However, in the presence of certain type of foods in food cans, tin can become active to iron. Such a change cannot be predicted by emf series.

The effect of film formation on the tendency of the metal to passivate solution cannot be predicted by emf series. For instance, titanium and aluminum are more negative than iron. However, in certain environments they form a film which makes their potential less active than iron. The effect of film formation on the tendency of the metals to corrode is kinetic and cannot be predicted by the thermodynamic emf series.

2.13.9 IR DROP (OHMIC DROP)

The *IR* drop is the ohmic voltage that results from the electric current flow in ionic electrolytes, such as dilute acids, salt water, etc. When the reference electrode for measurement of potential is placed in the electrolyte, an electrolytic resistance exists along the line between the test and the reference electrode. Because the current flows through an electrolyte resistance an ohmic voltage drop is also automatically included in the measurement of potential.

$$E = E_{\text{test}} - E_{\text{reference}}$$

IR drop (ohmic drop) is an unwanted quality which must be eliminated to obtain an accurate potential measurement. By Ohm's Law, $E = IR$.

The IR drop in a flash light battery, for instance, contribute to the internal resistance. For any measurement, the reference electrode is placed closed to the working electrode, if this is not done the measurement would be inaccurate as it would include an IR drop. This is because the potential which is measured will always include the potential difference across the electrolyte which is between the reference and working electrode. As E is measured as IR, the potential is called IR drop. The IR contribution must be subtracted to obtain the current emf of the cell.

2.13.9.1 Methods to Remove IR Drop

a. Extrapolation Method

The emf is measured with the reference electrode placed as close to the working electrode as

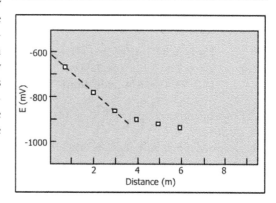

Figure 2.16   Extrapolation method

possible in the electrolyte. The electrode is moved at definite intervals of distance and potential (E) is measured. Emf is plotted as a function of distance and the curve is extrapolated to zero distance: Fig. 2.16. This method is applicable in electrolytes of high resistance, such as soils.

b. Current Interruption Method

When there is no current flow there cannot be any IR drop. However, when a current is flowing IR drop is included in the measurement. At a certain time, t, the current is interrupted so that $I = 0$, hence IR is also zero. The test electrode, therefore, shows a potential free from IR drop. Hence current interruption provides a good method for measurement of IR free cell potential. Unfortunately, depolarization may also occur when the current is interrupted. If the system is depolarized it is too negative at the anode area and too positive at the cathode. Various types of commercial electronic interrupters are available. The current can be interrupted by as much as 10 milliseconds of each half second. The potential must be read quickly, at the instant of interruption, by a fast response voltmeter such as an oscilloscope.

2.13.10 GALVANIC SERIES

Having examined the limitations of the emf series, the question arises, is there a better way to predict corrosion of metals and alloys than predicted by emf series. Fortunately, the limitations imposed by emf series are overcome by another series called '*galvanic series*' which will be now discussed (Table 2.5).

Table 2.5 Galvanic series in seawater at room temperature

1.	Magnesium
2.	Mg alloy AZ-31B
3.	Mg alloy HK-31A
4.	Mg alloy (hot-dip, die cast or plated)
5.	Beryllium (hot pressed)
6.	Al 7072 clad on 7075
7.	Al 2014-T3
8.	Al 1160-H14
9.	Al 7079-T6
10.	Cadmium (plated)
11.	Uranium
12.	Al 218 (die cast)
13.	Al 5052-0
14.	Al 5052-H12
15.	Al 5456-0, H353
16.	Al 5052-H32
17.	Al 1100-0
18.	Al 3003-H25
19.	Al 6061-T6
20.	Al A360 (die cast)
21.	Al 7075-T6
22.	Al 6061-0
23.	Indium
24.	Al 2014-0
25.	Al 2024-T4
26.	Al 5052-H16
27.	Tin (plated)
28.	Stainless steel 430 (active)
29.	Lead
30.	Steel 1010
31.	Iron (cast)
32.	Stainless steel 410 (active)
33.	Copper (plated, cast or wrought)
34.	Nickel (plated)
35.	Chromium (plated)
36.	Tantalum
37.	AM350 (active)
38.	Stainless steel 310 (active)
39.	Stainless steel 301 (active)
40.	Stainless steel 304 (active)
41.	Stainless steel 430 (active)
42.	Stainless steel 410 (active)
43.	Stainless steel 17-7PH (active)
44.	Tungsten
45.	Niobium (columbium) 1% Zr
46.	Brass, yellow, 268
47.	Uranium 8% Mo
48.	Brass, naval, 464
49.	Yellow brass
50.	Muntz metal 280
51.	Brass (plated)
52.	Nickel–silver (18% Ni)
53.	Stainless steel 316L (active)
54.	Bronze 220
55.	Copper 110
56.	Red brass
57.	Stainless steel 347 (active)
58.	Molybdenum, commercial pure
59.	Copper–nickel 715
60.	Admiralty brass
61.	Stainless steel 202 (active)
62.	Bronze, phosphor 534 (B-1)
63.	Monel 400
64.	Stainless steel 201 (active)
65.	Carpenter 20 (active)
66.	Stainless steel 321 (active)
67.	Stainless steel 316 (active)
68.	Stainless steel 309 (active)
69.	Stainless steel 17–7PH (passive)
70.	Silicone bronze 655
71.	Stainless steel 304 (passive)
72.	Stainless steel 301 (passive)
73.	Stainless steel 321 (passive)
74.	Stainless steel 201 (passive)
75.	Stainless steel 286 (passive)
76.	Stainless steel 316L (passive)
77.	AM355 (active)
78.	Stainless steel 202 (passive)
79.	Carpenter 20 (passive)
80.	AM355 (passive)
81.	A286 (passive)
82.	Titanium 5A1, 2.5 Sn
83.	Titanium 13 V, 11 Cr, 3 Al (annealed)
84.	Titanium 6 Al, 4 V (solution treated and aged)

(Contd)

Table 2.5   (*Contd*)

85.   Titanium 6 Al, 4 V (annealed)
86.   Titanium 8 Mn
87.   Titanium 13 V, 11 Cr, 3 Al (solution heat treated and aged)
88.   Titanium 75 A
89.   AM350 (passive)
90.   Silver
91.   Gold
92.   Graphite

The galvanic series is an arrangement of metals and alloys in order of their corrosion potentials in the environment. The potentials of the metals and alloys are measured in the desired environments, with the most noble (positive) at the top and the most active at the bottom. Table 2.5 shows a galvanic series of some commercial metals and alloys in seawater. The potentials are measured in seawater by means of a saturated calomel electrode and all potentials are expressed with reference to this electrode (Fig. 2.17).

2.13.11 CHARACTERISTICS OF GALVANIC SERIES (TABLE 2.5)

(1)   In the galvanic series, instead of potentials the relative positions of the metals and alloys are indicated.
(2)   The series is based on practical measurement of corrosion potential at equilibrium. The potential of a corroding metal in a given medium can be obtained by connecting the metal or alloy to the negative terminal of a voltmeter and the positive terminal to a reference electrode, generally, the calomel electrode.
(3)   The galvanic series indicates that alloys can be coupled without being corroded. For instance, alloys close to each other in the series can be safely coupled. As shown in the table, monel can be coupled to copper, or bronze, without any risk of galvanic corrosion. However, brass cannot be coupled

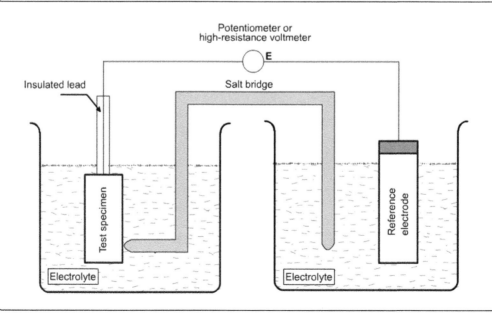

Figure 2.17   Laboratory technique for measuring the single electrode (corrosion) potential E_{corr} of metals and alloys in aqueous environments. Use of salt bridge

with tin, because the two are far away in the table and coupling them may cause serious galvanic corrosion.

(4) Some alloys exist in two places in the table. For instance, steel (18/8) exists in the passive state as well as in the active state. Steel in the passive state means steel with a film of oxide, which shifts the potential from active to passive. Joining of the two types (active and passive) may lead to serious corrosion. An example of stainless steel in the active state is if it is being continuously scraped.

(5) Metals and alloys in brackets can be conveniently joined with one another without any risk of corrosion.

Although galvanic series is widely used by designers, nevertheless, it suffers also from certain limitations, such as:

(1) Each environment requires a different galvanic series, for example, a galvanic series in static seawater cannot be used to predict galvanic corrosion in turbulently flowing seawater.

(2) Galvanic corrosion also depends on the extent of polarization of the metals in alloys and not only on how close they are in the galvanic series. Predictions based on their position in the galvanic series may not provide enough information on galvanic corrosion. For example, in galvanic corrosion the process of cathodic polarization predominates. For instance, titanium has a tendency to polarize cathodically in seawater. Any less resistant metal attached to titanium will not undergo corrosion as would be expected because of the cathodic polarization of titanium. In engineering applications, alloys rather than pure metals are used, however, the emf series has only a limited value. This series is of extreme importance for design engineers.

In spite of the limitations mentioned above, the galvanic series provides valuable information to engineers and scientists on the galvanic corrosion of metals and alloys in different environments.

2.14 POURBAIX DIAGRAMS (STABILITY DIAGRAMS)

2.14.1 INTRODUCTION

Potential–pH diagrams are also called Pourbaix diagrams after the name of their originator, Pourbaix (1963), a Belgium electrochemist and corrosion scientist. These diagrams represent the stability of a metal as a function of potential and pH. They are analogues to phase equilibrium diagrams, where the stability of various phases is shown as a function of temperature and percentage composition of the metal. At a particular temperature and composition a stable phase can be easily determined. Similarly, at a particular combination of pH and potential, a stable phase can be determined from the Pourbaix diagram. In such diagrams, the redox potential of the corroding system is plotted on a vertical axis and the pH on a horizontal axis. These diagrams are constructed from calculations based on Nernst equations and solubility data for metal and its species, such as Fe, Fe_2O_3, $Fe(OH)_2$, Fe_3O_4, etc. in equilibrium. A typical diagram for Fe–H_2O system is shown in Fig. 2.18.

2.14.2 CHARACTERISTICS OF A POURBAIX DIAGRAM

(1) pH is plotted on the horizontal axis and redox potential E *vs* SHE on the horizontal axis.

(2) The horizontal lines represent electron transfer reactions. They are pH-independent, but potential-dependent. These lines separate the regions of stability, e.g. Fe and Fe^{2+} in a potential–pH diagram for Fe–H_2O system. Variation of concentration of Fe^{2+} ($10^{-6} - 1$) leads to several parallel lines.

(3) The vertical lines are potential-independent but pH-dependent and not accompanied by any electron transfer, e.g. lines corresponding to the following reactions:

$$Fe^{2+} + 2H_2O \rightleftarrows Fe(OH)_2 + 2H^+$$

$$Fe^3 + 3H_2O \rightleftarrows Fe(OH)_3 + 3H^+$$

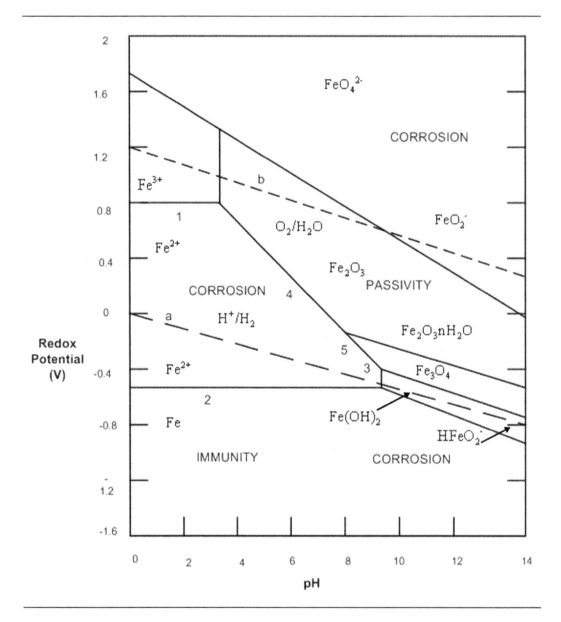

Figure 2.18 Potential–pH diagram for iron

(4) The sloping, straight lines give the redox potentials of a solution in equilibrium with hydrogen and oxygen, respectively. These equilibria indicate electron transfer as well as pH, e.g.

$$Fe_2O_3 + 6H^- + 2e \rightleftarrows 2Fe^{2+} + 3H_2O$$

The above reaction indicates both electron transfer and pH change.

(5) The concentration of all metal ions is assumed to be 10^{-6} mol per liter of solution. At lower concentration, corrosion should not occur.

(6) The diagram is computed for the equilibrium conditions at 25°C.

(7) The upper end of the redox potential axis is the noble end and the lower end, the active end, meaning that the oxidizing power increase with increasing potential.

(8) The hydrogen and oxygen lines are indicated in Pourbaix diagrams by dotted line.

It can be noted that Pourbaix diagrams may be constructed for all elements. The diagrams subdivide the potential–pH plots into regions of immunity, corrosion or passivation. These are very useful in prediction of tendency of metals to corrode. These diagrams, however, has several limitations which will be summarized at the end.

2.14.3 CALCULATIONS INVOLVED IN CONSTRUCTION OF POURBAIX DIAGRAMS

Numerous examples of applying Nernst equation to determine the potential and pH are given in this chapter. These equations are essential tools to calculate the redox potentials in the Pourbaix diagrams. For instance, consider Fig. 2.18, line 1 and let us determine the potential corresponding to reaction occurring in line 1. Here, Fe^{3+} is in equilibrium with Fe^{2+}. Applying Nernst equation, we obtain the reaction

Line 1

$$Fe^3 + e \rightleftarrows Fe^{2+}$$

$$E = E^\circ_{Fe^{3+}/Fe^{2+}} + \frac{RT}{f} \ln \frac{[a_{Fe^{3+}}]}{[a_{Fe^{2+}}]}$$

(one electron transfer)

$$E = 0.77 + 0.059 \log \left[\frac{a_{Fe^{3+}}}{a_{Fe^{2+}}} \right]$$

where $a = 10^{-6}$

Take $a = 0.6$ moles.
The log term drops out here as the term a becomes unity. Hence $E = 0.77\,V$, and we indicate this value as the potential axis.

Line 2

The equilibrium reaction is

$$Fe^{2+} + 2e \Rightarrow Fe$$

$$E = E^\circ_{Fe^{2+}/Fe} + \frac{0.059}{2} \log [a_{Fe^{2+}}]$$

$$E_{Fe^{2+}/Fe} = -0.44 + 0.030 \log [a_{Fe^{2+}}]$$

because $a(Fe) = 1$

but

$$a = 10^{-6}\,\text{mole/liter}$$

$$E_{Fe^{2+}/Fe} = -0.44 + 0.59[-6]$$

$$E = 0.62$$

This value is marked on the potential axis for line 2.

Line 3

The equilibrium reaction is

$$Fe + 2H_2O \rightleftarrows Fe(OH)_2 + 2H^+$$

There is no charge transfer in the above equilibrium system.

or

$$Fe^2 + 2OH^- \rightleftarrows Fe(OH)_2$$

Here the precipitation of $Fe(OH)_2$ is shown, and solubility product comes into play (K_{SP}), which can be described by

$$K_{SP} = [a_{Fe^{2+}}][a_{OH^-}]^2 = 10^{-14.71}$$

(Check table of solubility products)

$$\therefore \log [a_{Fe^{2+}}] + 2 \log [a_{OH^-}] = -14.71$$

but $\log [a_{OH^-}] = -2\,pH + 13.29$
Substituting for $a = 10^{-6}$

$$-6 = -2\,pH + 13.29$$

$$pH = 9.65$$

The corresponding pH value is indicated on the pH axis in the figure.

Line 4

The equilibrium reaction between Fe^{2+} and Fe_2O_3 is

$$Fe_2O_3 + 6H^+ + 2e = 2Fe^{2+} + 3H_2O$$

Note that there is some degree of potential-dependence shown by electron transfer as well some degree of pH-dependence, and the slope is finite as shown by the line. The expression is

$$E = E^{\circ}_{Fe_2O_3/Fe^{2+}} + \frac{0.0591}{2} \log \frac{\left[a_{Fe_2O_3}\right]}{\left[a_{Fe^{2+}}\right]^2}$$

$$\underset{\text{Reduced}}{+ \frac{0.0592}{2} \log \left[a_{H^+}\right]^6}$$

Multiplying $\left[\frac{0.059}{2}\right] \log \left[a_{H^{++}}\right]^6$ by 3, putting $a_{Fe_2O_3} = 1$, moving $a_{Fe^{++}}$ to numerator, and changing sign in the log term, we get

$$E = 0.73 - 0.059 \log \left[a_{Fe^{2+}}\right] - 0.177\text{pH}$$

The above expression shows the relativity between the pH and potential. Hence, the slope is now 0.777 which is indicated in the figure.

Line 5

It has also a finite scope and involves equilibrium between Fe^{2+} ions and Fe_3O_4. The equation for the reaction is

$$Fe_3O_4 + 8H + 2e = 3Fe^{2+} + 4H_2O$$

$$E^{\circ}_{Fe_3O_4/Fe^{2+}} = 0.98$$

Putting $E^{\circ} = 0.98$ for Fe_3O_4/Fe^{2+} in the Nernst equation for the above reaction and solving

$$E = 0.98 - 0.088 \log \left(a_{Fe^{2+}}\right) - 0.236$$

a slope of -0.236 is obtained, which is indicated in the figure. The slope shows a pH and potential dependence.

Similarly, we can calculate the potentials for other equilibrium reactions involving Fe_2O_3 and Fe_3O_4. The reaction is given by

$$3Fe_2O_3 + 2H^+ + 2e \rightleftharpoons 2Fe_3O_4 + H_2O$$

$$E^{\circ} = +0.221V_{SHE}$$

$$E = 0.221 + \frac{0.0551}{2} \log (a_{H^+})^2$$

$$= 0.221 - 0.059\,\text{pH}$$

Such calculations can be carried out for different metal–water systems to construct Pourbaix diagrams.

Other lines in the diagram:

- Hydrogen line (a)

$$H^+ + e = \frac{1}{2}H_2$$

or

$$H_2O + e = \frac{1}{2}H_2 + OH^-$$

- Oxygen line (b)

$$2H_2O = O_2 + 4H^+ + OH^-$$

or

$$4OH^- \rightleftharpoons O_2 + 2H_2O + 4e$$

Below hydrogen line (a) hydrogen is produced by a reduction of H^+ and H_2O, and above oxygen line (b), O_2 is produced by oxidation of H_2O and OH^-. Between the lines *a* and *b* water is stable.

Parallel lines identified by exponent of the activity of Fe^{2+} ions in solution ($a_{Fe^{2+}} = 10^0$, 10^{-2}, 10^{-4}, 10^{-6}). For instance line 8 corresponds to the formation of Fe_2O_3 from solution of $a_{Fe^{3+}} > a_{Fe^{2+}}$. Curves identified as 0, -2, -4, -6, corresponds to 10^0, 10^{-2}, 10^{-4}, 10^{-6}, respectively (Fig. 2.19).

2.14.4 REGIONS OF IMMUNITY, CORROSION AND PASSIVITY IN POURBAIX DIAGRAMS

Figure 2.20 shows the regions of immunity, corrosion and passivity which are described below.

2.14.4.1 Immunity Region

The region of immunity shown in the Pourbaix diagram for Fe–H_2O indicates that corrosion cannot occur in this region. For instance, at a point X in the diagram as the activity of Fe^{2+} would be very low ($\sim 10^{-10}$).

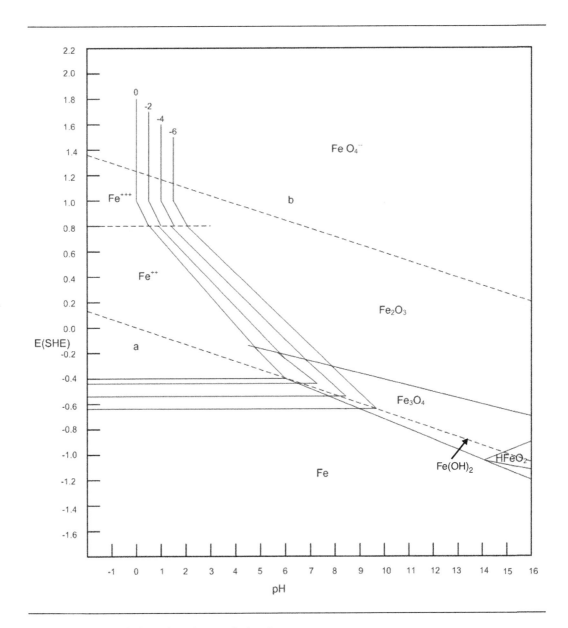

Figure 2.19 Simplified Pourbaix diagram for iron/water system

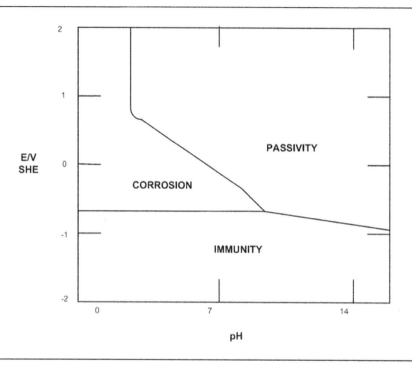

Figure 2.20 Pourbaix diagram for the iron–water system at 25°C showing nominal zones of immunity

2.14.4.2 Corrosion Region

As iron is transformed to soluble species, it is expected that iron would corrode.

2.14.4.3 Region of Passivation

An oxide species in contact with an aqueous solution along a boundary would not allow corrosion to proceed if it is impervious and highly adherent. The thin layer of oxide on the metal surface, such as Fe_3O_2 or Fe_3O_4 is highly protective under the above condition. Metals like aluminum and steel are known to resist corrosion because of development of oxide films in the air.

2.14.5 SOME EXAMPLES OF POURBAIX DIAGRAMS

2.14.5.1 Pourbaix Diagram for Aluminum

A Pourbaix diagram for aluminum and H_2O system is shown in Fig. 2.21. The pH varies (*x*-axis) from acidic at low pH to caustic at high pH. Line *a* is the hydrogen line below which water is no longer stable and decomposes into hydrogen and OH^- (alkalization). Line *b* is the oxygen line above which water decomposes into hydrogen, oxygen and H^+ (acidification). Water is stable between regions (*a*) and (*b*). In acidic conditions Al dissolves as Al^{+3}. In alkaline conditions aluminum dissolves as AlO_2.

$$Al_2O_3 + H_2O \rightarrow 2AlO_2^- + 2H^+$$

In neutral solutions (4–8 pH), the hydroxide is insoluble which makes aluminum surface passive. Aluminum dissolves both in acids and bases.

2.14.5.2 Pourbaix Diagram for Copper

A Pourbaix diagram for copper/water system is shown in Fig. 2.22. Copper ($E° = 0.337\,V$) is more noble than iron ($E°_{Fe} = -0.444\,V$), however, it is more stable in water (SHE) than iron. Copper is not passive in acid electrolytes. The oxide of copper, Cu^+ and Cu^{2+} are only

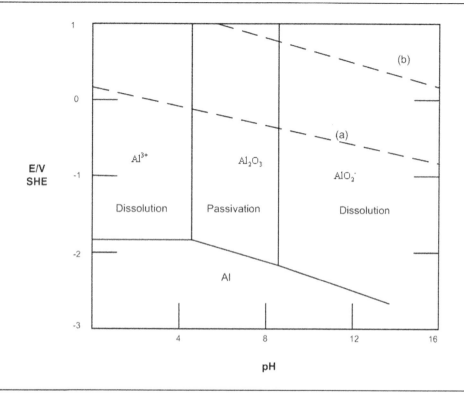

Figure 2.21 Pourbaix diagram for the aluminum–water system at 25°C

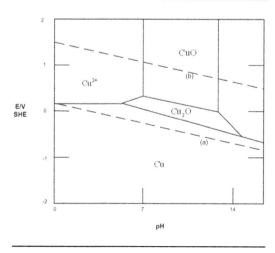

Figure 2.22 Pourbaix diagram for the copper–water system at 25°C

protective in weakly acidic and alkaline elec-trolytes. The region of immunity extends above the line (*a*) in the diagram which represents oxy-gen evolution. If the potential of copper is made more noble, it would corrode under mildly acidic and strongly acidic conditions. It would also cor-rode under strongly alkaline conditions in higher oxidizing potentials.

The scope of the chapter does not permit a broader treatment of Pourbaix diagrams. By iden-tifying the pH and potential ranges, it is possible to control the corrosion of metals in aqueous environment. Aluminum is highly active but it can be used with a minimum risk of corrosion because of its tendency to form protective oxide films which prevent the metal from corrosion. Although titanium is highly active with a very narrow immunity zone towards the bottom of the diagram, it has an excellent tendency for pas-sivation as shown by a highly extended passive zone which extends over the entire range of pH. Titanium, however, corrodes under reducing and highly oxidizing conditions as shown in the lower left region of the pH–potential diagram.

2.14.6 BENEFITS OF POURBAIX DIAGRAMS

Pourbaix diagrams offer a large volume of thermodynamic information in a very efficient and compact format. The information in the diagrams can be beneficially used to control corrosion of pure metals in the aqueous environment. By altering the pH and potential to the regions of immunity and passivation, corrosion can be controlled. For example, on increasing the pH of environment in moving to slightly alkaline regions, the corrosion of iron can be controlled. This can be achieved by water treatment. Similarly, changing the potential of iron to more negative values eliminate corrosion, this technique is called cathodic protection. Also, raising the potentials to more positive values reduces the corrosion by formation of stable films of oxides on the surface of transition metals. Steel in reinforced concrete does not corrode if an alkaline environment is maintained. On the contrary, an alkaline environment for aluminum is a disaster if the pH exceeds 8.0. The above example clearly demonstrate the merits of Pourbaix diagrams in prediction and control of corrosion. However, there are several limitations of these diagrams, which are summarized below:

1. These diagrams are purely based on thermodynamic data and do not provide any information on the reactions. The thermodynamic stability may not be achieved to the kinetics of the reaction. No information is provided in the rates of reaction.
2. Consideration is given only to equilibrium conditions in specified environment and factors, such as temperature and velocity are not considered which may seriously affect the corrosion rate.
3. The activity of species is arbitrarily selected as 10^{-6} g mol which is not realistic.
4. Pourbaix diagrams deal with pure metals which are not of much interest to the engineers.
5. All insoluble products are assumed to be protective which is not true, as porosity, thickness, and adherence to substrate are important factors, which control the protective ability of insoluble corrosion products.

Although the above disadvantages appear to be substantial, the advantages offered by the Pourbaix diagrams far outweigh their limitations.

QUESTIONS

A. MULTIPLE CHOICE QUESTIONS

Select one best answer:

1. The electrode potential of a metal is:

   [ ] The potential which exists at the interface between the metal and the electrolyte
   [ ] The potential between the anodic and cathodic areas of the metal
   [ ] The potential between two metals immersed completely in the same electrolyte
   [ ] The potential of a metal with respect to another metal, which is at a higher concentration

2. The double layer is formed as a result of

   [ ] attractive forces between negative charged metal surface and positive ions only
   [ ] repulsive forces between like positive ions only
   [ ] both attractive and repulsive forces between ions
   [ ] None of the above

3. The most acceptable method of obtaining standard electrode potential is by

   [ ] comparing the electrode potentials of a metal half cell with a hydrogen half cell
   [ ] comparing the electrode potential of a metal immersed in a solution of its ions at any concentration, with the hydrogen half cell

comparing the electrode potential of a metal with any standard electrode, such as Ag–AgCl or Calomel electrode

[ ] None of the above

4. A galvanic cell is formed

[ ] when two metals are immersed in solutions differing in concentration
[ ] when two different metals are immersed in one electrolyte
[ ] when two different metals are exposed to air
[ ] when two metals are brought close together and electrically insulated from one another

5. In *concentration* cells

[ ] the metal is in contact with two half cells having the same electrolyte but at different concentrations
[ ] the metal is in contact with two half cells, having the same electrolytes with the same concentration
[ ] no liquid junction is present
[ ] there is no migration of ions from one electrolyte to another electrolyte

6. If the free energy of a reversible process is negative, it implies that

[ ] the cell reaction is spontaneous
[ ] the cell reaction is not spontaneous
[ ] the cell reaction proceeds from right to left
[ ] no reaction takes place at all

7. The value of $(2.303RT)/F$ at 25°C varies with

[ ] temperature
[ ] the metal being considered
[ ] the melting temperature of the metal
[ ] None of the above

8. If $E > 0$, the cell reaction

[ ] proceeds in forward direction
[ ] proceeds in reverse direction

[ ] does not proceed
[ ] None of the above

9. The most common electrodes used for measurement of corrosion potentials are: (*Mark two correct answers*)

[ ] Ag–AgCl
[ ] Hg–Hg$_2$Cl$_2$
[ ] Cu–CuSO$_4$
[ ] Hydrogen electrode

10. A galvanic series is

[ ] a list of alloys arranged according to their corrosion potentials in a given environment
[ ] a list of metals and alloys according to their corrosion potentials in a given environment
[ ] a list of standard electrode potentials of alloys or metals arranged in order of their values
[ ] a grouping of metals and alloys based on their ability to get oxidized in a stated environment

B. REVIEW QUESTIONS

1. Distinguish between:

a) Metallic conduction and electrolytic conduction.
b) Standard electrode potential and corrosion potential.
c) Anode and cathode.
d) Electronic conduction and ionic conduction.

2. In the sign convention adopted by IUPAC:

a) What does the right-hand electrode indicate?
b) What does the left-hand electrode indicate?
c) How the cell potential, E_{cell}, is obtained?
d) What does the positive sign of the cell indicate?

3. If $E > 0$, in which direction will the cell reaction proceed, and conversely if $E < 0$, in which direction the reaction would proceed?

4. State which of the following statements are true:

   a) When two metals, e.g. Zn and Cd, are connected and placed in a solution containing both metal ions, the metal with the lower standard potential would corrode.

   b) Conversely, the metal with the higher potential would be deposited.

   c) The cell and cell reaction are written in opposite orders, for instance, for the cell Fe/Fe^{2+}(aq)/Cu^{2+}(aq)/Cu, the reaction is

   $$Fe^{2+} + Cu \rightarrow Cu^{2+} + Fe$$

   d) The cell potential is obtained by subtracting the electrode potential of the right-hand electrode from the left-hand electrode.

5. State the limitations of the emf series and the advantages of galvanic series for an engineer.

C. PROBLEMS

1. Devise electrochemical cells in which the following overall reactions can occur:

   a) $Zn(s) + Cu^{2+}(aq) \rightarrow Cu(s) + Zn^{2+}(aq)$

   b) $Ce^{+4}(aq) + Fe^{2+}(aq) \rightarrow Ce^{3+}(aq) + Fe^{3+}(aq)$

   c) $Ag^{+}(aq) + Cl^{-}(aq) \rightarrow AgCl(s)$

   d) $Zn(s) + 2Cl_2(g) \rightarrow ZnCl_2(aq)$

2. What is the mole fraction of NaCl in a solution containing 1.00 mole of solute in 1.00 kg of H_2O?

3. What is the molarity of a solution in which 1.00×10^2 g of NaOH is dissolved in 0.250 kg of H_2O?

4. What is the voltage (E_{cell}) of a cell comprising a zinc half cell (zinc in ZnSO$_4$) and a copper half cell (Cu in CuSO$_4$)? The metal concentrations of ZnSO$_4$ and CuSO$_4$

are 1 and 0.01, respectively. The activity coefficient for CuSO$_4$ is 0.047 and for ZnSO$_4$ is 0.70.

5. Calculate E for the half cell in which the reaction Cu^{++} (0.1 m) + 2e^- = Cu(s) takes place at 25°C.

6. Calculate the potential for each half cell and the total emf of the cell (E_{cell}) at 25°C:

   $$Pb \,|\, Pb^{2+} \ (0.0010 \ M)/Pt, Cl_2(1 \ atm)/$$

   $$Cl^-(0.10 \ M)$$

   $$E°Pb = Pb^{2+}/Pb° = -0.13 \ V$$

   $$E° \cdot (Cl_2 - Cl) = 1.358 \ V$$

7. Calculate the emf and the free energy of the cell given below:

   $$Fe^{2+} \cdot Fe \ and \ Ni^{2+} \cdot Ni$$

   (Obtain data from the literature.)

8. A piece of copper is immersed in an aqueous solution of KCl with a concentration of 1 kmol/m^3. The solubility product of CuCl at 25°C is (1.7×10^{-6}). Calculate the potential of the copper electrode.

9. A test piece of cadmium is placed (a) in flowing seawater, and (b) in stagnant seawater. Predict under which conditions cadmium would corrode. The following information is provided:
   pH = 7.0
   Concentration of cadmium ions:
   Cd(OH)$_2$ = 1.2×10^{-14} kmol/m^3 at 25°C
   in stagnant condition.
   Concentration of cadmium ions in flowing conditions:
   = 10^{-6} kmol/m^3
   $E°_{Cd} = -0.42$ V

10. Aluminum samples are exposed in Arabian Gulf water. It has been found that aluminum corrodes either as Al(OH)$_3$ or AlCl$_3$ in seawater. Show by calculations in which form aluminum corrodes or does it corrode in both forms. The concentration of Al^{+++} in AlCl$_3$ = 10^{-4}. The solubility product of Al(OH)$_2$ is 3.7×10^{-15}.

11. A piece of zinc measuring 2″ × 1.5″ is placed in a 0.002 molar solution of ZnCl$_2$. Show whether zinc would corrode in the given medium.

12. A piece of nickel measuring $6'' \times 4''$ is immersed in deaerated water having a pH of 8.0. The solubility product $Ni(OH)_2$ is 1.6×10^{-16}. Determine the potential of nickel in the given conditions and state whether nickel would corrode or not.

13. Calculate the emf of the following cell at 25°C:
    Fe^{3+} (a = 0.01), Fe^{2+} (a = 0.0001), Cu^+ (a = 0.01)Cu.
    Given: $E°$, $Fe^{3+}/Fe^{2+} = -0.771$ V and $E°$, $Cu^+/Cu = 0.5211$.

14. Find the potential of a cell where the reaction $Ni + Sn^{2+} \rightarrow Ni^{2+} + Sn$, proceeds. The concentration of Ni^{2+} is 1.3 and the concentration of Sn^{2+} is 1.0×10^{-4}. Predict if the reaction would proceed from right to left or left to right.

15. Write a balanced cell reaction and calculate the emf of the following cell:
    Pt/Sn^{2+} (a = 0.10), Sn^{4+} (a = 0.10), Fe^{3+}/Fe (a = 0.200).

16. The following is the reaction when iron corrodes in an acid:

    $$Fe + 2HCl(aq) + \frac{1}{2}O_2 \rightarrow FeCl_2(aq) + H_2O$$

    a) In what direction shall the reaction proceed if the activity of Fe^{2+} is 1, and $a(H^+) = 1$?

    b) If the temperature is maintained at 25°C, what activity of iron ($a_{Fe^{2+}}$) would be required to stop the corrosion in acid?

17. Calculate the voltage of the following cell:

    $$Zn + Cd^{2+} (a_{Cd^{2+}} = 0.2) = Zn^{++}$$

    $$(a_{Zn^{2+}} = 0.0004) + Cd.$$

18. The potential of an electrode is measured as -0.840 V relative to a 1 N Calomel electrode. What is the electrode potential on a standard hydrogen scale?

19. What pressure would be required to stop corrosion of zinc in deaerated water at 25°C? The major corrosion produced is $Zn(OH)_2$ and the solubility product is 1.8×10^{-14} at 20°C (Take pH = 7).

20. Calculate the theoretical tendency of cadmium to corrode when it is immersed in a solution of 0.001 M $CdCl_2$ (pH = 2.0).

21. For the cell H_2 (1 atm)·HCl·HgCl·Hg, $E° = 0.2220$ V at 298 K. If pH = 1.47, determine the emf of the cell.

22. Copper corrodes in an acid solution of pH = 2. Hydrogen is bubbled continuously in the solution. Calculate the maximum concentration of Cu^{2+} ions that would result.

23. A cell is composed of a pure copper and pure lead electrode immersed in solutions of their bivalent ions. For a 0.3 molar concentration of Cu^{2+}, the lead electrode is oxidized and shows a potential of 0.507 V. What would be the concentration of Pb ion at 25°C?

SUGGESTED BOOKS FOR READING

[1] Davis, J.P. ed. (2000). *Corrosion: Understanding the Basics.* 2nd ed. Ohio: Metals Park, USA.
[2] Revier, R.W. ed. (2000). *Uhlig's Corrosion Handbook.* New York: John Wiley.
[3] ASM Handbook (1987). *Corrosion.* Vol. 13, ASM, Ohio: Metals Park, USA.
[4] Stansbury, E.E. and Buchanan, R.A. (2000). *Fundamentals of Electrochemical Corrosion.* Ohio: ASM, USA.
[5] Piron, D.L. (1991). *The Electrochemistry of Corrosion,* Texas: NACE, USA.

KEYWORDS

Anode The region of the electrical cell where positive current flows into the electrolyte. Anode is the site where oxidation occurs. In a corrosion cell anode is the region which is dissolving. Cathode is the region of an electrical cell where positive electric current enters from the electrolyte. In a corrosion cell reduction reaction takes place at the cathode.

Electrochemical cell An electrochemical system comprising of an anode and a cathode in a metallic contact and immersed in an electrolyte.

Electrode potential It is the potential of an electrode in an electrolyte as measured against a reference electrode.

Electrolyte It is electrically conductive. It is usually a liquid containing ions that migrate in an electric field.

Free energy From the total heat content of a system (enthalpy), only a part can be converted to useful work. This part of total enthalpy is called free energy, G. Absolute value of free energy (G) cannot be measured, and only changes in free energy (ΔG) are measured.

Galvanic series A table of metals and alloys arranged according to their relative corrosion potential in a given environment.

Half cell An electrode immersed in a suitable electrolyte for measurement of potential.

IR drop Voltage drop caused by current flow in a resistor.

Reference electrode A stable electrode with a known and highly reproducible potential.

Reversible potential (*Equilibrium potential*) It is the potential of an electrode where the forward rate of reaction equals to the reverse rate of reaction.

Standard electrode potential (E°) It is reversible potential of an electrode measured against a standard hydrogen electrode (SHE) consisting of a platinum specimen immersed in a unit activity acid solution through which H_2 gas at 1 atm pressure is bubbled. The potential of the hydrogen electrode (half cell) is taken to be zero.

CHAPTER 3

CORROSION KINETICS

FARADAY'S LAWS OF ELECTROLYSIS AND ITS APPLICATION IN DETERMINING THE CORROSION RATE

The classical electrochemical work conducted by Michael Faraday in the nineteenth century produced two laws published in 1833 and 1834 named after him. The two laws can be summarized below.

3.1 THE LAWS

3.1.1 THE FIRST LAW

The mass of primary products formed at an electrode by electrolysis is directly proportional to the quantity of electricity passed. Thus:

$$m \propto It \text{ or } m = ZIt \tag{3.1}$$

where

I = current in amperes
t = time in seconds
m = mass of the primary product in grams
Z = constant of proportionality (electrochemical equivalent). It is the mass of a substance liberated by 1 ampere-second of a current (1 coulomb).

3.1.2 THE SECOND LAW

The masses of different primary products formed by equal amounts of electricity are proportional to the ratio of molar mass to the number of electrons involved with a particular reaction:

$$m_1 \propto \frac{M_1}{n_1} \propto Z_1 \tag{3.2}$$

$$m_2 \propto \frac{M_2}{n_2} \propto Z_2 \tag{3.3}$$

where

m_1, m_2 = masses of primary product in grams
M_1, M_2 = molar masses (g.mol^{-1})
n_1, n_2 = number of electrons
Z_1, Z_2 = electrochemical equivalent.

Combining the first law and the second law, as in equation (3.1)

$$m = ZIt$$

Substituting for Z, from equation (3.2) into (3.1)

$$m = k\frac{M}{n}It \tag{3.4}$$

or

$$m = \frac{1}{F} \cdot \frac{M}{n}It \tag{3.5}$$

where F = Faraday's constant. It is the quantity of electricity required to deposit the ratio of mass to the valency of any substance and expressed in coulombs per mole $(C (g\,equiv.)^{-1})$. It has a value of 96 485 coulombs per gram equivalent. This is sometimes written as 96 485 coulombs per mole of electrons.

3.1.3 APPLICATIONS OF FARADAY'S LAWS IN DETERMINATION OF CORROSION RATES OF METALS AND ALLOYS

Corrosion rate has dimensions of mass × reciprocal of time:

$$(g \cdot y^{-1} \text{ or } kg \cdot s^{-1})$$

In terms of loss of weight of a metal with time, from equation (3.5), we get

$$\frac{dw}{dt} = \frac{MI}{nF} \quad (I = \text{current}) \qquad (3.6)$$

The rate of corrosion is proportional to the current passed and to the molar mass. Dividing equation (3.5) by the exposed area of the metal in the alloy, we get

$$\frac{w}{At} = \frac{MI}{nFA} \qquad (3.7)$$

But, $\dfrac{I}{A}$ = current density (i). Then:

$$\frac{w}{At} = \frac{Mi}{nF} \quad (i = \text{current density}) \qquad (3.8)$$

The above equation has been successfully used to determine the rates of corrosion.

A very useful practical unit for representing the corrosion rate is milligrams per decimeter square per day (mg.dm^{-2}.day^{-1}) or mdd. Other practical units are millimeter per year (mmy^{-1}) and mils per year (mpy).

3.1.4 ILLUSTRATIVE EXAMPLES

Example 1

Steel corrodes in an aqueous solution, the corrosion current is measured as $0.1\,A \cdot cm^{-2}$. Calculate the rate of weight loss per unit area in units of mdd.

Solution:

For $Fe \rightarrow Fe^2 + 2e$

$$\frac{w}{At} = \frac{Mi}{nF} \qquad (3.9)$$

where

$M = 55.9\,g \cdot mol^{-1}$
$i = 0.1\,A \cdot cm^{-2}$
$n = 2$

Substituting the values in equation (3.9), we obtain

$$\frac{w}{At} = 2.897 \times 10^{-8}\,g\,cm^{-2}s^{-1}$$

Now converting g to mg ($\times 10^3$), we get

$$(2.897 \times 10^{-8}\,g\,cm^{-2}s^{-1})$$
$$= 2.897 \times 10^{-5}\,mg\,cm^{-2}s^{-1}$$

and converting from mg cm^{-2} s^{-1} to mg dm^{-2} s^{-1}, we get

$$(2.897 \times 10^{-5}\,mg\,cm^{-2}s^{-1})(1 \times 10^2\,cm^2/dm^2)$$

For converting the above expression to milligrams per decimeter square per day (mg dm^{-2} day^{-1}), we multiply by 24 h/day × 3600 s/h, and obtain the corrosion rate in the desired units:

$$(2.897 \times 10^{-5}\,mg\,cm^{-2}\,s^{-1})(24)(3600)$$
$$= 2.503\,mg\,dm^{-2}\,day^{-1}$$

Example 2

Iron is corroding in seawater at a current density of 1.69×10^{-4} A/cm^2. Determine the corrosion rate in

(a) mdd (milligrams per decimeter^2 day)
(b) ipy (inches per year)

Solution:
(a) Apply Faraday's law as before

$$\text{mdd} = 1.69 \times 10^{-4} \text{ A/cm}^2 \times 3600 \text{ s/h}$$

$$\times 24 \text{ h/day} \times 100 \text{ cm}^2/\text{dm}^2$$

$$\times \frac{55.85}{2} \text{ g/mol} \times 10^3 \text{ mg/g}$$

$$\times \frac{1}{96\,495} \text{ A} \cdot \text{s} \cdot \text{mole}$$

$$= 422.8 \text{ mg dm}^{-2} \text{ day}^{-1}$$

(b) Converting 422.8 mg dm^{-2} day^{-1} to inches per year (ipy) with the conversion factor [mdd $\times$ 0.00144/ρ], ρ = density

$$= 422.8 \times \frac{0.00144}{7.86}$$

$$= 0.077 \text{ ipy}$$

or 77 mpy, because 1 mil $= 1/1000$ of an inch.

Example 3

A sample of zinc anode corrodes uniformly with a current density of 4.27×10^{-7} A/cm^2 in an aqueous solution. What is the corrosion rate of zinc in mdd?

Solution:

$$\text{Zn} \rightarrow \text{Zn}^{+2} + 2e$$

$$\text{CR} = \left(4.27 \times 10^{-7} \text{ A/cm}^2\right)\left(1 \times 10^2 \text{ cm}^2/\text{dm}^2\right)$$

$$\left(24 \times 3600 \text{ s} \cdot \text{h/h} \cdot \text{day}\right)\left(\frac{65.38}{2} \text{ g/mol}\right)$$

$$\times 10^3 \text{mg/g} \left(\frac{1}{96\,495} \text{ A} \cdot \text{s} \cdot \text{mol}\right)$$

$$= 1.25 \text{ mdd}$$

Example 4

Penetration unit time can be obtained by dividing equation (3.8) by density of the alloy. The following equation can be used conveniently:

$$\text{Corrosion rate, } r = C \cdot \frac{Mi}{n\rho} \qquad (3.10)$$

where

ρ = density (g/cm^3)
i = current density (A/cm^2)
M = atomic weight (g $\cdot$ mol^{-1})
n = number of electrons involved
C = constant which includes F and any other conversion factor for units, for instance, $C = 0.129$ when corrosion rate is in mpy, 3.27 when in mm/year and 0.00327 when units are in mm^3/year.

For instant, the above relationship can be used to establish the equivalent of corrosion current of μA/cm^2 with the rate of corrosion for iron in mpy as shown below

$$1 \, \mu\text{A/cm}^2 = 0.129 \left[\frac{(55.8)(1)}{(2)(7.86)}\right] = 0.46 \, \text{mpy}$$

$$(3.10a)$$

Thus, for iron $1 \, \mu$A/cm^2 of current is equivalent to a corrosion rate of 0.46 mpy for iron. The equation can be extended to establish the equivalent in other units also.

The above example shows the correspondence between penetration rate and current density for a metal. A similar correspondence between the penetration rate and current density for alloys can be established. However, it would require the determination of equivalent weight (M/n) for the alloy.

Following is the relationship which is used to determine the equivalent weight of an alloy:

$$\text{Equivalent weight} = \frac{1}{\Sigma\left[\frac{f_i}{n_1/M_1}\right]} \qquad (3.10b)$$

where

f_i = mean fraction of an element present in the alloy
n_1 = electron exchanged
M_1 = atomic mass.

Example 5
Determine the corrosion rate of AISI 316 steel corresponding to $1\,\mu A/cm^2$ of current. Following is the composition of alloys:

Cr = 18%
Ni = 10%
Mo = 3%
Mn = 2%
Fe = balance

You may find the composition of alloys from ASM Handbook.

Solution:
Using equation (3.10a):

$$1\,\mu A/cm^2 = 0.128 \left[\frac{52.3}{(1)(7.19)}\right] 0.18$$

$$+ 0.128 \left[\frac{54.94}{(2)(7.45)}\right] 0.02$$

$$+ 0.128 \left[\frac{95.95}{(2)(10.1)}\right] 0.03$$

$$+ 0.128 \left[\frac{55.65}{(2)(7.86)}\right] 0.07\,\text{mpy}$$

$$= 0.514587\,\text{mpy}$$

Example 6
A sample of zinc corrodes uniformly with a current density of $4.2 \times 10^{-6}\,A/cm^2$ in an aqueous solution.

(a) What is the corrosion rate of zinc in $mg/dm^2/day$?
(b) What is the corrosion rate of zinc in mm/year?

Solution:
(a) Given current density, $i = 4.2 \times 10^{-6}\,A/cm^2$ = $(4.2\,\mu A/cm^2)$, we know that for zinc atomic weight, $M = 65.38\,g/mol$, density, $\rho = 7.1\,g/cm^3$, $n = 2$, $F = 96\,500$ coulombs/mole.

From the formula:

$$\text{Corrosion rate} = \frac{4.2 \times 10^{-6}\,A/cm^2}{2 \times 96\,500\,\text{coulombs/mol}}$$

$$\times \frac{100\,cm^2/1\,dm^2 \times 24\,h/1\,day}{2 \times 96\,500\,\text{coulombs/mol}}$$

$$\times \frac{3600\,s/1\,h \times 1000\,mg/1\,g}{2 \times 96\,500\,\text{coulombs/mol}}$$

$$\times \frac{65.38\,g/mol}{2 \times 96\,500\,\text{coulombs/mol}}$$

$$\Rightarrow \text{Corrosion rate of Zn} = 12.29\,mg/dm^2/day$$

(b) We can also use the relationship given below to determine the rate of corrosion in mm/year or other units by changing the constants. The constants for mm/year is 0.00327.

$$\text{Corrosion rate, } r = C \cdot \frac{Mi}{n\rho}$$

where ρ is the density in g/cm^3, i is the current density in $\mu A/cm^2$, and C is the constant = 0.00327 for mm/year.

$$\text{Corrosion rate} = \frac{0.00327 \times 65.38\,g/mol \times 4.2}{2 \times 7.13}$$

$$= 0.0629\,mm/year$$

Corrosion rate $= 0.0629\,mm/year$

The proportionality constant is 0.129 for mils per year (mpy).
We can also convert $12.29\,mg/dm^2/day$ to mm/year as below:

$$\Rightarrow 12.29\,mg/dm^2/day$$

$$= 12.29 \times \frac{1}{\rho} \times \frac{365\,days}{1\,year} \times \frac{100\,dm^2}{1\,m^2}$$

$$\times \frac{1000\,mm}{1\,m} \times \frac{1\,kg}{10^6\,mg}$$

$$= 12.29 \times \frac{1}{7130\,kg/m^3}$$

$$\times \frac{365 \times 100 \times 1000}{1\,000\,000}$$

$$= \frac{12.29 \times 365}{71\,300}$$

$$= 0.0629\,mm/year$$

$$\Rightarrow \text{Corrosion rate} = 0.0629\,mm/year$$

Example 7

AISI 316 steel has the following nominal composition:

Cr = 18%	n = 1	$\rho = 7.1 \, g/cm^2$
		At. wt. $= 52.01 \, g/mol$
Ni = 8%	n = 2	$\rho = 8.9 \, g/cm^2$
		At. wt. $= 58.68 \, g/mol$
Mo = 3%	n = 1	$\rho = 10.2 \, g/cm^2$
		At. wt. $= 95.95 \, g/mol$
Fe = 70%	n = 2	$\rho = 7.86 \, g/cm^2$
		At. wt. $= 55.85 \, g/mol$

Find the equivalence between the current density of $1 \, \mu A/cm^2$ and the corrosion rate (mpy).

Solution:
Find the equivalence between $i = 1 \, \mu A/cm^2$ and the corrosion rate.

$$\text{Corrosion rate} = C \cdot \frac{Mi}{n\rho}$$

where C is the constant for conversion depending on unit.

$$r = 0.129 \cdot \frac{Mi}{n\rho} \quad (C = 0.129 \text{ for units of mpy})$$

Since, we are given the composition of the steel 316, we will find the corrosion rate of steel as below:

Corrosion rate:

$$= 0.129 \left\{ \left(\frac{52.01}{1 \times 7.1} \right) 0.18 + \left(\frac{58.68}{2 \times 8.9} \right) 0.08 \right.$$

$$\left. + \left(\frac{95.95}{1 \times 10.2} \right) 0.03 + \left(\frac{55.85}{2 \times 7.86} \right) 0.70 \right\}$$

$$= 0.129 \, [1.318 + 0.263 + 0.282 + 2.48]$$

$$= 0.129 \times 4.343$$

$$= 0.55 \text{ mpy (mils/year)}$$

Example 8

Prove that for Al alloy 1100, the penetration rate of the alloy equivalent to $1 \, \mu A/cm^2$ is 0.43 mpy.

Solution:
For a penetration rate equivalent to $1 \, \mu A/cm^2$

$$\text{Corrosion rate} = C \cdot \frac{Mi}{n\rho}$$

Given:

$M = $ Equivalent weight $\times 3 = 8.99 \times 3$
$\qquad = 26.97$
Density $= \rho = 2.71 \, g/cm^3$
$n = 3$
$i = 1 \, \mu A/cm^2$

$$\Rightarrow \text{Corrosion rate} = \frac{0.128 \times 26.97 \times 1}{3 \times 2.71}$$

$$= 0.424 \text{ mpy}$$

$$= 0.43 \text{ mpy (mils/year)}$$

3.2 CORROSION KINETICS

Thermodynamics gives an indication of the tendency of electrode reactions to occur, whereas electrode kinetics addresses the rates of such reactions. The reactions of concern are mainly corrosion reactions, hence, it is more appropriate to call the kinetics of such reactions as corrosion kinetics. In order to understand the theory of aqueous corrosion, it is important to develop a complete understanding of the kinetics of reaction proceeding on an electrode surface in contact with an aqueous electrolyte. Methods which are used to study the rate of a reaction involve the determination of the amount of reactants remaining in products after a given time. In aqueous corrosion, it is very important to appreciate the nature of irreversible reactions which take place on the electrode surface during corrosion.

3.2.1 ANODIC AND CATHODIC REACTIONS

Consider a piece of iron corroding in an acid solution. The following are the basic reactions:

$$Fe \rightleftarrows Fe^{2+} + 2e \quad \text{(anodic)} \qquad (3.11)$$

$$H_2 \rightleftarrows 2H^+ + 2e \quad \text{(cathodic)} \qquad (3.12)$$

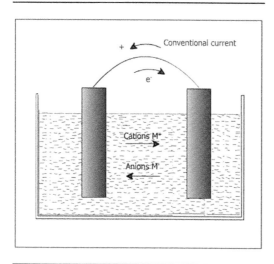

Figure 3.1   Anodic and cathodic processes in a corrosion cell

The overall reaction which takes place in the iron surface is

$$Fe + 2H^+ \rightarrow Fe^{2+} + H_2 \qquad (3.13)$$

The above process is illustrated in Fig. 3.1. The anodic reaction shown above involves the transfer of a metal atom from a metal lattice to the aqueous solution at the electrode/electrolyte interface. The reactions (3.11) and (3.12) shown above are called charge transfer reactions as they proceed by transfer of charge (electrons).

The metal reaction like $M \rightleftarrows M^{z+} + ze$ is reversible as the rate of forward reaction is equal to the rate of reversible reaction. An anodic reaction like $M \rightleftarrows M^{+z} + ze$, is, however, not a reversible reaction as the rate of forward reaction (i_f) is greater than the rate of reduction (i_r) and a net current i_A^{net} flows, $M \rightleftarrows M^{z+} + ze$, $i_A^{net} = i_f - |i_r|$.

The anodic reaction shown above, involves transfer of a metal atom from a metal lattice M to a metal cation having a positive charge M^{z+}. The water dipoles are attracted to the positive ions in solution and form a hydration or solvation sheath around each cation. The cations in solution are, therefore, hydrated or solvated.

The charge transfer reactions described above cannot proceed until a driving force is available.

Such a driving force is provided by the free energy of the reaction (ΔG^*).

3.2.2 ENERGY–DISTANCE PROFILES

These are plots of free energy (ΔG^*) *vs* the distance representing the progress of reaction through the double layer. The metal atoms are located in energy wells associated with lattice structure. A metal atom cannot be detached from the lattice and pass into the solution until it crosses an energy barrier, called activation energy. Figure 3.2 shows an energy–distance (from metal surface to the outer Helmholtz plane of double layer) profile. The solid circle in the figure represents the metal atom in the lattice and the circle with (+) charge, the cation in aqueous solution. The left-hand side of Fig. 3.2 is the lattice side and the right-hand side is the solution side. The hump in Fig. 3.2 represents the activation energy barrier, ΔG^*. The location of the hump is described by a symmetry factor β, and for all practical purposes it is taken as ($\beta = \frac{1}{2}$). The thermal energy of the ions in a metal crystal makes them vibrate with a frequency, generally of the order of 10^{12} per second. Any ion with sufficient thermal energy to reach to the top of the hump would vibrate with a characteristic frequency given by (f). An explanation of the

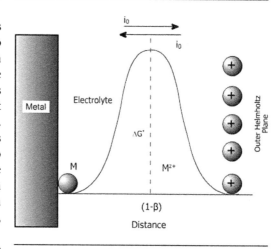

Figure 3.2   Energy–distance profile

reversible and irreversible processes illustrated by energy profiles is described below.

3.2.3 REVERSIBLE REACTIONS

Consider an electrode at equilibrium:

$$M \rightleftarrows M^{+z} + ze$$

The free energy of activation for the anodic (forward) and cathodic (reverse) reactions are located at the same level (Fig. 3.2). The dissolution and discharge reactions need the same energy of activation (ΔG^*). At equilibrium there is no net current, as $\vec{i} - \overleftarrow{i} = 0$. The electrode potential assumes its equilibrium value E_{rev}. Here, $\vec{r}_f = \overleftarrow{r}_r \cong i_0$, i_0 is called 'Exchange Current Density.' It is the current density associated with an electrode at equilibrium. Every reversible electrode has a characteristic exchange current density.

3.2.4 REVERSIBILITY AND EXCHANGE CURRENT

In order to understand reversibility, consider the following (Fig. 3.2):

The rate of reaction in forward direction, $\vec{r}_f$ is given by

$$\vec{r}_f = K_1 e^{-(\Delta G_A^*/RT)} \tag{3.14}$$

where K_1 is the constant depending on temperature, time and activity.

The rate of reverse process is

$$\overleftarrow{r}_r = K_2 e^{(\Delta G_C^*/RT)} \tag{3.15}$$

But if the system is at equilibrium $\vec{r}_f = \overleftarrow{r}_r$. Equating the two processes forward (anodic) and reverse (cathodic), we get

$$e^{-(\Delta G_A^* - \Delta G_C^*)/RT} = e^{zEF/RT} = \frac{K_2}{K_1} \tag{3.16}$$

Thus, each reversible process has a characteristic potential, called electrode potential. When the reaction is reversible, $\vec{i}_f \rightleftarrows \overleftarrow{i}_r = 0$, no net current flows. $\vec{i}_f = \overleftarrow{i}_r = i_{0(exchange)}$, i_0 is called the

exchange current density. There is no net transfer of charge as shown above. Each reversible electrode reaction has its own exchange current density. Consider a reaction, such as $Cu = Cu^{++} + 2e$. Although at equilibrium no net current will flow through the circuit but the interchange of Cu atoms and copper ions would take place at the electrode surface. Hence, a current would be associated with the anodic and cathodic partial reactions. An electrode will not achieve equilibrium electrode potential for $M^{z+} + ze \rightleftarrows M$, unless its i_0 is much greater than i_0 value of any other reversible reaction in the system. A high value of i_0 represents a high rate of reaction. The value of i_0 ranges between 10^{-1} and 10^{-5} A/cm^2 for different materials. The relationship between exchange reaction rate and current density can be derived from Faraday's law discussed earlier:

$$\vec{r}_f = \overleftarrow{r}_r = \frac{i_0 M}{nF} \tag{3.17}$$

where $\vec{r}_f$ is the rate of oxidation and $\overleftarrow{r}_r$ is the rate of reduction expressed in terms of current density (i_0). It is a convenient way of expressing $\vec{r}_f$ or $\overleftarrow{r}_r$ at equilibrium.

3.2.5 IRREVERSIBLE PROCESS

Figure 3.3 represents a situation when the electrode is irreversible, displaced from its reversible potential (E_{rev}). This irreversibility can be brought about by connecting the electrode to an external source of current. By connecting the electrode to the positive pole of an external source of current, the electrode is made the anode, and similarly it can be made a cathode by connecting to the negative terminal.

$$M \xrightarrow{i_f} M^{z+} + ze, \quad i_f > i_r$$

$$i_a^{net} = i_f - |i_r|, \quad i_a^{net} = \text{net anodic current density}$$

$$i_a^{net} = \vec{i}_f - \overleftarrow{i}_r, \quad M \underset{|i_a|}{\overrightarrow{\rightleftarrows}} M^{z+} + ze$$

The electrode is no more at equilibrium. The atom is, therefore, shifted from its equilibrium position to a new energy well in the direction of the positive potential. The magnitude of shift

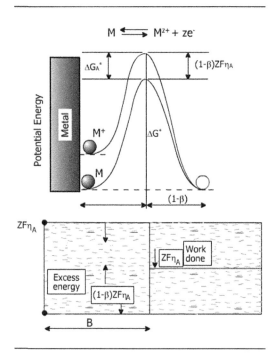

Figure 3.3 Energy–distance profile when the electrode is irreversible

of potential in the noble direction is called '*overpotential*'(η_A). It is given by $\eta_A = E - E_{rev}$, where subscript A is the anode, η is the over-potential, E is applied potential and E_{rev} is the reversible potential. The potential shifts in the noble direction, because $E > E_{rev}$ and η is positive. Consider now the electrode being connected to the negative pole of the external current source. The following reaction now proceeds in the reverse direction:

$$M \xleftarrow{i_r} M^{z+} + ze, \quad M \underset{|i_c|}{\overset{i_r}{\rightleftarrows}} M^{z+} + ze$$

The reverse process now prevails over the forward process:

$$i_r > i_f, \quad i_c^{net} = |i_r| - i_f \quad \text{or} \quad i_c^{net} = \overleftarrow{i_r} - \overrightarrow{i_f}$$

Cathodic reduction now prevails over anodic oxidation. The activation energy for the cathodic process ΔG_C^* is made more favorable than the rate of anodic process and the rate of transfer of charge by cathodic process is faster, $i_c^{net} = M \leftarrow M^{z+} + ze$, $\eta_C = E - E_e$, where $E < E_e$,

hence η_C must be negative. Thus the electrode potential moves to a more negative value.

Summarizing the discussions, on connecting the electrode at equilibrium to the positive terminal of an external current source, the electrode becomes the anode and the potential shifts in the noble direction; ΔG_A^* for the anodic process is made more favorable. On connecting the electrode to the negative terminal, the electrode becomes the cathode and the potential becomes more negative, ΔG_C^* for the cathodic process becomes more favorable.

3.3 HELMHOLTZ DOUBLE LAYER

When a metal dissolves in an aqueous solvent by release of cations (positively charged ion), it becomes negatively charged. As more and more ions are released, the metal surface becomes increasingly negatively charged. This process continues until an equilibrium is reached such that for any cation formed there must be a metal atom formed by the reverse process simultaneously.

$$M \rightleftarrows M^{z+} + ze$$

(Metal atom	(Hydrated ion
in lattice)	in solution)

At this point, the excess negative charge at the surface of the metal balances out the excess positive charge due to cations in solutions adjacent to the metal. Two layers having opposite charges, therefore, exist, one being negatively charged and the other positively charged (Fig. 3.4a). The separation of charges exists like in a capacitor. The two oppositely charged layers constitute the Helmholtz double layer. The double layer was compared to parallel plates by Helmholtz. A potential difference is thus created between the metal and solution. Under standard conditions, this potential difference is the standard electrode potential at the metal solution interface. Due to the separation of charges, a strong electric field in the space between the two charged layers is

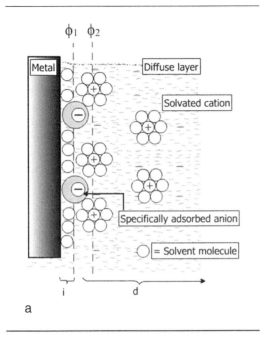

a

Figure 3.4a Electrical double layer showing inner Helmholtz, outer Helmhotz and diffuse layer ($\phi_1 =$ inner, $\phi_2 =$ outer)

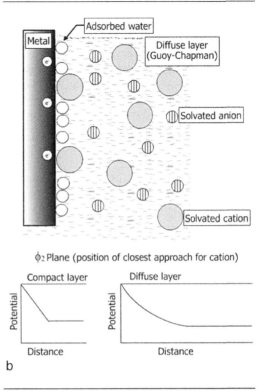

b

Figure 3.4b   Electrical double layer

set up. In the absence of external current, the electrode has a charged double layer, the capacitor C_{dl} is charged (Fig. 3.4b). The total impressed current, I_{total}, is divided into two parts, one crossing the double layer, through Faradic resistance (Faradic current I_F) on the other, I_{ch} (current of the charge), with a very small value. Faradic current is used in the electrode reaction, and I_{ch} is stored in the capacitor, C_{dl}. In a few seconds or less, the electrode charge reaches a steady value and it is proportional to the charge $Q = (I \times t)$ of the double layer (Fig. 3.4b), which is high enough to pull the ions across the double layer.

It is known that water is a dipolar molecule. The oxygen-end of a water molecule forms the negatively charged end and the hydrogen-end, the positively charged end. Due to the dipole nature, the water molecule is attracted towards the electrode and contributes to the potential difference across the double layer. The orientation of the water molecule depends on the charge on

the metal surface. If the metal has large negative charge, the H_2O molecules would be oriented with the positive ends (hydrogen) towards the metal, and negative ends (oxygen) towards the large positive charge. The water molecules are attracted towards the metal electrode and contribute to the potential difference. The water molecules form the first adsorbed layer on the metal surface. The cations which are hydrated and attracted towards the metal surface are limited in their approach to the metal surface because of the presence of water molecules on the metal surface. When electrostatic interaction operates, ions from the solution phase approach the electrode only as far as their solvation sheath allow. The surface arrays of these ions are thus cushioned from the electrode surface by a layer of solvent molecules. The line drawn through the center of such solvated cations at a distance of closest approach is called the '*outer Helmholtz plane.*' The anions are specifically adsorbed sometimes in the water molecules (solvent) (Fig. 3.4b).

The water molecules sometimes contain the specifically adsorbed anions. The water molecules form the inner Helmholtz layer. The line drawn through the center of these molecules is called the '*inner Helmholtz plane.*' The outer Helmholtz plane φ_2 (OHP) represents the locus of the electrical centers of the positive charges. This plane resides at a fixed distance from the metal due to the water molecules that are between the surface of the metal and ions. The outer Helmholtz plane (OHP) is significantly affected by hydrated cations M^{z+} (hydrated). In the simple model of the Helmholtz double layer developed earlier, the adsorption of dipoles was not considered. When a metal surface has a slight excess charge, the dipoles are adsorbed. This process contributes significantly to the potential difference across the double layer. Two factors seriously affect the potential difference across the double layer: (a) the magnitude of charges at the interface and (b) the presence of a layer of adsorbed dipole at the interface. The simple model presented above, however, takes no account of the presence of dipole layers, on the metal surface. It suggests that when the surface has excess negative charge, it would adsorb cations and when it is positively charged it would adsorb anions. However, it fails to take account of specific adsorption of ions (chemisorption) in contact with metal surface. The anions are held on the metal surface of the same charge either by chemical bonding or by electrostatic forces exerted by cations. Hence, they may be present in a solvent water molecule.

The model discussed above shows clearly two Helmholtz planes, the inner Helmholtz plane and the outer Helmholtz plane and it also shows how the anions are specifically absorbed. The Helmholtz double layer model is only applicable to a concentrated solution. Guoy and Chapman observed that the net charge in the compact double layer does not balance the charge on the metal surface. There is an additional region on the solution side of the layer, where either the cations are in excess of anions or anions are in excess of cations. These ions are distributed in a diffuse layer which extends from the outer Helmholtz plane to the bulk solution. Guoy and Chapman, therefore, proposed another charged layer, *Guoy–Chapman layer*, extending up to

1 μm from the outer Helmholtz plane in the bulk solution, the net charge being equal and opposite to that of the Helmholtz double layer. The size of the ions forming the OHP are such that the sufficient ions cannot fit these to neutralize the electrode, and the remaining charges which cannot fit that are held in increasing disorder. This less orderly arrangement constitutes the diffuse layer. As the concentration of charged ions in the Guoy–Chapman layer (Fig. 3.4b) increases, the thickness of the layer decreases. The ions are thus pushed down into the outer Helmholtz layer. It is analogous to a space charge layer in a semiconductor; where the thickness of diffusion layer decreases with an increase in the number of charge carriers. In case of concentrated solution, this layer is completely eliminated.

To summarize, the double layer consists of three constituents:

(a) An inner layer (inner Helmholtz layer) in which the potential changes linearly with the distance comprises the adsorbed water molecules and sometimes the specifically adsorbed anions.
(b) An outer Helmholtz layer. It comprises hydrated (solvated) cations and the potential varies linearly with the distance.
(c) An outer diffuse layer (Guoy–Chapman layer). It contains excess cations or anions distributed in a diffuse layer and the potential varies exponentially with the distance, *F*.

3.3.1 FACTORS AFFECTING EXCHANGE CURRENT DENSITY

A. **Forward reaction:** As described above, only those atoms which are energetically at unfavorable positions, such as at grain boundaries, dislocations, half planes, are able to detach themselves and participate in the reaction. Atoms are more easily pulled from the kink sites than terrace sites. The number of surface atoms available (N_s) in a given area can be calculated.
B. **Electrode composition:** It depends upon the composition of electrode (see Table 3.1). The exchange current density for Pt is 10^{-3} A/cm^2,

Table 3.1 Exchange current densities

Reaction	Electrode	Solution	i_0 (A/cm^2)
$2H^+ + 2e = H_2$	Al	2 N H$_2$SO$_4$	10^{-10}
$2H^+ + 2e = H_2$	Au	1 N HCl	10^{-6}
$2H^+ + 2e = H_2$	Cu	0.1 N HCl	2×10^{-7}
$2H^+ + 2e = H_2$	Fe	2 N H$_2$SO$_4$	10^{-6}
$2H^+ + 2e = H_2$	Hg	1 N HCl	2×10^{-12}
$2H^+ + 2e = H_2$	Hg	5 N HCl	4×10^{-11}
$2H^+ + 2e = H_2$	Ni	1 N HCl	4×10^{-6}
$2H^+ + 2e = H_2$	Pb	1 N HCl	2×10^{-12}
$2H^+ + 2e = H_2$	Pt	1 N HCl	10^{-3}
$2H^+ + 2e = H_2$	Pd	0.6 N HCl	2×10^{-4}
$2H^+ + 2e = H_2$	Sn	1 N HCl	10^{-8}
$O_2 + 4H^+ + 4e = 2H_2O$	Au	0.1 N NaOH	5×10^{-13}
$O_2 + 4H^+ + 4e = 2H_2O$	Pt	0.1 N NaOH	4×10^{-13}
$Fe^{+3} + e = Fe^{+2}$	Pt		2×10^{-13}
$Ni = Ni^{+2} + 2e$	Ni	0.5 N NiSO$_4$	10^{-6}

Source: Bockris, J. O'M. (1953). Parameters of electrode kinetics. *Electrochemical Constants*, NBS Circular 524, U. S. Government Printing Office, Washington, DC, 243–262.

whereas for mercury Hg it is 10^{-13} A/cm^2 in dilute acid.

C. **Surface Roughness:** Large surface areas provide a high exchange current.

D. **Impurities:** The exchange current density is reduced by the presence of trace impurities, such as As, S and Sb which are catalyst poisons.

Example 9

The exchange current density for $2H^+ + 2e \rightarrow H_2$ on mercury is 10^{-12} A/cm^2, but is nine orders of magnitude higher on platinum at 10^{-3} A/cm^2, which indicates Pt provides a favorable catalytic surface for the above reaction to occur rapidly on platinum. A large current density means a very stable potential which cannot be easily disturbed, hence, platinized platinum is made a standard hydrogen electrode, so that the potential remains stable.

It is well-known that most reactions occur very rapidly on the platinum surface. Platinized platinum has a larger surface area and consequently a higher exchange current density than pure platinum. It is, therefore, widely used as a catalyst because of the rapidity of the reactions which occurs on its surface.

Consider a forward anodic oxidation process:

$$M \rightarrow M^{z+} + ze$$

Note: The term anodic oxidation and anodic dissolution is used interchangeably in this text.

The total number of moles of atoms dissolving in a unit time, according to classical rate theory is given by

$$\text{Forward rate, } r_f = f\left(\alpha \frac{N_s}{N_0}\right)\exp\left[\frac{\Delta G^*}{RT}\right]$$

$$(r = \text{mol} \cdot \text{cm}^{-2} \cdot \text{s}^{-1}) \quad (3.18)$$

where

α = fraction of active atoms, which are chemically active

f = frequency factor (10^{12} s^{-1})

r = rate of forward reaction (mol/cm$^2 \cdot$ s)

The rate $\vec{r}$ can be converted to a current density term by multiplying $\vec{r}$ by zF

(zF = coulombs/mole) (z = No. of charges transferred)

$$r \left(\text{in mol/cm}^2 \cdot s \right) \times zF \left(\text{in coulomb/mol} \right)$$

As 1 coulomb (C) is equivalent to 1 ampere-second, we obtain A/cm^2 after multiplication, which is the unit of current density

$$\vec{i} = zF\vec{r} = \text{A/cm}^2$$

Equation (3.18) can now be rewritten as:

$$\vec{i_0} = zFf\alpha \left(\frac{N_s}{N_0} \right) \exp \left[-\frac{\Delta G^*}{RT} \right] \qquad (3.19)$$

where i_0 is the exchange current density.

3.4 REVERSE REACTION (CATHODIC REACTION)

Let us now consider the conditions under which a cation would be transformed back to a metallic

atom by the reverse process. Energy–distance profile for a reduction reaction is shown in Fig. 3.5. The rate of the reverse process is now given by i_r. Suppose C_s (mol/cm^2) is the concentration of metal cations, and V_L is the volume of this part of the Helmholtz layer, and if it is further assumed that all the $C_s V_L$ ions in the double layer are available for the reverse reaction $\overleftarrow{i_0}$, the magnitude of the cathodic partial current would be

$$\overleftarrow{i}_{0(\text{rev})} = zFf\, C_s V_L \exp \left[-\frac{\Delta G^*}{RT} \right]$$

where $C_s = 10^{-3}$ mol/cm^2, and $V_L = 10^{-8}$ cm^3 (typical values). At equilibrium $\vec{i_0} = \vec{i_f} = -i_{(\text{rev})}$.

The reverse current has a negative sign, however, i_0 is the magnitude of both forward and reverse process. As the cations are involved in the reverse process, N_s/N_0 (equation (3.18)) is replaced by $(C_s V_L)$.

$$\overleftarrow{i}_{(\text{rev})} = -zFfC_s V_L \exp \left(-\frac{\Delta G^*}{RT} \right)$$
$$\qquad (3.20)$$

$$\overleftarrow{i}_{(\text{rev})} = -zFf\alpha(C_s V_L) \exp \left[-\frac{\Delta G^*}{RT} \right]$$

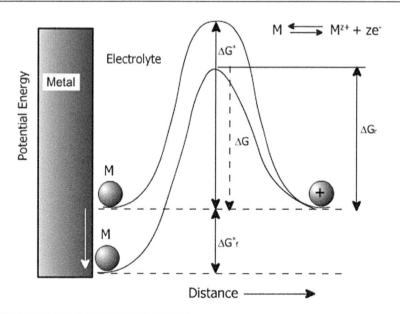

Figure 3.5 Energy–distance profile for a reduction reaction (r = reduction, f = forward reaction)

At equilibrium:

$$i_0 = \vec{i} = -\overleftarrow{i}_{(\mathrm{rev})} = zFf\alpha\left(\frac{N_\mathrm{s}}{N_0}\right)\exp\left(\frac{-\Delta G^*}{RT}\right)$$

$$= zFfC_\mathrm{s}\,V_\mathrm{L}\exp\left(\frac{-\Delta G^*}{RT}\right) \qquad (3.21)$$

Equation (3.21) shows that at i_0, the rate of dissolution is equal to the rate of deposition as discussed in Section 3.2.2.

3.5 DEPARTURE FROM EQUILIBRIUM [ACTIVATION OVER-POTENTIAL (η)]

When a metal electrode is in equilibrium, the partial current $\vec{i}_\mathrm{f}$ for forward reaction and partial current $\overleftarrow{i}_\mathrm{r}$ for reverse reaction are precisely equal and opposite. There is no net current flow.

If the potential drop across the double layer (Section 3.3) is altered by superimposing an external electromotive force (emf), the electrode is polarized and hence there is a deviation from the equilibrium condition. The extent of polarization is measured by the change in the potential drop (ΔE) across the double layer. The shift of potentials from their equilibrium value on application of an external current is called polarization. The magnitude of the deviation is termed 'over-voltage' (η) which is directly proportional to the current density. Due to polarization, an imbalance is introduced in the system. If $d(\Delta E)$ is positive, a net anodic current flows, $i_\mathrm{a}^{\mathrm{net}} = \vec{i}_\mathrm{f} \rightleftarrows \overleftarrow{i}_\mathrm{r}$, where $\vec{i}_\mathrm{f} \gg \overleftarrow{i}_\mathrm{r}$, and if $d(\Delta E)$ is negative, the electrons are pushed in the metal by connecting the electrode to the negative terminal of an external source of emf, a net cathodic current would flow, $i_\mathrm{c}^{\mathrm{net}}(\vec{i}_\mathrm{f} \rightleftarrows \overleftarrow{i}_\mathrm{r})$; here $\overleftarrow{i}_\mathrm{r} \gg \vec{i}_\mathrm{f}$.

To summarize, electrons are supplied to the surface and, hence, the surface potential (E) becomes negative to the equilibrium potential (E_e). On the other hand, in anodic polarization, electrons are removed from the metal surface which cause the potential (E) to become positive to the equilibrium potential, E_e. At the corrosion potential (E_corr), the rate of forward reaction (oxidation) $\vec{i}_\mathrm{f}$ equals the rate of reverse reaction, $\overleftarrow{i}_\mathrm{r}$ and hence the over-voltage ($\eta = 0$) (Fig. 3.6).

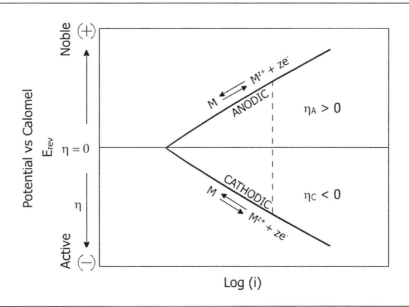

Figure 3.6 Polarization diagram (reversible electrode)

3.5.1 ACTIVATION POLARIZATION

Activation polarization can be a slow step in the electrical reaction for which an activation energy in the form of potential is required for the reaction to proceed. When a certain step in a half cell reaction controls the rate of electron flow, the reaction is said to be under *activation charge transfer* control and activation polarization occurs. For example, consider the reduction of hydrogen ions:

$$2H^+ + 2e^- \rightarrow H_2$$

Hydrogen evolution occurs in four major steps:

(1) 1st Step – H^+ is absorbed in the surface at the electrode.
(2) 2nd Step – $H^+ + e^- \rightarrow H_{ad}$, the species is reduced on the surface.
(3) 3rd Step – The two reduced species combine to form a hydrogen molecule $H_{ad} + H_{ad} \rightarrow H_2$.
(4) 4th Step – Hydrogen bubbles are formed by combination of hydrogen molecules.

The rate of hydrogen reduction is determined by the slowest of the steps. The rate, controlling step, varies with the metals, current density and environment. There is a critical activation energy needed to surmount the energy barrier associated with the slowest step. The rate of transformation is controlled by the magnitude of the energy barrier that an atom or ion must surmount to transform from metal to ion or from ion to metal. The energy that must be acquired is the activation energy ΔG^*.

The relationship between activation polarization and the rate of reaction is given by the Tafel equation:

$$\eta_A = \beta_a \log \frac{i_a}{i_0}$$

and

$$\eta_C = \beta_c \log \frac{i_c}{i_0}$$

where η is over-potential. A plot of overpolarization (η_{act}) vs $\log(I)$ is linear for both anodic and cathodic polarization. β_a and β_c are called the Tafel slopes and i_0 is the exchange current density (Fig. 3.7).

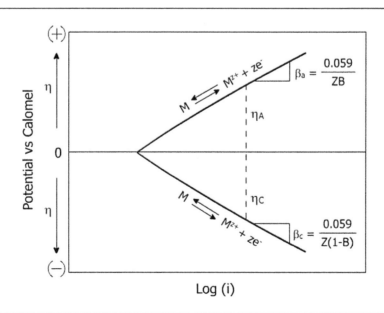

Figure 3.7   Polarization diagram of zinc in contact with its own ions

3.5.2 FACTORS AFFECTING ACTIVATION POLARIZATION

A. **Current Density:** Activation polarization increases with current density i, in accordance with the equation:

$$\eta = \pm \beta \log \frac{i}{i_0}$$

B. **Materials:** Activation polarization varies with one metal to another because of the specific effect of current density.
C. **Surface Roughness:** Activation polarization is high on a smooth surface compared to a shiny surface.
D. **Temperature:** Increased temperatures decrease polarization as less activation energy would be needed and the exchange current density would be increased.
E. **Pressure:** Hydrogen over-voltage increases rapidly with decreasing pressure.
F. **pH:** Over-voltage increase initially and decreases with increased pH value.
G. **Agitation:** It has no effect on activation polarization, because it is a charge transfer process involving electrons and not a mass transfer.
H. **Adsorption of Ions:** The hydrogen over-voltage is decreased by adsorption of anions and increased by adsorption of cations.

3.5.3 ANALYTICAL EXPRESSIONS FOR IRREVERSIBLE FORWARD AND REVERSE REACTION

It has been shown that when the rate of forward reaction $\vec{i_f}$ is equal to the rate of reverse reaction $\overleftarrow{i_r}$, there is no net accumulation of charges and no current flows in the external circuit, $M \rightleftarrows M^{z+} + ze$, The potential across the metal solution interface in the double layer is the equilibrium potential. The energy–distance profile for the reversible electrode is shown by broken lines (Fig. 3.5). The original position of the atoms is shown in the energy wells. By supplying an external emf, the atom has been pushed in a new position in a higher energy well. The over-voltage

required to shift the position of the atom from the reversible electrode is given by η.

In the forward reaction (anodic polarization), $M \rightarrow M^{z+} + ze$, the electrode is connected to the positive terminal external emf and the potential superimposed η_A is positive. It is greater than zero and the electrode polarizes in the noble direction. For an activated ion on the top of an energy hump, the magnitude of activation energy ΔG_A^* required is the total activation energy of the barrier $[\Delta G^* + (1-\beta)zF\eta_A]$ minus the over-potential $(zF\eta_A)$ applied. The magnitude of the total activation energy ΔG_A^* (the energy state of metal is increased) is given by:

$$\Delta G_A^* = \Delta G^* + [(1-\beta)zF\eta_A] - zF\eta_A \quad (3.22)$$

Hence

$$\Delta G_A^* = \Delta G - \beta zF\eta_A \quad (3.23)$$

But

$$\vec{i_f} = zFf\alpha\left(\frac{N_s}{N_0}\right)\exp\left[-\frac{\Delta G_A^*}{RT}\right] \quad (3.24)$$

After substitution

$$\vec{i_f} = zFf\alpha\left(\frac{N_s}{N_0}\right)\exp\left[-\frac{\Delta G^* - \beta zF\eta_A}{RT}\right] \quad (3.25)$$

The above expression can be written in terms of equation (3.23), hence

$$\vec{i_f} = zFf\alpha\frac{N_s}{N_0}\exp\left[-\frac{\Delta G^*}{RT}\right] \times \exp\left[\frac{\beta zF\eta_A}{RT}\right] \quad (3.26)$$

But $i_0 = zFf\alpha\dfrac{N_s}{N_0}\exp\left[-\dfrac{\Delta G^*}{RT}\right]$ by equation (3.19) (where i_0 is the exchange current density). Therefore,

$$\vec{i_f} = i_0 \exp\left[\frac{\beta zF\eta_A}{RT}\right] \quad (3.27)$$

Equation (3.27) suggests that as the anodic activation over-voltage increases, the rate of forward reaction also increases.

3.5.4 REVERSE REACTION (CATHODIC DEPOSITION)

If the reaction is now reversed, by connecting the electrode to the negative terminal of external emf:

$$M \xleftarrow[\overrightarrow{i_f}]{\overleftarrow{i_r}} M^z + ze, \quad |\overleftarrow{i_r}| \gg \overrightarrow{i_f}$$

Here, the rate of reverse reaction $\overleftarrow{i_r}$ is much higher than the rate of forward reaction $\overrightarrow{i_f}$. The over-potential η_C is less than zero. The activation energy for the forward reaction is higher than the reverse reaction. The total available energy for cathodic reaction is:

$$\Delta G_C^* = \Delta G^* + (1-\beta)zF\eta_C \qquad (3.28)$$

As shown earlier,

$$i_0 = zFf(C_s V_L)\exp\left[-\frac{\Delta G_C^*}{RT}\right] \text{(equation (3.24))}$$

Replacing ΔG_C^* in equation (3.24) by the expression in equation (3.28), we obtain:

$$\overleftarrow{i_r} = zFfC_s V_L \exp\left[-\left(\frac{\Delta G^* + (1-\beta)zF\eta_C}{RT}\right)\right] \qquad (3.29)$$

$$\overleftarrow{i_r} = -i_0 \exp\left[-\frac{(1-\beta)zf\,\eta_C}{RT}\right] \qquad (3.30)$$

Equation (3.30) shows that as the cathodic over-voltage is increased, the rate of cathodic reaction, i_r, is increased and the rate of forward reaction is decreased.

To sum up, the net current density under the two conditions when $\overrightarrow{i_f} \gg \overleftarrow{i_r}$ and $|\overleftarrow{i_r}| \gg \overrightarrow{i_f}$, is given by

$$i_{net} = |i_a| - |i_c| = i_0 \left\{\exp\left[\frac{\beta zF\eta_A}{RT}\right] - \exp\left[-\frac{(1-\beta)zF\eta_C}{RT}\right]\right\} \qquad (3.31)$$

which is the general form of current *vs* potential relationship. The above equation is called the Butler–Volmer equation.

Let us now consider the departure of the anodic and cathodic reactions from the equilibrium potential at large enough voltages, approximately >0.12 volt. Under the above conditions one of the two terms (partial current) becomes negligible, only one reaction would prevail, and the other would become negligible.

Consider anodic polarization ($\eta > 0$) only at voltages higher than 0.1 volt. The current (i_a^{net}) would be equal to the anodic partial current density and i_c would be negligible (the reverse cathodic reaction). Under this condition the right-hand expression of equation (3.31) would be eliminated, however, for the anodic reaction

$$i_a^{net} = i_0 \exp\left[\frac{\beta zF\eta_A}{RT}\right] \qquad (3.32)$$

which relates the partial anodic current density to the over-potential. By similar arguments, at a large enough cathodic polarization [η (negative)], the anodic partial current becomes negligible and the current density is

$$i_c^{net} = -|i_c| = i_0 \exp\left[-\frac{(1-\beta)zF\eta_C}{RT}\right] \qquad (3.33)$$

which shows that the anodic or cathodic current densities vary approximately as the exponential of the over-voltage.

When $|\overleftarrow{i_r}| \gg \overrightarrow{i_f}$; (for $\eta_C > 0.03$ V).

$$|i_c^{net}| = i_0 \exp\left[-\frac{(1-\beta)zF\eta_C}{RT}\right] \qquad (3.33a)$$

The expression derived above for i_c^{net} and i_a^{net} equations (3.32) and (3.33) can be written in terms of (η), the over-voltage. Taking logarithms of equations (3.32) and (3.33)

$$\ln \overrightarrow{i_a} - \ln i_0 = \frac{\beta zF\eta_A}{RT} \rightarrow \left(\text{when } \overrightarrow{i} \gg |\overleftarrow{i_r}|\right)$$

$$\eta_A = \frac{-\ln i_0 + \ln i_a}{\beta zF/RT}$$

$$\eta_A = \left(-\frac{RT}{\beta zF}\right)\ln\left(\frac{i_a}{i_0}\right)$$

$$\eta_A = \left(-\frac{RT}{\beta zF}\right)\ln i_0 + \left(\frac{RT}{\beta zF}\right)\ln i_a$$

For over-potential:

$$\eta = (>0), \eta_A = \left(-\frac{RT}{\beta zF}\right)\ln i_0 + \left(\frac{RT}{\beta zF}\right)\ln i_a$$

$$\eta_A = \left(\frac{RT}{\beta zF}\right)\ln\left(\frac{i_a}{i_0}\right) \tag{3.34}$$

Similarly for cathodic polarization, on taking logarithms,

$$\ln|\bar{i_c}| - \ln i_0 = -\left[\frac{(1-\beta)zF\eta_C}{RT}\right] \rightarrow \eta_C$$

$$= -\frac{RT}{(1-\beta)zF}\ln\left[\frac{|i_c|}{i_0}\right]$$

$$\eta_C = \frac{\ln i_0 - \ln|i_c|}{(1-\beta)zF/RT}$$

$$= \left[\frac{RT}{(1-\beta)zF}\right]\ln i_0 - \left[\frac{RT}{(1-\beta)zF}\right]\ln|i_c|$$

for $(\eta < 0)$

$$\eta_C = \left[\frac{RT}{(1-\beta)zF}\right]\ln i_0 = \left[\frac{RT}{(1-\beta)zF}\right]\ln|i_c|$$

$$\eta_C = \frac{-RT}{(1-\beta)zF}\ln\left(\frac{|i_c|}{i_0}\right)$$

$$\tag{3.35}$$

where

$\eta = (E - E_{rev})$
$\eta_A > 0$, anodic polarization
$\eta_C < 0$, cathodic polarization.

The experimental equations obtained by Tafel are very similar to equations (3.34) and (3.35), where the Tafel constant, b_a and b_c comprise of equation (3.32) by equation (3.34), and equation (3.33) by equation (3.35), are

$$b_a = \frac{2.3RT}{(1-\beta)zF}$$

$$b_c = \frac{2.3RT}{(1-\beta)zF}$$

The matter is further explained in Section 3.6.

3.6 TAFEL EQUATION

The relation between the over-voltage (η) and the reaction rate is extremely important. Consider anodic polarization only. At high anodic over-voltage, $i_r \gg i_f$, i_c is negligible and the current density (i_a^{net}) becomes equal to the anodic partial current density,

$$i_a^{net} = i_a = i_0 \exp\left[\frac{\beta zF\eta_A}{RT}\right] \quad \text{(equation (3.31))}$$

Taking logarithms, we get:

$$\ln i_a = \ln i_0 + \frac{\beta zF\eta_A}{RT} \rightarrow \eta_A$$

$$= \frac{RT}{\beta zF}\ln i_a - \frac{RT}{\beta zF}\ln i_0 \tag{3.36}$$

hence,

$$\eta_A = \frac{RT}{\beta zF}\ln\left(\frac{i_a}{i_0}\right) \tag{3.37}$$

On the other hand, at large enough cathodic polarization (η negative), the anodic partial current becomes negligible, hence

$$i_c^{net} = i_c = i_0 \exp\left[-\frac{(1-\beta)zF\eta_C}{RT}\right]$$

$$\text{(from equation (3.33))}$$

Taking logarithms, $\ln i_c = \ln i_0 + \dfrac{(1-\beta)zF\eta_C}{RT}$

$$\eta_C = \frac{RT}{(1-\beta)zF}\ln i_c - \frac{RT}{(1-\beta)zF}\ln i_0 \tag{3.38}$$

Equations (3.37) and (3.38) are Tafel equations. The above expressions can be written in a more generalized form of equation (Tafel equation)

$$\eta = a + b\log I \tag{3.39}$$

where a and b are constants and b is the Tafel slope. The left-hand term of equation (3.38) represents the constant a and the right-hand term constant b. The following are the empirical relationships between the current I and over-voltage η for the anodic and cathodic over-voltage as shown in equations (3.34) and (3.35).

$$\eta_A = b_a \log\left(\frac{i_a}{i_0}\right) \qquad (3.39a)$$

$$\eta_C = b_c \log\left(\frac{i_c}{i_0}\right) \qquad (3.39b)$$

From equation (3.36), constant a is $\left(-\dfrac{RT}{\beta zF}\right)$ $\log i_0$ and constant b, the slope is $\dfrac{RT}{\beta zF}$. It is more convenient to express these in the form of log to base of 10:

$$a = -\left(\frac{2.3RT}{\beta zF}\right) \log i_0 \qquad (3.39c)$$

$$b = \left(\frac{2.3RT}{\beta zF}\right) \qquad (3.39d)$$

In the expression (3.39c), appropriate values are inserted.

$T = 25 + 273 = 298\,^\circ K$

$R = 8.314\,J/deg.mol$

$F =$ Faraday's constant in coulombs/mole

2.303 is a conversion factor to convert natural log to the base 10

and $2.303 RT/F = 0.5916\,V$.

On inserting these values, the term in equation (3.39c) at 25°C becomes $\dfrac{0.059}{\beta z} \log i_0$.

Hence,

$$\frac{0.059}{\beta z} \log i_0 = a_a = -\left(\frac{2.3RT}{\beta zF}\right) \log i_0 \qquad (3.40)$$

The symmetry factor (β) represents the location of the energy hump. It ranges from 0 to 1 as shown in the energy profile diagram, however, to locate it precisely, it is extremely difficult. Generally, it is assumed that it is located in the center; here $\beta = 0.5$.

The slope $b = \left(\dfrac{2.3RT}{\beta zF}\right)$ (equation (3.39d)) inserting appropriate value again as in the previous case, $\dfrac{2.3RT}{\beta zF} = \dfrac{0.0592}{\beta z}$, hence,

$$\frac{0.0592}{\beta z} = b_a = \left(\frac{2.3RT}{\beta zF}\right) \qquad (3.41)$$

Similarly for cathodic polarization

$$a_c = \frac{0.059}{(1-\beta)z} \log i_0 \qquad (3.42)$$

$$b_c = -\frac{0.059}{(1-\beta)z} \qquad (3.43)$$

The Tafel equation is generally written as

$$\eta_A = \pm\beta \log \frac{i}{i_0} \quad (\eta_A = \text{activation polarization})$$

where

$\beta =$ constant, being positive for anode and negative for cathodic reaction

$i =$ the net rate of reaction

$i_0 =$ exchange density.

The situation is different if the over-voltage is low $\left(\eta \ll \dfrac{RT}{zF}\right)$; the electrode potential lower than 0.03 V. Under such conditions, the partial current densities are negligible. Equation:

$$i = |i_a| - |i_c|$$

$$= i_0 \exp\left[\frac{\beta zF\eta}{RT}\right] - \exp\left[-\frac{(1-\beta)zF\eta}{RT}\right]$$

can be expressed into a series:

$$e^x = 1 + x + \frac{x^2}{2^1} + \frac{x^3}{3^1} + \frac{x^{2n}}{n^1}$$

$$\rightarrow e^x = 1 + n$$

$$i = i_0 \left\{ \exp\left[\frac{\beta zF\eta}{RT}\right] - \exp\left[-\frac{(1-\beta)zF\eta}{RT}\right]\right\}$$

$$i = i_0 \left[+\frac{\beta zF\eta}{RT} - \frac{(1-\beta)zF\eta}{RT}\right]$$

$$\rightarrow i_a = \frac{i_0 zF}{RT}\eta$$

$$\rightarrow i_c = \frac{i_0 zF}{RT}\eta$$

Thus, the over-voltage is proportional to current density.

Consequently, polarization resistance can be defined as

$$R_P = \left(\frac{dn}{di}\right)_{i\rightarrow 0} = \frac{RT}{i_0 zF}$$

A low field approximate, $\eta < 0.002$, the hyperbolic sine function, approximates to a linear function so that

$$i_{net} = 2i_0 \sinh = \frac{zF\eta}{RT} = \frac{i_0 zF\eta}{RT} \qquad (3.44)$$

which is the lower field approximation for the Butler–Volmer equation.

3.7 Mixed Potential Theory and its Application

3.7.1 Multiple Electrodes

In a corrosion cell (electrolyte, anode, cathode and a metallic path), multiple reactions occur. For instance, when zinc corrodes in a dilute acid, the following reactions occur:

$$Zn \rightleftarrows Zn^{2+} + 2e \quad \text{(oxidation)}$$
$$\text{(Forward reaction, } \vec{i_f})$$

$$H_2 \rightleftarrows 2H^+ + 2e \quad \text{(reduction)}$$
$$\text{(Reverse reaction, } \overleftarrow{i_r})$$

The total reaction is $Zn + 2H^+ \rightarrow Zn^{2+} + H_2\uparrow$.

Accordingly, the metal constitutes a multi-electrode, because at least two different reactions occur on its surface simultaneously, one oxidation and one reduction.

The mixed potential theory partly mentioned earlier, is used with advantage to predict the rate of corrosion of metals and alloys in given environment. It was postulated by Wagner and Traud in 1938. It has two basic assumptions:

(a) Electrochemical reactions are composed of two or more partial anodic and cathodic reactions.
(b) There cannot be any accumulation of charges.

This theory was not applied until 1950, when Stern applied it in analysis of corrosion.

Consider a piece of zinc immersed in $ZnSO_4$. The reactions would be $Zn \rightarrow Zn^{2+} + 2e^-$ (oxidation) and $Zn^{2+} + 2e^- \rightarrow Zn$ (reduction). For the two reactions at rest an equilibrium potential would be observed which would be given by Nernst equation. Consider now an electrode, such as zinc, immersed in HCl. The following could be the possible reactions:

(a) $Zn \rightarrow Zn^{2+} + 2e^-$ (anodic)
(b) $2H + 2e^- \rightarrow H_2$ (cathodic)
(c) $Zn^{2+} + 2e^- \rightarrow Zn$ (cathodic)
(d) $H_2 \rightarrow 2H^+ + 2e^-$ (anodic)

Bubbles of hydrogen are observed from the surface of zinc electrode, and formation of bubbles of hydrogen is a cathodic reaction. Hydrogen is reduced and not oxidized. Similarly, zinc is oxidized and not reduced. Hence, only the two reactions (a) and (b) proceed. Under the condition of rest (no outside current), the potential of the electrode cannot be computed by the Nernst equation as it is not reversible. Also, the above electrode would not corrode in the absence of an external current. The potential assumed by the electrode under the above condition is the mixed potential and its value lies between the value of equilibrium potential of hydrogen and zinc. The value of the potential would depend on the metal and the environment. It is to be observed that the corrosion potential (E_{corr}) is not the equilibrium potential of either of the reactions, but some intermediate potential determined by the two partial anodic and cathodic reactions. Both the reactions

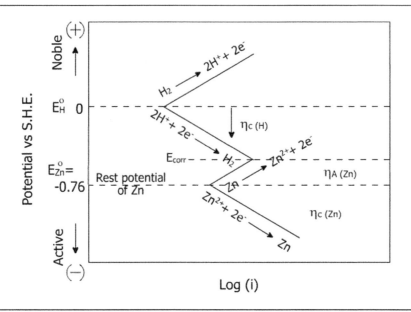

Figure 3.8 Diagram illustrating mixed potential theory

(a) and (b) are polarized to a common potential, the mixed potential (E_{mix}) (Fig. 3.8).

3.7.2 CONSTRUCTION OF A POLARIZATION DIAGRAM

A Tafel plot can be obtained by plotting the rate of corrosion in terms of current (I) *vs* the applied potential (η = over-voltage), either above or below the reversible potential (E_{rev}) (Fig. 3.6). The *x*-axis is the potential axis and *y*-axis is the current axis. Firstly, a zero over-potential point ($\eta = 0$) is located on the axis. This corresponds to the reversible potential of the system to be investigated, because at E_{rev}, $\eta = 0$. There is no driving force when $\eta = 0$, and so it provides a good starting point. At E_{rev},

$$M \rightleftarrows M^{z+} + ze$$

$$i_0 = \vec{i} - \overleftarrow{i}, \eta = 0$$

$$i^{net} = 0$$

For anodic polarization, a predetermined over-potential (η) is applied step-by-step in the noble direction. The upper curve, anodic polarization, corresponds to the reaction M $\rightleftarrows$ M^{z+} + ze and hence $i_a^{net} = \vec{i}_f - \overleftarrow{i}_r$ and $\eta_A > 0$. The variation of η with $|I|$ is linear. The departure from the equilibrium is shown by the over-potential (η).

Similarly, for cathodic polarization, starting from E_{rev}, the over-voltage (η), is increased in the more active cathodic direction (negative)

$$M \rightleftarrows M^{z+} + ze$$

hence,

$$I_{cathode}^{net} = \vec{i} - \overleftarrow{i}$$

$$\eta_C < 0$$

As η is increased, the rate of cathodic reduction (cathodic polarization) increases. The slope of the anodic portion of the curve is given by $\beta_a = -0.059/z\beta$ and that of the cathodic portion by $\beta_c = -0.059/z(1-\beta)$.

A linear relationship is observed in the Tafel plot. When the activity of the ion is unity, the reversible potential is the same as the standard potential measured against hydrogen. If zinc in contact with its ions at unity is polarized, it would be oxidized at more noble potentials: $M \rightleftarrows M^{z+} + ze$ and reduced at more active potentials $M^{z+} + ze \rightleftarrows M$.

$E_{rev}(Zn) = -0.763\,V_{SHE}\ (a_{Zn^{2+}} = 1)$. It is an example of a single electrode reaction when the metal is in contact with its ion in the solution. Unit activity of ions is 1 g ion per dm^3 or 1 g ion per liter.

A theoretical polarization diagram for the anodic and cathodic polarization of zinc is shown in Fig. 3.7. The linear Tafel relationship is illustrated in both diagrams.

The mixed potential diagram or *E vs I* diagram allows accurate prediction of corrosion rate to be made in specific environments. It also allows the prediction of current based on current density higher than the corrosion current (i_{corr}) and utilization of advanced technique for corrosion measurement, such as galvanostatic and potentiostatic techniques. These techniques will be described at the end of the chapter.

3.8 EVANS DIAGRAMS

The Evans diagram shows the electrode potential in volts in the ordinate and the reaction rate (ampere) in the abscissa (Fig. 3.8). Consider a base metal, such as iron or zinc, placed in acidic solution; the metal will dissolve at the same rate as hydrogen is evolved and the two reactions mutually polarize each other. This is shown in a very simple form by Evans diagram. In Evans diagram, either current or current density can be plotted against potential. If the ratio of anodic to cathodic areas is taken as unity, current density (i) may be used rather than current. A negligible resistance is assumed between the anode and cathode. As the change in the anodic and cathodic polarization has the same effect on corrosion current, the system is considered to be under mixed control.

Consider zinc corroding in dilute HCl. The following two reactions occur:

$$Zn \rightleftarrows Zn^{2+} + 2e \ (oxidation)$$

$$2H^+ + 2e \rightleftarrows H_2 \ (reduction)$$

$$E^\circ_{Zn} = -0.763\,V_{(SHE)}$$

A corroding electrode constitutes a polyelectrode because a minimum of two electrode reactions, oxidation and reduction, occur simultaneously. In case of a half cell, such as Fe/Fe^{2+}, H_2, Pt, parameters, such as current and individual electrode potentials, may be measured as a function of change in the external resistance. It is, however, impossible to measure the two parameters on the tiny anodic and cathodic areas of corroding metal. It is, however, possible to measure the mixed potential and the net current.

In case of zinc corroding in an acid, $Zn \underset{i_r}{\overset{i_f}{\rightleftarrows}} Zn^{2+} + 2e$, the following are the values of E° (potential, V) and i_0 (exchange current density):

$$E^\circ, Zn\,|\,Zn^{2+} = -0.76\,V$$

$$i_0, \left(Zn\,|\,Zn^{2+}\right) = 10^{-7}\,A/cm^2$$

and for $H_2 \rightleftarrows 2H^+ + 2e$, on $Zn = 10^{-10}\,A/cm^2 = 10^{-6}\,A/m^2$, $E^\circ_{H_2/H^+} = 0$ volts.

The difficulty in the measurement of individual current and potential is resolved by Evans diagram. The ratio of anodic to cathodic areas is taken as unity and current (I) is replaced by current density (i). On polarizing anodically, the potential is displaced in the positive direction, and on cathodic polarization, it is displaced in the negative direction. The intersection of anodic and cathodic curve gives E_{corr}, the corrosion potential, which can be measured experimentally. The potential corresponding to i_{corr} (corrosion current) is corrosion potential (E_{corr}). E_{corr} represents the contribution of both anode and cathode to polarization, hence, it represents a mixed potential. The rate of corrosion is given by i_{corr}, which represents the total current associated with the dissolution of zinc and evolution of hydrogen. The following are the reactions that take place as shown in Fig. 3.8:

(A) $Zn \underset{i_r}{\overset{i_f}{\rightleftarrows}} Zn^{2+} + 2e(1)$ or $Zn \rightarrow Zn^{++} + 2e$

and $Zn \underset{\overrightarrow{i_f}}{\overset{\overleftarrow{i_r}}{\rightleftarrows}} Zn^{++} + 2e(1a)$ or

$Zn^{2+} + 2e \rightarrow Zn(i_f \geq i_r)$

(B) $H_2 \underset{i_r}{\overset{i_f}{\rightleftarrows}} 2H^+ + 2e$ (2)   or $H_2 \rightarrow 2H^+ + 2e$

and $H_2 \underset{i_r}{\overset{i_f}{\rightleftarrows}} 2H^+ + 2e$ (2a) or $2H^+ + 2e \rightarrow$

$H_2 (i_r \geq i_f)$

(C) $Zn + 2H^+ \rightarrow Zn^{2+} + H_2$ (3) (total reaction)

Consider polarization (forward reaction) of zinc:

$$Zn \rightleftarrows Zn^{2+} + 2e \text{ (4)}$$

In the forward reaction, the oxidation rate of Zn^{2+} exceeds the reduction rate of zinc (equations (1) and (1a)). Similarly, the rate of reduction of hydrogen ions exceeds the rate of oxidation of hydrogen (equations (2) and (2a)).

$$H_2 \rightleftarrows 2H^+ + 2e \text{ (5) (cathodic reduction}$$
$$\text{of hydrogen)}$$

Equation (4) represents the anodic oxidation of zinc and equation (5), the cathodic reduction of hydrogen.

As observed above, four separate corrosion reactions are shown for zinc:

(1) $Zn \rightleftarrows Zn^{2+} + 2e$ (zinc oxidation) $(i_f \geq i_r)$
(2) $Zn^{2+} + 2e^- \rightleftarrows Zn$ (zinc reduction) $(i_r \geq i_f)$

For hydrogen:

(3) $H_2 \rightleftarrows 2H^+ + 2e$ (hydrogen oxidation)
    $(i_f \geq i_r)$
(4) $2H^+ + 2e^- \rightleftarrows H_2$ (hydrogen reduction)
    $(i_r \geq i_f)$

Of the four curves corresponding to the above reactions, only reactions (1) and (4) are of practical interest. Reaction (3) hydrogen oxidation, and reaction (2) zinc reduction, can be ignored because zinc can only be reduced at potentials more negative than -760 V and it continues to generate electrons with time. Hydrogen evolution can be clearly observed and it does not oxidize. Hence, only reactions (1) and (4) are of practical interest. The curves corresponding to the above reactions converge at a point of intersection called 'corrosion potential,' represented by

E_{corr}, which is the mixed potential of the system. The corresponding current at the current axis is called 'corrosion current,' represented by i_{corr}. At the corrosion potential, the main anodic and cathodic reactions are two different reactions, whereas, at an equilibrium potential, a single partial reaction occurs in both anodic and cathodic direction.

The corrosion rate given by i_{corr} is equal to the total anodic and cathodic rates of corrosion.

3.9 PREDICTION OF CORROSION TENDENCY ON THE BASIS OF MIXED POTENTIAL THEORY

According to the emf series, zinc has a more active potential than iron, $E°(Zn) = -0.76$ V, $E°(Fe) = -0.44$ V, hence zinc should corrode more easily than iron, but, surprisingly, iron corrodes faster than zinc, as explained below.

Figure 3.9 shows a comparison of the electrochemical parameters for iron and zinc in acid solution.

For iron:

$$Fe \rightarrow Fe^{2+} + 2e \quad \text{(anodic)}$$

$$2H^+ + 2e \rightarrow H_2 \quad \text{(cathodic)}$$

For zinc:

$$Zn^{2+} + 2e \rightarrow Zn \quad \text{(anodic)}$$

$$2H^+ + 2e \rightarrow H_2 \quad \text{(cathodic)}$$

$$E°_{Zn} = -0.763 \text{ V}, \ E_{H^+/H} = 0.00 \text{ V}$$

$$E°_{Fe} = -0.443 \text{ V}, \ E_{H^+/H} = 0.00 \text{ V}$$

The polarization diagram of zinc in acid is given by dashed lines and that of iron by solid line. It is observed that i_{corr} of zinc is less than that of iron, because of the lower i_0 value of hydrogen reduction on zinc (10^{-7} A/cm^2) compared to on iron (10^{-6} A/cm^2), so iron corrodes faster than zinc as shown in Fig 3.9.

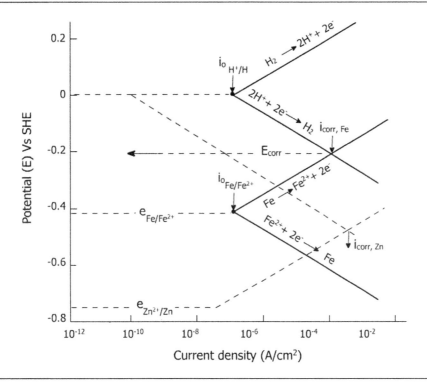

Figure 3.9 Comparison of electrochemical parameters for iron and zinc in solution

The reason for the success of Evans diagrams in corrosion is that they combine thermodynamic factors (E values) with kinetics factors (i values). The usefulness of corrosion kinetics in the study of corrosion rates is, therefore, obvious. The exchange current densities have been included in the polarization diagram by Stern, and such diagrams are called Stern diagrams. Evans diagrams do not include exchange current densities.

3.10 Application of Mixed Potential Theory

3.10.1 Effect of an Oxidizer

It is of interest to observe the effect of oxidizing metal ions on the corrosion rate of a metal in acid solutions, such as the effect of Fe^{3+} ions on the corrosion of zinc in hydrochloric acid. Log i vs E diagrams can be constructed to predict the effect of environment on the corrosion rate of metals.

Examine diagram (Fig. 3.10). When multiple reactions take place, E_{corr} and i_{corr} can be obtained by summing the currents at different potentials at which reactions occur (anodic and cathodic) to obtain total anodic and cathodic polarization curves. For instance, two partial anodic and two partial cathodic reactions are shown in Fig. 3.10. The total rate of oxidation and reduction is shown by broken line. Charge neutrality must be maintained.

Consider a metal, M, in HCl to which Fe^{3+} ions have been added. The corrosion rate of metal M in the absence of ferric salts is given by $i_{corr}(M)$. The corrosion increases from $i_{corr}(M)$ to $i_{corr}(M \rightarrow Fe^{3+}/Fe^{2+})$, on adding the oxidizer as shown in the diagram. The rate of hydrogen evolution is decreased from $i_{H_2} = i_{corr}(M)$ to $i_{H_2}(M \rightarrow Fe^{3+}/Fe^{2+})$ as shown in the figure. This is due to depolarization.

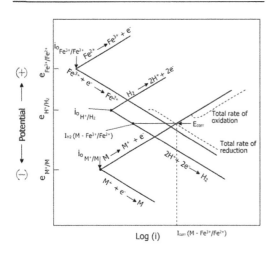

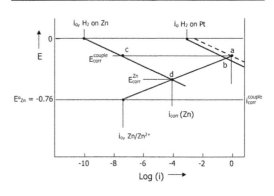

Figure 3.10 Effect of oxidizer on a metal in an air solution

The corrosion potential of the metal $E_{corr}(M)$ is shifted to a more noble value on addition of the oxidizing ions as shown by the E_{corr} arrowed on the figure obtained by the intersection of total anodic and cathodic curves. Thus, there are three consequences of adding an oxidizer:

(a) The corrosion rate of the metal is increased.
(b) The corrosion potential is shifted to a more noble direction.
(c) The rate of hydrogen evolution is decreased.

It is to be noted that the effect of Fe^{3+} ion is pronounced on the metal M, because of its high exchange current density. If the exchange current tendency is small, there would be little effect on the corrosion rate of metal M. The exchange current density of the oxidizing ions must be higher than the exchange current density of hydrogen on the metal surface, to have a significant effect.

3.10.2 COUPLING OF AN ACTIVE METAL TO A NOBLE METAL (FIG. 3.11)

Consider the coupling of zinc ($E° = -0.76\,V$) with platinum ($E° = 1.2\,V$) in a hydrochloric acid solution.

Platinum has a very noble potential and it does not oxidize in HCl. In fact its driving force is negative. Zinc dissolves in acid with the liberation of hydrogen. Platinum does not react. On connecting zinc with platinum electrically, it is observed that:

(a) The rate of hydrogen evolution on zinc (H on Zn) decreases.
(b) The rate of oxidation of zinc, $Zn^{2+} + 2e \rightarrow$ Zn, in the acid solution increases.
(c) The rate of hydrogen evolution on platinum surface (H on Pt) increases vigorously.

Consider first, the oxidation reaction of zinc ($Zn^{2+} + 2e \rightarrow$ Zn). The oxidation reaction polarizes zinc in the noble direction and the hydrogen reduction reaction ($2H^+ + 2e \rightarrow H_2$) in the active direction. The anodic and cathodic curves converge at a point d. The point of convergence yields i_{corr} for zinc. The potential corresponding to i_{corr} for zinc is E_{corr}. Now consider the oxidation of platinum. There is no oxidation of Pt in HCl, hence, no oxidation curve for platinum is shown in Fig. 3.11. However, the reduction of hydrogen on platinum is clearly shown by the cathodic polarization curve for platinum. It is observed that there is only one oxidation process ($Zn \rightarrow Zn^{2+} + 2e$) and two reduction processes $H_2 \rightarrow 2H^+ + 2e$ on platinum and $H_2 \rightarrow 2H^+ + 2e$ on zinc. The situation is analogous to Case 3.10.1 above. The total cathodic process is the sum of H_2 on Zn and H_2 on Pt, resulting in the dotted line.

Figure 3.11 Galvanic coupling of active to noble metal

As observed in Fig. 3.11, the i_{corr} of zinc (uncoupled) increases from 10^{-4} A/cm^2 to a very high value at a when coupled with platinum; consequently, on connecting zinc with platinum, the rate of corrosion of zinc is vigorously increased. As shown by the points of intersection, a and c, the rate of hydrogen evolution on zinc is drastically reduced (from d to c), whereas the rate of evolution of hydrogen on the platinum is vigorously increased. The current density at b is significantly higher than at c as shown in Fig. 3.11. Thus the rate of hydrogen evolution is decreased is shown by the reduction in the current from point d to point c. What does it all amount to? It is shown clearly that zinc is a very poor catalyst for reduction of hydrogen as shown by a very small exchange current density i_0(H$_2$ on Zn) $= 10^{-10}$ A/cm^2, whereas platinum is an excellent catalyst for reduction of hydrogen as shown by a very large exchange current density for reduction of hydrogen i_0(H$_2$ on Pt) $= 10^{-3}$ A/cm^2.

To summarize the effect of coupling of zinc to platinum, the following are the main points of interest:

(a) The rate of hydrogen evolution is decreased on zinc and increased on platinum.
(b) The rate of oxidation of zinc is increased significantly on coupling and zinc dissolves vigorously. Nothing happens to platinum.

(c) Platinum is an excellent catalyst for reduction of hydrogen and zinc is a poor catalyst.

3.10.3 EFFECT OF GALVANIC COUPLING (FIG. 3.12)

It is a common practice to use the potential of a metal in an emf series to predict its corrosion tendency. In case of galvanic couples, the difference of potential between the couples is taken as a measure of corrosion tendency of the couple. In general, the larger the difference between the thermodynamic potential of the metals forming a galvanic couple, the more severe would be the magnitude of galvanic corrosion. This is on the basis of thermodynamics. Consider coupling of zinc to gold and zinc to platinum. According to the thermodynamic approach, the difference between the potential of zinc ($E_{Zn}^{\circ} = -0.76$ V) and gold ($E_{Au}^{\circ} = 1.50$ V) is higher than the difference between the potential of the zinc and platinum ($E_{Zn}^{\circ} = -0.76$ V, $E_{Pt} = +1.2$ V). According to the thermodynamic approach, the Zn–Au couple should corrode faster than Zn–Pt couple. This is certainly not true and contrary to the experimental observation. The purely thermodynamic approach could not, therefore, be a good basis for prediction. Let us observe now the accuracy of prediction made on the basis

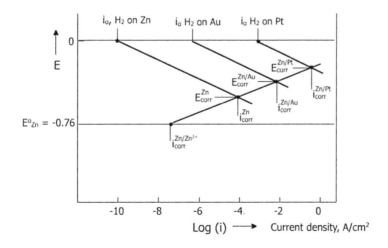

Figure 3.12   Effect of galvanic coupling of zinc with gold and platinum

of kinetics. Examine Fig. 3.12 showing the galvanic coupling of zinc with gold and platinum.

The oxidation of zinc polarizes the electrode in the noble (positive) direction and the cathodic reduction of hydrogen in the active (negative) direction. The intersection of the two curves gives i_{corr} for Zn. The exchange current density for hydrogen on zinc, Au and Pt is shown in the diagram. The exchange current density for zinc is shown in the first curve in Fig. 3.12. The oxidation curves for gold and platinum are not shown in the diagram as the above metals do not oxidize in HCl. The intersection of hydrogen reduction curve for gold with zinc oxidation curve yields the i_{corr} and E_{corr} of zinc coupled to gold (Au) and the intersection of the oxidation curve of zinc with platinum yields the i_{corr} and E_{corr} of zinc when coupled to platinum.

It is clearly observed that the highest value of i_{corr} is shown by Zn–Pt coupled. The i_{corr} value of Zn–Au couple is clearly very much lower than Zn–Pt couple. The hydrogen reduction reaction rate is the highest on a platinum surface $i_0(\text{H on Pt}) = 10^{-3}$ A/cm^2 followed by gold $i_0(\text{H on Au}) = 10^{-6}$ A/cm^2. The reduction rate of hydrogen is very low on zinc i_0 (H$^+$/H$_2$ on Zn) $= 10^{-10}$ A/cm^2. As observed in Fig. 3.12, the i_{corr} of Zn–Pt couple is higher than that of Zn–Au couple, hence, the corrosion rate of Zn–Pt

couple is higher than that of Zn–Au couple. This result in contrary to that obtained on the basis of only thermodynamic potentials but is true, due to the effect of kinetics. Gold is a poor catalyst of hydrogen evolution i_0 (H on Au) $= 10^{-6}$ A/cm^2, compared to i_0 (H on Pt) $= 10^{-3}$ A/cm^2.

Zinc when coupled to platinum, therefore, corrodes at a much higher rate than when it is coupled to gold. The above predictions are based on kinetics of reactions and are, therefore, more accurate and complete than the predictions based on thermodynamic potentials. The latter can be often misleading and if an accurate prediction is made, such as in case of active metal coupling, it may be more of a coincidence than the rule.

3.10.4 EFFECT OF AREA RATIO (FIG. 3.13)

The effect of anode–cathode area ratios has an important bearing on the rate of corrosion. This can be explained by a plot of log I vs E. In the plot (Fig. 3.13) current I is plotted vs E and not i (current density) to establish the effect of area ratio. To define the conditions, the reversible potential $E°$ of Zn (-0.760 V) and hydrogen (0.00 V) are located in the diagram. The values for hydrogen reduction on zinc, i_0 (H) on zinc (-1 cm^2),

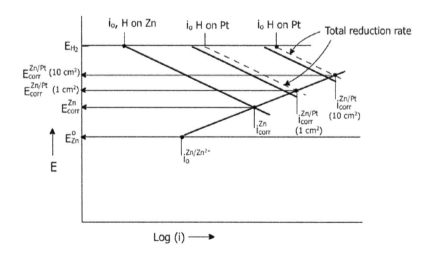

Figure 3.13 Effect of cathode anode area ratio on corrosion of zinc–platinum galvanic couples

hydrogen reduction on platinum, $i_0(H)$ on Pt $(10\,cm^2)$ are shown in Fig. 3.13. The intersection of hydrogen reduction curve (cathodic) on Zn and anodic oxidation curve of zinc gives i_{corr} and E_{corr} of zinc. The points of convergence of the cathodic reduction curves for Pt $(1\,cm^2)$ and $(10\,cm^2)$ with the zinc oxidation curve gives the i_{corr} value for Zn coupled to Pt $(1\,cm^2)$ and Zn coupled to Pt $(10\,cm^2)$, respectively. The i_{corr} of uncoupled zinc is lower than the i_{corr} of zinc coupled either to Pt $(1cm^2)$ or Pt $(10\,cm^2)$. The i_{corr} of Zn coupled to Pt $(10\,cm^2)$ is highest. The corrosion potential of coupled platinum shifts to more noble values as the area of platinum surface is increased from $1\,cm^2$ to $10\,cm^2$. The above observations prove something very important. As the area of the cathode increases, i_{corr} increases, hence, the rate of corrosion also increases.

In short, the smaller the anode to cathode ratio as in the case of Zn coupled to Pt $(10\,cm^2)$, the larger is the magnitude of corrosion. A valuable rule: Avoid a small anode to cathode area ratio to minimize the risk of serious galvanic corrosion.

3.10.5 EFFECT OF OXYGEN (FIG. 3.14)

Figures 3.14a and b show the reactions which occur when iron is placed in a deaerated acid and later oxygen is introduced into the solution. $E_{rev}(O_2)$ is the reversible potential of oxygen and $E_{rev}(H_2)$, the reversible potential of hydrogen, as shown in the above figures. The reversible potential of any metal is represented by $E_{rev}(M/M^{z+})$. In the deaerated condition, the following reactions take place:

$$Fe \overset{i}{\rightleftarrows} Fe^{2+} + 2e \text{ (anodic) } (i_f \geq i_r)$$

$$H_2 \rightleftarrows 2H^+ + 2e \text{ (cathodic) } (i_r \geq i_f)$$

The oxidation reaction polarizes iron in the anodic direction and the reduction reaction in the cathodic direction. The point of intersection gives i_{corr}. The magnitude of over-voltage from the reversible potential is given by η_C for cathodic polarization and by η_A for anodic polarization. Hydrogen is liberated and iron is corroded as Fe^{2+}.

Consider now an aerated condition. Oxygen is passed into the acid until it becomes saturated with oxygen. The reduction of oxygen is shown by the reaction, $1/2O_2 + 2H^+ + 2e \rightarrow H_2O$. In Fig. 3.14b, two partial reduction reactions are shown, i.e.

$$2H^+ + 2e \rightarrow H_2$$

$$\frac{1}{2}O_2 + 2H^+ + 2e \rightarrow H_2O$$

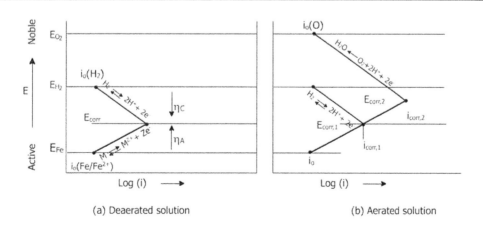

(a) Deaerated solution

(b) Aerated solution

Figure 3.14 Effect of aeration and deaeration

The summation of two partial reactions is shown in the figure. There are two cathodic processes and one anodic process. The exchange current density of oxygen on iron is very low (-10^{-10} A/cm^2). Even on a metal like platinum the exchange current density of oxygen is very low. The reason is that charge transfer is slowed down considerably due to film formation on the metal surface by oxygen. The metal electrode under oxidizing conditions becomes a less efficient catalyst for reduction. Oxygen reduction on platinum is approximately ten times slower than hydrogen reduction. The intersection of cathodic reduction curve of oxygen with the anodic polarization curve is given by i_{corr}. The intersection of cathodic reduction curve with the oxidation curve for iron is given by the appropriate i_{corr} (Fig. 3.14b).

It is observed that upon aeration of the acid solution, rate of reaction (i_{corr}) increases and the corrosion potential (E_{corr}) shifts to more noble value.

To summarize the results:

(a) Fe is oxidized to Fe^{2+}.
(b) Hydrogen is reduced.
(c) Oxygen is reduced.
(d) The rate of corrosion increases on aeration.
(e) The rate of corrosion decreases on deaeration.

3.11 CONCENTRATION POLARIZATION

In an earlier discussion pertaining to activation polarization it was assumed that dissolving metal ions move directly into solution and there is a plentiful supply of ions to be deposited during cathodic reduction. The above assumption is often not correct especially for oxygen reduction because O$_2$ takes time to diffuse in solution to the corroding interface, and metal ions take a definite time to cross the double layer, hence, in oxidation and reduction reaction there exists a metal ion concentration across the double layer. In a polyelectrode system, two separate electrode processes occur:

$$\text{M} \rightleftarrows \text{M}^{2+} + 2e \quad \text{(metal oxidation)}$$

$$\text{H}_2 \rightleftarrows 2\text{H}^+ + 2e \quad \text{(hydrogen reduction)}$$

or $\text{O}_2 + 2\text{H}_2\text{O} + 4e \rightleftarrows 4\text{OH}^-$

(oxygen reduction)

In order to achieve either an anodic or cathodic process, it is necessary to apply an over-voltage (η). If ΔG^* (activation energy) for the reaction is small, the over-voltage required is small and if it is large, the over-voltage required is also small. In the Evans diagram, the relationship between log I and η is observed to be linear for activation polarization:

$$\eta_a^A = b_a \log \left(\frac{i_a}{i_0} \right) \qquad \text{(i)}$$

and $\quad \eta_c^A = b_c \log \left(\frac{i_c}{i_0} \right) \qquad \text{(ii)}$

where A is activation polarization, subscript a and c represent anode and cathode, respectively. As described in detail, in Section 3.1, the Tafel equation is

$$\eta = a + b \log I \qquad \text{(iii)}$$

which indicates a linear relationship (Figs 3.6 and 3.7). As long as there is no concentration built up, an E vs log I plot shows a linearity and it is activation controlled (called activation polarization).

But if the current density is further increased there is a deviation from linearity. Here, the rate of diffusion is not fast enough to carry all the cations from outer Helmholtz plane (OHP) of the double layer by diffusion because the rate of generation of cations becomes very high at higher current densities. The concentration of ionic species is no longer at equilibrium and a concentration imbalance is created. The cations generated, therefore, accumulate and exert a repulsive force on the ions to be transported from the metal surface to OHP. A back emf is, therefore, created. This process, therefore, blocks the further progress of anodic dissolution. No further current can be carried by the cations, whatever the over-voltage is applied. Any amount of applied voltage at this stage has no bearing on the current, which has reached a limiting value. The current at this point is called 'limiting current, or i_L,' at this stage the current become independent of potential. The repulsive forces do not allow further oxidation and release of cations and the process becomes self-limited. It has been shown earlier that the rate of corrosion is controlled either by an anodic or a cathodic charge transfer process. However, if the cathodic reagent falls in short supply, mass transport becomes rate controlling. Mass transport control is much more noticeable for the cathodic oxygen reduction process (limiting O_2 diffusion rate in the solution).

Let us now consider a cathodic reaction. A limiting current density may be reached by a cathodic process, where the ions from bulk solution are unable to diffuse across the double layer to replace the ions that have been discharged.

At this point, infinite concentration polarization is reached and the current becomes independent of potential (Fig. 3.15). The over-potential is, therefore, the sum of two components: (a) activation polarization, and (b) concentration polarization:

$$\eta_{Total} = \eta_a^A + \eta_c^C$$

$$\eta_T = \beta_c \log \frac{i_c}{i_0} + 2.3 \frac{RT}{nF} \log \left(1 - \frac{i_c}{i_L}\right) \quad (3.45)$$

Concentration polarization over-potential (η_C) is obtained as shown below:

Consider a reduction process taking place on a metal surface. At the metal surface $x = 0$, $C_{x=0}$,

and in the bulk electrolyte $x = B$, $C = C_B$.

C_B = bulk electrolyte concentration

$C_{x=0} < C_B$, and $E_{x=0} < E_B$ (for reduction) (3.46)

where C_x is the concentration at the electrode surface, and

$C_{x=0} > C_B$ and $E_{x=0} > E_B$ (for oxidation) (3.47)

Applying Nernst equation

$$\eta_{conc} = E_{x=0} - E_B = \frac{RT}{nF} \ln \left[\frac{C_{x=0}}{C_B} \right] \quad (3.48)$$

From Fick's first law of diffusion

$$J = -D \frac{dc}{dx} \quad (3.49)$$

$$J = -D \frac{(C_{B=x=0})}{x} \quad (3.50)$$

$$J = \frac{i}{nF} = \frac{dQ}{dt} \quad \text{(in terms} \quad (3.51)$$
$$\text{of electrodes, Faraday's law)}$$

Where

$$i = -DnF \frac{(C_B - C_{x=0})}{x} \quad (3.52)$$

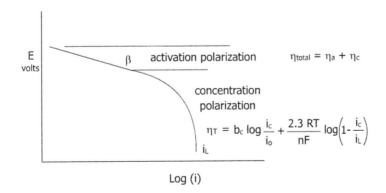

Figure 3.15 Combined activation and concentration polarization

Hence, when $C_{x=0}$, the concentration gradient is maximum, and $i = i_L$. Hence,

$$i_L = -\frac{DnFC_B}{x} \qquad (3.53)$$

where

D = thickness of diffused layer (mm)

η = number of electrodes

F = Faraday, 96 485 C(g equiv.)$^{-1}$

C = concentration of electrolyte (molar concentration).

Dividing $C_B - C_{x=0}$ by C_B from equations (3.52) and (3.53)

$$\frac{C_B - C_{x=0}}{C_B} = -i\frac{x}{DnF} \Big/ -i_L\frac{x}{DnF} \qquad (3.54)$$

$$1 - \frac{C_{x=0}}{C_B} = \frac{i}{i_L} \qquad (3.55)$$

$$\frac{C_{x=0}}{C_B} = 1 - \frac{i}{i_L} \qquad (3.56)$$

Replacing in the Nernst equation term

$$\eta_{conc} = \frac{RT}{nF} \ln\left(1 - \frac{i}{i_L}\right) \qquad (3.57)$$

where i_L is the limiting current density (A/cm^2), and η_{conc} is concentration over-potential.

3.12 EFFECT OF VARIOUS FACTORS ON CONCENTRATION POLARIZATION

Agitation. By agitation, the thickness of the diffusion layer is decreased, and the rate of diffusion of ions increases. There is no build up of any concentration gradient between the corroding surface and bulk electrolyte. The end result is a decrease in concentration polarization and an increase in the rate of corrosion. As the rate of agitation is increased,

the corrosion potential shifts in the noble direction because of a decrease in cathodic over-potential caused by concentration polarization.

Temperature. As the temperature rises, the thickness of diffusion layer is decreased and the corrosion current is increased.

Velocity. The higher the velocity, the less is concentration polarization. At a sufficiently high velocity, concentration polarization becomes zero because the ionic flux is now sufficient to maintain the surface concentration of ions at the electrode/electrolyte surface equal to the bulk concentration.

At high velocity, the driving force for the concentration polarization becomes zero. The process at high velocity becomes totally activation controlled and there is no concentration polarization at high velocity (Fig. 3.16).

Concentration of ionic species. Concentration polarization generally results from the depletion of ions in the vicinity of the cathode. When the concentration of species at the cathode reaches zero, $C_0 = 0$, the reaction is completely mass transport controlled. The higher is the concentration of species, the greater would be the concentration polarization.

Geometry. The geometry of fluid flow and the design of the cell (horizontal or vertical) affects concentration polarization.

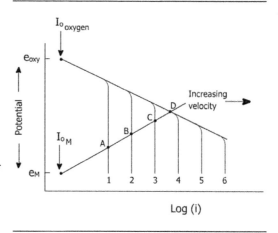

Figure 3.16 Effect of velocity on limiting current density, for $O_2 + 2H_2O + 4e \rightarrow 4OH^-$

3.13 RESISTANCE POLARIZATION (OHMIC POLARIZATION) (FIG. 3.17)

It is known that current is transported from the anode to the cathode by ions in the electrolyte and in the metallic path from the anode to cathode. Because of the normal high conductivity of metals, almost no resistance is offered to the current flow in the metallic path. However, resistance can be encountered if the distance between the anode and the cathode is appreciable.

The effect of ohmic polarization may be significant where this current flows from the anode to the cathode in an electrolyte, depending on the resistance (q) of the electrolyte.

Consider an electrolyte in seawater with a low resistivity or high conductivity represented by R_1. In the seawater, as electrolyte, the anodic and cathodic polarization curve would intersect at the point R_1, where the potentials of the anode and the cathode are polarized at the same value.

If the resistivity of the solution is high, a potential drop (IR) would result from the flow of the current, in the resistive solution and the potential of the anode and cathode would differ slightly in the case of seawater.

The reason is that the potential drop (IR) diminishes the driving force of activation polarization and the anodic and cathodic reactions are not polarized to the same potentials. This situation is represented by R_2 in Fig. 3.17, and further increasing the electrolyte resistance the magnitude of the ohmic drop would further increase as shown by R_3 in the figure. With an increasing resistance offered by the electrolytes, the magnitude of the corrosion current would decrease as shown by R_1, R_2 and R_3 in the figure.

Incidentally, increasing the solution resistance offers a good method of controlling corrosion since decreasing corrosion current means decreasing corrosion rate. Also, painting the metal surface inserts a high resistance into the corrosion circuit and is also illustrated by Fig. 3.17 (ohmic polarization).

3.14 MEASUREMENT OF CORROSION

3.14.1 CORROSION POTENTIAL AND CORROSION CURRENT

When a metal, M, corrodes in a solution, there must be at least one oxidation and one reduction process. What is measured is the sum total of all partial cathodic processes and partial anodic processes occurring during corrosion of a metal. An anodic curve represents the sum total of all partial oxidation processes and a cathodic curve, the sum total of all partial reduction processes. The point of intersection of anodic and cathodic polarization curves in an Evans diagram gives the mixed potential E_{corr} (corrosion potential), also called the compromise potential, or mixed potential, or free corrosion potential, and the corrosion current (i_{corr}).

An Evans diagram for a metal M is shown in Fig. 3.8. The lower limit of the potential in the diagram is given by the reversible potential of metal $\left(E_{M/M^{2+}}\right)$ and the upper limit by the reversible potential of hydrogen $\left(E_{H^+/H}\right)$.

The anodic process is

$$M \rightleftarrows M^{2+} + 2e$$

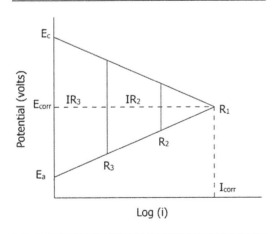

Figure 3.17   Effect of ohmic resistance and current

and the cathodic process is given by

$$H_2 \rightleftarrows 2H^+ + 2e$$

The corrosion rate is given by i_{corr} (corrosion current). The current cannot be measured between local anodes and local cathodes on freely corroding metal surface, however, this can be measured as shown below.

3.14.2 MEASUREMENT OF E_{corr} (CORROSION POTENTIAL)

The experimental arrangement for the measurement of corrosion potential is shown in Fig. 3.18a. The potential of the metal electrode (working electrode) is measured with respect to a standard Calomel electrode, which is non-polarizable. The reference electrode is kept in a separate container and it is connected electrically with the working electrode placed in a container in contact with the electrolyte via a salt bridge. A high impedance voltmeter is connected between the working electrode and the reference electrode. The negative terminal of the voltmeter is connected to the

working electrode and the positive to the reference electrode. The open circuit potential in the freely corroding state is shown by the voltmeter.

The corrosion potential is also referred to as the open circuit potential as the metal surface corrodes freely. It is called mixed corrosion potential as it represents the compromise potential of the anodes and the cathodes. At the rest potential no drive force is applied.

It is important to realize that a metal in an electrolyte will have a characteristic potential even if the salt bridge and voltmeter are removed.

3.14.3 MEASUREMENT OF CORROSION CURRENT (i_{corr})

Arrangements for polarization measurements are shown in Fig. 3.18b. In anodic polarization, an over-potential $(E-E_e)$ is impressed in the noble direction starting from the open circuit potential (natural corroding potential without any impressed current). The over-potential is a measure of how far the reaction is from the equilibrium where the over-voltage, η, is zero. If the potential is made more positive than the equilibrium potential (irreversible potential) then the rate of forward anodic reaction $\vec{i}_f$ is greater than the rate of reverse cathodic $\overleftarrow{i}_r$ reaction and metal dissolution continues. In the above case $\vec{i}_a > |\overleftarrow{i}_c|$. In cathodic polarization $\vec{i}_a < |\overleftarrow{i}_c|$. At small over-voltage both the anodic and cathodic reaction oppose each other, however, at a sufficiently large over-potential, only one reaction, either anodic or cathodic, takes place depending on which direction the potential is impressed.

Consider a freely corroding metal in an acid. The anodic partial process represented by $M \rightarrow M^{2+} + 2e$ intersects the cathodic partial process $H_2 \rightarrow 2H^+ + 2e$ at E_{corr}, the corrosion potential. The current corresponding to E_{corr} is i_{corr}. The current between the local (microscopic) anodes and cathodes cannot be obviously measured by conventional means, such as by placing an ammeter. At E_{corr} (freely corroding potential), the current is i_{corr} and the rate of forward process (i_f) is equal to the rate of reverse process $i_f = i_r$. What can, therefore, be done to measure the current? An experiment can be designed to measure

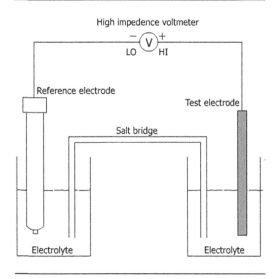

Figure 3.18a An experimental arrangement for making corrosion potential measurement

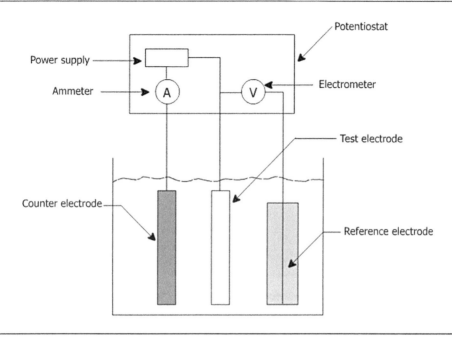

Figure 3.18b Circuitry associated with controlled potential measurements. A feedback circuit (not shown) automatically adjusts the voltage held between test electrode and reference electrode to any pre-set value

the current as well as the potential, which are the necessary electrochemical parameters. Impress a potential E_1 in the noble direction. Actual (theoretical) and measured (experimental) curves are shown in Fig. 3.18c. The corrosion potential (E_{corr}) at the intersection of the anodic and cathodic curves is shown in the actual diagram (Fig. 3.18c). This value is measured experimentally and plotted in the measured diagram. The potential lines E_1, E_2 and E_3 intersect the anodic curves at i_1, i_2 and i_3. As the overvoltage is shifted to E_3, the current becomes completely anodic, as shown by i_3. These three values are plotted in the measured diagram. On connecting the three points, a measured anodic polarization curve is obtained. In a similar manner but in an opposite direction, a cathodic measured polarization curve is obtained on increasing the potential from E_{corr} to E_1, E_2 and E_3 in the negative direction (active potential direction), points i_{c_1}, i_{c_2}, corresponding to cathodic currents similarly obtained in the measured polarization curve. The points are connected and a measured cathodic polarization curve is obtained. The portion of the curve near E_{corr} shows an asymptotic

behavior. The linear behavior is shown only above $50\,mV > E_{corr}$. If the linear portions of the measured anodic and cathodic polarization curves are extrapolated to E_{corr}, the point of intersection yields, i_{corr}, the corrosion current obtained from the measured diagram corresponds accurately with the i_{corr} in the actual diagram.

In the previous discussion, it was shown that a net anodic current is generated when the potential is raised from E_1 to E_3 in the noble direction. As the potential is raised from E_1 to E_3, an excess of electrons is being generated because of oxidation. A platinum electrode or a graphite electrode is connected with the working metal and a voltage is impressed on it by a power source. An ammeter is placed in the circuit to measure the current. In anodic polarization, the platinum electrode is the cathode and the working electrode is the anode, whereas in cathodic polarization it is the reverse. With this arrangement, the metal can be anodically and cathodically polarized and the net current can be measured.

In advanced corrosion measurement systems presently used, i_{corr} is automatically calculated

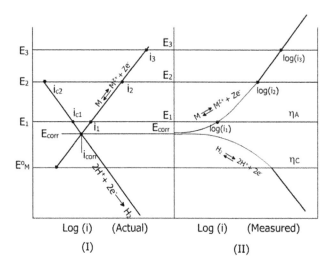

Figure 3.18c Actual and measured polarization curves for active metals

after the curve is recorded by the instrument upon the input of required parameters, such as the over-voltage (η), slopes b_a, b_c and scan rate.

3.15 DETERMINATION OF CORROSION RATES BY ELECTROCHEMICAL MEASUREMENTS

3.15.1 TAFEL EXTRAPOLATION METHOD

In this technique, the polarization curves for the anodic and cathodic reactions are obtained by applying potentials about 300 mV$_{SCE}$ well away from the corrosion potential and recording the current. Plotting the logarithms of current (log I) *vs* potential and extrapolating the currents in the two Tafel regions gives the corrosion potential and the corrosion current i_{corr}. A hypothetical Tafel plot is shown in Fig. 3.19.

Knowing i_{corr}, the rate of corrosion can be calculated in desired units by using Faraday's law. The modern techniques for measurement of corrosion rates are based on the classical work of Stern and Geary.

3.15.2 ADVANTAGES AND DISADVANTAGES OF TAFEL TECHNIQUES

(1) The specimen geometry requires a strict control to obtain a uniform current.
(2) The specimen is liable to be damaged by high current.
(3) The Tafel region is often obscured by concentration polarization and by the existence of more than one activation polarization process, e.g. as seen in Fig. 3.10.

For a complete description of Tafel plots refer to 'ASTM, Designation G3-74 (Re-approved 1981), *Annual Book of ASTM Standards*, Volume 03-02, 2001.'

3.16 POLARIZATION RESISTANCE (LINEAR POLARIZATION)

The theoretical relationship between a metal freely corroding potential (E), current density (i) and the rate of corrosion was developed by Stern.

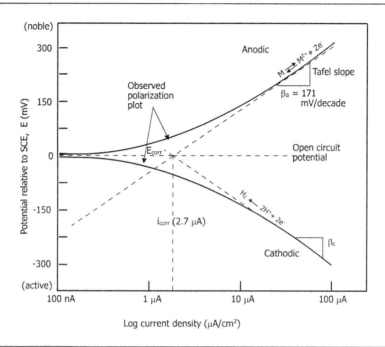

Figure 3.19   A hypothetical Tafel plot

Polarization resistance (R_p) of a corroding metal is defined using Ohm's Law as the slope of a potential (E) *vs* current density (log i) plot at the corrosion potential (E_{corr}). Here $R_p = (\Delta E)/(\Delta I)$ at $\Delta E = 0$. By measuring this slope, the rate of corrosion can be measured. The correlation between i_{corr} and slope (dI)/(dE) is given by

$$\frac{\Delta E}{\Delta I} = \frac{b_a b_c}{2.3 i_{corr}(b_a + b_c)} \qquad (3.58)$$

where b_a and b_c are Tafel slopes.

The potential–current density plot is approximately linear in a region of within $\pm 10\,mV$ of the corrosion potential. As the slope of the plot has the units of resistance, this technique is called 'Polarization Resistance' technique. The current measured in the external circuit equals the change in corrosion current, $\pm \Delta I$, caused by a small perturbation. As both anodic and cathodic reactions proceed in the vicinity of corrosion potential (E_{corr}), they are exponentially dependent upon the applied potential. Over a small potential range (20 mV), however, the exponentials are linearized (because mathematically Log $x \rightarrow x$ as $x \rightarrow 0$) and an approximate linear

potential–current relationship is obtained. This technique is quick and reliable.

3.17 THEORETICAL BACKGROUND (ELECTROCHEMICAL MEASUREMENTS)

In a corroding system:

$$M \rightleftharpoons M^{z+} + Ze^- \quad \text{(corroding metal)} \qquad (1)$$

$$X^{z+} + Ze^- \rightarrow \frac{1}{2}X_2 \quad \begin{array}{l}\text{(species in solution, such}\\\text{as hydrogen, oxygen)}\end{array} \qquad (2)$$

The reversible potential (E_{rev}) of the M in equation (1) is E_{rev}, M and that of X in equation (2) is E_{rev}, X. By increasing the over-voltage, a condition of irreversibility is created in the system. If E_{corr} is sufficiently removed from $E_{rev(M)}$, the reduction of M becomes insignificant compared to the rate of oxidation of M. $M \rightarrow M^{z+} + Ze^-$ and the rate of oxidation of X becomes insignificant compared to the rate of reduction of X (species in solution). The corrosion rate i_{corr} becomes equal to either $\vec{i}_M$ or $\overleftarrow{i}_X$ at E_{corr}.

If the potential of a metal M is held at the next negative value, and is then made more positive, the linear (Tafel) part of this right-hand graph will be ascended. But on reaching E_M°, an increasing current is added to the cathodic $2H^+ + 2e^- \rightarrow H_2$ line, but in the opposite direction, due to $M \rightarrow M^{z+} + Ze^-$ starting. This increasing anodic current becomes equal (and opposite) to the cathodic current when potential E_{corr} is reached and then the net (measured) current becomes zero. As the graph is of Log i (measured), the experimental line becomes asymptotic to the horizontal potential level of E_{corr}, because log 0 is ∞.

Thus the experimental line becomes curved and cannot follow the cathodic line $2H^+ + 2e^- \rightarrow H_2$ up to its origin at $E_{H_2 \, on \, m}^\circ$. Similarly, if the metal M were to be polarized to a very positive potential, starting from the top right point on the right-hand graph, it begins down a linear (Tafel) part of the curve, if made less positive, but deviation from linearity occurs as E_{corr} is approached, due to an increasing current in the opposite direction starting at potentials below $E_{H_2 \, on \, m}^\circ$.

Hence,

$$I_{measured} = \vec{i}_M - \overleftarrow{i}_X = 0 \text{ at } E_{corr}$$

where $I_{measured}$ is the net current.

To apply the linear polarization method for obtaining i_{corr}, an important condition is that the over-voltage must be small. Only perturbations of up to ± 30 mV are allowed. The b value may be determined in separate experiments or obtained from literature. If $b_a = b_c = 0.1$, the error will not be more than 20%. The polarization resistance plots are recorded from E_{corr} to, say ± 20 mV from E_{corr}.

For example, use a resistance of 0.1 ohm and let the applied potential on the instrument be $E = 30$ mV(SCE). The maximum current available is now 300 mA. Record E and current density i. Figure 3.20 shows a hypothetical linear polarization plot.

3.18 MODERN DEVELOPMENTS

The development of microprocessor based measurement systems have expanded the capability of corrosion scientists, relieved operator's tedium, and improved the accuracy and speed of the data required for experimenting. In these corrosion measurement systems, corrosion rates are calculated by using constants for areas, equivalent weight and density and Tafel slopes which are measured or keyed-in. The microprocessor examines data on both the anodic and cathodic sides of the first corrosion potential to find the semi-log straight line segment which will yield a Tafel constant. The straight line segment is extrapolated to intersect the corrosion potential and the value of current density is utilized in computing the corrosion rates. A typical output of a polarization resistance plot obtained by a microprocessor based on corrosion measurement system is shown in Fig. 3.21. Data from potentiodynamic polarization plots can be replotted as polarization resistance. Once i_{corr} is determined, the corrosion rates are calculated using Faraday's law as shown in Example 1.

Example 1
From a given plot of polarization resistance (Fig. 3.22), calculate the rate of corrosion.

Solution:
Data obtained:

$$R_p = 1.111 \times 10^2 \, \text{ohm} \cdot \text{cm}^2,$$

$$\beta_a = 100 \, \text{mV/decade}, \beta_c = 100 \, \text{mV/decade}$$

Calculations:

$$i_{corr} = \frac{1}{2.303 R_p} \left[\frac{\beta_a \beta_c}{\beta_a + \beta_c} \right]$$

$$= \frac{1}{2.303 \times 1.111 \times 10^2} \left[\frac{100 \times 100}{200} \right]$$

$$= 19.54 \times 10^{-5} \, \text{A} \cdot \text{cm}^{-2}$$

$$= 1.954 \times 10^{-4} \, \text{A} \cdot \text{cm}^{-2}$$

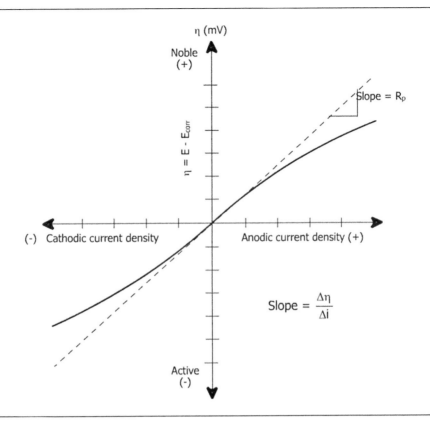

Figure 3.20   Hypothetical linear polarization plot

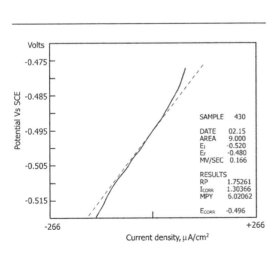

Figure 3.21   Typical polarization resistance plot

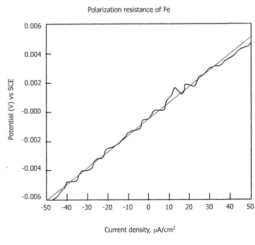

Figure 3.22   Output of a typical polarization resistance plot

Corrosion rate in mm/year can be found from i_{corr}. Here 0.00327 is a constant for mm/year.

$$r = 0.00327 \frac{Mi}{\eta\rho}$$

where

i = current density (μA/cm^2)

ρ = density (g/cm^3)

M, Eq. wt. = 27.92 g/equiv.

ρ = 7.87 g/cm^3

Corrosion rate, $\quad r = \dfrac{0.00327 \times 27.92 \times 195.4}{2 \times 7.87}$

$r = 1.13$ mm/year

3.19 KINETICS OF PASSIVITY

3.19.1 INTRODUCTION

Faraday published a theory in which he suggested that the metal surface of certain passive metals was oxidized and the oxide layer was very thin (*Phil. Mag.*, **9** (1836), p. 5). If a metal is converted to an oxide and the oxide which is formed is stable, the metal is considered passive as this oxide forms a barrier between the metal and the environment. For example, iron is not attacked in concentrated HNO$_3$, as a very thin film of an oxide is formed on its surface and causes loss of reactivity. A film of solid hydroxide or oxide may be precipitated from an aqueous solution if there are metal ions in the solution

$$Fe^{3+}(aq) + 3OH^-(aq) = Fe(OH)_3(s) \quad (1)$$

Later, the hydroxide converts to oxide by giving away the water molecule:

$$2Fe(OH)_3 = Fe_2O_3(s) + 3H_2O(s) \quad (2)$$

If there are no metal ions in solution the film may also be formed by chemical combination with adsorbed oxygen in solution

$$2Fe(s) + \frac{3}{2}O_2(ads) = Fe_2O_3(s) \quad (3)$$

Although the above reaction is possible, it is more likely that Fe$_2$O$_3$ (s) formed as a result of electrochemical oxidation at a sufficiently high potential followed by interaction of Fe^{3+} by OH$^-$ ions, the latter being formed by cathodic reaction O$_2$+2H$_2$O+4$e \rightarrow$ 4OH$^-$. Iron oxidizes as Fe(s) $\rightarrow$ Fe^{3+}(aq)+3e. As suggested by equation (1), the oxide formation is increased by increasing the activity of OH$^-$ ions. When a surface film of an oxide or hydroxide develops, corrosion is either eliminated or retarded. The passive films may be as thin as 2–10 nm, and they offer a limited electronic conductivity, and behave like semiconductors with metallic properties rather than the properties shown by bulk oxides. The films also allow a limited amount of conductivity of cations because of lattice defects and a slow anodic dissolution. Because of film formation, there is a reduction in the current density for cathodic reduction and an increase in the current density for an anodic polarization. The above factors lead to retardation of corrosion.

The important point is that once a film is formed, the corrosion rate sharply declines. The passivity on the metal surface which develops due to film formation on metal surface causes inhibition of the anodic dissolution process.

3.19.2 ACTIVE–PASSIVE METALS AND CONDITIONS FOR PASSIVATION

Transition metals, such as Fe, Cr, Ni and Ti, demonstrate an active–passive behavior in aqueous solutions. Such metals are called active–passive metals. The above metals exhibit *S*-shaped polarization curves which are characteristic of such metals. Consider, for instance, the case of 18-8 stainless steel placed in an aqueous solution of H$_2$SO$_4$. If the electrode potential is increased then the current density rises to a maximum, with the accompanying dissolution of the metal taking place in the active state. The current density associated with the dissolution process indicates the magnitude of corrosion. At a certain potential, the current density is drastically reduced as the metal becomes passivated because of the formation of a thick protective film. Iron shows passivity in acids containing $SO_4^{2-}, NO_3^-, CrO_4^{2-}$, etc. The above metals show a transition from an active state to a passive state and passivity occurs when the corrosion potential becomes more positive than the equilibrium potential of the metal and its metal oxide. The anodic dissolution behavior of a metal demonstrate an active–passive behavior as shown in Fig. 3.23. More details are shown in Fig. 3.24. The reversible potential of a hypothetical metal M is designated by $E_{(M/M^{z+})}$, oxygen by $E_{rev}(O)$. The anodic and cathodic polarization curves in the active region intersect at E_{corr}, the corrosion potential. The anodic curve shows a Tafel behavior with a slope b_a. Similarly, the cathodic curve can be obtained by deviation from E_{corr} in the active direction and the cathodic Tafel slope is given by b_c. The point of intersection gives E_{corr}. On polarizing the metal in the noble direction from E_{corr}, it is observed that as the potential increases, the rate of dissolution of the metal also increases. The highest rate of corrosion is achieved at a maximum current density, called critical current density ($i_{critical}$).

The lower portion of the anodic curve (nose of the curve) exhibits a Tafel relationship up to $i_{critical}$ which can be considered as the current required to generate sufficiently high concentration of metal cations such that the nucleation and growth of the surface firm can proceed. The potential corresponding to $i_{critical}$ is called the *primary passive potential* (E_{pp}) as it represents the transition of a metal from an active state to a passive state. Because of the onset of passivity, the current density (log i) starts to decrease beyond E_{pp} due to the oxide film formation on the metal surface. Beyond E_{pp} the current continues to decrease until at a certain value of potential, it drops to a value orders of magnitude lower than $i_{critical}$. The potential at which the current becomes virtually independent of potential and remains virtually stationary is called the *flade potential* (E_F). It represents the onset of full passivity on the metal surface due to film formation. The minimum current density required to maintain the metal in a passive state is called *passive current density* (i_p). It is an intrinsic property of oxidation.

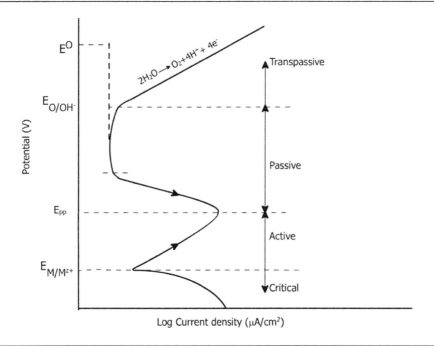

Figure 3.23 Polarization diagram for active–passive metal

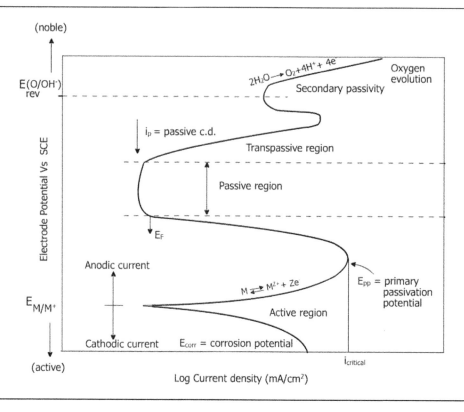

Figure 3.24   Anodic dissolution behavior of a metal as shown in Fig. 3.23 with more details

At i_p, the metal dissolution occurs at a constant rate and the oxide film begins to thicken. According to electric field theory, dissolution proceeds by transport of ionic species through the film under the influence of an electric field $[F = (\Delta E)/X]$. As the potential is increased in the noble direction, the film starts to get thicker. Hence, the electric field within it may remain constant. For instance, if the potential is increased from E_1 to E_2, the electric field $[F = (\Delta E)/X]$ would change. In order to maintain the electric field constant the film must get thicker. The thickness of the film proceeds by transport of cations M^{z+} outwards and combination of these cations with O^{-2} or OH^- ions at the film/solution interface. The dissolution rate in the passive region, therefore, remains constant. The process of dissolution is a chemical process, and it is not dependent on potential. The film which is dissolved is immediately replaced by a new film and a net balance is maintained between dissolution

and film reformation. The passivation potential is pH dependent. The E_{pH} equals the equilibrium potential, $E_{M_xO_y}$, for the formation of an oxide M_xO_y on the metal surface. For different oxides on iron FeO (wustite), Fe_3O_4 (magnetite) and αFe_2O_3 (hematite), the following values of the equilibrium potential have been calculated:

(a) $Fe + H_2O \rightarrow FeO + 2H^+ + 2e^-$
     $E^\circ_{eq}(Fe/FeO) = -0.04 - 0.059\,pH$

(b) $3Fe + 4H_2O \rightarrow Fe_3O_4 + 8H^+ + 8e^-$
     $E^\circ_{eq}(Fe/Fe_2O_3) = -0.09 - 0.059\,pH$

(c) $2Fe + 3H_2O \rightarrow \alpha Fe_2O_3 + 6H^+ + 6e^-$
     $E^\circ_{eq}(Fe/\alpha Fe_2O_3) = -0.05 - 0.059\,pH$

Experimentally it is found that at 25°C and pH = 6, $E_{pp} = 0.59 - 0.059$ pH. Once passivity has initiated, a small current is required to maintain it in a passive state. There is no change in current until a point is reached which marks the end of the passive region.

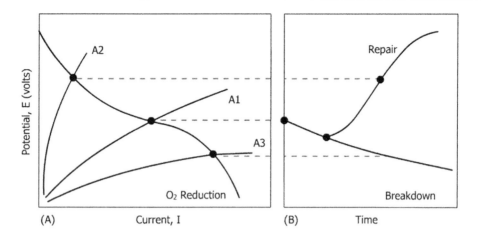

Figure 3.25 Film breakdown and repair mechanism

In general, surface films are heterogeneous and non-uniform, with inclusions which introduce stresses in these film. There are a number of pores in the film where it becomes less thick. The dissolution of metal starts at these points. The process of film repair takes place if the corrosion potential of the film exceeds the passivation potential E_{pp}.

Film repair would not take place if E_{corr} does not exceed E_p. The process of film repair is illustrated in Fig. 3.25A. If a film is repaired, the potential exceeds E_{pp} and moves up in the noble direction. If the repair does not take place the potential remains lower than E_{pp}. The polarization between a metal containing pores in an otherwise filmed surface is shown by curve (A). Curve (A) indicates that repair does not take place. A rise in potential with time often indicates the film repair, which is illustrated in Fig. 3.25B. The figure shows a rise of potential with time. The potential stabilizes finally at some value which exceeds the passivation potential.

The end of the passive region corresponds to the initial point of anodic evolution of oxygen. This happens above the reversible potential of oxygen ($E_{rev(O)}$).

$$2H_2O \rightarrow O_2 + 4H^+ + 4e \quad \text{(acid solution)}$$

$$4OH^- \rightarrow O_2 + 2H_2O + 4e \quad \text{(neutral solution)}$$

The evolution of oxygen on the filmed surface causes a sharp increase in the current. The basic polarization curve shown in Fig. 3.23 is highly generalized. Several deviations may occur from the basic polarization curve shown in Figs 3.23 and 3.24, depending on the metal and the environment.

The phenomena of breakdown of passive films at more noble values of potential leads to an accelerated rate of corrosion (transpassive corrosion). The potential at which the breakdown or rupture of protective film takes place and the current density rises sharply is called *transpassive potential* ($E_{transpassive}$). In Fig. 3.26, the effect of chromium

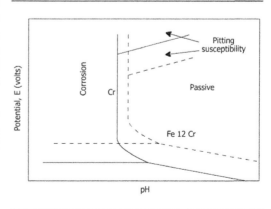

Figure 3.26 Effect of chromium on the transpassive region

Table 3.2 Electrochemical parameters of steels

Alloy	E_{pp} (V)SHE	$i_{critical}$ (A/m^2)	i_p (A/m^2)
Iron	+0.52 to +1.00	10^4	$10^{0.3}$
Fe–14 Cr	−0.10	10^3	$10^{-1.3}$
Fe–18 Cr	−0.15	$10^{2.5}$	$10^{-2.0}$
Fe–18 Cr–5 Mo	−0.20	$10^{0.3}$	$10^{-1.7}$
Fe–18 Cr–8 Ni	−0.20	10^1	$10^{-2.2}$

(Courtesy: John M. West, Basic corrosion and oxidation)

content on transpassivity is shown. Maximum anodic dissolution occurs at $i_{critical}$, the critical current density, just before E_{pp}, the passivation potential. In a variety of Fe–Cr alloys, E_{pp} is substantially reduced because of the formation of uniform protective films. The following are the values of E_{pp}, $i_{critical}$ and i_p, for steels containing additions of Cr, Mo and Ni (Table 3.2).

In the region of transpassivity, oxidation of Cr^{+3} to Cr^{+6} occurs in steels containing chromium:

$$Cr(III) + 4H_2O = HCrO_4^-(aq) + 7H^+ + 3e^-$$

The transpassive region increase with increasing chromium content. As the film dissolves, cation vacancies are created in the oxide surface and the conductivity of the film is increased.

3.20 DEFINITION OF IMPORTANT ELECTROCHEMICAL PARAMETERS FOR ACTIVE–PASSIVE METALS

Following are the definition of important terms related to potential and current as shown in the active–passive polarization diagrams (Fig. 3.24).

(a) **Equilibrium potential** (E_{eq} or $E_{M/M^{z+}}$). The potential of an electrode in an electrolyte when the forward rate of reaction

is balanced by the rate of reverse reaction ($M^{z+} + ze \rightleftarrows M$). It can be defined only with respect to a specific electrochemical reaction. This is also written as $E°$ and must not be confused with E_{corr}.

(b) **Passive potential** ($E_{passive}$). The potential of an electrode where a change from an active to a passive state occurs.

(c) **Flade potential** (E_F). The potential at which a metal changes from a passive state to an active state.

(d) **Transpassive potential** ($E_{transpassive}$). The potential corresponding to the end of passive region which corresponds to the initial point of anodic evolution of oxygen. This may correspond either to the breakdown (electrolysis) voltage of water, or, to the pitting potential.

(e) **Critical current density** ($i_{critical}$). The maximum current density observed in the active region for a metal or alloy that exhibits an active–passive behavior.

(f) **Passive current density** (i_p). The minimum current density required to maintain the thickness of the film in the passive range.

(g) **Pitting potential** (E_p). It is the potential at which there is a sudden increase in the current density due to breakdown of passive film on the metal surface in the anodic region.

3.20.1 ACTIVE–PASSIVE METALS VS ACTIVE METALS

It has been discussed earlier that when a metal is exposed to corrosive environments, two reactions take place:

$$\text{Oxidation: } M \rightleftarrows M^{z+} + Ze^-$$

Reduction:

$$O_2 + 2H_2O + 4e^- \rightleftarrows 4OH^- \text{ (in neutral solution)}$$

$$\left.\begin{array}{l} O_2 + 4H^+ + 4e \rightleftarrows 2H_2O \\ \text{or } 2H^+ + 2e^- \rightarrow H_2 \end{array}\right\} \text{ (in acid solution)}$$

Metals, like zinc, magnesium and aluminum, show a passive behavior in atmospheric corrosion.

Metals, like Fe, Cr, Ni and Ti, show a strong active–passive behavior. Whether passivity is achieved or not by the system depends on the cathodic reaction. Thus, cathodic reaction is a deciding factor in the establishment of passivity. A metal not showing any passivity will exhibit a linear E vs log i relationship. On the other hand, a metal exhibiting passivity would exhibit a non-linear anodic polarization. Anodic polarization curves for active–passive metals, such as Fe, Cr or Ni, show a highly polarized passive region (Fig. 3.24). The rate of corrosion depends upon the degree of polarization of the anode, contrary to the observation in active metals where the rate of corrosion depends on the degree of polarization of both the anode and the cathode.

3.20.2 ACTUAL (THEORETICAL) AND MEASURED POLARIZATION DIAGRAMS FOR ACTIVE–PASSIVE METALS

The actual polarization behavior is not generally obtained by anodic or cathodic polarization curves. At potentials more noble than 50 mV above E_{corr} and below 0.8–0.9 volts, the measured curves correspond closely with the actual polarization curves.

Locate $E_{M/M^{z+}}$ (reversible potential of the metal) and E_H in the actual diagram (Fig. 3.27). The diagram shows the exchange current density for metal oxidation i_0, M/M^{z+} and hydrogen reduction i_0 (H on M). The anodic oxidation and hydrogen reduction curves intersect at E_{corr} (corrosion potential) which is the first point on the measured diagram (Fig. 3.27). The starting point of the partial anodic oxidation ($M \rightleftarrows M^{z+} + ze$) and cathodic reduction ($H_2 \rightleftarrows 2H^+ + 2e$) are not shown in the measured diagram. As the over-voltage (η) is increased in the noble direction, the rate of dissolution of the metal is simultaneously increased. The dissolution kinetics observed initially is asymptotic, which becomes linear after 30–40 mV and shows a Tafel behavior. The slope is b_a is the same for both the actual and measured polarization diagrams. As the over-voltage is further increased, the curve bends back to the left to very low values of current. The onset of the reduction in the current density occurs at a particular current density called the *critical current density*, $i_{critical}$. It represents the onset of passivation due to the formation of a barrier film of Fe_2O_3 explained earlier. The current finally reaches a minimum value at a particular potential, the flade potential (E_F). Once it is reached the current does not show any response to any further increase in over-voltage until the reversible potential of oxygen is reached. The metal is in a passive state because of the formation of a barrier film of Fe_2O_3 which restricts the electrical conductivity. The current required to maintain the passive layer is called the passive current density (i_p). Both i_p and E_F are shown in the actual diagram. The measured diagram follows completely the kinetics shown in the actual diagram. The measured anodic polarization curve approximates the actual anodic dissolution behavior over a wide range of potential. However, in the measured diagram, the starting potential for the anodic and cathodic processes ($E_{M/M^{z+}}$ and E_H) are not shown. All the points in the actual diagram are identified in the measured diagrams.

3.21 MEASURED VS ACTUAL POLARIZATION BEHAVIOR OF ACTIVE–PASSIVE METALS

Consider three conditions which can arise when a cathodic process is superimposed on the curve. The three cases are discussed below to illustrate the polarization behavior of activation metals. The three curves representing different activation controlled process with different exchange current densities are shown in Fig. 3.27. The three reactions represent different rates of hydrogen reduction on the metal surface. The three exchange current densities associated with these reduction reactions are shown in Fig. 3.27 by $i_{0(H)_1}$, $i_{0(H)_2}$ and $i_{0(H)_3}$.

Case 1

The cathodic curve intersects the anodic polarization in the active region (Fig. 3.27). The point of

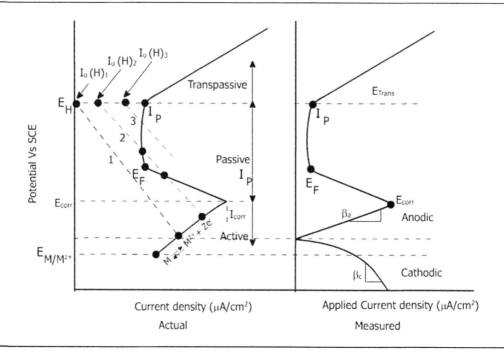

Figure 3.27 Three curves representing different activation controlled process with different exchange current densities

intersection of the cathodic partial process with the linear Tafel kinetics gives E_{corr}, the corrosion potential. It is the first point which is identified in the measured diagram. As the over-voltage is further increased in the noble direction, the measured curve follows the active–passive kinetics shown in the actual diagram. Initially, the curve in the measured diagram approaches asymptotically and later it follows the linear Tafel kinetics. The slope of the Tafel curve is given by b_a. As the potential is further increased in the more noble direction, the measured curve follows the characteristics of the actual curve. Both the actual and measured diagrams have the same slopes b_a, the only difference being that the measured curve approaches E_{corr} asymptotically. On deviation from E_{corr} in the more active (downwards) direction, the Tafel slope b_c can also be measured as shown in the measured diagram.

The above case is exemplified by titanium in deaerated H_2SO_4 or HCl or Fe in H_2SO_4. Titanium has a more active primary passivation potential (E_{pp}) than the reversible potential of

hydrogen ($E_{rev(H)}$). This example shows that a metal is stable in the active, active–passive or passive state. This condition is the least desirable because of the accompanying high rate of corrosion.

Case 2

The partial cathodic process represented by $i_0(H)_2$ shows a higher current density than the partial process (Fig. 3.28). The curve representing the partial cathodic process intersects the anodic polarization curve (actual) at three points, *a*, *b* and *c* (Fig. 3.28). The first point of intersection in the active region *c* corresponds to the corrosion potential (E_{corr}) which is the first point identified in the measured diagram. The other points of intersection are *a* and *b* at more noble potentials. It is, however, observed that although more and more over-voltage is applied in the noble direction, the cathodic process proceeds at a higher rate than the anodic process. The points of intersection *a* and *b* corresponds to a high magnitude of cathodic current rather than the anodic current

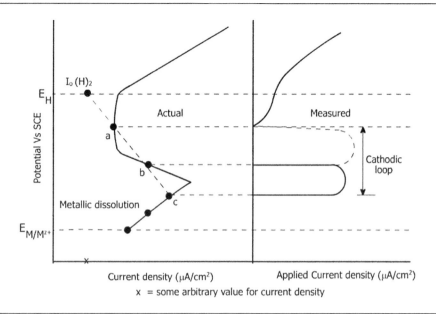

Figure 3.28   Actual and measured polarization curve for a cathodic curve intersecting the passive region of an anodic polarization curve

as shown in the actual diagram. The cathodic currents between the points *a* and *c* are identified as negative loops in the measured diagram. Hence between potentials *a* and *c*, cathodic loops are observed in the measured diagram. Below point *c*, the metal is in an active condition and above *a* it is in a passive condition. From point *a* onwards, the measured curve follows the characteristics of the actual curve in the passive region. The above case is typical for chromium in deaerated H_2SO_4 or stainless steel in H_2SO_4 containing oxides. Case 1 stated before is undesirable and Case 2 stated above is less desirable. Once the surface film becomes active it may not passivate again.

Case 3

In this case the exchange current $i_{0(3)}$ is higher than either of the two partial reduction processes represented by $i_{0(1)}$ and $i_{0(2)}$ (Fig. 3.29). The partial cathodic reduction curve represented by $i_{0(3)}$ intersects the anodic polarization curve in the middle of the passive range. The point of intersection identifies the corrosion potential (E_{corr}) and it is the first point in the measured polarization diagrams. As the over-voltage is increased in

the noble direction, the curve approaches E_{corr} asymptotically, followed by a quick shift in the passive range. The measured diagram shows the same characteristics as the actual diagram. There is no linear kinetics suggesting anodic dissolution observed in the measured diagram. The metal, therefore, simultaneously passivates. The above case is exemplified by Ti–Pt, Ti–Ni and Ti–Cu alloys. The above alloys do not undergo active dissolution and passivate instantaneously. Case 3 offers the best condition for passivity.

3.22 CONTROL OF PASSIVITY

Two general rules can be applied to control passivity. If corrosion is controlled by an activation control reduction process (I_C^A) an alloy which exhibits a very active primary potential must be selected. Conversely, if the reduction process is under diffusion control an alloy with

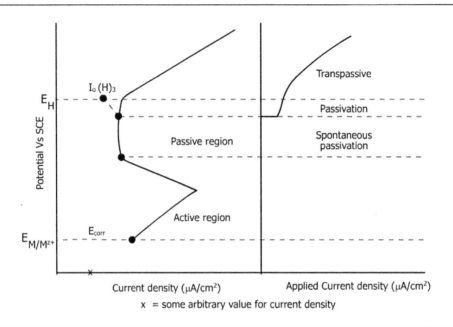

Figure 3.29 Actual and measured polarization curves for a cathodic curve intersecting the anodic polarization curve in the passive region

a smaller critical current density must be selected (Fig. 3.30). Thus, if activation polarization is the controlling factor, alloy 2 must be selected and if concentration polarization is rate controlling, alloy 1 must be selected. The following factors have a major bearing on the passivity.

3.22.1 ALLOYING ADDITION

Those alloying additions which decrease $i_{critical}$ are effective in increasing the passivating tendency. Consider alloying additions of Mo, Ni, Ta and Cb to Ti and Cr. The critical current density of Ti and Cr is reduced on addition of Mo, Ni, Ta or Cb. The potential of the above elements is active and their rate of corrosion is low. Generally, those alloying elements are useful which show low corrosion rates at the active potentials. Alloying with metals which passivate more readily than the base metal reduces $i_{critical}$ and induces passivity. Elements, like chromium and nickel, which have a lower $i_{critical}$ and $E_{passive}$ than iron, reduce the

$i_{critical}$ (critical current density) of iron (Fig. 3.31). Addition of up to 18% chromium reduces $i_{critical}$ iron. Similarly, addition of more than 70% nickel to copper reduces $i_{critical}$ and i_p.

3.22.2 CATHODIC REDUCTION

Passivity can also be increased by increasing the rate of cathodic reduction rather than changing the behavior of the anodic polarization curve. When the reduction process is made faster, the metal spontaneously passivates. Metals having a high hydrogen exchange current density tend to passivate spontaneously without any oxidizer being added. If alloying additions of noble metals are to be made to the matrix metal, it is preferable to select metals with active E_{pp}, such as titanium and chromium.

In highly oxidizing conditions, the addition of a noble metal causes an increase in the rate of corrosion of matrix metal as exemplified by the addition of platinum to chromium. This happens

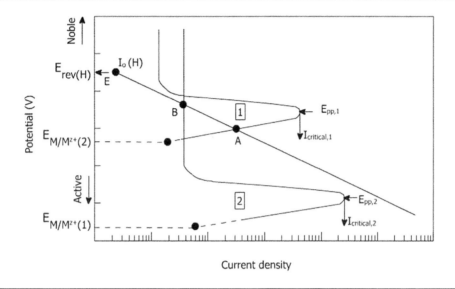

Figure 3.30   Behavior of active–passive alloys in a system containing an activation controlled cathodic reduction process

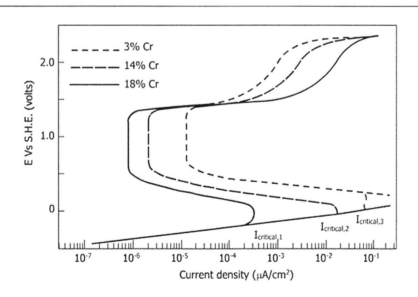

Figure 3.31   Effect of chromium content of anodic polarization curve of Fe–Cr alloys in 10% H_2SO_4 at 21°C

if the alloying element has an exchange current density greater than the matrix metal and the mixed potential of the metal is in the transpassive region. Chromium–platinum alloys have a low resistance to corrosion in oxidizing acids and a high resistance in non-oxidizing acid. Noble metal alloying additions are beneficial in only weak or moderately oxidizing acids. In highly oxidizing conditions the corrosion potential of chromium is very near the transpassive region. If platinum is added to titanium, the rate of corrosion of titanium is not increased because, unlike chromium, titanium does not exhibit a transpassive region. If the mixed potential of the metal is close to the transpassive region, the rate of corrosion is increased. The alloying metal must be selected carefully. The above effect is illustrated in Fig. 3.32. Suppose platinum is coupled to titanium in an acid solution. The point of intersection of hydrogen reduction reaction and titanium oxidation, give i_{corr} for titanium (uncoupled). There is no oxidation of Pt. The reduction curve for Pt is shown in Fig. 3.32. The total rate of reduction is shown by the broken line. It is observed that i_{corr} for the Ti–Pt couple is much lower than i_{corr} of titanium alone. Titanium spontaneously passivates and the rate of corrosion is reduced.

3.23 EFFECT OF ENVIRONMENT

A. EFFECT OF CHLORIDE ION CONCENTRATION

Chloride ions damage the protective films and cause the metal surface to be pitted. Stainless steel is subjected to serious pitting by stagnant water containing a high concentration of chloride ions. Steel pipes are subjected to pitting in brackish water and seawater. The higher the concentration of chloride the greater is the tendency of pitting. The effect of chloride concentration on the passivation of steel is shown in Fig. 3.33. Chloride ions break down the passivity and increase the rate of anodic dissolution.

B. EFFECT OF TEMPERATURE

An increase in temperature generally decreases the passive range and increases the critical current density ($i_{critical}$). An increase of temperature decreases polarization and enhances the dissolution kinetics.

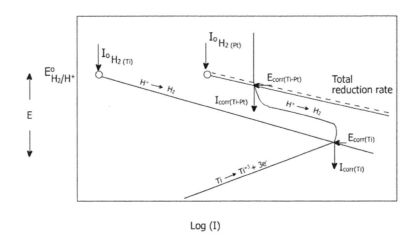

Log (I)

Figure 3.32 Spontaneous passivation of titanium by galvanically coupling to platinum

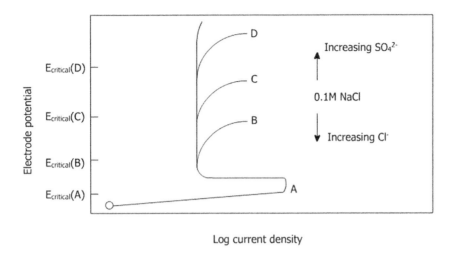

Figure 3.33 Schematic representation of the effect of aggressive ions (Cl^-) and inhibitive ions (SO_4^{2-}) on passivation of stainless steel

C. VELOCITY

When a metal is under cathodic diffusion control, agitation of the electrolyte increases the current density and the rate of corrosion increases up to a certain point. At a particular velocity (critical velocity), the rate becomes activation controlled rather than diffusion controlled, and, hence, the rate of corrosion becomes independent of velocity (Fig. 3.34). Beyond a certain point agitation has no effect. As $i_{critical}$ is exceeded, the metal attains passivity and corrodes only very slowly.

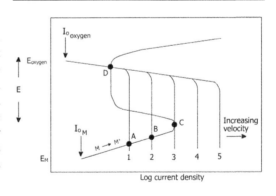

Figure 3.34 Effect of velocity on the electrochemical behavior of active–passive metals corroding under diffusion control

D. OXIDIZER CONCENTRATION

Addition of oxidizing agents, like Fe^{+2} or CrO_4^{2-} ions significantly affect the corrosion rate of metals which exhibit passivity. In the case of an active metal, the rate of corrosion is generally increased by an increase in the oxidizer concentration. Consider six curves (1–6) representing the oxidizer concentrations (Fig. 3.35). It would be observed that the oxidizer concentration having a concentration represented by curve 6, would allow passivity to be achieved quicker, because at this concentration, the lowest

value of i_{corr} is exhibited and the metal tends to passivate spontaneously. On increasing the oxidizer concentration represented by curve 6, the rate of corrosion is increased as i_{corr} is shifted in the transpassive region. Oxidizer concentration C_3 and C_4 produce a stable passive state that does not depend on oxidizer concentration. The corrosion rate, therefore, remain low and

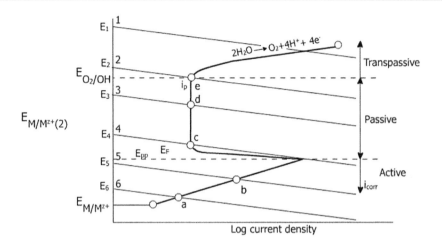

Figure 3.35 Effect of oxidizer concentration on electrochemical behavior of an active–passive metal

constant at concentration C_3. As the oxidizer concentration increases, the transpassive range is intersected resulting in a sudden increase in the rate of corrosion.

3.24 CONVERSION FACTORS

(a) For corrosion rates

 (1) To convert g/m²/day to mm/year, multiply by 365/d.
 (2) To convert mm/year to mdd, multiply by 27.4 d.
 (3) Inches per year to mils per year × 1000.

(4) mg dm^{-2} day^{-1} to mils per year × $(1.144)/\rho$.
(5) mils per year to mg dm^{-2} day^{-1} × $(1.144)/\rho$.
(6) mg dm^{-2} day^{-1} to mmy^{-1} × $(36.52)/\rho$ × mdd.
(7) mdd to ipy × $(0.365)/\rho$.
(8) gmd to mm/year × $(0.365)/\rho$.

Also remember:

$$1 \text{ mpy} = 0.0259 \text{ mm/year}$$
$$= 25.9 \text{ μm/year}$$
$$= 2.90 \text{ nm/year}$$
$$1 \text{ cm year}^{-1} = 0.0394 \text{ mpy}$$

Table 3.3 summarize the above conversion factors.

Table 3.3 Summary of conversion factors

	Unit	mdd	mm/year	mils/year	in/year
(1)	Milligrams per square decimeter per day	1	0.0365/d	1.144/d	0.00144/d
(2)	Millimeters per year (mm/year)	27.4 × d	1	39.4	0.0394
(3)	Mils per year	0.696d	0.254	1	0.001
(4)	Inches per year	696d	25.4	1000	1

Note: mdd = mg/dm²/day, d = day

(b) For current density

(1) To convert mA/cm^2 to A/m^2, multiply by 10.

(2) To convert $\mu A/cm^2$ to A/m^2, multiply by .010.

(3) To convert A/cm^2 to A/m^2, multiply by 10 000.

(c) Relationship between magnitude of current and rate of penetration of corrosion

The relationship between the magnitude of current (current density, $\mu A/cm^2$) and the rate of penetration of corrosion (mpy) is important. The rate of corrosion of different metals and alloys can be equated with $1\,\mu A/cm^2$ of current generated by a corroding metal as shown in Table 3.4. For instance, a conversion rate of 0.540 mpy for 316 stainless steel and 0.52 mpy for type 304 stainless steel corresponds to $1\,\mu A/cm^2$ of current. Such factors are called electrochemical conversion factors. They can be calculated for any desired alloy or metal as shown below.

Consider, for example, conversion factor for AISI type 316 steel (non-magnetite). The following is the nominal composition of the steel:

Cr – 18%
Ni – 10%
Mo – 3%
Mn – 2%
Fe – balance

$$1\,\mu A/cm^2 = 0.128\left(\frac{\text{at.wt. of the element}}{\text{valence} \times \text{density}}\right)$$

$$\times \text{percent of alloy} \times \text{mpy}$$

$$= \left(\frac{0.128(52.01)}{1(7.1)}\right)\frac{0.18}{Cr}$$

$$+ \left(\frac{0.128(58.69)}{(2)(8.9)}\right)\frac{0.10}{Ni}$$

$$+ \left(\frac{0.128(54.94)}{(23)(7.43)}\right)\frac{0.02}{Mn}$$

$$+ \left(\frac{0.128(95.95)}{(2)(10.2)}\right)\frac{0.03}{Mo}$$

$$+ \left(\frac{0.128(55.85)}{(2)(7.80)}\right)\frac{0.70}{Fe}(mpy)$$

$$= 0.169 + 0.0442 + 0.00945$$

$$+ 0.018 + 0.318$$

$$= 0.559\,mpy$$

$$1\,\mu A/cm^2 = 0.56\,mpy$$

3.25 ILLUSTRATIVE PROBLEMS

Problem 1
From an anodic polarization diagram, the value of i_{corr} recorded is $4.21 \times 10^{-7}\,A/cm^2$. Find the rate of corrosion in (a) mdd, and (b) in inches per year.

Solution:
Applying Faraday's law: $M = \dfrac{i}{nF} \times 1$

(a) $\dfrac{4.21 \times 10^{-7}\,A/cm^2}{96\,490\,A.s./mol} \times \dfrac{55.8\,g/mol}{2} \times \dfrac{1000\,mg}{g}$
[In S.I. units, atomic mass $= 55.8 \times 10^{-3}\,kg/mol$]
$= 1.21 \times 10^{-7}\,mg/cm^2/s$

(b) Convert to $mg/dm^2/day$
$1.21 \times 10^{-7}\,mg/cm^2/s \times 3600\,s/h \times 24\,h/day$
$\times 100\,cm^2/dm^2$
$= 1.052\,mdd$

(c) To get the answer in inches/year
$mdd \times \dfrac{0.00144}{7.87} = 1.044 \times 0.000183$
$= 1.924 \times 10^{-4}\,inch/year$

The above conversions can be done in S.I. units as shown below. It is suggested that S.I. units be used in all further problems.

Unit (Dimensions)	Conversion factor
Density, $kg\,m^{-3}$	$1\,g/cm^3 = 10^3\,kg/m^3$
Current Density, Am^{-2}	$1\,A/cm^2 = 10^4\,A/m^2$

The atomic mass (amu) is given in kg/mol and molality is given in mol/kg.

Table 3.4 Electrochemical and current density equivalence with corrosion rate

Metal/alloy	Element/oxidation state	Density (g/cm^3)	Equivalent weight	Penetration rate equivalence to 1 μA/cm^2 (mpy)
Pure metals				
Iron	Fe^{2+}	7.87	27.92	0.46
Nickel	Ni^{2+}	8.90	29.36	0.43
Copper	Cu^{2+}	8.96	31.77	0.46
Aluminum	Al^{3+}	2.70	8.99	0.43
Lead	Pb^{2+}	11.34	103.59	1.12
Zinc	Zn^{2+}	7.13	2.68	0.59
Tin	Sn^{2+}	7.3	59.34	1.05
Titanium	Ti^{2+}	3.51	23.95	0.69
Zirconium	Zr^{4+}	6.5	22.80	0.75
Aluminum alloys				
AA1100	Al^{3+}	2.71	8.99	0.43
AA5052	Al^{3+}, Mg^{2+}	2.68	9.05	0.44
AA6070	Al^{3+}, Mg^{2+}	2.71	8.98	0.43
AA6061	Al^{3+}, Mg^{2+}	2.70	9.01	0.43
AA7075	Al^{3+}, Mg^{2+}, Zn^{2+}, Cu^{2+}	2.80	9.55	0.44
Copper alloys				
CDA110	Cu^{2+}	8.96	31.77	0.46
CDA260	Cu^{2+}, Zn^{2+}	8.39	32.04	0.49
CDA280	Cu^{2+}, Zn^{2+}	8.39	32.11	0.49
CDA444	Cu^{2+}, Sn^{4+}	8.52	32.00	0.48
CDA687	Cu^{2+}, Zn^{2+}, Al^{3+}	8.33	30.29	0.47
Stainless steels				
034	Fe^{2+}, Cr^{3+}, Ni^{2+}	7.9	25.12	0.41
321	Fe^{2+}, Cr^{3+}, Ni^{2+}	7.9	25.13	0.41
309	Fe^{2+}, Cr^{3+}, Ni^{2+}	7.9	24.62	0.41
316	Fe^{2+}, Cr^{3+}, Ni^{2+}, Mo^{3+}	8.0	25.50	0.41
430	Fe^{2+}, Cr^{3+}	7.7	25.30	0.42
Nickel alloys				
200	Ni^{2+}	8.89	29.36	0.43
400	Ni^{2+}, Cu^{2+}	8.84	30.12	0.44
600	Ni^{2+}, Fe^{2+}, Cr^{3+}	8.51	26.41	0.40
825	Ni^{2+}, Fe^{2+}, Cr^{3+}, Mo^{3+}	8.14	25.52	0.50

Source: Adapted from Proposed Standard, ASTM G01.11.

Problem 2

An electrode has a potential of -0.80 V relative to a 1 N Ag Silver–Silver chloride electrode. What is the electrode potential on the hydrogen scale?

Solution:

$$AgCl + e \approx Ag + Cl^- \quad E° = -0.22 V$$

$$\gamma \pm 0.001 N K_{Cl} = 0.901 (\gamma = \text{activity coefficient})$$

$$[a_{Cl^-}] = (0.01)(0.901) = 9 \times 10^{-3}$$

Applying Nernst equation:

$$E_{Ag} = E° - 0.0592 \log[a_{Cl^-}] \quad n = 1$$

$$= 0.222 - 0.0592 \log[9 \times 10^{-3}]$$

$$= 0.343 V$$

The potential on hydrogen scale

$$-0.80 + 0.343 = -0.457 V$$

Problem 3

The potential of a platinum cathode at which hydrogen is evolved is -0.85 V relative to a saturated calomel electrode. If the pH of the aqueous electrolyte is 2.00, determine the hydrogen over-potential (η_C).

Solution:
$$E_{rev(H)} = -0.059 \times pH = -0.592 \times 2$$
$$= -0.11840 V$$

It is necessary to convert the potential given, -0.85 volt (SCE) to the hydrogen scale. The conversion factor is

$$SHE = SCE + 0.242$$

$$SHE = -0.85 + 0.242 \text{ (0.242 is the potential of standard calomel electrode)}$$

$$= -0.608 V \text{ (SHE)}$$

The over-potential

$$\eta_C = E - E_{rev} = -0.608 - (-0.11840)$$

$$= -0.4896 V$$

Problem 4

Consider the reaction $M \rightarrow M^{z+} + ze$ on an electrode surface. An over-potential of -0.155 V (η_C) is applied to the electrode. The exchange current density ($i_{0,M/M^{z+}}$) at 25°C is 5×10^{-7} A/cm^2. Determine the current density, i, if the change in the oxidation state is one unity.

Solution:

$$i_c = i_0 \exp\left[\frac{(1-\beta)zF\eta_C}{RT}\right]$$

$\beta = 0.5$ (symmetry factor)

$$Z = 2, \ F = 96\,500\frac{C}{mol}, \ \eta_C = 0.155 V$$

$R = 8.314$ (gas constant)

$T = 25 + 273 = 298K,$

$$i_c = i_0 \exp\left[-\frac{(1-\beta)(zF\eta_C)}{RT}\right]$$

$$= \left[-\frac{(1-0.5) \times 2 \times 96\,500 \times (-0.155)}{8.314 \times 298}\right] \times i_0$$

$$i_c = i_0 \exp[6.037]$$

$$i_c = 5 \times 10^{-7} \times \exp[6.037]$$

$$i_c = 2.093 \times 10^{-4} \text{ A/cm}^2$$

Problem 5

A nickel electrode is corroding in a deaerated electrolyte which has a pH of 3.0 and a concentration of nickel ions of 0.003 at 25°C.

(a) Determine the i_{corr} of nickel.
(b) Determine also E_{corr}.

The following data is provided

$$i_0(H \text{ on Ni}) = 10^{-6} \text{A/cm}^2$$

$$i_0(O \text{ on Ni}) = 10^{-14} \text{A/cm}^2$$

$$b_a = 0.04 \text{ V/decade}$$

$$b_c = -0.13 \text{ V/decade}$$

Solution:

Reaction:

Anodic: $Ni \rightarrow Ni^{2+} + 2e$
Cathodic: $H_2 \rightarrow 2H^+ + 2e$
Total: $Ni + 2H^+ \rightarrow Ni^{2+} + H_2 \quad E° = -0.25\,V$

(a) $E_{Ni} = E° + \dfrac{0.0592}{2} \log\left(a_{Ni}^{2+}\right)$

$= -0.25 + 0.0296 \log(0.003)$

$= -0.325\,V$

(1) $\eta_A = E_{corr} - E_{rev(Ni)} = b_a \log\left(\dfrac{i_{corr}}{i_0}\right)$

(2) $\eta_C = E_{corr} - E_{H_2} = b_c \log\left(\dfrac{i_{corr}}{i_0}\right)$

$E_{H_2} = -0.0592 \times pH$
$= -0.0592 \times 2$
$= -0.1184\,V$

$\eta_A = E_{corr}(-0.325) = -0.04 \log \times 10^{-7}$

$\qquad + 0.04 \log i_{corr} \qquad (1)$

$E_{corr} = -0.045 + 0.04 \log i_{corr}$

$\eta_C = E_{corr} - (-0.1184)$

$\quad = 0.13 \log i \times 10^{-6} - 0.13 \log i_{corr}$

$\eta_C = E_{corr} + 0.1184$

$\quad = 0.13 \log \times 10^{-6} \log i_{corr}$

$E_{corr} = -0.828 - 0.13 \log i_{corr} \qquad (2)$

Equating equations (1) and (2), we get

$-0.045 + 0.4 \log i_{corr} = -0.898 - 0.13 \log i_{corr}$

$i_{corr} = (-5.0176)^{10}$

$i_{corr} = 9.6 \times 10^{-6}\,A/cm^2$

(b) By substituting the value of i_{corr} either in equation (1) or (2), E_{corr} can be obtained

$E_{corr} = -0.045 + 0.04 \log 9.6 \times 10^{-6}$

$\quad = -0.246\,V(SHE)$

Problem 6

The rate of corrosion of a steel pipe in Arabian Gulf water is 3.5 gmd. The corrosion proceeds mainly by the reaction $4OH^- \rightleftarrows O_2 + 2H_2O + 4e^-$. The pH of water is 8.0. The pipe needs to be protected cathodically. Calculate:

(a) The voltage required for complete protection.
(b) The minimum initial current required for protection.

Solution:

(a) Find out the mass in $kg/m^2 \cdot s$

$$m = 3.5\,gmd \times \frac{1\,kg}{1000\,g} \times \frac{1\,day}{24\,h} \times \frac{1\,h}{3600\,s}$$

$$= 4.051 \times 10^{-8}\,kg/m^2 \cdot s$$

(b) Find i_{corr} from Faraday's law:

$$i_{corr} = \frac{m \times n \times F}{At.\ mass}$$

$$= \frac{4.051 \times 10^{-8} \times 96490\,(A.s./mol)}{0.055847\,kg/mol}$$

where

$m = $ mass (kg)
$n = $ number of moles of electrons
$F = $ Faraday (coulombs/mole)

$$i_{corr} = 0.140\,A/m^2$$

Taking $\beta_a = 0.08\,V/decade$

$i_0; \dfrac{Fe}{Fe^{2+}} = 10^{-4} A/m^2$

$K\,Fe(OH)_2 = 1.64 \times 10^{-14}\,mol/liter$

$\left[H^+\right] = 1 \times 10^{-8} \rightarrow \left[OH^-\right] = 1 \times 10^{-6}$

$\left[Fe^{2+}\right] = \dfrac{1.64 \times 10^{-4}}{(1 \times 10^{-6})^2} = 0.0164$

$a_{Fe} = 1.0(\gamma) \times 0.0164 = 0.0164$

$\eta_A = E_{corr} - E_{Fe^{2+}} = \beta_a \log \dfrac{i_{corr}}{i_0}$

$$E_{corr} = -0.493 + 0.08 \log\left(\frac{0.140}{10^{-4}}\right)$$

$E_{corr} = -0.241\,\mathrm{V}$, which is the required voltage.

(c) Since $\dfrac{\Delta E}{\beta_a} = \log\dfrac{(i_p)}{i_{corr}}$

$$\frac{-0.2412 + 0.493}{0.08} + \log 0.14 = \log i_p$$

$i_p = 195.94\,\mathrm{A/m^2}$, which is the minimum current density required for complete cathodic protection.

Problem 7

An iron pipe is used for transporting of $1\,\mathrm{N\,H_2SO_4}$ (pH $= 0.3$). The rate of flow of acid in the pipe is $0.3\,\mathrm{m/s}$ at $25^\circ\mathrm{C}$. The relationship between the limiting current density i_L and velocity is given by $i_L = V^{0.5}$. The following additional information is provided:

$$\beta_a = 0.100$$

$$\beta_c = -0.060$$

$$i_0\left(\frac{Fe}{Fe^{2+}}\right) = 10^{-5}\mathrm{A/m^2}$$

$$i_0(H) = 10^{-2}\mathrm{A/m^2}$$

If the surface acts as the cathode, determine:

(a) The corrosion potential of iron.
(b) The corrosion rate of iron in mm/year.
(c) The corrosion rate of iron in mdd.

Solution:

$$i_{corr} = (0.3)^{0.5} = 0.5478\,\mathrm{A/m^2}$$

$$\eta_c = E_{corr} - E_{H_2} - \beta_c \log\left[\frac{i_{corr}}{i_0 H}\right]$$

$$(E_{corr})_{Fe} = -0.0592 \times 3 - 0.0592 \times 3$$

$$-0.06 \log\left[\frac{0.5478}{1 \times 10^{-2}}\right]$$

$$= -0.122\,\mathrm{volt}$$

From Faraday's law:

$$m = \frac{i_{corr} \times At_{wt}}{n \times F}$$

$$m = \frac{0.5478\,\mathrm{A/m^2} \times 0.055847\,\mathrm{kg/mol}}{2 \times 96490(\mathrm{A.s./mol})}$$

$$m = 1.585 \times 10^{-7}\,\mathrm{kg/m^2 \cdot s}$$

The rate of corrosion in mm/year can be determine as below:

$$\mathrm{CR\,(mm/year)} = 1.585 \times 10^{-7}\,\mathrm{kg/m^2 \cdot s}$$

$$\times \frac{1}{7.87 \times 10^{-3}\,(\mathrm{kg/m^3})}$$

$$\times \frac{10^3\,\mathrm{mm}}{1\mathrm{m}} \times \frac{86400\,\mathrm{s}}{1\mathrm{day}}$$

$$\times \frac{365\,\mathrm{days}}{1\,\mathrm{year}}$$

$$= 1.585 \times 10^{-7}\,\mathrm{kg/m^2 \cdot s}$$

The rate of corrosion in mdd:

$$= 1.585 \times 10^{-7}\,\mathrm{kg/m^2 \cdot s} \times \frac{10^6\,\mathrm{mg}}{1\,\mathrm{kg}}$$

$$\times \frac{1\mathrm{m^2}}{100\,\mathrm{dm^2}} \times \frac{3600\,\mathrm{s}}{1\mathrm{h}} \times \frac{24\mathrm{h}}{1\mathrm{day}}$$

$$= 136.95\,\mathrm{mdd}$$

Problem 8

Calculate the concentration over-potential for silver depositing at a rate of $3 \times 10^{-3}\mathrm{g\,dm^{-2}}$ from a cyanide solution at $25^\circ\mathrm{C}$. The limiting current density is $3\,\mathrm{A/dm^2}$. The concentration over-potential is given by

$$\eta_{conc} = \frac{2.3RT}{zF} \log\left(1 - \frac{i}{i_L}\right)$$

Solution:
First to obtain i

$$i = \frac{3 \times 10^{-3}\mathrm{g.A.s}}{\mathrm{dm^2} \times 60\mathrm{s} \times 0.00118\mathrm{g}}$$

$$= 4.23\,\mathrm{Am^{-2}}$$

(0.00118 g of silver is liberated by 1 coulomb)

$$i_L = 3 \, A/dm^{-2} = 300 \, Am^{-2}$$

$$\eta = \eta_{conc} = 60 \log \left(1 - \frac{4.23}{300} \right)$$

$$= 60 \log[1 - 0.0141]$$

$$= 60 \log(0.986) = -0.37 \, mV$$

Problem 9

What volume of oxygen gas at STP must be consumed to produce the corrosion of 200 g of iron?

Solution:

$$Fe \rightarrow Fe^{2+} + 2e \quad \text{(one mole of Fe produces 2 moles of electrons)}$$

$$4OH^- \rightarrow O_2 + 2H_2O + 4e$$
$$(^1/_2 \, mole \, O_2 = 1 \, mole \, Fe)$$

$$mole \, O_2 \, gas = \frac{200 \, g}{55.85 \, gFe/mol \cdot Fe}$$

$$\times \frac{1/2 \, mol \cdot O_2}{1 \, mol \cdot Fe}$$

$$= 1.89 \, mol \, O_2$$

Using ideal gas equation

$$V = \frac{nRT}{p} = (8.314 \, J/mol)$$

$$\times \frac{8.314 \, J/mol \times (273k)}{(1 \, atm)(1 \, Pa/9.869) \times 10^{-6} \, atm}$$

$$= 0.042 \, J/Pa$$

$$= 0.042 \, J/Pa = 0.42 \, N \cdot m/Nm^2$$

$$= 0.0201 \, m^3$$

Problem 10

The maximum corrosion current density in a steel sheet coated with zinc and exposed to seawater is found to be $5 \, mA/m^2$. What thickness of the layer is necessary so that the coating may last for one year?

Solution:

Consider $1m^2$ of steel coated with zinc. Suppose the steel sheet is (X) meters (thick).

The mass lost is $= 7.13 \, mg/m^2 \times (1 \, m^2)(t_x) = (7.13 \, X) \, mg$.

The corrosion current can be determined as

$$5 \times 10^{-3} \, A/m^2 \times 1 \, m^2 = \frac{(7.13X) \, mg}{1 \, year} \times \frac{10^6 \, g}{mg}$$

$$\times \frac{0.6023 \times 10^{24} \, atoms}{65.38 \, g} \times \frac{2e}{atom}$$

$$\times \frac{0.16 \times 10^{-18} \, C}{e^-} \times \frac{1 \, year}{(365)(24)(3600)} \times \frac{1 \, A}{C/s}$$

$$= 7.50 \times 10^{-6} \, m$$

$$= 7.50 \, \mu m$$

Problem 11

An over-potential of 250 mV is applied to platinum in an aqueous and electrolyte. What would be the rate of hydrogen evolution if the exchange current of hydrogen on platinum is $1 \times 10^{-4} \, A/cm^2$ and $\beta = 0.5$?

Solution:

$$\eta = \frac{RT}{\beta zF} i_0 - \frac{RT}{\beta zF} i_c = \frac{RT}{\beta zF} \ln i_0 - \frac{RT}{\beta zF} \ln i_c$$

$$= \frac{0.059}{1 \times 0.5} \log i_0 - \frac{0.059}{1 \times 0.5} \log i_c$$

$$= 0.12 \log i_0 - 0.12 \log i_c$$

$$\log i_c = 0.25 = 0.12 \log \times (-4) - 0.12 \log i_c$$

$$\log i_c = \frac{-0.23}{0.12} = 1.2 \times 10^{-2} \, A/cm^2$$

QUESTIONS

A. MULTIPLE CHOICE QUESTIONS

Select one best answer: (*More than one answer may be correct in some question.*)

1. Which one of the following reactions is a reversible reaction?

   [ ] $M \rightarrow M^{z+} + ze$
   [ ] $H_2 \rightarrow 2H^+ + 2e$
   [ ] $M \rightarrow M^{3+} + 3e$
   [ ] $2H^+ + 2e \rightleftarrows H_2$

2. When a current flows as a result of impressing a voltage in the noble direction, the impressed current is

   [ ] net anodic
   [ ] net cathodic
   [ ] mixed current

3. The term over-voltage is given by

   [ ] $\eta = E - E^\circ$
   [ ] $\eta = E - E_{corr}$
   [ ] $\eta = E_{corr} - i_{corr}$
   [ ] $\eta = E^\circ - E$

4. When $\eta = 0$, and $E = E_{eq}$

   [ ] $i_a > -i_c > i_0$
   [ ] $i_a < i_c < i_0$
   [ ] $i_a = i_c = i_0$
   [ ] $i_a = * - i_0^* = i_0$

5. The polarization curve for the forward anodic reaction follows the general equation

$$i = i_0 \left\{ \exp \left(\left[\beta \frac{ZF}{RT} \eta_a \right] \right. \right.$$
$$\left. \left. - i_0 \exp \left[-(1-\beta) \frac{ZF}{RT} \eta_a \right] \right) \right\}$$

The above equation is called

[ ] Guoy–Chapman equation
[ ] Nernst equation
[ ] Tafel equation
[ ] Butler–Volmer equation

6. On the basis of corrosion kinetics iron corrodes faster than zinc. Which of the parameters given below is used for prediction?

[ ] The exchange current density (i_0 H on Fe)
[ ] The exchange current density (i_0(H) Zn/Zn^{2+})
[ ] E° Fe/Fe^{2+} and E° Zn/Zn^{2+}
[ ] i_{corr}(Fe) and i_{corr}(Zn)

7. One corroding metal (M) is connected to another corroding metal (N) in an acid electrolyte. Metal M has a relatively more noble potential E_{M/M^+} and i_{corr}(M) while metal N has a less noble potential E_{N/N^+} and corrodes at a rate of i_{corr}(N).

[ ] The corrosion rate of metal M is decreased whereas the corrosion rate of metal N is increased
[ ] The corrosion rate of metal M is increased whereas the corrosion rate of metal N is decreased
[ ] The corrosion rate of the couple M–N is less than the corrosion rate of metal M and N individually
[ ] The corrosion rate of metal N is higher than the corrosion rate of the couple

8. Consider the corrosion behavior of zinc (1 cm^2) coupled to platinum (1 cm^2) and platinum (10 cm^2) in an acid electrolyte.

[ ] Zinc coupled to platinum (10 cm^2) will show the highest rate of corrosion
[ ] Zinc coupled to platinum (1 cm^2) will show the highest rate of corrosion

[ ] Zinc uncoupled will show a higher rate of corrosion than zinc coupled with 1 cm^2 of platinum

[ ] Zinc uncoupled will show a higher rate of corrosion than zinc coupled with 10 cm^2 of platinum surface

9. Which of the following statements is not true about the linear polarization technique?

[ ] Measurements begin at -20 mV from OCP and end around $+100$ mV from OCP

[ ] The slope of the plot is $\Delta E / \Delta C$ is given in volts/amperes or mV/mA

[ ] This technique can be used to determine very low rate of corrosion

[ ] It can be used to monitor corrosion rate in process plant

10. Passivation occurs when

[ ] the corrosion potential (E_{corr}) becomes more positive than the potential corresponding to the equilibrium between the metal and one of its oxides/hydroxides

[ ] $E_{corr} < (E_{eq})$ M/M$_o$, M/M$_o$H

[ ] $E_{corr} > E_{passive}$

[ ] $E_{corr} = E_{eq}$, M/M$_o$

11. The most important criteria to compare passivity between two metals are

[ ] i_0

[ ] $i_{critical}$

[ ] $i_{passive}$

[ ] i_{corr}

12. Cathodic polarization takes place when: ($\eta = $ over-voltage)

[ ] $\eta > 0$

[ ] $\eta < 0$

[ ] $\eta = 0$

[ ] $\eta > E_{rev}$

13. Activation polarization is present

[ ] at low reaction rates

[ ] at high reaction rates

[ ] when the limiting current density is reached

[ ] when an oxidizing is introduced in the corroding system

14. In activation polarization

[ ] the current varies linearly with the applied potential

[ ] the current varies exponentially with the potential

[ ] up to 50 mV, a linear relationship between E and log I is observed followed by an exponential relationship

[ ] none of the above is observed in a plot indicating activation polarization

15. Concentration polarization occurs when

[ ] the concentration of electroactive species at the metal/electrode surface and the bulk solution is the same

[ ] a concentration gradient is built up between the electrode/electrolyte interface and the bulk solution

[ ] the rate of formation of an ion is balanced by the rate of its arrival by diffusion from the outer Helmholtz plane

[ ] the solution is continuously stirred or agitated

B. How and Why Questions

1. Explain why:

a) Only the atoms at the kink sites are chemically active.

b) Anodic polarization proceeds when the over-voltage applied is greater than zero.

c) No net current flows when rate of forward reaction (i_f) is the same as for the reverse reaction (i_r).

d) The rate of charge transfer by the cathodic process $\overleftarrow{i}$ is faster than by the anodic process when a negative voltage is impressed on the metal.

e) Cathodic polarization proceeds when the over-voltage is less than zero.

2. Answer the following questions with regard to the energy vs distance profiles.

a) What renders the dissolutions of a metal M more rapid when the reaction is irreversible and why the reverse process does not take place?

b) What causes a decrease in the free energy of activation required for the forward process of dissolution?

c) The general form of an anodic dissolution process is given by

$$\overrightarrow{i} = i_0 \exp\left[\frac{\beta z \eta F}{RT}\right]$$

$$- i_0 \exp\left[\frac{-(1-\beta)z\eta F}{RT}\right]$$

If the irreversibility of the electrode process is high ($\eta = 0.05$ V), why the second term from the above equation is dropped?

3. Answer the following question with regard to polarization.

a) What is the numerical value of a and b?

b) What is indicated by $\left(\dfrac{2.3RT}{\beta zF}\right)$

c) What is the significance of $-\dfrac{2.3RT}{(1-\beta)zF}$? Why the polarization term η is multiplied by zF? What are the units for zF?

d) Why the current density is a linear function of over-potential (η) in activation polarization?

4. a) What is the significance of $i_a^{net} = i_f - |i_r|$?

b) Consider the reaction $M \underset{i_r}{\overset{i_f}{\rightleftharpoons}} M^z + ze$.

If $\quad \Delta G_A^* = \Delta G^* + [(1-\beta)F\eta_A] - zF\eta_A$,

show that $i_f = i_0 \exp\left[\frac{\beta z F \eta_A}{RT}\right]$.

5. Answer the following questions:

a) Why anodic polarization occurs when $\eta_A > 0$ and cathodic polarization takes place when $\eta_C < 0$?

b) Why the relationship between E vs $\log I$ is exponential if concentration polarization occurs?

c) Why activation polarization takes place at low applied over-voltages?

d) Why activation polarization is not affected by agitation?

6. a) If $i_A^{net} = i_f = |i_r|$ write the full expression for anodic activation.

b) When $\eta_A > 0.03$ V, write an expression for i_a^{net}.

c) If $i_0^{net} = |i_r| - i_f$, write an expression for cathodic activation.

d) If $i_a = i_0 \exp\left[\frac{\beta z F \eta_A}{RT}\right]$ write an expression for η_A.

e) As asked in Problem (d), write an expression for in terms of η_C (consult energy distance profiles in the text).

7. Why iron corrodes faster than zinc in an acid electrolyte, according to kinetics? What is the basis of this prediction and why it is different from the prediction based on thermodynamic potentials?

8. When zinc is galvanically coupled to platinum:

a) Why the rate of corrosion of zinc coupled with platinum increases? (Consult the diagram in the text.)

b) Why the rate of corrosion of zinc coupled to Pt (1 cm^2) is lower than the rate of corrosion of zinc coupled to platinum (10 cm^2).

c) Why zinc coupled to gold corrodes faster than zinc coupled to platinum?

9. What is the effect of the following on concentration polarization?

a) Agitation

b) Temperature

c) Velocity

d) Concentration of species

10. State the reasons for the following questions on concentration polarization:

a) For small shifts in potential, charge transfer is completely controlling and the over-potential is purely an activation over-potential.

b) For larger shifts in potential, the current is less than expected.

c) If sufficiently large shift, the current becomes independent of potential.

11. A metal surface exposed to an aqueous electrolyte corrodes. The metal has several tiny anodic and cathodic areas.

a) Why it is not possible to measure the current between a local anode and a local cathode?

b) How the corrosion potential is measured experimentally?

c) When a net anodic reaction is generated, electrons are released. How are these electrons consumed? What is the method which is used to collect these electrons.

d) Why an auxiliary electrode is used in the setup for measurement of corrosion current? What is its function?

e) What are the functions of the reference, working and the auxiliary electrodes?

12. Explain why:

a) It is essential to polarize the electrode sufficiently away (at least 100 mV) either in the anodic direction or cathodic direction in the linear polarization method.

b) Tafel relationship is dependent on activation control method and not on diffusion control.

c) A luggin probe is used in the polarization cell and the reference electrode is not generally introduced directly to the electrolyte.

d) The Tafel method is applied to systems containing one reduction reaction only.

13. Answer the following questions related to passivity of metal:

a) Why zinc does not show an active–passive behavior whereas iron shows such a behavior?

b) Why the current starts to recede as soon as the critical current density ($i_{critical}$) is reached as observed in a S-shaped anodic polarization diagram for active–passive metals?

c) What is the function of passive current density (i_p)?

d) What is the significance of transpassive region? What reaction generally takes place in this region?

e) What reaction is likely to take place above the reversible electrode potential of oxygen?

f) What would be the effect of addition of 10–15% chromium on the Flade potential (E_p).

14. Distinguish between the following:

a) Aeration and deaeration.

b) i_0 and $i_{critical}$.

c) E_F and E_{pp}.

d) Outer Helmholtz plane and the diffused layer.

e) Active region and passive region in a polarization plot of an active–passive metal.

C. CONCEPTUAL QUESTIONS

1. Explain clearly the difference between activation polarization and concentration polarization. Under what conditions would activation polarization change to concentration polarization?

2. Explain clearly the difference between the inner Helmholtz plane, outer Helmholtz plane and diffuse layer in electrical double layer.

3. Explain on the basis of kinetics why zinc corrodes more rapidly in an aerated solution of HCl than in a deaerated solution.

4. Explain the effect of the following on the anodic polarization curve for active–passive metal:

   a) Effect of adding an oxidizer.
   b) Effect of chloride in concentration.
   c) Effect of temperature.
   d) Effect of addition of inhibitors.

5. Explain how the passivating tendency can be increased by the following. Give examples.

   a) Increasing the rate of cathodic reduction.
   b) Reducing the critical current density.
   c) Noble metal addition.

D. PROBLEMS

1. The potential of a cathode at which hydrogen is generated is -0.80 volts with respect to a standard Calomel electrode. If the pH $= 4$, determine the hydrogen over-potential.

2. An electrode has a potential of -0.70 V with respect to a silver–silver chloride in 0.01 N KCl at 25°C. Determine the potential on the hydrogen scale.

3. If the anodic over-voltage of an electrode is 0.12 V, and the exchange current density i_0, Fe/Fe^{2+} $= 10^{-6}$ A/cm^2, calculate the anodic current density.

4. From the following data on the corrosion of a nickel electrode in an acid electrolyte (pH$=2.0$), calculate

   a) i_{corr} of Ni.
   b) E_{corr} of Ni.

   Data:

   1) i_0(H on Ni)$=10^{-6}$ A/cm^2
   2) i_0(H)$=10^{-14}$ A/cm^2
   3) $b_a=0.04$
   4) $b_c=-0.13$
   5) $a_{Ni^{++}}$ at 25°C

5. A steel tank used for the storage of water of pH $= 7$ corrodes at the rate of 10 mg/dm^2/day. It is desired to protect the tank cathodically by using an external current. Estimate the minimum current density (A/m^2) and the voltage required for complete cathodic protection of the storage tank. Assume that corrosion occurs by oxygen depolarization. The following constants are provided:

$$\beta_a = 0.06 \, \text{V/decade}$$

$$i_0, \frac{\text{Fe}}{\text{Fe}^{2+}} = 10^{-2} \, \text{A/cm}^2$$

$$K, \text{Fe(OH)}_2 = 1.64 \times 10^{-14} \, \text{mol/l}$$

6. A bimetallic platinum copper couple is immersed in an acid solution at 25°C and oxygen is passed into the solution rapidly. Calculate the anodic current density on the copper assuming that the corrosion rate of uncoupled copper in the same solution can be neglected and that the area of platinum and copper are 10 and 1 cm^2, respectively. The solubility of oxygen in the solution is 1.4 μmol/cm^3. The diffusion coefficient of oxygen is 1.75×10^{-5} cm^2·s^{-1}. The thickness of the diffusion layer is 0.05 cm. The Faraday's constant is 96 500 C.

7. Convert 0.122 mg/mm^2/second to

   a) mg/dm^2/day
   b) mg/m^2/day
   c) mm/year

8. A tin immersed in seawater shows a current density of 2.45×10^{-6} A/cm^2. What is the rate of corrosion in mdd?

9. In a polarization resistance experiment, an applied over-voltage of 10 mV results in a current density of 5 mA.

   a) What is the current density for corrosion if $b_a=0.06$ V and $b_c=-0.12$ V?
   b) Using Faraday's law, calculate the rate of corrosion in mdd.

10. In a linear polarization experiment, a current increase of 10 μA for a voltage increment of 4 mV is recorded. The area of the specimen

is 10 cm². Calculate (a) the polarization resistance, and (b) the rate of corrosion. Assume $b_a = 0.01$ V, $b_c = 0.02$ V/decade.

11. A metal M has been anodically polarized to 0.50 mV during anodic polarization. The following values for electrochemical parameters have been obtained:

$$b_a = 0.20\,V,\ b_c = -0.2\,V,\ \text{and}$$

$$I = 20\,mA/m^2.$$

Determine the current density.

12. What would be the rate of corrosion of copper in aerated seawater (pH = 7)? The solubility product for $Cu(OH)_2 = 5.6 \times 5 \times 10^{-20}$ mol/l.

13. The value of i_{corr} obtained for iron corroding in an aqueous solution (electrolyte) by the Tafel extrapolation technique is 3.74×10^{-4} A/m². Find the rate of corrosion in (a) mm/year and (b) mdd.

14. From the following data, calculate the thickness of the diffusion layer:

$$i_L = 12\,A/dm^2,\ c = 1\,mol/l,$$

$$D = 10^{-5}/cm^2/s,\ \eta = 2.$$

15. In a corrosion cell involving chromium, which forms Cr^{3+}, an electrical current of 20 mA is measured. How many atoms per second are oxidized at the anode?

16. A piece of steel corrodes in seawater. The corrosion current density is 0.2 mA/cm². Calculate the rate of weight loss in mdd units.

17. If zinc surface is corroding at a current density of 2×10^{-5} and current of 0.2 mA/cm², what thickness of metal would corrode in 8 months.

18. Determine the potential of an electrode on hydrogen scale which has a potential of 0.87 relative to an Ag/AgCl electrode.

19. The corrosion rate of iron in deaerated HCl is 40 mdd. Calculate the E_{corr} of iron with respect to 0.1 N Calomel electrode ($b_a = 0.1$)

20. A mild steel cylindrical tank 1 m high and 50 cm dia contains aerated water to the 60 cm level and shows a loss in weight due to corrosion of 304 g after 6 weeks. Calculate the corrosion current density involved.

SUGGESTED READING

[1] Davis, J.R. ed. (2000). *Understanding the Basics*, Materials Park, Ohio: ASM International, USA.

[2] Stansbury, E.E. and Buchanan, R.A. (2000). *Fundamentals of Electrochemical Corrosion*, Materials Park, Ohio: ASM, USA.

[3] Bockris, J.O.M and Reddy, A.K.N. (1973). *Modern Electrochemistry*, New York: Plenum Press.

[4] Pourbaix, M. (1974). *Atlas of Electrochemical Equilibria*, NACE, 1974.

[5] Vetter, K.J. (1967). *Electrochemical Kinetics*, New York: Academic Press, USA.

[6] Stern, M. and Geory, A.L. (1957). Electrochemical polarization I: A theoretical analysis of the shape of polarization curves, *J. Electrochem Soc*, **104**, 56–63.

[7] Stern, M. and Geory, A.L. (1957). Electrochemical polarization II, *J. Electrochem Soc*, **104**, 559.

[8] Stern, M. (1957). Electrochemical polarization III, *J. Electrochem Soc*, **104**, 645.

KEYWORDS

Activation polarization Polarization of an electrode controlled by a slow step in reaction sequence of steps at the metal/electrolyte interface. There is a critical activation energy needed to surmount the energy barrier associated with the slowest step.

Anodic polarization The shift of the potential of an electrode in a positive direction by an external current.

Concentration polarization (Diffusion or transport over-potential) It is the change of potential of an electrode caused by concentration change near the electrode/electrolyte interface. The concentration changes are caused by diffusion of ionic species in the electrolyte.

Corrosion potential It is the potential of a corroding surface in an electrolyte with reference to a reference electrode. It represents the mutual polarization of the potentials of the anodic and cathodic reactions which constitute the overall corrosion reaction.

Corrosion rate It is the rate showing slow or fast corrosion proceeds on a metallic surface. Corrosion rate is commonly expressed in terms of millimeters per year (mm/year), mils per year (1 mil = 1/1000 of an inch), or milligrams per decimeter square per day (mdd).

Critical current density The maximum current density exhibited in the active region of a metal/alloy system which exhibits an active/passive behavior.

Exchange current density It is the rate of exchange of electrons (expressed as electrical current) when an electrode reaches equilibrium at the equilibrium potential. At the equilibrium potential, the rate of forward reaction $\xrightarrow{r_f}$ (anodic) balances the rate of reverse reaction $\xleftarrow{}{r_r}$.

Faraday It is the quantity of electrical charge required to bring a change of one electrochemical equivalent ($F = 96\,500\,C$).

Flade Potential It is the potential signifying the onset on passivity in an active/passive metal/alloy system.

Helmholtz double layer Ions at the metal/electrolyte interface create separation of changes due to repulsive forces. The separation of negative and positive charges like in a capacitor constitute a double layer. It was introduced by Helmholtz, hence it is known as Helmholtz double layer.

Mixed polarization The combined potential of a specimen where two or more electrochemical reactions proceed on the surface of the specimen.

Passive current density The current density at which the metal or alloy is pushed in the passive region.

Polarization The shift of the potential of an electrode from its equilibrium potential ($\eta = c$).

Transpassive region The region of anodic polarization above the passive potential range, which shows a sudden increase in the current density due to breakdown of passivity.

Types of Corrosion: Materials and Environments

4.1 Introduction

A wide spectrum of corrosion problems are encountered in industry as a result of combination of materials, environments and service conditions. Corrosion may not have a deleterious effect on a material immediately but it affects the strength, mechanical operations, physical appearance and it may lead to serious operational problems. Corrosion may manifest itself as a cosmetic problem only, but it can be very serious if deterioration of critical components is involved. Serious corrosion problems, such as the pitting of condenser tubes in heat exchangers, degradation of electronic components in aircrafts and corrosion fatigue of propellers can lead to catastrophic failures. When catastrophic failures occur, the cost in terms of lives, equipment, and time is very high. While evaluating the long range performance of materials, it is essential for an engineer to consider the effects of corrosion along with other characteristics, such as strength and formability. In order to be able to select suitable materials, it is important for an engineer to understand the nature of corrosion, types of corrosion and methods of prevention of different types of corrosion. The mechanism of corrosion has already been discussed at length in Chapter 3, and so is not repeated here.

Environment plays a very important part in corrosion. The severity of corrosion varies considerably from one place to another. For instance, the rate of corrosion of steel will not be the same in Dhahran and Riyadh,[1] because the former has a sea-coastal (marine) environment and the later, a typical desert environment. Generally, the variation in severity of corrosion at different geographical locations can be attributed to the variation in moisture, temperature and air-borne substances found in the atmosphere. The marine environment is considered the most corrosive of natural environments. The environmental areas are classified on the basis of the degree of atmospheric contamination. Air pollutants are found in liquid, solid and gaseous forms.

The familiarity of environment and the type of corrosion is very important for design engineers. The engineer should recognize the potential hazards of corrosion and should be familiar with the methods used to mitigate various types of corrosion attacks. This chapter, therefore, will be devoted mostly to the study of localized corrosion. The scope of localized corrosion is very extensive and the literature is abundant on the theoretical aspect of localized corrosion. The main object of this chapter is to familiarize engineers with the type of localized corrosion so that they may identify the various forms of corrosion and suggest remedial measures. Contradictory mechanisms will, therefore, be left to a minimum. Each form

---

[1] Both are located in Saudi Arabia. Dhahran has a marine environment, and Riyadh has a desert environment.

of corrosion will be discussed in the following manner:

(1) Definition.
(2) Environment.
(3) Mechanism.
(4) Examples from industry.
(5) Methods of prevention.

4.2 UNIFORM CORROSION

4.2.1 DEFINITION

It is the uniform thinning of a metal without any localized attack. Corrosion does not penetrate very deep inside. The most familiar example is the rusting of steel in air.

4.2.2 ENVIRONMENT

(1) Dry atmosphere.
(2) Damp atmosphere.
(3) Wet atmosphere.
(4) Acids (HCl, $HClO_4$, H_3PO_4).
(5) Atmospheric contaminants.
(6) Process water containing hydrogen sulfide.
(7) Brines.
(8) Industrial atmosphere.
(9) Hydrocarbon containing wet hydrogen sulfide.

4.2.3 EFFECT OF POLLUTANTS

Corrosion can proceed in a dry environment without any moisture if traces of sulfur compounds or H_2S or other pollutants are present in the air. Tarnishing of silver in dry air in the presence of H_2S traces is an example of dry corrosion. Industrial atmospheres contain SO_2 as the major contaminant.

The rate of corrosion in the presence of SO_2 increases in the presence of moisture. The sulfur dioxide released in the atmosphere reacts with the rust formed on the metal surface as shown in reaction(4.3). Much of the SO_2 is converted to

SO_3 in the upper atmosphere. The reaction is:

$$SO_3 + H_2O \rightarrow H_2SO_4 \text{ (Sulfuric acid)} \qquad (4.1)$$

$$SO_2 + H_2O \rightarrow H_2SO_3 \text{ (Sulfurous acid)} \qquad (4.2)$$

$$SO_2 + 2Fe_2O_3 \rightarrow FeSO_4 + Fe_3O_4 \qquad (4.3)$$

The $FeSO_4$ formed accelerates corrosion. Once rusting has started, corrosion cannot be stopped even after SO_2 is removed from the air. Iron corrodes faster than any other engineering material in an industrial and marine atmosphere (Fig. 4.1). Other contaminants are nitrogen compounds, H_2S and also dust particles. CO_2 does not play a significant role in uniform corrosion. Sulfur compounds are abundant in an atmosphere where petroleum industry is located.

4.2.4 EFFECT OF HUMIDITY

Corrosion can be caused in the atmosphere when about 70% of the humidity is present, as this is the value in equilibrium with saturated NaCl solution and NaCl is commonly present on surfaces. In the presence of such humidity, an invisible thin film of moisture is formed on the surface of a metal. The thin film of moisture acts as an electrolyte for the passage of current. Structures which are exposed to open air, are affected by damp environments. Beyond 80% relative humidity, a sharp increase in the rate of corrosion is observed. Each metal has a critical value of relative humidity beyond which the rate of corrosion increases significantly.

4.2.5 WATER LAYERS

If visible water layers are formed on the metal surface, corrosion initiates. Splashing of seawater, rain and drops of dew provide the wet environment. The water layer on the metal surface acts as an electrolyte and provides a passage for the flow of current, similar to the situation in a corrosion cell.

4.2.6 DEW FORMATION

If the dew becomes acidic, due to the presence of SO_2, it increases the rate of corrosion.

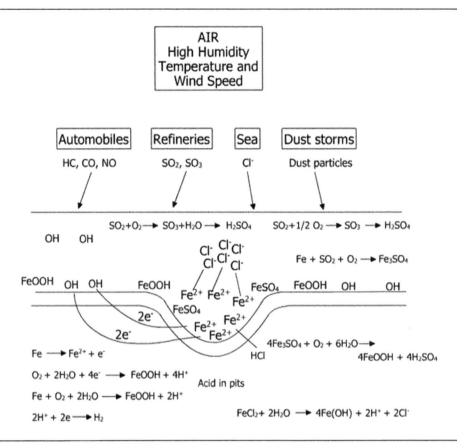

Figure 4.1   Corrosion of iron in marine atmosphere

Automobiles left open in the air may be subject to corrosion through acidic dew formation.

4.2.7 CORROSION PRODUCT

If the corrosion product on the metal surface is microporous, it can condense the moisture, below the critical value. Corrosion proceeds rapidly in such a case, even if the moisture content is below the critical limit.

4.2.8 MECHANISM OF UNIFORM CORROSION

Corrosion mechanism in aqueous solution has been amply demonstrated. In atmospheric corrosion which also exemplifies uniform corrosion, a very thin layer of electrolyte is present. It is probably best demonstrated by putting a small drop of seawater on a piece of steel. On comparing the atmospheric corrosion with aqueous corrosion, the following differences are observed:

On a metal surface exposed to atmosphere, only a limited quantity of water and dissolved ions are present, whereas the access to oxygen present in the air is unlimited [1]. Corrosion products are formed close to the metal surface, unlike the case in aqueous corrosion, and they may prevent further corrosion by acting as a physical barrier between the metal surface and environment, particularly if they are insoluble as in the case of copper or lead. The following is a simplified mechanism of aqueous corrosion of iron (Fig. 4.2):

At the anodic areas, anodic reaction takes place:

$$Fe \rightarrow Fe^{++} + 2e \qquad (4.4)$$

Aqueous Corrosion

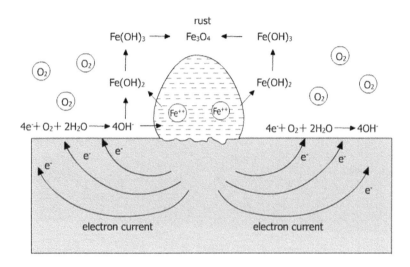

Figure 4.2   Aqueous corrosion of iron

At the cathodic areas, reduction of oxygen takes place:

$$O_2 + 2H_2O + 4e \rightarrow 4OH^- \qquad (4.5)$$

The OH ions react with the Fe^{++} ions produced at the anode:

$$Fe^{2+} + 2OH^- \rightarrow Fe(OH)_2 \qquad (4.6)$$

With more access to oxygen in the air, $Fe(OH)_2$ oxidizes to $Fe(OH)_3$ and later it loses its water:

$$4Fe(OH)_2 + O_2 + 2H_2O \rightarrow 4Fe(OH)_3 \qquad (4.7)$$

Ferrous hydroxide is converted to hydrated ferric oxide or rust by oxygen:

$$4Fe(OH)_2 + O_2 \rightarrow 2Fe_2O_3 \cdot H_2O + 2H_2O \qquad (4.8)$$

Rust ($Fe_2O_3 \cdot H_2O$) is formed halfway between the drop center and the periphery which is alkaline. The electrons flow from the anode (drop center) to cathode (periphery) in the metallic circuit. The current flow is shown in Fig. 4.2. The ferrous ions on the surface of iron are soluble whereas those in solution are oxidized by oxygen

to insoluble hydrated oxides of ferric called *rust*. The rust is formed away from the corroding site. The corrosion rate is very high if the ferrous ion is oxidized to ferric oxide rapidly. $Fe(OH)_3$ is insoluble and if it forms away from a metal surface, the corrosion reaction speeds up as equilibrium is to be maintained by supplying more ferrous ions (Fe^{++}) from the surface. If, however, $Fe(OH)_3$ is formed on the surface of a metal very rapidly, the corrosion is prevented (a passive film).

If SO_2 is present as a pollutant, in air, $FeSO_4$ is produced (equations (4.1–4.3)). The corrosion of iron is significantly affected by the presence of soluble sulfate ion in solution. The sulfate ion continues to attack iron and the surface becomes uneven and even pitted. In this case, layers of porous rust are formed. As no protection is provided by the porous rust to the metal, corrosion continues to take place. The effect of SO_2, SO_3 and Cl^- ions is illustrated in Fig. 4.3.

4.2.9 EXAMPLES OF UNIFORM CORROSION

(1) Tarnishing of silver ware.
(2) Tarnishing of electrical contacts.

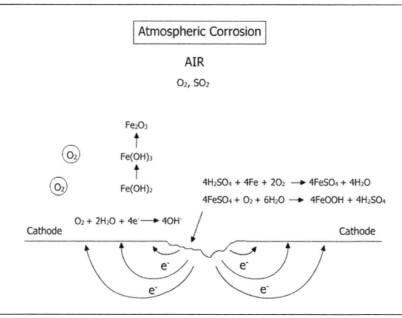

Figure 4.3 Effect of SO₂ and humidity on metallic corrosion. Reaction occur in a very thin (invisible) aqueous layer

(3) Rusting of steels in open air.
(4) Corrosion of offshore drilling platforms.
(5) Corrosion of galvanized steel stairways.
(6) Failure of distillation columns.
(7) Corrosion of electronic components.
(8) Corrosion of underground pipes (composite asphalt coated).
(9) Corrosion of automobile bodies.
(10) Corrosion of heat exchanger tubes.
(11) Corrosion of structural steels.

4.2.10 FAILURE CASE HISTORIES

4.2.10.1 Case 1 – Failure of a Distillation Column Wall

A distillation column wall, originally 20.0 mm thick failed in a humid atmosphere. The operating temperature varied between 0 and 15°C (32–59°F). After a service of three years, the column failed.

(a) Environment Ninety percent relative humidity. Conditions: condensed moisture and oxygen.

The pressure was 0.1 MPa. The insulation was wetted by exposure to air.

(b) Micro and macro examinations A macro examination with a magnifying lens indicated severe corrosion of the external surface. The surface was thinned to the point of perforation. Micro examination was not considered necessary.

(c) Remarks The operating temperature of the base of the distillation tower is about 60°C, whereas the top operating temperature is 0°C. Because of the leakage of water in the insulation material and ice formation, moisture penetrated below the insulation surface. As a result of wetting of insulation, the corrosion rate substantially accelerated. The situation could have been worse, had the temperature range been 60–80°C.

(d) Remedy Coat the entire column of steel with epoxy-phenolic coating. Cathodically protective pigments should not be used in the coating system, particularly in the hotter section as the polarity of zinc can be changed at temperatures above 60°C, the zinc becomes cathodic and iron becomes anodic. Iron corrodes and zinc is protected.

Table 4.1   Galvanic corrosion chart

Magnesium	Anodic (least noble) corroded
Magnesium alloys	
Zinc	
Beryllium	
Aluminum 110, 3003, 3004, 5052, 6053	
Cadmium	
Aluminum 2017, 2024, 2117	
Mild steel 1018, wrought iron	
HSLA steel, cast iron	
Chrome iron (active)	
430 Stainless (active)	
302, 303, 321, 347, 410, 416 Stainless steel (active)	
Ni-resist	
316, 317 Stainless (active)	
Carpenter 20Cb-3 stainless (active)	
Aluminum bronze (CA687)	
Hastelloy C (active) Inconel 625 (active titanium (active))	
Lead/tin solder	
Lead	
Tin	
Inconel 600 (active)	
Nickel (active)	
60% Ni–15% Cr (active)	
80% Ni–20% Cr (active)	
Hastelloy B (active)	
Naval brass (CA464), yellow brass (CA268)	
Red brass (CA230), admiralty brass (CA443)	
Copper (CA102)	
Manganese bronze (CA675), Tin bronze (CA903, 905)	
410, 416 Stainless (passive) phosphor bronze (CA521, 524)	
Silicon bronze (CA651, 655)	
Nickel silver (CA732, 735, 745, 752, 754, 757, 765, 770, 794)	
Cupro nickel 90–10	
Cupro nickel 80–20	
430 Stainless (passive)	
Cupro nickel 70–30	
Nickel aluminum bronze (CA630, 632)	
Monel 400, K500	
Silver solder	
Nickel (passive)	
60% Ni–15% Cr (passive)	
302, 303, 304, 321, 347 Stainless (passive)	
316, 317 Stainless (passive)	
Carpenter 20Cb–3 Stainless (passive), Incoloy 825 (passive)	
Silver	
Titanium (passive), Hastelloy C and C276 (passive)	
Graphite	
Ziconium	
Gold	
Platinum	Cathodic (most noble) protected

4.3.3 MECHANISM OF GALVANIC CORROSION

To understand the mechanism of galvanic corrosion, caused by joining of two metals differing in potential, such as iron and copper, consider a galvanic cell shown in Fig. 4.3. For the formation of a galvanic cell, the following components are required:

(1) A cathode.
(2) An anode.
(3) An electrolyte.
(4) A metallic path for the electron current.

In the case of copper and steel, copper has a more positive potential according to the emf series, hence, it acts as a cathode. On the other hand, iron has a negative potential in the emf series ($-0.440\,\text{V}$), hence, it is the anode. As a matter of principle, in a galvanic cell, the more noble metal always becomes the cathode and the less noble always the anode. Moisture acts as an electrolyte and the metal surface provides a metallic path for the electron current to travel. Thus, when a piece of copper is joined to iron, all qualifications required for the formation of a galvanic cell are fulfilled and galvanic corrosion proceeds (Fig. 4.4).

The positive ions (Fe^{++}) flow from the anode (iron) to cathode (copper) through the electrolyte, which is water. Iron, therefore, corrodes. The hydrogen ions (H^+) are discharged at the copper cathode, and ultimately hydrogen is released. The Fe^{++} ions travel towards the cathode and OH^- towards the anode. They combine to form insoluble iron hydroxide, $Fe(OH)_2$. Positive electricity (conventional current) flows from the cathode ($+$) to the anode ($-$) through the external metallic path. In the electrolyte the electric current flows from anode to cathode by positive ions (cations). From the cathode to the anode, it is carried by negative ions (anions). In the external circuit, the current (conventional current) is actually carried by electrons from anode to cathode. The electrons, after being released by the anodic dissolution of iron, participate in the reduction process, such as

$$2H^+ + 2e \rightarrow H_2 \text{ or } Cu^{2+} + 2e \rightarrow Cu \quad (4.9)$$

The analogy of the galvanic cell described above, with a copper and iron galvanic couple, illustrates why iron corrodes and copper does not corrode. Naturally, copper does not corrode because it acts as a cathode whereas iron corrodes and generates a more negative potential and it acts as an anode.

Consider a steel pipe of 4″ OD joined to a copper pipe of the same diameter and exposed to soil containing some moisture. The steel pipe would become the anode and, therefore, corrode.

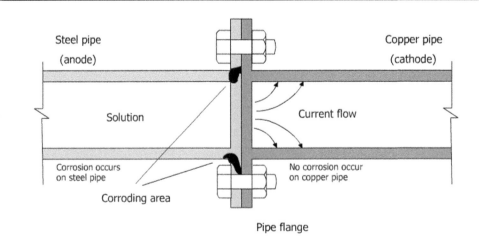

Figure 4.4 Formation of galvanic cell by joining of two dissimilar metals

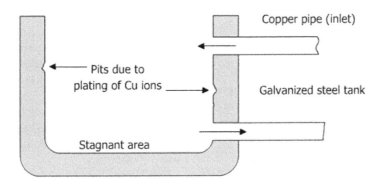

Figure 4.5   Galvanic corrosion in a hot water tank

Figure 4.5 shows a galvanized steel tank with a copper inlet pipe and an aluminum tank with an aluminum water inlet pipe. The steel tank would corrode as shown in the figure. Copper ions in the tank which are leached from the copper pipe would deposit on the wall of the tank and form a galvanic cell, hence the galvanized steel tank would corrode as shown in Fig. 4.5.

To summarize, a more active metal in potential series tends to corrode, whereas a less active or more noble metal does not corrode.

4.3.4 FACTORS AFFECTING GALVANIC CORROSION

The following factors significantly affect the magnitude of galvanic corrosion:

A. Position of metals in the galvanic series.
B. The nature of the environment.
C. Area, distance and geometric effects.

A. Position of Metals in the Galvanic Series

As mentioned earlier, the further apart the metals are in the galvanic series, the greater is the chance for galvanic corrosion. The magnitude of galvanic corrosion primarily depends on how much potential difference exists between two metals. For a particular environment, the metals selected should be close to each other in the galvanic series to minimize galvanic corrosion. Active metals should not be joined with passive metals. Thus, aluminum should not be joined to steel, as aluminum being more active would tend to corrode.

B. The Nature of Environment

Due consideration must be given to the environment that surrounds the metal. For instance, water containing copper ions, like seawater, is likely to form galvanic cells on a steel surface of the tank. If the water in contact with steel is either acidic or contains salt, the galvanic reaction is accelerated because of the increased ionization of the electrolyte. In marine environments, galvanic corrosion may be accelerated due to increased conductivity of the electrolyte. In cold climates, galvanic corrosion of buried material is reduced because of the increased resistivity of soil. In warm climates, on the other hand, it is the reverse because of the decreased resistivity of the soil.

C. Area, Distance and Geometric Effects

Effect of Area

The anode to cathode area ratio is extremely important as the magnitude of galvanic corrosion is seriously affected by it. The area ratio can be unfavorable as well as favorable.

Unfavorable Area Ratio

The area ratio of the anode to cathode plays a dominant role in galvanic corrosion. As a given amount of current flows in a galvanic couple, the current density at the anode or cathode controls the rate of corrosion. For a given amount of current, the metal with the smallest area has the largest current density and, hence, is more damaged if corrosion occurs at it. For similar reasons, the current density at a large metal is very small. The rate of corrosion increases with the ratio of cathodic to anodic areas (Fig. 4.6). Take the example of steel plates joined by aluminum bolts (Fig. 4.6). Aluminum has a smaller anodic area and steel, a larger cathodic area. Aluminum is more active in the galvanic series than steel. The current density on aluminum is, therefore, extremely large and serious galvanic corrosion of aluminum takes place. This shows the end result of an unfavorable anode/cathode ratio. The other ratio, large anode/small cathode, would only slightly accelerate the rate of galvanic corrosion.

Effect of Distance

It is a known principle that the solution conductivity varies inversely with the length of the conduction path. Most corrosion damage is caused by current which cover short paths. Hence, the greatest galvanic damage is likely to be encountered near the junction of the two metals and the severity would be decreased with increased length. If two different metals are far away from each other, there would be no risk of galvanic corrosion, because of very little current flow. This is possible in designs, like oil rigs and other complex structures requiring a very large variety of material.

The effect of area and distance may be best understood by the examples of utility lines in a large building. Consider, for instance, the copper tubes transporting water and natural gas. Coated carbon steel pipelines are laid in the same trench in a soil of low resistivity. In the trench, a corrosion cell would be formed if the pipes touch (metal-to-metal contact), or if they are bonded together somewhere (for electrical earthing requirements). Copper acts as cathode and steel acts as anode in an electrolyte of soil and galvanic corrosion would initiate. Now consider the area effect. Being coated, steel pipe would have a small anodic area at the sites of coating defects, whereas copper pipes would have a large cathodic area. Due to the small anode area and large cathode areas, galvanic corrosion would initiate and leaks in the carbon pipe would soon start. The problem also illustrates the effect of distance (Figs 4.7a and b). If the above utility pipe were in two different trenches, a sufficient distance between them would not have allowed the galvanic current to flow.

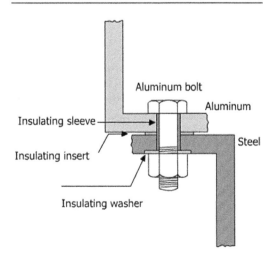

Figure 4.6 Avoidance of galvanic corrosion

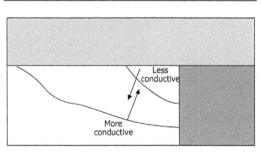

Figure 4.7a Representation of galvanic corrosion in less conductive and more conductive solution

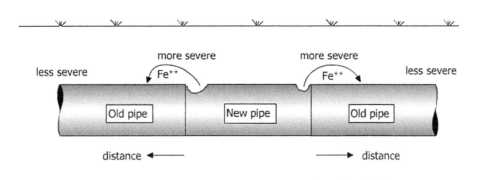

Figure 4.7b Dissimilar metal couple mechanism

Galvanic attack would be restricted for two dissimilar metals in contact with soil of high pH and low in carbon dioxide.

Effect of Geometry

Geometry of components and their design also influence galvanic corrosion. As current does not flow around the corners, the geometry of the circuit affects the degree of galvanic corrosion. Polarization may be affected by a break in the continuity of the current.

Component design is also a factor in galvanic corrosion as the current circuit geometry affect the magnitude of galvanic corrosion and the polarization process. Any obstacle to polarization would accelerate galvanic corrosion.

4.3.5 APPLICATION OF PRINCIPLES OF GALVANIC CORROSION

A. Non-metallic Conductors

Many non-metallic materials are cathodic to metals and alloys. For example, impervious graphite used in heat exchanger applications is noble to more active metals. The nature of non-metallic conductors must be known before their application. Graphite packing around a steel pump shaft can cause galvanic corrosion of the steel shaft if it is wet.

B. Metallic Coatings

Two types of metallic coatings are generally used, noble and sacrificial type. Zinc coating is an example of the sacrificial type. Zinc corrodes eventually and it protects the steel substrate both by its barrier effect and also by providing electrons ($Zn \rightarrow Zn^{2+} + 2e$) into the steel which prevent Fe^{++} ions from escaping from the steel (cathodic protection – see below). Noble coatings act as a barrier only between the metal substrate and the environment. Nickel, silver, copper, lead and chromium are called *noble* metal coatings. Formation of pores and damage to the noble coating can cause galvanic corrosion of the substrate (Fig. 4.8), as there is no sacrificial cathodic protection of the substrate.

C. Cathodic Protection

As mentioned above, a positive use of the principle of galvanic corrosion is cathodic protection. In a sacrificial system of cathodic protection, anodes of active metals, like Zn, Mg and Al, are used for protection of steel structures. The sacrificial galvanic anodes provide protection to the less active metals, like steel because they corrode and release electrons. The electrons which are released by the

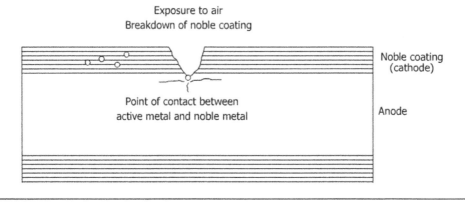

Figure 4.8 Initiation of galvanic corrosion by breakdown of coating

corroding metals enter the steel structures, which become cathodic and, therefore, do not corrode. This system of cathodic protection is based on galvanic corrosion, however, in this case a beneficial use is the result of galvanic corrosion.

4.3.6 EXAMPLE OF FAILURES OF GALVANIC CORROSION

Numerous examples of galvanic corrosion are encountered as it is one of the most frequently encountered forms of corrosion. The following are common examples:

(1) Galvanic corrosion of aluminum shielding in buried telephone cables.
(2) Galvanic corrosion of steel pipe with brass fittings.
(3) Galvanic corrosion of the body of the ship in contact with brass or bronze propellers.
(4) Galvanic corrosion on a gannet wheel where the steel securing the bolts is in contact with the magnesium wheel.
(5) Galvanic corrosion between the tubes and the tube sheet in heat exchangers.
(6) Aluminum conduit buried in steel reinforced concrete.
(7) Galvanic corrosion of steel coated with copper due to the defects in copper coating.
(8) Galvanic corrosion of the Statue of Liberty.

(9) Galvanic corrosion inside horizontal stabilizers in aircrafts.

4.3.7 FAILURE CASE HISTORIES

A. Case 1 – Failure of Aluminum Alloy Spacers by Galvanic Attack

Description of the problem Several rebuilt hydraulic actuators had been in storage for three years. At each joint, there was an aluminum alloy spacer and a vellum gasket. The mounting flanges of the steel actuators were nickel plated. While assembling the actuators, a lubricant containing molybdenum sulfide had been applied to the gasket to serve as a sealant. The galvanic attack occurred on the aluminum alloy spacer [2].

Identification The three major components: aluminum spacers, vellum gasket and actuators housing were examined in a laboratory. The following were the conclusions of the examination:

(a) Corrosion had penetrated to a depth of $64\,\mu m$ in some areas of the aluminum spacers which had been stained badly.
(b) The vellum gasket was electronically conductive.

(c) A potential of 0.1 V was measured between the aluminum alloy spacers and the actuator housing.
(d) No potential was detected after breaking contact between aluminum spacer and the actuator housing.
(e) Deposits of molybdenum sulfide were found on the aluminum surface of aluminum spacers.

Conclusion Galvanic corrosion was the reason for failure. Molybdenum disulfide acted as an electrolyte between the aluminum spacers and the nickel plated steel actuator housing. In order to remedy the situation, dry vellum gaskets were used and the use of molybdenum disulfide was discontinued.

B. Case 2 – Failure of Hot Water Galvanized Pipes in a Housing Area

Description of the problem In a housing society built for an aluminum company, the household plumbing was severely affected. All hot water pipes were galvanized steel. The temperature of the water inside the pipe was about 80°C. Every morning, the color of the water becomes red and it gave an offensive odor. The residents of the houses could not use this water for washing purpose as it stained their hands [3].

Identification Upon metallographic examination of the internal surface of the pipe it was found that the pipe was seriously corroded. The corrosion was of uniform type. A chemical analysis of the water showed the leaching of iron from the pipe.

Conclusion It was concluded that iron was severely corroding. The amount of iron increased overnight when the water was relatively unused. In the morning, when the taps were opened, a substantial amount of iron leached in the water rendering it unsuitable for use. The reason was the reversal of polarity which occurs in hot water, zinc became the cathode and iron the anode. The zinc coating did not, therefore, protect the steel pipe which continued to corrode by becoming anode.

Remedy

(1) The operating temperature of the boiler was lowered. The water temperature in the pipe was not allowed to exceed 60°C.
(2) The water was treated with sodium, hexametaphosphate, before it entered the piping system of the house.
(3) Normally this problem is avoided by using copper for hot water pipes and tanks.

4.3.8 METHODS OF PREVENTION OF GALVANIC CORROSION

(1) Select metals, close together, as far as possible, in the galvanic series.
(2) Do not have the area of the more active metal smaller than the area of the less active metal.
(3) If dissimilar metals are to be used, insulate them.
(4) Use inhibitors in aqueous systems whenever applicable and eliminate cathodic depolarizers.
(5) Apply coatings with judgment. Do not coat the anodic member of the couple as it would reduce the anodic area and severe attack would occur at the inevitable defect points in the coating. Therefore, if coating is to be done, coat the more noble of the two metals in the couple which prevents electrons being consumed in a cathodic reaction such as $2H^+ + 2e \rightarrow H_2$, which is likely to be corrosion rate controlling.
(6) Avoid joining materials by threaded joints.
(7) Use a third metal active to both the metals in the couple.
(8) Sacrificial material, such as zinc or magnesium, may be introduced into this assembly. For instance, zinc anodes are used in cast iron water boxes of copper alloy water-cooled heat exchangers.
(9) In designing the components, use replaceable parts so that only the corroded parts could be replaced instead of the whole assembly.

Above all, understand materials compatibility which is the key to control galvanic corrosion.

4.4 DEZINCIFICATION

4.4.1 INTRODUCTION

It is a form of corrosion in which zinc is selectively attacked in zinc-containing alloys, like brasses. It mainly occurs in alloys containing less than 85% copper. De-alloying and selective leaching are broader terms which refer to the corrosion of one or more constituent of a solid solution alloy.

Dezincification is a form of de-alloying. As the phenomenon was first observed in brass in which zinc separated by dissolution from copper, the term *dezincification* is still used.

Ordinary brass consists of about 30% zinc and 70% copper. Dezincification can be observed by naked eyes, because the alloy changes in color from yellow to red.

4.4.2 TYPES OF ATTACKS

Two types of dezincification are commonly observed:

(1) Uniform type (layer type) (Fig. 4.9)
(2) Plug type (Fig. 4.10)

The two types of attack can be observed in Figs 4.9 and 4.10. In the uniform type of dezincification, the active area is leached out over a broad area of the surface and it is not localized to a certain point of the surface. On the other hand, the plug type of attack is localized, at a certain point on the surface and the surrounding area remains unaffected. Dezincification can occur on grain boundaries, such as α–β brasses (Fig. 4.9).

4.4.3 ENVIRONMENT

Dezincification generally takes place in water under stagnant conditions. Copper–zinc alloys containing more than 15% zinc are susceptible to dezincification.

4.4.4 MECHANISM

There is a disagreement between the workers on the mechanism of dezincification. One group contends that firstly the entire alloy is dissolved and later one of its constituent is re-plated from the solution which leached the alloy. This is

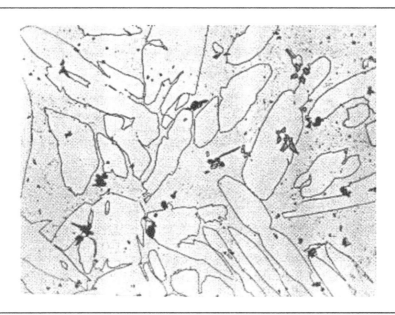

Figure 4.9 Dezincification of α–β brass (500 x 300)

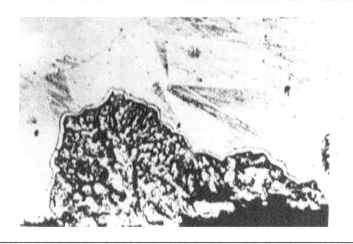

Figure 4.10 Dezincification plugs in 70-30 Brass exposed to NaCl. Evidence of two mechanisms operating simultaneously can be observed in the picture. (From Vernik, R.D. Jr. and Heiderbach, R.H. Jr. (1972). In Localised Corrosion, Cause of Metal Failure, STP 516, *ASM*. Reproduced by kind permission of ASM, Metals Park, Ohio, USA)

the basis of the dissolution and redeposition mechanism.

$$Zn \rightarrow Zn^{++} + 2e$$

$$2H^+ + 2e \rightarrow H_2$$

$$Zn \rightarrow Zn^{2+} + 2e$$

$$CuCl + e \rightarrow Cu + Cl^- \quad \text{(I) dissolution}$$
$$(4.10)$$

$$M \rightarrow M^{n+} + ne$$

$$H_2 \rightarrow 2H^0$$

$$H^+ + e \rightarrow H$$

$$Cu \rightarrow Cu^{++} + 2e \quad \text{(II) dissolution} \quad (4.11)$$

$$Cu^{+2} + Zn \rightarrow Cu + Zn^{+2} \quad \text{(III) plating}$$
$$(4.12)$$

Copper is deposited as a fine copper dust which readily dissolves once more in any electrolyte. Zinc is leached out of the brass leaving behind a highly porous mass. The three steps involved above are:

Step I Dissolution of Cu and Zn
(equations I and II)
Step II Zinc stays in solution
Step III Copper plates back (equation III)

It has been shown that zinc has a high tendency to dissolve, whereas copper has a high tendency to plate. The $E°$ for zinc is $-0.763\,V$, whereas the $E°$ for copper is $0.337\,V$, which shows the above tendencies. Zinc replaces copper in the solution because copper is far nobler than zinc in the EMF series. As copper is redeposited as a porous mass, the strength of brass is significantly lowered. The alloy is mechanically weakened to the extent that a slight increase in the load causes the alloy to rupture. Dezincification causes only a slight change in the appearance and does not affect the dimensions.

Contrary to the opinion of one group of workers, the other group believes that one species is dissolved selectively from the alloy, and a porous residue of the more noble metal is left [4]. A third group believe that both the mechanisms operate. The evidence of that both mechanisms may operate is shown in Fig. 4.10. In the figure, a dezincified region is shown below the metal surface. The perturbations shown above are deposits of copper. The evidence that both mechanisms may operate at the same time, is shown in the above figure.

Use has been made of electrochemical hysteresis methods to evaluate the mechanism of dezincification. On the basis of the information generated by electrochemical methods, pH–potential diagrams have been constructed

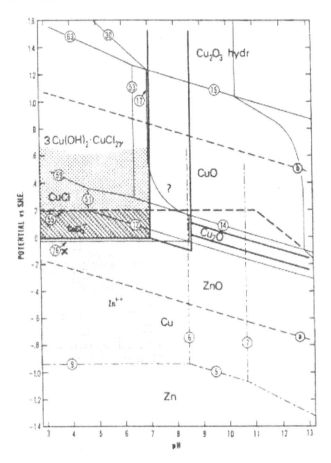

Small dots indicate the domain in which selective removal of zinc is expected insolutions
free of copper ions. Larger dots indicate the domain in which both copper and zinc dissolve.
Crosshatching indicates the region in which copper is expected to deposit.

Figure 4.11 Potential–pH diagram of 70-30 Cu–Zn in NaCl Solution. Figure shows the domains of copper leaching. (From Vernik, R.D. Jr. and Heiderbach, R.H. Jr. (1972). In Localised Corrosion, Cause of Metal Failure, STP 516, *ASM*. Reproduced by kind permission of ASM, Metals Park, Ohio, USA)

(Fig. 4.11). From these diagrams, the domains of selective leaching of zinc, and domains of dissolution of both zinc and copper can be observed. The regions where copper deposition takes place are also shown in the figure. From the diagram it is observed that at potentials between 0.000 and $+0.200\,V_{SHE}$, both zinc and copper dissolve, but in stagnant conditions copper may deposit on the specimen without any evidence of selective leaching.

Above $+0.200\,V_{SHE}$, both alloys dissolve but there is no redeposition of copper. If the potential

is held above $+0.200\,V_{SHE}$ for a long period of time and then lowered below $+0.200\,V_{SHE}$, deposition of copper would take place ($CuCl + e \rightarrow Cu + Cl^{-}$).

In spite of the development and evolution of new techniques, such as microprobe analysis and advances in the electrochemical methods, it is not clearly known which mechanism is operative. It, however, appears on the basis of the literature published that both mechanisms may operate, i.e. selective leaching or dissolution of both components and re-plating of one of the component, or selective dissolution of one species only leaving a porous residue of the more noble species.

4.4.5 EXAMPLES OF DEZINCIFICATION

(a) **Layer Type**
Uniform layer type dezincification occurs in tooth of gear wheel. It also occurs on the inner surface of admiralty brass heat exchanger tubes when exposed to water at pH = 8.0 and temperature range 31–49°C.
(b) **Plug Type**
It is found particularly in α-brass heat exchanger pipes. If the heat exchanger is not cleaned and dried, differential aeration cells are formed in which the brass dissolves. The corroded region is filled with the re-precipitated copper.

4.4.6 CASE HISTORY

Failure of copper alloy (C27000) inner cooler tubes for air compressors.

Description of the problem Yellow brass tubes (65% copper) in air compressor showed leaking in cooling water after 17 years of service.

Conditions The cooling water was chlorinated well-water. The water was recirculated.

Physical examination A visual examination showed a thick layer of porous brittle copper on the inside surface. A plug type deposit had penetrated deep in the well of the tube. At many points the wall of the tube was completely damaged [5].

Analysis of the tubes showed that they were fabricated from copper alloy C27000. It was observed that the alloy before dezincification contained 35% zinc. The analysis showed that only a trace of the original 35% zinc remained.

It is obvious that most of the zinc was lost as shown by the analysis of the base metal and the brittle layer. This was due to the leaching of zinc by dezincification.

Recommendations Arsenic addition in the range 0.02–0.06%, provides high resistance to

dezincification, and it should, therefore, be added to the tubes. A small amount of Mg is also required with the As. High copper alloys (copper above 85%) are immune to dezincification and they can be used safely. If dezincification is very severe, the use of more expensive copper–nickel 70-30 is recommended.

4.4.7 DE-ALLOYING

Dezincification is a special case of de-alloying. The phenomenon of de-alloying by selective leaching can occur in various materials (as shown in Table 4.2). Selective leaching (de-alloying) of some important materials is summarized below.

(a) Graphitic Corrosion

Gray cast iron sometimes show the effect of selective leaching out of iron in mild corrosive environments. The surface layer of the iron becomes like graphite and it can be easily cut with a knife. Because of the attack, the iron or steel matrix is dissolved and an interlocking nobler graphite network is left. The graphite becomes cathodic to iron and a galvanic corrosion cell is formed. Iron is dissolved and a porous mass of voids and complex iron oxides is left behind. This graphitized cast iron loses its strength and other metallic properties (Fig. 4.12), but to a casual view it looks dirty but unchanged in shape, which can lead to dangerous situations.

Graphite corrosion does not occur in nodular and malleable cast iron. One common mistake in the books is to use the term '*graphitization*' rather than graphitic corrosion. Graphitization occurs when a low alloy steel is subjected to high temperature for an extended time period. Graphitization results from the decomposition of pearlite into ferrite and carbon, whereas in graphitic corrosion the gray cast iron is selectively attacked. The presence of graphite is necessary for leaching to take place.

(b) De-aluminification

Copper alloys containing more than 8% Al may be subjected to the preferential dissolution of aluminum component of the alloy. The α-phase

Table 4.2 Alloys subjected to leaching

Alloy	Environment	Element removed
Brasses	Many waters, especially under stagnation conditions	Zinc (dezincification)
Gray iron	Soils, many waters	Iron (graphitic corrosion)
Aluminum bronzes	Hydrofluoric acid, acids containing chloride ions	Aluminum
Silicon bronzes	Not reported	Silicon
Copper nickels	High heat flux and low water velocity (in refinery condenser tubes)	Nickel (de-nickelification)
Monels	Hydrofluoric and other acids	Copper in some acids, and nickel in others
Alloys of gold or platinum with nickel, copper or silver	Nitric, chromic and sulfuric acids	Nickel, copper or silver (parting)
High-nickel alloys	Molten salts	Chromium, iron, molybdenum and tungsten
Cobalt–tungsten–chromium alloys	Not reported	Cobalt
Medium-carbon and high-carbon steels	Oxidizing atmospheres, hydrogen at high temperatures	Carbon (decarburization)
Iron–chromium alloys	High-temperature oxidizing atmospheres	Chromium, which forms a protective film
Nickel–molybdenum alloys	Oxygen at high temperature	Molybdenum

of aluminum–bronze is attacked and a porous residue of copper is left behind.

(c) De-nickelification

Although, not common, the de-alloying of nickel in 70-30 Cu–Ni alloy has been observed under low flow conditions.

4.4.8 TECHNIQUES FOR EVALUATION OF DEZINCIFICATION AND DE-ALLOYING

Since the phenomenon of de-alloying is extended over a long period of time, accelerated techniques have been in demand, but still no single

reliable technique acceptable to researchers has been developed.

The technique of superimposing an electrochemical hysteresis curve on the pH–potential Pourbaix diagram has shown a fair degree of promise. As shown in the text related to dezincification, this technique provides a basis for the prediction of the tendency for de-alloying as a function of potential. From a knowledge of potential, it can be predicted whether zinc is being leached or copper is being deposited or copper is being leached.

However, more work is needed to improve the existing technique and to develop a new technique to make such predictions.

4.4.9 PREVENTION

(1) Use copper alloys with copper content above 85%.

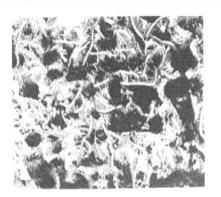

A pump impeller section showing
graphite corrosion attack

Figure 4.12 Graphite corrosion in gray iron pipe.
(From Ports, R.D. (1987). Nalco Chemical Company)

(2) Use brass alloys with tin, arsenic or antimony
addition.
(3) Avoid environments where the solution
becomes stagnant and deposits accumulate
on the metal surface.

4.5 CREVICE CORROSION

This is a localized form of corrosion, caused by
the deposition of dirt, dust, mud and deposits on
a metallic surface or by the existence of voids,
gaps and cavities between adjoining surfaces.
An important condition is the formation of a
differential aeration cell for crevice corrosion to
occur. This phenomenon limits the use, particularly of steels, in marine environment, chemical
and petrochemical industries.

4.5.1 CAUSES

(a) Presence of narrow spaces between metal-to-
metal or non-metal to metal components.
(b) Presence of cracks, cavities and other defects
on metals.
(c) Deposition of barnacles, biofouling organisms and similar deposits.
(d) Deposition of dirt, mud or other deposits on
a metal surface.

4.5.2 MATERIALS AND ENVIRONMENT

The conventional steels, like SS 304 and SS 316,
can be subject to crevice corrosion in chloride containing environments, such as brackish
water and seawater. Water chemistry plays a very
important role.

Factors affecting crevice corrosion (Fig. 4.13):

(a) Crevice type.
(b) Alloy composition.
(c) Passive film characteristics.
(d) Geometry of crevice.
(e) Bulk composition of media.
(f) Bulk environment.
(g) Mass transfer in and out of crevice.
(h) Oxygen.

(a) Crevice Type

Crevice type means whether the crevice is between
metal-to-metal, metal to non-metal or a marine
growth, like barnacles or other marine biofouling
organisms, on the metal surface. It is important to know whether factors affecting crevice
are man-made or natural in order to select
appropriate methods for prevention.

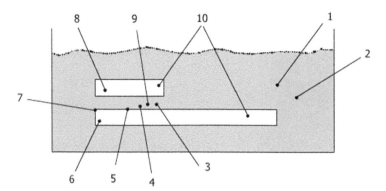

1. Bulk solution composition

2. Bulk solution environment

3. Mass transport in and out of crevice
   - convection

4. Crevice solution

5. Electrochemical reactions
   - metal dissolution
   - O₂ reduction
   - H₂ evolution

6. Alloy composition
   - major constituents
   - minor constituents
   - impurities

7. Passive film characteristics
   - passive current
   - film stability

8. Crevice type
   - metal/metal
   - metal/non-metal
   - metal/marine growth

9. Crevice geometry
   - gap
   - depth

10. Total geometry
    - exterior to interior crevice area ratio
    - number of crevices

Figure 4.13   Factors affecting crevice corrosion. (Oldfield, J.W. and Sutton, W.H. (1978). *Br. Corros. Jr.*, **13**, (1). Reproduced by kind permission of British Corrosion Journal)

(b) Alloy Composition

It is important to know whether or not the alloy is resistant to crevice corrosion. For instance, work on the various grades of steels, such as SS 304, carpenter alloy, Incoloy (Alloy 825), Hastelloy (Alloy g), and Inconel (625) showed that the later two alloys were highly resistant to crevice corrosion in ambient and elevated temperature seawater [6]. The alloying elements in various grade of steel affect both the electrochemical and chemical processes, such as hydrolysis, passive film formation, passive current density and metal dissolution. Hence, the effect of major alloying elements, such as Fe, Cr, Ni and Mo and

Fe–Cr–Ni–Mo class of steels on crevice corrosion initiation is important. It has been shown that iron, chromium, nickel and Mo improve the resistance of steels to crevice corrosion [7]. Crevice corrosion is not usually observed with 6% molybdenum. The effect of chromium can be harmful if it increases the acidity in the crevice by hydrolysis. At a higher chromium concentration, it generally increases the stability of the passive film, lowers the pH below that required for crevice to initiate, and reduces the passive current density. The rate of entry of chromium ions in the crevice is, therefore, minimized, hence hydrolysis which causes acidic conditions in the crevice is minimized by chromium.

(c) Passive Film Characteristics

The type of passive film formed is important, as the breakdown of a passive film results in the onset of crevice corrosion.

The improved performance of certain steels can be attributed to enrichment of the surface film by chromium. The quality of surface films affects the susceptibility of steels to crevice corrosion. The formation of passive films is dependent on oxygen and hence, its concentration affects the magnitude of crevice corrosion. It has been reported that as the seawater temperature is raised from ambient to 70°C, the resistance of steel types 304 and 317 is increased. This has been attributed to the enrichment of the passive film by chromium which increases its stability.

(d) Geometry of Crevice

The magnitude of crevice corrosion also depends on the depth of the crevice, width of the gap, number of crevices and ratio of exterior to interior crevice. It has been shown for types 316 and 304 stainless steel that smaller the gap, the less is the predicted time for initiation of crevice corrosion (Fig. 4.14). The reason is that when the ratio of crevice solution volume to creviced area is small, the acidity is increased and the critical value for initiation of crevice is achieved rapidly. The ratio of the bold area to the creviced area also affects crevice corrosion (Fig. 4.15). Generally, the larger is the bold area (cathodic) and smaller the creviced area (anodic), the larger is the probability of crevice corrosion. This has been shown by work on types 304 and 316 stainless steel and

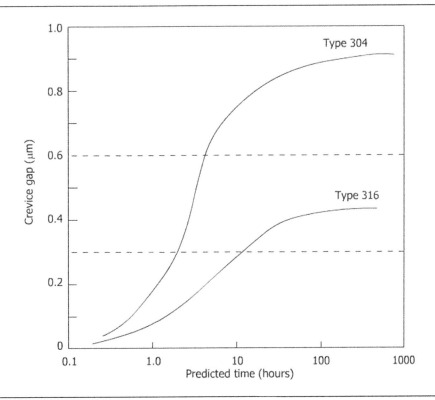

Figure 4.14 Effect of crevice gap on predicted times to onset of crevice corrosion for Type 304 and Type 316 stainless steel. (Kain, R.M. (1979). NACE Paper No. 230, *Corrosion*, March) [10]

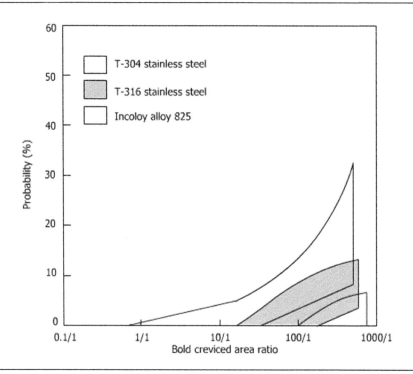

Figure 4.15 Probability of crevice corrosion initiation in one month. (From Andreason D.B. (1974). R & D INCO, *Paper presented on ASTM-ASM Symposium in Pitting Corrosion*, Detroit, MI, October 23. Reproduced by kind permission of ASM, Metals Park, Ohio, USA)

Incoloy alloy 825 [7]. If the depth is increased, acidity is also increased in the crevice.

(e) Effect of Temperature

The rate of crevice corrosion propagation of stainless steel alloys, such as types 304 and 316, is decreased in natural seawater, when the temperature is increased from ambient to 70°C [8,9]. The tendency for initiation of crevice corrosion is not affected by increase in temperature. The above trend may be attributed to the decreased solubility of oxygen with increased temperature and changes in the nature of the passive film which is formed.

(f) Bulk Solution Composition

The breakdown of a passive film on a metal surface largely depends on the aggressiveness of the electrolyte. All chloride containing solutions are highly aggressive and contribute to onset of crevice corrosion. Seawater and brackish water are high aggressive and promote crevice corrosion of steels.

(g) Mass Transfer in and out of Crevice

There are three forms of mass transport: migration, diffusion and convection. Most of the current is carried by migration and diffusion. The effect of bulk chloride concentration on predicted time for onset of crevice corrosion is shown in Fig. 4.16. The prediction for onset is shortened with increased chloride concentration.

The process of bringing chloride ions increases the concentration of chloride ions in the small crevice, hence, the aggressiveness of the electrolyte inside the crevice is increased.

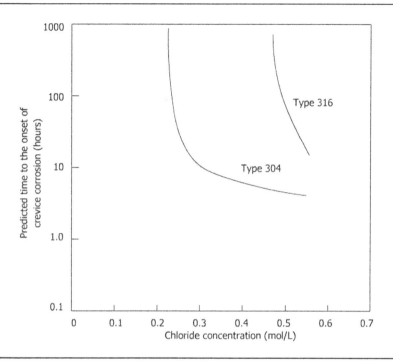

Figure 4.16 Effect of bulk chloride concentration on predicted initiation of crevice corrosion for Type 304 and Type 316 for a given set of severe crevice conditions. (From Kain, R.M. Tech. Paper No. FP-37, LCCT, Corrosion/80, NACE, Chicago. Reproduced by kind permission of NACE, Int., Texas, USA)

The solution at a stage becomes highly acidic and the pH is lowered.

(h) Oxygen

The onset of crevice corrosion is strongly linked with the nature of the passive film on the metal surface. If the passive film is very stable, crevice corrosion is blocked. Oxygen is essential for the formation of passive film on the metal surface. Hence, it has an important influence on the onset of crevice corrosion. As oxygen is consumed within the crevice, a potential difference is set up between the creviced areas and the boldly exposed areas open to oxygen. A differential concentration is, therefore, set up which accelerates oxygen. However, as the temperature is increased, oxygen solubility decreases which retards the crevice process. The exact relationship between oxygen and temperature is not understood. Thermal insulation in process plants provide water and oxygen for crevice corrosion to develop.

4.5.3 MECHANISM OF CREVICE CORROSION

Most textbooks give only an over-simplified picture of the mechanism of crevice corrosion, hence, one may not appreciate its complexity. In the last fifty years, a significant progress on the understanding of the mechanism of crevice corrosion has been made. The phenomenon is extremely complex, and a unified mechanism does not exist presently. Most of the mechanism is based on certain type of concentration cells [10,11]. The following is a summary of some of the concentration cell mechanisms, which have contributed to the understanding of the crevice corrosion.

(a) Metal ion concentration cells [11]. According to this old theory, a difference in metal ions exists between the crevice and outside, hence, a corrosion cell is formed. The area with low metal concentration becomes the anode and other, the cathode. Anodic dissolution at the anode, therefore, initiates.

(b) A high concentration of oxygen on the surface outside the crevice and a low oxygen concentration inside a crevice creates a differential aeration cell, which initiates crevice corrosion. A unified mechanism of crevice corrosion was given by Fontana and Greene [12]. A unified mechanism which is rather over-simplified is given below (Fig. 4.17):

(1) The site at which a crevice is formed, becomes the anode and the site outside the crevice, the cathode. The reason for the above can be attributed to the formation of differential metal ion, oxygen concentration or active–passive cells. The following reactions take place:

Anode (in the crevice)

$$M \rightarrow M^{++} + 2e \text{ (M represents}$$
$$\text{a metal)} \quad (4.13)$$

Cathode

$$\frac{1}{2}O_2 + 2H_2O + 2e \rightarrow 4OH^- \text{ (oxygen}$$
$$\text{reduction outside the crevice)}$$
$$(4.14)$$

(2) After sometime, the oxygen in the crevice is consumed, but the concentration of oxygen at the cathode remains unchanged, hence, the reaction continues unabated.

(3) Within the crevice, the following processes continue to occur:

$$Cr \rightarrow Cr^{+++} + 3e \text{ (chromium}$$
$$\text{contained in the stainless steel)}$$
$$(4.15)$$

$$Fe \rightarrow Fe^{++} + 2e \quad (4.16)$$

To preserve electroneutrality, the chloride ions are attracted by Cr^{+++} or Fe^{++} ions and metallic chlorides are formed:

$$Cr^{+++} + 3Cl^- \rightarrow CrCl_3 \quad (4.17)$$

$$Fe^{++} + 2Cl^- \rightarrow FeCl_2 \quad (4.18)$$

With the formation of metallic chlorides, the process of anodic dissolution continues, and the cavity becomes deeper and deeper.

(4) Hydrolysis of these chlorides takes place immediately which results in the production of acid conditions in the pit. Hydrolysis increases the level of acidity in the crevice. The geometry of a crevice

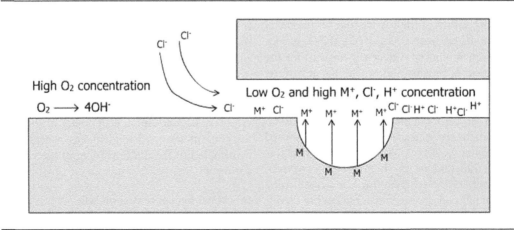

Figure 4.17 Mechanism of crevice corrosion. (From France, M.W.D. (1987). Localized Corrosion, cause of metal failure, ASM STP 516. Reproduced by kind permission of ASTM, Philadelphia, USA)

limits the exchange of solution between the structure and the crevice in the bulk, thus creating acid conditions in the pit. The above is generalized by:

$$M^+ + Cl^- + HOH \rightarrow MOH + HCl \tag{4.19}$$

In case of 18-8 steel:

$$CrCl_3 + 3HOH \rightarrow Cr(OH)_3 + 3HCl \tag{4.20}$$

$$FeCl_2 + 2HOH \rightarrow Fe(OH)_2 + 2HCl \tag{4.21}$$

It can be observed that acid is produced and hence acid conditions are produced inside the crevice. The pH may attain a value of as low as 1.0 inside the crevice. Once the acid conditions are generated, the process continues until the reaction is terminated. The mechanism described above is self-generative and once it starts, it continues.

The above mechanism is, however, purely qualitative and does not provide an explanation of the following:

(a) Why crevice corrosion takes place even in non-aggressive environments?
(b) What is the critical concentration of chloride ions necessary to induce crevice corrosion?
(c) The major emphasis is on the formation of a differential aeration cell, whereas other differential cells also may affect crevice corrosion.
(d) The relationship of time, chloride concentration and passivity is not explained clearly.

The mechanism is, however, far more complex than given by a unified mechanism. This phenomenon is highly unpredictable because of many of design, metallurgical and environmental factors associated with it. It is, therefore, essential to identify the parameters before any reliable mechanism could be given. The development of mathematical models have been a useful step in this direction [13].

4.5.4 DEVELOPMENT OF A MATHEMATICAL MODEL

An outstanding contribution has been made by Oldfield and Sutton [13]. They have developed a mathematical model of crevice corrosion in which observations, such as the fall in pH, localized breakdown of passive films and the onset of crevice corrosion in stainless steels, is presented. A summary of the model developed by Oldfield and Sutton is given below. For calculation and detailed discussion, reference must be made to the original paper. In the model developed, crevice corrosion is considered a four stage mechanism. The four stages of crevice corrosion suggested are shown in Fig. 4.18.

(1) Depletion of oxygen in crevice due to consumption in the cathodic reaction.
(2) Increase of acidity in the pit by the process of hydrolysis.
(3) Breakdown of the passive film on the surface at a critical value of pH.
(4) Propagation of crevice corrosion with further hydrolysis and production of acidity.

(a) First Stage

The following is the initial reaction when stainless steel is placed in a oxygenated neutral chloride solution:

$$\text{Anode: } M \rightarrow M^{2+} + ze \tag{4.22}$$

$$\text{Cathode: } H_2O + 2e + \frac{1}{2}O_2 \rightarrow 2OH^- \tag{4.23}$$

Overall reaction:

$$2M^{2+} + \frac{1}{2}O_2 + 2H_2O + 4e \rightarrow 2M(OH)_2 \tag{4.24}$$

Due to passivation, the potential of the specimen shifts into the passive region where the small $I_{passive}$ is equal to the cathodic current. The film becomes thicker by the driving force created by the cathodic reaction. If steel containing a crevice is now placed in seawater, the reaction would occur all over the surface. However,

STAGE - I	Depletion of oxygen in the crevice solution
STAGE - II	Increase in acidity and chloride content of the crevice solution
STAGE - III	Permanent breakdown on the passive film and then onset of rapid corrosion
STAGE - IV	Propagation of crevice corrosion

Figure 4.18   The four stages of crevice corrosion

if the crevice is very small, oxygen diffusing into it is slower than its removal by the cathodic reaction [14]. The solution in the crevice becomes slowly depleted in oxygen (Fig. 4.19). In the first stage:

(a) Oxygen is depleted and the solution inside becomes deoxygenated.
(b) The metal ion concentration is increased.

(b) Second Stage

In this stage, the cathodic reduction of oxygen proceeds outside the crevice and slow dissolution of the metal takes place inside the crevice. The concentration of metal cations produced by the reaction $M \rightarrow M^{z+} + ze$ inside the crevice increased until the solubility of one of the hydroxides is exceeded. The entry of metal ion through

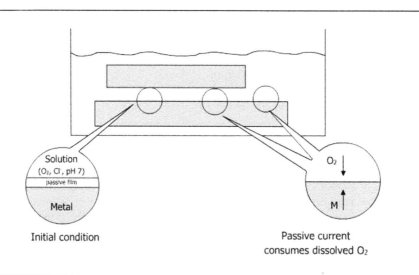

Figure 4.19   Stage I of crevice corrosion. (From Kain, R.M. and Lee, T.S. (1979). Technical Paper No. FP 40, *Corrosion*. Reproduced by kind permission of NACE, Int., Texas, USA)

the passive film has two effects: (1) chloride ions migrate from the bulk solution to the crevice to maintain charge neutrality and (2) hydrolysis of metal chloride immediately takes place which causes an increase in acidity inside the crevice (Fig. 4.20).

$$M^{n+} + HOH \rightarrow M(OH)^{(n-1)} + H^+ \text{(hydrolysis)} \tag{4.25a}$$

$$M^{n+}(Cl^-) + HOH \rightarrow MOH + H^+Cl^- \tag{4.25b}$$

or

$$M^{n+}(Cl^-)_2 + 2H_2O \rightarrow M(OH)_2 + 2H^+ + Cl^- \tag{4.25c}$$

The acidity within the crevice in increased and the pH inside the crevice is reduced. The cations (M^{n+}) are moved out of the crevice and the anions (Cl^-) are moved inside the crevice.

The aggressiveness of the solution inside the crevice is increased.

(c) Third Stage

In the third stage (Fig. 4.21), accelerated corrosion takes place due to the breakdown of the passive film, because the solution inside the crevice in the second stage is highly aggressive. The concentration of solution at which the passive film breaks down is called 'critical crevice solution.'

(d) Fourth Stage

In the final state (fourth stage), the crevice corrosion continues to propagate. The propagation is terminated when the metal is perforated. Rapid dissolution of the alloy inside the crevice continues. The process is autocatalytic; once it starts, it continues until termination.

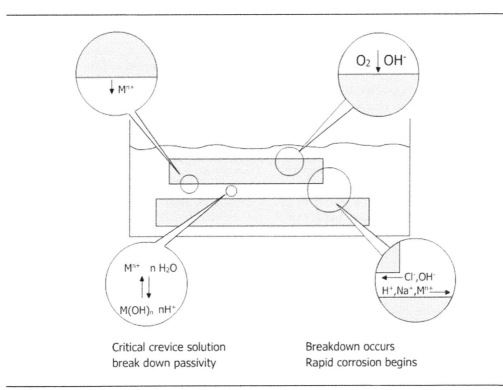

Critical crevice solution
break down passivity

Breakdown occurs
Rapid corrosion begins

Figure 4.20 Stage II of crevice corrosion. (From Kain, R.M. and Lee, T.S. (1979). Technical Paper No. FP 40, *Corrosion*. Reproduced by kind permission of NACE, Int., Texas, USA)

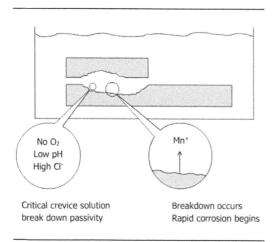

Critical crevice solution break down passivity

Breakdown occurs Rapid corrosion begins

Figure 4.21 Stage III of crevice corrosion. (From Kain, R.M. and Lee, T.S. (1979). Technical Paper No. FP 40, *Corrosion*. Reproduced by kind permission of NACE, Int., USA)

4.5.5 SOME IMPORTANT QUESTIONS AND ANSWERS ON THE MATHEMATICAL MODEL TO PROVIDE A BASIC UNDERSTANDING

(1) Why pH fall is so important in the model?
The passive current generates metals ions within the crevice, which hydrolyze and generate hydrogen ions. This is indicated by a fall in pH. Hence, fall of pH is very important to understand what is happening within the crevice.

(2) What can be predicted by a plot of crevice solution composition and the change of pH?
The change in the concentration of metal ion can be predicted in the crevice. For instance, as pH falls, the concentration of Cl^- within the crevice increases. These changes show the exit of positive ions from the crevice and entry of negative ions into the crevice.

(3) At what pH value generally is the passive film assumed to be broken?
For most stainless steels, a pH between 1 and 2 indicates the breakdown of a passive film.

(4) What are the three important parameters which form a basis for determination of crevice corrosion susceptibility of the alloy?

They are: critical crevice solution (CCS), passive current (I_p) and composition of the alloy.

(5) What value of passive current for stainless steel is assumed?
From 10^{-1} to 10 $A \cdot cm^{-2}$.

4.5.6 CASE HISTORIES [15]

(a) Case 1

Development of leaks at the rolled joint of the pipe and pipe bottom in a heat exchanger.

Description of the problem A heat exchanger was made from a pipe of 35-29 stainless steel, 25/30 mm diam. The oil at the pipe was heated externally from 90 to 170°C by superheated steam of 8–10 atm.

Investigation A physical examination of the pipe showed that the surface was pitted all around the rolled joint very close to the steam chamber. Also grooves were noticeable in the vicinity of this spot.

Identification of localized corrosion The attack was due to crevice corrosion. The rolled joint was not very tight, and, therefore, the steam condensate penetrated into the gap. A differential aeration cell was formed due to the depletion of oxygen in the crevice formed by the rolled joint because of its not being tight.

Prevention
(1) Inconel 625 could be used as an alternate alloy in such a situation. Alloy 20 Cr–25 Ni–4.5 Mo–1.5 Cu could also provide a useful service.
(2) Minimize the gap by a change in design of the rolled joint.

(b) Case 2

Crevice corrosion of a tubing in a hydraulic oil cooler.

Problem Leakage occurred in horizontal heat exchanger tubes in an electric power plant after 18 months of service.

Service conditions

(a) Coolant – river water.
(b) Tube dimensions: 9.5 mm, 0.65 mm wall thickness.
(c) Material – Cu–10% Ni.

Investigation Visual examination revealed nodules on the inner surface and holes through the nodule. On metallographic examination of a pit, a high rate of attack was observed. On examining a nodule, it was found that corrosion had penetrated 65%. A greenish residue of copper carbonate hydroxide [$CuCO_3$, $Cu(OH)_2$] was observed. The formation of nodules is very characteristic of crevice corrosion. They are generally isolated from each other. The nodules are generally black from inside and rusted on the outside. Microbial corrosion is very likely in such situations, enhancing crevice corrosion.

Conclusion The tubes failed by crevice corrosion. Deposits on the surface of the tubes were formed from the dirt in the river water.

Prevention and remedial measures The following measurements were adopted:

(1) The cooling water supply was changed by a cleaner water not containing dirt.
(2) The tubings were replaced. Even those nodules which were not leaking, were also replaced.

4.5.7 PREVENTION OF CREVICE CORROSION

(1) Use welded joints in preference to bolted or riveted joints.
(2) Seal crevices by using non-corrosive materials.
(3) Eliminate or minimize crevice corrosion at the design stage.
(4) Minimize contact between metals and plastic, fabrics and debris.
(5) Avoid contact with hygroscopic materials.
(6) Avoid sharp corners, edges and pockets where dirt or debris could be collected.
(7) In critical areas, use weld overlays with highly corrosion resistant alloy.

(8) Use alloys resistant to crevice corrosion, such as titanium or Inconel. Increased Mo contents (up to 4.5%) in austenitic stainless steels reduce the susceptibility to crevice corrosion. Use appropriate alloy after prescribed service tests for a specific application.
(9) Apply cathodic protection to stainless steels by connecting to adjacent mild steelstructure.
(10) For seawater service, maintain a high velocity to keep the solids in suspension.
(11) For better performance of steels in seawater, allow intermittent exposure to air to allow the removal of protective films.
(12) Use inhibiting paste, wherever possible.
(13) Paint the cathodic surface.
(14) Remove deposits from time to time.
(15) Take precautions against microbial corrosion, which creates crevices and is very damaging to low Mo stainless steels.

4.6 PITTING CORROSION

4.6.1 DEFINITION

It is a form of localized corrosion of a metal surface where small areas corrode preferentially leading to the formation of cavities or pits, and the bulk of the surface remains unattacked. Metals which form passive films, such as aluminum and steels, are more susceptible to this form of corrosion. It is the most insidious form of corrosion. It causes failure by penetration with only a small percent weight-loss of the entire structure. It is a major type of failure in chemical processing industry. The destructive nature of pitting is illustrated by the fact that usually the entire system must be replaced.

4.6.2 ENVIRONMENT

Generally, the most conducive environment for pitting is the marine environment. Ions, such as Cl^-, Br^- and I^-, in appreciable concentrations tend to cause pitting of steel. Thiosulfate ions also induce pitting of steels.

Aluminum also pits in an environment that cause the pitting of steel. If traces of Cu^{2+} are present in water, or Fe^{+3} ions are in water, copper or iron would be deposited on aluminum metal surface and pitting would be initiated. Oxidizing metal ions with chloride, such as cupric, ferric and mercuric, cause severe pitting. Presence of dust or dirt particles in water may also lead to pitting corrosion in copper pipes transporting seawater. With soft water, pitting in copper occurs in the hottest part of the system, whereas with hard waters, pitting occurs in the coldest part of the system.

4.6.3 CONDITIONS

The most important condition is that the metal must be in a passive state for pitting to occur. Passive state means the presence of a film on a metal surface. Steel and aluminum have a tendency to become passive, however, metals which become passive by film formation have a high resistance to uniform corrosion. The process of pitting destroys this protective film at certain sites resulting in the loss of passivity and initiation of pits on the metal surface. It may be recalled that passivity is a phenomenon which leads to a loss of chemical reactivity. Metals, such as iron, chromium, nickel, titanium, aluminum and also copper, tend to become passive in certain environments.

The following are the conditions for pitting to occur:

(1) Breaks in the films or other defects, such as a lack of homogeneity in the film on the metal surface.
(2) The presence of halogen ions, such as Cl^-, Br^- and I^-, and even $S_2O_3^-$.
(3) Stagnant conditions in service. Steel pumps in seawater serve for a good number of years as long as they keep on working. When taken out of service even for a short period stainless steel pumps tend to pit. For this reason, stainless steel pumps must run for a few minutes several times per week even if not required for pumping duty.

Sites which are most susceptible to pitting are grain boundaries. Steels having small grain size are more susceptible to pitting than steels with large grain sizes.

4.6.4 SHAPES

Studies on stainless steels have shown that sulfide inclusions are the most probable sites for pit nucleation. Pits may grow in several forms, such as circular, square, pyramidical and hexagonal. However, it is the depth of the pit which matters more than the shape of the pit. Crystallographic pits may be observed in certain alloys, such as Al 5052 associated with very low dissolution rates. Pits generally grow downwards from horizontal surface. Several months or sometimes even years may be needed before the pits become visible. The period intervening between their initiation and becoming visible is called *induction period*, which depends upon a particular metal and environment.

4.6.5 MECHANISM OF PITTING

In order for pitting to take place, the formation of an anode is a prerequisite. With the formation of an anode, a local corrosion cell is developed. The anode may be formed as a result of

(a) lack of homogeneity at the metal corrosive interface. Lack of homogeneity on the metal surface is caused by impurities, grain boundaries, niches, rough surface, etc. The difference in the environments can cause, for instance, the formation of concentration cells on the metal surface.
(b) destruction of a passive film. The destruction of a passive film results in the formation of a small anode. Therefore, a breakthrough of the protective film on the metal surface results in several anodic sites and the surrounding surface acts as a cathode. Thus, an unfavorable area ratio results.
(c) deposit of debris or solids on the metal surface. This generally leads to the formation of anodic and cathodic sites.
(d) formation of an active–passive cell with a large potential difference. The formation of a small anode on the passive steel surface, for

instance, leads to the formation of the above cell.

4.6.5.1 Conditions

(1) The passive metal surrounding the anode is not subject to pitting as it forms the cathode and it is the site for reduction of oxygen.
(2) The corrosion products which are formed at the anode cannot spread on to the cathode areas. Therefore, corrosion penetrates the metal rather than spread, and pitting is initiated.
(3) There is a certain potential characteristic of a passive metal, below which pitting cannot initiate. This is called pitting potential, E_p.

With the above considerations, it should be possible to suggest a simple mechanism for pitting corrosion.

4.6.6 PITTING PROCESSES

(1) The formation of anodic sites by disruption of the protective passive film on the metal surface. The metal dissolution takes place at the anode. Pitting is, therefore, initiated. The anodic metal dissolution reaction is represented by

$$M \rightarrow M^{n+} + ne \qquad (4.26)$$

This is balanced by the cathodic reaction of oxygen on the adjacent surface.

$$O_2 + 2H_2O + 4e \rightarrow 4OH^- \qquad (4.27)$$

Initially, the whole surface is in contact with the electrolyte containing oxygen so that oxygen reduction as shown above takes place (Fig. 4.22).

(2) Due to the continuing metal dissolution, an excess of positive ions M^+ is accumulated in the anodic area. The process is self-stimulating and self-propagating. Conditions are produced within a pit which are necessary for continuing the activity of the pit. To maintain charge neutrality negative ions (anions), like chloride, migrate

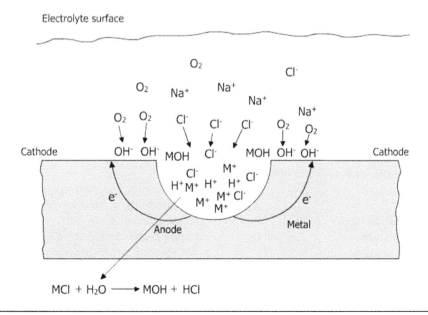

Figure 4.22   Schematic of an active corrosion pit on a metal in a chloride solution

from the electrolyte (consider, for example, seawater or a 5% NaCl solution)

$$M^+Cl^- + H_2O \rightarrow MOH + H^+ + Cl^-$$

$$(4.28)$$

OH$^-$ ions also migrate (but more slowly) to neutralize the positive charges. The reaction of M$^+$Cl$^-$ with H$_2$O resulting in the formation of M$^+$OH$^-$ + H$^+$Cl$^-$. This process is called *hydrolysis* (equation (4.28)).

(3) The presence of H$^+$ ions and chloride content, prevents repassivation. The above process generates free acid and the pH value at the bottom of the pit is substantially lowered. It has been measured between 1.5 and 1.0. The pH value depends on the type of steel-values for stainless steel pits which are lower than for mild steel and have a pit pH of 4. This is due to the solubility product effects and depends on the presence of chromium. In stainless steels, the pH is reduced by hydrolysis of Cr^{3+} and Fe^{2+} as well as by accumulation of chloride.

(4) The increase in the rate of dissolution at the anode increases the rate of migration of the chloride ions and the reaction becomes time dependent and continues, resulting in the formation of more and more M$^+$Cl$^-$ and, therefore, generation of more and more H$^+$Cl$^-$ by hydrolysis (equation (4.28)).

(5) The process continues until the metal is perforated. The process is autocatalytic and it increases with time resulting in more and more metal dissolution.

(6) Finally, the metal is perforated and the reaction is terminated.

As shown above, basically three processes are involved:

(a) pitting initiation
(b) pitting propagation
(c) pitting termination.

Step (1) of the mechanism shown above, describes the initiation of pitting, steps (2–5) describe the propagation of pitting, and step (6) termination of pitting.

4.6.7 MAIN STEPS INVOLVED IN THE PITTING OF CARBON STEEL AND STAINLESS STEEL

The following are the main steps involved in the pitting of carbon steel and stainless steels.

According to Wranglen [16], sulfide inclusions are responsible for the initiation of attack in both carbon steel and stainless steels. The following reactions occur in the pit interior (Fig. 4.23):

$$Fe \rightarrow Fe^{2+} + 2e \qquad (4.29)$$

$$Fe^{2+} + H_2O \rightarrow FeOH^+ + H^+ \qquad (4.30)$$

$$MnS + 2H^+ \rightarrow H_2S + Mn^{+2} \qquad (4.31)$$

The above reactions (equations (4.29–4.31)) are anodic. Reaction (4.30) results in the decrease of pH due to the production of H$^+$ ion which reacts with the inclusion (MnS) and results in the formation of H$_2$S and Mn^{+2}, as shown in equation (4.31). Hydrogen is evolved as H$^+$ accepts electrons released by the anodic reaction. S^{2-} and HS$^-$ stimulate the attack. Because of an excess of positive ions as shown by equation (4.29), the concentration of the Cl$^-$ ions increases with time. As the process continues with time, the pit depth becomes more and more.

Rust is formed at the pit mouth. It may be either Fe$_3$O$_4$ or (FeOOH). Further oxidation of Fe^{2+} and FeOH may take place at the pit mouth:

$$2FeOH^+ + \frac{1}{2}O_2 + 2H^+ \rightarrow 2Fe(OH)^{2+} + H_2O$$

$$(4.32)$$

$$2Fe^{2+} + \frac{1}{2}O_2 + 2H^+ \rightarrow 2Fe^{3+} + H_2O$$

$$(4.33)$$

The rust formed does not allow the mixing of the products of the anodes and the cathode (anolyte and catholyte).

The pH of the solution falls at the pit mouth as H$^+$ is formed by hydrolysis of FeOH$^+$ or Fe^{3+} formed (equations (4.34) and (4.35)):

$$FeOH^{2+} + H_2O \rightarrow Fe(OH)_2^+ + H^+ \qquad (4.34)$$

$$Fe^{3+} + H_2O \rightarrow FeOH^{2+} + H^+ \qquad (4.35)$$

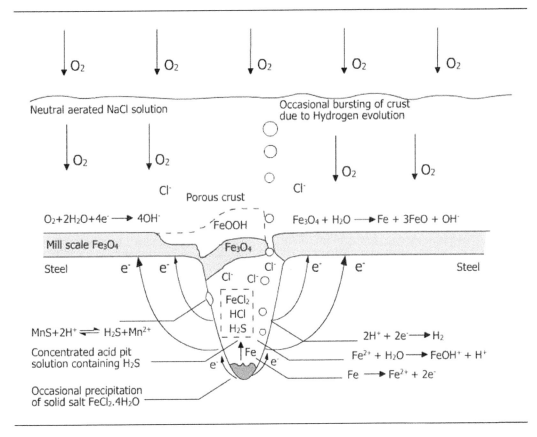

Figure 4.23 Electrochemical reactions that occur when a pit is initiated at sufide inclusion in a carbon steel. (From Wranglen, A. (1974). *Corrosion Science*, **14**, 331)

The process continues, the pH falls, and the reaction becomes autocatalytic.

Outside the pit, the surface is cathodically protected, the main reactions being the reduction of oxygen and the formation of rust. The surface adjacent to the pit acts as the cathode as shown earlier, and, therefore, it does not pit. The main cathodic reactions are:

$$O_2 + 2H_2O + 4e \rightarrow 4OH^- \qquad (4.36)$$

$$3FeOOH + e \rightarrow Fe_3O_4 + H_2O + OH^- \qquad (4.37)$$

Production of OH^- ions increases the pH of the cathodic area. Because the electrons are consumed at the surface adjacent to the pit, the surface around the pit is also cathodically protected to a certain extent.

4.6.8 MECHANISM OF PITTING CORROSION OF ALUMINUM

One important mechanism suggests the following steps which lead to pitting [16,17]:

(1) Absorption of the halide on the oxide film (at the oxide solution interface):

$$Cl^- \text{ (in bulk)} = Cl^- \text{ (adsorbed on } Al_2O_3, \\ nH_2O \text{ sites)} \qquad (4.38)$$

(2) Formation of basic hydroxychloride aluminum salt which separates from the lattice and goes into solution:

$$Al^{+++} \text{(in } Al_2O_3, nH_2 \text{ lattice)} \\ + 4Cl^- \text{(adsorbed)} = AlCl_4^- \\ \text{in solution of lower pH} \qquad (4.39a)$$

$$Al^{+++} \text{(in } Al_2O_3, \ nH_2O \text{ lattice)} + 2Cl^-$$
$$+ 2OH^- = Al(OH)_2Cl_2^-$$
$$\text{(readily soluble)} \quad (4.39b)$$

$Al(OH)_2Cl_2^-$ is formed by:

$$Al(OH)_3 \rightleftarrows Al(OH)_2^+ + OH^-$$

$$Al(OH)_2^+ + Cl^- \rightarrow Al(OH)_2Cl$$

(3) The thinning of oxide films: The oxide film is thinned to the extent that aluminum ions can pass from the metal to the solution interface. The chloride ion gets entry by one of the following possible methods:

   (a) By competitive absorption of halide ions.
   (b) Penetration of halide in the oxide film.
   (c) Diffusion of halide through the oxide film and attack on the metal.
   (d) Formation of soluble complexes with the halide ions [17].

For a complete understanding of the mechanism of corrosion of aluminum, refer to literature cited at the end of the chapter.

4.6.9 ELECTROCHEMICAL PARAMETERS OF PITTING

The classical techniques involving exposure to a corrosive medium and metallographic examination by optical microscopy have been supplemented by time-saving electrochemical techniques in recent years. One such electrochemical technique is the polarization hysteresis technique used for observing the anodic polarization behavior of steels, aluminum and their alloys.

The anodic polarization scan is commenced from the open circuit potential of aluminum and continued in the noble direction at a slow scan rate. The scan is reversed on reaching a predetermined current value and continued in a reverse direction until the loop is closed at a more active potential and the current drops to a minimum value. The plot obtained is called a cyclic polarization curve, as shown in Fig. 4.24. The diagram is conveniently divided in three regions as below:

(1) **Region I:** This region represents immunity to pitting and at a potential more negative than E_{pp} (protection potential) pitting is not expected to propagate.
(2) **Region II:** In this region, new pits do not initiate, but the pits already initiated in Region III continue to propagate. This region is, therefore, the propagation region for pits.
(3) **Region III:** In this region, pits initiate and propagation take place on and above a certain potential, called breakdown or critical pitting potential (E_p). The critical pitting potential signifies the onset of pitting.

Corrosion potential is the potential at which the rate of oxidation equals the rate of reduction. It is a potential that a corroding metal exhibits under specific conditions of time, temperature, aeration, velocity, etc. The pitting potential, the protection potential and corrosion potential, can be read from the polarization hysteresis diagram. The pitting resistance of aluminum alloys may be compared and predicted by the magnitude of the above values and the relative position of the pitting potential (E_p) compared to the corrosion potential. For instance, if the value of pitting potential (E_p) is positive to the corrosion potential (E_c), the resistance of the alloy to pitting is high. Generally, the more negative the value of E_c compared to E_p, the high is the pitting resistance. There is a disagreement between the workers on the application of the above parameters for prediction of pitting resistance of aluminum alloys, however, it is useful to understand the significance of these parameters.

In contrast to the situation for aluminum, electrochemical studies on the pitting of stainless steels are fraught with contradictions because of the effect of experimental factors. Several investigations have used breakthrough potential or breakdown potential, as a measure of pitting susceptibility. As pointed out earlier, the current increases rapidly at the breakdown potential (Fig. 4.24), this is considered as an indication of a breakdown of the passive, corrosion-resistant steel film and initiation of pitting.

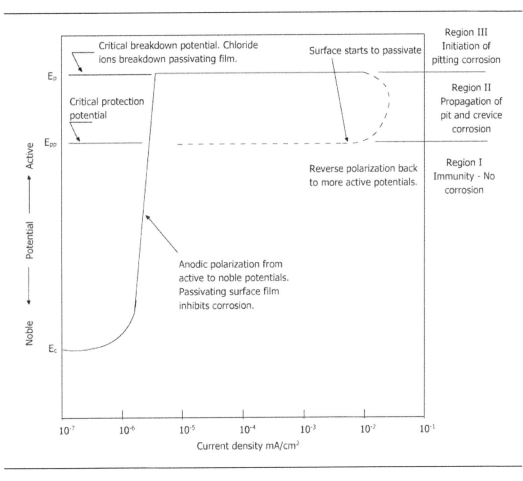

Figure 4.24 Shape of the polarization curve for a typical stainless steel in a solution showing how corrosion behavior changes as the passivating surface film forms

Figure 4.25 shows the effect of chloride ion concentration on the polarization behavior of steel 304. The effect of various factors, such as concentration, heat treatment and composition of the solution, can be determined by breakdown potentials. For instance, in Fig. 4.24, an increase in the chloride concentration from 1000 ppm to 40 000 ppm decreases the breakdown potential from $+400\,mV_{SCE}$ to about $-10\,mV_{SCE}$. It is to be pointed out that below E_p, the metal is in the passive state and pitting does not occur. Above E_p the passivity breaks down and pitting occurs. The most accurate method of measuring E_p is by the potentiostatic or potentiodynamic technique. In this technique, the potential is held at an assigned value which can be changed gradually at a constant rate, to plot potential against current. The potentiostat keeps the potential difference between specimen and reference electrode at a set value, by applying any necessary voltage to an auxiliary electrode and the voltage and current are recorded.

4.6.10 FACTORS AFFECTING THE BREAKDOWN POTENTIAL (PITTING POTENTIAL)

(a) **Chloride ions concentration:** A certain concentration of chloride ions is necessary to initiate pitting of steel and aluminum alloys.

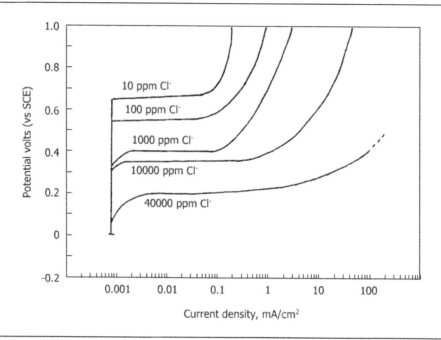

Figure 4.25 Potentiodynamic polarization curves of type 304 stainless steel in various chloride ion concentration using NaCl as the salt

(b) **Alloying elements**: The effect of certain alloying elements on the breakdown potential is important. Generally, the alloying elements which have a beneficial effect shift E_p to a more positive value, whereas the elements having detrimental effects shift E_p to more negative value. Generally, Mo, Cr, Ni, V, Si, N, Ag and Re improve pitting resistance of stainless steels.

(c) **Effect of electrolyte composition**: Ions, such as $SO_4^{2-}, OH^-, ClO^-, CO_3^{2-}$ and MoO_4^{2-} shift E_p to more positive value. Nitrates and chromates when added to a solution containing Cl^-, inhibit pitting. In the case of 18-8 stainless steels, the inhibitor efficiency of ions decreases in the order: $OH^- > NO_3^-$ Acetate $> SO_4^{2-} > ClO_4^-$.

(d) **Effect of pH**: In the acid range of pH (pH value less than 6), the pitting potential remains unaffected by pH changes as shown by 18-8 stainless steel in 0.1 NaCl. In the alkaline range, however, the pitting potential shifts in the noble direction. The pitting potentials obtained in this region are generally non-reproducible and general dissolution take place in a limited range. Thus, in the alkaline region, the dissolution pattern is changed from pitting to uniform corrosion. It is to be observed from the effect of pH on critical potential, that pitting is stopped by addition of hydroxyl (OH^-) ions.

As shown above, the measurement of pitting potential (E_p) can be utilized to predict the tendency of steels to pitting in chloride containing environment.

(e) **Effect of temperature**: It has been shown that an increase of temperature by 10°C shifts the critical pitting potential (E_p) by 30 mV for certain grades of steels. This relationship is, however, differs for other grades of steels. Generally a rise in temperature shifts the critical pitting potential (E_p) in the more active direction, which leads to acceleration of pitting. Critical pitting potential (E_p) is synonymous with the breakdown potential (E_b). There is a critical pitting temperature below which pitting will not occur.

(f) **Effect of heat treatment and cold working:** It is not possible to make a general statement on the effect of heat treatment on the pitting potential of steels as different types of steels respond differently to cold working or hot working. The nature of alloying elements, their concentration and microstructural changes that they introduce, affects the response of various grades of steels to cold working and hot working. For instance, the effect of cold working on the value of E_p for 18 Cr–10 Ni, 25 Cr–25 Ni, steel is negligible. In case of 18 Cr–2 Ni–Ti steel, alloyed with 3% Si, the pitting resistance is appreciably decreased.

(g) **Induction time (T):** This is the time required for the first pit to initiate. The induction period of pitting varies from one steel to another and depends on the environment. Mainly, the induction time depends upon the concentration of Cl^- ion in the solution. The higher the concentration of chloride, the less the induction time [18,19].

It has been observed that the induction time decreases with increasing potential or the induction time increases with increasing potential. Studies on the induction time of 18 Cr–8 Ni steel have shown that the rate of breakdown of oxide film by the chloride ions resulting in initiation of pit is proportional to the chloride ion concentration [18]. It has also been observed that E_{pit} (current density inside pit) decreases with decreasing potential, decreasing chloride concentration and increasing temperature for steel [20,21].

4.6.11 FACTORS AFFECTING PITTING

(1) **Surface defects,** such as inclusions, second phases and surface heterogeneity, have been identified with initiation of pits. It has been established that sulfide inclusions are mainly responsible for pitting of carbon and stainless steels.

(2) **Degree of cold work.** Severe cold working increases pitting susceptibility as shown by austenitic steels in ferric chloride solution.

(3) **Surface finish.** A smooth surface does not allow accumulation of impurities and stagnancy to be built up on the metallic surface, thus minimizing the risk for the formation of differential aeration cells and minimizing corrosion and pitting. However, the pits which are formed on a smooth and shiny surface are deeper and larger compared to a rough surface.

(4) **Sensitizing temperature.** The temperature zone which promotes intergranular corrosion of steels also promotes pitting, because of the weakening of grain boundaries.

(5) **Velocity.** It is a matter of common observation that a stainless steel pump handling seawater gives a good service if it is used to run continuously, but it would tend to pit if stopped because of the formation of differential aeration cells. Pits are often associated with stagnant conditions. High velocity may cause passivation of the steel surface by greater control of the oxygen with the steel surface, thus enabling the integrity of the passive film to be maintained.

(6) **Environmental contamination.** An environment contaminated by dust particles would be more conducive to pitting. Dust particles on metals absorb moisture and lead to the formation of differential aeration cells. Salt particles in a marine environment when deposited on a metal surface lead to intensive pitting of equipment and appliances.

4.6.12 INFLUENCE OF ALLOYING ELEMENTS ON THE PITTING OF ALUMINUM

(1) **Chromium.** It is usually added to Al–Mg alloy and Al–Mg–Zn alloys in amounts of 0.1–0.3%. It has a beneficial influence on pitting resistance in seawater.

(2) **Copper.** It reduces the corrosion resistance of aluminum alloys and increases the susceptibility to pitting. The corrosion potential of Al–Cu alloys become cathodic in direct proportion to the amount of Cu in solid solution.

(3) **Iron.** It exists in the form of $FeAl_3$ and promotes pitting of aluminum alloy. At

low copper levels (<0.05%), addition of 1.25% Mn (e.g. AA-3003), minimizes the detrimental effect of Fe due to the formation of $Mn(Fe)Al_6$.

(4) **Magnesium**. It provides substantial strengthening with good ductility in addition to excellent corrosion resistance. The aluminum–magnesium family (5xxx) have the lowest pitting probability and penetration rates. Alloy 5052 containing 2.5% Mg is found to possess an excellent resistance to pitting in seawater. In the presence of Si, Mg_2Si is formed which leads to slight enhancement of pitting and may lead to intergranular corrosion. Pitting is generally combined with intergranular corrosion in Al–Mg alloys.

(5) **Manganese**. Up to 1.25% Mn can be added to AA 1099 without increasing pitting probability. It is normally used in heat treatable alloy AA 5182 in concentration ranging from 0.20 to 0.50%. It reduces the pitting probability of 99.5 or 99.8% aluminum silicon. It does not increase the pitting probability of AA 1099 in water. It has a less pronounced effect on pitting than Fe. At a higher level (0.3%) it can have an adverse effect.

(6) **Zinc**. Up to 1.0% zinc can be added to AA 1099 without increasing the pitting probability. At higher levels, its influence would depend upon other constituents, such as Mg and Cu, and the microstructure produced.

(7) **Scandium**. Aluminum alloys with 0.15 to 0.6% scandium have been reported in recent years. Scandium addition up to 0.6% has a dramatic influence on strength and it does not adversely affect the resistance of Al–2.5 Mg alloys to pitting.

4.6.13 ALLOY ADDITION IN IRON AND STEEL

Alloying additions of chromium, nickel, molybdenum, increase the pitting resistance, whereas silicon, sulfur, carbon and nitrogen, decrease the resistance to pitting. Addition of titanium increase resistance of steels only in solutions of $FeCl_3$ and not in other mediums.

4.6.14 PITTING OF COPPER

Copper tubes (alloy C-106) are widely used in water distribution system. One of the serious problems encountered in these tubes containing a small carbon residue, is pitting. The following mechanism of pitting has been suggested.

A. Anodic Dissolution

$$Cu \rightarrow Cu^+ + e \qquad (4.40)$$

Copper ions (positive charged) combined with chloride ions present in water to form CuCl:

$$Cu^+ + Cl^- \rightarrow CuCl \qquad (4.41)$$

Cuprous chloride is not stable in near-neutral pH ranges and hydrolyzes to form cuprous oxide which is precipitated on the metal surface.

$$2CuCl + H_2O = Cu_2O + 2HCl \qquad (4.42)$$

B. Cathodic Reduction

The main cathodic reaction is the reduction of oxygen on the surface adjacent to the pit:

$$O_2 + 2H_2O + 4e \rightarrow 4OH^- \qquad (4.43)$$

If the water is acidic, the OH^- produced in equation (4.43) is removed. If bicarbonate is present in the water, OH^- is also removed. Pitting occurs in copper tubes at a certain potential, the *critical pitting potential*. The pH inside the pit is substantially lowered by the production of HCl as shown in reaction (4.42) (Fig. 4.25).

4.6.15 THEORIES OF PITTING

The pitting corrosion process is a highly complex process with a sequence of steps. The following are the steps:

(1) Breakdown of passivity
(2) Early pit growth

(3) Late pit growth
(4) Repassivation of pits.

The primary step in pitting is the breaking of the passive layer on the metal surface which is explained by

(A) ion penetration [22]
(B) adsorption [23,24]
(C) film breakdown [25–27]
(D) repassivation mechanism [28,29]
(E) point defect model [30].

A. Ion Penetration Theory

It is based on the assumption that the tendency of the chloride ions to penetrate and break the passive layer is because of its small diameter. The penetration of the aggressive anions, however, is not an isolated process as it is combined with the transport of O^{2-} to the electrolyte to maintain charge neutrality. The transport of O^{2-} is, however, a slow process and, therefore, if the above assumption is taken to be true, the charge neutrality would not be maintained. It has also been shown that NO_3^- has an identical rate of penetration to Cl^-, However, pitting is not caused by the former. Also, it has been shown by some studies that the thickness of the passive layer does not affect pitting and hence the rate of penetration of Cl^- ion cannot be a decisive factor. It is known that SO_4^{2-}, ClO_4^- and SCN^- cause pitting, whereas according to the penetration theory they should not cause pitting because of their relatively larger diameter compared to Cl^- ion.

B. The Adsorption Theory

This theory is based on the assumption that aggressive anions which are absorbed at the energetically favored sites on the passive layer cause the nucleation of pits. In order to overcome the repulsive forces between anions, there should be an exchange of OH^- or O^{2-} of the passive oxide layer by the aggressive anions. Experiments have shown that the passive layer (oxide layer) is removed and replaced by an adsorbed salt layer containing the aggressive anions. Where O^{2-} is absorbed the metal is passivated and where Cl^- is adsorbed, the

passivity breaks down resulting in the initiation of pits. It has been observed that the breakdown of a passive surface is accompanied by high current densities. If values of currents are plotted against potential after a pre-determined time, it would be observed that at the instant a pit initiates, the current at a certain value of potential shoots to a very high value. This high value of current indicates the initiation of a pit on a passive surface.

This theory, like other theories on pitting, has its own points of weakness. For instance, the displacement of a doubly charged negative ion by a singly charged anion should be theoretically favored by a shift of the potential to more negative values and not to move positive values of potential as suggested by the adsorption theory. If the above suggestions were true, PO_4^{3-} should be more aggressive than Cl^-, which is contradictory to the observations.

C. Film Breakdown Theory

This theory assumes a direct access of the aggressive anions to the metal surface by breaking the passive film. In order to break the passive layer, a high concentration of aggressive anions is necessary which could be caused by high concentration of the aggressive anions in the bulk solution. The high concentration stabilizes the adsorption of salt layer which prevents the repassivation of pits. Pitting, therefore, is likely to occur at a certain value of concentration (critical concentration). It has been experimentally shown that below a certain concentration, pitting does not initiate.

It is very difficult to decide which of the two mechanisms, adsorption or film breakdown, explains the mechanism more accurately. As discussed above, the concentration of aggressive anions and the effect of electrolytic composition on pitting seem to suggest that the adsorption theory is more appropriate to explain the phenomenon observed in pitting.

If inhibiting anions are absorbed in place of aggressive anions, pitting would not be initiated because the oxide layer would not be broken as it would be by aggressive anions. This observation can be best explained by the film breakdown theory; the pitting does not initiate until the passive layer is broken. It is, therefore, not possible to distinguish between the merits of the two theories clearly.

D. Repassivation Theory

In repassivation, the aggressive ions at the pit are replaced by a passive layer – at more negative potentials. Pitting would, therefore, not initiate. In the process of repassivation, the layer of aggressive anions is replaced by a passive layer. The aggressive ions are removed by diffusion from the pit to the electrolyte. Repassivation is expected only if the pitting potential becomes more than the flade potential (the potential in a polarization curve of an active–passive metal at which the current density is minimum). Repassivation is prevented if the concentration of chloride exceeds 1 M.

In spite of a substantial progress made in the field of pitting corrosion, no single theory exists which can explain all the known observations on pitting.

E. Point Defect Model

This model assumes the transport of both anions (oxygen ions) and cations (metal) ions or their respective vacancies. The anions diffuse to the metal surface with a resultant thickening of the film. Cation diffusion cause film dissolution and metal vacancies are created at the metal/film interface. When the rate of production of metal vacancies in the metal exceeds their rate of migration, the vacancies pile up at the metal/film interface and result in void craters. This is the process of pit incubation. At a certain critical size, the passive film suffers local collapse and the incubation period ends. The collapsed site dissolves much faster than at any other place and leads to pit growth.

There are other mechanism, such as that of Pickering [31], however, discussion of all of them is beyond the scope of this chapter.

The theories described above are summaries of the work done by various investigators. As pitting is an extremely important form of localized corrosion, it is necessary for engineering students to have some ideas about the mechanism

of pitting beyond the basic material provided in most of the books on corrosion. For a better and more detailed understanding of the mechanism of pitting, references are provided at the end of the chapter.

Although the theories outlined above are conflicting, there are certain points which are generally agreed upon by most of the investigators. These are:

(a) Pits initiate above a certain critical potential called breakdown, rupture or pitting potential.
(b) The breakdown potential is a function of halide ion concentration.
(c) The breakdown of a passive film takes place at a highly localized site.
(d) There exists an induction time for a pit to appear on the surface.
(e) A passive surface is essential for pitting to take place.
(f) The induction time is potential dependent. Pitting does not appear at values negative to the pitting potential.
(g) The two processes, breakdown of a passive film, and adsorption of chloride ions, play a major role in the initiation of pits. Whether pits are initiated by film breakdown or by adsorption of chloride or by formation of complex of chloride is not clear, however, all the three factors appear to affect the process of pitting.

Some other salient features of the pitting mechanism suggested by various workers are:

(a) Three or four halide ions are jointly absorbed on the oxide film surface around a lattice cation and a transitional complex is formed which possesses high energy, and separates from the oxide. Once the complex is formed, the process is repeated until the passive layer is completely destroyed.
(b) Pits develop on the site where oxygen absorbed on the metal surface is displaced by Cl^- ions. The breakdown potential (pitting potential) is the potential at which the aggressive anions produces a reversible displacement of the oxygen from the metal surface.

(c) The bond between the metal and oxygen is weakened at many points and exchange of chloride ions with oxygen takes place at these points.
(d) Voids are created by accumulation of vacancies at the metal/film surface and at a certain critical void size, the film suffers damage.

4.6.16 CASE HISTORIES

A. Case 1

A seawater pumping plant had been designed to provide seawater for cooling purpose to a thermal power plant (TPP), a turbo-blower station (TBS) and other plants, such as refrigeration, oxygen, rolling mills and refractory plants, in a steel mill. Pitting was observed to be the major problem in welds of the outer casing of the main pump and in impellers.

Conditions The seawater contained 36 020 ppm TDS, 24 200 ppm chloride and 3 076 ppm SO_4 ions. The average pH was 8.0.

Material The casing material was stainless steel (304 L).

Inspection After three years of operation, the inside of the casing was examined. The surface was found to be covered by a thin dark brown layer of uniform thickness over the surface. Removal of the layer showed a large number of pits on the weld metal. On analyzing the problem, it was found that the following factors caused pitting:

(a) High contents of dissolved salt
(b) Suspended solids like sand, silt, etc.
(c) Seepage of seawater through concrete.

Remedial measures The following remedial measures were adopted:

(1) The pH of the water was maintained at 12.5.
(2) Seepage of the seawater through the concrete was stopped.
(3) The casing was cathodically protected by galvanic anodes.

B. Case 2

An impeller was made of steel, ASTM designation A296, Grade CF-8, having the following composition:

C = 0.06, Si = 1.47, Mn = 0.98, P = 0.02, S = 0.019, Ni = 8.88, Cl = 18.81, Balance Fe.

The impellers were inspected after three years of operation. Pitting was observed to be the major problem.

Conditions The same as for the casing in Case 1 above.

Material 18-8 Cr–Ni steel.

Remedial measures The following measures were recommended:

(1) Steel 18-8 with 2% Mo be used instead of 18-8 Cr–Ni steel.
(2) The pH of the seawater be maintained at 11.00.

C. Case 3

Pitting of Type 321 stainless steel aircraft fresh water tanks caused by retained metal cleaning solution.

Conditions One tank had been in service for 32 h and the other for 10 h. The tanks contained fresh water.

Service history The sodium hypochlorite solution used for cleaning of the tank was three times the prescribed strength. After sterilizing, a small amount of cleaning solution remained in the bottom of the tank.

Observations Metallographic examination showed a large number of pits, most of the pits were about 6–13 mm (1/4–1/2 in) from the weld bead at the outlet.

Remedial measures It was recommended that

(1) the cleaning solution of hypochlorite be made according to the recommended strength.

(2) the drainage after cleaning be thoroughly done and no amount of cleaning solution be retained in the vessel.

(3) welding be properly controlled and with an inert gas to avoid the formation of a black scale.

(4) the design of the tank be changed to achieve 100% drainage of the liquid.

4.6.17 PREVENTION OF PITTING CORROSION

(1) Use materials with appropriate alloying elements designed to minimize pitting susceptibility, e.g. molybdenum in stainless steel.

(2) Provide a uniform surface through proper cleaning, heat treating and surface finishing.

(3) Reduce the concentration of aggressive species in the test medium, such as chlorides, sulfates, etc.

(4) Use inhibitors to minimize the effect of pitting, wherever possible.

(5) Make the surface of the specimen smooth and shiny and do not allow any impurities to deposit on the surface.

(6) Minimize the effect of external factors on those design features that lead to the localized attack, such as the presence of crevices, sharp corners, etc.

(7) Apply cathodic protection, wherever possible.

(8) Coat the metals to avoid the risk of pitting.

(9) Do not allow the potential to reach the critical value.

(10) Add anions, such as OH^- or NO_3^- to chloride environment.

(11) Operate at a lower temperature, if service conditions permit.

4.6.18 EVALUATION OF PITTING DAMAGE

Conventional weight-loss techniques cannot be used to evaluate pitting because of its localized nature. What matters in pitting is the depth of the deepest pit, which is an indicator of the resistance of a material to pitting. Although the depth is related to the sample size, it is often used as a criteria of pitting resistance in laboratory tests. The prediction should always be made on a combination of results obtained from number of pits per mm^2 and average pit depth.

The development of pit depths follows a time function:

$$D = Ktc \qquad (4.44)$$

where

D is depth of the deepest pit

K is an alloy surface area and environment-dependent constant

t is time and

c is an environment-dependent parameter often close to *a*.

For alloys showing a good resistance to pitting with t in years and D in mm, K can be typically close to 0.75. Estimation of the life time of material on the basis of controlled laboratory tests is of little use. However, by extreme value statistical treatment of the data quantitatively, the life and time of materials can be satisfactorily predicted [32,33].

4.7 INTERGRANULAR CORROSION

Intergranular corrosion refers to preferential corrosion along the grain boundaries. Grains are 'crystals' usually on a microscopic scale, that constitute the microstructure of the metal and alloys. By analogy, they are like the grains of sand which constitute a sandstone.

4.7.1 DEFINITION

It has been defined commonly as a form of localized attack on the grain boundaries of a metal or alloy in corrosive media, which results in the loss of strength and ductility. The localized attack may lead to dislodgment of the grain. It works inwards between the grains and causes more loss of strength than the same total destruction of metal uniformly distributed over the whole surface. The attack is distributed over all the grain boundaries cutting the surface. Intergranular corrosion is less dangerous than stress corrosion, which occurs when stress acts continuously or cyclically, in a corrosive environment, producing cracks following mostly intergranular paths. The difference between the two, i.e. intergranular corrosion and stress corrosion cracking is important. There are many materials which are susceptible to intergranular corrosion and not to stress corrosion cracking. The attack is very common on stainless steel, nickel and aluminum alloys.

Metals are crystalline materials. They consist of grains. As corrosion proceeds along the grain boundaries, the grains become weaker particularly at the grain boundaries and they eventually disintegrate (Fig. 4.26).

All metals and alloys are joined together by grain boundaries. The intergranular corrosion of steels, brasses, bronzes and aluminum alloys containing copper is of particular interest to engineers. Because of the importance of steels, the largest amount of work reported in the literature is on steels. It would be, therefore, appropriate also here to review the phenomenon of intergranular corrosion with a particular emphasis on steels. It would be appropriate, hence, to review the types of steels and some important technical terms related to intergranular corrosion.

4.7.2 TYPES OF STEELS

On the basis of their structure and hardening response, stainless steel may be divided into a number of classes which are summarized in Table 4.3.

A. Austenitic Stainless Steels

Austenitic stainless steels have a face centered cubic structure that is attained by the addition of further alloying addition to chromium and iron, such as nickel, manganese, carbon and nitrogen. Nickel is added in concentration ranges of 3.5–3.7%, Mn in the range of 1.15%, nitrogen in the range of 0.1–0.4%, and carbon from 0.02 to 1.0%. The following are the main characteristics:

(1) **Strength**. Increase in strength obtained by cold working. For maximum work hardening low chromium and nickel concentrations are recommended.

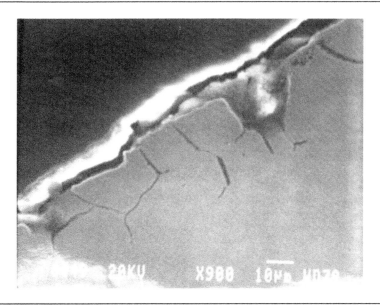

Figure 4.26 SEM micrograph of grain boundary corrosion in HAZ of 316L SS Weldment (×900)

Table 4.3 Classification of stainless steels

Steel type	Strength range (MN/m^2)	Characteristic properties
Austenitic	200–500	Non-hardenable except by cold working. Very high toughness. Very good formability. Creep resistant. Non-magnetic. Very good corrosion resistance. High cost.
Ferritic	250–500	Good formability. Inferior corrosion resistance to austenitic steels. High temperature oxidation resistance. Lower cost than austenitic steels.
Martensitic	400–2000	High strength and wear resistance. Inferior toughness.
Precipitation hardened	500–1640	High strength. High toughness. Optimum creep properties. High cost.

(2) **Toughness.** Very high impact toughness. They are used for cryogenic applications.

(3) **Corrosion resistance.** Excellent resistance to atmospheric corrosion and scaling at elevated temperatures. Corrosion resistance in acids and chloride solution is increased by addition of molybdenum. Addition of titanium and niobium increases resistance to intergranular attack. Scaling resistance is increased by addition of chromium and silicon.

(4) **Creep resistance.** For high temperature application, steels stabilized with titanium or niobium are used.

(5) **Weldability.** To avoid intergranular attack steels containing niobium or titanium are used. Austenitic steels are easily welded by inert gas welding.

On the whole, austenitic chromium nickel steels are used with the advantage for pipings, pressure vessels and other numerous applications, because they posses a unique combination of properties summarized above. The composition and mechanical properties are shown in Tables 4.4 (a–d).

B. Ferritic Stainless Steels

They usually contain 16–30% chromium, although it is possible to have a ferritic structure with lower chromium contents. AISI Type 409 and AISI Type 430 ferritic steels are used for vehicle mufflers and automobile trims, respectively. Ferritic steels have BCC structure. The following are the main characteristics.

(1) **Strength.** They possess relatively lower strength than austenite steels. They are used in either the annealed or cold work condition.

(2) **Toughness.** Impact toughness decreases with increasing strength and carbon content. Optimum toughness may be obtained by tempering above 650°C. Upon heating in the range (399–482°C), for prolonged periods, the notch toughness is reduced.

(3) **Corrosion resistance.** They do not offer a good resistance to atmospheric corrosion as chromium nickel austenitic steels. Their resistance to corrosion is increased by addition of molybdenum and tantalum.

(4) **Formability.** The formability is not as good as that of austenitic steels.

(5) **Weldability.** Is not as easy as for austenitic steels. Post-welding treatment is necessary to avoid cracking.

C. Martensitic Stainless Steels

They contain 11–16% chromium as alloying element, although it is possible to develop

Table 4.4a   The composition and mechanical properties of steels – Austenitic stainless steel

Cr	Ni	Mo	C max	Other elements	AISI Type	Effect on properties
18	8	–	0.15		302	Basic steel
17	7	–	0.15		301	Lower Cr and Ni for greater work hardening
18	8	–	0.15	S or Se 0.15 min	303	S added for enhanced machinability
19	10	–	0.08		304	Lower C for improved ductility and weldability
19	10	–	0.03		304L	Lower C for welding of thicker sections
18	11.5	–	0.12		305	Higher Ni for increased formability
23	13.5	–	0.20		309	Higher Cr and Ni for corrosion and scaling resistance
25	20.5	–	0.25		310	Highest Cr and Ni to increase scaling resistance
24.5	20.5	–	0.25	Si 2.75	314	Si higher to increase scaling resistance
17	12	–	0.08		316	Mo added for more corrosion resistance
18	10.5	–	0.08	$Ti = 5 \times C$ min	321	Ti added to avoid intergranular corrosion
18	11	–	0.08	$Nb + Ta = 10 \times C$ min	347	Nb, Ta added to avoid intergranular corrosion
18	3.5–6.0	–	0.15	Mn 5.5×10.0 N 0.25 max		Substitution of nickel by manganese gives similar corrosion resistance at lower cost. Strength is approximately 50% higher but weldability is impaired

martensitic structure with even higher chromium levels. The martensitic structure of the steels is obtained by rapid quenching from the austenite range.

(1) **Strength**. Steels with medium strength and good toughness contain 0.15–0.35% C. Steels with very high strength and wear resistance may contain up to 0.12% C.
(2) **Toughness**. Optimum toughness is obtained at temperature of 650°C or above.
(3) **Corrosion resistance**. They are the least resistant members of the steel family. Resistance to corrosion is increased by tempering in the range 360–650°C. The corrosion resistance is increased with increasing chromium content.
(4) **Weldability**. Post-treatment and pre-treatment is necessary to avoid cracking. Welding of martensitic steels is not generally recommended.

D. Duplex and Precipitation Hardening Steels

There are some stainless steels in addition to the above three classes, which exhibit a wide range of high strengths because of the formation of carbide

Table 4.4b The composition and mechanical properties of steels – Ferritic stainless steels

| Composition (wt%) | | | | | | Condition | Mechanical properties | | | | Remarks |
C	Mn	Si	Cr	Mo	Nb (or Ta)		0.2% Proof strength (Mn/m^2)	Tensile strength (Mn/m^2)	Reduction of area (%)	Impact toughness (Joules)	
0.08 max	1.0 max	1.0 max	12.0 14.0	–	–	Annealed cold worked	240 520	450 600	60	50	Good formability
0.12 max	1.0 max	1.0 max	14.0 18.0	–	–	Annealed cold worked	300 550	520 650	60	–	Free machining grades containing 0.15% sulfur available
0.12 max	1.0 max	1.0 max	16.0 18.0	0.75 1.25	$5 \times C$	Annealed	350	550	60	–	Improved corrosion resistance and weldability
0.20	1.5 max	1.0 max	23.0 27.0	–	–	Annealed	375	575	45	5	Excellent resistance to high temperature corrosion and scaling

Table 4.4c The composition and mechanical properties of steels – Martensitic stainless steels

	Composition (wt%)						Typical mechanical properties			Remarks
C	Mn	Si	Cr	Ni	Mo		0.2% Proof strength (Mn/m²)	Tensile strength (Mn/m²)	Impact toughness (Joules)	
0.15 max	1.0 max	1.0 max	11.5–13.5	–	–		400–1020	630–1350	50–100	For high strength tempering temperatures of 150–180°C are employed. For optimum toughness tempering temperatures of 650–750°C are recommended.
0.15 max	1.0 max	1.0 max	11.5–13.5	1.25–2.50	–		720–1050	850–1400	60	
0.15–0.20	1.0 max	1.0 max	12.0–14.0	–	–		1350	1600	15	High strength and wear resistance.
0.20 max	1.0 max	1.0 max	15.0–17.0	1.25–2.50	–		650–1100	875–1400	50	High strength with toughness
0.60–0.75	1.0 max	1.0 max	16.0–18.0	–	0.75 max		1150	1825	6	
0.75–0.95	1.0 max	1.0 max	16.0–18.0	–	0.75 max		1900	1950	4	Very high strength and wear resistance.
0.95–1.20	1.0 max	1.0 max	16.0–18.0	–	0.75 max		1920	1980	3	Very low toughness.

Table 4.4d The composition and mechanical properties of steels – Precipitation hardened stainless steels

Steel designation	Composition (wt%)										Mechanical properties			Remarks
	C	Cr	Ni	Mo	V	Ti	Nb	Al	Cu	Co	0.2% Proof strength (Mn/m^2)	Tensile strength (Mn/m^2)	Impact toughness (Joules)	
17-14 Cu Mo	0.12	17	14	2.5	–	–	0.25	0.45	3.0	–	300	600	35	Good high temperature strength and creep resistance
FV 520 B	0.07 max	14	6	1.5	–	–	0.5	–	1.5	–	1050	1300	40	Good high temperature strength
15-5 PH	0.07 max	15	5	–	–	–	0.4	–	3.5	–	1170	1300	20	High strength
16-6 PH	0.08	16	6	–	–	0.07	–	–	–	–	1275	1310	–	Good high temperature strength
AFC 77	0.15	14	–	5	0.2	–	–	–	–	14	1480	2000	12	Good creep resistance
17-7 PH	0.09	17 max	7	–	–	–	–	1.0	–	–	1520	1620	8	Good forgeability
PH 14-8 Mo	0.05 max	14	8	2.5	–	–	–	1.0	–	–	1520	1620	–	Good toughness
Custom 455	0.03	12	9	–	–	1.2	0.3	–	2.3	–	1635	1690	14	Good corrosion and oxidation resistance

and intermetallic compounds. Elements, like aluminum, molybdenum, copper and titanium, are added to obtain the intermetallic strengthening effect (increase of strength by the formation of an intermetallic compound). Elements, like titanium, molybdenum and vanadium, are added to obtain the carbide strengthening effect. In these steels the strengths is derived from the formation of intermediate precipitates, such as Ni_3Al, Ni_3Ti, Ni_3Mo and $CuAl_3$. Their high temperature strength is superior to that of austenitic steels.

(1) **Strength**. Very high strength.
(2) **Toughness**. The toughness is superior to that of martensitic steels. The toughness decreases with increased strength.
(3) **Corrosion resistance**. Good resistance to stress corrosion cracking. Superior to the martensitic stainless steels, inferior to ferritic and austenitic steels.
(4) **Weldability**. They are welded by inert gas welding technique. These steels may be used in aged conditions or as-rolled condition. In the as-rolled condition, they are strengthened by niobium carbide precipitation and further hardening is obtained on aging by the precipitation of copper.

4.7.3 TECHNICAL TERMS

(A) **Sensitization**. When steels, such as austenitic type, are slowly cooled through the range 550–850°C, they frequently become susceptible to intergranular attack in a corrosive medium.
(B) **Stabilized steel**. Steels containing stabilizers, such as titanium and niobium, which form carbides that are more stable than chromium carbides, are called stabilized stainless steels.
(C) **Chromium depletion**. When the concentration of chromium adjacent to the grain boundaries becomes less compared to its concentration away from the grain boundaries, the phenomenon is termed *chromium depletion*.
(D) **Quenching**. It consists of heating the alloy in the temperature range of 1950–2000°F (1056–1110°C) and cooling rapidly by quenching in water.

(E) **Weld decay zone**. It is a band in the parent plate removed from the weld.

4.7.4 WHAT HAPPENS WHEN THE STEEL IS SENSITIZED

When stainless steel, for example 18-8 steel, is cooled slowly through 550–850°C, it becomes susceptible to intergranular corrosion when exposed to a corrosion medium. The following observations characterize the intergranular attack:

(1) Sensitization occur only in the temperature range 550–850°C because of the diffusion of the carbon to the grain boundaries. The sensitization band is usually 1/8–1/4 in thick adjacent but slightly removed from the weld.
(2) Fast cooling rate through the sensitizing temperature range does not provide adequate time for the process of sensitizing to occur.
(3) The degree of sensitization increases with increasing carbon contents and decreasing chromium contents.
(4) Chromium carbide precipitate at the grain boundary. The precipitation of carbide is a time–temperature dependent phenomenon. The carbide precipitated is mainly $Cr_{23}C_6$, so that a single atom ties up almost four chromium atoms.

4.7.5 EFFECT OF VARIOUS FACTORS ON THE SENSITIZATION OF STEELS

A. Effect of Carbon

Carbon plays a very crucial role in the precipitation of carbides and a decrease of carbon content of steels restricts the formation of harmful carbides, such as chromium carbide, which is primarily responsible for the depletion of chromium adjacent to the grain boundaries. The effect of carbon is illustrated in Fig. 4.27. The ordinate is the temperature, and the abscissa the time of annealing (slow cooling in the furnace). Two types of steels are shown, one steel containing

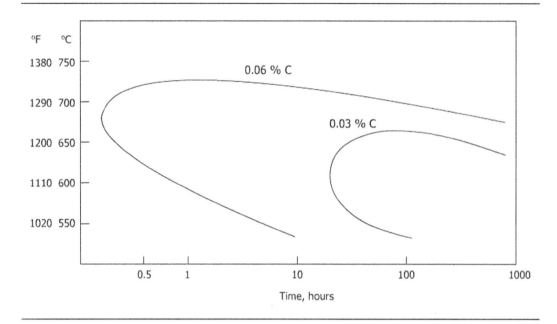

Figure 4.27 Comparison of time–temperature sensitization curves for 0.06 and 0.03 carbon austenitic stainless steel. (From Sandvik, Steel works)

0.06% C and another steel containing 0.03% C. It can be observed that above 725°C and below 400°C, these steels do not become sensitized. Steel containing 0.06% C becomes sensitized at 600°C in less than an hour, whereas the steel containing 0.03% C takes about 25 h to become sensitized. The above observations show the effect of time and temperature on the precipitation of carbides.

B. Effect of Alloying Elements

(1) **Nickel**. The susceptibility to sensitization decreases with increased nickel content because the carbon solubility decreases with increased nickel content.
(2) **Molybdenum**. The susceptibility of steels to intergranular attack decreases with increased Mo content. The Mo also increases the resistance to pitting and crevice corrosion.
(3) **Chromium**. It generally increases the resistance to sensitization.
(4) **Nitrogen**. The effect of nitrogen depends upon the amount contained in a steel. Nitrogen is an austenite stabilizer.

Steels containing 18% chromium, 9–12% Ni and 0.02–0.03% C which are sensitized at 500°C for 100 h showed a maximum resistance sensitization at 0.04% nitrogen.

(5) **Titanium and Niobium**. These are added to give stabilization against the harmful precipitates, such as chromium carbides. These elements combine with carbon and nitrogen and do not allow carbon and nitrogen to precipitate as carbides and nitrides which lead to intergranular corrosion. The optimum temperature for formation of TiC (titanium carbide) is 900–950°C.
(6) **Silicon**. It is added to increase corrosion resistance in sulfuric acid.
(7) **Sulfur and Selenium**. These are added to increase machinability, however, the corrosion resistance is decreased.

4.7.6 CARBIDES

Carbides play a very important role in intergranular attack. Intergranular attack does not occur without the formation of chromium carbides. The carbides are present in dendritic shapes. All the carbon in the alloy does not precipitate carbide as exemplified by $Cr_{23}C_6$ in which only half of the amount of carbon present in the steel is tied up as carbide.

4.7.7 SIMPLIFIED MECHANISM OF INTERGRANULAR CORROSION

Various theories and models have been proposed to account for the intergranular corrosion of steels. These are summarized below. For in-depth understanding of the theories, references are cited at the end of the chapter. The theories are reviewed here to provide a basic understanding of the mechanism of intergranular corrosion to engineers.

The following three steps lead to the intergranular corrosion of austenitic steels:

(1) Sensitization in the temperature range 550–900°C.
(2) Diffusion of carbon to a grain boundary and formation of carbide. This leads to depletion of chromium content along the grain boundary.
(3) Weakening of the grain boundary and disintegration of the grains.

4.7.7.1 Preferred Site

Preferential attack at the grain boundaries occurs because of the higher free energy of the grain boundary region. This is enhanced by the segregation of grain boundary impurities and by the precipitation of second phase particles which may be noble and lower the resistance of the matrix. The grain boundaries are the preferred location of localized attack compared to the matrix.

4.7.7.2 Conditions Leading to Attack

A majority of intergranular attacks on stainless steels can be attributed to the metallurgical changes which occur in the temperature range of 427–816°C due to slow cooling, or to prolonged welding of steel. The steels become sensitive to intergranular corrosion in this temperature range (sensitization temperature range) because of the grain boundary precipitation of chromium carbides ($Cr_{23}C_6$).

4.7.7.3 The Attack

Carbon has a strong tendency to form carbides with chromium present in the steel ($Cr_{23}C_6$) in the sensitizing temperature zone during slow cooling from annealing treatment, stress relieving in the sensitizing temperature zone or during fabrication procedures involving prolonged welding and flame cutting. Hence, chromium carbide is precipitated at the grain boundary and the zone next to the grain boundary becomes depleted of chromium, thus making this zone anodic to the remainder of the surface. In austenitic steels, a minimum of 12% chromium is required to resist corrosion, however, depletion of chromium below this level causes weakening of the grain boundary and may lead ultimately to dislodging of grains. As localized attack progresses through the grains, the attack is more often called '*intergranular attack.*' It must be noted that it is not the chromium carbide which is attacked but a zone adjacent to the grain boundary which is depleted of chromium and is attacked. The term '*weld decay*' is of historic origin when the mechanism of intergranular corrosion was not understood. The welders called the localized corrosion attack adjacent to the welded zone [heat affected zone (HAZ)], as weld decay. The heat affected zone (a narrow zone) becomes susceptible to intergranular attack because it reached the sensitization temperature range during welding (400–900°C), whereas the temperature of the weld region is much higher. Hence, the region adjacent to the weld fails by intergranular corrosion, commonly called '*weld decay.*' The rest of the base metal remains unaffected.

4.7.7.4 Effect of Time and Temperature

Time and temperature effects are important in welding. For instance, electric arc welding produces intense heating in a shorter time compared to gas welding. The metal will be in a sensitizing zone for a longer time if gas welding is used, hence, a greater amount of carbide would precipitate. Thus, steel which has been welded by gas flame would be more sensitive to attack by intergranular corrosion. It was observed from a time–temperature sensitization diagram (Fig. 4.27) that a stainless steel (SS 304) with 0.05% carbon content remained in the sensitizing zone for 10 s. The minimum time required to cause susceptibility to weld decay was 40 s. Therefore, theoretically, this alloy should not have undergone weld decay (intergranular attack). This alloy was, however, subjected to a rapid weld decay attack in nitric acid in the heat affected zone, contrary to the theoretical assumptions. To explain the above contradiction, it was suggested by Tedmon [35] that the nucleation of chromium carbide may initiate at a higher temperature, and the growth of carbide may take place at a lower temperature. This explains why weld decay may occur even after an exposure period of 10 s. The time required for sensitization includes the period of initiation as well as the period of growth and propagation.

A substantial loss of mechanical strength accompanied by a loss in corrosion resistance occur. Figure 4.28 shows the chromium carbide, grain boundary and low chromium areas. The grains of austenitic steels contain 18% chromium. Figure 4.29a shows the chromium depleted zone

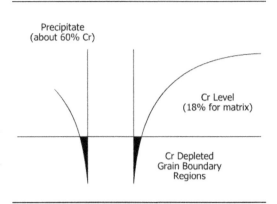

Precipitate
(about 60% Cr)

Cr Level
(18% for matrix)

Cr Depleted
Grain Boundary
Regions

Figure 4.29a Cr depleted zone around a grain boundary precipitate of secondary carbide

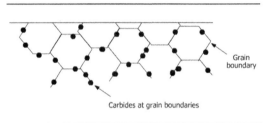

Grain boundary

Carbides at grain boundaries

Figure 4.29b Sketch showing the occurrence of carbides at grain boundaries

around a grain boundary precipitate of a secondary carbide. Figure 4.29b shows carbides at grain boundaries due to Cr depletion.

4.7.7.5 Method of Prevention

The following are the methods of prevention of austenitic nickel chromium stainless steels from intergranular corrosion:

(a) Purchase and use stainless steel in the annealed condition in which there is no harmful precipitate. This only applies when the steel is not to be exposed to the sensitizing temperature.

(b) Select low carbon grade steel with a maximum of 0.03% C, such as 304 L. This would prevent the formation of harmful chromium carbide during fabrication.

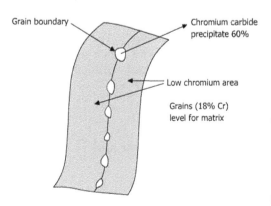

Grain boundary

Chromium carbide
precipitate 60%

Low chromium area

Grains (18% Cr)
level for matrix

Figure 4.28 Grain boundary in sensitized type 304 stainless steel

(c) Select stabilized grades of stainless steels, such as type 347 (niobium stabilized) or steel type 321 (titanium stabilized). Stabilized grades of steels do not allow the depletion of chromium to occur adjacent to the grain boundary, as carbon has a greater affinity to combine with either titanium or niobium than with chromium, hence chromium carbide, the harmful precipitate is not formed. It is possible to reclaim steel which suffers intergranular corrosion by heating above 1000°C followed by water quenching.

(d) Solution treatment at 1121°C re-dissolves the carbides into the matrix, followed by rapid quenching to prevent re-precipitation of the carbides.

(e) Cold working of the steel prior to sensitization. The application of cold work causes an increase in the precipitation of chromium carbide on dislocations, and reduces the amount of grain boundary carbide.

(f) Modification of analysis of steel to produce delta-ferrite. The precipitation of carbides at the delta-ferrite/austenite interface reduces the severity of grain boundary precipitation.

The success of the remedial measures would depend upon the selection of materials, design and cost. The second remedy can be rejected because it may not be practical for large welded structures. There are, therefore, two options: use of low carbon steels or ferritic steels. The main disadvantage of low carbon steels is their lower strength. Remedial measure (C) is also not practical because the stabilized grades suffer from another localized form of corrosion called 'knife-line attack,' discussed in Section 4.7.9 below. Ferritic steels may be used with advantage. The ferritic steels may become sensitive to weld decay while cooling from 900°C, hence, they can be conveniently annealed by a torch in the temperature range 600–800°C to avoid sensitization. There are, however, other limitations, such as hydrogen embrittlement (diffusion of hydrogen in the metal and loss of mechanical strength), which may restrict the use of ferritic steels. The fifth remedy could be applied only in the context of the mechanical properties.

4.7.8 THE CHROMIUM DEPLETION THEORY

The above theory was developed by Bain, *et al* [34]. According to the assumptions made in this theory a passive film is not formed on the surface of the material, if the chromium content is less than 12%. It means that if the amount of chromium in the austenitic steel is less than 12%, it would be subjected to localized corrosion. As a result of sensitization between the temperature range of 550–900°C, carbon diffuses towards the grain boundaries and forms chromium carbide, which is insoluble and precipitates out of solid solution if the carbon content is higher than 0.02%. The bulk alloy contains only 18% chromium. However, on analyzing carbides it has been observed that the chromium content in carbide can be as high as 94% which suggests that chromium is depleted from the areas of the alloy matrix immediately adjacent to the carbides (Fig. 4.30). The corrosion resistance of the steel which is primarily due to chromium (optimum concentration for corrosion resistance is 12%) is, therefore, lowered in the region adjacent to the grain boundary.

When steel is in the temperature range (500–900°C), the diffusion rate of chromium from the matrix to the depleted area in the grain boundary is too slow to replenish the amount of chromium depleted. The rate of recovery of austenitic steels from intergranular corrosion, however, is greater at a higher temperature than at a lower temperature. As chromium is depleted from an area adjacent to grain boundary, it becomes mechanically weaker and is a preferred site for corrosion attack. Grain boundary attack is used synonymously with intergranular corrosion.

A. Positive Points of the Theory

According to the above theory, the susceptibility of the steels to intergranular corrosion should decrease with:

(a) decreased carbon content,
(b) increased chromium content.

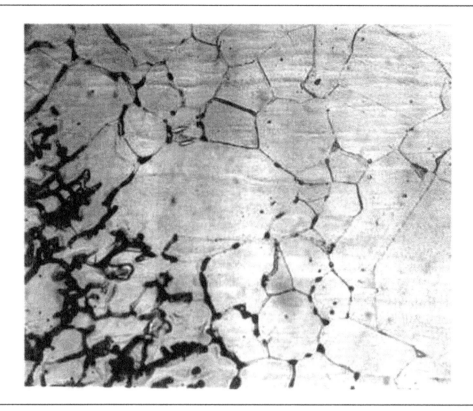

Figure 4.30   Grain boundaries in 316 L stainless steel etched in acidified $CuCl_2$ (x 500)

Both these facts have been found to be true and verified by several investigators.

(c) The existence of chromium depleted regions has been experimentally verified electrochemically and by microanalysis using electron microprobes.

Proof

Sensitized steels were maintained at potentials between 0 and +350 mV in 2N H_2SO_4 and the products of dissolution were analyzed. On analysis of the products, it was found that the chromium content decreased from 18.6% at 0 to 13.7% at +350 mV. It has been shown by different tests on steels that a decrease in chromium content is accompanied by an increase in passive current density (the density required to achieve passivation). Intergranular corrosion was

observed to take place at +350 mV; hence it was concluded that chromium depletion precedes intergranular corrosion.

B. Quantitative Model

A quantitative model was developed by Strawstrom, Hillert and Tedmon *et al.* [35,36]. This further strengthened the chromium depletion theory. Figure 4.31 shows a comparison of $Cr-C-Cr_{23}C_6$ chromium carbide equilibrium data for 304 stainless steel with experimentally obtained corrosion data for several temperatures. The bulk carbon concentration is shown on the abscissa. The chromium content in equilibrium with $Cr_{23}C_6$ in the steel is shown as the ordinate.

If 12% chromium is assumed to be the minimum quantity to form a passive film, it is observed

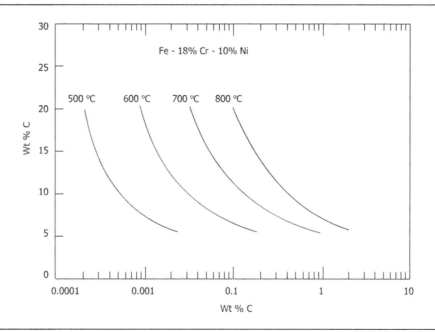

Figure 4.31   Comparison of Cr–C–Cr$_{23}$C$_6$ equilibrium data for type 304 stainless steel with data for several temperatures

that if the alloy contains more than 0.15% carbon it would become sensitized at 800°C. For the alloy annealed at 700°C, the carbon content from the plot is determined to be 0.08 and 0.02% at 600°C. If the carbon contents exceed the levels indicated above, sensitization is likely to occur. It is not possible to sensitize 304 stainless steel at 500°C, even after a prolonged exposure period, because carbide does not precipitate. It is, however, possible to sensitize the samples below 600°C, if the alloy has been equilibrated at a high temperature where carbide first precipitated. A sample of type 304 stainless steel containing 0.075% carbon was annealed at 700°C for one hour and quenched in water. When this sample was put in a test solution of boiling acid copper sulfate, no intergranular attack was observed although the sample contained precipitates of chromium carbides. The answer is simple. The chromium content was below the required limit of 12–15%, to cause intergranular corrosion. This proves the point made earlier in the chromium depletion theory that an optimum concentration of 12–15% of chromium is necessary to cause intergranular attack which is now proved.

The same sample was slowly annealed for several hours at 500°C and water quenched. This treatment caused an increase in chromium carbide content. This sample was exposed again in the acid copper sulfate solution and rapid grain boundary attack occurred. The carbide was already precipitated on exposing the sample to 700°C as shown above, hence, the Cr–C–Cr$_{23}$C$_6$ equilibrium is rapidly attained even at 550°C. The amount of maximum allowed carbon at 550°C was determined to be 0.01%. As this amount of carbon contained in the alloy exceeds 0.075%, intergranular attack was observed. Thus, evidence is provided in favor of chromium depletion theory by the above work. Why would sensitized stainless steel held at sensitizing temperature for a very long period of time not be susceptible to intergranular corrosion? Because sufficient time is provided for the diffusion of chromium from the grain matrix to the depleted region. Of course, thousands of hours may be required to achieve this. Strawstrom and Hillert [36] obtained good agreement between the theoretical and experimental results. The chromium depletion model has been widely accepted.

C. Objections Against the Chromium Depletion Theory

(1) Lack of direct proof of the impoverished areas. With the rapid progress in electron microprobe analysis, the objection has been largely resolved. The difficulty in direct detection is due to the narrowness of the depleted area in the atomic dimensions, which can be up to $2\,\mu m$. The strongest evidence in favor of chromium depletion has been provided by anodic polarization studies mentioned in the text to follow.

(2) After sensitizing a 302 stainless steel containing 0.10% C for 150 h at 650°C, only 45% of the carbon precipitates as carbide and recovery from the intergranular attack is possible. The above observation, if correct, means that depletion does not occur as suggested, as only 45% of the carbon is precipitated. Also, the process of self-healing should not occur while the carbon particles are still growing. The above objections have been largely resolved. According to Baumel [37], the precipitation of carbide at the grain boundary stops at every stages and the process of healing can start while the precipitation of carbon occurs in the interior of the austenitic grains. Recovery from intergranular attack is, therefore, possible after sensitizing the alloy for 150 h at 650°C. Thus, the objection raised by Stickler and Vinckier [38] is not valid.

Also, the theoretically calculated annealing time for developing a chromium depleted zone is in good agreement with the experimental annealing time.

4.7.9 STRAIN THEORY [39]

On the basis of chromium depletion theory, a uniform attack occurs on both sides of the grain boundary carbide. Therefore, the observation that the knife-line attack (knife-line attack occurs in a very narrow band in the parent metal immediately adjacent to the weld), occurs only on one side of the carbide, austenite interface cannot be explained by the chromium depletion theory. The knife-line attack arises as a result of strains at the interface. The strains arise from a distorted lattice adjacent to the carbide precipitate.

It has been suggested that the chromium carbide $(Cr_{23}C_6)$ formed at the grain boundary is only in registry with one grain and not with other grains. The magnitude of the knife-line attack is found proportional to the magnitude of mis-orientation at the carbide austenite interface.

It was found by Streicher [40] that the rate of grain boundary attack on a particular grain was dependent on the orientation of that grain. It was confirmed also by other workers that the density of carbide precipitated was dependent on the mis-orientation between the grains. An explanation of the healing effect was offered by this theory. It was postulated that the knife-line attack does not occur above 800°C because the total strain reduces due to the precipitation carbides at this temperature. The above explanation, however, was not very convincing. It was through the work of Stickler and Vinckier [38] that sufficient evidence was provided in support of the strain theory. It was shown that in intergranular corrosion, healing only occurred in the absence of a continuous grain boundary film of carbide and the growth of carbide particles was essential for the process of healing to occur. The knife-line attack does not propagate in the absence of a continuous grain boundary film even if the grains are not in registry, because of the healing effect. It was pointed out earlier that the magnitude of the intergranular corrosion depends on the de-registry of the grain accordingly to the strain theory. Thus, the latter finding gave a reasonable explanation of the healing effect. The later observations also support the chromium depletion theory.

4.7.10 THE ELECTROCHEMICAL THEORY

It was shown that a piece of $M_{23}C_6$ (metallic carbide), when attached to 18-8 steel (annealed) and immersed in nitric acid, developed intergranular attack. From the above experiment, it was concluded that a potential difference existed between the metallic carbide and stainless steel. It was shown that $M_{23}C_6$ was more noble than the steel matrix. It was further suggested that the intergranular attack was accelerated in the presence of residual stresses. The evidence described above shows that:

(1) A corrosion cell is formed between an anode and cathode. The metallic carbides being more noble act as cathode and the grain matrix surface as anode. It was suggested that whether the carbide is noble or the steel (boiling 65% HNO_3) would depend on the electrolyte used. The Huey solution attacked $M_{23}C_6$, whereas the Streicher solution attacked the grain matrix.
(2) An emf is set up between the metallic carbide and the grain surface in the presence of an electrolyte and intergranular attack takes place.

Objections

(a) The electrochemical theory was questioned by Baumel, *et al.* [37]. They suggested that localized corrosion could not be confined to a very narrow zone ($2 \mu m$) and the corrosion should extend to the matrix.
(b) The potential measurement of 18-9 steel, platinum, $M_{23}C_6$, and copper in Strauss solution (boiling 16% H_2SO_4 + 5.7 % $CuSO_4$ + Cu) showed that the potentials of the four materials is nearly the same within an accuracy of $\pm 1\%$. This work was in contradiction of the results obtained on the effect of electrolyte on potential of $M_{23}C_6$ and austenite.

From the above description, it is clear that the electrochemical theory [41] is controversial and it does not present a better picture of intergranular corrosion than predicted by the chromium depletion theory.

4.7.11 SOLUTE SEGREGATION THEORY [42–45]

This theory was developed by Aust [43] and suggested the presence of the following as the primary cause for the intergranular corrosion of non-sensitized austenitic steels:

(1) Continuous grain boundary path of a second phase.

(2) Soluble impurity segregates resulting from solute vacancy interaction. The studies were conducted on annealed material.

This model is concerned primarily with intergranular attack on non-sensitized steel and only secondarily with carbide forming sensitized steel. According to the theory, the resistance to intergranular attack is improved if discontinuous carbides are precipitated. This can be achieved by heating between 800 and 900°C followed by water quenching. By this treatment, chromium carbides would be precipitated in a discontinuous form and phosphorus and silicon would be precipitated. Although the dissolution rate of carbides is increased, the portion of the material in between the carbides would be low in phosphorous and silicon, hence the rate of corrosion would be decreased. This improvement is attributed to the incorporation of a major part of the segregating solute into the carbide phase. It is a general opinion that grain boundary segregation of elements, such as silicon and phosphorus, is the cause of intergranular corrosion but the exact mechanism is not known. It has been further suggested that the driving force for the attack is provided by the potential difference which exists between the grain matrix and the grain boundary. Experiments on hardness measurements showed the boundaries to be harder due to the presence of solute segregates. Intergranular attack was observed on alloys in a highly oxidizing solution, like HNO_3 + K_2CrO_4.

The evidence that intergranular attack occurs on non-sensitized steels in nitric acid containing oxidizing ions, like Cr^{+3}, is strong evidence in favor of grain boundary segregation of impurity solute. The steel does not remain passive in the oxidizing solution of nitric acid and potassium dichromate as the protective film possibly breaks down. It would be difficult to otherwise explain why intergranular attack is observed on an alloy which is not subjected to sensitization (formation of metallic carbide in the range of 550–850°C). Grain boundary segregation may increase the electronic conductivity of the oxide, at the boundary. It increases the attack because Cr^{+3} is oxidized to Cr^{+6} which is soluble, whereas Cr^{+3} is not soluble. A film of oxide which is not soluble will prevent intergranular attack.

The attack takes place at the grain boundaries because they are high energy regions.

The de-adsorption of impurities from the grain boundaries at elevated temperatures above 1100°C causes a reduction of the rate of intergranular attack which is in conformity with the solute segregation theory.

Objections

(1) It does not seriously consider the importance of oxidizing power of the corroding environment. In nitric acid solution, the insoluble Cr^{+3} ions can be transformed to soluble Cr^{+6} state. The stainless steel does not remain passive in this solution and general corrosion along with localized intergranular attack is likely to take place. Under such conditions, the presence of solute segregate is not, therefore, necessary. The above observation, therefore, contradicts the solute segregation theory.

(2) The postulate states that the presence of a continuous second phase leads to intergranular attack, does not seem to be valid because of the observation that steel 316 in which sigma phase precipitated at the grain boundary showed no acceleration of intergranular corrosion until the solution was made highly oxidizing.

Generally speaking, the solute segregation theory is only valid for non-sensitized steel and attempts to extend it to the sensitized steels have not been successful because the tests were carried out in highly oxidizing solution in which general corrosion as well as intergranular takes place as discussed above.

4.7.12 SENSITIZATION OF HIGH NICKEL ALLOYS

The nickel base alloys are typically single-phase, multi-component and contain iron and chromium. Carbon is present in the range of 0.03–0.08 wt%. From several investigations, it is clear that chromium increases the corrosion resistance of nickel base alloys by virtue of formation of a passive film containing chromium. Inconel 825 becomes susceptible to intergranular corrosion between 650 and 750°C in the Huey test. The reason of sensitization is stated to be the depletion of chromium at the grain boundary because of the formation of $M_{23}C_6$. It was further shown that nickel base alloys, such as alloys containing 38–46% nickel, 19–25% chromium, 2–3% molybdenum, 1–3% copper, 0–5% C, 0.6–1.2% titanium and balance iron, could be sensitized in 5 h at 680°C. By the above treatment, titanium carbide was precipitated instead of chromium carbide and the chromium distribution was not affected. The alloys, therefore, showed a good resistance to intergranular attack because of the addition of titanium.

Experimental work related to chromium–carbide–carbon equilibrium on a nickel base alloy (15% chromium, 10% iron, balance Ni), showed that the carbon solubility in nickel base alloys is lesser than in austenitic steels. Nickel-based alloys containing chromium can be sensitized in the temperature range 500–700°C.

In high nickel alloys containing molybdenum, such as 15% Cr, 15% Mo, 4% W, 5% Fe, 0.06% C and balance Ni, intergranular attack is generally attributed to the depletion of molybdenum and chromium, because of the formation of M_6C (molybdenum carbides and chromium carbides ($M_{23}C_6$).

Sufficient evidence is not available to present a more acceptable mechanism of intergranular corrosion of nickel base alloys.

4.7.13 SENSITIZATION OF FERRITIC STAINLESS STEELS

Sensitization in ferritic stainless steels is introduced by high temperature heat treatment (above 925°C) and relieved by heating for a short time between 650 and 815°C, which is opposite to the observations on austenitic steels. Annealing sensitized ferritic steel at 788°C for several minutes will eliminate intergranular attack. The susceptibility is reduced by the addition of titanium or niobium. The presence of carbon or nitrogen is necessary to cause sensitization.

It was postulated by Houdermont and Tofante [46,47] that austenite is formed at the grain boundaries during heating in the sensitization range because of the higher solubility of carbon in austenite than in ferrite. At 927°C, there is a transfer of carbon from ferrite to austenite, and ferrite is enriched in chromium. The grain boundary austenite becomes rich in carbon. During cooling, austenite is decomposed to give chromium carbide, and these carbides are responsible for the intergranular attack.

In another theory, Bond [48] postulated that:

(1) The ferritic steels become immune to intergranular corrosion if the carbon or nitrogen content is made very low.
(2) Intergranular corrosion in alloys containing a very high amount of carbon and nitrogen becomes less severe than observed in alloys containing a smaller amount.

He demonstrated that on increasing the carbon content of an alloy (containing 17% Cr and 0.0031% C) to 0.012% C, and raising the nitrogen content from 0.0095 to 0.022%, intergranular corrosion was observed.

If the carbon or nitrogen content is increased to sufficiently high values as suggested [49], the austenite phase is extended; hence, when the alloy is cooled martensite precipitates. As martensite contains most of the carbon and nitrogen in solution, the precipitation of carbides (chromium-rich) and nitrides is reduced. The magnitude of sensitization would be consequently reduced in alloys containing sufficiently high amounts of these elements compared to alloys containing lower amounts of these elements.

The above model is based on the chromium depletion theory described earlier. The mechanism proposed by Hodges [50] is based on the following observations on high purity ferritic steels.

(1) Alloys containing 17–26% chromium, C = 0.001–0.002%, N = 0.004–0.009%, when water quenched from 1000°C, were not found to be sensitive to intergranular attack. On the contrary, the same sample when air cooled, showed sensitivity to. intergranular attack.
(2) He further observed that alloys containing molybdenum were more sensitive to

intergranular attack than molybdenum-free ferritic steels when furnace cooled.

These results were explained in terms of chromium depletion theory. Because the alloys contained small amounts of carbon and nitrogen, diffusion occurred over long distances to the grain boundary to form chromium carbides. In conventional low purity alloys, the distances involved for diffusion are smaller. Thus, when the alloy is water-quenched sufficient time is not allowed for diffusion of either carbon or nitrogen to form carbides or nitrides; hence the intergranular attack does not occur under this condition. The immunity from intergranular attack, therefore, depends on how much time is available for carbon to diffuse to the grain boundaries which depends on the type of heat treatment, i.e. annealing, normalizing or quenching. Thermal treatment which produces susceptibility in ferritic alloys are those which are used to minimize it in the austenitic range. The following table (Table 4.5) illustrates the point. It can be observed that heat treatments beneficial for austenitic steel 304 is harmful for ferritic 430, and the heat treatment beneficial for ferritic (430) is harmful for austenitic 304.

4.7.14 SENSITIZATION OF DUPLEX STEELS

Steels which contain both the body-centered cubic alpha-phase and the face-centered cubic gamma-phase, are called 'Duplex Steels.' Austenitic ferritic steels containing about 5% or more delta-ferrite show a higher resistance than

Table 4.5 Heat treatment of ferritic stainless steels

Heat treatment	Corrosion rate	
	Ferritic 430	Austenitic 304
1300°F, water-quenched (1 h)	40	400
2000°F, water-quenched (1 h)	400	10

the austenitic alloys only with the same carbon content, because the islands of ferrite minimize the formation of continuous grain boundary precipitate in austenitic grain boundaries. They can be either low carbon low chromium (0.03% C, <20% Cr) or higher carbon high chromium alloys (C > 0.03%, Cr > 20%). The low carbon, low chromium alloys are typified by alloys, such as 308 L; the steel has a matrix of austenite and the amount of ferrite is between 5 and 10%. The other class is typified by alloys, such as AISI 326, which has a ferrite matrix and contains 40–50% austenite.

Chromium-rich duplex alloys, such as 326 and Uranus, are highly resistant to intergranular attack because:

(a) the alloys offer more resistance to intergranular corrosion by virtue of their higher chromium contents which cause the formation of a well adherent and stable chromium-rich oxide film,

(b) a better resistance to sensitization is offered because one particular heat treatment affects only one phase; a heat treatment which sensitizes the ferrite phase has no adverse effect on the austenite phase.

Alloys, such as type 329 (Carpenter 7 Mo), is used in the form of annealed heat exchanger tubing. The other important alloys developed is Sandvik 3 RE60 which is in the form of annealed tubing.

Stabilizing elements are added to duplex steels to improve intergranular resistance. These alloys are quite resistant, if a network of continuous carbides is not formed.

4.7.15 KNIFE-LINE ATTACK

This phenomenon is restricted to the stabilized grades of steel, such as 321 and 347. The knife-line attack occurs immediately adjacent to the weld and shows as a thin line of intergranular corrosion. It results from intergranular corrosion like the weld decay. It may be noted that weld decay develops at some distance away from the weld. The following is the mechanism suggested for the knife-line attack of stabilized steels.

During the process of welding, the metal at the weld pool and the base metal interface is held at a temperature above that required for the precipitation of either niobium carbide or titanium carbide (1220°C). During the process of cooling, the metal at the weld pool-metal interface passes through the temperature range at which the precipitation of carbide of niobium or titanium occurs. At room temperature, the heat affected area becomes rich in carbon and on reheating this area (550–780°C), the chromium carbide precipitates faster than the niobium or titanium carbide, which accounts for the knife-line attack. The precipitation of carbide leaves a narrow band susceptible to intergranular corrosion adjacent to the fusion line.

It is also been suggested that eutectic carbides of niobium and titanium are formed at the grain boundary during cooling (600–900°C). The carbides dissolve and cause intergranular corrosion. One good method of prevention against knife-line attack is to heat the stabilized steel after welding to 1061.5°C followed by cooling to 891°C.

4.7.16 INTERGRANULAR ATTACK WITHOUT SENSITIZATION: EFFECTS OF ENVIRONMENT

It has been observed that even those steels which are not susceptible to intergranular attack undergo intergranular attack in the presence of strong oxidizing agents, like CrO_4^{2-}, MnO_4^-, Fe^{3+}, Ce^{4+}, V^{5+}, and high temperature water containing $FeCl_2$, $CuCl_2$ or KOH. It has been shown already that intergranular attack can be avoided by adopting the following means:

(a) Reducing carbon content to 0.2%.
(b) Solution treatment.
(c) Addition of stabilizers, like niobium and titanium.

In spite of these measures, commercial iron–chromium–nickel alloys are found susceptible to intergranular attack under highly oxidizing condition. What could be the reason? Ions, like Cr^{+6}, depolarize the cathodic reaction and increase the rate of anodic dissolution. The grain boundaries being regions of high energy are preferentially attacked. It is also likely that the elements segregated at grain boundaries induce a driving force for corrosion because of a difference of potential between the grain boundary and the matrix. Armaijo [51,52] who showed that the elements responsible for the above behavior were phosphorus and silicon.

The attack on non-sensitized alloys can be attributed to the following factors:

(1) Surface heterogeneity of grain boundaries.
(2) Impurity segregation at the grain boundaries. This point was proved by hardness measurement in which alloys containing phosphorous showed increased hardness due to grain boundary segregation. No grain boundary hardening effect is observed in alloys of high purity.
(3) Increase in the dissolution rate in the transpassive region due to high chromium contents and decrease due to low chromium content [53].
(4) Acceleration of dissolution by lower silicon concentration and retarding of dissolution by higher silicon concentration [54].
(5) The change of character of Cr^{+3} from being protective to being soluble Cr^{+6} at more noble potentials. The impurities segregates affect the flow of electrons and the rate of corrosion. Dissolution would depend on how conductive are the impurities at the grain boundaries.

It is not very clear which of the above factors contributes more to the acceleration of intergranular attack on non-sensitized steels. It is clear that the rate of intergranular attack is affected by all factors stated above, however, the effect of each factor cannot be quantitatively ascertained.

4.7.17 FAILURE CASE HISTORIES

A. Case 1

Failure of pipeline system for the feed of a swimming pool with seawater [55].

Material   Welded 18-8 stainless steel pipe.

Environment   Seawater.

Analysis of failure   In order to analyze the failure, test pieces of the pipes were prepared and subjected to metallographic examination. The metallographic examination showed macroscopic welding defects in the transveric welding seam. The welding did not penetrate the transversal joints of stainless steel pipes. Corrosion was observed adjacent to the longitudinal weld. Deep cavities were observed on the transverse section of the weld about 3 mm from the weld metal corresponding to the external points of the attack. A detailed microscopic examination of the hole showed intergranular attack and grain boundary carbides (Fig. 4.32). Further optical studies showed weld edge misalignment. It was also observed that the weld metal did not penetrate sufficiently at some points of the seam.

Prevention

(1) Selection of stabilized steels for the piping systems, such as AISI 321 and 347.
(2) Eliminating welding defects as far as possible by adopting proper technique of welding.
(3) Using ferritic stainless steels, wherever possible.

B. Case 2

Failure of Type 304 stainless steel fused salt pot [55].

Material   Stainless steel type 304.

Environment   Molten eutectic mixture of sodium, potassium and lithium chloride at melt temperatures ranging from 500 to 600°C. The pot used was a weld cylinder with 3 mm thick steel walls and 305 mm in height.

Analysis of problem   Six specimens were prepared for evaluation and examination. All specimens exhibited carbide precipitation in the grain boundaries. Specimens taken about 100 mm from the melt level showed no attack as these regions were not exposed to the sensitizing temperature. Similarly, specimens taken about 45 mm

Figure 4.32 Inter-crystalline attack

from the melt also showed no evidence of inter-granular cracking as in these regions also the sensitization temperature was not reached. Specimens taken from the regions which were in direct contact with the fused salt and were exposed to temperature of 500–650 °C and showed intense grain boundary attack. Welding followed by exposure to a temperature in the sensitizing range caused intense intergranular attack.

Prevention

Stabilized grade steel

Stainless steel 347, stabilized grade, could be used with advantage. After welding, the steel should be stress relieved to about 900°C for 2 h and rapidly cooled in order to minimize the stresses. Also, Hastelloy (70 Ni–17 Mo–7 Cr–5 Fe) could be used with advantage. It should, however, be solution treated to 1120°C and cooled rapidly.

Ferritic steels

(1) The most important method is to control the amount of interstitial carbon and nitrogen to less than 50 ppm and 150 ppm, respectively. Lowering the carbon to 0.03% does not appreciably help, because of the lower solubility of carbon in ferrite. It has been shown for 18 Cr–2 Mo alloys that the maximum level of C + N is 60–80 ppm, and for 26 Cr–1 Mo steel, the level required is 150 ppm. Use ferritic steels in annealed conditions (788°C).

(2) Addition of stabilizing elements. Both titanium and niobium can be used as stabilizing elements. Both elements minimize intergranular attack of ferrite steels, hence, titanium stabilized alloys are now recommended for applications in nitric acid environments. The amount of stabilizing elements, titanium or niobium, required for immunity to intergranular corrosion by $CuSO_4$–16% H_2SO_4,

a test for 18 Cr–2 Mo alloys, is given by $Ti + Nb = 0.2 + 4 (C + N)$. These limits are valid for carbon contents ranging from 0.02 to 0.05%.

Duplex stainless steels

(1) Avoid the formation of harmful phases by rapid cooling through 900–700°C. These phases, however, affect the mechanical properties mainly and do not significantly affect intergranular corrosion.
(2) Keep the carbon lower than 0.03%. As the duplex stainless steels are not as widely used as other grades of steels, sufficient information on preventive methods is not available.

4.7.18 Test of Intergranular Corrosion

The following tests are used to check the susceptibility of steels to stress corrosion cracking:

(1) **Huey test**. It is the nitric acid test described in ASTM A262-68 practice.
(2) **Acid ferric sulfate test**. It is described in ASTM designation A262-68.
(3) **Strauss test**. It consists of exposing a sample to a boiling solution of 6 wt% $CuSO_4$ + 16 wt% H_2SO_4 for a period of 72 h.
(4) **Streicher test**. It consists of electrolytically etching in a polished sample (3% grid paper) in a room temperature solution consisting of 100 g of $H_2C_2O_4$, $2H_2O$ dissolved in 900 ml of distilled water. The current required is 0.1 A/cm^2 for 15 min. It is described as ASTM designation A262-68 Practice A.

4.7.19 Standard Test Methods

■ ASTM A262 – practices for detecting susceptibility to intergranular attack in austenitic stainless steel.
■ ASTM G28 – test methods for detecting susceptibility to intergranular attack

in wrought, nickel-rich, chromium-bearing alloys.
■ ASTM G34 – test method for exfoliation corrosion susceptibility in 2xxx and 7xxx series aluminum alloys (EXCO Test).
■ ASTM G66 – test method for visual assessment of exfoliation corrosion susceptibility of 5xxx series aluminum alloys (Asset Test).
■ ASTM G67 – test method for determining the susceptibility to intergranular corrosion of 5xxx series aluminum alloys by mass loss after exposure to nitric acid (Namlt Test).

4.8 Stress Corrosion Cracking and Hydrogen Damage

4.8.1 Introduction

Stress corrosion is the failure of a metal resulting from the conjoint action of stress and chemical attack. It is a phenomenon associated with a combination of static tensile stress, environment and in some systems, a metallurgical condition which leads to component failure due to the initiation and propagation of a high aspect ratio crack. It is characterized by fine cracks which lead to failure of components are potentially the structure concerned. Stress corrosion cracking is abbreviated as SCC. The failures are more often sudden and unpredictable which may occur after as little as few months or years of previously satisfactory service.

4.8.2 Requirements for SCC

The following conditions are necessary for SCC to occur:

(1) A susceptible metal.
(2) A specific environment.
(3) A tensile or residual stress.

Figure 4.33 shows a stress corrosion triangle.

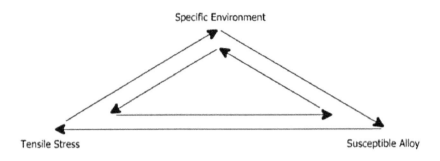

Figure 4.33 The stress corrosion triangle

4.8.3 HISTORICAL

The development of SCC can be historically divided into three periods; the first period between 1930s and 1950s, being the identification period, the second period between 1960s and 1970s, being the mechanistic period in which important basic mechanisms of SCC were explained and various alloys were evaluated for application in selected environments and the third period being the application and development period of the eighties.

4.8.4 APPLICATION OF FRACTURE MECHANICS

One important characteristic of fracture in metals is the occurrence of plastic deformation at the tip of the crack. The fracture toughness is proportional to the energy consumed in plastic deformation, which is difficult to measure accurately.

The fracture toughness of most materials is determined by a parameter called 'stress intensity factor,' K_i. It is a measure of the concentration of stresses at the tip of the crack. It is given by

$$K_i = y\sigma_y\sqrt{\pi a} \text{ (units : MPa}\sqrt{\text{m})} \quad (4.45)$$

where

$\sigma_y =$ nominal applied stress
$a =$ crack length
$y =$ correction factor.

Failure occurs when the stress intensity factor reaches a critical value, K_{1c}. This critical value is called '*fracture toughness.*' The relation between fracture stress to fracture toughness is given by:

$$\sigma_f = \frac{K_{1c}}{y\sqrt{\pi a}} \quad (4.46)$$

where

$K_{1c} =$ fracture toughness (psi$\sqrt{\text{in}}$ or MPa$\sqrt{\text{m}}$)
$\sigma_f =$ nominal stress at fracture (psi or MPa)
$a =$ crack length (one-half of the crack length depending on geometry)
$y =$ correction factor.

For an edge crack, the crack length is a, for a center crack the crack length is $2a$. Most of the past research of stress corrosion cracking utilizes the fracture mechanics approach in which the relation between the initially applied stress, stress intensity factor K_i and the time to failure of pre-cracked specimen is observed. The time to failure increases as K_i decreases from the level required for the fracture to occur (K_{1c}, fracture toughness) to a threshold level (K_{1SCC}) below which SCC does not occur.

If the stress intensity factor K_i is plotted against the time to failure as shown in Fig. 4.34, it is observed that below a certain value of K_i (stress intensity factor), failure does not occur, or there is a threshold limit for a failure to occur. The value of K_i below which failure does not occur is, therefore, designated K_{1SCC}.

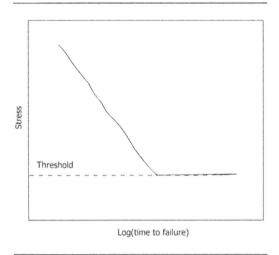

Figure 4.34 A typical curve showing failure time as a function of stress for normal fatigue (no corrosion)

4.8.5 TYPES OF SCC

Various types of SCC are distinguished as below:

(a) Chloride stress corrosion cracking. It occurs in austenitic steels under tensile stress in the presence of oxygen, chloride ion and high temperature.

(b) Caustic stress corrosion cracking. Cracking of steels in caustic environments where the hydrogen concentration is high, for instance, cracking of Inconel tubes in alkaline solutions.

(c) Sulfide stress corrosion cracking. Cracking of steels in hydrogen sulfide environment as encountered in oil drilling industry.

(d) Seasonal cracking. The term is now obsolete. It had a historical significance only. It refers only to SCC of brass in ammoniacal environment, but still occasionally occurs in refrigeration plant using ammonia refrigerant.

4.8.6 INTERGRANULAR AND TRANSGRANULAR CRACKING

Intergranular cracking. If the cracking proceeds along the grain boundaries, it is called intergranular cracking (Fig. 4.35a).

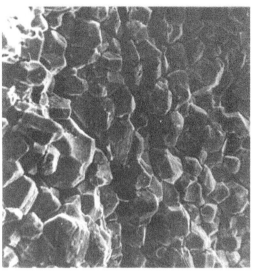

(a) Intergranular Cracking

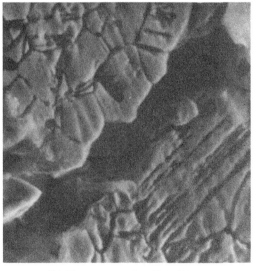

(b) Transgranular Cracking

Figure 4.35 Micrographs of intergranular and transgranular cracking

Transgranular cracking. The crack proceeds across the grain. It is the most common type in concentrated chloride environment (Fig. 4.35b).

4.8.7 IMPORTANCE OF STRESSES

The tensile stresses, a key requirement for SCC to occur, are '*static*' and they may be residual, thermal or applied. It is generally agreed that residual stresses are dangerous, because of their greater magnitude, which may sometimes approach yield strength, such as in fusion welds. For SCC to be caused alone by applied stress, it must be of very high magnitude. Stress corrosion can be caused by either residual, thermal or applied or a combination of all. Cracking caused by cyclic stresses is called '*corrosion fatigue*,' however, the two have different fracture mechanisms. Compressive residual stresses are used either to prevent SCC or corrosion fatigue or to delay their onset. Table 4.6 shows the sources for stress for SCC.

The question, 'what is the threshold stress for a certain alloy for SCC to occur,' has no definite answer because the magnitude of stress required would vary with environment. In literature, the values are quoted under specific conditions of stress and temperature and environment.

4.8.8 SCC SITES

Stress corrosion cracking is a deleterious phenomenon which occurs under a tensile stress, either residual or applied in a corrosive environment. The cracks are initiated and propagated by the combined effect of stresses and the environment. The mechanism of stress corrosion cracking is highly complex and despite extensive research, it is not conclusively understood. However, various important factors which lead to SCC are given below. A simplified diagram of SCC is shown in Fig. 4.36. The SCC cracks can be both intergranular or transgranular, depending on the alloy, stress conditions and the environment.

Stress corrosion cracking initiate and propagate without any outside evidence of corrosion. The failures can take place without any previous warning. They often initiate at pre-existing flaws or flaws formed during the service period of the component. The following are generally the sites for crack initiation.

(a) **Surface discontinuities.** Cracks may initiate at surface irregularities, such as grooves, laps or defects arising from fabrication process (Fig. 4.37).

(b) **Corrosion pits.** SCC can also initiate at the pits which are formed on the surface due to breakdown of passivity by chloride ions. Pits are formed at a certain value of potential, the critical pitting potential. In many cases, stress corrosion cracks have been observed to initiate at the base of the pits by intergranular corrosion. The electrochemistry at the base of the pit is the controlling factor in the initiation of cracking at pit sites.

(c) **Grain boundaries.** Intergranular corrosion resulting from the sensitization by impurities, such as phosphorus or sulfur, at the grain boundary which makes the grain boundaries very reactive to SCC. Dissolution of the slip planes is caused by the breakdown of the protective film and initiates SCC.

Table 4.6 Sources of stress for SCC

Sources	Applications
Residual	Rapid heating and quenching
Welding	Thermal expansion
Shearing, punching, cutting	Vibrations
Bending, crimping, riveting	Rotation
Machining (lathe/mill/drill)	Bolting (flanged joints)
Heat treatment	Pressure (internal or external)
Straightening, breaking, deep drawing	Structural loading

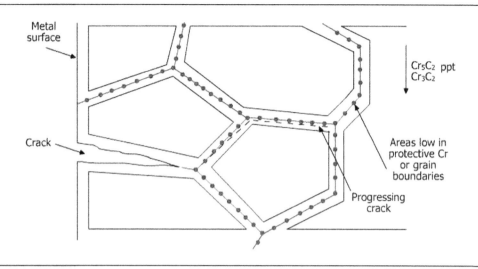

Figure 4.36 A schematic sketch showing crack propogation along grain boundaries (intergranular stress corrosion cracking)

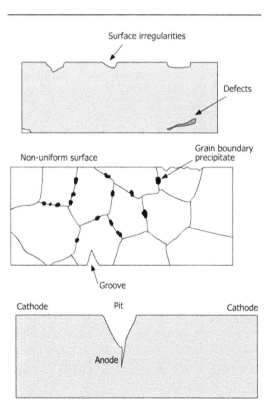

Figure 4.37 Initiation of crack at the pit

4.8.9 Examples of Typical Engineering Materials which Undergo SCC

(1) High strength steel in water.
(2) High strength aluminum alloys in chloride solutions.
(3) Copper alloys in ammonical solutions.
(4) Mild steels in hydroxide and nitrate solutions.
(5) Austenitic steels in hot chloride solution and hydroxide solution.
(6) Titanium alloys in chloride solutions and hot solid chloride.
(7) High nickel alloys in high purity steam.

Table 4.7 shows stress corrosion cracking environments.

4.8.10 Characteristics of SCC

(1) Stress, either residual or applied, is required. There is a simultaneous action of stress and corrosion.

Table 4.7 Stress corrosion cracking environments

Metals	Environment
Carbon steel	NaOH solution
	$NaOH + Na_2SiO_3$ solutions
	Nitric acid (concentrate)
	Ammonia ⎤
	Calcium ⎬ Nitrate solutions
	Sodium ⎦
	HCN solutions
	$HCN + SnCl_2 + AsCl_2 + CHCl_2$ solutions
	CH_3COOH solutions
	NH_4Cl solutions
	$CaCl_2$ solutions
	$FeCl_2$ solutions
	H_2S solutions
	$NH_4 CNS$ solutions
	Hydrofluoric acid
	$H_2SO_4 - HNO_3$ solutions
	Mixture of antimony chloride, hydrochloric acid and aluminum chloride in hydrocarbon
	MEA (mono-ethanolamine), DEA, etc. (amines)
Low alloy steels	H_2S solutions
	May also be cracked by media listed for carbon steel
Ferritic and martensitic stainless steels (Fe–Cr)	Seawater
	H_2S solutions
	$MgCl_2$ solutions
	NaCl solutions
	$NaCl + H_2O_2$ solutions
	NaOH solutions
	NH_3 solutions
	HNO_3
	H_2SO_4
	$H_2SO_4 + HNO_3$ solutions
	High temperature and high pressure water (high purity)
	May also be cracked by media listed for carbon steel
	NH_3 ⎤
	Ba
	Co
	Ca
	Fe
	Mg
	Hg ⎬ Chloride solutions
	K
	Li
	Na
	Vinyl
	Ethyl
	Methyl ⎦

(Contd)

Table 4.7 *(Contd)*

Metal	Environment
Austenitic stainless steel (Fe–Cr–Ni)	Seawater
	$NaCl + H_2O_2$ solutions
	Dichlorethane
	CCl_4 solutions
	Chloroform
	Aniline chlorohydrate
	Ortho-dichlorobenzene
	Oxychlorobenzene
	Aluminum sulfate
	Na_2SO_4
	Sodium aluminate
	Ammonium carbonate
	KOH solutions
	NaOH solutions
	Molten NaOH
Aluminum alloys	
Al–Zn	Air
Al–Mg	$NaCl + H_2O_2$ solutions
	NaCl solutions
Al–Mg	Seawater
Al–Cu–Mg	
Al–Mg–Zn	
Al–Zn–Cu	NaCl solutions
	$NaCl + H_2O_2$ solutions
Al–Zn–Mg–Mn	Seawater
Al–Zn–Mg–Cu–Mn	
Titanium alloys	HNO_3 (fuming red)
	HCl
	Molten NaCl
	Seawater
	Salt water
	Molten N_2O_4
	Trichloroethylene
	Acetic acid
	HCN
	HF
	Fluosilicic acid
	H_2SO_3
	Naphthenic acid
	Polythionic acid
	NH_3
	Calcium ⎤
	Mercurous ⎬ Nitrate solutions
	Sodium ⎦

Table 4.7 *(Contd)*

Metal	Environment
	Nitric acid
	Formaldehyde
	Steam (260°C)
	$H_2SO_4 + CuSO_4$ solutions
	NaOH + sulfide solutions
	H_2SO_4 + chloride solutions
	Na_2CO_3 + 0.1% NaCl solutions
	Conc. boiler water
	$NaHCO_3 + NH_3 + NaCl$ solutions
	Nickel alloys (Monel, etc.)
Nickel alloys (Monel, etc.)	HF
	Molten NaOH
	Hg salts
	Hydrofluosilicic acid
	Chromic acid
Copper alloys (mainly brass and bronze)	NH_3
	Ammonia compounds
	H_2S solutions
	Moist SO_2
	HNO_3 fumes
	Some amines

(2) Generally all alloys are susceptible to SCC, however, there are a few pure metals which have been observed to undergo SCC, such as 99.999% Cu and high purity iron.

(3) SCC of a specific alloy is caused by only a few chemical species in the environment.

(4) There is a period called '*induction period*' which is necessary to produce crack initiation, similar to the induction period required for producing pitting.

(5) Conditions of cracking are specific to alloy and environment. An alloy may be corroded in one corrosive medium while it may not under SCC. All the specific environments for a particular alloy are not known.

(6) The mode of cracking may be intergranular or transgranular. The transition from intergranular to transgranular depends upon factors, such as heat treatment, corrosive medium, stress level and temperature.

(7) The rate of attack is very rapid at the crack tip and very low at sides of the crack.

(8) There is a particular corrosion reaction critical for SCC to occur.

(9) SCC cracks are microscopically brittle in appearance.

(10) The fracture mode of an alloy in SCC is always different from its fracture mode in a plain strain fracture.

(11) For some systems, there appears to be a threshold value of stress below which SCC does not occur.

(12) There may exist a critical potential below which SCC does not occur.

4.8.11 ENVIRONMENTAL AND MATERIALS FACTORS IN STRESS CORROSION CRACKING

The cracking environments are specific because not all environments promote stress

Table 4.8 Common environmental stainless steel alloy systems susceptible to SCC

- Concentration of chloride (evaporation)
- Elevated temperature
- pH > 2
- Oxygen
- Time

corrosion cracking. The specific environments which promote SCC of alloys are shown in Table 4.8. Alloys which are inherently corrosion resistant, such as austenitic steels which develop protective films, require aggressive ions to promote SCC, whereas alloys of no inherent corrosion resistance, such as carbon steels, require an environment which is partially passive. Thus, for cracking of Mg–7% Al, a mixture of CrO_4^{-2} and Cl^- ions is required, and no SCC would be caused by either of the species alone.

In order for SCC to occur, it is important to maintain a balance between passivity and reactivity. If either of the above is not met, there would be no SCC but localized corrosion or general corrosion on the one hand, and complete passivity on the other hand.

Cracking is only achieved in certain specific environments and not in all environments. It is, therefore, not very practical to suggest specific solution requirements for cracking to occur.

4.8.12 MATERIALS CHEMISTRY AND MICROSTRUCTURE

The effect of materials chemistry and microstructure of materials in SCC and the interrelationship between the two is highly complex. The composition of the alloy has a significant bearing on the properties of the passive films and phase distribution. For example, a high amount of carbon in steels tends to form chromium carbide which causes sensitization of steel and leads to intergranular corrosion. Similarly, impurity elements in steels segregates and affects the corrosion dissolution process.

The effect of materials chemistry and microstructure on intergranular corrosion can be classified in two categories: (1) grain boundary precipitation and (2) grain boundary segregation.

A. Grain Boundary Precipitation

This is best illustrated by the formation of chromium carbide ($Cr_{23}C_6$) on the grain boundary and depletion of chromium adjacent to the grain boundary in stainless steels, such as AISI 304. The grain boundary region is subjected to corrosion attack called intergranular attack. Carbide precipitation also takes place in nickel alloys, such as alloy 600. In the case of aluminum alloys intermetallic compounds, such as Mg_2Si, are precipitated. For instance, the intermetallic compound which is precipitated in aluminum alloy (5086-H34) containing more than 3% Mg is beta phase (Mg_2Al_3) which is highly anodic to the aluminum magnesium matrix.

B. Grain Boundary Segregation

Impurities, such as phosphorus, sulfur, carbon and silicon, segregate at the grain boundary and cracking contribute to the SCC steels and nickel base alloys.

Examples of grain boundary precipitation

(1) Precipitation of chromium carbide in stainless steel in temperature range 500–800°C.
(2) Chromium carbide precipitation in nickel base alloys, such as alloy 600.

4.8.13 THERMODYNAMIC REQUIREMENTS FOR ANODICALLY ASSISTED DISSOLUTION

(a) The dissolution or oxidation of the metal must be thermodynamically possible.
(b) The protective film formed on the metal surface should be thermodynamically stable.
(c) On the basis of (a) and (b), it has been suggested that a critical potential exists at which SCC occurs. The values of critical

potential for SCC of alloys are shown in Fig. 4.38. The critical values are experimentally obtained from the anodic polarization curves as discussed in Chapter 3. Zones for intergranular and transgranular cracking of alloys 600 and 800 are shown in Figs 4.39 and 4.40. The zone of intergranular cracking of AISI Type 304 stainless steel in 10% NaOH at 288°C is shown in Fig. 4.41. It is observed that intergranular corrosion occurs over a much wider range compared to transgranular cracking.

In zone 1, the metal is in a transition state from active to passive formation (Fig. 4.38). The conditions of film formation on crack wall and film dissolution at the crack tip exist in this zone. An active surface is necessary for the process of dissolution to occur, such as active surface provided by the anodic zones formed in the material. Similar conditions also exist in zone 2. The above conditions are ideal for transgranular cracking which takes place in the two zones, 1 and 2.

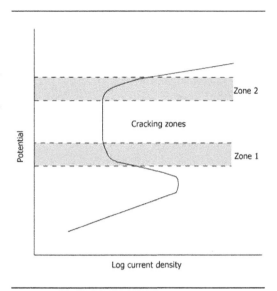

Figure 4.38 Potentiokinetic polarization curve and electrode potential values at which stress corrosion cracking appears. (From Jones, R.H. Battelle Pacific Northwest Laboratory, Ricker, R.E. N.B.S., *Metals Handbook*, **13**, ASM. Reproduced by kind permission of ASM, Metals Park, Ohio, USA)

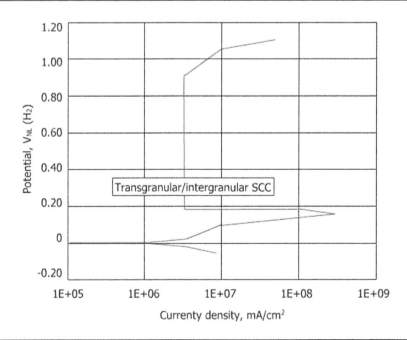

Figure 4.39 Alloy 600 in 10% NaOH solution at 288°C showing transgranular/intergranular SCC

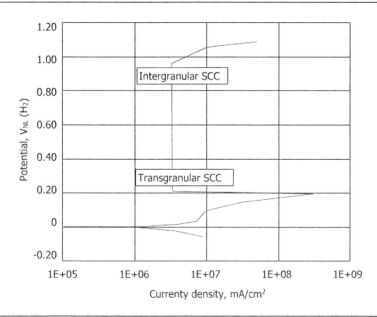

Figure 4.40 Alloy 800 in 10% NaOH solution at 288°C showing regions of intergranular and transgranular SCC. (From Jones, R.H. and Ricker, R.E. (1987). *Metals Handbook*, **13**, Corrosion. Reproduced by kind permission of ASM, Metals Park, Ohio, USA)

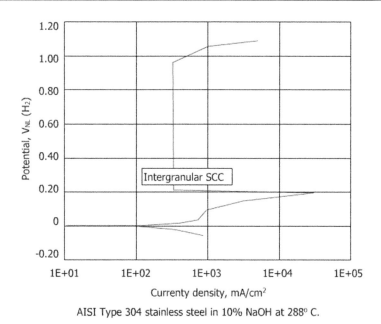

AISI Type 304 stainless steel in 10% NaOH at 288° C.

Figure 4.41 Electrode potential values at which intergranular and transgranular stress corrosion cracking appears. (From Jones, R.H. and Ricker, R.E (1987). *Metals Handbook*, **13**. Reproduced by kind permission of ASM, Metals Park, Ohio, USA) [64]

Intergranular cracking takes place over a much wider range of potential because of a larger difference of potential between the segregation of impurities at the grain boundary and the metal matrix.

4.8.14 SOURCES OF ANODIC ZONES

Following are the sources of formation of anodic zones which assist dissolution:

(a) Composition and microstructural differences.
(b) Grain boundaries and sub-boundaries.
(c) Stress disorder effects.
(d) Local rupture of surface films due to stress.
(e) Heterogeneous effects of stress, or prior cold work.

4.8.15 THE INFLUENCE OF STRESS ON RATES OF LOCALIZED CORROSION

Localized corrosion as a result of stress or deformation may arise from one of the following factors:

(a) Rupture or damage of surface film.
(b) Changes in the metallurgical characteristics of the alloy.
(c) Change in the polarization characteristics.
(d) Change in the electrode potential.

4.8.16 CHARACTERISTICS OF STRESS AND PLASTIC DEFORMATION

(a) The internal energy of the materials is increased by plastic deformation.
(b) The amount of plastic deformation varies from grain to grain and even within the grain.
(c) Fabricating processes also generate residual stresses. The residual stress influences the susceptibility to SCC very significantly.

(d) Grain orientation is affected by plastic deformation.

4.8.17 MECHANISM OF STRESS CORROSION CRACKING

The first recognized case of environmentally induced corrosion was that of seasonal cracking of brass in the rainy season. It was a classical example of SCC of brass in ammoniacal environment. Soon upon being realized as a major mechanism, it became the subject of intense research and a brief review of literature would show that thousands of papers have been written on the subject. The most intriguing is the mechanism of stress corrosion cracking and stress corrosion crack propagation. A full discussion of the mechanism is beyond the scope of this chapter. An online search would reveal thousands of articles and update references. Four basic categories of mechanisms have been proposed:

(a) Mechano–electrochemical model [56].
(b) Film rupture model [57–59].
(c) Embrittlement model [60].
(d) Adsorption model [61,62].

(a) Mechano–Electrochemical Model

This suggests that there are pre-existing paths in an alloy which become intrinsically susceptible to anodic dissolution. For instance, a grain boundary precipitate anodic to the grain boundary would provide an active path for localized corrosion to proceed. Similarly, if a more noble constituent is precipitated along a grain boundary, the impoverished zone adjacent to the precipitate would provide an active path for localized corrosion. Also, the removal of the protective film at the crack tip by plastic deformation would facilitate the onset of localized corrosion. For instance, the precipitation of $CuAl_2$ for an Al–4 Cu alloy depletes of copper along the grain boundaries and provides active paths for localized corrosion. Application of cathodic protection stops crack propagation and its removal restarts the process, thus establishing conjoint action of mechanical and electrochemical processes.

(b) Film Rupture Model

This is based on a repetitive cycle comprising (a) localized film disruption, (b) localized attack at the point of disruption and (c) film repair. Plastic strain plays a major role. This mechanism has several variations.

(c) Embrittlement Model

This is based on the postulate that an electrochemical process embrittles the materials in the vicinity of a corroding surface. This mechanism was based on a study of high strength martensitic steels in chloride media [63].

(d) Adsorption Model

Adsorption of damaging species causes weakening of cohesive bonds between surface metal atoms by specific damaging species. The surface energy associated with crack formation is lowered by adsorption of species and the probability of a metal forming a crack under tensile stress is increased. A universal theory does not exist and several theories have to be examined for results and analyzed to reach a plausible explanation.

As observed in the above discussions, it appears that there is not a single but two or three different mechanisms which operate. The mechanism of crack propagation falls into two basic categories, the dissolution model and the mechanical fracture model.

4.8.18 EXISTING MECHANISMS OF STRESS CORROSION CRACKING

On the basis of the progress made in the understanding of the SCC phenomenon in recent years, the existing mechanisms of SCC can be divided into three categories: (1) pre-existing active path mechanisms, (2) strain assisted active path mechanism and (3) adsorption related mechanisms [64–66]. These subdivisions are only from a point of view of understanding of one continuous process with only shift of emphasis from one mechanism to another. The above mechanisms are summarized below.

4.8.19 PRE-EXISTING ACTIVE PATH MECHANISM [67–69]

(a) Mechanism

If solute segregation or precipitation occurs at the grain boundary, such as the segregation of impurities like sulfur and phosphorus, and the precipitation of chromium carbide as discussed under intergranular corrosion, the electrochemical properties of the matrix and segregate are changed. The area adjacent to the grain boundary is depleted of by one of the alloying elements and it is preferentially attacked. Under such conditions, localized galvanic cells are created with the segregate being generally anodic to the matrix of the grain (Fig. 4.42). The polarity can also reverse in some situations. Such structural features lead to intergranular corrosion. The changes brought about by change in the grain boundary structure, produce compositional difference and provide active paths for SCC to occur. This mechanism is predominant in cases where SCC is governed by electrochemical or metallurgical factors rather than stress.

(b) Evidence in Support

(1) The cracking susceptibility can be altered by altering the ratio of anode to cathode area at the grain boundary. For instance, the susceptibility to SCC is reduced if the grain boundary regions become less anodic with respect to the grain matrix, i.e. the difference of potential between the active area of the grain boundary and the grain matrix is reduced by alloying.

(2) Steel samples are etched when immersed in a cracking environment at predetermined potentials. Corrosion attack takes place only in certain selective regions, i.e. upon pre-existing paths determined by the changes in the grain boundary structure. In nitrate solutions, the grain boundary attack only penetrates to a depth of few millimeter in the

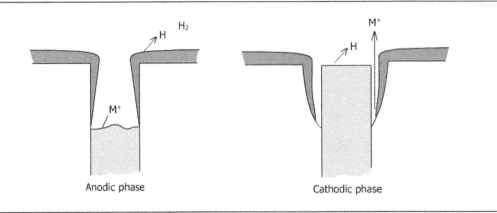

Figure 4.42 Galvanic cell mechanism. (From Perkins, R.N. (1972). *Br. Corros. Jr.*, **7**, 15. Reproduced by kind permission of British Corrosion Journal, London, UK)

absence of a stress. However, on applying anodic polarization, the steel is virtually destroyed by intergranular corrosion. The stress and applied potential both appear to contribute to the dissolution of the active path [70].

(3) The low susceptibility of low carbon steels can be explained by the change in the distribution of carbon which occurs after a deformation processes. Carbon is not generally present at the ferrite grain boundaries [71]. The carbon distribution is intimately related to the active path, and it appears to fix the site of the active path.

(4) Pure iron containing about 0.005% impurity, is not susceptible to SCC. It becomes susceptible when it is carburized [72].

(5) Aluminum alloys, such as Al–Mg, Al–Cu and Al–Zn–Mg, are susceptible to intergranular corrosion in the absence of stress after appropriate heat treatment [73]. If the heat treatment produces precipitates, such as $MgAl_3$, $MgZn_2$ and $CuAl_2$, intergranular corrosion and SCC become very severe. Galvanic cells are set up as a result of differences in the electrochemical properties between the precipitates and the grain matrix. The cracking susceptibility can be altered by altering the grain boundary precipitate volume (anode/cathode area ratio).

(6) A precipitate-free zone exists next to the grain boundaries in Al–Zn–Mg alloys [74]. It has been observed that the susceptibility

of alloy having a narrow precipitate-free zone is more than the alloy having a wide precipitate-free zone. It has been observed that deformation occurs more readily in the precipitate-free zone and causes the oxide film to rupture. Although the later is disputed, it appears that the precipitate-free zone influences SCC by its effect on deformation at the crack tip. This shows that mechanical effect is also important but it only plays a secondary role in SCC of aluminum alloys.

4.8.20 STRAIN GENERATED ACTIVE PATH MECHANISMS [74–76]

(a) Mechanism

In contrast to the active paths mechanism, there are many systems where strain is the controlling influence, e.g. high strength steels in chloride solutions or tin alloys in methanol. Such instances are best explained by strain-generated active path mechanisms. It is based on the idea of strain-induced rupture of the protective film. Various theories have been proposed which relate crack propagation to dissolution of the crack tip and the existence of strain/stress conditions existing in that region. These theories depend on the existence of strain/stress regions for crack propagation and hence, they are classified as a strain-generated active path mechanism. The term '*active path*' has been explained earlier.

The most outstanding mechanism which has found a wide acceptance is the film rupture mechanism. This mechanism has been extensively studied in stress corrosion cracking of alpha-brass in ammoniacal environment, although it was originally proposed for caustic cracking of boiler steel. Here are some salient features of the mechanism. The items are illustrated in Fig. 4.43.

(1) The theory assumes the existence of a passivating film on a metal surface. The existence of such films has been experimentally verified.
(2) The passivating film is ruptured by plastic strain in the underlying metal.
(3) After the film is ruptured, the bare metal is exposed to the environment.
(4) The processes of disruptive strain (disruption of protective film) and film formation (due to repassivation), alternate with each other.

(5) The rate of cracking is controlled by the rate of film growth. A process of competitive adsorption between species promoting passivity, and species promoting Cl^-, takes place. If repassivation occurs too quickly, only a small amount of corrosion would take place and the crack would grow.
(6) The crack propagates by successive dissolutions of the crack tip when it successively becomes bare due to the rupture of the oxide film due to plastic strain in the underlying metal.

In some other theories, it has been suggested that the crack tip always remains covered with an oxide and the film is only periodically ruptured by emergence of slip steps. It has also been suggested that the crack tip remains bare because the rate of rupture of the oxide film is higher than the rate of repassivation of the film. In general, the rate of attack is determined by stress (applied or residual), electrochemical potential, total strain rate and specific ions and effect of solute segregates.

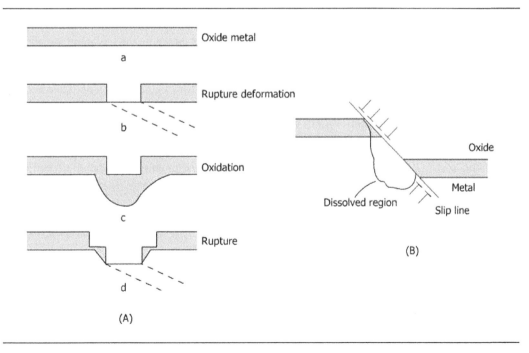

Figure 4.43 Strain generated active path mechanisms (A) Film rupture model (B) slip step dissolution model. (From Sedriks, A.J., Slattery, P.W. and Pugh, E.M. (1969). *Trans. Am. Soc. Met.*, **62**, 1238. Reproduced by kind permission of ASM, Metals Park, Ohio, USA)

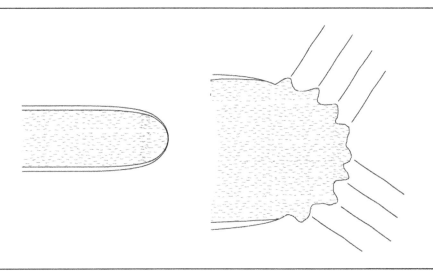

Figure 4.44 Schematic representation of crack propogation by the film rupture model. (From Staehle, R.W. (1971). in *Stress Corrosion Cracking in Alloys*, NATO, p. 223)

A generalized mechanism based on a slip rupture model is shown in Fig. 4.44.

The formation of passive film is most important, because SCC takes place commonly in alloys that are covered by a highly protective film, such as aluminum or steel alloys. Under a tensile strain, the slip plane breaks the protective film as shown in Fig. 4.45a, a small part of the film undergoes dissolution as shown in (b) and later repassivation takes place as in (c). As pointed out earlier, if repassivation occurs too rapidly, corrosion attack would be too small and the crack would propagate slowly. On the other hand, if repassivation occurs very slowly, excessive metal dissolution occurs on the crack tip and sides. This widens and blunts the crack tip, and the crack growth is arrested. The greatest damage is caused by moderate repassivation rates.

(b) Evidence in Support

(1) Ellipsometric studies on the effect of increase of zinc content on the film growth in ammoniacal solution confirm the film rupture mechanism [77].
(2) It has been shown that addition of arsenic to brass increases its resistance to SCC.

This is because arsenic addition directly influences the film forming characteristics of the brass.
(3) It has been shown that the preferential growth of film occurs along grain boundaries, as shown by the preferential oxidation of copper at the grain boundary by $FeCl_3$ in copper–gold (Cu_3Ag) alloys. The removal of copper leaves gold in a spongy state and the grain boundary is attacked due to the rupture of the film. The film growth, therefore, is a structural dependence process, and its rupture leads to crack propagation [78].
(4) A strain rate which produces a bare metal at a faster rate than the film formation rate is sufficient to cause deformation. The currents associated with straining electrodes are higher than at a static surface as shown by Hoar [79].
(5) Generally, the segregation of solute on dislocation results in localized corrosion. Under such conditions the corrosion progresses along the active paths. If such segregates are not present, corrosion takes place because of moving dislocations which generate chemical activity similar to that produced by segregation leading to stress corrosion cracking.

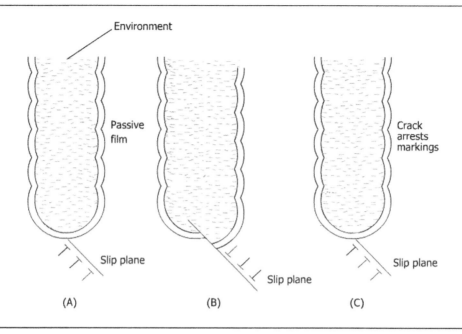

Figure 4.45 Schematic of film rupture model (a) illustrates the main feature of the model – the protective film is ruptured at the crack tip by localized slip, permitting propagation by anodic dissolution. The crack tip is depicted in greater details in (b) which illustrates the view that crack advance results from a number of independent film rupture and transient dissolution events. (From Staehle, R.W. (1979). *Stress Corrosion Cracking in Alloys*, NATO, p. 223)

(6) The observation that the rate of crack propagation observed is greater than that accounted for by electrochemical observation, has been supported by the evidence that crack propagation may be discontinuous. It has also been suggested that the discrepancy in the rate of crack propagation actually observed and accounted by electrochemical process, is due to the crack formation by mechanical factors and crack growth by electrochemical factors. The mechanical crack propagation is followed by electrochemical crack growth.

It has been stated earlier that active paths are generated by formation of segregates, precipitates at the grain boundary. There exists a difference of potential between the segregates, precipitates and the grain matrix and hence, galvanic cells which are formed by the precipitates/segregates and metal matrix provide favorable sites for continuous paths and the active corrosion to proceed.

Active paths for SCC may also be generated by the rupture of an oxide film or by the dissolution of slip plane.

To conclude, it is observed that in mechanisms based on a pre-existing active path, the susceptibility to SCC at one end is predominantly controlled by localized corrosion where stress is not necessary, such as carbon, steel in NO_3^-, and Al–Zn–Mg in Cl^-, brass in NH_3 solution, to the other limit where the stress influences the propagation of a crack from an intergranular notch, such as in Mg–Al alloys in CrO_4^-, Cl^- and high strength steel in water.

4.8.21 ADSORPTION RELATED PHENOMENON [80–85]

(a) Mechanism

This model is based on the assumption that the adsorption of environmental species lowers the interatomic bond strength and the stress required for cleavage fracture (a brittle fracture in which the crystallographic planes are separated). There are several features which identify a cleavage fracture, such as river patterns, tongues, herring bone (Fig. 4.46). The theoretical fracture stress required to pull apart two layers of atoms of spacing b is given by [80]:

$$\partial_{Fr} = \left(\frac{E\gamma}{b}\right)^{1/2} \qquad (4.47)$$

where

$E =$ modulus of elasticity
$\gamma =$ surface energy
$b =$ spacing between atoms.

The above theory implies that if the surface energy term γ is reduced, then ∂_{Fr} will also be reduced. This led Orwan [80] to suggest that delayed fracture of glass occurs by adsorption of environmental species which lowers the surface energy, and hence the stress required to cause fracture. The same principles were later adopted to account for the SCC of metals. If environmental species are present and they are adsorbed at the crack tip, the bond strength is reduced, the surface energy is effectively lowered, and fracture takes place. The species may also diffuse into the metal and become absorbed in some region in advance of crack tip where the stress/strain conditions are favorable for the nucleation of a crack. In the latter case, hydrogen is the only known species which diffuses and causes SCC referred to as hydrogen embrittlement. The latter is, however, considered as a particular case of SCC.

Other mechanisms suggest that hydrogen atoms are formed by the reduction of hydrogen ions within the crack. These hydrogen atoms cause the weakening of the bonds beneath the surface of the crack tip. This can also be achieved by formation of metal hydrides which are known to be brittle.

The formation of hydrogen gas in small quantities formed by H atoms diffusion through the metal lattice leads to the development of enormous pressures in micro-cavities where $H + H$ combine to H_2 gas. This leads to the rupture of the metals. Pressures as high as 3000–20 000 atmospheres (0.3 to 2 GPa) are developed.

The adsorption mechanism, however, fails to explain the phenomenon of SCC of metals where considerable plastic deformation occurs. SCC progresses without any significant plastic deformation, but localized plastic deformation occurs at the crack tip, and under such conditions the surface energy term is significantly lower than the plastic work term, hence, the reduction of surface energy would be insignificant and it would exert only a negligible effect on fracture stress (∂_{Fr}). Equation (4.47) can be modified to equation (4.48):

$$\partial_{Fr} = \frac{[E(s - p)]^{1/4}}{b} \qquad (4.48)$$

where

$s =$ specific surface energy
$p =$ plastic work term.

In the above condition, $p > s$, hence, fracture would only be slightly easier with the lowering of the surface energy. The effect of lowering of surface energy is, therefore, not significant.

Figure 4.46 Typical cleavage fracture in iron

It can be concluded that specific adsorption does not account properly for SCC of metals where plastic deformation is associated with fracture.

(b) Grain-size Relationship

It has been shown that coarse grain size materials are more susceptible to SCC than fine grain size material. A patch-type of relationship connecting grain diameters and the stress required to initiate a SCC crack is well-known: $\partial_{Fr} = \partial_0 + Kd^{-\frac{1}{2}}$, where ∂_0 and K are constants. K is related to the surface energy associated with formation of new surface:

$$K = \frac{(\partial \pi G \gamma)}{1 - v} \qquad (4.49)$$

where

γ = surface energy
v = Poissons ratio
G = modulus of rigidity.

Measurement of the relation between the grain size and stress corrosion fracture, allows the surface energy to be calculated. A lowering of surface energy was observed for mild steel in nitrate solution [62]. The surface energies so calculated [80] were found to be lower than in other conditions. Hence, it was deduced that surface energy is lowered by adsorption of active species during stress corrosion cracking. This supported the adsorption mechanism. However, later it was found that the yield stress and work hardened flow stress after '*plastic*' deformation are also dependent on grain size similar to the surface energy. It is, therefore, observed that grain size is related to plastic flow as it is to the lowering of surface energy. If plastic deformation accompanies stress corrosion cracking, the plastic strain term (γ_p) becomes more significant than (γ_s), the surface energy lowering, hence adsorption would be insignificant. The adsorption theory appears to be more valid where the γ_p term is not much higher than γ_s term.

(c) Evidence of Support

(1) Studies on Al–Mg alloys have shown that the process of adsorption assists the mechanical part of crack propagation.

(2) The failure of engineering materials by adsorption of hydrogen causing embrittlement of metal in advance of the crack tip.

(3) The stress corrosion cracking of α-titanium alloys occurs by nucleation of hydride (by the interaction of absorbed hydrogen with the metal).

(4) The failure of high strength steels by hydrogen adsorption lends strong support to hydrogen adsorption mechanism; only specific environments promote stress corrosion cracking, and this goes in favor of an adsorption mechanism.

(5) Lack of systematic explanation of the environmental aspect of cracking. Specific environments are more consistent with the adsorption theory.

(6) Failure of materials by absorbed hydrogen provides strong evidence in favor of adsorption theory. The cracking of high strength steels is not dependent on specific environment and failure would occur as long as a source of hydrogen is available. The hydrogen lowers the surface energy by adsorption and lowers the cohesive forces of the lattice. The lowering of cohesive forces causes the lowering of the stress required to cause SCC of the metal. This particular mode of cracking by hydrogen is called '*hydrogen embrittlement.*' A sufficient reduction in the hydrogen gas pressure surrounding the specimen containing a crack at a given intensity would cause arrest of the crack but a subsequent increase in the pressure would restart the crack. That cracking results from the formation of a hydride phase [85] due to the adsorption of hydrogen by titanium is supported by a strong evidence. The SCC of α-titanium alloys constitutes an example of slow strain rate hydrogen embrittlement. Other researchers believe that SCC in titanium alloys results from dissolution. Thus, adsorption theory accounts well for hydrogen induced cracking (HIC).

(d) Limitations

(a) The adsorption mechanism accounts mostly for metals where $\gamma_p \gg \gamma_s$ and it does not account for the SCC of metals that undergo significant plastic deformation.
(b) It does not explain how the crack maintains a sharp tip in a normally ductile material because it does not include a provision for limiting deformation in the plastic zone.
(c) The discontinuous nature of cracks observed in some instances is not explained.

4.8.22 EFFECT OF ENVIRONMENTAL FACTORS ON SCC

(a) Specific Ions

Not all environments promote SCC and hence, the environments causing SCC are specific. However, all environments that cause SCC are not fully known. Electrical conductivity is a very important characteristic and if a particular ion of certain electrical conductivity promotes SCC, other ions of similar electrical conductivity should also promote SCC. The specificity of ions can be explained better by an adsorption mechanism than other mechanisms. The propagation of SCC requires the reaction which occurs at the crack tip to proceed at a much faster rate than any dissolution process. A balance between passivity and reactivity has, therefore, to be maintained. For those alloys which develop a protective film, an aggressive ion is required to promote SCC. Other metals with a low corrosion resistance, such as carbon steel or Al–Mg alloy, require a partially passivating balance between chemical reactivity and passivity and if this balance is not maintained, the result would be general corrosion or pitting rather than SCC. The evidence for the above is provided by anodic polarization of mild steel in $(NH_4)_2 CO_3$ at 75°C by potentiodynamic method [86] (Fig. 4.47).

It can be observed from Fig. 4.47 that between -600 and -800 mV, the specimens suffered pitting and between -500 and -600 mV cracking

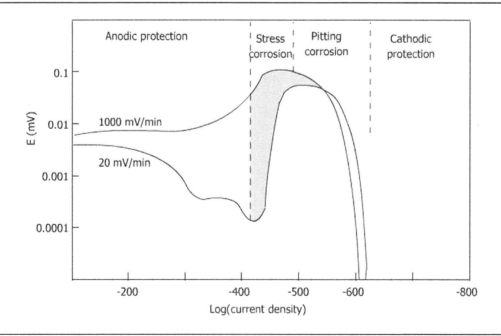

Figure 4.47 Fast and slow sweep rate anodic polarization curves for steel in $(NH_4)_2 CO_3$ at 75°C. (Perkins, R.N. (1972). *Br. Corros. Journal*, **7**, January. By kind permission of British Corrosion Journal, London, UK)

and below -850 mV, the steel remained immune to cracking. These observations demonstrate the phenomenon of passivity and reactivity. The occurrence of SCC requires the existence of a critical balance between the two. If the balance is not maintained, there would be either pitting and general corrosion or complete passivity.

The above observations are supported by electrochemical studies of steel in $MgCl_2$ solution [87]. In Fig. 4.48, current *vs* time curves are shown. Three types of curves are observed. Curve 1 is indicative of the fact that the film breaks down by emergence of slip steps as discussed earlier. The surface re-passivates so rapidly that no time is given for crack propagation to occur. In curve 3, passivation occurs too slowly which is a favorable situation for pitting in curve 2; there is the possibility of crack propagation because of a passivation time which allows the crack propagation and a critical balance is maintained between reactivity and passivity. Propagation occurs because more fresh metal surface is generated than can be re-passivated. The above observations support the electrochemical mechanism of SCC. Electrochemical studies are also used to distinguish between SCC and HE (hydrogen embrittlement).

(b) Effect of Oxygen

It is commonly believed that oxygen is essential for SCC to take place. This opinion has, however, been contradicted, and it has been suggested that oxygen only acts as a reducible species, a function which could well be served by Fe^{+3} and Cr^{6+} rather than oxygen. It has been argued that the hydrogen bubbling only indicates a reduction process and it does not suggest hydrogen embrittlement. It is suggested that hydrogen enters steel when it is not escaping from its surface.

(c) Role of Hydrogen

The role of hydrogen in SCC is very crucial and yet more controversial. The effect of hydrogen has been mostly explained in terms of hydrogen embrittlement, which causes a loss of ductility in the metal. The following observations are in support of a hydrogen embrittlement mechanism:

(1) Escape of hydrogen bubbles from cracks.
(2) Fractographic studies.
(3) Stimulation of cracking by hydrogen embrittlement.

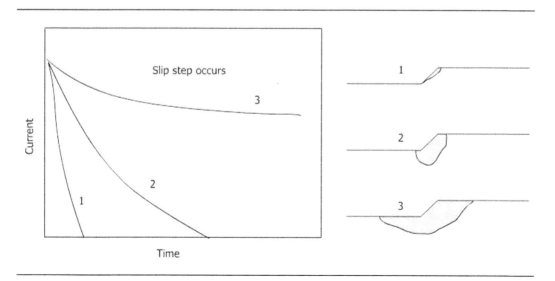

Figure 4.48 Schematic illustrations of the relationship between reaction current transient and the amount of reaction occurring at an emergent slip step. (From Staehle, R.W., Royuela, J.J., Raredon, T. *et al.* (1970). *Corrosion*, **26**, 451. By kind permission of NACE, Houston, Texas, USA)[87]

In order to account for the failures of high strength steel by hydrogen two mechanisms have been suggested: (a) Active path corrosion (APC), anodic dissolution and (b) hydrogen embrittlement (HE). The active path mechanism which accounts for cracks by providing active paths for propagation of cracks to proceed in SCC differs from hydrogen embrittlement where adsorption of hydrogen at cathodic sites is followed by embrittlement of the material. The two processes, SCC and hydrogen embrittlement (HE), must be differentiated from each other, which for some systems is extremely difficult. To distinguish the two, the relationship between time to failure (TF) and potential is studied. In the case of hydrogen embrittlement, the time to failure decreases as cathodic current is applied and increases with anodic polarization. The reverse is true for the anodic dissolution mechanism – (active path mechanism) when SCC occurs (Fig. 4.49).

4.8.23 STRESS CORROSION CRACKING OF STAINLESS STEEL

In the following section, an attempt is being made to summarize the practical information regarding stress corrosion cracking (SCC) of various types of steels. Table 4.9a lists primary conditions which can cause SCC of stainless steels. Table 4.9b shows some typical results on the SCC of steels

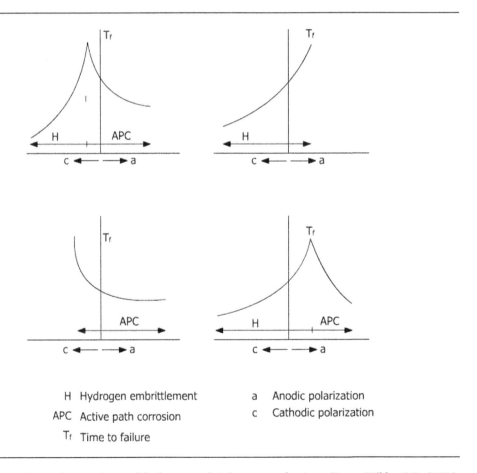

H   Hydrogen embrittlement
APC   Active path corrosion
T_f Time to failure

a   Anodic polarization
c   Cathodic polarization

Figure 4.49 Active path corrosion and hydrogen embrittlement mechanism. (From Wilde, B.E. (1971). *Corrosion*, **27**, 326. By kind permission of NACE, Houston, Texas, USA)

Table 4.9a Primary conditions which cause stress corrosion cracking of stainless steels

- Susceptibility of the alloys
- Metallurgical conditions
- Damaging environment
- Time

Table 4.9b Results of stress corrosion cracking on various stainless steels and high alloys exposed at 100°C to 100 ppm chloride

Cracked	Resisted cracking
304, 304 L, 316, 316 L, 347, 310, 202	329, 430, 446, Fe–Ni-Cr 20, Fe–Cr–Fe Alloy 600, Fe–Ni–Cr 825

Table 4.11 Effect of various elements on SCC susceptibility of austenitic steels

Element	Effect
Ni (>8%)	Harmful
C	Beneficial
Si	Beneficial
Cb	Beneficial
Ti	Beneficial
V	Beneficial
P	Harmful
Mo	Harmful
Cr (>16%)	Harmful
Cu	Harmful

Several factors affect the SCC of austenitic steels in chloride solutions. The effect of some important factors are given below.

(a) Alloy Composition

The nominal compositions of selected AISI standard grades of stainless steels are given in Table 4.4a. The highest susceptibility to SCC is shown by austenitic steels containing 8% nickel. Alloys containing less than 8% or more than 8% nickel exhibit a higher resistance (Fig. 4.50). The beneficial effect of a higher nickel content is shown by Incoloy, alloy 800, which fails in boiling 45% $MgCl_2$, but which is resistant to the environments in which alloys of lower nickel content, steel 304 (10% Ni) and steel 316 (12% Ni), fail by stress corrosion cracking. Resistance to SCC in most environments is obtained by adding about 30% nickel. Nickel is useful in resisting corrosion in mineral acids. Nickel also stabilizes the austenitic phase [88].

Chromium is an essential element for forming a passive film, whereas other elements assist in stabilizing the film. The passive film is not very effective up to 10.5% chromium content, but as the chromium content is increased between 17 and 20%, the passivating film becomes very stable. Austenitic steels generally contain 18–20% chromium. Whereas the corrosion resistance of austenitic steel is increased by addition

and high alloys exposed to 100 ppm chloride. On top of the table is the 300 series stainless steels in annealed condition in chloride environment, as they represent the most common situation because of their wide application. Unfortunately, the environment–alloy interaction is least understood. The SCC of austenitic stainless steel in chloride has been the subject of considerable research. The factors responsible for the SCC of austenitic steels in chlorides are given in Table 4.10. The effect of various elements on SCC susceptibility is shown in Table 4.11.

Table 4.10 Contributing factors to SCC in chloride solution

No.	Contributing factors
1.	Concentration of chloride
2.	Elevated temperature
3.	pH > 2
4.	Oxygen
5.	Time

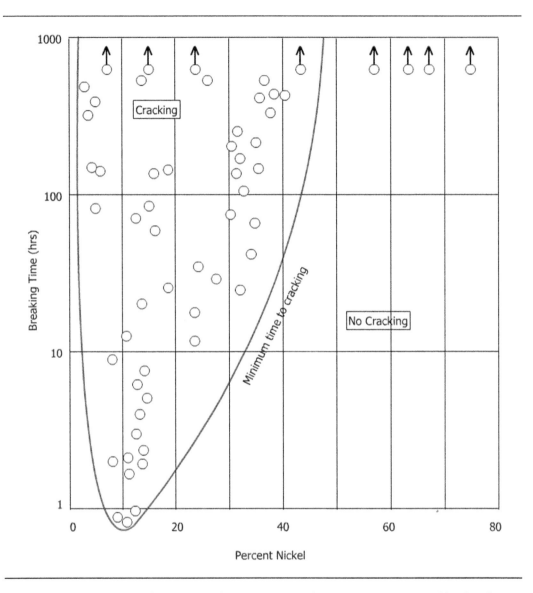

Figure 4.50 Breaking time of iron–nickel–chromium wires in boiling 45% magnesium chloride solutions. (From Copson, H.R. (1959). *Phy. Metallurgy of Stress Corrosion Fracture*, Interscience Publishers) [88]

of chromium, its resistance, particularly to SCC, is slightly lowered at levels higher than 10%. Steel 304 contains 18–20% of Ni and steel 316 has 16–18% Cr. Chloride SCC occurs generally at pH levels above 2.0 provided the other conditions for cracking also exist. As the pH value increases towards the alkaline side of the pH scale, the tendency for SCC is reduced. At high temperature and when free caustic is present in a concentrated form, steels 316, 304 and other 300 series suffer caustic cracking. The type of cracking will be separately dealt with. The beneficial and detrimental effects of various elements on SCC of austenitic stainless steels are shown in Table 4.11.

It is to be noted that the harmful or beneficial effect is only with regard to stress corrosion cracking and not to any other form of corrosion.

(b) Effect of Temperature

The susceptibility of austenitic steels to SCC increases with temperature. Mostly failure occurs by concentration of chloride by evaporation. In normal testing of SCC in chloride solution, heat has to be applied to such a degree as to cause evaporation of chloride or alternate wetting and drying to concentrate the chloride. It is to be remembered that stainless steels do not crack in strong chloride environments at ambient temperature.

Dana and Warren [89] conducted SCC tests on cracking of hot stainless steel pipe under wet chloride-bearing thermal insulation. It was shown by the experiment that at 90% the yield strength, SCC occurs only at temperatures above 60°C (Fig. 4.51). It is observed from the figure that the specimen exposed to 1.00 ppm chloride took 6 months to crack at 60°C, whereas it took only one day to crack at 100°C. Types 304 and 316 steels have been reported to be susceptible to SCC in water containing 20–60 ppm chloride above 60°C.

(c) Stress Levels

It is to be noted that SCC would never occur in the absence of either corrosive environment or stress. SCC can occur both by residual and tensile stress and never by surface compressive stress, although the later may be used to prevent SCC. SCC can occur from stress close to the yield point, down to as low as 0.1% of the yield strength [90]. For many systems, a '*threshold*' or limiting stress has been observed. Stresses as high as 4000–7000 psi are built up by corrosion products as shown by the SCC tests on steel 347. Whereas the role of stress is extremely

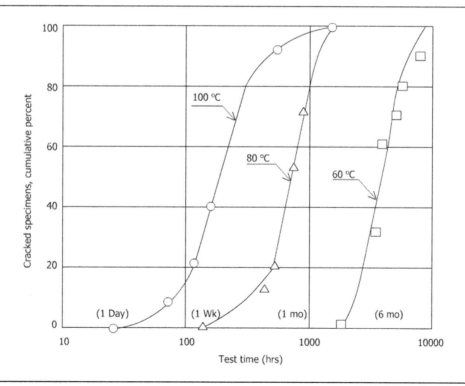

Figure 4.51   Effect of temperature on the time for cracking of Type 304 stainless steel specimens exposed to water containing 100 ppm chloride. (From Dana, A.W. and Warren, D. (1957). Bulletin, DuPont Engg. Dept., *ASTM Bulletin No. 225*, October. Reproduced by kind permission of ASTM, Int., Ohio, USA) [89]

important, it is to be considered that 'neither the residual nor the external stress are prerequisites for SCC because the corrosion process in chloride media can generate stresses of sufficient magnitude to cause SCC.' Stress corrosion cracking can occur in unnotched austenitic steel specimens which are stress-free when exposed to 42% $MgCl_2$ at 135°C. Some of the stresses have been shown in Section 4.4 as intergranular cracking.

concentrated by chemisorption, by leaching or by migration to pitting sites. If the temperature is ambient, no SCC would occur even in the presence of 10 000 ppm chloride. SCC of austenitic steels would not occur in seawater if it is cathodically protected. The time for SCC of 304 steel in chloride-containing water at 100°C is reduced significantly on an increase in the concentration of chloride as shown in Fig. 4.52 [91].

(d) Chloride Concentration

It is impossible to specify a chloride concentration at which SCC occurs. Stress corrosion can occur at a wide range of concentration, such as ranging from 10 000 to 0.02 ppm Cl^- because chloride in small concentration can be

(e) Electrochemical Potential

Intergranular SCC occurs over a wide range of potentials. The process of film formation and film breakdown occurs at a certain value of potential termed '*critical potential.*' Transgranular type SCC occurs in zone 1 and zone 2, whereas

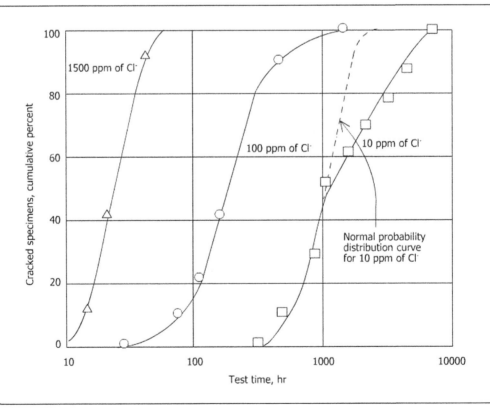

Figure 4.52 Effect of chloride concentration on the time for cracking of Type 304 stainless steel specimens exposed at 100°C to chloride bearing water. (From Warren, D. (1969). *Proc. 15th Annual Purdue Industrial Waste Conf.*, Purdue Univ., May 1–19) [91]

intergranular corrosion occurs over a wider range of potential between the two zones [92]. It is to be noticed that in zone 1, there is a transition from active to passive state and hence, film formation takes place on the side of the crack and film rupture on the crack tip. Similar conditions are observed in zone 2, however, it is observed that zone 2 is above the critical pitting potential (E_p) and hence, cracking is initiated by pitting in this zone. Because of chemical heterogeneity at the grain boundary and increased activity of the grain boundary with respect to the bulk material, intergranular SCC occurs over a wide range of potentials. The pitting potential of steels in 1000 ppm chloride solution is as shown in Table 4.12.

(f) Oxygen

The role of oxygen is very important and yet not fully understood. It is generally believed that oxygen is essential for SCC of austenitic steels in chloride medium. Oxygen reduction is the primary cathodic reaction in the SCC process. SCC can also occur in solutions where hydrogen undergoes the cathodic reduction. It is also argued that it is not the presence of oxygen but the presence of a species which can be reduced which is important. The oxygen content in boiling $MgCl_2$ test is estimated to be 0.3 ppm [93].

(g) Effect of Anions and Cations

It is known from the experimental work that chlorides, bromides and fluorides cause SCC

Table 4.12 Pitting potential of steels in 1000 ppm chloride solution

Steel type	Potential (V) *vs* SCE
430	−0.05
304	−0.22
316	−0.33
216	−0.95

of austenitic steels. It has been reported that iodides inhibit the SCC of steel. Similarly, silicates, phosphates, carbonates, iodides and sulfites are effective inhibitiors. The sulfites inhibit corrosion of stainless steels by removing oxygen from the system [94].

4.8.24 STRESS CORROSION CRACKING OF FERRITIC STEELS

The simplest stainless steel contains only iron and chromium. Chromium stabilizes the ferrite phase. Ferrite has a body centered cubic structure, it is magnetic, high in yield strength and low in ductility. Ferrite shows a very low solubility for carbon and nitrogen. The ferritic steel, AISI Type 446, is used for high-temperature applications and Types 430 and 434 for corrosion applications, such as automotive trim.

(a) Stress Corrosion Cracking of Ferritic Steels in Chloride Media

Ferritic steels, Types 430 and 434, are resistant to SCC in $MgCl_2$ at 140°C. High purity ferritic stainless steels are subjected to SCC in boiling 30% sodium hydroxide in tests exceeding 1000 h and in 42% $MgCl_2$ in a sensitized condition. Types 430 and 446 stainless steels are subject to chloride SCC in the welded conditions. In the presence of high residual levels of copper (0.37%) and nickel (1.5%), the alloys become susceptible to SCC in 42% $MgCl_2$ solution.

Conventional ferritic stainless steels containing the same amount of carbon and chromium as austenitic steels undergo sensitization more rapidly than the austenitic steels. The ferritic steels generally become sensitized when quenched from 927–1149 °C. Sensitization is eliminated or minimized by annealing between 732 and 843°C for a sufficient length of time. If the steel is held for a sufficiently long period of time, chromium diffuses back to the grain boundaries and sensitization is eliminated.

4.8.25 STRESS CORROSION CRACKING OF MARTENSITIC STEELS

It is possible to obtain austenite at elevated temperature with low chromium and relatively high chromium content. Fast cooling of austenite transforms it into martensite which has body centered tetragonal structure. Martensite is strong and brittle and it can be tempered to a desired level of toughness and strength. The corrosion resistance of martensite is, however, not as high as that of ferrite and austenitic steels.

Martensitic steels are generally susceptible to SCC in a wide range of environments. The martensite steels, such as grade AISI 400 types, are susceptible to hydrogen embrittlement when stressed or exposed to sulfide or chloride environments. The specific anions to cause failure are not as important as hydrogen.

Figure 4.53 shows the polarization curve for a modified 12% Cr martensitic steel in 3% NaCl at 25°C [95,96]. It is observed in the figure that in aerated conditions, pitting occurs as the corrosion potential is noble to (higher than) the pitting potential of the steel. Pitting is facilitated as the anodic current density is increased. If a cathodic current is applied, it is observed that there is an increase in the time of failure. It is also observed from the figure that if the corrosion potential (E_c) is more active (lower) than pitting potential (E_p), pitting is not likely to occur. The only cathodic reaction is the reduction of water which lowers the pH. At cathodic current densities higher than $25\,\mu A/cm^2$, however, hydrogen evolution is the major cathodic reaction and the time to failure decreases because of the absorption of hydrogen. There is a great deal of evidence to show that cathodically generated hydrogen can cause embrittlement of high strength steel. In a high strength, low-alloyed steel, with a martensitic structure, tempering at 400°C eliminates the susceptibility to SCC by hydrogen embrittlement.

The best method to distinguish between SCC and hydrogen embrittlement or hydrogen induced cracking (HIC) is by cathodic polarization.

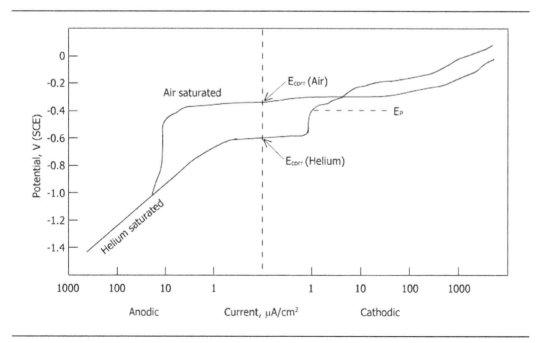

Figure 4.53 Polarization curves for a modified 12% Cr martensitic stainless steel in 3% NaCl at 25°C. (B.E. Wilde, The theory of stress corrosion cracking in alloys, NATO Scientific Committee Research Evaluation Conference, 1971. Reproduced by kind permission of British Corrosion Journal)

By increasing cathodic polarization, SCC is prevented but hydrogen cracking is accelerated.

4.8.26 HYDROGEN DAMAGE (HIGH TEMPERATURE HYDROGEN ATTACK)

The role of hydrogen in the embrittlement of martensitic steel has already been discussed above. However, embrittlement is not the only way in which materials are damaged by hydrogen. Steels are also damaged by hydrogen blistering at high temperatures. Thus, there are three categories of hydrogen damage (Fig. 4.54)

(a) High temperature hydrogen attack (hydrogen damage)
(b) Hydrogen blistering
(c) Hydrogen embrittlement.

A comparison of three types of attack is shown in Table 4.13.

(a) High Temperature Hydrogen Attack

This type of attack requires the presence of atomic hydrogen because of the inability of the molecular hydrogen to permeate steel at atmospheric temperatures. At temperatures above 230°C and hydrogen partial pressure above 100 psi (7 kg/cm^2), atomic hydrogen reacts with the carbon component in the steel to form methane.

$$Fe_3C + 4H \rightarrow 3Fe + CH_4 \qquad (4.50)$$

The removal of carbon as shown by equation (4.50) causes a loss of strength. Accumulation of methane inside builds up high internal pressure inside the steels and creates fissures preferentially at the grain boundary or non-metallic inclusions. Since neither molecular hydrogen nor methane is capable of diffusion through the steel, so these gases accumulate. Therefore, hydrogen attack at high temperature results mainly from the formation of fissures by methane and by decarburization of steel. The steel after hydrogen attack may also be found to contain blisters in addition to the fissures. These blisters, however, differ from the low temperature blisters in that they contain methane (CH_4) instead of hydrogen.

Use of stabilized grades of steel is one preventive method. Prevention of blisters is difficult to avoid by coatings or linings because of the high permeability of hydrogen.

(b) Hydrogen Blistering (Hydrogen Induced Cracking)

This is caused by the atomic hydrogen diffusing into a steel and being trapped at a non-metallic inclusion or at a grain boundary to produce

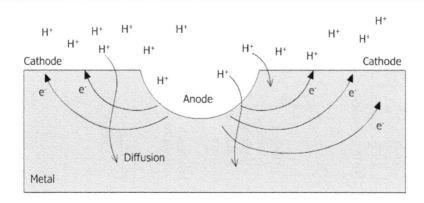

Figure 4.54 Illustration of Hydrogen diffusion

Table 4.13 Classification of hydrogen damages

No.	Type of attack	Environment	Source of atomic hydrogen	Type of metal deterioration	Method of prevention
1.	Hydrogen attack	Temperature > 450°F (230°C). H_2 Pressure > 100 psi (7 kg/cm^2· G)	Equilibrium dissociation of H_2 molecules: $H_2 \rightarrow 2H^0$	Decarburization and fissuring: possibly methane blisters. Serious loss of strength.	Use of adequately resistant alloys.
2.	Hydrogen blistering (hydrogen induced crack, HIC)	Temperature 212°F (100°C) in the presence of moisture and usually H_2S. Promoted by cyanides.	Generated by corrosion: $H^+ + e \rightarrow H$	Blisters when defects are shallow from the surface. Cracks parallel to the surface, when defects are deep.	1. Protective linings. 2. Use of adequate materials (anti-HIC steel). 3. Chemical treatment of corrosive medium with water, polysulfides or inhibitors.
3.	Hydrogen embrittlement	High strength steel in environments same as above.	Same as above.	Severe loss of ductility at low strain rates and delayed fracture	Same as hydrogen blistering.
4.	Hydrogen cracking	Nearly atmospheric temperature by rapid cooling of high strength equipment operating in conditions similar to those in hydrogen attack.	As in hydrogen attack.	Severe loss of ductility at low strain rates and hydrogen assisted crack growth.	1) Use of adequate materials. 2) Hydrogen degassing. 3) Appropriate startup procedures.

molecular hydrogen. As a result of formation of molecular hydrogen inside the non-metallic inclusions or grain boundaries, a high pressure is localized at the inclusions or grain boundaries until the bulging occurs, producing blisters or cracks. These cracks are parallel to the surface along the original laminations generated at various depths. These are finally connected together. Stepwise cracking occurs when short blisters at varying depths within the steel link together to form a series of steps. A schematic of hydrogen induced cracking in steel is shown in Fig. 4.55. Details are illustrated in Figs. 4.56 and 4.57.

Blistering is a very common name and it is often confused with high temperature hydrogen attack described earlier. It is, therefore, recommended not to use the name '*blistering.*' Hydrogen induced cracking (HIC) can be used as an alternate name to avoid confusion.

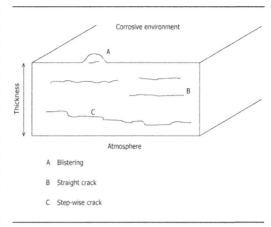

Figure 4.55   Hydrogen induced cracking

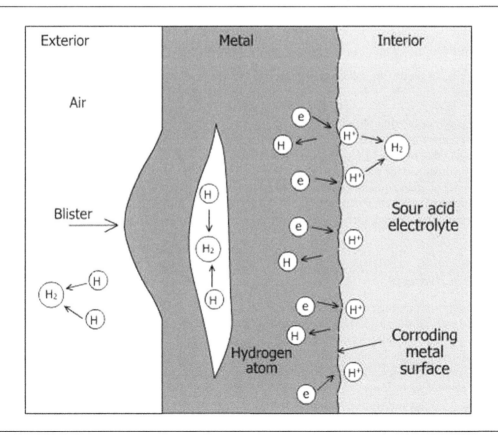

Figure 4.56   Hydrogen blistering in the wall of a container

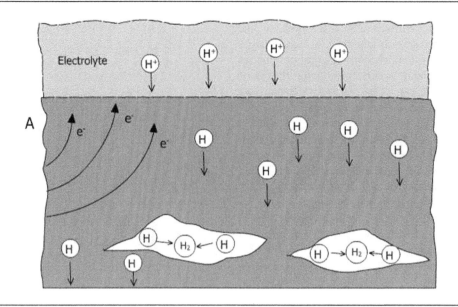

Figure 4.57 Blister formation

Conditions

For the HIC to occur, the following conditions must occur:

(a) The presence of water phase.
(b) The presence of atomic hydrogen.
(c) An agent that retards the formation of molecular hydrogen at the surface.
(d) Presence of grain boundaries or inclusions.
(e) Maintenance of an active surface.
(f) Discontinuity in metal, such as slag, inclusion and/or void.

Sources of Hydrogen

In order to protect the high strength non-stainless steels against corrosion, a good surface treatment must be provided. Incidently, surface treatment like pickling or electroplating, are likely to introduce hydrogen in the steel. In high strength steels, a small amount of hydrogen may cause serious cracking. Hydrogen absorption may also take place during manufacture, fabrication, welding and heat treatment.

Mechanism of Hydrogen Formation

The formation of atomic hydrogen is exemplified, for instance, by the following reactions:

$$H_2S \rightarrow 2H^+ + S^{2-} \qquad (4.51)$$

$$Fe + 2H^+ \rightarrow Fe^{++} + 2H^0 \qquad (4.52)$$

$$2H^0 \rightarrow H_2 \qquad (4.53)$$

Such a reaction takes place in the presence of hydrogen sulfide. Free hydrogen ions are produced at the cathode during the formation of iron sulfide scale as well as during the dissolution of this scale.

Prevention

(a) Changing the corrosive environment.
(b) Coating or lining.
(c) Using steel resistance to hydrogen induces cracking, such as steels containing Cu or cobalt.

(c) Hydrogen Embrittlement

Once hydrogen has been absorbed by a material, its effect, regardless of the source from where it has been absorbed, is the same. Gaseous hydrogen and hydrogen released from a cathodic reaction differ from each other in the following respects:

(1) Cathodic hydrogen is adsorbed on the surface as atomic hydrogen (reduced), whereas gaseous hydrogen is adsorbed in the molecular form and it then dissociates to form atomic hydrogen.
(2) The internal pressure produced by the gaseous hydrogen is much lower than produced by cathodic hydrogen, due to the log term in the Nernst equation which converts an E value into an exponent on the hydrogen pressure.

Hydrogen embrittlement occurs during the plastic deformation of alloys in contact with hydrogen gas and is strain rate dependent. Alloys, like ferritic steel, nickel-base alloys, and titanium, show highest degradation in properties when the hydrogen pressure is very high and strain rate is low.

Hydrogen embrittlement is a phenomenon whereby hydrogen is absorbed in the metal (diffuses), exerts local stresses, and leads to embrittlement of material, such as high strength steels.

Examples of Embrittlement

(1) In plating operations.
(2) In pickling operations.
(3) In cleaning of high strength steels in chloride or fluoride solution.
(4) Manufacturing and fabrication processes.

Materials Most Susceptible

Iron, titanium, zirconium, martensitic steels, high strength aluminum alloys.

Identification

Hydrogen embrittlement results in a brittle fracture throughout the embrittled material as a result of hydrogen adsorption unless the strength of the remaining material is less than the load applied. Later, instantaneous final fracture occurs. The failure by hydrogen embrittlement is mostly intergranular. The fractured surface has, therefore, a crystalline appearance.

Mechanism

No definite mechanism of hydrogen embrittlement has been suggested. It has been considered sufficient to identify hydrogen as a cause for cracking. It is a general opinion that impurity segregations at the grain boundary act as poisons, and increase the adsorption of cathodic hydrogen at these sites.

It is widely believed that in BCC materials hydrogen embrittlement is caused by the interaction of hydrogen with defects in the structure. Such defects are vacancies, dislocations, grain boundaries, interface, voids, etc. Hydrogen is trapped in these defects and growth of a crack is facilitated. A large number of such defects interact with hydrogen and the combined trapping results in a significant loss of ductility.

Difference between SCC and Hydrogen Embrittlement

It is possible to distinguish between SCC and hydrogen embrittlement by applied currents. If on applying a current, a specimen becomes more anodic and cracking is accelerated, the attack is SCC, whereas, if cracking is accelerated in the opposite direction and hydrogen evolution is observed, the attack is hydrogen embrittlement. The statement is, however, oversimplified. It has been discussed earlier.

The following are the differences between SCC and hydrogen embrittlement from a fractographic point of view:

(1) SCC begins at the surface, whereas hydrogen embrittlement begins internally.
(2) The magnitude of corrosion is higher at the origin of SCC than observed with hydrogen embrittlement.

Prevention of Hydrogen Embrittlement

(1) Select materials resistant to hydrogen embrittlement for elevated temperature applications.

(2) In pickling or plating operations, submit the material to low temperature aging at temperature between 161 and 370°C to eliminate the effects of hydrogen embrittlement. Chlorides and fluorides are known to cause hydrogen embrittlement. There are recommended concentrations of such solutions for use in processing so that damage may be avoided. Hydrogen embrittlement of titanium can be avoided if the ratio of HNO_3/HF exceeds 10. Cleaning and plating solutions should be carefully screened to avoid hydrogen embrittlement problem, special care being needed when cleaning welding tints off welds.

(3) Use of inhibitors. Inhibitors may be added to pickling baths to minimize corrosion.

(4) Baking. It is recommended that high strength steel be subjected to baking at low temperature after plating. Temperatures between 200 and 300°F are generally employed.

(5) Change of design. Avoid sharp corners, as they act as stress raisers. Also eliminate sites for crevice corrosion.

(6) Heat treatment. Temper at a higher temperature for a long period of time. This would lower the strength.

(7) Surface preparation technique. Use techniques which introduce compressive stresses on the surface.

(8) Substituting alloys. Alloys with nickel or molybdenum reduce susceptibility but do not eliminate it.

(9) Use low hydrogen welding rods.

4.8.27 CAUSTIC CORROSION

Austenitic nickel–chromium stainless steels and mild steel are subject to stress corrosion cracking in caustic soda (caustic cracking) at elevated temperatures. The phenomenon, 'caustic cracking' is mostly encountered in boilers. Caustic is added as an additive to boiler water in order to preserve the thin film of magnetic iron oxide by raising the pH. Caustic addition creates problems only when it becomes concentrated. It can become concentrated by one of the mechanisms summarized below. Four instants in the life of a steam bubble are shown in Fig. 4.58.

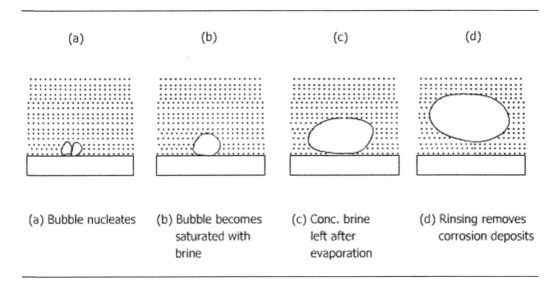

(a)   (b)   (c)   (d)

(a) Bubble nucleates

(b) Bubble becomes saturated with brine

(c) Conc. brine left after evaporation

(d) Rinsing removes corrosion deposits

Figure 4.58   Four instants in the life of a steam bubble

(a) Departure from Nucleate Boiling (DNB) [97]

This refers to the condition in which bubbles of steam nucleate at points on the metal surface. As these bubbles form, minute concentrations of boiler water develop at the metal surface at the bubble/water interface. As the bubbles separate, the water redissolves sodium hydroxide. The rate of bubble formation exceeds the rinsing rate at the onset of the departure from nucleate boiling (DNB) and, hence, caustic as well as other solids begin to concentrate.

(b) Under Deposits

Many times steam is formed under the insulating deposits, which leaves a rich caustic residue.

(c) Evaporation at Waterline

It is possible also for the caustic to concentrate at the waterline. Generally, the waterline area is always most sensitive to corrosion. Two types of failures are common in boilers and both are related to the effect of concentration of caustic. One which forms discontinuous microcracks and results in the bursting of tubes is called 'hydrogen damage,' and the other which results in the formation of continuous microcracks leading to intergranular corrosion is called 'caustic cracking.' Both are briefly discussed below:

(1) Mechanism of Hydrogen Damage by Concentrated Caustic Solutions

Caustic solution acts by dissolving the magnetic iron oxide

$$4NaOH + Fe_3O_4 = 2NaFeO_2 + Na_2FeO_2 + 2H_2O \qquad (4.54)$$

The protective coating of magnetic iron oxide is thus destroyed. After the coating is destroyed, water reacts with iron to evolve hydrogen:

$$3Fe + 4H_2O = Fe_3O_4 + 4H_2 \qquad (4.55)$$

Also, the caustic soda may react with iron to produce hydrogen:

$$Fe + 2NaOH = Na_2FeO_2 + H_2 \qquad (4.56)$$

If atomic hydrogen is produced it diffuses to the metal at inclusions and grain boundaries to form molecular hydrogen, or alternatively

$$Fe_3C + 4H \rightarrow CH_4 + 3Fe \qquad (4.57)$$

It forms CH_4 as shown in the equation.

Molecular hydrogen cannot diffuse into steel and, therefore, it accumulates on the grain boundary. Due to enormous pressures exerted by hydrogen, discontinuous microcracks are formed in the grain boundaries. The metal strength continues to decrease by the pressure by methane or hydrogen until damage occurs. A steel tube in a boiler would burst under the pressure.

(2) Caustic Cracking

In the presence of sufficient tensile stress and traces of silicon, hot caustic solutions can induce SCC of boiler steels. This phenomenon is not called 'caustic embrittlement,' as no loss of ductility occurs in caustic cracking. Tensile stress and caustic concentration cause the formation of continuous intergranular cracks in the metal. As the cracks progress, the strength of the metal is exceeded and fracture occurs.

Improved water treatment practices and improved boiler designs have minimized the problem.

4.8.28 SULFIDE STRESS CORROSION CRACKING

This phenomenon is generally encountered in high strength steels of Rockwell hardness above 22 in a sour oil field environment. It is a special case of hydrogen stress cracking, also called hydrogen embrittlement. Factors responsible for sulfide stress corrosion cracking (SCC) are:

(1) Notches, pits, irregularities, inclusions on the metal surface.

(2) Hydrogen diffusion.
(3) Tensile stress.
(4) Environment. H_2S concentration. The partial pressure of H_2S for a 80 ksi yield strength stress is 0.01 atm, and 0.0001 atm for a 130 ksi strength steel.
(5) pH below 6.0.
(6) Chloride has a significant effect on 12% Cr steels and little effect on low alloy steels.
(7) Highest sensitivity is observed at 20°C.

Metallurgical factors

(1) Yield strength. Resistance to SCC decreases with increased strength.
(2) Cold work. It decreases the resistance to SCC.
(3) Hardness. Susceptibility to SCC increases with hardness.

(4) Microstructure. It varies from one alloy system to another but it has a direct influence on SCC.

Mechanism The following steps lead to the formation of FeS (ferrous sulfide), which causes SCC (Fig. 4.59):

$$\text{Anode: Fe} \rightarrow Fe^{2+} + 2e \qquad (4.58)$$

$$\text{Cathode: } H_2S + H_2O \leftrightarrows H^+ + HS^- + H_2O \qquad (4.59a)$$

$$HS^- + 2H_2O \leftrightarrows 2H^+ + HS^- + H_2O \qquad (4.59b)$$

$$\text{Combination: } 2e^- + 2H^+ + Fe^{2+} + S^{2-} \leftrightarrows 2H^0 + FeS \qquad (4.60)$$

$$\text{Net: Fe} + H_2S \xrightarrow{\text{HOH}} FeS + 2H^0 \qquad (4.61)$$

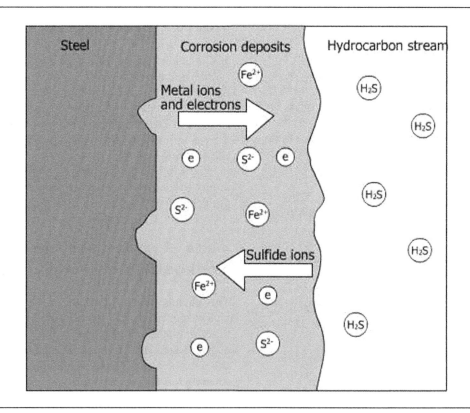

Figure 4.59 Mechanism for sulfide stress corrosion cracking

Prevention

(1) Use alloys less susceptible to SCC.
(2) Use steels of HRC below 22.
(3) Avoid the sensitizing temperature.
(4) Minimize tensile stresses in the system.

4.8.29 FAILURE CASE HISTORIES

A. Case 1

In 1985, the concrete roof of a swimming pool collapsed. The roof was supported by stainless steel rods.

Environment Chlorine-based disinfectants were used in the pool. The temperature was maintained at about 1°C above water temperature. High levels of humidity were observed.

Cause Stress corrosion cracking was established to be a major cause. Chlorine-containing compounds were transferred via the pool atmosphere to surface away from the pool itself resulting in the production of a highly corrosive film on the stainless steel rods which supported the roof structure. The failure occurred around 30°C in highly stressed components which were not frequently cleaned. Such components included brackets, rods, bars, wire ropes, fasteners, etc.

Prevention The following remedies were recommended by the consultants:

(1) A better design to resist localized corrosion and elimination of possible side of crevices.
(2) A rigorous inspection.
(3) Effective management including maintenance.
(4) Reduction of temperature around highly stressed components.
(5) Regular cleaning of stressed components.
(6) Recirculation of pool.
(7) Lower humidity and chloride buildup.

B. Case 2

Stress corrosion cracking of a U-bend waste fuel heat exchanger tube.

Environment Shell side – hot waste fuel. Tube side – cooling water.

Material Type 316 stainless steel U-bend exchanger tube. Temperature of tube side. Service life is 5 years at 21°C.

Cause of failure Failure analysis showed that 316 SS heat exchanger tubes suffered chloride stress corrosion cracking in straight sections as well as U-bend of heat exchanger tubes. The root cause was attributed to a blockage in cooling water which activated the hot oil at 21°C to raise the temperature of the water containing about 800 ppm chloride. The problem was resolved by free circulation of the cooling water.

C. Case 3

Chloride stress corrosion cracking of a 304L SS, 28" diam buried pipeline carrying CO_2 feedstock, after seven years of service life.

Environment The pipeline ran parallel to a roadway which was frequently de-iced with salts during winter period. The adjacent soil of the pipe contained 150–260 ppm of chloride.

Cause A metallurgical history of the pipe showed that it was exposed to tensile stresses, thermal stresses and residual stresses from weldment. The temperature reached about 45°C. Surface topography suggested intergranular cracking.

Prevention SCC occurred because of the leaching of chlorides from the adjacent soil which was de-iced by salt in winters. The ingress of chloride may be stopped either by some insulation or by changing the location of the pipe well away from the roadway.

D. Case 4

Failure of landing gears. (*Courtesy: S. J. Keccham, Naval Development Center, Westminister, PA, USA.*)

A main landing gear shock strut piston assembly failed catastrophically while on the ground, separating both axle stubs and lower portion of the shock strut piston from the gear.

The aircraft had 559 flight hours and no records of overweight.

Material The steel used was air-melt 4340 and not vacuum melt.

Observations The cracks near the origin indicated the failure to be caused by intergranular cracking or hydrogen embrittlement followed by fracture. The axle bore showed active pitting. The failure occurred by intergranular stress corrosion cracking.

Recommendations It was recommended to use vacuum melt steel. Air-melt steel is more susceptible to stress corrosion cracking than vacuum melt steel. To avoid hydrogen embrittlement, cadmium plating in vacuum ("*ion plating*" vacuum equipment) was also recommended.

E. Case 5

Failure of 17-4 PH Bolts. (*Courtesy: James A. Stanley, Proceedings of the First Joint Aerospace and Marine Corrosion Technology Seminar, Los Angeles, CA, p. 30, 1960.*)

On a launch vehicle, high strength 17-4 PH steel bolts were used on an aluminum body oxidizer valve. This martensite steel was heat treated (H-900 condition) to a minimum strength of 190 000 psi. On exposure to marine atmosphere at Cape Kennedy, the bolts were found to be cracking.

Observations The failure was not caused by SCC, shown by laboratory tests, but by hydrogen induced cracking (HIC). The bolts were used to hold down aluminum forging of alloy 7075-T6. Because of galvanic corrosion, due to contact of steel with aluminum, hydrogen was being released into steel, which caused HIC.

Recommendations

(1) Protection by over-aging. Over-aging at 51°C for one hour was recommended to eliminate HIC.
(2) Vacuum deposited coating to a thickness of not less than 0.5 mil.
(3) Anodizing of aluminum alloy 7075-T6 which is in contact with steel.
(4) Application of lubricants, greases and paints.

4.8.30 PREVENTION OF SCC

The following general methods are recommended to minimize stress corrosion cracking of austenitic steels:

(1) Eliminate chloride salts as far as possible in acid environments.
(2) Remove oxygen or use oxygen scavengers in the case of chloride-containing environment.
(3) Convert the tensile stresses to compressive stresses by shot peening.
(4) Avoid concentration of chloride or hydroxide by proper equipment design. Allow complete drainage in equipment. Prevent designs that would concentrate chlorides or caustics by vaporization.
(5) Reduce stresses by stress relief of the steel. The stress relief of non-stabilized grades or low carbon stainless steels renders them more susceptible to intergranular cracking than if they are not heat treated.
(6) Adopt a good design practice. Highly stressed areas can be minimized by adopting a good design practice. For instance, forged elbows are preferred over mitreed elbows, and forged tees are preferred over direct stub-in joints. Avoid poor welding workmanship.
(7) Select a more crack-resistant alloy if other preventive measures fail to work. Use high nickel alloys or alloys containing very low levels of nitrogen and other impurities, if present. For instance, purified 16 Ni–20 Cr is not susceptible to cracking. Choose one of the lower nickel duplex stainless steels or a ferritic steel.
(8) Use coatings to provide a barrier between the steel and the environment. Coatings have been successfully used to prevent access of moisture on the outside of stainless steel lines, operating at 49–260°C in the chloride-laden atmosphere of the Arabian Gulf.
(9) Use inhibitors. High concentration of phosphate have been successfully used.
(10) Cathodic protection. Impressed current cathodic protection system has been

successfully used to prevent SCC of steels. This system has been used to protect steel in seawater.

(11) Lower the temperature if service conditions permit. For instance, keep below the critical temperature (85°C) for SCC to occur in chloride media. Whenever possible, maintain high velocity to prevent any deposition of debris.

(12) Modify the environment, if possible, by changing pH or reducing oxygen content as in heat exchangers or fluids.

In addition, the following measures may be taken to prevent SCC of martensitic, or ferritic steels.

(a) Do not allow hydrogen to be generated by any source as hydrogen in these steels cause hydrogen damage.

(b) If cathodic protection is being applied take care that no over-protection occurs, as a slight generation of hydrogen would lead to hydrogen damage.

(c) Reduce the hardness of martensitic or precipitation hardened steels to values lower than Rockwell C-400.

(d) Take precautions against nitrogen or carbon pick-up during mill processing or during the welding of fabricated equipment.

All of the methods given are not ideal and they must be used judiciously taking into consideration the past experience, materials compatibility, environments and the service life.

(13) Reduce tensile stresses by shot peening. Shot peening introduces surface compressive stresses. Shot peening counter-balances tensile stresses.

4.9 CORROSION FATIGUE

4.9.1 INTRODUCTION

Fatigue failures are as old as the Industrial Revolution and they have been responsible for some of the major engineering catastrophes, for example the Comet aircraft in the past. Mechanisms of mechanical failure in aircraft have been investigated over the past fifty years [98] by Royal Aerospace Establishment. Fatigue is indicated to be the most frequent of the failure types compared to other modes. Fatigue failure is a major mode of failure of components in automobiles, transport, petroleum, petrochemical, shipping and construction industries.

Corrosion fatigue is a process in which a metal fractures by fatigue prematurely under conditions of simultaneous corrosion and repeated cyclic loading at lower stress levels than would be otherwise required in the absence of a corrosive environment. Metals and alloys will crack in the absence of corrosion if they are subject to high cyclic stress for a number of cycles. The number of cycles for failure decreases as the stress is increased. Below a certain stress the metal will last indefinitely. This level is termed as '*Endurance Limit*' of the material.

If, however, the material under cyclic stress is subject to a corrosive environment, the endurance limit of the material is sharply reduced. The premature failure of a material from the exposure to the combined action of corrosion and cyclic stress is called '*Corrosion Fatigue.*'

This type of failure is generally encountered in pump shafts, heat exchanger tubes, rotors, steam turbine blades, aircraft wheels, boiler and steel equipment. Some major examples where corrosion fatigue is encountered are given below:

(1) Ships
(2) Offshore platforms
(3) Drilling rigs
(4) Navigation tuners
(5) Aircrafts
(6) Submarine pipelines
(7) Communication equipments
(8) Heat exchanger tubes
(9) Pump shafts
(10) Alternate dry and wet zones.

4.9.2 CHARACTERISTICS OF CORROSION FATIGUE

(a) The main distinguishing feature of corrosion fatigue cracks is the presence of several cracks near the fracture. In a normal fatigue failure, usually one crack is present.

(b) The corrosion fatigue cracks run parallel to each other and are aligned perpendicular to the direction of principal stress.

(c) Corrosion fatigue cracks in carbon steels often propagate generally from the base of corrosion pits.

(d) Generally, the fatigue corrosion cracks in certain steels are transgranular in nature (Fig. 4.60).

(e) The striations formed in corrosion fatigue failure are less pronounced than in normal fatigue failure, as shown by electron microscopy. Fatigue striations are shown in Fig. 4.61.

(f) The cracks generally initiate at the surface where the stress is maximum.

(g) The fractured surface is dull in appearance and may contain corrosion products.

A good indication of the effect of corrosion on the fatigue strength of materials is obtained by the conventional *S–N* curves (stress *vs* number of cycles to failure). A corrosive environment promotes crack initiation and shortens the fatigue life. Figure 4.62 shows a typical *S–N* fatigue curve. The endurance limit is represented by the asymptotic portion of curve. It is defined as the stress below which no failure will occur for a given number of cycles. The effect of corrosive environment on the fatigue life is clearly shown in Fig. 4.63. It is observed from the figure that fatigue life is drastically reduced in the environments shown. It is also observed that the fatigue life in 0.1 M NaCl is much less than in distilled water. Most important, no endurance limit is shown by the *S–N* curves in either distilled water or in NaCl.

The variables affecting corrosion fatigue may be classified into three categories: (a) mechanical

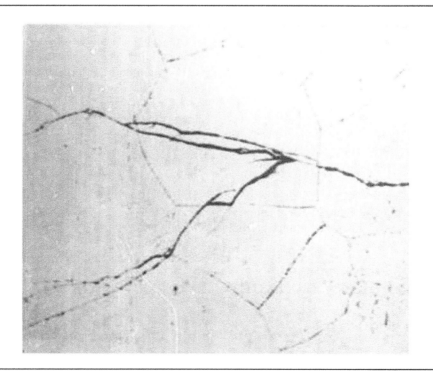

Figure 4.60   Transgranular cracks in corrosion fatigue

(b) metallurgical and (c) environmental, as shown below:

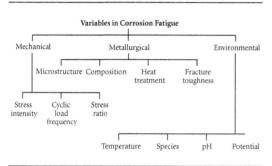

Figure 4.61 Fatigue striation in a 7178 aluminum alloy. (From Pelloux, M.N. (1965). In *Metals Engineering Quarterly*, ASM, November)

4.9.3 EFFECT OF VARIABLES ON FATIGUE AND CORROSION FATIGUE

A. Mechanical Factors

1. Stress Intensity

This factor is of a fundamental importance in the prediction of brittle fracture using linear fracture mechanics (LEFM) principles. It is a function of both crack geometry and associated loading.

The analytical expression given below can be used to express the relationship of the stress intensity factor to crack growth:

$$\frac{da}{dN} = C(\Delta K)^n \qquad (4.62)$$

where ΔK is stress intensity factor range $(K_{max} - K_{min})$, C and n are constants, the value of n varies between 2 and 4, depending on temperature, environment and frequency. The above equation is called Paris equation. The constants are called Paris constants. They can be found from Table 4.14.

An expression similar to equation (4.63) can be applied to fatigue data for a range of materials in various environments. A schematic diagram of stress intensity, ΔK vs fatigue crack growth rate (da/dN) is shown in Fig. 4.64. The exponent varies with the environments. The corrosion fatigue crack propagates in both inert and aggressive environments as shown.

(a) In some cases, the presence of air and environment causes a da/dN vs ΔK plot to move to a higher da/dN value at a given ΔK.

(b) In some cases, corrosion fatigue increases da/dN more at low ΔK than at high ΔK, so there is a reduction of exponent n in the Paris equation.

(c) In the intermediate ΔK range, the Paris law is approximated [98]. In this region, it is reported that the growth of fatigue cracks is by ductile striation which is associated with corrosion fatigue.

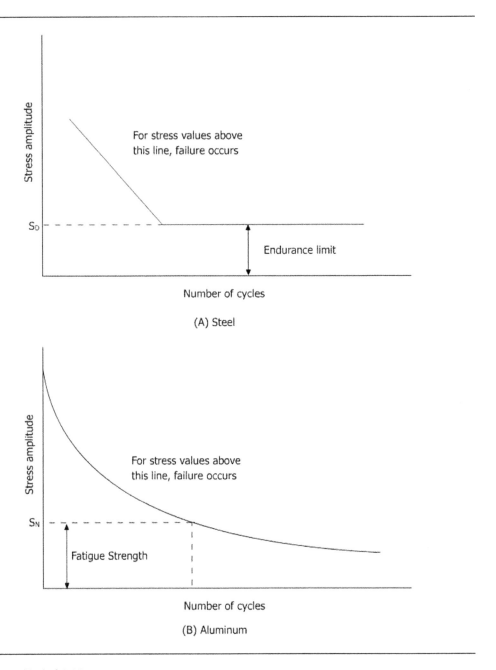

Figure 4.62 Typical *S–N* curves

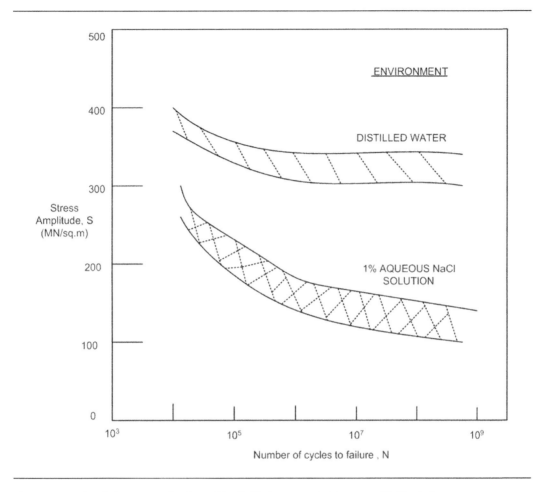

Figure 4.63 *S–N* fatigue curves for Cr steel in distilled water and 1% aqueous NaCl solution

A schematic representation of fatigue crack growth is shown in Fig. 4.64. It has basically three regions:

Region A. The crack growth rates are very low. The region begins with a threshold value of stress intensity, below which crack propagation does not occur. This region continues until the slope becomes constant.

Region B. Shows a linear relationship between log ΔK and log da/dN. Equation (4.62) shows a power law dependence. It can be expressed as

$$\log \frac{da}{dN} = \log[C(\Delta K)^n] \qquad (4.63)$$

On simplification it yields

$$\log \frac{da}{dN} = n \log \Delta K + \log C \qquad (4.64)$$

Equation (4.64) shows a straight line relationship between (da/dN) and ΔK, with slope n and intercepts C. Region B shows the steady-state rate of crack growth. The value of n is 3 for steels and in the range of 3–4 for aluminum alloys.

Region C. It exhibits a steep gradient and the crack growth rate accelerates. In Region C, the value of $K_{\max}$ is attained. Region B is the most important region

Table 4.14 Paris constant and Paris exponent values

Alloy	Environment	$\log_{10} C$	n
5086-H116	Air	−10.75	3.53
	Seawater, free corrosion	−10.47	3.41
	Seawater – 0.75 V	−10.09	3.16
	Seawater – 1.3 V	−10.82	3.69
	Seawater – 1.4 V	−11.13	3.76
5086-H117	Air	−10.78	3.53
	Seawater, free corrosion	−9.09	2.26
	Seawater – 0.75 V	−9.55	3.05
	Seawater – 1.3 V	−10.50	3.30
	Seawater – 1.3 V	−11.43	3.95
	Seawater – 1.4 V	−11.47	3.95
5456-H116	Air	−10.67	3.53
	Air	−10.90	3.68
	Seawater, free corrosion	−11.16	4.20
	Seawater – 0.75 V	−10.95	3.79
	Seawater – 1.3 V	−10.77	4.25
	Seawater – 1.3 V	−10.95	3.70
	Seawater – 1.4 V	−13.96	5.10
5456-H117	Air	−11.01	3.69
	Seawater, free corrosion	−9.85	2.97
	Seawater – 0.75 V	−9.43	2.74
	Seawater – 1.3 V	−11.81	4.11
	Seawater – 1.4 V	−15.02	6.31
	Seawater – 1.5 V	−15.66	6.86
5456-H116	Air	−9.91	3.0 ($\Delta K < 28$ MN m$^{-3.2}$)
	Seawater – 0.75 V	−12.91	3.0 ($\Delta K < 28$ MN m$^{-3.2}$)
		−9.91	3.0 ($\Delta K < 28$ MN m$^{-3.2}$)
		−12.91	3.0 ($\Delta K < 28$ MN m$^{-3.2}$)
	Seawater – 0.95 V	−11.3	4.4
	Seawater – 1.3 V	−9.91	3.0
		−12.91	5.0
2219-T851	Argon	−10.07	2.73
	H_2O	−10.00	2.73
	H_2O	−10.00	2.73
	H_2O	−9.95	2.73
	H_2O	−9.70	2.73
	H_2O	−9.65	2.73
	H_2O	−9.60	2.73

(Contd)

Table 4.14   *(Contd)*

Alloy	Environment	$\log_{10}C$	n
HY130, 10 Ni–Cr–Mo–V	Air	−9.58	2
10 Ni–Cr–Mo–V	3% NaCl	−9.38	2
10 Ni–Cr–Mo–V	3% NaCl	−9.24	2
12 Ni–5 Cr–3 Mo	3% NaCl	−9.11	2
4340	Air	−9.49	2.56
	3% NaCl	−9.10	2.56
	3% NaCl	−8.52	2.56
2 Ni–5 Cr– 5 Mo	Air	−9.58	2
(maraging steel)	3% NaCl, 10 Hz	−9.44	2
	3% NaCl, 1 Hz	−9.39	2
	3% NaCl, 0.1 Hz	−9.08	2
	3% NaCl, 1 Hz square wave	−9.59	2
AISI 4145 Cr–Mo steel	Air 15 Hz	−10.96	2.8
	0.1 N H_2SO_4 1.7 Hz	−9.51	2.02
	0.1 N H_2SO_4 0.7 Hz	−8.74	1.7
AISI 4340	Dry argon, 20 Hz	−9.82	2
	Water vapor 10 Hz	−9.70	2
	Water vapor 4 Hz	−9.56	2
	Water vapor 2 Hz	−9.35	2
	Water vapor 1 Hz	−9.12	2
	Water vapor 0.5 Hz	−9.0	2
	Water vapor 0.1 Hz	−8.35	2
3.5% Ni–Cr–Mo–V	Air	−11.86	3.31
	H2, 5 psi, R = 0	−11.42	3.18
	H2, 5 psi, R = 0.4	−12.41	3.94

for evaluating the life of engineering structures.

Materials which are affected by environment are characterized by high crack growth rate. The effect of environment is generally to enhance crack growth rates.

2. Stress Ratio

Stress ratio is defined as the ratio of minimum stress to maximum stress. The rate of corrosion fatigue crack propagation is increased by an increase in the stress ratio (Fig. 4.65).

$$R = K_{\min}/K_{\max} \qquad (4.65)$$

The situation is rather complex. A higher load ratio results in a higher mean load, which causes an accelerated interaction between the environment and the crack tip. This would, however, vary from one material system to another. A uniaxial stress seriously shortens the corrosion fatigue life. The extension of corrosion fatigue relative to dry air increased from a factor of four at R (0.11 to 0.24) to as much as 20–30 fold at high

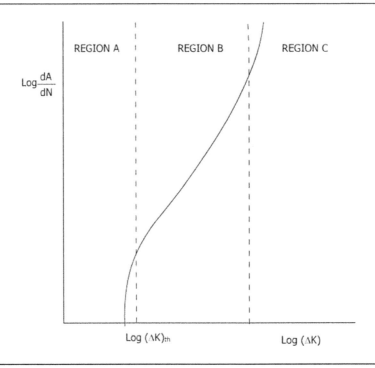

Figure 4.64 A schematic diagram of fatigue crack growth

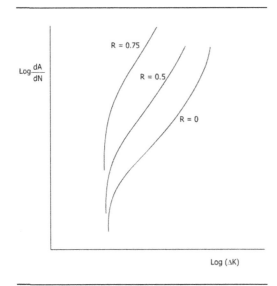

Figure 4.65 Effect of crack growth rate on varying *R* ratio

R (0.61 to 0.71). R is specified to quantify the amount by which a mean stress value is removed from zero. Aluminum alloys are sensitive to variations of R, whereas its effect on steel is limited. A negative value of K_{min} indicates closure of the tip. If K_{min} is negative, $R = \Delta K = K_{max}$, $R = 0$.

3. Cyclic Stress

The cyclic load frequency seriously affects corrosion fatigue. The service life of a component depends upon:

(a) Number of stress cycles required to initiate cracks.
(b) The number of stress cycles required for cracks to grow to a critical size before final fracture occurs.

Local stress is not important in defining the conditions of crack growth. However, by defining it in terms of stress intensity factor, it is possible to predict how much time will be taken for a crack of a sub-critical size to reach a critical value before fracture occurs.

Interaction between the material and environment depends on the rate of transfer of the environmental species to the metal. The time available for interaction is, therefore, important. The more is the time available, the greater the interaction.

If sufficient time is available, the corrosion species will have sufficient time to reach the diffusion boundary layer, and cause failure by corrosion fatigue. If the time available is short, due to faster loading rates, sufficient time would not be available to the species to reach the boundary layer and, hence, the damage would be purely mechanical in nature.

4. Stress Amplitude

In general, a low amplitude of cyclic stress favors a greater contact of the environment with the metal and favors a longer fatigue life. In case of high amplitude, there is hardly an environmental interaction. In case of high frequency, there may not be an interaction between the metal and the environment.

5. Effect of Strain Rate

New surface continues to be generated as the fatigue crack continues to grow, and the strain rate at the tip of the growing crack is high. The surface at the tip is influenced by passivation, adsorption and anodic dissolution. At low frequency, the corrosive environment is in contact for a long time period to allow the crack tip–environment interaction. As discussed earlier, at high frequency, sufficient time is not available for crack tip–environment interaction.

6. Mean Stress

If the mean tensile stress is high, the level of stress intensity at the crack tip increases and the crack remains open for a longer period of time. Also the crack has a wider opening under this condition. The corrosion fatigue damage increases because the crack tip remains in contact with the environment and also because of interaction with a larger volume of environment.

7. Stress Waveshapes

The effect of environment is affected by the shape of the cyclic stress. Exposure of the metal to an environment at the peak stress accelerates corrosion fatigue failure. The effect of waveform is important on the crack growth. Certain waveforms have a more pronounced effect than others. The effect of triangular, square, saw tooth and trapezoidal waveforms on corrosion fatigue has been demonstrated. It has been generally observed that maximum effect occurs with sinusoidal, triangular and positive saw tooth waveforms. The square and negative saw tooth wave have no effect, because of very short rise time [99]. Different types of loading cycles are shown in Fig. 4.66. It is not possible to generalize that the effect of load waveform on corrosion fatigue at this stage, as the phenomenon is not clearly understood because of lack of corrosion fatigue data. In the case of stainless steels, the kinetics of reaction depends largely on the rupture of the oxide film on it, which is controlled by the strain rate. The effect of stress waveshapes is, however, not clear and more investigations are required to establish the effect of shape of the stress waves on corrosion fatigue. The effect of waveform is particular to a specific material environment system.

8. Stress Corrosion Cracking

Stress corrosion cracking can also occur in a system susceptible to stress corrosion cracking above K_{1SCC}. The adverse effect of environment takes place at loads below K_{1SCC} only during the increasing load portion of the stress cycle. Crack growth is cycle dependent. Crack growth propagation occurs below the threshold for the time-dependent stress corrosion. The time-dependent crack growth can be attributed to the constant or the rising portion of load cycle.

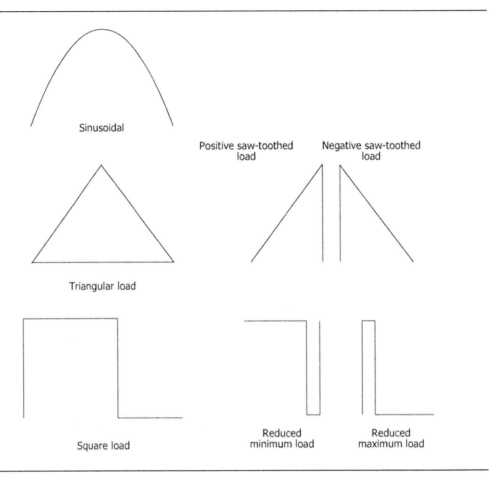

Figure 4.66 Different types of loading cycles. (Barsom, M.J. (1971). In *Corrosion Fatigue*, Int. Conf., June 14–18, NACE)

9. Environmental Factors

The aggressivity of the environment is intimately related to fatigue life. Increased concentration of a corrodant generally decreases the corrosion fatigue resistance of metals and alloys. For instance, in seawater, chemical, physical and biological factors affect the resistance of materials to corrosion. The effect of environment is shown in Fig. 4.67. It is to be observed that the corrosion fatigue limit in salt water is lower (30 ksi) than in fresh water (40 ksi).

The corrosion fatigue resistance of high strength alloys is also affected by relative humidity and condensation conditions in the environment.

10. Temperature

Temperature affects the rate of fatigue crack propagation and, hence, the fatigue life. Increase of temperature increases the rate of transport of the active species to the tip of the crack, and accelerates the propagation of the crack. It also increases the corrosion process by lowering the hydrogen over-voltage. The fatigue crack growth rate of a high strength steel in 3.5% NaCl increases with a rise in the temperature of sodium chloride.

11. Electrode Potential

Application of a constant electrode potential can either decrease or increase corrosion fatigue.

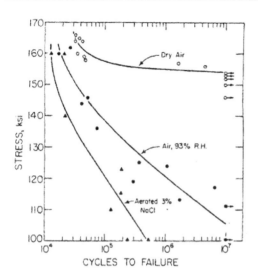

Fatigue life of 4140 steel R_c 52, in moist and dry air and in aerated 3% NaCl, 25°C.

Figure 4.67 Effect of environment on corrosion fatigue. (From Lee, H.H. and Uhlig, H.H. (1972). *Met. Trans*, **3**, 351, November)

Whether the corrosion fatigue would be accelerated or slowed down would depend on the environment, material (passive or active) and the direction in which the potential is applied (active or noble). Generally, fatigue life increases with more active potential, however, the potential applied to achieve cathodic protection is not sensitive to applied stress. The crack propagation rates are sensitive to changes in potential. The potential changes can either decrease or increase the crack propagation rates. The increase or decrease would depend on the applied potential in the negative or positive direction (cathodic or anodic), the nature of reactive species in the environments, and the mechanism of interaction with the specimen.

Moderate cathodic protection can improve a corrosion fatigue performance of low strength steels. In the case of high strength alloy, cathodic protection may accelerate fatigue crack initiation. Steel specimens, AISI 4040 (R_c 20), in 3% NaCl were stressed 10% above and below the fatigue limit. The specimens were maintained at a series of constant potentials by a potentiostat. The fatigue life increased as the applied potential in the active direction (cathodic) increased as shown by failure tests (Fig. 4.68). At an applied stress of 53×10^3 psi below the fatigue limit failure did not occur within 10^7 cycles below -0.48 V(SHE). At 65×10^3 psi, maximum life was observed below -0.48 V(SHE). It shows that the protection potential in either case does not change with the applied stress. For specimens with a higher hardness value, R_c 52, tests at an applied stress of 141 ksi showed increased life for more active potentials, but the maximum life did not reach more than 10^7 cycles (Fig. 4.69). A maximum life of 7×10^6 cycles was observed at more active potentials from -0.55 to -0.85 V. The life is almost 100 times the life at corrosion potential. Hydrogen is, however, evolved at more active value, which causes damage by hydrogen embrittlement and shortens the life. In conclusion, cathodic protection is effective in 3% NaCl for steels between hardness values of R_c 20 and R_c 44 in 3% NaCl.

Anodic polarization of steels, aluminum alloys and copper alloys, decrease their resistance to fatigue crack initiation, whereas cathodic polarization increases the resistance.

4.9.4 METAL ENVIRONMENT INTERACTION

When a metal is subjected to continuously applied cyclic stress, the extent to which fatigue crack initiation and propagation is influenced would depend on:

(1) The thermodynamic tendency of the metal and the environment to react.
(2) The reaction kinetics, the type of reaction and reaction produced.

In aqueous systems, the following processes may take place:

(a) Adsorption of atoms and molecules on the surface of the atom.
(b) Film formation at the metal–electrolyte interface.
(c) Dissolution of the metal.

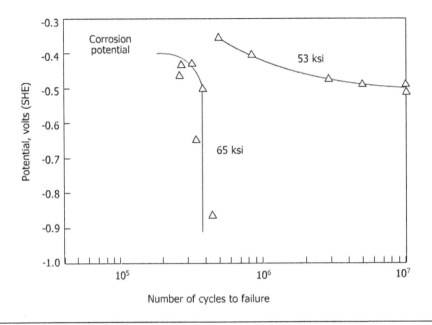

Figure 4.68 Effect of applied potential on 414 steel R_c 20 stressed above and below the fatigue limit, on fatigue life in aerated 3% NaCl. (From Lee. H.H. and Uhlig, H.H. (1972). *Met. Trans. ASM*, **3**, November, 2949–2957. Reproduced by kind permission of ASM, Metals Park, Ohio, USA)

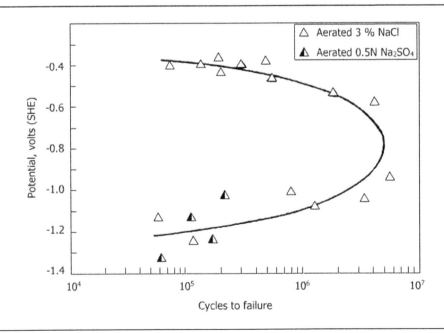

Figure 4.69 Effect of applied potential of 4140 steel R_c 52 stressed at 141 ksi (below the fatigue limit) on fatigue life in aerated 3% NaCl and aerated 0.5N Na_2SO_4, 25°C. (Lee, H.H. and Uhlig, H.H. (1972). *Met. Trans.*, 2949, November. Reproduced by kind permission of ASM, Metals Park, Ohio, USA)

However, as the crack progresses, the effect of environment also alters as forces such as mechanical, become very predominant. The effects of adsorption which cause embrittlement may, for instance, increase with an increase in the stress intensity. The effect of environment on crack initiation and crack growth varies from one metal to another. The fatigue life in the presence of an aggressive environment at a given stress value generally decreases. The corrosion fatigue is highly dependent on particular metal environment combination. The following can be the environment variables:

(1) Temperature.
(2) Type of environment.
(3) Concentration of reactive species.
(4) Composition of electrolyte.
(5) pH.
(6) Viscosity of electrolyte.
(7) Coatings on metal surface or inhibitors added to electrolyte.

A basic knowledge of the principles involving the mechanism of fatigue crack initiation and propagation is essential to the understanding of corrosion fatigue behavior of different metals.

4.9.5 INITIATION OF FATIGUE CRACKS

The following are the possible sites for crack nucleation:

(1) Discontinuities in metal, such as near the surface.
(2) Inclusions or second-phase particles.
(3) Scratches on a metal surface.
(4) Sites of pitting or intergranular corrosion.
(5) Twin boundaries.

The cracks are initiated generally on the surface, however, initiation from subsurface is possible in the presence of surface defects. Intrusions may also develop into cracks. During the loading part of the cycle, slip occurs in a plane favorably oriented. The surface created by slip may oxidize during the unloading period. The first cyclic step may create either an intrusion or extrusion as shown in Fig. 4.70. By continued deformation

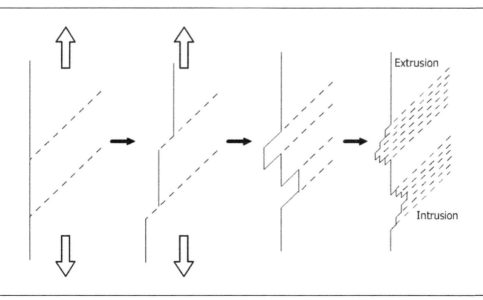

Figure 4.70 Slip band intrusion and extrusion. (From Fuchs, H.O. and Stephans, R.I. (1982). In *Metal Fatigue in Engineering*, New York: John Wiley & Sons Inc., p. 28. Reproduced by kind permission of John Wiley, New York)

in subsequent cycles an intrusion may grow and form a crack [100–102].

4.9.6 FATIGUE CRACK PROPAGATION

The following is a summary of fatigue crack propagation:

(a) A crack initiates in a crystallographic shear mode. It penetrates a few tenths of a millimeter. There is a large effect of microstructure, stress ratio and environment [103].

(b) The crack propagates in a direction normal to the stress axis (Stage 1). The stress concentration at the crack tip causes local deformation in a zone in front of the crack. The plastic zone increases in size as a result of crack growth. It continues to grow until it reaches the thickness of the specimen.

When the size of the plastic zone becomes nearly equal to the thickness of the specimen, plain strain conditions do not exist any more. In the plain strain condition, the plane of fracture instability is normal to the axis of the principal tensile stress. There is zero strain in the direction normal to both the axis of applied tensile stress and the direction of crack growth. The crack propagates in a direction perpendicular to the tensile stress (Stage 2). A shear decohesion also contributes to crack propagation [104].

The crack undergoes rotation and the final rupture occurs in a plane stress mode (the plane of fracture instability is inclined to 45° to the axis of principal tensile stress, Stage 3). The crack propagates during each cycle. The crack growth during stage 3 is only a few nanometers per cycle. In the second stage, the rate of crack growth depends upon the square of fourth power of stress intensity range:

$$\frac{da}{dN} = \Delta K^4 \tag{4.66}$$

The striations observed on the fracture surface are related to the load cycle. Each striation represents one load cycle. At high ΔK values, striations become less significant.

4.9.7 CORROSION FATIGUE CRACK GROWTH

Corrosion fatigue crack growth for many systems occurs only above a certain threshold stress intensity factor ΔK_{th}. The growth increases linearly with increasing ΔK and above ΔK_{th}. It enters a region where da/dN is linearly dependent upon log ΔK. The rate becomes very rapid as K_{max} approaches K_{1C}. Time-dependent stress corrosion cracking is shown in Fig. 4.71. In case of SCC, crack growth does not occur below a certain stress intensity factor (K_{1SCC}). Cycle dependent fatigue crack growth is shown in Fig. 4.72. In an aggressive environment, corrosion fatigue crack growth may be quite different from the pure fatigue curve because of the metal–environment interaction.

The fatigue crack growth response of materials in an environment depends on K_1 at K levels approaching K_c or K_{1c} (fracture toughness and plain strain fracture toughness, respectively) at high stresses and at levels approaching threshold at the lower end, with an intermediate region depending on some power of K. The environment-assisted fatigue growth can be represented by three patterns of behavior. In general, three stages are encountered. In region (1), at low ΔK value, the crack growth rate is extremely dependent on stress intensity and the curves become almost parallel to the crack growth rate. The corresponding stress intensity appears to be environment-dependent and it is denoted by ΔK_{ICF} in analogy to K_{1SCC}. Below K_{ICF}, the corrosion fatigue crack growth is negligible. In region (2), the fatigue crack growth rate depends strongly on environment. Most results on corrosion fatigue studies have been reported in this region. It may be noted that acceleration of corrosion fatigue crack growth in region (2) may also occur around or below K_{1SCC}, which shows that stress corrosion cracking and corrosion fatigue are two different phenomenon. Hence, a fatigue crack which accelerates in region (2) is called

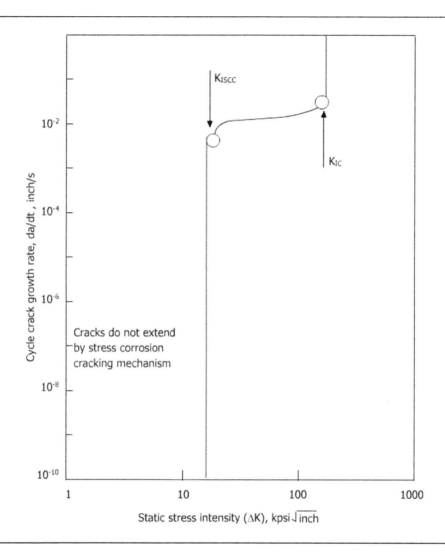

Figure 4.71 Time-dependent stress corrosion cracking. (From Gallaghar, J.P. and Wei, R.P. (1971). *Int. Corrosion Fatigue Conf.*, June 14–18, NACE. Reproduced by kind permission of NACE, Houston, Texas, USA)

'*True Corrosion Fatigue (TCF)*.' Both processes may take place in region (2). Stress corrosion cracking may occur at a stress intensity threshold called K_{1SCC} for a particular metal environment system. The effects of air fatigue and stress corrosion cracking on corrosion fatigue become apparent in *dn/di vs* log ΔK curve and their shape may change depending on which of the above is predominant. Cyclic-dependent fatigue crack propagation occurs below the threshold for stress corrosion. Type A (Fig. 4.73) shows the combined effect of fatigue and corrosion and it, therefore, represents true corrosion fatigue (TCF). This system is typified by Al–H_2O system.

As observed, there is a reduction in upper stress threshold for crack growth due to the corrosive environment. The crack growth is also accelerated. The environmental effect is diminished as K approaches K_c or K_{1c}, because of mechanical–chemical interactions. There is a

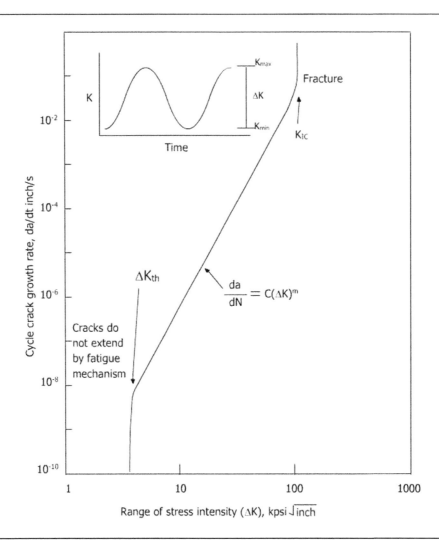

Figure 4.72 Cycle-dependent fatigue crack growth. (From Gallagher, J.O. and Wei, R.P. (1971). In *Proc. Int. Corrosion Conf.*, June 14–18, NACE. Reproduced by kind permission of NACE, Houston, Texas, USA)

combined interaction of cyclic load and aggressive environment in the region where the rate of crack growth is not very rapid because the mechanical factors which lead to cracking are more dominant than chemical factors. The second type [B] represents stress corrosion fatigue. There is no environmental interaction below K_{1SCC}. Here the sustained load contributes significantly to the cyclic crack growth. Above K_{1SCC}, the stress intensity, cyclic frequency and waveform shape are major contributors. It is typified by the hydrogen/steel system. The environmental effects become very strong above a threshold for stress corrosion cracking and negligible below it. A mixed behavior is shown by type [C]. It is typified by a broad range of metal–environment systems. In type [C], type [A] behavior is observed at K levels below the threshold K level, and type [B] behavior above the threshold K level. It shows a mixed pattern of SCC and CF.

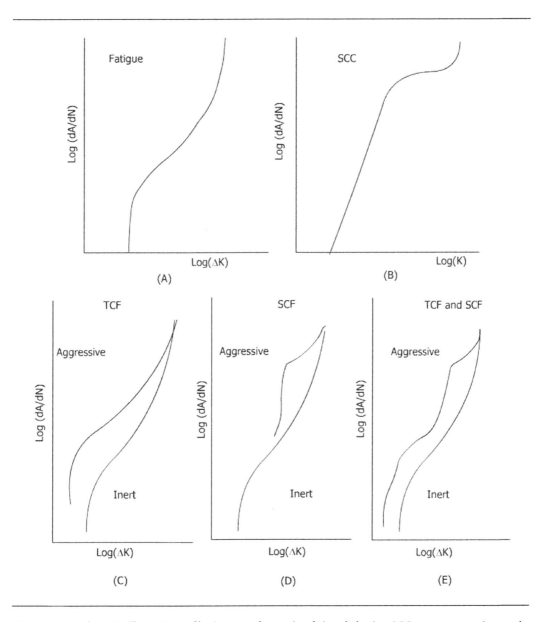

Figure 4.73 Schematic illustrations of basic types of corrosion fatigue behavior, SCC = stress corrosion cracking, TCF = true corrosion fatigue, SCF = stress corrosion fatigue. (From McEvily, A.J. and Wei, R.O. (1971). In *Proc. Int. Fatigue Conf.*, June 14–18, NACE. Reproduced by kind permission of NACE, Houston, Texas, USA)

4.9.8 THEORIES OF CORROSION FATIGUE CRACK INITIATION AND PROPAGATION

The theories of corrosion fatigue are based on one or more of the following principles. Extensive analytical studies have been made [105–108].

(1) Stress concentration at the base of pits. The crack nucleation is related to pits formed by corrosive attack.
(2) Preferential dissolution of deformed material which acts as a local anode with the undeformed material acting as a local cathode.

(3) Rupture of the protective film on the metal surface in the presence of aggressive ions by cyclic deformation.

(4) Supply of reactants and removal of products from crack tip.

(5) Reduction of surface energy due to adsorption of species.

(6) Diffusion of a chemical species ahead of crack tip.

1. Stress Concentration at the Base of Pits

The role of pitting is not very clear. Materials which develop pitting are found to be susceptible to corrosion fatigue, however, corrosion fatigue is also observed without pitting. Pitting is not a prerequisite to cracking as shown by low carbon steel in sodium chloride solutions. It is possible that pits may have been developed after cracking. The role of pitting was found to be crucial in fatigue crack initiation of martensitic 13% Cr stainless steels. There is sufficient evidence to suggest that fatigue cracks are initiated by pits. Pitting plays a crucial role in corrosion fatigue. However, in view of the varying nature of results on various metal–environment systems, the effect of pitting on the fatigue crack growth cannot be generalized.

2. Electrochemical–Mechanical Attack

There is a possibility that fatigue cracks advance by an electrochemical mechanism. The distorted metal acts as the anode whereas the undistorted metal acts as cathode, and a galvanic cell is, therefore, set up which provides the required driving force for the crack to advance. The active sites are provided by plastic deformation at the crack tip.

3. Film Rupture Theory

It has been proposed that corrosion fatigue proceeds by rupture of the surface film. There is considerable evidence to suggest that the rupture of a protective film leads to initiation of fatigue crack. Figure 4.74 shows a schematic of a film rupture model. The film is ruptured at the crack tip by slip resulting in an increase in the plastic strain. The surface, after the film rupture, becomes very active and dissolves anodically in the presence of corrosive media. The crack advances by the conjoint action of film rupture and anodic dissolution. That potential drop continues throughout the length of an alternating stress experiment which suggests that the protective film is destroyed.

Several investigators do not support the film rupture theory. For instance, corrosion fatigue occurs in acid solutions, where there is no possibility for film formation. The observations given below do not support the film rupture theory:

(a) There is a critical rate of corrosion which is associated with corrosion fatigue.

(b) There is no fundamental shift of equilibrium potential of steel, as shown by the effect of cathodic protection on corrosion fatigue of steel. The potential is always shifted by film formation.

(c) Emergent slip pattern is observed by metallographic studies for various grades of steels.

All these findings show that the film rupture theory cannot be accepted as a general mechanism.

As shown by metallographic studies, slip bands are formed by emergence of slip steps or intrusion–extrusion mechanism (Fig. 4.74). The intrusion–extrusion pairs are larger in the presence of a corrosive media. These observations suggest that the emergent slip bands are attacked by a corrosive medium which causes local stress intensification leading to premature failure. At preferentially attacked slip band steps, dislocations are unlocked and it becomes easier for the metal to be deformed (Fig. 4.75). This has been observed for carbon steels in aqueous media. The slip bands produce numerous sites for crack initiation. The density of slip bands is increased by corrosion.

Investigations on polycrystalline copper suggest that copper fails in an intergranular manner by corrosion fatigue. The dissolution of the grain boundaries is enhanced by deformation with respect to the slip bands. In the case of steels, however, corrosion fatigue failures are always transgranular.

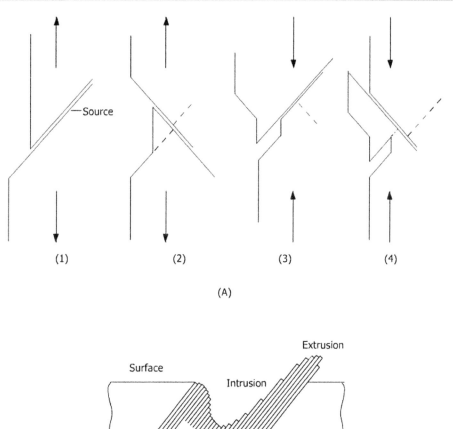

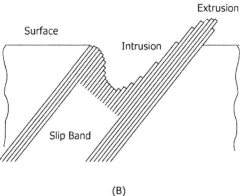

Figure 4.74 Schematic example of mechanisms leading to a fatigue crack initiation (a) extrusion and intrusion in slip region due to cyclic stress (b) slip band intrusions and extrusions prior to crack initiation (Note the stress concentration is formed on the surface in this process). (From Fuchs, H.O. and Stephans, R.I. (1980). *Metal Fatigue in Engineering*, New York: John Wiley & Sons Inc., p. 28. By kind permission of John Wiley, New York, USA)

In the case of aluminum alloys, failure occurs by hydrogen embrittlement according to mechanisms suggested under *Hydrogen Damage*. Hydrogen embrittles the regions in the vicinity of crack tip. The corrosion reaction continuously supplies hydrogen to the growing tip of the crack. It has been also suggested that the surface films on aluminum prevent hydrogen from escaping from the surface alloy and lead to the accumulation of dislocation debris. The situation may enhance the formation of cavities and lead to the propagation of cracking.

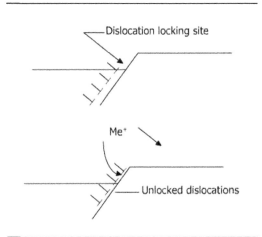

Figure 4.75 Schematic of simple model for corrosion enhanced crack nucleation at emerging slip step. (From Duquette, D.J. and Uhlig, H.H. (1969). *Trans. ASM*, **62**, 839. Reproduced by kind permission of ASM, Metals Park, Ohio, USA)

4. Adsorption Theory

It is postulated that specific ions are absorbed and interact with strained bonds at the surface of the crack tip, thus reducing the bond strength, and permitting continued brittle fracture. This theory has been supported by observations in SCC. By chemisorption of the environmental species on the crack tip, the local fracture stress of the metal lattice is reduced. The theory has been applied to hydrogen embrittlement and liquid metal embrittlement. The adsorption phenomenon may be used to interpret the crack propagation mechanism of alloys which fail by hydrogen embrittlement, such as aluminum alloy 7075.

4.9.9 CASE HISTORIES

A. Case 1

Two admiralty brass heat exchanger tubes from a cooler in a refinery unit showed cracks.

Examination Both tubes showed cracks extending circumferentially above 180°C on the tension side of the U-bend. Tube 1 showed a relatively smooth surface, whereas Tube 2 showed buildup of corrosion products. Tube 1 showed blunt transgranular cracking with minimum branching propagating from inside the tube. By EDX-ray analysis, the presence of copper and zinc ions, and some small amounts of chloride, sulfur, silicon and tin were observed.

Conclusion The tubes failed by corrosion fatigue.

Recommendation Admiralty brass tube was recommended because of its good corrosion performance. The tubes may be annealed to reduce residual stresses.

B. Case 2

A number of cabin bolts from elevators had fractures.

Examination Longitudinal sections from four bolts were examined. The bolts had a sudden transition in cross section and coarse grinding grooves, both of which are liable to decrease the fatigue strength. Longitudinal sections of bolts were prepared from the bolts and examined under microscope. A small crack (0.1 to 0.2 mm deep) was observed in one of the bolts. The other bolts were broken open at the point of cracking. In some bolts, narrow crescent-shaped cracks were observed.

Some of the bolts had been normalized and some were quenched and tempered, therefore, the strength of the bolts fell into three groups: (a) 47, (b) 60–62 and (c) 65–71 kg/mm. Two bolts from group (a), two from the group (b) and four from group (c) had fractured.

Conclusion The fatigue fracture was due to faulty material.

4.9.10 PREVENTION OF CORROSION FATIGUE

The aim of the protective measures is to prevent corrosion fatigue cracks from initiating rather than stopping them from propagation once they have been initiated.

(1) The fatigue life in corrosive environments can be increased by introducing compressive stresses. The compressive stresses can be introduced by shot-peening, shot-blasting, nitriding and carburizing.

(2) Cathodic protection has been used to prevent fatigue crack initiation. It can be achieved by attacking a more active metal than that being protected or by the impressed current technique. The use of electrodeposited coatings, such as zinc coating or cadmium coating, have been found to offer a better protection than the hot dipped zinc and cadmium coatings, because of the compressive stresses introduced during electrodeposition.

(3) Application of nickel coating has been found to be effective in preventing corrosion fatigue. Use has been made of non-metallic coatings, like epoxy, rubber and enamel, to increase fatigue life, hence, once the coating is mechanically damaged, the fatigue resistance is adversely affected due to acceleration of underneath the coating.

(4) The fatigue life of carbon steels and stainless steels is known to be improved by the formation of a passive layer on the surface. The passive layers can be formed by a technique known as '*anodic protection*.' In anodic protection, the steel is polarized by raising its potential to the passive region. The fatigue life of carbon steels and stainless steels may be improved by this technique in oxidizing environment. This technique is generally used to improve the fatigue life of steels in an oxidizing environment.

(5) The composition of an alloy can be modified to increase its resistance to corrosion fatigue. The resistance of stainless steels can be improved by increasing the Ni content. The resistance of mild steel can be improved by addition of 1–2% titanium. Such modifications in compositions are not the ideal solutions as the cost factor may sometimes be prohibitive for making such modifications.

(6) The corrosion fatigue life can also be improved by environmental modification, such as adding inhibitors to the environment. The fatigue resistance of steel in a chloride–sulfate solution is increased by adding sodium dichromate.

Such inhibitors act by forming protective films on the surface of the metal being subjected to corrosion fatigue. The selection of a right inhibitor and its effective concentration is important for prevention of corrosion fatigue. The price is very expensive as large amounts of inhibitors are required.

4.9.11 Why to Study Corrosion Fatigue

There is a great demand for high static strength materials and a desire for higher structural performance from these materials. The demand for high strength-to-weight ratio of aerospace materials is increasing and resistance to fatigue is an important criteria. There is a need to improve the performance of aircraft alloys and of powder metallurgy products.

Aggressive environments have a deleterious effect on the fatigue life, hence, an understanding of corrosion fatigue and of fatigue–environment interaction is extremely important for selection of materials resistant to corrosion fatigue.

This is a localized form of corrosion, caused by the deposition of dirt, dust, mud and deposits on a metallic surface or by the existence of voids, gaps and cavities between adjoining surfaces. An important condition is the formation of a differential aeration cell for crevice corrosion to occur. This phenomenon limits the use, particularly of steels, in marine environment, chemical and petrochemical industries.

4.10 Fretting Corrosion

Fretting is a phenomenon of wear which occurs between two mating surfaces subjected to cyclic relative motion of extremely small amplitude of vibrations. Fretting appears as pits or grooves surrounded by corrosion products. The deterioration of material by the conjoint action of fretting and corrosion is called '*Fretting Corrosion*.' Fretting is usually accompanied by corrosion in a corrosive environment. It occurs in bolted parts, engine components and other machineries.

4.10.1 EXAMPLES OF FRETTING DAMAGE

(1) Fretting of the blade roots of tube blades.
(2) Loosening of the wheels from axles. For example, the coming out of railroad car wheel from the axle due to vibrations.
(3) Fretting of electrical contacts.
(4) Fretting damage of riveted joints.
(5) Surgical implantations.

4.10.2 FACTORS AFFECTING FRETTING

1. **Contact load**
   Wear is a linear function of load and fretting would, therefore, increase with increased load as long as the amplitude is not reduced.
2. **Amplitude**
   No measurable threshold amplitude exists below which fretting does not occur. An upper threshold limit, however, exists above which a rapid increase in the rate of wear exists [109]. Amplitude oscillations as low as 3 or 4 nm are sufficient.
3. **Number of cycles**
   The degree of fretting increases with the number of cycles. The appearance of surface changes with the number of cycles. An incubation period is reported to exist during which the damage is negligible. This period is accompanied by a steady-state period, during which the fretting rate is generally constant. In the final stage, the rate of fretting wear is increased.
4. **Temperature**
   The effect of temperature depends on the type of oxide that is produced. If a protective, adherent, compact oxide is formed which prevents the metal-to-metal contact, fretting wear is decreased. For example, a thick layer of oxide is formed at 650°C on titanium surface.

The damage by fretting is, therefore, reduced at this temperature. The crucial factor is not the temperature by itself, but the effect of temperature on the formation of oxide on a metal surface. The nature and type of the oxide is the deciding factor.

5. **Relative humidity**
   The effect of humidity on fretting is opposite to the effect of general corrosion where an increase in humidity causes an increase in the rate of corrosion, and an increase in dryness causes a decrease in corrosion. Fretting corrosion is increased in dry air rather than decreased for metals which form rust in air. In case of fretting, in dry air, the debris which is formed as a consequence of wear on the metal surface is not removed from the surface and, therefore, prevents direct contact between two metallic surfaces. If the air is humid, debris becomes more mobile and it may escape from the metal surface, providing sites for metal-to-metal contact.

4.10.3 MECHANISM OF FRETTING CORROSION

Although the mechanism of fretting corrosion is only partially understood, it is well-known that fretting proceeds in three stages:

(a) The first stage is the metallic contact between two surfaces. The surfaces must be in close contact with each other. The contact occurs at few sites, called *asperities* (surface protrusions). Fretting can be produced by very small movements, as little as 10^{-8} cm.
(b) The second stage is oxidation and debris generation. There is a considerable disagreement between the workers on whether the metal is oxidized prior to its removal or after its removal. It is possible that both processes may occur, each process being controlled by conditions which lead to fretting. In either case, the debris is produced as a result of oxidation.
(c) Initiation of cracks at low stresses below the fatigue limit.

4.10.4 CHARACTERISTICS OF EACH STAGE

A. Adhesion

In order for the metals to be in physical contact with each other, there must be no protective oxide layer. The breakdown of the protective layer is essential for the onset of fretting. The asperities are bonded together at adhesion sites created by the relative slip of the surfaces. Fretting may occur at amplitude as small as 10^{-8} cm.

If two metals in intimate contact are similar, protective films on both shall be disrupted, however, if one metal is soft and the other is hard, the layer on the soft metal will be destroyed, and on the other metal it would not be disrupted.

B. Generation of Debris

According to one school of thought, the metal is oxidized before it is removed from the surface and according to other oxidation precedes the removal of the metal from the surface. This is no clear explanation for the two opposite views. The material removed from the metal surface due to fretting is called *debris*. The debris produced by low carbon steel consists of mainly ferric oxide, Fe_2O_3. The debris can also contain unoxidized particles in the case of non-ferrous metals. The composition of the debris differs from one metal to another metal. If the particles of oxide becomes embedded in the softer material, the rate of wear is reduced and hence fretting is minimized. Loose particles increase the rate of wear and hence fretting proceeds at a high rate.

C. Crack Initiation

Cracks grow in a direction perpendicular to the applied stress at the fretting area. Some of the cracks may not propagate at all at low stresses because the impact of stress on a fretted surface extends to a shallow depth only. The propagation of cracks is either restrained or prevented by the presence of favorable compressive stresses. The stage of crack initiation is called *fretting fatigue* [110]. Crack propagation at higher stresses is of practical importance as it can lead to failure of components, such as shafts and axles. The crack originates at the boundary of a fretted zone and propagates. During propagation, if a corrosion medium contacts the crack, corrosion fatigue also contributes to the crack propagation. Outside the sphere of the surface contact stress, the crack propagates as a fatigue crack, and upon fracture, a characteristic lip is observed.

The general mechanism of fretting corrosion is shown in Figs 4.76 and 4.77.

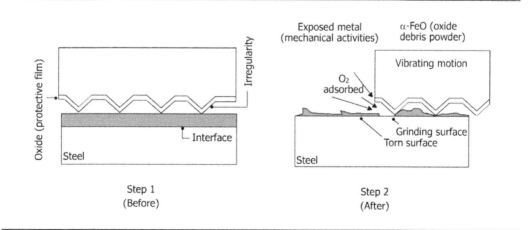

Figure 4.76 The mechanism of fretting corrosion at a steel surface (schematic view)

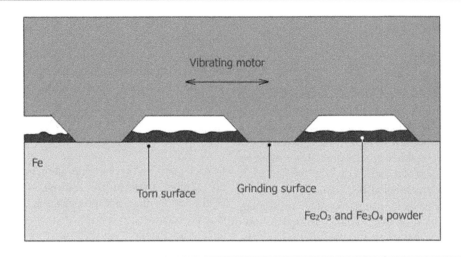

Figure 4.77 Sketch illustrating the mechanism of fretting corrosion

4.10.5 CASE HISTORY

Fretting failure of raceways on 52100 steel rings of an automotive front wheel bearing [111].

Description of problem Fretting and pitting on the raceway of the ball bearing is shown in Fig. 4.78. The inner and outer rings were made of cold drawn 52100 steel tubing.

Examination On physical examination, it was discovered that serious fretting of the raceway in the ball contact area has occurred. Fretting and pitting occurred at spacings equivalent to the spacings of the balls in the retainer (Fig. 4.78). Examination of the inner raceway showed lesser attack.

Conclusion The failure was caused by fretting. During transportation of vehicle by sea, the body of the vehicle was continuously vibrating without any rotation of the bearing.

Recommendations Sufficient preventive measures were not taken during the transportation of the above vehicle. The rolling elements should be taken off during packing and replaced by wooden packing. Vibrations should be eliminated, as far as possible, during transportation of vehicles by sea.

4.10.6 PREVENTION

(1) Increase the magnitude of load at the mating surfaces to minimize the occurrence of slip.
(2) Keep the amplitude below the level at which fretting occurs, if known for a particular system. There is lower threshold limit below which fretting does not occur.
(3) Decrease the load at the bearing surfaces, however, it may not always work.
(4) Use materials which develop a highly resistant oxide film, at high temperature to minimize the adverse effect of temperature on fretting.
(5) Use gaskets to absorb vibration.
(6) Increase the hardness of the two contacting metals, if possible, by shot-peening. Compressing stresses are developed during shot-peening, which resist and increase fretting resistance.
(7) Use low viscosity lubricating oils.
(8) Use materials to resist fretting corrosion (Tables 4.15 and 4.16).
(9) Use thick resin bonded coatings.
(10) Separate the surfaces, for instance, by interleaving sheets of wrapping paper between stacked sheets.

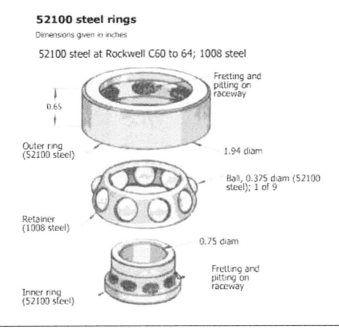

Figure 4.78   Fretting and pitting on raceway

Table 4.15   Fretting resistance of various materials

Poor	Average	Good
Aluminum on cast iron	Cast iron on cast iron	Laminated plastic on gold plate
Aluminum on stainless steel	Copper on cast iron	Hard tool steel on tool steel
Magnesium on cast iron	Brass on cast iron	Cold-rolled steel on cold-rolled steel
Cast iron on chrome plate	Zinc on cast iron	Cast iron on cast iron with phosphate coating
Laminated plastic on cast iron	Cast iron on silver plate	Cast iron on cast iron with coating of rubber cement
Bakelite on cast iron	Cast iron on copper plate	Cast iron on cast iron with coating of tungsten sulfide
Hard tool steel on stainless steel	Cast iron on amalgamated copper plate	Cast iron on cast iron with rubber gasket
Chrome plate on chrome plate	Cast iron on cast iron with rough surfaces	Cast iron on cast iron with Molykote lubricant (MoS_2)
Cast iron on tin plate	Magnesium on copper plate	Cast iron on stainless steel with molykote lubricant
Cast iron on cast iron with coating of shellac	Zirconium on zirconium	–

Source: McDowell, J. R. (1952). *ASTM Special Technical Publication*, No.144, American Society for Testing Materials, Philadelphia, 24.

Table 4.16 Fretting resistance of various materials

Combination	Fretting resistance
Aluminum on cast iron	Poor
Aluminum on stainless steel	Poor
Bakelite on cast iron	Poor
Cast iron on cast iron, with shellac coating	Poor
Cast iron on chromium plating	Poor
Cast iron on tin plating	Poor
Chromium plating on chromium plating	Poor
Hard tool steel on stainless steel	Poor
Laminated plastic on cast iron	Poor
Magnesium on cast iron	Poor
Brass on cast iron	Average
Cast iron on amalgamated copper plate	Average
Cast iron on cast iron	Average
Cast iron on cast iron, rough surface	Average
Cast iron on copper plating	Average
Cast iron on silver plating	Average
Copper on cast iron	Average
Magnesium on copper plating	Average
Zinc on cast iron	Average
Zirconium on cast iron	Average
Cast iron on cast iron with coating of rubber cement	Good
Cast iron on cast iron with Molykote lubricant	Good
Cast iron on cast iron with phosphate conversion coating	Good
Cast iron on cast iron with rubber gasket	Good
Cast iron on cast iron with tungsten sulfide coating	Good
Cast iron on stainless steel with Molykote lubricant	Good
Cold-rolled steel on cold-rolled steel	Good
Hard tool steel on tool steel	Good
Laminated plastic on gold plating	Good

Source: McDowell, J. R. (1952). In *Symposium in Fretting Corrosion*, STP 144, ASM.

4.11 EROSION–CORROSION AND CAVITATION DAMAGE

4.11.1 INTRODUCTION

The problem of cavitation corrosion has plagued engineers continuously in a variety of disciplines, ranging from civil engineers to rocket engineers. Cavitation occurs wherever the local pressure in a flow field falls below a certain critical value.

Cavitation corrosion is a form of localized corrosion combined with mechanical damage, that occurs in a rapidly moving liquids and takes the form of areas or patches of pitted or roughened surface. Many of the terms related to cavitation damage need to be distinguished from each other:

Cavitation corrosion It is a conjoint action of corrosion and cavitation.

Cavitation damage   It is the degradation of a solid body caused by cavitation. The damage appears in the form of loss of material, change in appearance, surface deformation or changes in properties. It takes place when the velocity becomes so high that its static pressure is lower than the vapor pressure of liquid.

Cavitation erosion   It is a continuous loss of material by the impact of cavitation under the influence of erosion.

Generally, cavitation includes the corrosion as well as the mechanical damage.

4.11.2 CHARACTERIZATION OF CAVITATION DAMAGE

The following are the characteristics of damage caused by cavitation corrosion:

(a) Honeycombed surface of the metal.
(b) Absence of corrosion product on the surface.
(c) Occurrence of the attack in very sharply defined areas with a very sharp boundary between the affected and the unaffected metal.
(d) Difference in the intensity of attack on two sides of the exposed surface.

4.11.3 ENVIRONMENTAL FACTORS AFFECTING CAVITATION DAMAGE

(1) Amount of entrained air.
(2) Presence of dust particles. The dust particles act as nuclei for cavity formation.
(3) Temperature. There is a critical temperature above which the intensity of attack is decreased.
(4) Corrosiveness of the media. For instance, the attack in salt water is at least 50% more than observed in distilled water. The intensity would, therefore, vary from one medium to another.
(5) Selection of materials. Materials, like 18-8 steels and titanium, are resistant to cavitation damage.

(6) Operation conditions. For instance, rotary pumps can be operated at the highest head pressure to avoid crevice corrosion.

4.11.4 MECHANISMS OF CAVITATION

Under certain conditions, for example, in pumps and at the back of propellers, a partial vacuum is created. The water boils at ambient temperature at this reduced pressure, in the form of bubbles which collapse when they move to a position of increased pressure. The pressure exerted by the collapsing bubbles are very intense. Collapsing bubbles can produce pressure as high as 60 000 psi. If the process is repeated too often, the protective oxide layer is completely destroyed by the implosion of bubbles. The resistance of the metal to cavitation depends on its ability to form a compact, dense and adherent film. Brittle films are not normally protective and are subjected to cavitation damage.

The collapse of bubbles is accompanied by vibrations and noise. The above phenomenon can be observed also during boiling, when vapor-filled bubbles are convected away from regions of high temperature into cooler regions, where condensation of vapors and bubble collapse is accompanied by a crackling sound. Many times large vapor-filled cavities develop which remain attached to the solid surface. The metal, which is subjected to a tensile force greater than its cohesive strength, is split open and forms such cavities. When such cavities form on surface, such as the blades of pumps, turbines, hydrofoils, etc. a marked change occurs in the flow field. Small bubbles are formed on the surface of these cavities and they collapse. The process is repeated continually. Physical damage always occurs if the collapse is adjacent to a solid boundary. Estimates have shown that 2×10^6 cavities may collapse over a small area in a matter of seconds. Shock waves with pressures as high as $200 \, \text{MN/m}^2$ are produced. Such high forces are sufficient to cause plastic deformation. Failed pump parts show slip lines, which is the evidence of plastic deformation brought about by the collapse of cavities [112].

Several mechanisms have been proposed. The various stages of bubble collapse mechanisms are shown in Fig. 4.79 [112].

The theory of bubble dynamics is well developed and impulsive pressures sufficiently high to cause metal failure can be predicted from various models of bubble collapse.

The schematics of various mechanisms are shown in Fig. 4.79. The mechanism of spherical collapse, non-symmetrical collapse and for vital collapse are shown [113]. Readers interested in the above mechanism should refer to the references cited at the end of this section.

4.11.5 EXAMPLES

Cavitation occurs wherever irregular water flow takes place, particularly if high pressures and vacuum are involved in the flow systems. Cavitation is a serious problem in desalination plants. In multi-stage flash distillation units, cavitation occurs in:

(a) suction of brine cycle, blowdown and distillate pumps,
(b) the drain opening of the last heat recovery stage (HRS),
(c) the brine recirculation control valve,
(d) heat reject pipe.

Cavitation is common in hydraulic equipment, ship propellers, impellers, inlets to heat exchanger tubes and steam turbines.

Cavitation can lead to the shutdown of a desalination plant. If there is cavitation in an ejector condensate pump, it will fail to reach the discharge pressure and the required amount of condensate would not be extracted. One of the requirements of pumping liquid is that the pressures in any point in the suction arm should never be reduced to the vapor pressure of the liquid as this causes boiling (at reduced pressure). Too low a pressure at the pump suction must always be avoided so that cavitation is not caused. Cavitation in pumps is noticed by a sudden increase in the noise level and its inability to reach discharge pressure [114].

A plot of performance and noise in a turbomachine as a function of cavitation is shown in Fig. 4.80.

4.11.6 CASE HISTORY

A cylinder lining from a diesel motor had suffered extensive damage on cooling water side (Fig. 4.81) [115].

Visual examination Heavy pitting was observed on the cooling water side. The outside wall was also found to be coated with Fe_3O_4, SiO_2 and CaO.

Identification The attack was cavitation. This attack is generally encountered in diesel motor cylinders. The attack is observed when a body vibrating in a liquid attains high values of amplitude and frequency and the sluggish liquid cannot keep in pace with body. Implosion of bubbles on the point of maximum vibration occur and high shock wave pressures are produced on the metal surface. It may be pointed out that at the moment of reversal, the cylinder is subjected to bending vibrations by the sideways, pressure of the piston and the cooling water may not be unison with the cylinder vibrations.

Remedy

(1) Reduce the piston pressure.
(2) Use a thicker wall lining to reduce the amplitude.
(3) Add a suitable oil to the cooling water so that a film is formed on the cylinder surface, but note that an oil film may reduce heat transfer.

Some interesting applications The scales which develop on the surface of the pipes in geothermal plants are extremely hard and in many instances, they approximate the strength of the pipe. These scales are extremely difficult to remove. Cavitation cleaning systems have been used to remove such scales. Cavitation nozzles at pressure drops up to 100 MPa have been tested. The focal length of the bubble can be adjusted to cut the required scale thickness.

1. Spherical collapse and rebound

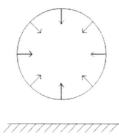
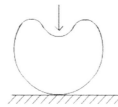

Cavitation bubble collapse commences.

Motion arrested at minumum volume.

Bubble rebounds due to non-condensable gas. A pressure wave propagates outward.

2. Non-symmetrical collapse

A cavitation bubble near a boundary begins to collapse.

Due to the presence of the boundary the bubble indents.

A high velocity jet is formed that strikes the boundary.

3. Final toroidal collapse

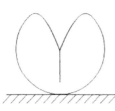

A jet strikes the boundary but causes little or no damage.

The cavity collapses further in the form of a torust until minimum volume is reached.

The bubble rebounds due to noncondensable gas. A pressure wave propagates outward.

Figure 4.79 Schematic of various bubble collapse mechanisms. (From Arndt, R. E. (1977). Paper No. 91, *Corrosion*, NACE) [113]

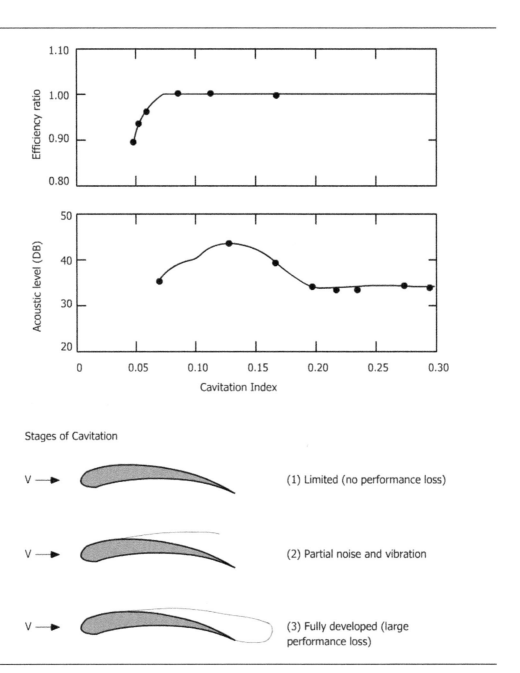

Figure 4.80 Performance and noise in a turbo-machine as a function of cavitation index. (From Arndt, R.E. (1977). Paper No. 91, *Corrosion*. Reproduced by kind permission of NACE, Houston, Texas, USA) [113]

Microstructure of specimen
taken from the cylinder liner

Figure 4.81   Failure of cylinder liner from a diesel motor. (From Kauczor, E. (1979). *Case Histories in Failure Analysis*, p. 232, ASM. Reproduced by kind permission of ASM, Metals Park, Ohio, USA)

In dental hygiene, extensive use has been made of cavitation cleaning for the removal of calcerous deposits. Water is fed to the tip of the tool at 35 cm^3/min under pressure. The vibrating tip of the tool used for cleaning induces cavitation in the water flow, and removal of the scale is achieved by the cavitation cleaning technique.

4.11.7 PREVENTION OF CAVITATION DAMAGE

Cavitation damage can be reduced by:

(1) Using materials resistant to cavitation damage (Table 4.17) [116].
(2) Changes in design, for instance, by changing the diameter of the pipe, the flow geometry can be affected. The flow pattern can be changed from turbulent to lamellar by increasing the pipe diameter. Smooth finishes on pump impellers and propellers reduce the cavitation damage.
(3) Reducing the amount of entrained air, if the process of protective film formation is not adversely affected.
(4) Avoiding dust particles and other impurities.
(5) Optimizing operation conditions, for instance, operate rotary pumps at highest pressures. Cavitation indicates that there is insufficient net positive suction head (NPSH) (NPSH is a measure of the suction head to prevent vaporization of the lowest pressure point in a pump). A good design can eliminate the cavitation in impeller vanes by providing sufficient NPSH and positive pressure at the suction inlet.
(6) Adding inhibitors, like chromates and nitrates. In order to prevent the cavitation damage of diesel engine cylinder liners, fNa$_2$ CrO$_4$, is generally added to cooling

Table 4.17 Resistance of metals to cavitation corrosion [116]

Group I	Most resistant. Subject to little or no damage. Useful under extremely severe conditions.
	Stellite hard-facing alloys
	Titanium alloys
	Austenitic and precipitation-hardening stainless steels
	Nickel–chromium alloys, such as Inconel alloys 625 and 718
	Nickel–molybdenum–chromium alloys, such as Hastelloy C
Group II	These metals are commonly used where a high order of resistance to cavitation damage is required. They are subject to some metal loss under the most severe conditions of cavitation.
	Nickel–copper–aluminum alloy Monel K-500
	Nickel–copper alloy Monel 400
	Copper alloy C95500 (nickel–aluminum bronze, cast)
	Copper alloy C95700 (nickel–aluminum–manganese bronze, cast)
Group III	These metals have some degree of cavitation resistance. They are generally limited to low-speed low-performance applications.
	Copper alloy C71500 (copper–nickel, 30% Ni)
	Copper alloys C92200 and C92300 (leaded tin bronzes M and G, cast)
	Manganese bronze, cast
	Austenitic nickel cast irons
Group IV	These metals normally are not used in applications where cavitation damage may occur, except in cathodically inhibited solutions or when protected by elastomeric coatings.
	Carbon and low alloy steels
	Cast irons
	Aluminum and aluminum alloys

Note: Applies to normal cavitation–erosion intensities, at which corrosion resistance has a substantial influence on the resistance to damage.

water, but chromates are undesirable for health reasons.

(7) Cathodic protection application. Hydrogen bubbles produced by cathodic protection mitigate the effect of imploding cavitation bubbles. Sacrificial anodes, such as zinc or magnesium, can also be used.

(8) Minimizing opportunities for cavities to be formed. It could be accomplished by keeping the liquid temperature and pressure high and by introducing air bubbles.

(9) Making the surface very smooth.

(10) Coating with resilient materials, like rubber.

QUESTIONS

UNIFORM CORROSION

Multiple Choice Questions

Mark one correct answer from the following:

1. The major component affecting corrosion in industrial atmosphere is

a) carbon particles

b) dust particles

c) SO_2

2. In an industrial atmosphere with high humidity and SO_2 content

   a) the corrosion product $[Fe(OH)_2]$ being formed is protective
   b) the corrosion product is formed away from the corroding site
   c) the corrosion product is formed at the corroding site

3. The following protective measures can be very effective in preventing uniform corrosion:

   a) Select a thinner material
   b) Prevent the material from atmosphere by covering
   c) Use an epoxy coating
   d) Give a corrosion allowance

4. If a metal undergoes uniform corrosion it becomes

   a) thinner
   b) thicker
   c) perforated

5. The following is the example of uniform corrosion:

   a) Cracking of a pipe in the soil
   b) Perforation of a water pipe
   c) Rusting of a steel tank in air
   d) Leakage of water pipe by corrosion

Galvanic Corrosion

Multiple Choice Questions

Mark one correct answer from the following:

1. Galvanic corrosion is restricted when

   a) corrosion products form between the two metals
   b) pH of the electrolyte is reduced
   c) electrolytic conductivity is high
   d) polarization of less active metal takes place with difficulty

2. Referring to the galvanic series of some commercial metals and alloys in seawater, mark the condition which would lead to minimum corrosion by galvanic coupling.

   a) Coupling of 18-8 steel active with chromium stainless steel, 13% Cr (active)
   b) 70-30 brass with pure copper
   c) Silver and copper
   d) Magnesium and aluminum plates

3. Which of the following shows an unfavorable area ratio?

   a) Copper rivets in copper plate
   b) Steel rivets in copper plate
   c) Aluminum rivets in aluminum plate
   d) Brass rivets in copper plate

4. Which of the following would be most effective in preventing galvanic corrosion?

   a) Avoiding unfavorable area ratio
   b) Adding inhibitors
   c) Using welded joints of the same alloy as the alloy to be welded
   d) Increasing temperature at the point of contact of the two metals

5. Which of the following pairs of metals would show the highest rate of corrosion in seawater?

   a) Copper and steel
   b) Copper and zinc
   c) Copper and aluminum
   d) Copper and brass

How and Why Questions

1. In the following couples, which one would form the anode and which one the cathode?

   a) Copper and iron
   b) Stainless steel and brass
   c) Active and passive steel
   d) Zinc and aluminum

2. In the galvanic series why active steel is placed far away from the passive steel and not bracketed together?

3. Explain why:

   a) The magnitude of galvanic corrosion is maximum at the junction of the two metals

   b) The magnitude of galvanic corrosion decreases as the distance from the junction of the two metals increases

   c) A new pipe joined to an old pipe corrodes

   d) A metallic couple with an unfavorable area ratio (small anode/large cathode) corrodes faster than a couple with large anode and small cathode

   e) Galvanic attack is accelerated in the presence of NaCl and humidity

4. Explain why:

   a) If steel rivets are fixed in a copper plate, the rivets are completely corroded after sometime

   b) Galvanized steel pipes are not suitable for hot water system above 80°C

   c) The polarization of the reduction reaction predominates

   d) The aggressiveness of the environment determines the extent of galvanic corrosion

   e) When dissimilar metals are to be coated, the more noble must be coated

   f) Water heaters with copper water inlet tube corrode

   g) Welded joints are better than riveted joints

Review Questions

1. State the mechanism of galvanic corrosion for a steel and copper couple immersed in 1 N solution of aerated NaCl.

2. State three main differences between the emf series and galvanic series.

3. What is the effect of area ratio and distance on the magnitude of galvanic corrosion?

4. State the mechanism of atmospheric corrosion of steel in an industrial atmosphere containing SO_2, C and chloride particles.

PITTING CORROSION

Multiple Choice Questions

Mark one correct answer from the following:

1. The most important condition for pitting to take place is

   a) the metal must be pure
   b) the metal must be in a passive state
   c) the metal must be in an active state
   d) the metal can be in any state

2. Pitting can take place under the following conditions:

   a) Water in contact with stainless steel and flowing at a high velocity
   b) Stagnant conditions in service
   c) Presence of NO_3^- and CrO_4^- ions in water
   d) Deaerated water

3. The following reaction occurs in the pit cavity:

   a) $2H_2O + O_2 + 4e \rightarrow 4OH^-$
   b) $Fe^{++} + 2e \rightarrow Fe$
   c) $Fe \rightarrow Fe^{++} + 2e$
   d) $2H^+ + 2e \rightarrow H_2$

4. On the outside of the pit cavity, the main reaction is

   a) anodic polarization
   b) cathodic reduction of hydrogen
   c) cathodic reduction of oxygen

5. Which one of the following processes occurs mainly around the pit on aluminum?

   a) Cathodic protection by the anodic reaction
   b) Passivation by alkali formed in the cathode reaction
   c) Oxidation of aluminum
   d) Deposition of active metals

6. The following methods are used to prevent pitting:

   a) Adding inhibitors containing Fe^{++} ions or Hg^{++} ions
   b) Operations at the highest temperature
   c) Increase of oxygen concentration

d) Elimination of chloride from the environment, wherever possible

How and Why Questions

1. Explain why:

   a) Stainless steels exposed to water containing Fe^{3+}, Cu^{2+} or Hg^{2+} develop pits within hours.
   b) Addition of 3% $NaNO_3$ to 10% $FeCl_3$, inhibits the pitting of 18-8 steel.
   c) A passive surface is necessary for the onset of pitting.
   d) Oxygen is a prerequisite for pitting.
   e) A polished surface has a lesser tendency to develop pits than a rough surface.
   f) Addition of 2% Mo to steel increases its resistance to pitting.

2. Explain why:

   a) Acidity is developed within a pit.
   b) The outside of a pit cavity is cathodically protected.
   c) Pitting reaction is sustained even after oxygen is consumed.
   d) The pitting process is autocatalytic.
   e) Pitting takes place in contact with water under stagnant conditions.

3. Explain how:

   a) The surface adjacent to a pit is protected from corrosion.
   b) The resistance of an alloy to pitting can be predicted.
   c) Pitting can be differentiated from fretting corrosion.
   d) Pitting can be differentiated from crevice corrosion.
   e) Pitting can be prevented by cathodic prevention.
   f) The resistance of steels to pitting can be increased.
   g) Pitting damage can be evaluated microscopically.
   h) Pitting attack is decreased by an increase of velocity of corroding fluid.

Review Questions

1. Explain the mechanism of pitting when it is initiated at sulfide inclusion in a carbon steel.
2. Explain the reactions which take place during the pitting of aluminum

   a) at the mouth
   b) inside the pit mouth
   c) outside the pit.

3. How can you distinguish between the detrimental effect and beneficial effect of an alloying element with respect to the shift of corrosion potential either in the positive or in the negative direction?
4. State briefly the main characteristics of adsorption theory as applied to aluminum and steel. Mention one item of evidence against the theory.

INTERGRANULAR CORROSION

Multiple Choice Questions

Mark one correct answer from the following:

1. Addition of molybdenum is particularly made steels

   a) to increase the tensile strength
   b) to increase its fluidity
   c) to increase the resistance of stainless steels (ferrite and austenite) to pitting and crevice corrosion

2. Intergranular corrosion is

   a) restricted to stainless steels
   b) not restricted to stainless steel but includes nickel-base alloys also
   c) restricted to alloys containing low carbon content

3. Sensitization of austenitic steel would not occur on cooling fast through 580–850°C

   a) if the crosssection of the specimen is small
   b) the amount of carbon is increased

c) if it is reheated after quenching between 550 and 850°C temperature range

d) if it is slowly cooled through 800°C, quenched and reheated to 235°C

4. Titanium and niobium carbides are formed at temperatures above 1219°C, whereas chromium carbides are formed in the range 787–1219°C. Steels 321 and 347 contain either chromium or titanium carbides as received from the mill. Intergranular corrosion would develop in these steels if

a) material is heated below 400°C

b) metal is heated to 1310°C, quenched in water and reheated for 30 min to 500°C and cooled

c) the material is heated after welding to above 1055°C and cooled rapidly

d) metal is heated between 505 and 760°C.

5. Which of the following is true about carbides?

a) They are present in the grain boundaries in the form of a continuous film

b) They utilize all the carbon present in the alloy for their formation

c) The carbides are generally represented by $M_{23}C_6$, where $M = Cr$ and Fe

d) The composition of carbides is fixed

6. Which of the following assumptions regarding the chromium depletion theory for stainless steels has more evidence in support of it?

a) The local chromium–carbon–carbide equilibrium is reached at the grain boundary and the chromium content at the grain boundary is thermodynamically determined by the equilibrium

b) The chromium content in the vicinity of the carbide particles is negligible and the variation in the degree of sensitivity results from changes in the morphology of distribution of particle along the grain boundary

c) Heat treatment does not influence distribution and precipitation of carbide and, hence, the degree of sensitization

7. Sensitization in ferritic stainless steels can be eliminated by

a) lowering the carbon content to 50 ppm

b) lowering chromium to less than 12%

c) lowering the amount of interstitial carbon and nitrogen to 50 ppm and 150 ppm, respectively

8. Which of the two stainless steels, ferritic or austenitic, would sensitize more under the given conditions?

a) Ferritic steel when quenched from 800°C or higher

b) Ferritic steels on slow cooling from 800°C

c) Austenitic steels when slowly cooled through 500°C

9. Knife-line attack in stainless steels is caused by one of the following:

a) The base metal (18-8 steel) adjacent to the fusion line is heated in the temperature range in which the steel is sensitized

b) Subsequent welding passes during welding pre-heat the area adjacent to the fusion line and the temperature range to cause chromium carbide to be formed

c) Heating stabilized steel in the temperature range 510–760°C

10. Which of the methods given below can be used for prevention of intergranular corrosion of ferritic stainless steels?

a) Reduction of interstitial carbon and nitrogen content

b) Heating at 925°C and rapid quenching

c) Increasing the nickel content

How and Why Questions

1. Why the precipitation of carbide results in the depletion of chromium in the grain boundary,

and why it becomes sensitive to the attack of corrosion?

2. Are the carbides stoichiometric or they can be represented by $M_{23}C_6$?

3. For 304 stainless steel, the maximum carbon allowed for 700°C anneal is 0.08%, and at 600°C it is 0.02%. Why the alloy would be susceptible to intergranular attack if the above carbon amounts are exceeded? Assume that 12% chromium is required to form a passive film.

4. Why ferritic stainless steel resists corrosion if it is held at 788°C for a sufficient period of time?

5. Why intergranular attack is not prevented in ferritic stainless steel even if carbon is reduced to 0.3%?

6. Although carbon reacts with titanium and niobium which are added as stabilizers, yet intergranular corrosion is not observed in their presence. Explain.

7. What is the major difference between knife-line attack and weld decay as far as the location of the attack with respect to the weld is concerned?

8. Why stainless steels which are stabilized undergo knife-line attack? If the zone on which knife-line attack has occurred is reheated between 950–1450°C, would you expect the steel to be sensitive to intergranular attack?

Conceptual Questions

1. Differentiate clearly between the austenitic, ferritic, martensitic, duplex and precipitation hardened stainless steel. List two examples of each.

2. State four salient characteristics of the model of grain boundary attack based on chromium depletion theory.

3. What would happen if austenitic stainless steel is held at the sensitizing temperature for, say 2000 h? Would it be still susceptible to inter-granular attack? If yes, why. If no, explain the reason.

4. Differentiate clearly between knife-line attack and weld decay in terms of mechanism.

STRESS CORROSION CRACKING

Multiple Choice Questions

1. Which of the following is true about stress corrosion cracking?

   a) The stress required to cause cracking may be below the macroscopic yield stress
   b) It is not necessary to apply a tensile stress for SCC to occur
   c) It can take place in vacuum
   d) It cannot take place in the absence of a thermal stress

2. Select one correct statement:

   a) Specific ions are necessary for metals and alloys
   b) An environment that causes SCC of one alloy would also causes SCC of another alloy
   c) As specific ions form a film on the metal surface, hence, those ions which form films may be adequate to cause SCC

3. State which of the following is characteristic of stress corrosion cracking:

   a) The failure by SCC is always ductile
   b) The failure by SCC is dependent on the metallurgical condition of the alloy
   c) Cracks generally become visible very soon (in hours) and take a very long time to propagate

4. The following variables may affect SCC (two):

   a) Temperature
   b) Impurity
   c) Mechanism of crack propagation

5. In the film rupture mechanism, it is implied that

   a) the stress acts to open the crack and ruptures the protective film

b) the crack advances by preferential dissolution of the grain boundaries

c) the crack tip remains bare, because the
rate of film rupture at the crack tip is less
than the rate of repassivation

6. By the adsorption of specific active species on
the metal surface

a) the interatomic bond strength is increased

b) the stress required for cleavage fracture is
decreased

c) the cracks propagates in a discontinuous
manner

7. The process of stress corrosion cracking can
be adequately explained by

a) mechanistic chemical model

b) a broad spectrum of mechanism, such
as the theories based on active path,
strain assisted and adsorption related
phenomenon

c) the film rupture mechanism

8. In the stress corrosion cracking of austenitic
stainless steels

a) the greatest susceptibility to SCC is exhibited at a nickel content of 8%

b) the least resistance is shown by alloys
of a higher nickel content

c) nickel content has no particular effect on
SCC of austenitic stainless steels

9. The stress levels required for SCC of steels is

a) 0.1% of the yield stress

b) vary from one metal to another

c) close to the yield stress

10. Ferritic stainless steels like austenitic stainless
steels are

a) subjected to SCC in chloride media, such
as boiling $MgCl_2$ at 140°C

b) only subjected to cracking in 5% NaOH
at 25°C

c) do not crack in chloride media

11. Addition of a small amount of Ni to ferritic
stainless steel

a) increases its resistance to corrosion

b) decreases its resistance to corrosion to
SCC

c) has no effect

12. Which of the following similarities are common between stress corrosion cracking and
hydrogen embrittlement? (Mark two correct
answers.)

a) A general increase in sensitivity to the
degradation with the increased strength
of the alloy

b) Discontinuous crack propagation

c) Liberation of hydrogen upon cathodic
polarization

d) Appearance of blisters on the surface of
material which has cracked

13. The following are the reasons for concentration of caustic which lead to caustic cracking:

a) Departure from nucleate boiling in which
case the rate of bubble formations exceed
the rinsing rate

b) Concentration of sodium hydroxide at
the water level

c) Formation of steam over the insulating
deposits

d) Adding a higher concentration of caustic
than the concentration prescribed

14. By the term caustic embrittlement, it is
understood that

a) there is a loss of ductility

b) there is no loss of ductility

c) the failure occurs by weakening of the
grain boundaries leading to cracking
without any loss of ductility

Conceptual Questions

1. Briefly mention at least six important characteristics of stress corrosion cracking.

2. List three different sources of residual
stresses.

3. Explain the main difference between the pre-existing active path, strained generated active paths and specific adsorption theories.
4. State three major factors which cause the concentration of caustic. Explain briefly the mechanism of caustic cracking in boilers.
5. Why localized corrosion is initiated by the rupture of a protective film on steels, and how it assists SCC?
6. Specific anions are not important for SCC of martensitic steels, as long as hydrogen is evolved. Substantiate this statement.
7. State briefly how the rupture of a protective film assists in SCC of steels.
8. How a local galvanic cell is formed by precipitation of carbides at the grain boundary?
9. Why low carbon stainless steels are not as much susceptible to SCC as high carbon steel?
10. Outline the main difference between hydrogen attack at $200°C$ and blistering caused by hydrogen.
11. What would happen if hydrogen is adsorbed on an active metal surface?
12. If by applying a cathodic current, the time to failure decreases and increases by anodic polarization, which mechanism of SCC is suggested? Active path mechanism or hydrogen embrittlement?
13. Explain the main difference between the pre-existing active path, strained generated active paths and specific adsorption theories.
14. State three major factors which cause the concentration of caustic. Explain briefly the mechanism of caustic cracking in boilers.

c) chlorides, bromides and iodides
d) deposits, debris on a metallic surface

2. The main reason for crevice corrosion is

a) the formation of differential, aeration cells
b) the difference of potential between two contacting surfaces
c) development of stress cells on the metal
d) polarization of the anodic and cathodic areas

3. When the ratio of the crevice solution to crevice area is small

a) the critical value for initiation of crevice corrosion is rapidly achieved
b) the critical value for initiation of crevice corrosion is delayed
c) the critical value has no affect on the initiation of crevice corrosion

4. The acid produced in the crevice is by

a) Oxidation of Fe^{++}
b) Reduction of Fe^{++}
c) Hydrolysis of M^{++}
d) Hydrolysis of M^+Cl^-

5. To prevent the crevice corrosion of 18-8 steel, the following measure is very important:

a) Decreasing the velocity of water
b) Painting the anodic surface
c) Avoiding contact with non-hygroscopic material
d) Using welded joints in preference to bolted joints

CREVICE CORROSION

Multiple Choice Questions

Mark one correct answer from the following:

1. Crevice corrosion is caused by

a) metallic contact between two different metals
b) high water velocity

Review Questions

1. What are the three important metallurgical factors which affect crevice corrosion? Explain the reasons.
2. Explain the mechanism of crevice corrosion based on concentrations of metal-ion cell and differential aeration cell.
3. Explain the mechanism of crevice corrosion.

How and Why Questions

1. How is oxygen depleted in the crevice?
2. Why does the surface adjacent to the crevice not become the anode?
3. Why hydrolysis causes a fall in pH in the crevice solution?
4. How is acidity produced in a crevice?

FRETTING CORROSION

Multiple Choice Questions

Mark one correct answer from the following:

1. Select the best fretting resistant material from the following:

   a) Hard tool steel on tool steel
   b) Cast iron on cast iron with phosphate coating
   c) Aluminum on cast iron
   d) Chrome plate on chrome plate

2. Fretting corrosion occurs at contact areas between material under load and subjected to

   a) attack by environment
   b) static load
   c) vibration and slip
   d) cavitation damage

3. Which one of the following is not a requirement for fretting corrosion to occur?

   a) The interface must be under load
   b) Relative motion between contacting surface
   c) Sufficient relative motion to produce slip
   d) Presence of a fatigue crack

4. Which one of the following is not an example of fretting corrosion?

   a) Fretting of blade roots of turbine blades
   b) Bolted tie plates on railroad rails
   c) Fretting of electrical contacts
   d) Failure of a ship propeller

5. Which of the following is true about the oxidation-wear mechanism?

   a) A thin layer of a protective film which is formed on the metal surface is ruptured and oxide debris is produced
   b) When metals are placed in contact, the oxide layer is broken at all points
   c) No frictional forces are involved in the theory based on oxidation-wear concept
   d) Cold welding or fusion occurs at the interface between the contacting surface

How and Why Questions

1. What are the four important factors which affect fretting corrosion?
2. Why film rupture causes fretting corrosion?
3. What is the essential difference between the wear-oxidation and oxidation-wear mechanism?
4. How is the debris which is formed as a result of oxidation affects fretting?

Review Questions

1. State the main points of disagreement and agreement between the theory of wear-oxidation and oxidation-wear.
2. What is the essential difference between fretting wear, fretting fatigue and fretting corrosion? State two examples of each.
3. State four important factors which affect fretting.
4. State four methods of prevention of fretting corrosion.

CAVITATION DAMAGE

Multiple Choice Questions

Mark one correct answer from the following:

1. Cavitation corrosion damage on metals may

   a) be the result of chemical changes in the environment
   b) occur in slowly moving liquids

c) be limited sometimes by inhibitor

d) be easily eliminated by cathodic protection

2. Cavitation damage is caused by

  a) degradation of a solid body resulting from exposure to cavitation

  b) loss of original material from a solid surface

  c) collapse of cavities within a liquid when subjected to rapid pressure changes

  d) loss of mechanical strength of the metal

3. Which of the following are the examples of cavitation corrosion?

  a) Cracking of 18-8 steel pipe transporting brackish water

  b) Vibration in the blades of turbines

  c) Loosening of fly wheels from shafts

  d) Failure of a ship propellers

4. Cavitation corrosion may be prevented by

  a) coating of materials

  b) allowing dust particles to interact with the surface

  c) applying cathodic protection

  d) reducing the size of cavities

5. Cavitation in a pump can be observed by

  a) failure to reach the discharge pressure

  b) quiet operation

  c) regular power consumption

6. A classical example of cavitation damage is shown by

  a) a bolted tie plate on railroad rails

  b) interface between press fitted ball bearing shaft

  c) pump impeller

  d) boiler tubes

How, Why and What Questions

1. How are extremely high pressures created on the surface of metals by cavitation?

2. Why is the magnitude of cavitation corrosion in salt water greater than in distilled water?

3. What is the most important factor which leads to the cavitation damage of pumps?

4. How does cathodic protection prevents cavitation damage?

Review Questions

1. State clearly the difference between cavitation corrosion and cavitation damage.

2. Briefly describe the mechanism of cavitation damage.

DEZINCIFICATION

Multiple Choice Questions

Mark one correct answer from the following:

1. In dezincification of brass

  a) copper is selectively attacked

  b) zinc is selectively attacked in preference to copper

  c) the attack is confined to both zinc and copper

  d) zinc is selectively attacked only if its concentration is more than 35%

2. Plug type of dezincification is observed

  a) in admiralty brass when exposed to water (pH = 8.0)

  b) in brasses with a high zinc content in alkaline environment

  c) in heat exchanger tube of 90-10 Cu–Ni alloys

  d) in tooth of a gear wheel

3. Which one of the following steps does not represent the mechanism of dezincification of a 70-30 brass?

  a) Oxidation of copper

  b) Plating of copper ions

  c) Reduction of H_2O

  d) Anodic dissolution of zinc

4. Dezincification can be minimized by

  a) addition of oxygen
  b) addition of 0.2% titanium
  c) addition of 0.1% arsenic
  d) using brasses with a zinc content higher than 45%

How, Why and What Questions

1. What is the main cause of dezincification of α-brass?
2. What is the essential difference between the plug type and layer type of dezincification?
3. Why zinc is always leached in seawater in preference to copper?

Review Questions

1. State the essential difference between the two main theories of dezincification.
2. What is de-alloying? Is it appropriate to call dezincification as de-alloying?
3. State four important factors which affect dezincification. What alloys you would suggest to be used for seawater services?

CORROSION FATIGUE

Multiple Choice Questions

Mark one correct answer from the following:

1. In corrosion fatigue

  a) the number of cycles for failure increase as the stress is increased
  b) there is always a greater effect of environmental factors than mechanical factors
  c) the endurance limit of a material is sharply reduced
  d) the surface remains bright after fracture

2. The following are the metallurgical factors which affect corrosion fatigue:

  a) Stress intensity
  b) Potential

  c) Composition of alloy
  d) Cyclic loading frequency

3. A plot of log (ΔK) *vs* log (da/dN)

  a) can be integrated in conjunction with stress intensity solution to predict life of a material
  b) cannot be used for corrosion fatigue studies as it is purely used for design and evaluation of engineering structures
  c) cannot be used to mitigate corrosion fatigue
  d) region II of the plot is not affected by environment

4. One of the following factors does not influence corrosion fatigue:

  a) Load frequency
  b) Stress ratio
  c) Environmental composition
  d) Alloy composition
  e) Loading parameter

5. Corrosion fatigue proceeds more rapidly

  a) at higher frequencies
  b) at low frequencies
  c) at low stress ratios
  d) at very high frequencies

6. Low cycle fatigue region is characterized by

  a) $N < 10^4$ cycles
  b) $N < 10^3$ cycles
  c) $N > 10^2$ cycles
  d) $N > 10^5$ cycles

7. According to film rupture theory

  a) it is not necessary that a film be present as a metal surface
  b) the surface becomes active only after the protective film is ruptured
  c) there is no fall of potential while the film is being destroyed because all parts of the film are not ruptured
  d) there is a fall of surface potential when the film is being destroyed

8. Which of the following elements responsible for corrosion fatigue? (*Mark YES or NO*)

 a) The basic material
 b) Mechanical deformation
 c) pH
 d) Passivity

9. The following are the electrochemical factors which contribute to corrosion fatigue: (*Mark YES or NO*)

 a) Potential
 b) Current density
 c) Film formation
 d) Passivity

10. In an aggressive environment, cyclic frequency

 a) affects the fatigue strength significantly
 b) does not affect fatigue strength
 c) affects the fatigue strength only in highly acidic medium

11. The fatigue life in corrosive environments can be increased by

 a) introducing tensile stresses
 b) use of coatings, such as zinc and cadmium coatings
 c) application of anodic current
 d) increasing the stress ratio

How, Why and What Questions

1. How does corrosion affect fatigue life?
2. How can corrosion of certain steels be prevented by cathodic protection?
3. How does film rupture accelerate the propagation of fatigue crack growth?
4. Why is fatigue life of a material drastically reduced in corrosive environment?
5. Why is the rate of corrosion fatigue not appreciably accelerated at a high frequency?
6. What is the difference between stress corrosion cracking and corrosion fatigue?
7. How is the rate of corrosion fatigue increased by a higher stress ratio?

8. How is crack growth related to the kinetics of mass transport?
9. Why is the rate of corrosion fatigue crack propagation increased by high stress ratios?

Review Questions

1. State the effect of stress ratio and stress intensity range on corrosion fatigue.
2. Describe the characteristics of a $\log(\Delta K)$ *vs* $\log(da/dN)$ curve. Specify the characteristics of each region.
3. Does the above curve take into consideration the effect of varying stress ratio?
4. State the effect of electrode potential on corrosion fatigue crack growth.
5. State the difference between a SCC crack and a corrosion fatigue crack.
6. What is the basis of film rupture mechanism? Give one piece of evidence in support of the above theory.
7. Differentiate clearly between fatigue, corrosion fatigue and stress corrosion cracking.
8. State clearly the effect of stress ratio, stress amplitude and mean stress on fatigue crack propagation.
9. State the factors which play an important role in corrosion fatigue crack propagation.

REFERENCES
Uniform Corrosion

[1] Ahmad, Z. and Afzal, M. (1995). Brackish water corrosion of glass lined water heaters in sea-coastal environment. *AJSE*, **20**, (4 A), October.

Galvanic Corrosion

[2] Ashbaugh, W.G. (1986). *Corrosion Failures in Metals - Handbook*, Vol.11, 9th ed. ASM.
[3] Ahmad, Z. (1973). A technical report submitted to Sakaii Housing Corporation. Ministry of Works, Iran.

Dezincification

[4] Vernick, E.D. and Jr. Heldersback, R.H. (1972). In ASTM-STP, 516, 303, ASTM.
[5] Ashbough, W.G. (1986). *Corrosion*, **11**, ASM, 178.

Crevice Corrosion

[6] Kain, R.M. (1980). Technical Paper No. F-P-37, Laque Center, Wrightsville, March.
[7] Jackson, R.P. and Van Rooyen, D. (1974). *Localized Corrosion*, ASTM-STP, 516, 210.
[8] Brigham, R.J. (1974). *Corrosion*, 396, November.
[9] Anderson, D.B. (1978).Statistical aspects of crevice corrosion. Paper TL-26, T-O.P. (Rand, D.) INCO.
[10] Kain, R.M. (1979). *Corrosion*, Paper No. 230, March.
[11] McKay, R.J. (1922). *Transactions TESOA*, **41**, 201.
[12] Fontana, M.G. and Greene, N.D. (1987). *Corrosion Engineering*, 3rd ed. New York: McGraw-Hill.
[13] Oldfield, J.W. and Sutton, W.H. (1978). *B. Corros. J.*, **13**, 3.
[14] Nauman, F.K. and Spies, F. (1978). Eisenforschung, Max Planck Institute, Dusseldorf.
[15] ASM(1988). *Metals Handbook*, **13**, 632.

Pitting Corrosion

[16] Wranglen, G. (1969). *Corr. Sc.*, **9**, 585.
[17] Foroulis, Z.A. and Thubrikar, M.J. (1975). *J. Electrochem. Sc. and Tech.*, **122**, 1296.
[18] Engell, H.J. and Stolica, N.K. (1959). *Corr. Sc.*, **9**, 455.
[19] Engell, H.J. and Stolica, N.D. (1959). *Z. Phys. Chem. N. F.*, **20**, 113.
[20] Broli, Aa., Holton, H. and Andreassen, J.B. (1976). *Werkstoffe und Korrosion*, 27, 497–504.
[21] Broli, A., Holton, H. and Midjo, M. (1973). *Br. Corros. J.*, **8**, 173.
[22] Hoar, T.P., Mears, D.C. and Rothwall, G. D. (1965). *Corr. Sci.*, **5**, 279.
[23] Kolotyrkin, Y.M. (1964). *Corrosion*, **19**, 216t.
[24] Hoar, T.P. and Jacobs, W.R. (1967). *Nature*, **216**, 1299.
[25] Vetter, K.J. and Strehblow, H.H. (1970). *Ber. Bunsenges. Phys. Chem.*, **74**, 1024.
[26] Sato, N., Kudo, K. and Noda, T. (1971). *Electrochem. Acta.*, **16**, 1909.
[27] Sato, N. (1971). *Electrochem. Acta.*, **16**, 1683.

[28] Strehblow, H.H. (1984). *Werkstoffe und Korrosion*, **35**, 437.
[29] Strehblow, H.H. (1976). *Werkstoffe und Korrosion*, **27**, 729.
[30] Chao, C.Y., Lin, L.F. and McDonald, D.D. (1981). *J. Electrochem. Sc. and Tech.*, **8**, 1187.
[31] Pickering, H.W. and Frankenthal, R.P. (1972). *J. Electrochem. Soc.*, **119**, 129.
[32] Kowaka, M., Tsuge, H., Akashi, M., Masamura, K. and Ishimota, H. (1984). Applications of statistics of extremes, *Japan Society of Corrosion Engineers*, Marzen Pub. Co.
[33] Makabe, H. (1978). How to use Weibull probability, papers, *JIS*, Tokyo (in Japanese).

Intergranular Corrosion

[34] Bain, E.C., Aborn, R.H. and Rutherford, J.B. (1933). *Trans. Am. Soc. Steel Treating*, **21**, 481.
[35] Tedmon, C.S. Jr. and Vermilyea, D.A. (1971). *Corrosion*, **9**, 376.
[36] Strawstrom, C. and Hillert, M. (1969). *J. Iron and Steel Institute*, **77**, 207.
[37] Baumel, A., Buhler, H., Schuller, E., Schwaab, H., Schwenk, P., Ternes, W. and Zitter, H. (1964). *Corros. Sc.*, **4**, 89.
[38] Stickler, R. and Vinckier, A. (1961). *Trans. Am. Soc. Metals*, **54**, 362.
[39] Kinzell, A.B. (1952). *J. Metals*, **4**, 469.
[40] Streicher, M.A. (1959). *J. Electrochem. Soc.*, **161**, 106.
[41] Stickler, R. and Vinckier, A. (1963). *Corros. Sc.*, **3**, 1.
[42] Aust, K.T., Armaijo, J.S. and Westbrook, J.H. (1966). *Trans. Am. Soc. Metals*, **59**, 544.
[43] Aust, K.T., Armaijo, J.S. and Westbrook, J.H. (1967). *Trans. Am. Soc. Metals*, **60**, 360.
[44] Aust, K.T., Armaijo, J.S. and Westbrook, J.H. (1968). *Trans. Am. Soc. Metals*, **61**, 270.
[45] Hanneman, R.F. (1968). *Ser. Metall*, **2**, 235.
[46] Houdermont, E. and Tofante, W. (1952). *Stahl U. Eisen*, **72**, 539.
[47] Columbier, L. and Hochmann, J. (1965). *Stainless and Heat Resisting Steels*, New York: St. Martins Press, p. 168.
[48] Bond, A.P. (1969). *Trans. Met. Soc. AIME*, **245**, 2127.
[49] McGuiire, M.F. and Trolano, A.R. (1973). *Corrosion*, **29**, 7, 268.
[50] Hodges, R.J. (1971). *Corrosion*, **27**, 119.
[51] Armaijo, J.S. (1968). *Corrosion*, **24**.
[52] Armaijo, J.S. (1968). *Corrosion*, **21**.
[53] Edeleanu, C. and Snowdenn, P.P. (1985). *J. Iron and Steel Institute*, **186**, 406.
[54] Wulpi, D.J. (1990). *Understanding How Components Fail*, ASM.
[55] ASM(1986). *Failure Analysis and Prevention Metals Handbook*, Vol.11, p. 180.

Stress Corrosion Cracking

[56] Hoar, T.P. and Hines, J.G. (1956). In *Stress Corrosion Cracking and Embrittlement*. New York: John Wiley, p. 107.

[57] Champion, F.A. (1948). In *Symposium on Internal Stresses in Metals and Alloys*. Inst. Of Metals, London, 469.

[58] Logan, H.L. (1950). *J. Res. N. B. S.*, **48**, 99.

[59] Forty, A.J. (1959). In *Physical Metallurgy of Stress Corrosion Cracking*, New York: Interscience, p. 99.

[60] Keating, F.W. (1948). *Symp. On Internal Stresses in Metals and Alloys*, I. O. M. London, 311.

[61] Sandoz, G. (1972). *Stress corrosion cracking in high strength steels and in titanium and aluminum alloys*, Naval Research Lab, 131.

[62] Petch, N.J. (1956). *Phil. Mag.*, **1**, 331.

[63] Ulhig, H.H. (1952). *J. Res. N. B. S.*, **48**, 99.

[64] Jones, R.H. (1987). *Metals Handbook*, **13**, ASM, 145.

[65] Wearmouth, N.R., Dean, C.P. and Parkins, R.N. (1973). *Corrosion*, **29**, 251.

[66] Parkins, R.N. (1974). *5th Symposium*, American Gas Association, Catalog No. L-30174.

[67] Dix, E.P. (1940). *Trans. Alml.*, **11**, 137.

[68] Mear, R.B., Brown, R.H. and Dix, H. (1944). *Symposium on SCC of Metals, ASTM/AIME, Phil.*, 329.

[69] Dix, E.H. (1940). *Trans. AIME*, **11**, 137.

[70] Parkins, R.N. (1980). *Corros. Sc.*, **20**, 197.

[71] Humpries, M.J. and Parkins, R.N. (1969). *Proc. 4th Int. Cong. Met. Corros.*, Australia: Sydney.

[72] Bakish, R. and Robertson, W.D. (1956). *J. Electrochem. Soc.*, **103**, 320.

[73] Sprowls, D.O. and Browny, R.H. (1969). In *Fundamental Aspects of SCC Conference*, Houston: NACE.

[74] Parr, S.W. and Straub, F.G. (1928). *Univ. Ill. Bull.*, 177.

[75] McEviley, A.J. and Bond, A.P. (1965). *J. Electrochem. Soc.*, **112**, 131.

[76] Pugh, E.N. (1971). Theory of Stress Corrosion Cracking, NATO.

[77] Green, J.A.S., Mengelberg, H.D. and Yollan, H.T. (1970). *J. Electrochem. Soc.*, **117**, 433.

[78] Robertson, W.D. and Bakish, R. (1956). *Stress Corrosion Cracking and Embrittlement*. New York: John Wiley, p. 32.

[79] Hoar, T.P. and West, J.M. (1962). *Proc. Royal Soc.*, **A.268**, 304.

[80] Orwan, E. (1944). *Nature*, **154**, 341.

[81] Paleh, N.J. and Stables, P. (1952). *Nature*, **169**, 842.

[82] Paleh, N.J. (1956). *Phil. Mag.*, **1**, 331.

[83] Coleman, E.G., Weinstein, D. and Kostokar, W. (1961). *Acta Metallurgia*, **9**, 491.

[84] Heady, R.B. (1977). *Corrosion*, **33**, 441.

[85] Scully, J.C. and Powell, D.T. (1970). *Corros. Sci.*, **10**, 719.

[86] Perkins, R.N. and Fessler, R.R. (1978). *Mater. Eng. Appl.*, **80**.

[87] Staehle, R.W., Royuela, J.J., Raredon, T.L., Morin Serrate, E. and Farrar, R.V. (1970). *Corrosion*, **26**, 451.

[88] Copson, H.R. (1959). *Physical Metallurgy of Stress Corrosion Cracking*, Rhodin, T.N., ed. New York: Interscience, 247–272.

[89] Dana, A.W. and Warren, D. (1957). DuPont Engineering Department, *ASTM Bulletin*, 225, October.

[90] Hoar, T.P. and Hines, J.G. (1956). *J. Iron and Steel*, **184**, 166–172.

[91] Warren, D. (1960). *Proc. 15th Annual Purdue Industrial Waste Conference*, Purdue University Mag., 1–19.

[92] Bruce Craig, (1988). In Corrosion, *Metals Handbook*, **13**, ASM, 151.

[93] Engle, J.P., Floyd, G.L. and Rosenes, R.B. (1959). *Corrosion*, **15**, (2), 69–72.

[94] Snowden, P.P. (1961). *Nuclear Engineering*, **6**, 409–411.

[95] Wilde, B.E. (1971). Theory of Stress Corrosion Cracking of Alloys, NATO, SCREC.

[96] Parkins, R.N. (1972). *Br. Corros. J.*, **7**, 15.

[97] Port, R.D. (1984). *Corrosion-84*, New Orleans: NACE.

Corrosion Fatigue

[98] Griffith, A. (1920). *Phil. Trans. Royal Soc. Lond.*, 163.

[99] Meyers, M. and Chawla, K.K. (1984). *Mechanical Metallurgy*, Prentice Hall.

[100] Inglis, C.E. (1913). *Proc. Inst. Naval Arch.*, **55**, 163.

[101] Irwin, G.R. (1957). *J. App. Mech.*, **24**, 361.

[102] Mitchell, M.R. Fundamentals of modern fatigue analysis for design, *Fatigue and Micro-structure*, ASM, Ohio: Metals Park, p. 385.

[103] Feltner, C.E. and Landgraf, R.W. (1969). *ASMF*, Paper No. 69, DE-59, New York: ASME.

[104] Coffin, L.F. (1954). *Trans. ASME*, **76**, 31.

[105] Sander, B.I. (1972). Fundamentals of cyclic stress and strain. University of Wisconsin-Madison, Wisconsin.

[106] Paris, P.C. and Erdogan, F. (1963). *J. Basic Eng. Trans. ASME*, **85**, 528.

[107] Erdogan, F. (1967). Crack propagation theories, NASA-CR-901.

[108] Forman, R.G., Kearney, V.E. and Engle, R.N. (1967). *J. Basic Eng. Trans. ASME*, **89**, 459.

Fretting Corrosion

[109] Pugh, E.N., Bertocci, U. and Ricker, R.E. (1987). *N. B. S.*

[110] Lindley, T.C., and Nix, K.J. (1982). *ASTM, Std. 853*, 340.

[111] Widner, R.L. (1990). In *ASM Metals Handbook*, Vol. 11, 420, Ohio: Metals Park, USA.

Cavitation Corrosion

[112] Hammilt, F.O. (1987). In *Metals Handbook Corrosion*, **13**, 9th ed., p. 163.

[113] Arndt, R.E. (1977). *Corrosion*, Paper No. 91.

[114] Khan, A.H. (1956). *Desalination Processes and Multistage Flash Distillation Plants*, Elsevier, Oxford, UK.

[115] Unterweiser, P.M., ed. (1979). Case Histories in Failure Analysis, ASM, 232.

[116] Tuthill, A.H. and Schoollmaker, C.M. (1965). *Ocean Science and Ocean Engineering Conference*, Marine Tech., Washington: SOC.

SUGGESTED READING FOR GALVANIC CORROSION

[117] Dillon, C.P., ed. (1982). *Forms of Corrosion, Recognition and Prevention*, Chapter 3, p. 45, NACE.

[118] Hack, H.P. (1988). *Galvanic Corrosion*, STP, 978, ASM.

[119] Bradford, S.A. (1998). *Practical Self-study Guide to Corrosion Control*, Casti Publishers.

[120] Fontana, M.G. (1986). *Corrosion Engineering*, 3rd ed. McGraw-Hill, pp. 41–51.

[121] Jones, D.A. (1996). *Principles and Prevention of Corrosion*, 2nd ed. Prentice Hall, p. 168.

WEBSITE FOR GALVANIC CORROSION

[122] http://www.corrosion-doctor.com

SUGGESTED READING FOR FRETTING CORROSION

An excellent review of fretting corrosion is given by Waterhouse, R.B. in:

[123] *Fretting Corrosion* (1972). Pergamon Press, pp. 111–113.

[124] *Theories of Fretting Processes in Fretting and Fatigue* (1981). Applied Science, pp. 217–211.

KEYWORDS

Galvanic Corrosion

Anode/cathode ratio It is the ratio of the anodic area to the cathodic area.

Bimetallic corrosion Corrosion of two metallics in electrical contact. The term galvanic corrosion is used generally in place of bimetallic.

Cathode The electrode of a galvanic cell where positive current flows from the solution to the electrode.

Concentration cell A galvanic cell in which the emf is due to the difference in the concentration of reactive constituents of the electrolyte.

Differential aeration cell A corrosion cell formed by the difference in the concentration of oxygen in the electrolyte in contact with a metal.

Galvanic series It is a list of metal and alloys based on their relative potentials in a specified environment. The environment generally used is seawater.

Galvanic couple A pair of dissimilar metals in contact with an electrolyte.

Galvanic current The current which flows between two dissimilar metals in contact with an electrolyte.

Galvanic corrosion The corrosion caused by a galvanic cell.

Galvanic potential The potential arising from two or more electrochemical reactions preceding simultaneously on the metal surface. It is also called 'galvanic cell potential.'

Pitting Corrosion

Active–passive metal A metal which shows transition from an active state to a passive state, e.g. stainless steel, titanium.

Passive A metal characterized by low corrosion rate in a certain potential range whose oxidation is the predominant reaction.

Passivity It is the reduction in the anodic rate of reaction because of the formation of a protective film of oxide.

Pits It is the result of localized corrosion confined to a small area in the metal surface, in the form of a small cavity.

Pitting factor The ratio of the depth of the deepest pit divided by the average penetration of pits on the metal surface.

Pitting frequency The number of pits per unit area.

Pitting probability $P = N_p/N \times 100$, where N_p is the number of specimens in the pit, and N is the total number of specimens.

Pitting potential (E_p). Also called critical pitting potential. It is the most negative potential required to initiate pitting on a metal surface which is in the region of the passive potential. At this potential, the current density shoots up sharply during anodic polarization.

Passive–active cell (Active–Passive) A corrosion cell in which the area of the metal surface in the active state is the anode and the area in the passive state is the cathode.

Protection potential (E_{pp}). It is the potential below which no pitting is expected to initiate or propagate.

Intergranular Corrosion

Austenitic A solid solution of one or more elements in a face centered cubic iron.

Austenitizing Heating a steel into the austenitic temperature range so that its structure becomes austenitic.

Carbide A compound of iron and carbon Fe_3C (cementite) is a hard and brittle substance.

Concentration cell A corrosion cell formed by a difference in the concentration of some constituents in the electrolyte.

Chromium depletion Sensitization of 18-8 stainless steel results in the precipitation of chromium carbide at grain boundaries with depletion of chromium in regions adjacent to the grain boundary.

Diffusion (Self) It is the migration of atoms in a material by thermal random-walk.

Diffusion (Interstitial) The migration of interstitial atoms in a matrix lattice.

Duplex steels Alloys containing both the body centered cubic alpha-phase and the face centered gamma-phase are known as duplex steels. They may be low carbon (less than 0.03%) low chromium alloys (less than 20% Cr) with a continuous austenitic matrix or high chromium (>20%) alloys with a ferrite matrix.

Dislocation A crystalline imperfection in which a lattice distortion is centered around a line.

Ferritic A solid solution of one or more elements in a BCC iron (α-phase).

Ferritic steel Binary iron chromium alloys containing 12–30% Cr. Their structure remains ferritic under normal heat treatment condition.

Grain boundary A surface imperfection which separates grains of different orientation in a polycrystalline aggregate.

Heat Affected Zone (HAZ) The portion of a metal whose microstructure is altered by heat, such as during welding or heat treating.

Intergranular corrosion Preferential corrosion occurring at grain boundaries or at regions adjacent to the grain boundaries.

Inter-crystalline corrosion Same as intergranular corrosion.

Knife-line attack Intergranular corrosion of an austenitic steel (stabilized grade) extending only a few grains away from the fusion line of a weld during sensitization of steel.

Martensitic stainless steel Essentially Fe–Cr alloys containing 12–17% Cr with 0.15–1.0% C so that a martensitic structure is produced by quenching from the austenite phase region. A martensitic structure is body-centered tetragonal.

Non-sensitized steels Steels which are specially heat treated so that they are not susceptible to intergranular corrosion are called non-sensitized steels.

Precipitation hardening Hardening caused by the precipitation of a constituent from a supersaturated solution.

Precipitation hardening steels These are steels with high toughness, high strength and having optimum creep properties. Aluminum, molybdenum, copper and titanium additions are made for intermetallic strengthening.

Residual Stress Stress that is within a metal as a result of plastic deformation.

Precipitation of carbides At the grain boundaries of stainless steels, chromium reacts with carbon to precipitate chromium carbides. Carbides of Ni, molybdenum, titanium and niobium, are well-known. Chromium carbide is represented by $Cr_{23}C_6$.

Sensitization When austenitic steels, such as AISI Type 304 are cooled slowly through the range 550–580°C, they are frequently susceptible to severe intergranular attack when exposed to a corrosive environment. This phenomenon has been termed '*sensitization*' which means the metal is sensitive to grain boundary corrosion.

Sigma-phase In stainless steels containing Mo, stainless steel types, such as 316 and 317, and in high nickel alloys containing Mo, heating to temperatures in the range 700–850°C causes the precipitation of sigma-phase.

Solution-heat treatment Heating an alloy to a temperature which causes one of the constituents or more to enter into solid solution. A solid solution is a single, solid crystalline phase containing two or more species.

Streicher test This is a test to determine the susceptibility of Fe–Ni–Cr alloys to intergranular corrosion. The specimen is exposed in a solution of 50 wt% H_2SO_4 + 25 g/liter $CuSO_4$ for 120 h.

Stabilized steels Steels containing elements which form carbides that are more stable than chromium carbides. Niobium and titanium are used as stabilizing elements. Steel AISI Types 321 and 347 are stabilized grades of steels.

Transpassive region The region of an anodic polarization curve above the passive potential range which is characterized by a significant increase in current density with an increase in the potential to the noble range.

Weld decay This is the intergranular attack which occurs in stainless steels or certain nickel-base alloys as a result of sensitization of the steel in the sensitizing temperature range. It occurs a short distance away from the weld bead.

Stress Corrosion Cracking

Active path (pre-existing) Structure features, such as impurity, segregates, precipitates, which exist on the grain boundaries and cause a local galvanic cell to be set up. These cells provide active paths for the propagation of cracks.

Active path (strain generated) If the protective film is disrupted, the exposed metal is attacked until the disrupted film is again formed. As a result of disruption of the protective films, the metal becomes active. The plastic strain in the underlying metal is responsible for disrupting the film. This type of disruption provides a strain generated active path.

Adsorption This is the surface retention of solid, liquid or gas molecules, atoms or ions by a solid or liquid.

Anodic zone Anodic zones are produced as a result of microstructural differences, compositional differences, grain boundaries and sub-grain boundary or heterogeneous effects of stress or prior cold work.

Brittle fracture Brittle fracture proceeds along characteristic crystallographic planes called cleavage planes and has a rapid crack propagation. Most brittle fractures are transgranular. They proceed with very little plastic deformation.

Cleavage. This is the splitting of a crystal in a crystallographic plane. A cleavage fracture is marked by characteristic pattern, such as river markings, herring bones, tongues, etc.

Ductile fracture A fracture or tearing of metal associated with gross plastic deformation.

Film (breakdown and repassivation) This is the rupture of a film and the reformation of a film on the metal surface. An emergent slip step breaks the thin protective film. The film may reform again on the surface except a small area. The latter is called 'repassivation.'

Film rupture This is the breakdown of a protective film on a metal surface.

Plane strain This is a stress condition in which the strain is zero in a direction normal to both the axis of the applied tensile stress and direction of crack growth (parallel to the crack). This condition is achieved with thick specimens.

Plane strain fracture toughness (K_{1c}) The minimum stress intensity required to cause catastrophic failure.

Plane stress This is a stress condition in which the stress in the thickness direction is zero. This stress arises as a result of strain gradient. This condition is achieved in thin specimens.

Residual stress The stress that is confined in a material as a result of plastic deformation.

Slow strain rate technique The specimen is pulled in uniaxial tension at a controlled very slow strain rate in a given test environment to detect its susceptibility to SCC.

Slip Plastic deformation by irreversible shear displacement of one part of crystal with respect to another in a definite crystallographic direction and on a specific crystallographic plane.

Strain rate This is determined from the specimen gauge length during tensile test. In evaluation of SCC, a constant strain rate of 10^{-6}sec^{-1} is applied to a tensile specimen in a given environment. It represents the rate of deformation of a metal.

Stress intensity It is denoted by symbol K and it denotes the intensity of the applied stress at the tip of the crack. K is a function of nominal stress, crack length and component geometry.

Stress intensity factor (K_1)

$$K_1 = K \times \partial \times \sqrt{C}$$

where

> K_1 is stress intensity factor derived from crack length
> C is crack length
> M is stress across the narrow section of the specimen
> k is constant.

It is also reported as $K_1 = \partial \sqrt{ay}$, where a is the crack length and y is the correction factor.

Stress intensity factor for SCC (K_{1-SCC}). The value of stress intensity above which cracks grow because of stress corrosion cracking.

Stress raisers Discontinuities in the structure which produce local increases in the stress.

Stress relief This occurs when the alloys after welding are subjected to stress relief by appropriate heat treatment in order to reduce the residual stresses. This is also called 'post-weld heat treatment.'

Stress level The magnitude of stress required to cause stress corrosion cracking.

Threshold stress intensity factor ΔK_{th}. The value of ΔK_i below which crack propagation does not occur.

Transgranular cracking This is the cracking which takes place across the grains.

Hydrogen Damage

Caustic embrittlement It is the phenomenon of stress corrosion cracking of steels in caustic soda at elevated temperatures.

Hydrogen blistering The formation of blisters on the metal surface due to excessive internal pressure of hydrogen.

Hydrogen embrittlement The ductility of a metal is reduced and it cracks due to absorption of hydrogen.

Hydride formation The degradation of mechanical properties and cracking of metals, such as titanium, resulting from hydride formation due to pick-up of hydrogen.

Hydrogen Induced Cracking (HIC) This is characterized by the brittle fracture of a ductile alloy normally in the presence of hydrogen (same as hydrogen embrittlement).

Hydrogen sulfide cracking This refers to cracking of high strength steels in a sour gas environment encountered generally in oil drilling.

Hydrogen trapping This is the binding of hydrogen atoms to impurities and structural defects. Hydrogen may be trapped at mobile dislocations or grain boundaries.

Seasonal cracking This refers to the cracking of brass in ammoniacal environment.

Cavitation Damage

Cavitation The formation and collapse of innumerable tiny cavities within a liquid subjected to intense pressure changes.

Cavitation corrosion This is the deterioration of a metal subjected to a conjoint action of cavitation and corrosion.

Cavitation damage This is the damage to a material caused by cavitation.

Cavitation erosion This is the gradual loss of material from a solid surface as a result of continuous exposure of a material to cavitation.

Erosion Removal of surface material by individual impact of solid and liquid particles.

Protective film This is generally a continuous film of oxide on a metal surface which acts as a barrier between the metal and the environment.

Shockwave A wave of the same nature as a sound wave but of much greater intensity.

Dezincification

De-alloying It is the selective leaching of one or more component of an alloy.

De-aluminification Selective leaching of aluminum in aluminum containing alloys.

De-nickelification Selective dissolution of nickel from nickel containing alloys.

Dezincification Selective leaching of zinc from copper–zinc alloys.

Graphitic corrosion This is the deterioration of cast iron in which the iron-rich matrix is selectively leached or converted to corrosion products and a weak residue of graphite flakes is left behind.

Leaching Dissolution of a component of an alloy or dissolution of a metal by leaching agents, such as sulfuric acid or sodium hydroxide, etc.

Corrosion Fatigue

Anodic polarization Polarizing a metal in a positive direction by applying an external EMF.

Axial strain A linear strain parallel to the longitudinal axis of the specimen.

Brittle fracture The separation of a solid accompanied by either little or no plastic deformation.

Cathodic polarization Polarizing a metal surface in an active (positive) direction by application of an external EMF.

Corrosion fatigue Cracking produced by the conjoint action of repeated or fluctuating stress and a corrosive environment.

Crack size A linear measure of a principal plane dimension of a crack.

Crack extension An increase of crack length.

Crack opening displacement This is the displacement at the tip of the crack (CTOD).

Cyclic loading This is a periodic or non-periodic fluctuating loading applied to a test specimen (also called fatigue loading).

Ductile fracture Tearing away of a metal accompanied by appreciable plastic deformation.

Constant amplitude loading A type of loading in fatigue in which all peak and valley loads are equal.

Fatigue life The number of loading cycles that a specimen can sustain before failure.

Fatigue limit This is the limiting value of medium fatigue strength.

Fatigue crack growth The rate of crack growth under constant amplitude fatigue loading (da/dN).

Fracture toughness This is a measure of the resistance of a material to crack extension ($MPa\sqrt{m}$).

Fracture toughness (plane strain) K_{1C} The crack extension resistance under conditions of crack tip plane strain.

Loading amplitude (or alternate loading). This is one-half of the range of a cycle.

Loading ratio

$$R = \frac{\text{Minimum load}}{\text{Maximum load}} = \frac{\sigma_{\min}}{\sigma_{\max}} \ (\sigma \text{ is given in MPa})$$

Mean load

$$\delta_{m} = \frac{\delta_{\max} + \delta_{\min}}{2} \ (\sigma \text{ is given in MPa})$$

Plane strain A stress condition in linear elastic fracture mechanics in which there is zero strain in the direction normal to the axis of applied tensile stress and direction of crack growth. It is achieved in thick plate, along a direction parallel to the plate.

Plane stress A stress condition in linear elastic fracture mechanics in which a stress in the thickness direction is zero. It is achieved in loading thin steel along a direction parallel to the surface of a sheet.

Plane stress fracture toughness, K_C The value of stress intensity at which crack propagation becomes rapid in sections thinner than those in which plane strain conditions occur.

Stress intensity factor is given as $K_a = \sigma\sqrt{\pi a f(a/w)}$, whereas $a/w \to 0$, $f(a/w) \to 1.0$. In the above expression $f(a/w)$ is the correction factor which depends on the width (w) and flow size (a). The critical value of stress intensity, K_{1C}, is commonly known as plane stress fracture toughness. It is expressed in units of MPa.

S–N diagram A plot of stress amplitude against the number of cycles to failure.

Stress corrosion cracking A process of cracking which occurs by the action of a corrodent and a stress.

Stress concentration factor It is the ratio of greatest stress in the region of a notch or discontinuity to the corresponding nominal stress.

Stress intensity (K_i) It represents the intensification of applied stress of the tip of a crack of known size and shape. The range of stress intensity factor is given by

$$\Delta K = K_{max} - K_{min} \, (MPa\sqrt{m}).$$

Stress ratio (*A* **or** *R*) It is the ratio of minimum stress (S_{min}) to maximum stress (S_{max}).

Subscripts for loading conditions:

K_C = plane stress fracture toughness.
K_1 = stress intensity factor.
K_{1C} = plane strain fracture toughness.
K_{1SCC} = threshold intensity for stress corrosion cracking.
K_{th} = threshold intensity for SCC.
ΔK = the range of stress intensity factor during a fatigue cycle.

CHAPTER 5

CATHODIC PROTECTION

5.1 INTRODUCTION

Cathodic protection is a proven corrosion control method for protection of underground and undersea metallic structures, such as oil and gas pipelines, cables, utility lines and structural foundations. Cathodic protection is now widely applied in the protection of oil drilling platforms, dockyards, jetties, ships, submarines, condenser tubes in heat exchangers, bridges and decks, civil and military aircraft and ground transportation systems.

The designing of cathodic protection systems is rather complex, however, it is based on simple electrochemical principles described earlier in Chapter 2. Corrosion current flows between the local action anodes and cathodes due to the existence of a potential difference between the two (Fig. 5.1). As shown in Fig. 5.2, electrons released in an anodic reaction are consumed in the cathodic reaction. If we supply additional electrons to a metallic structure, more electrons would be available for a cathodic reaction which would cause the rate of cathodic reaction to increase and that of anodic reaction to decrease, which would eventually minimize or eliminate corrosion. This is basically the objective of cathodic protection. The additional electrons are supplied by direct electric current. On application of direct current, the potential of the cathode shifts to the potential of the anodic area. If sufficient direct current is applied, the potential difference between the anode and cathode is eliminated and corrosion would eventually cease to occur.

As the cathodic current increases (more transfer of electrons), the cathodic reaction polarizes in the direction of local action anode potential, thus reducing further the potential difference between the anodes and cathodes. Complete cathodic protection is achieved when the metallic structure becomes cathode (more negative). The severity of corrosion is directly proportional to the magnitude of the difference of potential between the anode and the cathode, hence by eliminating this difference, corrosion may be eliminated.

5.2 BASIS OF CATHODIC PROTECTION

Figure 5.3 illustrates the simple principle of cathodic protection. On application of an external current, the difference of potential between the cathodes and anodes on the structure decreases. Corrosion stops when potential of cathode becomes equal to the potential of anode. The anode would become more negative and the cathode more positive. Cathodic protection is, therefore, achieved by supplying an external negative current to the corroding metal to make the surface acquire the same potential to eliminate the anodic areas. The anodic areas are eliminated by transfer of electrons. After a sufficient current flow, the potential of anodic areas would become negative enough for corrosion to stop.

(a) There must be an anode, a cathode, an electrolyte and a metallic path for the transfer of electrons.
(b) A source of DC current to supply electrons.
(c) Sufficient direct current should be applied to eliminate the potential difference between the anode and the cathode.

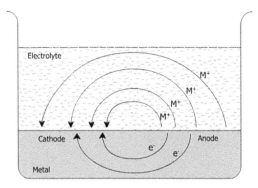

Figure 5.1 An electrochemical cell. Corrosion cell between two areas on single metal surface. Current flows because of a potential difference that exists between anode and cathode. Anions (e.g. Fe^{++}) leave at the anode which corrodes and are accepted at the cathode where corrosion is prevented. Electrons are insoluble in aqueous solutions and move only in the metal

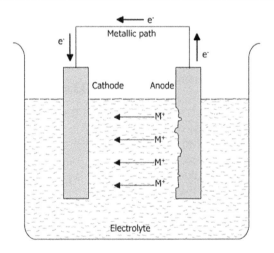

Figure 5.2 Simple electrochemical cell

5.3 Working of Cathodic Protection System

Figures 5.4a and b show how, in principle, a cathodic protection system works. Figure 5.4a shows a buried pipeline with anodic and cathodic areas prior to the application of the cathodic current. Figure 5.4b shows the same pipeline after the cathodic protection.

5.4 Factors Leading to Corrosion of Underground Metallic Structures

The corrosion encountered by metals in aqueous solutions is always electrochemical in nature. It occurs because of the formation of anodic and cathodic areas and the flow of electrons through

Potential Difference

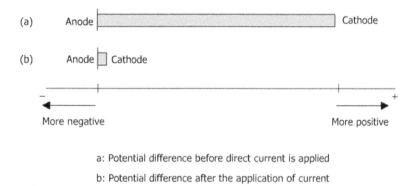

(a)   Anode ⬛⬛⬛⬛⬛⬛⬛⬛⬛ Cathode

(b)   Anode ⬜ Cathode

More negative   More positive

a: Potential difference before direct current is applied

b: Potential difference after the application of current

Figure 5.3   Effect of external current in the metallic structure

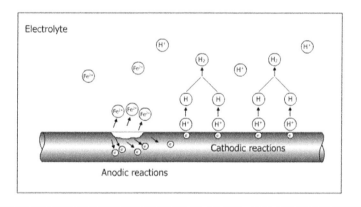

Figure 5.4a   Anodic and cathodic reactions on a metal surface

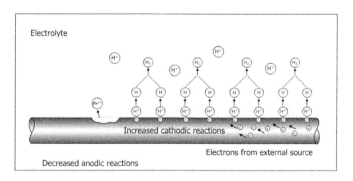

Figure 5.4b   Increased cathodic reaction and decreased anodic reaction (insignificant) caused by introducing electrons from an external source

the metallic path. Structures, such as pipelines buried in the ground are affected by the presence of concentration cells, galvanic cells and stray currents. Some typical cells are described below.

(a) **Concentration cell**

A concentration cell is formed by the differences in concentration of salts, degree of aeration, soil resistivity and degree of stress to which the metals are subjected. The above differences cause the formation of anodic and cathodic areas on the surface which lead to corrosion.

(b) **Galvanic cell**

A galvanic cell is formed when two metals differing in potential are joined together. For instance, if copper is joined to aluminum, aluminum would corrode because it has a more negative potential (-1.66 V) than copper ($+0.521$ V). Copper being less active becomes the cathode and aluminum becomes the anode. But if iron is joined to aluminum, the iron corrodes (in seawater), due to the passive film on aluminum which causes it to behave like a nobler metal than iron (but not nobler than copper). The formation of such galvanic cells often leads to the corrosion of underground buried structures. A steel plate with copper rivets

would form a galvanic cell. Several examples of galvanic cells are given in Chapter 4. A galvanic cell is basically a corrosion cell.

5.4.1 EXAMPLES OF CONCENTRATION CELLS

(a) Differential Aeration Cell

Suppose, a pipeline is buried in a completely uniform soil and that some areas of the line have a free supply of oxygen and other areas have a restricted supply. The part of the pipe buried in the soil with a free supply of oxygen (high oxygen content) would form the cathode and the part with a less supply of oxygen or poorly aerated forms that anode (Fig. 5.5). The current (Fe^{++} ions) would flow in the soil from anode to cathode resulting in the corrosion of the pipe end buried in poorly aerated soil. Such a cell is commonly called a *differential aeration cell.*

(b) Dissimilar Electrode Cells

Joining of a new pipe to an old pipe results in the corrosion of the new pipe, as the new pipe becomes anodic to the old pipe. The old pipe by virtue of film formation has a less active potential

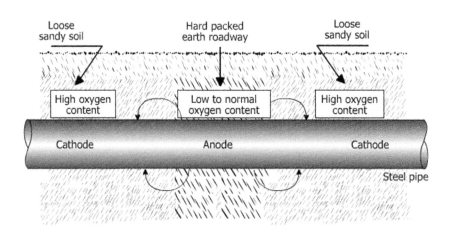

Figure 5.5 Formation of differential oxygen cells. (From CORRINTEC, USA, Presentation at KFUPM Dhahran, Saudi Arabia, March 5, 1982)

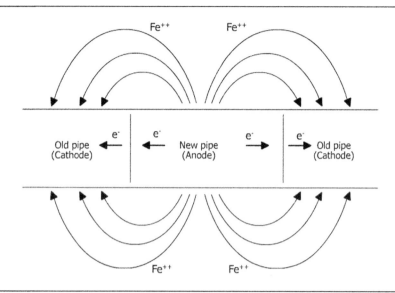

Figure 5.6 Dissimilar electrode cell formation by joining of an old pipe with a new pipe

than the new pipe and it, therefore, becomes the cathode (Fig. 5.6). Corrosion also occurs at the coating flaws when the new pipe is connected to the main with a bonded coupling.

(c) Stress Cells

Anodic areas may be formed in regions of high stresses and cathodic areas in regions of low stresses as shown in Fig. 5.7. Bright steel at threads

and areas of stress at welds form the anode and corrosion occurs at these areas.

(d) Different Types of Soils

Suppose a pipeline is buried in two types of soils, one being a clay soil and the other sandy soil. The area of pipe buried in the clay soil would form the anode, whereas the area buried in the sandy soil would form the cathode, because of

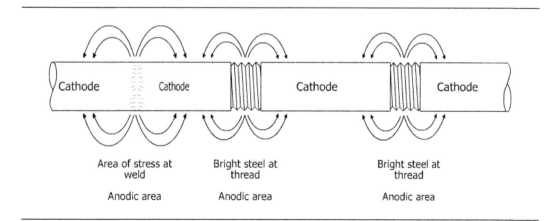

Figure 5.7 Stress cell. Corrosion caused by areas of stress and areas of bright steel on an underground pipeline

excess of oxygen and higher porosity. The pipe in the clay soil, therefore, corrodes. Corrosion caused by different types of soils is shown in Fig. 5.8.

(e) Different Moisture Contents

If a pipe is buried in two types of soil having different moisture contents, the area of the pipe in contact with wet soil (high moisture) would corrode, whereas the pipe in contact with the dry soil would not corrode. The area of the pipe in contact with the dry soil becomes the cathode and the area in contact with the wet soil, the anode. Differential concentration cell is shown in Fig. 5.9.

At the anode: $Fe \rightarrow Fe^{++} + 2e^-$

At the cathode: $O_2 + 2H_2O + 4e^- \rightarrow 4OH^-$

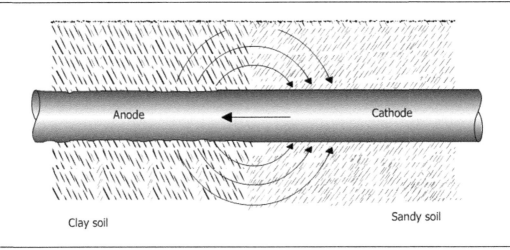

Figure 5.8   Corrosion caused by different types of soils

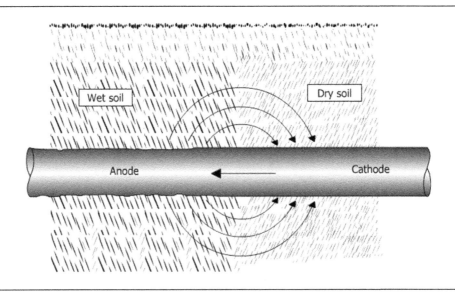

Figure 5.9   Differential concentration cell

5.4.2 GALVANIC CELLS

This type of corrosion arises by two dissimilar metals in contact with each other. In pipe lines it can occur when replacing an old section of the pipe by a new section. The same problem arises when the new lines are laid adjacent to the existing lines and connected across by means of branches. Any dissimilarity within the two metals would bring about this type of corrosion. The more active metal in this situation becomes the anode and the less active, the cathode. For example, if zinc is connected with steel, zinc would corrode as it would be anodic (more active potential), and by sending electrons to the cathode (steel) it would protect the steel. Any metal with a more active potential would always form the anode and the one with less active potential the cathode. The conventional current in the soil always flows from anode to cathode.

Figure 5.10 shows a cast iron box containing copper tubes. Electrons would flow in the metallic path from the cast iron box to the copper tubes as the cast iron has a more negative potential and so would be the anode, whereas copper has a noble potential would be the cathode. Eventually the box would be corroded and the copper tubes protected.

By installing a magnesium anode, a hot water steel tank may be protected against corrosion as shown in Fig. 5.11. In this case magnesium being very active would become the anode and corrode.

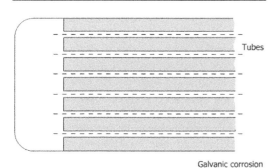

Galvanic corrosion

Figure 5.10   Current flows in the metallic path from case iron box into copper based tubes

5.5 ELECTRICAL BASIS OF CATHODIC PROTECTION

Let us suppose a pipe is buried in a soil. A difference of potential exists between the anode and the cathode, i.e.

$$\Delta E = (E_a - E_c)$$

ΔE is the difference in potential. If $R_a + R_c$ is controlling resistance, and $E_a - E_c$ has a finite value of circuit resistance, the corrosion current (I_c) would flow. If complete freedom from corrosion is desired, I_c must be made zero.

$$I_c = \frac{E_a - E_c}{R_a + R_c} = 0 \left(I = \frac{E}{R} \right) \quad \text{Ohms law}$$

where

I_c = corrosion current (A)
E_a = cathode potential
E_c = anode potential
R_a = anode resistance (ohms)
R_c = cathode resistance.

The corrosion current can be made zero by making $R_a + R_c$ equal to infinity, for example by pointing. This can also be achieved by equalizing the potential difference between E_c and E_a or making $(E_c - E_a)$ equal to zero. The structure can be made cathode (negative) by supplying an electric current from outside until its potential becomes equal to the potential of the anode, and the difference between E_c and E_a completely disappears.

To further illustrate the principle, the equivalent circuit (a) of a cathodically protected metal is shown in Fig. 5.12A. Complete cathodically protected metal is shown in Fig. 5.12B. Here

$$E_A - E_C = I_1 R_A + R_C(I_1 + I_2)$$

$$I_1 = \frac{E_A - (E_C + R_C I_2)}{R_A + R_C}$$

where

E_A = open circuit anode potential
E_C = open circuit cathode potential
R_A = effective anode resistance

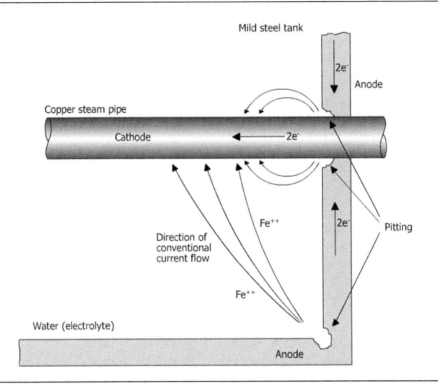

Figure 5.11 Galvanic corrosion in a hot water tank

R_C = effective cathode resistance
I_1 = current from anodic area
I_2 = current from the external anode.

If corrosion is not to occur, I_1 must be zero, therefore, $(E_C + R_C I_2) = E_A$. It means that sufficient current must flow through R_C for the potential of the anode to be equal to the potential of the cathode (open circuit potential).

The above relationship suggests that enough current must be provided by an external anode to flow through the cathode resistance R_C to make the cathode potential (E_C) equal to the anode potential (E_A). When this condition, $E_A = (E_C + R_C I_2)$, is achieved, no corrosion would occur and the structure would be cathodically protected. Figure 5.12 illustrates the complete and incomplete protection. In part(b) current is not flowing to all anodic areas, whereas in part(c) current is supplied to all anodic areas. Figure 5.12A shows

partial protection and Fig. 5.12B shows complete protection, $E_A = (E_C + R_C I_2)$.

5.5.1 SOURCES OF CURRENT

It has been shown above that there must be a source of current to supply electrons to the areas of the metal which is corroding. In a metal buried in ground, anodic areas corrode by release of electrons and if an equal number of electrons are not introduced from an external source, the metal would continue to corrode. An external anode which supplies such current is called *auxiliary anode* in the electrochemical cell and referred to as anode in a cathodic protection system. Electrodes of graphite, cast iron, platinum and titanium act as conductors of electricity and supply the desired current to the structure to be protected. The conductors are energized by a DC source. The rate of consumption of anode electrodes

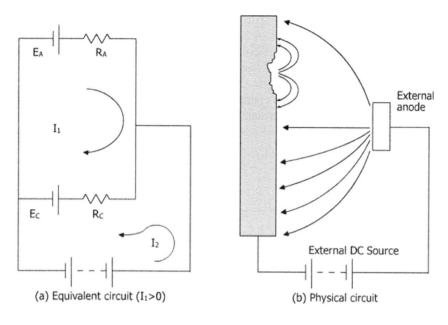

(a) Equivalent circuit ($I_1>0$)   (b) Physical circuit

(A) Partial protection

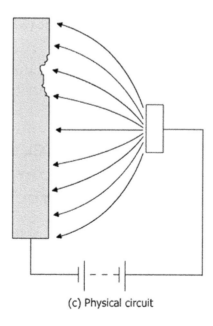

(c) Physical circuit

(B) Complete protection

Figure 5.12   Equivalent circuit of cathodically protected metals. (From Spencer, K.A. (1960). The Protection of Gas Plant and Equipment from Corrosion. *Joint Symp. Corrosion Group of the Society of Chemical Industry,* Sept 22–23)

depends on the material used. The more expensive the anode, such as titanium, the lesser the rate of consumption. The above anodes are called *impressed current anodes*. The other types are *sacrificial anodes* which corrode in the soil and generate electric current required for protection of the structure. These anodes are not energized by an outside DC source. Details of galvanic and impressed current anodes are provided later in this chapter.

A structure without cathodic protection is analogous to an electrochemical cell, and is shown in Fig. 5.13a.

In a cathodic protection system, the current (electrons) leaves the anode and enters both the anodic and cathodic areas of the buried metal or by analogy to the electrochemical cell, the current from the auxiliary anode (AA) enters both the anodic and cathodic areas of corrosion, A and C, where A = anode and C = cathode, returning to the source of DC current (Fig. 5.13b). Naturally, when the cathodic areas are polarized to the open circuit potential of the anode, all the metal surface is at the same potential and the local active current would no longer flow. The metal does not corrode as long as the external current is maintained. The electrodes A and C would achieve the same potential if they are polarized by the external current supplied by AA. The potential of the cathode would shift close to the potential of the anode, and after some time both A and C would acquire the same potential, thus, by supplying an external current it is possible to prevent corrosion. As the two electrodes, A and C, become more negative, the method is called *cathodic protection*. Cathodic protection is, therefore, a method of preventing the corrosion of a metallic structure by making it a cathode with respect to the auxiliary anode (anode supplying the current).

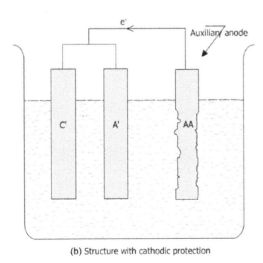

(a) Structure without cathodic protection
A = anodic area of structure
C = cathodic area of structure

(b) Structure with cathodic protection

Figure 5.13 Diagram illustrating the principle of cathodic protection

5.6 ELECTROCHEMICAL THEORY OF CATHODIC PROTECTION

In Fig. 5.14a, $E_{o,c}$ represents the reversible potential of the cathode and $E_{o,a}$ represents the reversible potential of the anode. β_a represents the Tafel slope of the linear portion of anodic curve (slope = 0.12 V/decade), and β_c the cathodic slope. E_{corr} represents the corrosion potential, where the rate of oxidation equals the rate of reduction. I_{corr} is the corrosion current corresponding to corrosion potential $(100\,\mu\text{A/cm}^2)$ (Figs 5.14a and b). Extrapolation of the linear portion of the anodic curve to the equilibrium potential of the anode, gives the exchange current density for the anodic reaction $(I_{o,c})$. I_o is defined

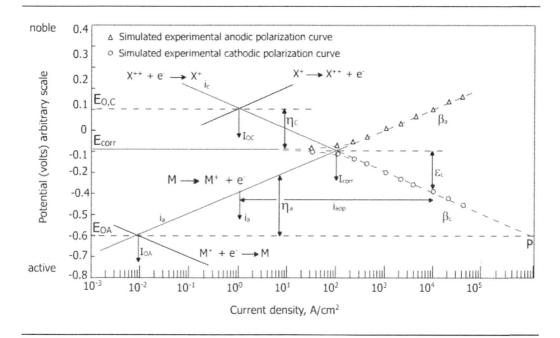

Figure 5.14a Anodic and cathodic Tafel curves for corrosion of metal, M, showing the principle of cathodic protection. (From Jones, D.A. (1981). *Principles of Measurement and Prevention of Buried Metal Corrosion by Electrochemical Polarization.* In Underground Corrosion, STP 741, ASTM. Reproduced by the kind permission of ASTM, Philadelphia, PA, USA)

as equal and opposite ('exchange') current per unit area on a reversible electrode at equilibrium potential (reversible potential). I_o is a kinetic factor and has a significant influence on the rate of corrosion. For example, pure zinc has no reaction in reducing acids (e.g. H_2SO_4), whereas impure zinc evolves hydrogen. The I_o (H^+/H_2) on pure zinc is much lower than I_o (H^+/H_2) on commercial zinc. The values of $I_{o,a}$ and $I_{o,c}$ in the diagram are 10^{-2} and $1.0\,\mu A/cm^2$, respectively.

The relation between the current and overvoltage is given by the Tafel equation discussed in Chapter 3.

$$\eta = \pm\beta\log\frac{i}{i_o}$$

$$Mg \rightarrow Mg^{2+} + 2e \quad \text{(Tafel equation)}$$

$$M^+ + e \rightarrow M$$

Here η denotes the over-voltage of an anode (the difference between the actual potential of the electrode at certain value of applied current

and the equilibrium potential E_o). If the cathode, for instance, is polarized by applying an external current, the cathodic process would be accelerated ($M^+ + e \rightarrow M$) and the anodic process ($M \rightarrow M^+ + e$) would be retarded. Similarly, an anodic process could be accelerated by supplying an external current to the anode and the cathodic process suppressed. In both cases, however, charge would be conserved and the current applied would be equal to the difference between the i_c and i_a. $i_{applied} = i_c - i_a$ (where i_c is current density at the cathode and i_a is current density at the anode).

If i_a is made smaller than i_c, $i_{applied}$ would be equal to i_c and only the cathode would be polarized. Under such conditions, a linear Tafel behavior would be observed. On applying an electrical current ($I_{applied}$) equal to $10^4\,\mu A/cm^2$, and continuously polarizing the cathode, the corrosion current would be reduced from $100\,\mu A/cm^2$ at I_{corr} to $1\,\mu A/cm^2$ at I_a in Fig. 5.14a which is a 99% reduction in the rate of corrosion of M (Fig. 5.14a). By continuously increasing the

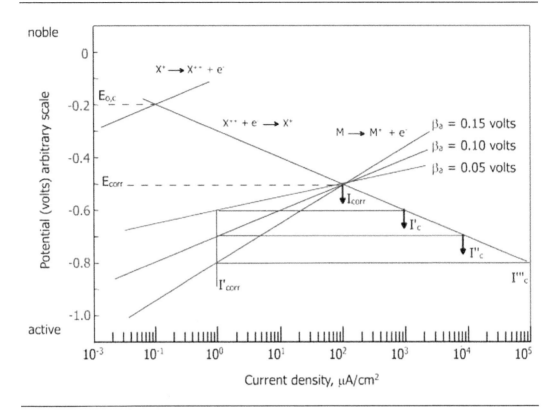

Figure 5.14 b Effect of anodic Tafel slope on polarization and current necessary for cathodic protection of three different metals with (hypothetically) identical E_{corr} and I_{corr}, but different Tafel β slopes. (From Jones, D.A. (1981). *Principles of Measurement and Prevention of Buried Metal Corrosion by Electrochemical Polarization.* In Underground Corrosion, STP 741, ASTM. Jones, D.A. (1971). *Corrosion Science.* **11**, 439. Reproduced by the kind permission of ASTM, Philadelphia, PA, USA)

applied current, the value of I_{corr} can be made negligible and corrosion can be stopped. In order to reduce the current density to zero, an applied current density of 10^6 μA/cm^2 would be required (point P in Fig. 5.14a).

Figure 5.14b (for three different metals) shows how the amount of corrosion is controlled by the anodic Tafel slope, β_a. The lower the value of β_a, the greater is the reduction in the rate of corrosion for a given cathodic polarization. Lower β_a means lower polarization required to achieve the required degree of protection. The Tafel slope β_a gives the measure of the amount of cathodic polarization required to reduce the anodic reaction by one order of magnitude. In order to reduce I_{corr} from 10^2 μA/cm^2 to 1 μA/cm^2, a

potential change of 100 mV (0.1 V) is required for $\beta_a = 0.05$. To achieve the same amount of reduction in I_{corr} a potential change of 0.2 V and 0.3 V would be required, respectively, for $\beta_a = 0.1$ V and $\beta_a = 0.15$ V. Thus, lesser polarization is needed with a lower value of β_a. It is, however, not necessary to reduce the rate of corrosion to zero at $E_{o,a}$ (Fig. 5.14a) as this would require a large amount of current which might not be economically justified. It is only sufficient to reduce the corrosion to a negligible amount which would depend on the number of years the cathodic protection structure is to be designed for.

By cathodic polarization the potential of the cathode ($E_{o,c}$) becomes nearly equal to the equilibrium potential of anode ($E_{o,a}$) and the metal surface attains a uniform potential, hence, corrosion is prevented. As long as the value of E_{corr} is brought very close to the value of $E_{o,a}$ by applying an external current, corrosion is prevented, hence, it is not necessary to reduce the corrosion rate to zero completely.

5.7 ANODIC POLARIZATION

Imagine what would happen if the structure is now polarized in the opposite direction. It would amount to polarizing the potential of the anode to that of cathode in the positive direction. Theoretically, such a practice should result in creating corrosion rather than protection. But for some metals, positive polarization forms a protective oxide/hydroxide surface film and this phenomenon of passivation for a limited number of metals results in retardation of corrosion. By this method called *anodic protection*, it is possible to passivate active–passive metals. Metals, such as iron, chromium and nickel are passivated by anodic polarization, which leads to retardation of corrosion. The potential of this must, however, be maintained in the region of passivity by a potentiostat. Anodic protection is widely applied in transport of acids and corrosives in containers and other applications.

5.8 CATHODIC PROTECTION SYSTEMS

Two types of cathodic protection systems exist:

(a) Galvanic anode system or Sacrificial anode system.
(b) Impressed current anode system.

In the galvanic or impressed current system, the metallic structure is made the cathode (negative) by connecting it to galvanic anodes, which are more negative than the metallic structure to be protected. In this system, the current is generated by the corrosion of active metals, such as magnesium, zinc and also aluminum, which are

galvanic anodes:

$$Al \rightarrow Al^{+++} + 3e$$
$$Mg \rightarrow Mg^{++} + 2e$$
$$Zn \rightarrow Zn^{++} + 2e$$

The anodes of the above materials are utilized as sources of electrons which are released when the anodes are buried in the soil corrode. The electrons released pass through the metallic connection between anode and steel, and thus enter the structure to be protected.

A suitable anode is buried adjacent to and level with the invert (lowest part) of a pipeline. A connection is made between the anode and the pipeline. The anode, generally magnesium or zinc, is connected to the pipeline or any buried metallic structure by an insulated cable. A schematic diagram of a galvanic anode cathodic protection system is shown in Fig. 5.15a. The figure shows a carbon steel pipe (A), magnesium anode (B), chemical backfill (C) surrounding the anode, wires connecting the carbon steel pipe to the anode, the soil (E) and test station (F). The details of test station are shown in Fig. 5.15b. The resistive component of a galvanic circuit is shown in Fig. 5.16.

The copper wire connection provides a passage for flow of electrons to the pipe to be protected. The electrons are released by the consumption of Mg anode in accordance with the anodic reaction, $Mg \rightarrow Mg^{++} + 2e$. The outer circuit is completed by the passage of electrons from the pipe (cathode) to the anode (Mg anode) through the copper wire (D). The pipe continues to be protected as long as it receives a regular supply of electrons from the anode. A typical anode installation in detail is shown in Fig. 5.17. The figure shows galvanic anodes (A) connected by a test station (F) and separated from each other by a distance of 8 feet. The test station provides a connection between the anode lead wire and the structure via the test panel. The details of the surface box housing test station are shown in the figure. The surface box is sometimes buried below the ground level. The anodes are connected to the pipe via a central control test panel. For measurement of pipe-to-soil potential and currents from the magnesium anode ground-bed, test stations

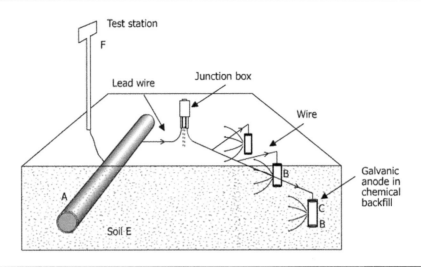

Figure 5.15 a Typical vertical galvanic anode in soil (arrows show the direction of convention positive current flow)

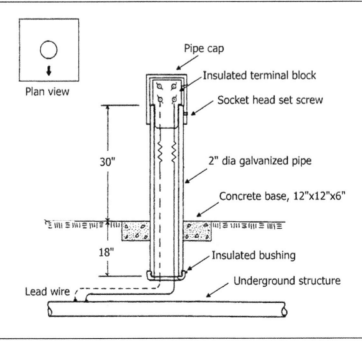

Figure 5.15 b Details of a test station

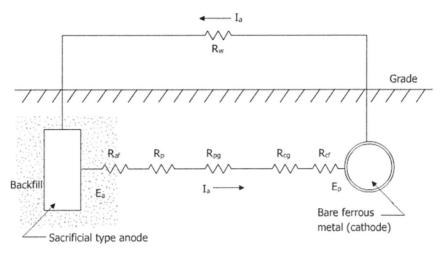

E_a = potential of anode (half cell potential)
E_p = potential of pipe (half cell potential)
R_w = resistance of connecting wire (metallic resistance)
R_{af} = anode film resistance
R_p = backfill resistance
R_{pg} = resistance of anode backfill to ground
R_{cg} = resistance of cathode to ground (this value can become the major
       component if pipe is coated
R_{cf} = cathodic film resistance
I_a = anode current

Figure 5.16   Resistive components of a galvanic circuit

are utilized. A special backfill, such as hydrated gypsum, bentonite and clay is placed around the anode, to ensure low resistance contact to the local soil.

The anodic installation is often designed for ten years but may last much longer if current demand is low. The potential of the pipe must be continuously monitored and the value should not be allowed to fall below -0.85 V ($CuSO_4$ reference electrode used). A 70 lb Mg anode practically gives a current of more than 300 mA in a soil of average resistivity of 2000 ohms-cm. Bare steel sometimes requires about 15 mA/ft^2. A single anode can protect about 2 square feet of the pipe. By applying a coating, the current requirement is reduced to 0.5 μA/cm^2, hence one Mg anode can protect up to 6000 square feet

of the pipe surface. A potential value of 1.11 V is obtainable from the magnesium anode. This subject will be discussed in details later in this chapter.

The following are the advantages and the disadvantages of the galvanic anode system:

(a) Advantages

1. It requires no external source, which might fail.
2. It is economical.
3. It can be easily installed.
4. It can be easily maintained.
5. It can be used in areas where the soil resistivity is low.

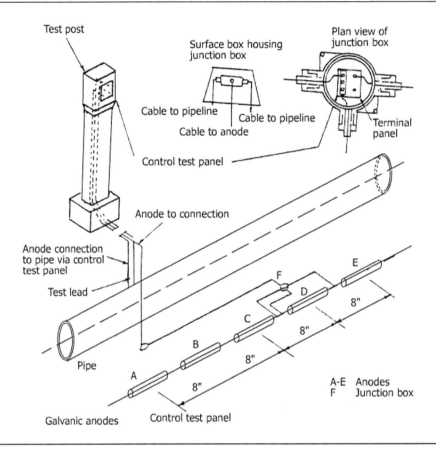

Figure 5.17 Typical galvanic anode installation layout and test points. (By kind permission of Cheveron Corp., USA)

6. Lesser interference with the other metallic structures is caused because of a relatively low current output.

7. The current is evenly distributed.

(b) Disadvantages

1. It has limited applications compared to impressed current.

2. Driving voltage is fixed and cannot be manipulated, except by selecting Mg instead of Zn for example.

3. The cost of protection is high for bare systems (uncoated structures).

4. As no above-ground equipment is used, it is difficult to trace the protected system, unless contact posts are provided.

The following are the advantages and disadvantages of impressed current anode systems:

(a) Advantages

1. Rectifiers available in unlimited current output.

2. May be designed for long lives.

3. More economical.

4. Possibility of variation of current to suit the changes in the system.

(b) Disadvantages

1. External power is essential.

2. More complicated system for installation.

3. Less economical for smaller jobs.

4. Limited to use below a soil resistivity of 3000 ohms-cm.

5.9 COMPONENTS OF GALVANIC SYSTEMS

5.9.1 ANODES

As mentioned earlier, magnesium and zinc anodes are customarily used in this system. The magnesium anodes are most popular because of their high current output. They have the following advantages:

(a) Favorable position in the emf series (very active).
(b) Non-polarizing.
(c) High current output. An average current output of 500–600 A-h/lb can be obtained assuming the efficiency to be 65%. Magnesium anodes are usually applied in soils of resistivity lower than 3000 ohms-cm. A standard 17 lb magnesium anode will produce 1 A of current for one year or 0.1 A for ten years, and so on. The current output, however, depends on the soil resistivity. With a resistivity of 1000 ohm-cm, the current produced would be 0.1 A. If the resistance is very high, the produced current would be lower. To obtain a maximum efficiency the anodes are surrounded with a mixture of gypsum and bentonite called as *backfill*. The composition of a typical backfill is given in Table 5.1.

The mixture must be 100% capable of passing through a 20 mesh screen and 50% through a 10 mesh screen.

5.9.2 MAGNESIUM ANODES

The details of a magnesium anode are shown in Fig. 5.18. The open circuit potential (no current

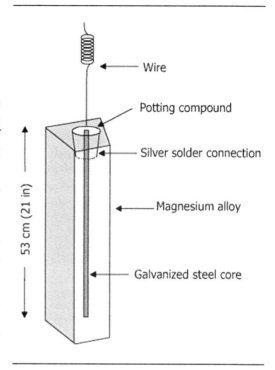

Figure 5.18 Typicl 14.5 kg magnesium galvanic anode

draw) of a standard alloy magnesium anode is −1.55 V with respect to Cu–CuSO$_4$ reference electrode. The open circuit potential of a high manganese magnesium anode is −1.75 V with respect to a copper–copper sulfate half cell. If iron is polarized to −0.85 V, the driving potential of standard alloy magnesium would be (1.55 − 0.85 = 0.70 V) and that of a high manganese magnesium anode is (1.75 − 0.85 = 0.90 V). The composition of some magnesium anodes is shown in Tables 5.2a and 5.2b.

At 100% efficiency, the output is limited by local corrosion cells. To avoid this, magnesium is alloyed with Al and Zn. Practically, an efficiency of 60–70% can only be obtained. Galvomag anodes are used for seawater service. A typical composition is shown in Table 5.3a. The consumption rate is 17.52 lb/A-year. The compositions of high purity and Galvomag Mg anodes are also shown in Table 5.3a.

Table 5.1 Composition of backfill

Element	Percentage
Ground hydrated gypsum	75
Powdered wyoming bentonite (local brand)	20
Anhydrous sodium sulfate	5

Table 5.2a Composition of standard magnesium anodes

Element	Percentage
Aluminum	5.30–6.7
Manganese	0.15–min
Zinc	2.50–3.5
Silicon	0.30–max
Copper	0.05–max
Nickel	0.003–max
Iron	0.003–max
Other impurities	0.300–max
Magnesium	Balance

Table 5.2b High manganese magnesium anodes

Element	Percentage
Aluminum	0.010–max
Manganese	0.50–1.31
Copper	0.02–max
Iron	0.03–max
Nickel	0.001–max
Other impurities	0.3–max
Magnesium	Balance

Table 5.3a Specifications of galvomag Mg anodes

Specification	Galvomag
Specifications	
Cu	0.02
Al	0.01 max
Si	–
Fe	0.03
Mn	0.5–1.3
Ni	0.001
Zn	–
Sn	0.01 max
Pb	0.01 max
Mg	Remainder
Efficiency	50%
Potential	−1.70 V Ag/AgCl
Capacity – Ampere hours	1230 per kg 560 per lb

AgCl is a reference electrode

5.9.3 ZINC ANODES

Zinc anodes are frequently used for protection of submarine pipelines. They are commercially available in weights from 5 to 60 lb. Prepared backfill should be used for anodes if they are to be installed in the earth. They have a driving potential of −1.10 V compared to a Cu–CuSO$_4$ reference electrode. The details of zinc anodes are shown in Fig. 5.19.

Specifications of some zinc anodes are given below. Suppliers specifications must, however, be consulted before application. Some commercial specifications and details of zinc anodes are appended to this chapter.

The composition and characteristics of zinc anodes are shown in Tables 5.3b and c.

(a) Characteristics of zinc anodes

(1) Corrosion products insulate the anodes and the anodes are, therefore, installed below the water table in soils with no free carbonate or phosphate so that passivity does not occur.
(2) Zinc is not generally polarized anodically and the theoretical current output can be as high as 372 A-h/lb for a 99.99% pure zinc.

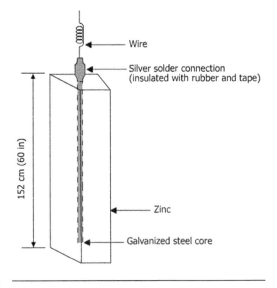

Figure 5.19   Typical 13.6 kg (30 lb) zinc anode

Table 5.3b Zinc anode for soil or fresh water use

Element	Percentage
Aluminum	None
Cadmium	0.004 min
Iron	0.0015 max
Lead	0.006 max
Copper	0.005 max
Impurities (Total)	0.010
Zinc	99.99

Table 5.3c Zinc anode for sea-water use

Element	Percentage
Aluminum	0.3 max
Cadmium	0.06 max
Iron	0.003 max
Lead	0.006 max
Copper	0.005 max
Impurities (Total)	0.014
Zinc	Remainder

Table 5.4a Elements added in various classes of aluminum anodes

Type	Additives
I	3.0–7.0% Zn
II	3.5–9.0% Zn, 0.10–0.5% Sn
III	7.0% Zn, 1% Sn
IV	0.5% Sn
V	0.45% Zn, 0.045% Hg
VI	0.1–0.40% Zn, 6.0–8.0% Mg and 0.08–0.15% Hg
VII	1–5% Zn, 0.1–0.05% Mn and 0.06–0.15% Hg

Table 5.4b Solution potential of aluminum anodes

Type	Open circuit potential $(E_{Cu-CuSO_4})$	Working potential (Driving potential)	Consumption rate (kg/A-year)
I	−1.06	−1.00	4.7–3.6
II	−1.10	−1.15–1.25	4.3–3.6
III	−1.35	−1.10	3.3–2.9
VI	−1.43	−0.93–1.30	–
VII	−1.10	−0.93–1.13	–

(3) Zinc can operate up to 95% efficiency and a current output of 335 A-h/lb can be obtained ($0.90 \times 372 = 335$ A-h/lb).

(4) The open circuit potential of zinc is generally −1.10 V with respect to a copper–copper sulfate reference electrode.

(5) Based on a polarization potential of −0.85 V for steel, the driving potential of zinc is 0.25 V.

5.9.4 ALUMINUM ANODES

These are mostly employed for seawater applications. The base metal contains 98–99% of aluminum. Elements commonly added in different types of aluminum anodes are shown in Table 5.4a. Table 5.4b shows the solution potential of various classes of anodes.

(a) Characteristics of aluminum anodes

(1) The cost is low and they are light in weight.

(2) The corrosion products do not contaminate the water.

(3) The rate of consumption varies between 7 and 9 lb/A-year. The efficiency varies between 87 and 95%.

(4) The anodes are easily passivated and must be rinsed with NaCl to reactivate. Backfill must be used with aluminum anodes.

(5) The consumption of Galvalum–Aluminum anodes (Fe = 0.08, Si = 0.11–0.21, Zn = 0.35–0.50 and Hg = 0.035–0.40) is 1285 A/h/lb (Table 5.4c). It is a very popular anode.

Table 5.4c Composition of galvalum–aluminum anodes

Galvalum*–aluminum	
Specification:	
Fe	0.08 max
Si	0.1 max
Zn	0.35–0.50
Al	Remainder
Efficiency	90%
Potential	−1.05 V, Ag/AgCl
Capacity – Ampere hours	2830 per kg
	1285 per lb

*There are three types; Galvalum I, Galvalum II and Galvalum III

5.9.5 INSTALLATION OF GALVANIC ANODES

Galvanic anode installations are simple compared to impressed current installations. Figure 5.20 shows a single packaged anode installation. The common 17 pound packaged magnesium anodes are commonly used for a packaged installation. When several magnesium or zinc anodes are to be installed at a single location, the anodes are connected to a header wire which is directly connected to the pipeline. Packaged anodes in multiple may be installed as shown in Fig. 5.20. The anode line may be parallel or perpendicular to the pipeline. Zinc anodes are generally kept at a distance of 5 ft from the pipelines, if they are in parallel lines. Magnesium anodes are generally 15–20 ft from the pipelines.

If space is not available and soil resistivity conditions are very low, galvanic anodes are

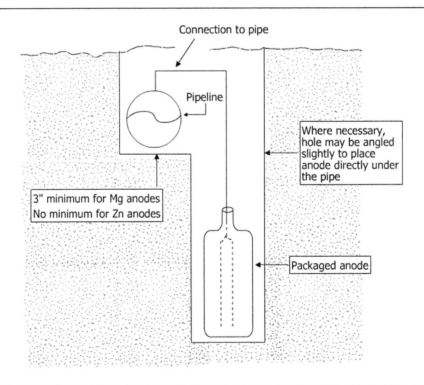

Figure 5.20 Galvanic anode installation for packaged anodes. (From Source: Peabody, A.W. *Control of Pipeline Corrosion*, NACE)

Table 5.5 Anode spacing factors

No. of anodes in parallel	Adjusting factors			
	Anode spacing (feet)			
	5	10	15	20
2	1.839	1.920	1.946	1.964
3	2.455	2.705	2.795	2.848
4	3.036	3.455	3.625	3.714
5	3.589	4.188	4.429	4.563
6	4.125	4.902	5.223	5.411
7	4.652	5.598	6.000	6.232
8	5.152	6.277	6.768	7.036
9	5.670	6.964	7.536	7.875
10	6.161	7.643	8.304	8.679

installed below the pipe. Very deep holes are required to place multiple anodes. The anodes are safer in deep soils as they are not affected by variation in the moisture contents and soil conditions. Due to seasonal changes, non-packaged anodes are frequently used. In this case, the anode and the backfill is not installed as single unit, but separately. Galvanic anode may also be installed horizontally. Table 5.5 shows the anode spacing factors.

5.9.6 CALCULATIONS OF CURRENT OUTPUT OF MAGNESIUM ANODES

The current output of galvanic anodes is affected by the resistivity of the soil. A higher current would be required for a low resistivity soil. The geometry of the anode is also important, for instance, the longer the length of the anode, the higher the current output would be. Lastly, the higher the potential of the alloy, the higher would be the current output. The efficiency of anodes is a major factor even under the best conditions. The Mg anodes, for instance, does not have an efficiency higher than 50%. The current output is an important information which must be known.

The current output of a magnesium alloy can be determined by the following relationship

$$I_{Mg} = \frac{150\,000\,FY}{\rho} \quad \text{(for an uncoated pipeline)}$$

$$(5.1)$$

$$I_{Mg} = \frac{120\,000\,FY}{\rho} \quad \text{(for a coated pipeline)}$$

$$(5.2)$$

where:

I_{Mg} = current output of magnesium anode
ρ = soil resistivity, ohm-cm
F = factor from Table 5.6
Y = correction factor for pipe-to-soil potential value (Table 5.7)

Example 1

Calculate the current output for a 32 pound packaged magnesium anode buried in a 1750 ohm-cm soil for the protection of a bare pipeline where the expected resultant pipe-to-soil potential is −0.85 V.

Solution:

The given data for a 32 pound anode is: $F = 1.06$ from Table 5.6. From Table 5.7, the Y factor for a pipe-to-soil potential of −0.85–1.00. Pipe-to-soil potential is the potential between the pipe and the surrounding soil.

Putting the data in equation (5.1), one obtains

$$I_{Mg} = \frac{150\,000 \times 1.06 \times 1}{1750} = 90.9\,\text{mA}$$

For multiple anodes, the current output obtained for a single anode should be multiplied by an adjusting factor (Table 5.5).

Extending the above example further for a 4 parallel anodes, 10 ft apart, one obtains:

3.455 (adjustment factor – Table 5.5) × 90.9 mA
= 310.50 mA

Similar calculations can be made for the output of magnesium anodes used with the coated pipelines. Equation (5.2) would be used under such conditions.

Table 5.6 'F' factor for different anode weights

Anode weight (pounds)	Details	Factor 'F'
Standard Anodes:		
3	(Packaged)	0.53
5	(Packaged)	0.60
9	(Packaged)	0.71
17	(Packaged)	1.00
32	(Packaged)	1.06
50	(Packaged – anode dimensions 8″ diam × 16″)	1.09
50	(Packaged – anode dimensions 5″ × 5″ × 3.1″)	1.29
Long Anodes:		
9	(2.75″ × 2.75″ × 26″ backfill 6″ × 31″)	1.01
10	(1.5″ × 1.5″ × 72″ backfill 4″ × 78″)	1.71
18	(2″ × 2″ × 72″ backfill 5″ × 78″)	1.81
20	(2.5″ × 2.5″ × 60″ backfill 5″ × 66″)	1.60
40	(3.75″ × 3.75″ × 60″ backfill 6.5″ × 66″)	1.72
42	(3″ × 3″ × 72″ backfill 6″ × 78″)	1.90
Extra Long Anodes:		
15	(1.6″ diam × 10′ backfilled to 6″ diam)	2.61
20	(1.3″ × 20′ backfilled to 6″ diam)	4.28
25	(2″ diam × 10′ backfilled to 8″ diam)	2.81

Table 5.7 'Y' correction factors

P/S	Magnesium	Zinc
−0.70	1.14	1.60
−0.80	1.07	1.20
−0.85	1.00	1.00
−0.90	0.93	0.80
−1.00	0.79	0.40
−1.10	0.64	0.00
−1.20	0.50	0.00

Example 2 – Current Output of Zn Anodes

A bare pipeline is to be protected by zinc anode. A 32 pound packaged is to be used for protection. The anode is buried in a 750 ohm-cm soil with an expected pipe-to-soil potential of −0.85 V. Determine the current output.

$$I_{Zn} = \frac{50\,000 \times F \times Y}{\rho}$$

Putting the data from the tables, as in the case of Mg anodes:

$$I_{Zn} = \frac{50\,000 \times 1.06 \times 1}{700} = 70.7\,\text{mA}$$

For a four parallel anodes, 10 ft apart, the estimated output would be: (*The factor for 4 parallel anodes, 10 ft apart is 3.445*) (Table 5.5)

$$I = 70.7 \times 3.455 = 243.92\,\text{mA}$$

5.9.7 CALCULATIONS OF NUMBER OF ANODES REQUIRED AND SPACING

The number of anodes required for a pipe or any bare structure can be conveniently estimated as per example given below.

Suppose 10 000 ft of a bare 4 inch pipeline is to be protected and if the resistance of soil is 1000 ohms per cubic centimeter, the pipe requires $1 \, mA/ft^2$ for protection. The anode output curve shows 100 mA per anode in this type of soil. The number of anodes can be easily calculated as shown below.

Calculations

(a) First calculate the area of the pipe:

$$Area = \frac{4 \times 3.14 \times 10\,000}{12} = 10\,500 \, ft^2$$

$$Total \; current \; requirement = 10\,500 \times 1.0$$

$$= 10\,500 \, mA$$

Number of anodes required

$$= \frac{Total \; current \; required}{Current \; output \; of \; a \; single \; anode}$$

$$= \frac{10\,500 \, mA}{100 \, mA/anode} = 95 \; anodes$$

(b) Total current requirement $=$

Protected area $(m^2) \times$
Current density (A/m^2)

or

Protected area $(ft^2) \times$
Current density (mA/ft^2)

(c) Total weight of anode material required:

$$\frac{Current \; (A) \times Design \; life \; (years) \times 8760}{Capacity \; A\text{-}h/kg}$$

5.10 IMPRESSED CURRENT SYSTEM

In contrast to the galvanic anode system, the flow of current from the anode to the cathode is forced from a DC source in the impressed current system. Thus, whereas the current is provided by the corrosion of the electrode in the anodic galvanic system, the electrode acts as a conductor and hardly corrodes in the impressed system and the AC input is transformed and rectified to a varying DC voltage. A transformer rectifier is the most important component of the system.

Figure 5.21 shows a schematic diagram of an impressed current cathodic protection system. The direct terminal supply is obtained from a transformer rectifier (T/R unit) designed to step down normal alternating mains voltage and then rectify this to the direct current. This output is adjustable over a wide range to suit requirements. An AC power supply with an ammeter is connected to the rectifier. A switch box is installed in the AC circuit which contains a magnetic or thermal circuit breaker to protect the rectifier against overloads. The AC goes to the rectifier unit. It passes through the primary coil (P), the resulting magnetic field continuously expands, contracts and reverses direction. The changing field induces AC in the secondary coil, that is proportional to the turns ratio between the two coils.

$$\frac{Primary \; turns}{Secondary \; turns} = \frac{Primary \; volts}{Secondary \; volts}$$

The rectifier cells are shown from B to E. Their function is to allow the current to flow in one direction and to block it in the opposite direction. When AC is applied to the AC terminal of the rectifier, DC appears at the output. The negative terminal of the rectifier assembly is connected to the pipeline (G) and the positive to the anode bed. The five anodes are represented by numbers 1 to 5. The anodes are surrounded by a mixture of coke breeze. The anodes are energized by the rectifier and they are made of silicon cast iron, steel scrap, platinized titanium, graphite or lead silver.

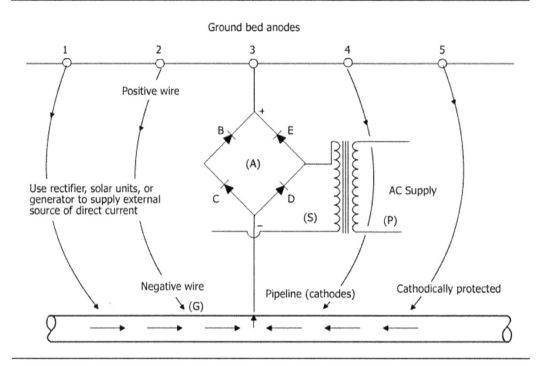

Figure 5.21 Schematic diagram of impressed current cathodic protection system

The electrons from the rectifier enter the pipe or any structure to be protected. The anodes are placed in a trench, which is termed a ground-bed. As a matter of practice, the anodes are placed in the lowest resistivity soil.

Details of a typical impressed current installation is shown in Fig. 5.22. The AC power unit is connected to the rectifier (R). The negative terminal is attached to the structure (S) and positive terminal to the ground-bed. The anodes are silicon cast iron and represented in the figure by A_1 to A_4. The individual anode leads are brought into a junction box (J) and a shunt is inserted in each anode lead. The junction box allows the measurement of current output of individual anodes by determining the IR drop. The anodes can be connected directly to the positive terminal of the rectifier by a ring main cable system shown in the figure or via a control box (direct connecting system) to monitor the current output above ground.

5.10.1 ADVANTAGES AND DISADVANTAGES OF IMPRESSED CURRENT SYSTEM

(a) **Advantages**

(a) One installation can protect a large area of metal.

(b) The system can be used for a wide variety of voltage and current requirement.

(c) Schemes can be designed for life in excess of 20 years, if required.

(d) Current requirements and potentials can be easily adjusted to the varying needs of protection.

(e) Can be applied to a wide range of structures.

(f) Requires generally a small total number of anodes.

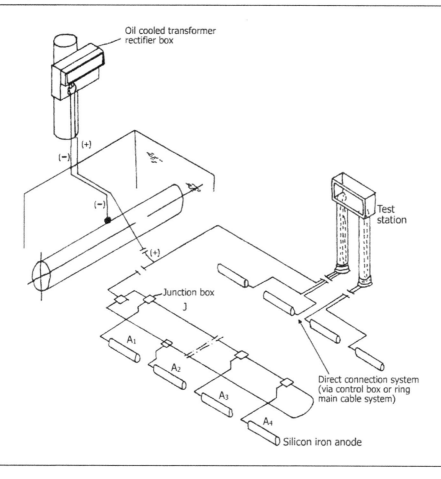

Oil cooled transformer
rectifier box

(+)

(−)

(−)

(+)

Test
station

Junction box
J

A₁

A₂

A₃

A₄

Silicon iron anode

Direct connection system
(via control box or ring
main cable system)

Figure 5.22 Details of typical power impressed ground-bed installation

(g) Requires simple controls which can be automated.

(h) The use is less restricted by soil resistivity.

(i) A large area can be protected by one installation.

(b) Disadvantages

(a) Possibility of interference effects on other buried structures.

(b) Regular maintenance is essential.

(c) Electrical power cost is high.

(d) Power failures can cause serious problems, and faults may go unnoticed for long times.

5.11 COMPONENTS OF IMPRESSED CURRENT SYSTEMS AND AC/DC SYSTEMS

5.11.1 POWER SOURCE

Rectifiers are commonly used as the source of power. Although rectifiers are most commonly used as a source of DC power for the impressed current systems, other external power sources, such as wind generators, thermoelectric generators and wind-driven generator units can also be employed if conventional AC power is not available.

5.11.2 RECTIFIERS

The rectifier units consist of a transformer, rectifier stacks, meter, switch and the transformer tap connections. The rectifiers convert AC to DC current. The transformer is used to step down the supply of voltage to that required for operation of the rectifier stacks. There are two types of rectifiers:

(a) Selenium (Oil and Air-cooled) Rectifier

The selenium air-cooled rectifiers have lower initial cost than the oil immersed units, but have poor ventilation. The oil immersed rectifiers are less vulnerable to air and dust than the air-cooled type. In either case, the structure to be protected is connected to the (−) negative terminal of the rectifier. The installation is normally at 8 miles interval for 36″ pipe and 12 miles for 24″ pipe.

(b) Silicon (Oil and Air-cooled) Rectifier

The silicon rectifiers (oil and water-cooled) have longer lives and higher efficiency compared to selenium rectifiers. The oil-cooled types are not susceptible to damage by dust and dirt. They are 20–50% smaller in size than the selenium rectifiers.

5.11.3 COMPONENTS OF A RECTIFIER

The three-phase bridge is the most common circuit for rectifiers operated from a three-phase AC power line. Each phase of a three-phase AC current is spaced 120 electrical degrees apart and therefore the voltage of each secondary winding reaches its peak at different times.

Figure 5.23 shows the operation of a single-phase bridge rectifier. The direction of flow reverses 60 times per second for 60 cycles AC. In a positive half-cycle (diagram A), current originates at T_1 on the secondary winding. It is blocked by D_3 (silicon diode). The current, therefore, flows through direction D_1, follows the path (3) and through diode D_4 it enters the negative terminal T_2. In the next half-cycle (1/120th) of a second later, polarities at T_1 and T_2 are reversed (see diagram B). The current is blocked by diode D_4 and flows through D_2, follows the path (3) through D_3 in the same direction as before. The load R_L thus receives energy in the form of pulses at 120 per second.

Although three-phase rectifiers are used as mentioned before, each single bridge shares a pair of diodes with one of the other bridges. The three-phase bridge is like three single-phase bridges, with each bridge sharing a pair of diodes with one of the other bridges.

A rectifier consists of three important components: circuit breaker, transformer and rectifying elements (stacks) (Fig. 5.24). Brief details are given below.

(a) **Circuit breaker**

These are basically switches with an internal mechanism which opens the switch when the current exceeds a prescribed designed limit. They also serve as 'on and off' switches. There are two types of switches: (1) magnetic and (2) thermal. The circuit breaker protects equipment from over loading.

In the magnetic type, a coil is woven around a brass tube and a magnetic field is set up by a current flowing in the coil. The magnetic slug is held at one end of a tube by a spring. The magnetic field attracts the slug, but at or below the rated current the slug does not move. At overload, the magnetic field pulls the slug into the coil. When the slug is drawn to the opposite end of the tube, the circuit is completed for the trip mechanism and the breaker switch trips. The movement of the magnetic flux is slowed down and a time delay is provided. The breaker can trip on to 101–125% of the rated current. Overloads of ten times the rated currents can be sustained. The dropping is very fast when the overload is ten times.

In thermal magnetic breakers, the thermal tripping is caused by the flowing current through the resistor close to the bimetallic strip. When the current exceeds the rated value, the bimetallic element trips the

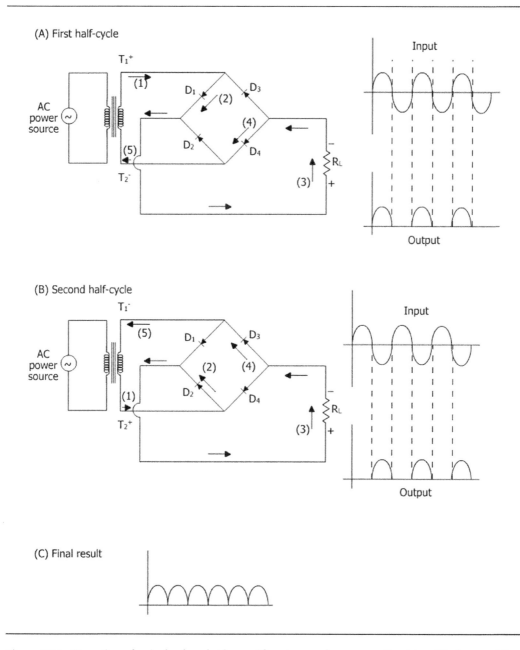

Figure 5.23   Operation of a single phase bridge rectifier. Arrows show conventional (postitive) current flow direction

breaker and a long time delay is involved before the breaker can be closed.

(b) Transformer

This consists of two coils of wire wound around an iron core. The coils are not connected electrically, but the core provides a magnetic link between them. AC voltage is applied to one coil (primary), the changing magnetic field crosses to the other coil (secondary) and induces a voltage in it. The changing field induces the AC voltage in the secondary coil that is proportional to the

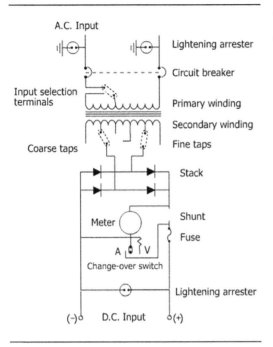

Figure 5.24 Components of a rectifier

turns ratio between the two coils.

$$\frac{\text{Primary turns}}{\text{Secondary turns}} = \frac{\text{Primary volts}}{\text{Secondary volts}}$$

(c) **Rectifier cells**

The change of AC power to DC is done by rectifying elements. They act like check valves by offering low resistance to current flow in one direction and high resistance in the other direction. The function of the rectifying element is to allow the current to flow readily in one direction and to block current flow in the opposite direction. The Selenium cell is the most common rectifier cell. Selenium is applied to one side of an aluminum base plate which has been nickel plated. A thin metallic layer is applied over the selenium layer. This layer acts as counter electrode. It collects the current and provides low resistance to the contact surface. These cells may be arranged in stacks or parallel to produce the desired voltage and current rating.

5.11.4 RECTIFIER EFFICIENCY

This is the ratio between the DC power output and AC power input. Rectifiers are used as a source of DC power. Rectifiers convert the AC current (60 cycles) to DC current through rectifier operated at maximum efficiency at the full rated loads.

Overall rectifier efficiency

$$= \frac{\text{DC power output}}{\text{AC power input}} \times 100\%$$

An efficiency filter can be used to minimize the ripples.

5.11.5 CABLES

Cable conductors to the anodes are made from copper or aluminum. These conductors must, therefore, be insulated. High-density polyethylene has good properties with respect to abrasion and high-temperature while maintaining excellent dielectric properties. Technical information about cables is provided by the manufacturers.

5.11.6 INSULATION AND ELECTRICAL CONTINUITY

The structure to be protected must be free from all interconnected metal work in order to limit the flow of electric current on the structure. This is made possible by electrical insulation. It is the condition of being electrically isolated from other metallic structures. The term insulation is used interchangeably with isolation. The use of isolated flanges and joints is common in pipelines.

5.11.7 INSULATION OF INSULATING FLANGES

Insulating flanges are installed to isolate the cathodically protected buried pipelines from above-ground pipelines. An insulating kit consists of a non-conducting gasket separating the flange faces, an insulating sleeve and a washer for the bolt (Fig. 5.25a). Details of an insulated flange

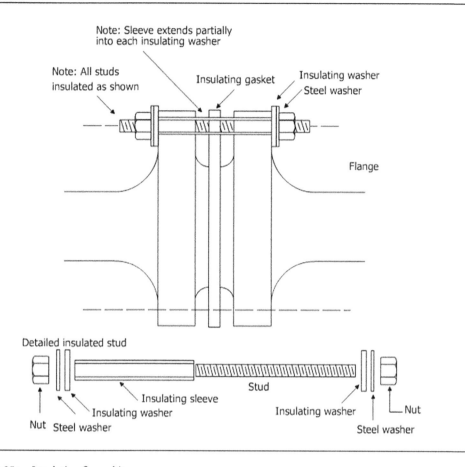

Figure 5.25a Insulating flange kit

are shown in Fig. 5.25b. Insulated flange sets are assembled prior to installation.

5.11.8 ISOLATION JOINT

It is a flangeless factory assembled pipe-length which incorporates an insulating coupling.

5.12 TYPES OF GROUND-BEDS

There are two types of ground-beds:

(1) Close ground-bed system (represented by distributed anode system)
(2) Remote ground-bed system.

5.12.1 CLOSE GROUND-BED SYSTEM

In a situation where only a limited amount of pipeline length is to be protected, distributed anodes are used. In congested areas, such as in tank farms, pipeline terminals or pump stations, close ground-beds or anodes have been used effectively. It is apparent that the density for the current flowing from the anode becomes less and less with distance. The highest potential drop is observed in the earth where the current density is highest. Most of the resistance to remote earth for an anode is, therefore, confined within the first few feet. In the close system, the positive potential gradient of the earth is used with advantage. The earth in the vicinity of the

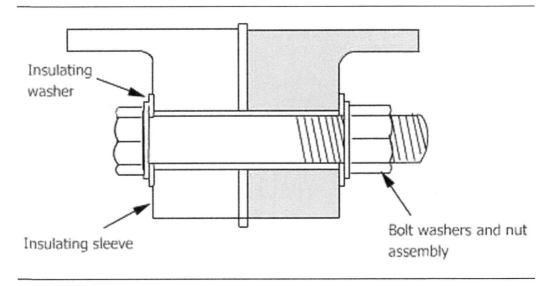

Figure 5.25b Details of insulated flange 12″ B.S.T.E.

anode is positive with respect to the remote earth. The positive earth is, therefore, very close to the anode, if the anodes are placed very close to the pipeline. The area of influence (area where voltage drop is caused by the pipeline) of the pipeline would pass through the earth which is at positive potential as compared to the remote earth (outside the area of influence with zero voltage drop). Consequently, a limited area nearest to the anode will attain cathodic protection because of the sufficient potential difference between the soil and the pipeline brought about by close anodes. In the above system, the potential of the local earth is changed rather than changing the potential of the pipe. The magnitude of the earth potential change depends on the voltage impressed on the anode and not on the current discharged from the anode. The potential to earth must be known so that the magnitude of earth potential change necessary to bring the potential of the pipe to −0.85 V is determined. The current requirements of the power source are designed after collecting information on the resistance of the parallel anodes and resistance of the cables and the back emf between pipeline and anode. In this system the anodes are required to be put close together. If the distance to be protected is very long, sufficient voltage for protection would not be obtained and,

therefore, closer space between anodes may be required.

5.12.2 REMOTE GROUND-BED

Basically, there are two types of anode ground-beds, close and remote. The terms, close and remote, are related to the area of influence in the electrolyte around the anode. In the case of a remote anode ground-bed surrounding the pipe, the pipeline is negative with respect to the remote earth as opposed to the close ground-bed system, where the local earth is made positive with respect to the pipe. The anode is so located that only a small area of structure is influenced by it. The length of pipeline that can be protected by close ground-bed anode depends on the voltage impressed by the anode. Only limited areas of structure are protected.

In the remote ground-bed, the current flowing to the pipeline from a DC source causes a voltage drop in the soil adjacent to the pipe. The ground-bed is considered remote if there is no overlapping between the area of influence surrounding the pipe and the area of influence surrounding the ground-bed (Fig. 5.26). The potential gradient readings continue to decrease

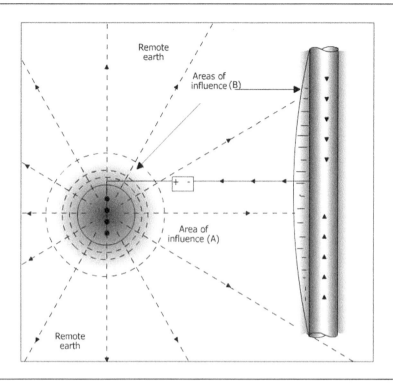

Figure 5.26   Two areas of influence caused by a remote ground-bed

until at a point of is no further decrease, which is the remote location. At this position, the current density and potential gradient approach zero. Large sections of structure may be protected, the only limiting factor being the resistance of the pipe itself (Fig. 5.26). The area of influence ends where there is no change in the soil potential. Beyond this point the earth is remote. When the current enters the remote earth there is no more resistance from the soil. Remote earth is considered as an infinite conductor for all practical purposes. Under such conditions, the current flows from the ground-bed, through the mass of the earth and then to the pipeline which is to be protected.

5.12.3 GROUND-BED DESIGN

The basic requirement of a good cathodic protected system is the protection of the structure at a minimal total annual cost over a protected period of life. It is not always the ground-bed with the

least resistivity that is most suitable, but rather the ground-bed that fits into the economics of the design.

(a) Vertical Impressed Current Types

These are very commonly used for the impressed current system. A typical vertical anode installation is shown in Fig. 5.27. This type of ground-bed has a low resistivity which is an outstanding advantage. Almost all of the current flow is from the anode to the backfill by direct contact so the consumption of the material is mostly at the edge of the backfill. Vertical anodes are installed with a carbonaceous backfill.

(b) Horizontal Type

Anodes are installed in a continuous or non-continuous coke breeze bed in a horizontal ditch.

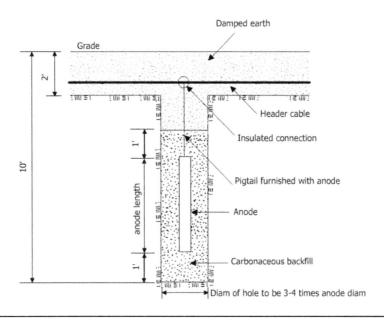

Figure 5.27 Typical vertical anode installation for impressed current cathodic protection system

These are used because of the near presence of the other structures and also because of the limitations of the soil depth. A typical installation is shown in Fig. 5.28. The anodes should always be placed in moist soil or in hygroscopic backfill so that it does not dry out during the dry season.

(c) Deep Vertical Anodes

A deep ground-bed is defined as a ground-bed in which the anodes are installed vertically in a drilled hole at a depth of 50 ft or more. A typical deep anode installation is shown in Fig. 5.29.

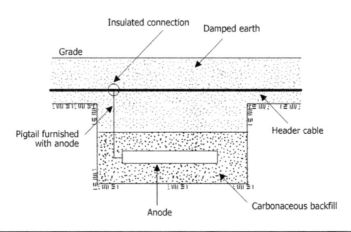

Figure 5.28 Typical horizontal anode installation for impressed current cathodic protection system

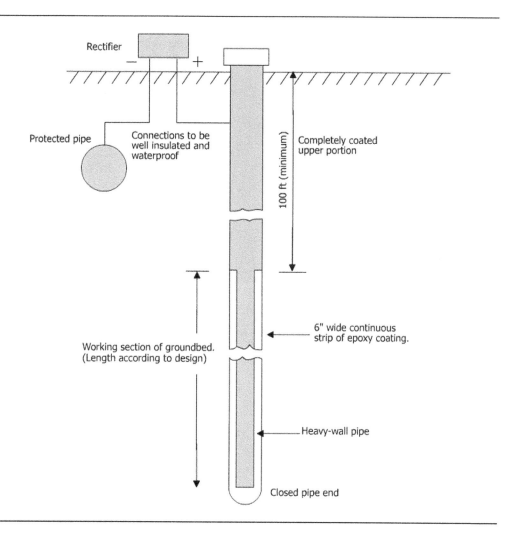

Rectifier

−   +

Protected pipe

Connections to be
well insulated and
waterproof

100 ft (minimum)

Completely coated
upper portion

6" wide continuous
strip of epoxy coating.

Working section of groundbed.
(Length according to design)

Heavy-wall pipe

Closed pipe end

Figure 5.29   A heavy-wall deep impressed current anode ground-bed 'Casing method.' (From Peabody, *Control of Pipeline Corrosion*, NACE. Reproduced by kind permission of NACE, Int., Texas, USA)

Such anodes are used in areas where the resistivity of the soil is very high, such as in deserts. They are also suitable for areas where otherwise a large ground-bed electrically remote is required to keep the resistance to minimum. If the surrounding deep soil has a low resistivity, excellent distribution of current is obtained. The designs vary according to soil condition.

5.13 MAJOR IMPRESSED CURRENT ANODES

The following are the major impressed current anodes:

5.13.1 HIGH SILICON CAST IRON

Silicon iron anodes are composed of iron as the base metal with about 15% silicon and 1% carbon, additionally alloyed with chromium (5%), manganese (1%) and molybdenum (2%). The maximum current output is 50 A/m^2 and the rate of consumption is between 90 and 250 g/A/year. Anodes containing Mo are used in high-temperature media.

A typical analysis of high silicon cast iron anode is shown below in Table 5.8. It is generally used for onshore cathodic protection applications.

Table 5.8 Analysis of a typical silicon cast iron anode

Element	Percentage
Silicon	14.35 min
Carbon	0.85 max
Manganese	0.65 max
Iron	Remainder

5.13.2 METAL SCRAP

It has the advantage of being cheap and abundantly available. The rate of consumption of mild steel scrap pipes, rails and cast iron scrap castings varies. For mild steel scrap the rate of consumption is 6.6–9.0 kg/A/year, and for cast iron the rate is 0.9–9.0 kg/A/year. Steel in the form of old railroad line, pipes and structural sections is used. The rate of consumption of steel scrap is generally uniform. The material is mostly available in the form of long and thin section and depending whether these sections are installed horizontally or vertically, they may encounter soil strata with different resistivities resulting in non-uniform corrosion. Cast iron has the advantage of being thick in section and of such form that any one piece will be in soil of more or less uniform resistivity. Also, a graphite surface is left exposed as the outer iron is consumed. The remaining iron in the form of a graphite, therefore, acts as a graphite anode.

5.13.3 GRAPHITE ANODES

These have the advantages of long-life corrosion protection, low maintenance cost and high efficiency. The typical anode current density is between 10.8 and 40.0 A/m^2 (1.4 A/ft^2). The rate of consumption is between 0.225 and 0.45 kg (0.5 and 1.0 lb) per year. These are generally cylindrical in shape, although other forms are available.

5.13.4 PLATINIZED TITANIUM

These anodes are used for salt water or fresh water where the conductivity is very low. Titanium develops an adherent oxide layer of high electrical resistance. The oxide layer prevents corrosion by acting as a barrier. Titanium acts as an inert support for the platinum. Platinum can withstand very high current density and it is generally applied to a small area only. The platinum layer is normally 2.5 microns in thickness and it has an estimated life expectancy of 10 years. Titanium sheets, 1–2 mm thick with a platinum coating of 2.5–5.0 μm, can be loaded to 10 A/dm^2 or over a period of years. Rod anodes of 10–25 mm diameter are used frequently for protection of vessels, pipes, condensers, heat oil terminals, etc. Current densities up to 50 A/ft^2 (540 A/m^2) can be obtained. The anode should, however, be used at a low voltage.

5.13.5 LEAD ANODES

Lead anodes are made of various lead alloys, such as Pb–1Ag–6Sb and Pb–1Ag–5Sb–1Sn. The density of a lead anode is around 11.0 to 11.2 g/cm^3.

Pb-1Ag-6Sb has a capacity of 160–220 A/m^2 and a consumption rate of 90 g/A/year or 0.009 kg/A/year at a current density of 10 A/ft^2 (108 A/m^2). The other anode containing 10% Sn and 5% antimony has a capacity of 500 A/dm^2 and the rate of consumption 0.3 to 0.8 kg/year. This alloy has good mechanical properties, and can be extended to any shape. Lead–silver or lead–platinum anode with a diameter of 7.5 cm, length 75 cm and weighing 36 kg or with a diameter of 5 cm, length 180 cm and weighing about 45 kg are used in the form of round anodes to protect corrosion of marine structures. These are also used for protection of ships.

5.14 PROTECTION OF SUBMARINE PIPELINES

Pipelines in seawater are protected by so-called bracelets (annular anodes) as shown in Fig. 5.30. In marine structures, corrosion is at maximum at a small distance below the water line and decreases with depth. Corrosion is less severe in mud. For protection of bare steel in seawater, an initial current density of 15 mA/ft^2 (161 mA/m^2) is required, and this decreases after some time

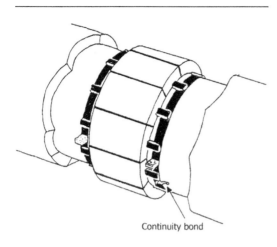

Continuity bond

Figure 5.30   Bracelet type of anode for marine application. (From B.K.L. Cathodic Protection Division, Birmingham, England)

to $5\,mA/ft^2$ or $(43\,mA/m^2)$. In the impressed current system non-consumable graphite anodes are required, whereas in the galvanic system a magnesium anode is the best material. Zinc and aluminum anodes are also used as galvanic anodes, but the cost is high.

The protective current densities for steel structures in various seawaters are given in Table 5.9a. The current requirement of steel is affected by the degree of hardness as shown in Table 5.9b. The current required for protection of steels in various environments is shown in Table 5.9c.

Complete protection of buried steel or iron may require 0.75–$5.0\,mA$ of current per square foot of the surface. On a well-coated line, the current may be as low as 0.01–$0.2\,mA/ft^2$.

Table 5.9a   Protective current requirements in different seawater

Area	Protective current requirements (A/m^2)
Gulf of Mexico	80–150
Nigeria	85
Alaska	250
Arabian Gulf	65–85
North Sea	90–150
U.S. West Coast	24

Table 5.9b   Degree of hardness *vs* the protective current density

Carbonate, hardness (Degree of hardness)	Protective current density (mA/m^2)
<2	250–320
2–10	100–150
>10	70–120

Table 5.9c   Recommended current densities

Environment	Current density (mA/ft^2)	Current density (mA/m^2)
Soil	0.75–5.0	40–58
Fresh water	1–3	11–32
Seawater	4–5	43–64
Moving seawater	1–3	11–32
Sea mud	1–3	11–32

The potential necessary to protect buried steel is $-0.85\,V$, however, in the presence of sulfates, reducing bacteria a minimum potential of $-0.95\,V$ with respect to copper sulfate electrode would be necessary. Approximately 15–$100\,mA/ft^2$ current is needed for protection of bare steel in sluggish water. In rapidly moving water, 1–$10\,mA/ft^2$ for bare steel in a soil would be necessary. Current requirements in various environments can be found abundantly in the literature as well as cathodic protection specifications. For submarine pipeline, a current density of $5\,mA/ft^2$ is required.

5.15 DESIGN PARAMETERS IN CATHODIC PROTECTION

The basic design requirement for cathodic protection is the choice of current density per square foot or square meter of the surface area to be protected. The choice of current density can vary from something in the order of $100\,mA/ft^2$ for a bare structure in water to as low as $0.0001\,mA/ft^2$ for well-coated pipes or structures of high resistivity. To estimate the current requirements, knowledge of soil resistivity is of a primary importance. The following are the major characteristics of soils:

(1) Sandy-type soils are less corrosive than non-homogeneous soils.
(2) Homogeneous soils are less corrosive than heterogeneous soils.
(3) Well aerated soils are less corrosive than sparsely aerated soils. The more aerated soils tend to be brown in color.
(4) Soils low in organic matter are less corrosive than soils with high amount of organic matter.
(5) High acid and high alkaline soils (high pH) are more corrosive.
(6) Soils containing sulfate reducing bacteria are more corrosive than soils free from this bacteria.
(7) Soils having low electrical resistivity are more corrosive than soils having high electrical resistivity.

The soils can be classified as below with respect to the corrosivity of steel and iron (Table 5.10).

5.16 CURRENT REQUIREMENTS

Generally, the quantity of current varies with the type of soil and the quality of coating. The current requirement for an uncoated structure may be seventy times the current requirement for a coated structure. For instance, 10 miles of well-coated pipelines (24″) may require about 8 A for protection compared to 560 A required by a non-coated line.

Table 5.10 Classification of soils according to corrosivity of steel and iron

Soil Resistivity	Tendency to Corrode
0–1000 ohm-cm	Very corrosive
1000–3000 ohm-cm	Corrosive
3000–5000 ohm-cm	Mildly corrosive
10 000 and above	Non-corrosive

It is necessary to experimentally conduct a current requirement test for the structures to establish the magnitude of power requirement for the system to be installed.

As a rough guide, the following current densities would be required to ensure anode life for 10 years in soils of different resistivities.

800 ohm-cm and below 300 mA
800–200 ohm-cm, 180 mA
2000–3000 ohm-cm, 90 mA

5.17 BACKFILL

Galvanic anodes are surrounded by a backfill which is usually a mixture of gypsum, bentonite, and clay. Table 5.11 shows a typical composition of a backfill.

The backfill for galvanic anodes serves the following purposes:

(1) It isolates the anode from the surrounding soil and protects the anode from the effect of chemicals contained in the soil.
(2) It provides a lower anode-to-earth resistance because of its low resistivity.
(3) It provides a higher current output because of the low resistivity of the surrounding soil.

For the impressed current anodes, the standard material is coke. The physical and chemical analysis of the coke breeze is given in Tables 5.12a and 5.12b, respectively.

Backfill should be installed very dry around the anodes except in desert conditions. The coke breeze provides a low resistance between anode and earth and a longer life for the impressed current anode. Because the greater part of the current passes from the anode to the backfill particles, the anode is consumed at a slower rate. Table 5.13 shows the amount of backfill required.

Table 5.11 Composition of backfill

Soil Resistivity (ohm-cm)	Gypsum (%)	Bentonite (%)
Below 3000	50	50
3000 and above	75	20

Table 5.12a   Physical analysis of coke breeze

1% maximum to remain in No. 2 mesh screen	(0.525 in)
85% minimum to remain in No. 3 mesh screen	(0.263 in)
12% maximum to remain in No. 6 mesh screen	(0.131 in)

Table 5.12b   Chemical analysis of coke breeze

Constituent	Percentage
Moisture	14.70
Volatile matter	3.14
Fixed carbon	76.66
Ash	1.50
Sulfur	4.00

Table 5.13   Amount of backfill required

Anode size (in)	Hole size (in)	Backfill (lb)
1 × 60	6 × 80	62
1.5 × 60	6 × 84	61
2 × 60	8 × 84	108
3 × 60	10 × 84	180

5.18 MEASUREMENTS IN CATHODIC PROTECTION

5.18.1 SOIL RESISTIVITY MEASUREMENTS

Soil resistivity measurement is the first important step in the design of a cathodic protection system as the current requirement would differ from one soil resistivity to another for the system. It may, however, be pointed out that there is no single method available to determine precisely the degree of corrosivity caused by soils. Soil resistivity only provides a rough guide to the corrosivity of the soils. There are several methods available to determine the soil resistivity and a few important methods are given below.

(a) Four-pin Direct Current Method (Wenner's Method)

On the site, four stainless steel pins are buried in the ground and spaced in a straight line. The spacing between the pins represents the depth to which the resistivity is measured. By increasing pin spacing, the resistivity to a greater depth is measured. The instruments used for measurement are easily available. Instruments, such as a Vibroground, are often employed for the measurement of soil resistivity (Fig. 5.31a). This method is based on voltage drop. Figure 5.31b shows two outside pins C_1 and C_2, and the two inside pins P_1 and P_2. The outer pins are connected to a DC source (12 V). In the meter all instruments, such as the power supply, the current meters, the switch, etc. are contained in one single case. A desired amount of current is made to flow in the earth between the two

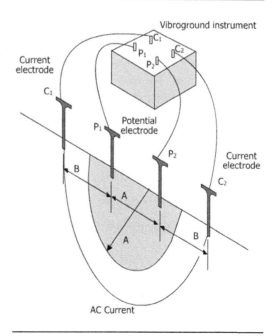

Figure 5.31a   Vibroground method for determination of soil resistivity

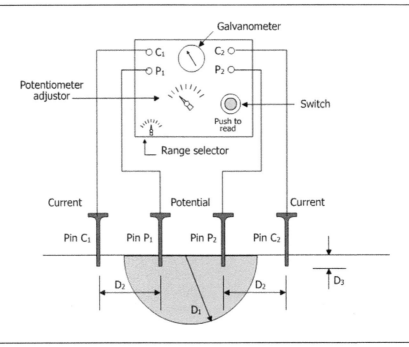

Figure 5.31b Four-pin (Wenner) resistivity measurement

outside pins C_1 and C_2, and the voltage drop is measured between the two inner pins P_1 and P_2. The millivoltmeter is used to measure the voltage drop in the earth between the inner pins P_1 and P_2, located along the current path between C_1 and C_2. The existing voltage between the two potential pins P_1 and P_2 is recorded before any current is applied. The test current is turned on and the voltage drop between the two potential pins is recorded. Resistivity is calculated by using the relationship

$$\rho = \frac{V_{on} - V_{off}}{I} \times \text{Spacing of pins (ft)}$$
$$\times 191.5 \text{ (Factor) (Voltage in mV)}$$

Illustrative problem
Calculate soil resistivity from the following data given:

Pin spacing – 3 ft
Impressed voltage – 10 V

Current reading – 8 mA
Voltage across the inside pins – 50 mV with no current flowing
Voltage across the inside pins – 500 mV with current flowing.

Solution:

$$\rho = \frac{191.5 \times D \times \Delta V}{I}$$

$$\rightarrow \frac{191 \times 3 \times (500 - 50)}{8}$$

$$= 32\,231 \text{ ohm-cm}$$

where

$$D = \text{distance between pins}$$
$$\Delta V = \text{voltage drop}$$

Similarly, if $D = 10'6''$

$$\Delta V = 412 \text{ mV}$$

$$I = 267 \text{ mA}$$

$$\rho = \frac{191 \times 10.5 \times 412}{267}$$

$$= 3094 \text{ ohm-cm}$$

(b) Soil Box Method

In the soil box method, no multiplying factor is necessary for calculation of resistivity as the dimensions are made such. A typical soil box has dimensions of $1''(D) \times 1.5''(W) \times 8.5''(L)$. It is constructed from a non-conducting material like plastic. The box contains the current terminals and potential terminals. The end plate of the box acts as current terminal and the inside contact plates as potential terminals. The box is filled with required soil which is packed firmly. The potential change is divided by the current to obtain the desired resistivity quickly. The resistivity of soil or water can also be measured by a soil box which has four terminals. The measurements can be conveniently made in the laboratory. The soil box method is a replica of a four pin method mentioned earlier. A typical soil box is shown in Fig. 5.32. In the figure shown, C_1 and C_2 are current terminals and P_1 and P_2 potential terminals. A DC source is connected to C_1 and C_2 terminals and the voltmeter is connected to the terminals P_1 and P_2 (potential terminals). A measured current is passed through the soil sample and the voltage drop is read across the P_1 and P_2 pins. Knowing the voltage drop and the current introduced, the resistivity of the soil can be conveniently measured.

The box can be used either with AC instrument, such as Nilssons or Vibroground or with Miller meter. When using Vibroground, the C_1 and C_2 terminals are connected to the end plates of the box and the P_1 and P_2 terminals to the intermediate terminals of the box.

5.18.2 SIGNIFICANCE OF SOIL RESISTIVITY

Soil resistivity gives an indication of the corrosivity of soil. Some typical approximations are given as below:

Soil resistivity (ohms-cm)	Corrosion
0–900	Severe corrosion
901–2300	Severe corrosion
2301–5000	Moderate corrosion
5000–10 000	Mild corrosion
10 000 and above	Very mild corrosion

The above is only a rough guide to predict corrosion. A soil resistivity survey is required to determine the current requirement for a given pipeline. Soil resistivity may be very high in cold areas, such as Alaska, and virtually no cathodic protection may be required for coated pipes. On the contrary, in tropical areas near the sea shores,

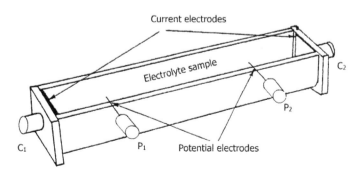

Figure 5.32 Typical soil box

Table 5.14 Characteristics of corrosive and non-corrosive soils

Corrosive	Non-corrosive
(1) Brackish water	Low moisture content
(2) Poorly aerated	Well aerated
(3) High acidity or alkaline pH	Low acidity or neutral pH or slightly acidic
(4) Sulfides present	Sulfates present
(5) Black or gray color	Red or brown color
(6) Anaerobic microorganism	Aerobic microorganism
(7) Soil with high salt contents	Dry soil

in particular, the soil resistivity may be low and cathodic protection for the pipes would be essential. Soil resistivity measurements will, however, depend upon dampness of soils and the prevalent weather conditions. The following are the characteristics of corrosive and non-corrosive soils (Table 5.14).

5.19 PIPE-TO-SOIL POTENTIAL

In order to determine the extent of corrosion of a buried pipe in soil, it is essential to determine its pipe-to-soil potential. It is to be realized that a potential difference existing on a pipe surface can cause corrosion, similar to the situation in a dry battery cell in which zinc is the anode and corrodes. The pipe-to-soil potential would indicate whether the pipe is corroding or it is fully protected.

The normal procedure for measurement of the structure-to-soil potential is to connect the negative terminal of a high resistance voltmeter (cm/V) to the pipe and the positive terminal to a standard copper/copper sulfate half cell (reference electrode) which is placed near the pipe (Fig. 5.33a). The Cu–CuSO$_4$ half cell acts as the cathode and the steel pipe as the anode. The difference of potential between the Cu–CuSO$_4$ half cell and the steel pipe is generally -0.55 V (pipe negative), when all connections have been made. The reading of the voltmeter indicates the structure-to-soil potential. The reference Cu–CuSO$_4$ is shown in Fig. 5.33b and connections are shown in Fig. 5.33c.

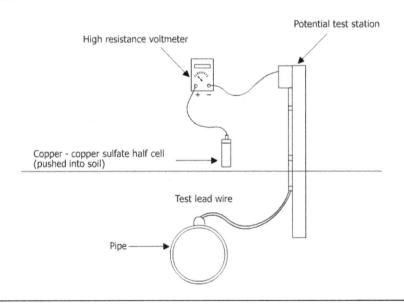

Figure 5.33a Typical pipe-to-soil potential measurement

A. Half-cell components

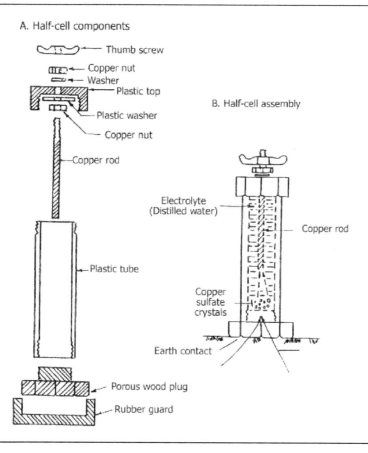

Thumb screw
Copper nut
Washer
Plastic top
Plastic washer
Copper nut
Copper rod

B. Half-cell assembly

Plastic tube

Electrolyte
(Distilled water)

Copper rod

Copper
sulfate
crystals

Earth contact

Porous wood plug
Rubber guard

Figure 5.33b Copper–Copper sulfate half cell components

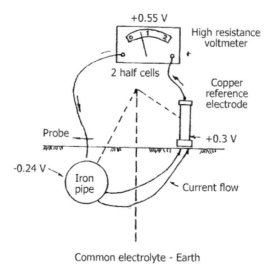

+0.55 V

High resistance
voltmeter

2 half cells

Copper
reference
electrode

Probe

+0.3 V

-0.24 V

Iron
pipe

Current flow

Common electrolyte - Earth

Figure 5.33c Copper–Copper sulfate half cell used as a reference electrode

5.20 Factors Affecting Pipe-to-Soil Potential

5.20.1 Placement of Reference Electrode for Measurement of Pipe-to-Soil Potential

One of the important points in the measurement of pipe-to-soil potential is the placement of the reference Cu–CuSO$_4$ electrode. Two locations are normally used, one near the structure and the other electrically remote. As a matter of principle, the reference electrode should be placed in such a way that the circuit resistance remains at the minimum. If a pipe is fully coated, the reference electrode may be placed in either remote or on the pipe position and it would not make any difference as the coating resistance makes up most of the resistance between the pipe and the soil. The potential of a cathodically protected pipe is measured by placing the reference electrode at least five or six pipe diameters away from the structure along the surface of the earth. In the case of an unprotected system, it may be essential to take measurements of pipe-to-soil potential with the reference electrode directly over the pipe and also at a position remote to the pipe. At the close position the potential of a small segment of the pipe is only indicated. Suppose an electrode is placed at a distance of 3 feet from a 12 inches diameter uncoated pipe, it would only survey a segment of six feet long as the electrode surveys a distance equal to twice its distance from the pipe. For a pipe length of 100 feet, the electrode must be placed at a distance of at least 50 feet. As a rule of thumb, the circuit resistance must be kept to minimum. For a bare structure, the reference electrode must be placed at a position electrically remote from the test section. This can be done by continuously increasing the distance and observing the readings in the voltmeter which becomes more and more negative. When the successive reading becomes more negative, the electrode is not outside the structure-to-earth resistance. Once, a point is reached at which there

is no further increase in the negative reading, the remote position is reached which implies that the electrode is outside the influence of the structure-to-earth resistance. This is the desired location for placing the reference electrode. At this position, there is no further change in IR drop (potential drop). The remote earth does not offer any resistance to current flow. A remote location is outside the potential gradient of the anode bed and the pipe.

5.20.2 Electrical Contacts

Extreme caution must be taken to maintain a good electrical contact between the probe and the structure, and between the soil and the reference electrode.

5.20.3 Instruments

Only high resistance voltmeters must be used to avoid errors in reading of potential. The voltmeter should have a minimum sensitivity of 50 000 ohms/V.

5.21 Potential Survey

It has been stated earlier that the tendency for a metal to corrode can be predicted by its potential in a particular environment. The potential surveys are made to assess the magnitude of corrosion of pipelines and detect special areas and spots where the degree of corrosion is severe; the later being termed as a hot spot. A general idea of corrosion of a pipeline can be obtained from the average pipe/soil potential. Newer pipelines show generally a lower negative potential than the older and coated pipelines. For example, a new pipeline may show an average potential of -1.65 V when compared to a potential of -1.2 V shown by an older pipeline. It is, however, important to detect the areas on the pipelines which are subjected to severe corrosion (hot spots).

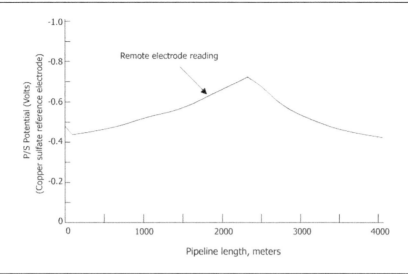

Figure 5.34 Potential survey – remote electrode

Figure 5.34 shows a plot of pipeline length *vs* pipe-to-soil potentials taken with the electrode in the remote position. The peak in the curve indicates the points towards which current is flowing to the pipe and discharging to the soil. It should be noted that the higher the value of *P/S* potential, the more would be the magnitude of corrosion. The peak, therefore, represent the hot spot or the areas where corrosion is active.

5.21.1 POTENTIAL SURVEY MEASUREMENTS

The distribution of potential along the survey of earth above the pipeline indicates the location of corroding areas. Different types of soil encountered by the pipe affect the potential of the pipe. The changes in soil resistivity also induce potential differences. The surface potential surveys are made to determine the anodic and cathodic areas on the pipe. The structure-to-soil potentials do not give a qualitative measurement of corrosion, however, they are very useful in prediction of corrosion when used in conjunction with other data,

such as soil resistivity. Several methods are used for potential measurement, such as the one electrode and the two electrodes methods. The two electrodes method is described below.

The two electrodes method for measuring the surface potential is shown in Fig. 5.35. Two $Cu/CuSO_4$ electrodes (a) and (b) are placed on a wooden clipboard and connected to each other through a resistance voltmeter. One of the electrodes is in rear position and the other in the forward. The figure also shows the positions along which the electrodes are marked 1, 2 and 3. The anodic and cathodic areas are also shown in the figure. The IR (potential) drops at each location is measured between the two electrodes. The two electrodes leap-frog method is shown in Fig. 5.36. The survey is started at position 1 and the potential is recorded. Electrode 'A' is left at position 1 and electrode 'B' is moved at position 2. The potential difference is recorded and also the polarity of electrode is noted. This procedure is continued along the whole length of the pipeline. For instance, consider the potential at position 1 to be −0.72 V. If the potential drop at position 2 is 0.065 V

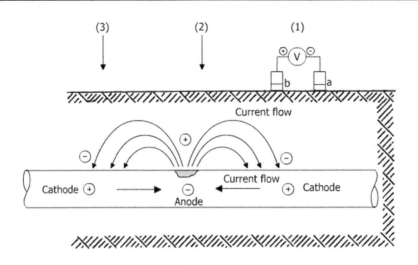

Figure 5.35 Surface potential method for determining corrosion involves electrodes (a) and (b) connected through a voltmeter (V). Potential is measured between the two electrodes

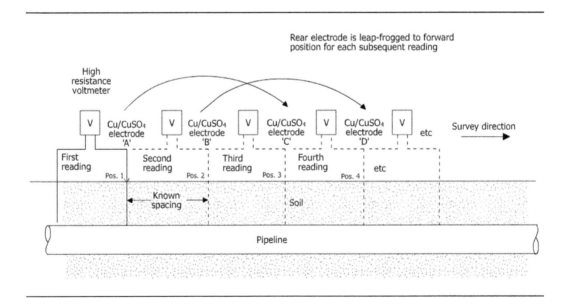

Figure 5.36 Two electrodes method of measurement of potential. (By kind permission of Cheveron, USA)

and the polarity is positive, then the pipe-to-soil potential would be −0.785 V. If the polarity of forward electrode is negative, the potential drop would be subtracted to get the final pipe-soil potential.

5.22 MEASUREMENT OF CURRENT FLOW

It is possible to measure the current flow in a pipeline or any buried structure by determining the voltage drop for a given length, if the resistance is known. Take, for instance, the current flow in a 6″ OD steel pipeline. Assume the length to be 500 feet. As the voltage drops are often small, low resistance voltmeters may be used. For example, a voltmeter with an internal resistance of 100 ohms/V on a 2 mV scale and external resistance of 2 ohm would be suitable for this measurement. Refer to Fig. 5.37 for the measurement technique.

Before determination of current flow, the true voltage between the tests points A and B and the resistance of the steel must be known. The latter information can be determined from standard tables for steel pipes.

Voltage drop

= Voltmeter reading between A and B

= 1.5 mV

True voltage between A and B =

Voltmeter reading

$$= \frac{\text{Voltmeter resistance} + \text{External resistance}}{\text{Voltmeter resistance}}$$

$$= \frac{1.5(2.0 + 2.0)}{2.0} = 3.0 \, \text{mV}$$

Thus the true voltage is 3 mV (Cu–CuSO$_4$).

The next step is to determine the resistances of the pipe. The resistance of a standard weight pipe is obtained by substituting in a simple equation:

$$R = \frac{215.8 \, L}{1\,000\,000 \, W}$$

where

R = resistance of the pipe per given length in ohms

W = weight of pipe per linear foot (consult tables)

L = length of the test pipe in feet (given)

The weight of a 6 inch steel pipe is 19 pounds per linear feet. The test length is 500 feet. The resistance of the pipe is, therefore,

$$R = \frac{215.8 \times 500}{19 \times 1\,000\,000}$$

$$= \frac{0.000216 \times 500}{19}$$

$$= 0.0005684 \, \text{ohms}$$

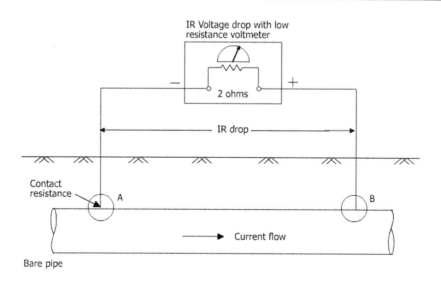

Figure 5.37 Measurement of current flow

By applying Ohm's law, $I = E/R$, the magnitude of current flow can be determined:

$$I = \frac{3.0}{0.0005684} = 0.5275\,A$$

As shown in the figure, the polarity indicates that the flow of current is from left to right. In the previous example, the magnitude of the current already flowing in a pipeline is determined. The amount of current flow may also be determined by passing a known amount of battery current and measuring the voltage drop between two points. The current is determined from Ohm's law ($E = 1/R$).

5.22.1 ZERO RESISTANCE AMMETER

The zero resistance ammeter can be used to determine the exact amount of current that would flow between two points of different voltages in a given circuit.

The zero resistance ammeter is useful for measurement of current flow in low resistance and low current circuits. If an ordinary ammeter is used to read current in such a circuit, an appreciable margin of error would be introduced by its internal resistance, the true voltage would not be shown. The basic advantage of this instrument is its ability to read test current accurately without introducing any resistance. For instance, if the circuit has a total resistance of 0.1 ohm, and an ammeter of 0.2 ohm resistance is introduced, the current flow would be reduced by 67% by a 0.2 ohm ammeter. Thus, the margin of error would be 67%. By not introducing any resistance in the circuit the meter is able to read true current. It is designed to work on a null principle.

The arrangement for measurement of current by zero resistance ammeter (ZRA) is shown in principle in Fig. 5.38. According to the arrangement, the open circuit potential between the pipe and the anode balanced by a potentiometer circuit. Sufficient current is allowed to flow from a battery through a variable resistor, until the IR drop balances the open circuit potential. At this point, the galvanometer registers zero current. The true current between the pipe and anode when bridged is then shown by the ammeter. The modern method employs using AC audio frequency energy for measurement.

5.23 REFERENCE ELECTRODE

The emf of a cell is comprised of its two half cells. For instance, the emf of a Daniel cell is the sum of the two half cells, i.e. Zn in $ZnSO_4$ and Cu in $CuSO_4$. The resulting emf is 1.1 V. In order to determine the half cell potential of zinc in any media, zinc is connected to the negative end of a voltmeter and a reference electrode of a known potential to the positive end. The difference of potential is directly read from the voltmeter.

The protective potentials for steel in seawater at 25°C and soils with respect to commonly used reference electrodes are shown in Table 5.15. Several reference electrodes can be used for measurement of half cell potential as shown in Chapter 2.

The most common electrode used for cathodic protection measurements is the copper–copper sulfate electrode. It contains a copper rod, a saturated solution of $CuSO_4$ containing the copper rod and a porous plug for contact with the electrolyte. The container is non-conducting. The space within the tube is filled with a saturated solution of copper sulfate and should contain an excess of crystals to insure that the solution would be saturated all the time. A rubber cap may be placed over the wooden plug when the electrode is not in use.

An increase in the temperature causes an increase in concentration of the solution and a decrease in temperature causes a decrease in concentration of the solution. The temperature correction is 0.5 millivolts per degree change (°F). Temperature potential correction charts may be used for correction.

Example

Determine the true half cell potential of a steel structure to a copper–copper sulfate reference electrode when the outside temperature is 90°F.

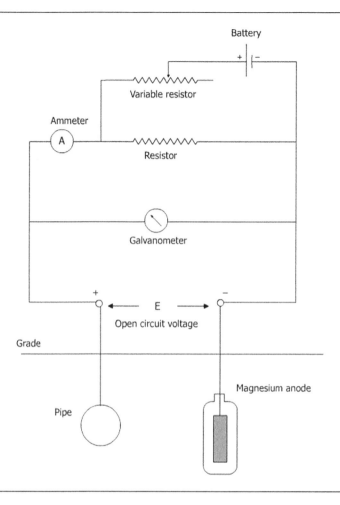

Figure 5.38   Measurement of current flow by the zero resistance ammeter method

The pipe-to-soil resistance potential reading is −0.70 V.

(a) Temperature correction +0.316 + (+0.007) = +0.323 V (Correction factor from the chart).
(b) Half cell potential of the steel structure without correction:

$$= -0.70 - (+0.316) = 0.384\,V$$

(0.316 V is the potential of Cu–CuSO$_4$ electrode.)
And with correction:

$$= -0.70 - (+0.323) = -1.023\,V.$$

5.24 COATING RESISTANCE TESTS

Suppose the coating resistance is to be measured on coated 24″ diameter steel pipe, the procedure is simple. Pass a known amount of current in the given section of the test pipe and note the voltage drop. By applying Ohm's law, determine the resistance (R) of coating. An experimental setup is shown in Fig. 5.39. Between the insulated flanges, three voltmeters are fixed at a predetermined spacing. A is the test section of a pipe, B is the on and off switch, C is the power source, D is an ammeter and E, F, G are the voltmeters. H is a copper–copper sulfate reference electrode.

Table 5.15 Protective potentials for steel in seawater and soils (aerated)

Reference electrode	Value for E_{prot} Fe in aerated soil or seawater
Copper/Copper sulfate (CSE) Cu/CuSO$_4$	−0.85 V
Saturated calomel (SCE) Hg/Hg$_2$Cl$_2$(s)/KCl (Saturated)	−0.80 V
Silver chloride Ag/AgCl/Seawater	−0.80 V
Standard hydrogen electrode (SHE) Pt, H$_2$/H$^+$($a_H^+ = 1$)	−0.53 V
Zinc Zn/Seawater	+0.20 V

A known amount of current is passed and the values of potential 'on' and 'off' are obtained by opening and closing the switch which controls the circuitry. The value of ΔV is obtained from V(on) and V(off). The amount of current passed is already known. The ΔV values of the voltmeters are averaged. By applying Ohm's law, the coating resistance is determined. Here is an example.

Example

(a) Length of the pipe = 2 miles, area = πDL, OD = 24″, soil resistivity = 10 000 ohms

The amount of current passed = 0.096 A, $\Delta V_E = 0.77$ V, $\Delta V_F = 0.75$, $\Delta V_G = 0.77$

Average voltage drop $(\Delta V_E + \Delta V_F + \Delta V_G)$ $\Delta = 0.75$ V.

The resistance by Ohm's law is $R = \dfrac{E}{I} =$

$\dfrac{0.75 \text{ V}}{0.096 \text{ A}} = 7.8125$ ohms

The area of the above pipe = 24 × π × 2 × 5280 = 66 316 ft^2

The effective coating resistance would, therefore, be

7.8125 ohms × 66 316

  = 51 792.7 ohms   per average square foot (approximately)

The soil resistivity affects the coating resistance significantly. The pipeline resistance is the combined resistance of coating and the resistance of the pipeline to remote earth.

In the example shown above, the resistance was determined in 1000 ohm-cm soil. The resistance can be changed from one soil resistivity to another soil resistivity. For instance, if a resistance of 0.4 ohms obtained in a soil resistivity of 10 000 ohms, the resistance is 10 000 ohm, would be 10 000/1000 × 0.4 = 4 ohms.

(b) If now it is assumed that the test section was bare and in 1000 ohm-cm soil, the resistances to earth of the above line would be lower. Let us assume it is found to be 0.0040 ohm in a soil having a resistivity of 1000 ohm-cm. If the average soil resistivity is 100 000 ohm, the resistance would be

$$0.0040 \times \frac{100\ 000}{1000}$$

$$= 0.40 \text{ ohms} \quad \text{(This is the resistance of the bare pipe)}$$

The difference between the resistance of the bare and coated pipe is 0.40 ohm. The difference is added to the resistance obtained above, i.e. 7.8125 and multiplied by the area of the pipe to obtain the effective coating resistance of the pipe per average square foot. In the example above, the effective average coating resistance would be 7.8125 + 3.60 = 11.412 × 66 316 = 756 002 ohm per average square foot. Area of pipe = 518 093 ohms per average square foot. The resistance of the steel pipes of various dimensions can be estimated from Table 5.16. It is based on the relationship

$$R = \frac{16.061 \times \text{Resistivity in micro-ohms}}{\text{Weight per foot}}$$

$$= \text{Resistance of one foot of pipe in micro-ohms}$$

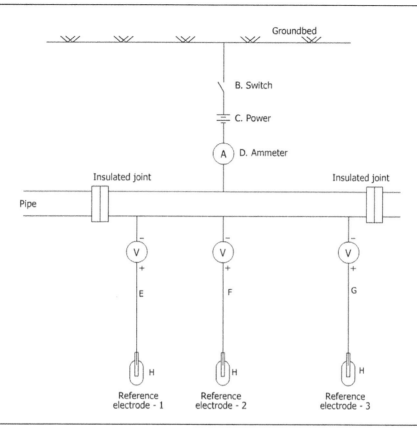

Figure 5.39 Determination of coating resistance. (From TEXACO Cathodic Protection – Design and application school, Texaco Houston Research Center, Training Manual. Reproduced by kind permission of Cheveron, Houston Research Center, USA)

5.25 CURRENT REQUIREMENT TESTS

Basically sufficient current is supplied from a power source to lower the potential of a structure to −0.85 V. At this potential all cathodic areas are polarized to the open circuit potential of the anodic areas, hence no corrosion would occur. A temporary ground-bed is made and a known amount of current is forced in the structure to determine the total current required for cathodic protection.

Protection of steel in an aggressive soil is achieved if the metal–soil potential is more negative than −0.85 V or −0.95 V for anaerobic soil conditions. In a bare pipeline or a structure, sufficient current is introduced in the system until a protective potential is achieved. To protect a structure from corrosion, say, a pipeline, a current requirement test is crucial, because if the required current to achieve a pipe-to-soil potential of −0.85 V is not provided, the pipe would not be protected as desired. If a bare pipe is buried in a soil of known resistivity and its characteristics are well-known, a selected current density can be given to the structure, without the necessity of current requirement test. In such instances experience is the best guide. In areas where experience lacks, the situation is different and current density required for protection can be determined experimentally by applying a range of current densities and selecting the most desirable current density to achieve cathodic protection.

The '*current drain*' test is commonly employed to determine the current requirement for a coated pipeline. A temporary drain point is set to

Table 5.16 Steel pipe resistance[1]

Pipe size (inches)	Outside diameter (inches)	Wall thickness (inches)	Weight per foot (pounds)	Resistance of one foot [2] in (ohms $\times 10^6$) (millionths of an ohm)
2	2.375	0.154	3.65	79.2
4	4.5	0.237	10.8	26.8
6	6.625	0.280	19.0	15.2
8	8.625	0.322	28.6	10.1
10	10.75	0.365	40.5	7.13
12	12.75	0.375	49.6	5.82
14	14.00	0.375	54.6	5.29
16	16.00	0.375	62.6	4.61
18	18.00	0.375	70.6	4.09
20	20.00	0.375	78.6	3.68
22	22.00	0.375	86.6	3.34
24	24.00	0.375	94.6	3.06
26	26.00	0.375	102.6	2.82
28	28.00	0.375	110.6	2.62
30	30.00	0.375	118.7	2.44
32	32.00	0.375	126.6	2.28
34	34.00	0.375	134.6	2.15
36	36.00	0.375	142.6	2.03

(1) Based on steel density of 489 pounds per cubic foot and steel resistivity of 18 micro-ohm-cm as stated in the text

(2) $R = \dfrac{16.061 \times \text{Resistivity in micro-ohm-cm}}{\text{Weight per foot}}$ resistance of one foot of pipe in micro-ohms

determine how much current would be required to protect the line. The site must be selected where the pipeline is exposed or a valve box is located so that a connection may be made. The test current can be supplied either by a storage battery (100 A) or a welding machine. A machine with an output voltage of 40 V is sufficient for the test. The temporary ground-bed usually consists of scrap length of pipes, driven steel rods or aluminum foil. If desired, actual anode could be used instead of scrap length of pipes and left on the location. They can be later utilized in the permanent ground-bed. For final installation, currents from 2–3 A up to full ratings of the machine can be utilized. The procedure is continued until the quantity of current required to protect the entire line of test section is determined. The ground-bed must be between 100 and 50 ft of the pipeline. If the temporary installation

(pipeline + ground-bed) is operated, for instance, at 25 A and a sufficient protection is obtained, it signifies that a rectifier, when set at this value, would also protect the pipeline. It is to be remembered that the temporary ground-bed described above has a high resistance, therefore, a high voltage is required from the welding machine. When a rectifier is installed with a proper ground-bed (low resistance), the voltage used would not be as high. The temporary ground-bed also would be consumed in a few days under test conditions because of high power consumption, however, it has no bearing on the protective current density determined for the protection of the pipe by the above test method. The method is illustrated in Fig. 5.40. Current is drained at point A by the welding machine. Two points, such as B and C are located on the pipeline test section after a polarization run of 3 or 4 hours. The two points should

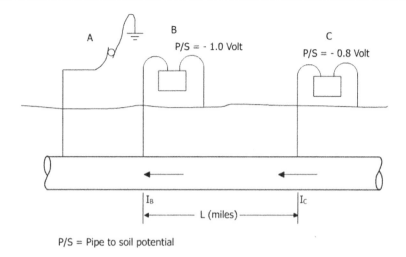

Figure 5.40   Current drainage method

have potentials of 1.0 and 0.8 V, respectively. The section between the two points will be at a potential of about 0.9 V, which is the potential required for protecting the line. The line current is measured between each of these points. The difference between the line currents at the two points would be the current picked up. If two points having P/S of −1.0 V and −0.8 V are not located, readings are taken which approximate the above values. Potential and current are adjusted to obtain the desired quantity. A static potential of −0.6 V is assumed. On the above basis, P/S and current curves are drawn as straight lines, and the points where the values are −1.0 and −0.8 are located in Fig. 5.41. If the distance between the two points is L miles, the average current required to bring them to 0.9 V line $I_B − I_C$, and $(I_B − I_C)/L$ would be average current, in amperes per mile. Once the two points are located, the current value at those points is determined. The total current requirement to raise the potential in a range −1.0 to −0.8 V can, therefore, be estimated.

It also possible to obtain data that would give information on the current requirement while the coating resistance tests are being made. Such a setup is shown in Fig. 5.42. In the setup a current interrupter is provided (automatic switch). The switch is 'on' position for 30 seconds and in 'off' position for 15 seconds. The data is taken at

every specified length's interval for calculation of coating resistance.

The values of ΔV (voltage 'on' and voltage 'off') and ΔI (current 'on' and current 'off') are measured. The average pipe potential change, which is mostly the voltage drop across the pipe coating due to the current increment (arithmetic difference between ΔI value) is determined. From the data, the current required to protect the coated pipe as well as the resistance of the coating is estimated. The estimate is preliminary only.

5.26 CURRENT DENSITY REQUIREMENT FOR A BARE STRUCTURE

In case of a bare structure, or a pipe, a current requirement test is made by placing the electrode at a position remote to the structure. The setup for a current requirement test of a bare structure is shown in Fig. 5.43. The most important condition is that the structure must be fully polarized to a potential −0.85 V for complete cathodic protection. It is to be noted on the current off position, the potential must show −0.85 V immediately. This value should not be reached gradually on

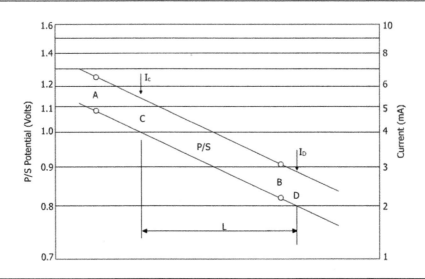

Figure 5.41 Measurement of pipe-to-soil *P/S* potential and line current. The difference in the current values for the two points gives the current demand for the distance *L*

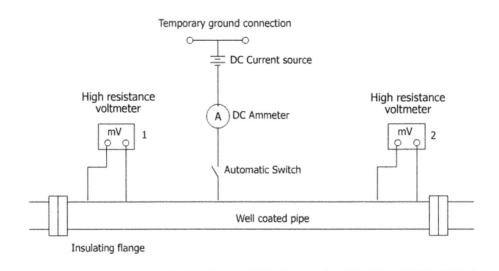

Figure 5.42 Current requirement test using close Cu/CuSO$_4$ electrodes. (From TEXACO Cathodic Protection – Design and application school, Texaco Houston Research Center, Training Manual. Reproduced by kind permission of Cheveron, USA)

opening the '*current on*' switch. It may take a very long time in several instances to achieve polarization. To shorten the waiting period for achievement of complete polarization (−0.85 V), the curve can be extrapolated. Figure 5.44 shows a polarization curve. It is a plot of pipe potential to remote earth *vs* time. The immediate objective is to determine the amount of current for the test section to achieve a potential of −0.85 V.

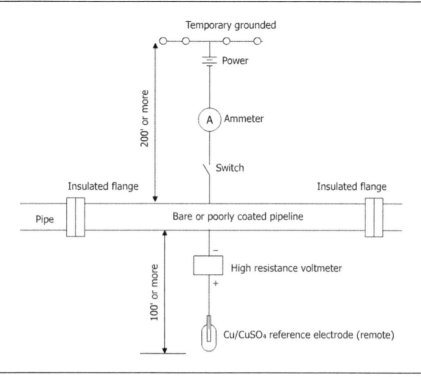

Figure 5.43 Cathodic protection of a bare structure. (From TEXACO Cathodic Protection – Design and application school, Texaco Houston Research Center, Training Manual. Reproduced by kind permission of Cheveron, USA)

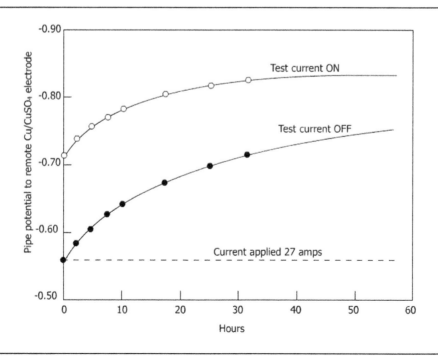

Figure 5.44 Polarization curve. (From TEXACO Cathodic Protection – Design and application school, Texaco Houston Research Center, Training Manual. Reproduced by kind permission of Cheveron, USA)

Referring to the polarization plot shown in Fig. 5.44, the following are the salient characteristics:

(1) The broken line in the plot shows the pipe-to-soil potential before the application of cathodic protection, in the diagram it is shown as 0.560 V.
(2) Variation of pipe potential with time read in current 'off' position (curve II).
(3) Variation of pipe potential with time read in current 'on' position (curve III). The curve extrapolated to read maximum potential, −0.838 V.
(4) The value of current recorded to achieve maximum potential is recorded. Assume the current value as 25 A (I_A).

Example

Assume that the test section of bare pipe, 1 mile long, 6–5/8″ OD, is to be protected, and the approximate current requirement is 25 A. An approximate quantity of current is introduced in this pipe or buried structure. This approximate value is based on the experience of cathodic protection engineer. Later this approximate quantity is modified on the basis of the results obtained from the polarization plot. The following steps show clearly, how the exact current requirement is worked out:

(a) Determine the total change of voltage (ΔE) from the original value of pipe potential (−0.560 V)

$$\Delta E = -0.838 - (-0.560) = -0.278 \text{ V}$$

(b) Calculate the average potential change per ampere for the test section

$$\frac{\Delta E}{I} = \frac{0.278}{25} = 0.0112 \frac{\text{V}}{\text{A}}$$

(c) The additional current required to raise the potential from −0.838 to −0.850 is

$$-0.850 \text{ V} - (-0.838) = -0.01 \text{ V}$$

$$\frac{0.01}{0.0112} = 0.822 \frac{\text{V}}{\text{A}}$$

(d) The total current required would, therefore, be 25 + 0.822 = 25.822, or say 26.0 A. The field test data can be modified by this method.

(e) The current approximation for a temporary DC power source on the basis of 2.50 mA/ft is calculated as below for a bare line

$$= \left[\frac{6-5/8 \text{ in}}{12} \right] (\pi) \left[5280 \frac{\text{ft}}{\text{mile}} \right] \left(0.0025 \frac{\text{A}}{\text{ft}^2} \right)$$

$$= \left[\frac{0.552}{12} \right] (3.41\pi) \left[5280 \frac{\text{ft}}{\text{mile}} \right] \left(0.0025 \frac{\text{A}}{\text{ft}^2} \right)$$

$$= 0.25 \text{ A}$$

5.26.1 CURRENT DENSITY REQUIREMENTS

Table 5.17 shows typical current requirements of uncoated steel in various environments.

Table 5.17 Typical current density requirements for cathodic protection of uncoated steel

Environment	Current density, (mA/ft²)	
	AFM 88–9	Gerrard
Neutral soil	0.4 to 1.5	0.4 to 1.5
Well-aerated neutral soil	2 to 3	2 to 3
Wet soil	1 to 6	2.5 to 6
Highly acidic soil	3 to 15	5 to 15
Soil supporting active sulfate reducing bacteria	6 to 42	up to 42
Heated soil	3 to 25	5 to 25
Stationary fresh water	1 to 6	5
Moving fresh water	9 to 25	5 to 6
Turbulent fresh water containing dissolved oxygen	5 to 15	5 to 15
Seawater	3 to 10	5 to 25

5.27 STRAY CURRENT CORROSION

5.27.1 INTRODUCTION

In a cathodic protection system, the conventional current (e.g. Zn^{++} ions) flows from the ground-bed through the earth towards the metallic structure. If a current encounters a metallic structure on its way it is picked up by the metallic structure, transmitted to other parts of the structure and finally discharged from the structure through the earth and returns to the cathode. The following are the two routes:

(a) Ground-bed (Anode) – Earth – Metallic structure – Earth – Cathode
(b) Ground-bed (Anode) – Earth – Cathode.

If an underground metallic structure is present, the point at which the current is discharged from the metallic structure to the ground becomes the anode and, therefore, corrodes (forming Fe^{++} ions). The undesired current which enters the metallic structure on its way to the cathode is called stray current and the corrosion caused by stray current is called *stray current corrosion*. This type of corrosion is associated with DC transit systems, such as in the case of electric trains on super-grid contacts which cross the electrical lines. There are numerous other sources of stray currents, such as welding machines, elevators, electroplating machines, etc. Speaking simply, stray currents are uncontrolled currents which originate mostly from DC systems and cause corrosion at the point of leakage from the system (site of exit of Fe^{++} ions). There are two main categories into which these currents can be divided:

(A) Static type

   (1) Cathodic protection rectifier
   (2) Railroad signal batteries

(B) Dynamic type

   (1) DC equipment in mines
   (2) Electric railway generating equipment.

The part of the pipe receiving the current from the rails becomes the cathode and the part of the pipe from which the current leaves the pipe becomes the anode. It has been shown earlier that anode is the area of exit of Fe^{++} ions (corrosion site).

5.27.2 MAJOR SOURCES OF STRAY CURRENTS

(a) DC Transmit Systems, such as Electric Trains

Stray currents create a great source of difficulty in DC transit systems. In such transit systems, the overhead feeder is connected to the positive bus of DC substations. The load current which is required to operate the trains is expected to return via the tracks connected to the negative bus. For illustration, one train and a substation of a transit system is shown in Fig. 5.45. A part of the load current may enter the earth as the tracks are not completely insulated from the earth. Load current would enter where the tracks are positive and take a path back to the substation. In the current pick areas, any underground bare pipes or other metallic structures would pick up the current and they would be cathodically protected at no cost to the other. The current after being picked up by a pipeline would be carried in the neighborhood of the DC substation where it would be discharged as Fe^{++} ions to the earth. The discharge area of the pipe would, therefore, become the anode and the pipe would seriously corrode. Thus, in a current pickup area the pipe would be protected and in a current discharge area it would corrode. If all negative return was to be carried by the tracks completely insulated, this problem would not exist. In DC systems, the negative side of the generator is grounded because of the contact of rails or structures with the earth. In the example given of an electric railway system the positive side of the generator is connected to the trolley contact wire through the feeder cables and the negative side to the rail.

The return path of the current is shown in Fig. 5.45. The current flows back to the generator, either through the rails or an earth path parallel to rails. There is strong possibility of such current

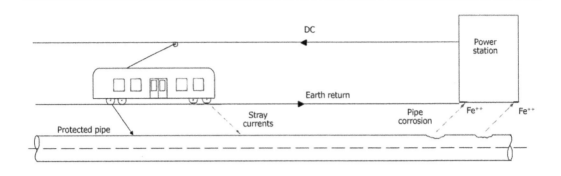

Figure 5.45 Showing how a current flowing between a train and the distant generating station tends to split itself between the track and a pipeline

being picked up by a nearby metallic structures. From the structure the current would discharge to the earth, returning to the rails and finally to the negative terminal of the generator. The underground metallic structure would corrode at the point where the current discharges back to the ground before its entry to the rail and its return to negative terminal of the generator.

Any underground metallic structure would corrode at the point of exit of Fe^{++} ions. To prevent this undesirable stray current corrosion a metallic bond, such as a bond cable between the pipeline and the negative bus of the DC substation, is installed as shown in Fig. 5.46. The current is then drained off by the metallic bond and all the surface of the secondary pipes becomes completely cathodic. The situation here is rather over-simplified, as there may be hundreds of substations serving the system depending on the traffic load and the load may vary during the 24 hours period. In certain instances, a bond connected may not be useful as the direction of flow of current in the bond may reverse and the current may flow to the pipelines rather than to the negative return. In order to handle this problem, rectifier discs may be inserted in the circuit so as to prevent the reversal of the current flow.

(b) Welding Generators

During the electric arc welding of ships and fabrication of large structures, large currents are involved and they escape to earth where they are picked up by the buried metallic structures. While the ships are being repaired in the dock basin, welding generators are placed on shore and the ground DC lines are taken to the ship. Any current returning from welding electrode through the water to the shore would destroy paint work and cause damage to the hull of the ship. This happens particularly where the earth circuit is closed. The stray current on the ground would be picked up by underground metallic structures and cause corrosion at the point of exit. In such instances, the DC generator should be placed on ship rather than on the shore to minimize the magnitude of stray currents.

5.28 CATHODIC PROTECTION INTERFERENCES

For cathodic protection of an underground structure, current is directly injected into the soil in the neighborhood of the pipe or structure to be protected. High current density, therefore, exists close to the ground-bed. If another metallic structure, such as a water pipe or a cable, happens to be present in the immediate neighborhood of the cathodically protected structure, it would pick up some current at one point and discharge it via the soil at other point on the metallic structure.

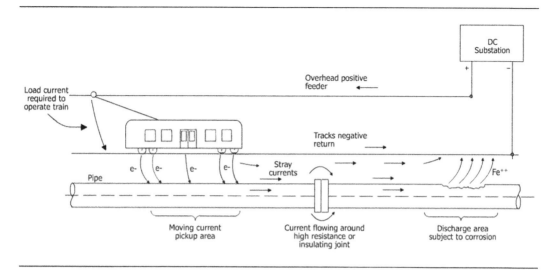

Figure 5.46 Stray current corrosion caused by DC transit system. (From Peabody, A.W. (1967). *Control of Pipeline Corrosion*, NACE. Reproduced by kind permission of NACE, Int., USA)

Any water pipe or cable in the vicinity would corrode as a result of discharge of positive current from it (Fe^{++} ions). This is an unwanted corrosion and is referred to as cathodic protection interference. It refers to an undesirable current discharge from a pipeline or structure in the vicinity of a cathodically protected pipeline or structure.

5.28.1 EXAMPLES OF INTERFERENCE

A foreign (unprotected) pipeline may either cross a protected pipeline or passes close to the ground-bed of the protected line. Such situations cause varying degrees of interference. Suppose a foreign pipeline does not cross the cathodically protected line but passes very close to a ground-bed of cathodically protected line. This is an example of interference caused by radial current flow. When a foreign pipeline lies across the normal flow lines of current, there would be a little difference in the potential between the path nearest to the anode and nearest to the cathode as only small amount of current would be interrupted. If, however, the foreign pipeline runs parallel to flow lines of current, there would be

a large amount of current pickup which would be discharged to the protected structure. This is because current prefers a low resistance path (metal path) rather than higher resistance (via soil). A tank close to an anode bed may pickup current at one point and discharges it as Fe^{++} ions from another point (Fig. 5.47). The discharge area would undergo heavy damage and corrode due to the consequent loss of Fe.

In areas of positive soil potential, the foreign pipeline would pickup the current. This current would leave the foreign pipeline in remote areas in order to reach the protected pipeline and flow back to the rectifier to complete the circuit. Because of several areas of Fe^{++} current discharge, the foreign pipeline would corrode.

However, if a foreign pipeline is coated, it would not be affected to the same degree as a bare pipeline or structure, although it may be in the region of potential gradients. To ascertain the damage, potential readings can be taken at the points of crossing of the pipes. If the potential of the foreign pipeline becomes positive, it would indicate corrosion caused by Fe^{++} current discharge. In order to prevent this, the bare pipelines may be coated in the region of crossing or a metallic bonding may be introduced between the two as illustrated in Fig. 5.48.

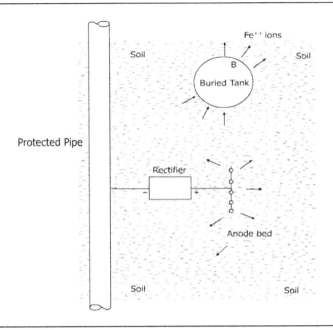

Figure 5.47 Damage caused by current flow. Corrosion occurs at position 2 due to discharge of Fe^{++} ions

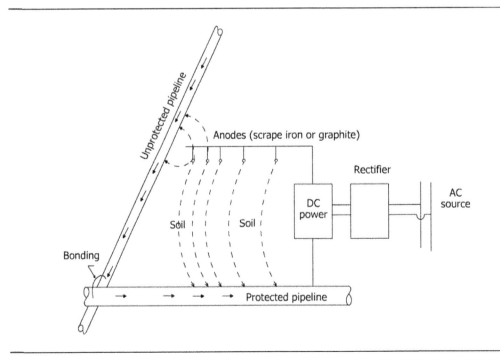

Figure 5.48 Showing current impressed by a DC source being picked up by an unprotected line and returning to the protected line via a bonding link

The following is a summary of preventive methods which could be used to minimize the interference:

(1) The current output of the main rectifier may be reduced.
(2) The ground-bed may be re-sited, if necessary. This is applicable if a foreign pipeline passes close to the ground-bed.
(3) Installation of a crossing bond between the pipes. A bond between the two points of crossing is installed and the amount of current flow is controlled by a resistor.
(4) Installation of magnesium anodes on the corroding structure.
(5) Isolation of the anodic section of the structures and installation of continuity bonds across the anodic section.
(6) Coating the metal/electrolyte interface, or the contact surfaces. In cases of corrosion involving a foreign line, the foreign line must be coated.

5.29 DESIGN CHARTS

Figure 5.49 shows a typical horizontal anode design chart where the number of anodes is shown on the abscissa and the resistance of the anode at different spacings on the ordinate axis. The following basic information is required for the construction of a design chart:

(a) Anode size
(b) Size of coke breeze column
(c) Resistivity of coke breeze
(d) Anode spacing
(e) Soil resistivity

The resistance of a given number of anodes in a soil of 100 ohm-cm at a specified spacing can

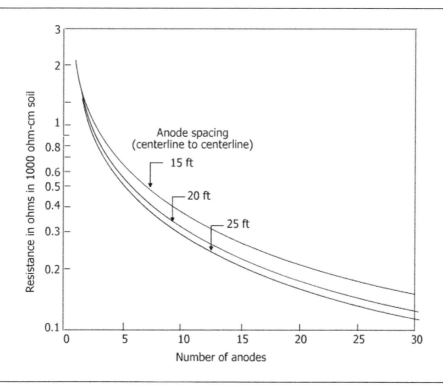

Figure 5.49 Typical horizontal anode design chart for impressed current ground beds. (From TEXACO Cathodic Protection – Design and application school, Texaco Houston Research Center, Training Manual. Reproduced by kind permission of Cheveron, USA)

be directly noted from the chart. It is to be noted that separate charts need preparing for each anode size, size of coke breeze column and resistivity of the coke breeze. Resistance can be conveniently converted from one soil resistivity to another soil resistivity.

Suppose it is desired to obtain resistance of 20 anodes in parallel at 20 ft spacing in a soil of 3000 ohm-cm, the following is a step-wise procedure.

(1) Obtain the parallel resistance of 20 anodes at 20 ft spacing in 1000 ohm soil directly from the chart.
(2) Determine the resistance in a soil of 3000 ohm by conversion. Suppose the anode resistance determined is 0.50 ohm. It would be

$$0.50 \times \left(\frac{3000}{1000}\right) = 1.5 \, ohm\text{-}cm$$

(3) Add to this resistance (1.5 ohm-cm) the internal resistance of the anode and divide by the number of anodes in parallel (twenty in this case).

The above procedure can be used to determine the anode to earth resistance. Design charts can be constructed for vertical anodes and horizontal anode ground-beds for both impressed and galvanic anode systems. The development of design curves is discussed in Section 5.31.2.

5.30 GROUND-BED DESIGN

The ground-bed design is an essential component of a cathodic protection system. The following are the salient points which need consideration while selecting a site for the ground-bed installation:

(1) Presence of metallic structures in the vicinity of the ground-bed.
(2) Location of the site with respect to the pipeline.
(3) Source of power.
(4) Ready access.

Once the current requirement of a structure is established, the ground-bed is designed to keep the resistances as low as possible. Such an arrangement allows a small voltage for the purpose of driving the required protective current. By lowering the resistances, the cost of power consumption is also reduced. The simplified relation given below can be used to determine the resistance of a single or vertical anode to earth.

$$R_V = 0.00521\frac{\rho}{L} \times \left[2.3 \log \frac{8L}{d} - 1\right]$$

It is known as D'Wights equation for single vertical anode, where

R_V = resistance of vertical anode (ohms) to earth
L = anode length in feet
ρ = soil resistivity (ohm-cm)
d = anode diameter (in feet)

A more common form of equation for multiple vertical anodes is:

$$R_V = 0.0052\frac{\rho}{NL}\left[\left(2.303 \log \frac{8L}{d} - 1\right)\right.$$
$$\left. + \left(\frac{2L}{S} \times 2.303 \log 656N\right)\right]$$

It is called Sundae equation.

H. B. D'Wights equation can also be used to determine the resistance to earth for a single horizontal anode.

$$R_h = \frac{0.00521\rho}{L}\left(2.303 \log \frac{4L}{D}\right.$$
$$\left. + 2.303 \log \frac{L}{H} + \frac{2h}{L} - 2\right)$$

where

L = anode length in feet
R_h = resistance of horizontal anode
d = anode diameter (including the backfill)
h = depth from surface to center of the anode (ft).

For multiple anodes, multiply R_h by the adjusting factors for the parallel anodes given in standard tables.

The following relationship is widely used to determine the resistances of the anode backfill:

$$R_V = \frac{0.0171\rho}{L}\left(2.303 \log \frac{8L}{d} - 1\right)$$

Illustrative Examples

Calculate the resistance to earth of a 3 in × 60 in long vertical anode in 1000 ohm soil. The anode length is 5 ft and diameter is 0.25 in. What would be its resistance in a 5000 ohm-cm soil?

Solution:

Using the formula mentioned above (D'Wights equation)

$$R = \frac{0.00521\rho}{L}\left(2.303 \log \frac{8L}{d} - 1\right)$$

and inserting the appropriate values

$$R = \frac{0.00521 \times 1000}{5}\left(2.303 \log 8 \times \frac{5}{0.25} - 1\right)$$

$$= 4.24 \, \text{ohms}$$

In a soil of 5000 ohm-cm, the resistance would be

$$4.24 \times \frac{5000}{1000} = 21.20 \, \text{ohms}$$

With the help of the above formula, it can also be shown that a long slender anode has a lower resistance than a short square anode, given the same weight.

5.31 SOIL RESISTIVITY AND PIPE-TO-SOIL POTENTIAL SURVEY

5.31.1 SOIL RESISTIVITY SURVEY

Conducting soil resistivity surveys is a primary step in designing of a cathodic protection system for pipelines. The methods of determination of soil resistivity have been described in an entire section. Pipelines in low resistivity soils would require a greater amount of current for protection than pipelines in a high resistivity soils, because of a higher magnitude of corrosion in the former. Hence, low soil resistivity areas are selected to install the anode ground-bed.

Figure 5.50 shows a typical plot of soil resistivity survey. The peaks in the plot indicate the areas of high soil resistivity and the valleys, the

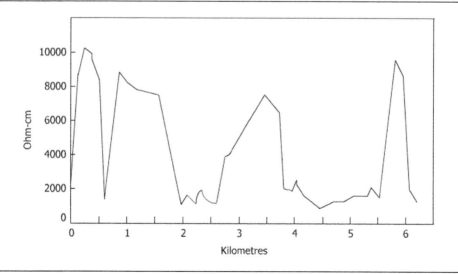

Figure 5.50   Soil resistivity data

areas of low soil resistivity, the preferred areas for installation of anode ground-beds.

The basic idea of the survey is to detect anodic and cathodic areas on the pipe for cathodic protection. The potential readings differ for coated and uncoated pipelines and are influenced by galvanic corrosion and external current. Therefore, no generalized statement can be made on the value of potentials above which the areas are anodic and below which are cathodic. More negative potentials indicate the anodic areas in general and these areas need to be protected. The following are the guidelines for determining the pipe-soil potential:

(1) Potentials for newer pipelines are more negative than the other ones.
(2) Potentials are more negative in neutral and acidic soils than those in alkaline environments.
(3) Regions of more negative potential correspond to locations of low soil resistivity (anodic areas) in uncoated pipelines.
(4) Typical values of potentials along uncoated pipelines are, in general, in the range of $-0.5\,V - -0.6\,V$ and $-0.65\,V - -0.75\,V$

along wet coated pipelines. The peaks in the potential profile represent locations liable to corrode.

Determining the degree of cathodic protection: A reading less negative than $-1.2\,V$ *vs* Cu/CuSO$_4$ would indicate adequate C. P. and a reading more positive than $-1.2\,V$ would indicate inadequate C. P.

5.31.2 COMBINED PLOTS (COMPOSITE PLOTS)

Composite plots of pipe-to-soil potential and soil resistivity surveys can be constructed to identify the hot spots (corroding areas determined by both soil resistivity and pipe-to-soil potential surveys). The corrosive areas in the composite figure are shown by shaded areas where the peaks in the pipe-to-soil correspond with the valleys in the soil resistivity survey (Figs 5.50–5.52). A low soil resistivity (say, 1000 ohm-cm) and a high negative value of pipe-to-soil potential (say, $-1.3\,V$) would constitute a corrosion area.

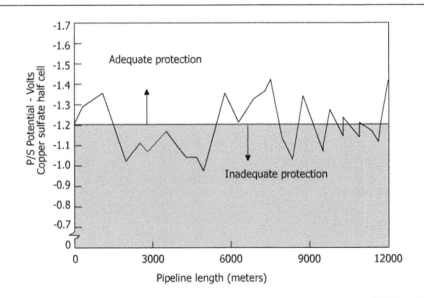

Figure 5.51 Pipe-to-soil potential survey. (From TEXACO Cathodic Protection – Design and application school, Texaco Houston Research Center, Training Manual. Reproduced by kind permission of Texaco, Houston, USA)

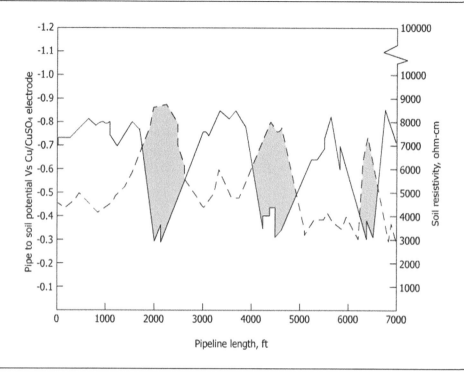

Figure 5.52   Composite data – pipe-to-soil potential and soil resistivity surveys. Shaded areas represent hot areas which need protection. (From TEXACO Cathodic Protection – Design and application school, Texaco Houston Research Center, Training Manual. Reproduced by kind permission of Texaco, Houston, USA)

5.32 CALCULATIONS IN CATHODIC PROTECTION DESIGN

5.32.1 DESIGN CURVE CALCULATIONS

It has been stated earlier that the resistance of a number of anodes can be determined from design charts. The design charts are made for definite sizes of anodes, in a known soil resistivity. The following calculations show how the design curves are constructed.

(A) Determination of resistance of the anodes to outer edge of the backfill (internal resistance)
     In order to determine the internal resistance, the following data would be required:

(a) Resistivity of the backfill
(b) Dimensions of anode

(c) Dimensions of the backfill

Suppose the following data is given:

$a = 40$ ohm-cm   (backfill resistivity)
$b = 3'' \times 5'$ (anode dimensions)
$c = 6'' \times 7'$ (backfill dimensions)

Solution:
Applying Sundae equation, determine first the resistance of vertical anode to outer edges of the backfill (R_1).

$$R_{V_a} = \frac{0.00521}{L}\rho\left(2.3\log\frac{8L}{d} - 1\right)$$

$$R_{V_a} = \frac{(0.00521)(40)}{5}\left[2.3\log\frac{40}{0.25} - 1\right]$$

$$= 0.166\,\text{ohms}$$

(use anode dimensions and backfill resistivity)

$$R_{V_{bf}} = \frac{(0.00521)(40)}{7} \left[2.3 \log \frac{56}{0.5} - 1 \right]$$

$$= 0.107 \text{ ohms}$$

$$R_1 = R_{V_a} - R_{V_{bf}}$$

$$= 0.166 - 0.107 \text{ ohms}$$

$$= 0.059 \text{ ohms}$$

(B) Calculation of total resistance of one anode to earth

Data required:

R_I = resistance of anode to outer edge of backfill (internal resistance)
ρ = resistivity of soil (1000 ohm-cm)
d = dimension of backfill column
   (6″ × 7′) as in (A)
R_2 = resistance of backfill column to soil

Solution:

$$R_{V(total)} = R_1 + R_2 \text{ (internal resistance} + \text{backfill to soil)}$$

R_1 — has already been determined in (A) to be 0.059 ohms

R_2 — resistance of backfill to soil is now to be determined, which is

$$R_2 = \frac{(0.00521)(1000)}{7} \left(2.3 \log \frac{56}{0.5} - 1 \right)$$

$$= (0.74)(2.3 \log 112 - 1)$$

$$= (0.74)(2.3 \times 2.04 - 1)$$

$$= (0.74)(3.692)$$

$$= 2.76 \text{ ohms}$$

R_1 — has already been determined in (A) to be 0.059 ohms

As $R_{V(total)} = R_1 + R_2 = R_T$

$$R_T = R_{V(total)} = 0.059 + 0.76$$

$$= 2.822 \text{ ohms}$$

(C) A design curve is to be developed for 10, 15, 25 and 30 anode beds at a 10 ft spacing in 1000 ohm-cm soil resistivity.
Required data:

(1) ρ = 1000 ohm-cm   (soil resistivity)
(2) N = 10, 15, 20, 25, 30   (number of anodes)
(3) S = 10 ft   (anode spacing)
(4) d = 6″ × 7′   (backfill dimensions)

$$R = \frac{0.00521}{NL} \rho \left\{ \left(2.3 \log \frac{8L}{d} - 1 \right) \right. $$
$$\left. + \left(\frac{2L}{S} 2.3 \log 0.656 \, N \right) \right\}$$

Workout R for 10, 15, 20, 25 and 30 anodes from the given data.

(1) For $N = 10$

$$R = \frac{(0.00521)(1000)}{10(7)} \left\{ \left(2.3 \log \frac{8 \times 7}{0.50} - 1 \right) \right. $$
$$\left. + \left(\frac{2 \times 7}{10} 2.3 \log 0.656(10) \right) \right\}$$

$$= 0.47216$$

(2) For $N = 15$

$$R = \frac{(0.00521)(1000)}{15(7)} \left\{ \left(2.3 \log \frac{8 \times 7}{0.50} - 1 \right) \right. $$
$$\left. + \left(\frac{2 \times 7}{10} 2.3 \log 0.656(15) \right) \right\}$$

$$= 0.343$$

(3) For $N = 20$

$$R = \frac{(0.00521)(1000)}{20(7)} \left\{ \left(2.3 \log \frac{8 \times 7}{0.50} - 1 \right) \right. $$
$$\left. + \left(\frac{2 \times 7}{10} 2.3 \log 0.656(20) \right) \right\}$$

$$= 0.272$$

(4) For $N = 25$

$$R = \frac{(0.00521)(1000)}{(25)(7)} \left\{ \left(2.3 \log \frac{8 \times 7}{0.50} - 1 \right) \right.$$

$$\left. + \left(\frac{2 \times 7}{10} 2.3 \log 0.656(25) \right) \right\}$$

$$= 0.227$$

(5) For $N = 30$

$$R = \frac{(0.00521)(1000)}{(30)(7)} \left\{ \left(2.3 \log \frac{8 \times 7}{0.50} - 1 \right) \right.$$

$$\left. + \left(\frac{2 \times 7}{10} 2.3 \log 0.656(30) \right) \right\}$$

$$= 0.1955$$

From the above calculations tabulate the results as below:

Number of anodes	R (ohms)
10	0.47216
15	0.343
20	0.272
25	0.227
30	0.1955

Repeat the calculations for the following anode spacing

$S = 10$

$S = 15$

$S = 20$

$S = 25$

For $S = 15$, $N = 10$

(1) R_{10} for $S = 10$

$$\frac{(0.00521)(1000)}{10(7)} \left\{ \left(2.3 \log \frac{8 \times 7}{0.5} - 1 \right) \right.$$

$$\left. + \left(\frac{2 \times 7}{10} 2.3 \log 0.656(10) \right) \right\}$$

$$= 0.471 \, \Omega$$

(2) R_{10} for $S = 15$

$$\frac{(0.00521)(1000)}{10(7)} \left\{ \left(2.3 \log \frac{8 \times 7}{0.5} - 1 \right) \right.$$

$$\left. + \left(\frac{2 \times 7}{15} 2.3 \log 0.656(10) \right) \right\}$$

$$= 0.471 \, \Omega$$

(3) R_{10} for $S = 20$

$$\frac{(0.00521)(1000)}{10(7)} \left\{ \left(2.3 \log \frac{8 \times 7}{0.5} - 1 \right) \right.$$

$$\left. + \left(\frac{2 \times 7}{20} 2.3 \log 0.656(10) \right) \right\}$$

$$= 0.3742 \, \Omega$$

(4) R_{10} for $S = 25$

$$\frac{(0.00521)(1000)}{10(7)} \left\{ \left(2.3 \log \frac{8 \times 7}{0.5} - 1 \right) \right.$$

$$\left. + \left(\frac{2 \times 7}{25} 2.3 \log 0.656(10) \right) \right\}$$

$$= 0.355 \, \Omega$$

(5) R_{15} for $S = 10$

$$\frac{(0.00521)(1000)}{10(7)} \left\{ \left(2.3 \log \frac{8 \times 7}{0.5} - 1 \right) \right.$$

$$\left. + \left(\frac{2 \times 7}{10} 2.3 \log 0.656(15) \right) \right\}$$

$$= 0.343 \, \Omega$$

(6) R_{15} for $S = 15$

$$\frac{(0.00521)(1000)}{15(7)} \left\{ \left(2.3 \log \frac{8 \times 7}{0.5} - 1 \right) \right.$$

$$\left. + \left(\frac{2 \times 7}{15} 2.3 \log 0.656(15) \right) \right\}$$

$$= 0.290 \, \Omega$$

(7) R_{15} for $S = 20$

$$\frac{(0.00521)(1000)}{15(7)} \left\{ \left(2.3 \log \frac{8 \times 7}{0.5} - 1 \right) \right.$$

$$\left. + \left(\frac{2 \times 7}{20} 2.3 \log 0.656(15) \right) \right\}$$

$$= 0.2635 \, \Omega$$

(8) R_{15} for $S = 25$

$$\frac{(0.00521)(1000)}{15(7)} \left\{ \left(2.3 \log \frac{8 \times 7}{0.5} - 1 \right) \right.$$
$$\left. + \left(\frac{2 \times 7}{25} 2.3 \log 0.656(15) \right) \right\}$$
$$= 0.248 \, \Omega$$

(9) R_{20} for $S = 10$

$$\frac{(0.00521)(1000)}{20(7)} \left\{ \left(2.3 \log \frac{8 \times 7}{0.5} - 1 \right) \right.$$
$$\left. + \left(\frac{2 \times 7}{10} 2.3 \log 0.656(20) \right) \right\}$$
$$= 0.272 \, \Omega$$

(10) R_{20} for $S = 15$

$$\frac{(0.00521)(1000)}{20(7)} \left\{ \left(2.3 \log \frac{8 \times 7}{0.5} - 1 \right) \right.$$
$$\left. + \left(\frac{2 \times 7}{15} 2.3 \log 0.656(20) \right) \right\}$$
$$= 0.2275 \, \Omega$$

(11) R_{20} for $S = 20$

$$\frac{(0.00521)(1000)}{20(7)} \left\{ \left(2.3 \log \frac{8 \times 7}{0.5} - 1 \right) \right.$$
$$\left. + \left(\frac{2 \times 7}{20} 2.3 \log 0.656(20) \right) \right\}$$
$$= 0.205 \, \Omega$$

(12) R_{20} for $S = 25$

$$\frac{(0.00521)(1000)}{20(7)} \left\{ \left(2.3 \log \frac{8 \times 7}{0.5} - 1 \right) \right.$$
$$\left. + \left(\frac{2 \times 7}{25} 2.3 \log 0.656(20) \right) \right\}$$
$$= 0.192 \, \Omega$$

For R = 25 and S = 10, 15, 20 and 25:

(13) R_{25} for $S = 10$

$$\frac{(0.00521)(1000)}{25(7)} \left\{ \left(2.3 \log \frac{8 \times 7}{0.5} - 1 \right) \right.$$
$$\left. + \left(\frac{2 \times 7}{10} 2.3 \log 0.656(25) \right) \right\}$$
$$= 0.229 \, \Omega$$

(14) R_{25} for $S = 15$

$$\frac{(0.00521)(1000)}{25(7)} \left\{ \left(2.3 \log \frac{8 \times 7}{0.5} - 1 \right) \right.$$
$$\left. + \left(\frac{2 \times 7}{15} 2.3 \log 0.656(25) \right) \right\}$$
$$= 0.19 \, \Omega$$

(15) R_{25} for $S = 20$

$$\frac{(0.00521)(1000)}{25(7)} \left\{ \left(2.3 \log \frac{8 \times 7}{0.5} - 1 \right) \right.$$
$$\left. + \left(\frac{2 \times 7}{20} 2.3 \log 0.656(25) \right) \right\}$$
$$= 0.17 \, \Omega$$

(16) R_{25} for $S = 25$

$$\frac{(0.00521)(1000)}{25(7)} \left\{ \left(2.3 \log \frac{8 \times 7}{0.5} - 1 \right) \right.$$
$$\left. + \left(\frac{2 \times 7}{25} 2.3 \log 0.656(25) \right) \right\}$$
$$= 0.158 \, \Omega$$

From the above results tabulate the calculations in the following form:

No. of anodes	Anode spacing			
	10	15	20	25
R_{10}	0.472	0.407	0.37428	0.355
R_{15}	0.343	0.29	0.2635	0.248
R_{20}	0.272	0.2275	0.205	0.192
R_{25}	0.229	0.19	0.17	0.158

Plot the data and obtain the number of anodes (N) *vs* resistance curves for a soil of 1000 ohm-cm resistivity and backfill dimensions of 0.5 ft diameter and 7 ft long.

The following example shows how the design curve can be used to obtain the resistance of impressed current and galvanic anodes. Separate charts are constructed for impressed current ground-beds and horizontal ground-beds. With each chart the following design information is provided:

(1) Anode dimensions
(2) Backfill dimensions
(3) Anode resistance
(4) Soil resistivity
(5) Spacing of anodes
(6) Internal resistance of anode
(7) Type of anode

The charts can be prepared for special applications by utilizing the data and working out the resistance according to the method given above.

5.32.2 CURRENT REQUIREMENTS

Example to illustrate the total current requirements for a bare 1 mile section of a 10(3/4") OD pipe. The current density required to protect the pipe is 50 mA/ft^2.

Solution:
For a bare steel structure, a potential of -0.85 V on it provides a reasonable degree of cathodic protection. The structure is polarized by a known amount of current until a potential of -0.85 V$_{Cu-CuSO_4}$ is acquired by the pipe.

Step 1 – Estimation of current requirement

$$\left[\frac{10.75}{12} \right] \times (3.14) \times 5280 \, \frac{ft}{mile} \left(0.005 \, \frac{A}{ft^2} \right)$$
$$= 0.895 \times 3.14 \times 5280 \times 0.005 \, A$$
$$= 74.2 \, A$$

(An approximate current density on the basis of experience is first selected.)

Apply a current of approximately 74.2 A to the pipe. In order to ensure the accuracy of the estimation, a pipe potential *vs* time polarization curve is plotted as shown in Fig. 5.52. By extrapolation of the polarization curve, the maximum potential achieved by the predetermined current is noted. If, for instance, by applying 74.2 A of current, the potential achieved is -0.8386, the additional current needed to achieve -0.85 is worked out and added to the approximated current to obtain an accurate value of current requirements. For example, the amount of current approximated raised the potential of the structure to -0.838 rather than -0.85 V, which is the required potential to achieve cathodic protection. If the voltage before cathodic protection is applied is -0.6 V, then the total voltage change is

$$\Delta E = -0.838 - (-0.6) = -0.238 \, V$$

The average potential change per ampere is

$$\frac{\Delta E}{\Delta I} = \frac{0.238}{74.2} = 0.0032 \frac{V}{A}$$

Amount of current required to raise the potential from $-0.85 - (-0.838) = 0.01$ V.

Therefore, additional amount of current required:

$$= \frac{0.01 V}{0.0032 V/A} = 3.125 \, A$$

Total current $= 74.2 + 3.125 = 77.325$ A.

Additional current needed to raise the potential from -0.838 V to -0.85 V is 0.892 A. The total current required is, therefore,

$$74.2 \, A + 0.892 \, A = 75.092 \, A$$

which is not too far away from the approximation (74.2 A).

5.32.3 DETERMINATION OF COATING RESISTANCE OF A PIPE

Determine the coating resistance of 4(6/8)" OD pipe, 4 miles long.

Solution:

Data provided:

Location	Volts (ON)	Volts (OFF)	DF
1	−1.10	−0.83	0.27
2	−1.45	−1.19	0.26
3	−1.30	−1.00	0.30

The current measured is 0.05 A.

(A) Calculate the average value of ΔE

$$\Delta E = \frac{0.27 + 0.26 + 0.30}{3}$$

$$= \frac{0.83}{3}$$

$$= 0.276\,\text{V}$$

(B) Calculate the area of the pipe

$$\text{Area} = \pi DL = \frac{4.625}{12} \times (\pi)(4)(5280)$$

$$= 25559.6\,\text{ft}^2$$

(C) Calculate the resistance, R

$$R = \frac{\Delta E}{I}$$

$$= \frac{0.276\,\text{V}}{0.05\,\text{A}}$$

$$= 5.52\,\text{ohms}$$

The resistance of the coating for one average square feet is $(5.52\,\text{ohms})(25559.6) = 141088.99\,\text{ohms·ft}^2$.

Illustrative Problem 1

From the design chart for 3″ diam × 5′ long anode, determine the resistance of 25 graphite anodes at 10 ft spacing in a 3000 ohms-cm soil.

Solution:
(1) Suppose the resistance of 25 anodes at 10 ft spacing in 1000 ohm-cm soil is 0.26 ohms from a design chart.
(2) The resistance in 3000 ohm-cm soil

$$R_{(3000)} = 0.26 \times \frac{3000}{1000} = 0.78\,\text{ohms}$$

Illustrative Problem 2

Design an impressed current system to protect a coated pipeline 4 mile long, 6(5/8)″ OD in a soil of 2000 ohm-cm resistivity. Graphite anodes 3″ × 5′ are to be used. The back voltage between the pipeline and ground-bed is 3.0 V.

Data

(1) In order to cause a potential shift (ΔV) of 0.2 V, (−0.65 to −0.85), 0.13 A of current is applied as required by the current requirement test (extrapolation method).
(2) The coating has 2% holidays.
(3) A current density of 3 mA/ft² is to be applied.
(4) Resistivity of anode bed = 2000 ohm-cm.

The following values are to be calculated:

(1) Current requirement.
(2) Pipe-soil resistance.
(3) Maximum allowed circuit resistance.
(4) Wire resistance.
(5) Anode bed resistance.
(6) Anode bed size.
(7) Weight of the backfill and its volume.

Solution:
Current required

(1) From the data given above, surface area of the pipe

$$= \left[\frac{6.6}{12}\right](\pi)(4)(5280) = 36\,474\,\text{ft}^2$$

(2) Percentage of uncoated pipeline = 2%.
(3) Current density required, 0.003 A/ft². Total current needed = $36474 \times 0.02 \times 0.003 = 2.188$ A. A rectifier, possibly 12 V, 4 A may be installed. However, an 18 V, 6 A rectifier is recommended to make allowances for future requirements.
(4) Pipe-to-soil resistance from the given data:

$I =$ current applied for voltage drop
$\quad = 0.13$ A
$\Delta E =$ voltage drop $= 0.2$ V
$$R = \frac{\Delta E}{I} = \frac{0.2}{0.13} = 1.5\,\text{ohms}.$$

(5) Calculations of maximum allowable circuit resistance in this case

Back voltage
= 3.0 V (provided in the data)
Rectifier voltage
= 18 V (available from the rectifier)
$\Delta E = 18 - 3$
= 15 V (maximum rectifier output)
$I = 6$ A (maximum current output)
Maximum allowable circuit resistance:

$$R = \frac{\Delta E}{I} = \frac{15}{6} = 2.5 \text{ ohms.}$$

(6) Calculation of resistance of wire

Wire selected (gauge of wire) = 6
Resistance of the wire
= 0.410×10^{-3} ohms (from tables)
Length = 60 ft
Additional length for safety
= 10% of the original = 6 ft
Total length of wire = 66 ft
Resistance of the wire
= 0.259×10^{-3} ohms

$$(0.259 \times 10^{-3})66 \text{ ft} = 0.017 \text{ ohms.}$$

(7) Anode bed resistance

Maximum circuit resistance = 2.5 ohms
Pipeline-soil resistance = 1.5 ohms
Wire resistance = 0.017 ohms
$R_{(\text{anode bed})}$ = maximum circuit
resistance − pipe to soil resistance
− wire resistance
= $2.5 - 1.5 - 0.017 = 0.983$ ohm

in a soil of 2000 ohm-cm.
The resistance in a soil of 1000 ohm-cm shall be

$$0.983 \times \frac{1000}{2000} = 0.491 \text{ ohms.}$$

(8) Calculation of the anode bed size
Suppose 25 anodes are placed at 10 feet spacing.

The resistance of 25 anodes from the design chart (suppose) = 0.26 ohms in 1000 ohm-cm soil.

The resistance of 25 anodes (anode bed) in a soil of 2000 ohm-cm

$$= 0.26 \times \frac{2000}{1000} = 0.520 \text{ ohms.}$$

Internal resistance (from data)
= 0.520 ohm/25 = 0.020 ohm.
The header cable resistance (assume)
= 0.017 ohm
Total anode bed resistance = Resistance of 25 anodes bed + internal resistance + header wire resistance.
Total anode bed resistance:

$$= 0.52 + \underset{\substack{\text{(internal} \\ \text{resistance)}}}{0.020} + \underset{\substack{\text{(header-wire} \\ \text{resistance)}}}{0.017}$$
$$= 0.557 \text{ ohms}$$

(9) Total resistance of circuit

Pipe-to-soil resistance
= 1.5 ohms (given in data)
Anode bed resistance = 0.557 (Step 8)
Lead wire resistance = 0.017 (Step 6)
Total = $1.5 + 0.557 + 0.017 = 2.077$ ohms
It is obviously less than the maximum permissible resistance of 2.5 ohms (as shown in Step 5).

(10) Weight of backfill

Anode dimension (given) = $3'' \times 5'$
= 0.245 ft^3
Backfill dimension = $8'' \times 7' = 2.446$ ft^3
Number of anodes is 25 at a spacing
of 15 ft

Backfill volume $= 25\,(2.45 - 0.246)$

$= 55.1 + 20\%$ for

over-design

$= 66.12\,\text{ft}^3$

Weight of backfill at $70\,\text{lb/ft}^3$

$= 70 \times 66.12 = 4628\,\text{lb/ft}^3$

Illustrative Problem 3

Calculate the expected current output of a single 48 lb Galvamog magnesium alloy anode. The size of the backfill package is $10'' \times 40''$. The steel has been polarized to a potential of -0.85 V. The resistivity of soil is 2000 ohm-cm. The solution potential of Galvamog is -1.75 V. Calculate the life of 48 lb magnesium anode.

Solution:

(a) First determine the resistance of the anode from the given data:

$$d = 10'' = \frac{10}{12}\,\text{ft} = 0.83\,\text{ft}$$

$$L = 40'' = \frac{40}{12}\,\text{ft} = 3.33\,\text{ft}$$

$$R = \left[\frac{0.00521}{L} \times \rho\right]\left[\ln\frac{8L}{d} - 1\right]$$

$$= \frac{0.00521}{3.33}(2000)\left[\ln\frac{8 \times 3.33}{0.83} - 1\right]$$

$$= 7.72\,\text{ohms}$$

(b) The driving potential
$= -1.75 - (-0.85) - (-0.10) = 0.8\,\text{V}$.

(c) According to Ohm's law:

$$I = \frac{E}{R}$$

$$I = \frac{0.8}{7.72} = 0.103\,\text{A}$$

The life expectancy is determined by

$$\frac{(\text{Weight of anode})(0.116)(0.50^*)(0.85^{**})}{0.103}$$

$$= \frac{(48)(0.116)(0.50)(0.85)}{0.103}$$

$$= 22.97\,\text{years}$$

*consumption rate $= \text{A/year/lb}$ (inverse of theoretical consumption rate)

**efficiency of anode

Illustrative Example 4

The interior of a tank is to be protected. The tank contains 5000 barrels of salt water. The following data is provided:

(1) Water level $= 25.0\,\text{ft}$
(2) Height of tank $= 30.0\,\text{ft}$
(3) Diameter of tank $= 40\,\text{ft}$
(4) Water level is maintained at 25 ft
(5) Current density $= 4\,\text{mA/ft}^2$
(6) Resistivity of water $= 10\,\text{ohm-cm}$
(7) Length of 2 AWG wire $= 130\,\text{ft}$

Resistance $= 0.162 \times 10^{-3}\,\text{ohms/ft}$

From the above data, calculate the following:

(1) Current requirement.
(2) Current output per anode if $3'' \times 5'$ graphite anodes are used and the anode current density is $2.0\,\text{A/ft}^2$.
(3) Number of anodes required.
(4) Anode to electrolyte resistance.
(5) Resistance of lead wire (R_w) 130 ft long.
(6) Resistance of anode.
(7) Potential drop in the circuit.
(8) Size of the rectifier to be selected.

Solution:
(1) Firstly, the area of the tank in contact with water is calculated (wetted area, A_w)

$$A_w = \frac{\pi d^2}{4} = \pi d L$$

$$= \frac{\pi(40)^2}{4} + \pi(40)(25)$$

$$= 1256 + 3140$$

$$= 4396\,\text{ft}^2$$

The current requirement is:

$$I = 4396 \times 0.004 = 17.6\,\text{A}.$$

(2) Current output of each anode.
From the previous calculation we have observed that the current requirement is 13.2 A.
The area of the anode surface is (A_a)

$$A_a = \frac{d\pi L}{12} = \frac{3\pi}{12} \times 5$$

$$= 3.92\,\text{ft}^2(\sim 4\,\text{ft}^2)$$

The current output is, therefore,

$$C_a = Cd_a \times A_a \; (C_a = \text{current output of anode}, A_a = \text{area of anode})$$

$$Cd_a = 2.0\,\frac{\text{A}}{\text{ft}^2} \times 3.92$$

$$= -8.0\,\frac{\text{A}}{\text{anode}} \; (Cd_a = \text{current density of anode})$$

As the total current requirement is 17.6 A, the number of anodes required is:

$$\frac{17.6}{8} \simeq 2\,\text{anodes}$$

(3) Anode to electrolyte resistance use D'Wight equation as before:

$$R = \frac{0.00521}{L}\rho\left[\ln\frac{8L}{d} - 1\right]$$

$$= \frac{0.00521}{5}(10)\left[\ln\frac{40}{0.25} - 1\right]$$

$$= 0.0425\,\text{ohms}$$

(4) Calculation of resistance of lead wires
The resistance of the lead wires is the estimated average length × resistance of 2 AWG wire. The known resistance is 0.162 × 10^{-3} ohms/ft.

$$\text{Resistance} = 130 \times 0.162 \times 10^{-3}\,\text{ohms}$$

$$= 0.021\,\text{ohms}.$$

(5) The resistance of the anode. We have now all the necessary information to calculate the resistance of the two anodes.

The resistance of the circuit is given by

$$\frac{1}{R} = \frac{4}{R_w + R_a}$$

Here

R_w = resistance of the wire which is calculated to be 0.02 ohms

R_a = resistance of the anode to electrolyte which has been determined to be 0.0425 ohms

Therefore,

$$\frac{1}{R} = \frac{4}{0.021 + 0.0425}$$

$$= \frac{4}{0.0635}$$

$$= 62.992$$

$$R = 0.016\,\text{ohms}$$

(6) The potential drop is determined by Ohms' law:

$$E = I \times R$$

$$= 17.6 \times 0.016$$

$$= 0.282\,\text{V}$$

The size of the rectifier to be used must be a 30 V–30 A circuit rectifier would be quite suitable.

Illustrative Example 5
Calculate the current output and life expectancy of a Galvomag anode from the following data:

Backfill package size
$= 8'' \times 32.5'' \; (0.65' \times 2.71')$
Potential of the polarized structure $= -0.95\,\text{V}$
Soil resistivity $= 2000\,\text{ohm-cm}$
Solution potential of Galvomag anode
$= -1.75\,\text{V}$
Weight $= 48\,\text{lbs}$
Polarization potential $= -0.10\,\text{V}$

Solution:

Apply the following relationship to obtain the resistance:

$$R = \frac{0.00521}{L}\rho\left[\ln\frac{8L}{d} - 1\right]$$

Inserting the value

$$R = \frac{0.00521}{2.71}(2000)\left[\ln\frac{21.7}{0.67} - 1\right]$$

$$= 9.53\,\text{ohms}$$

The driving potential (E_D)

$$E_D = -1.75 - (-0.95) - (-0.1)$$

$$E_D = 0.7\,\text{V}$$

Determine the current I, by using $E = IR$

$$I = \frac{0.700}{9.53} = 0.073\,\text{A}$$

$$\text{Life expectancy} = \frac{(48)(0.116)(0.50)(0.85)}{0.73}$$

$$= 32\,\text{years}$$

5.33 IMPORTANT FORMULAE IN CATHODIC PROTECTION CALCULATIONS

(1) Series Circuit

$$R = R_1 + R_2$$

(2) Parallel Circuit

$$\frac{1}{R} = \frac{1}{R_1} + \frac{1}{R_2} = \frac{R_1 + R_2}{(R_1)(R_2)}$$

(3) $\rho = \dfrac{RA}{L}$

where

ρ = resistivity in ohm-cm
R = resistance
L = length between two points
A = area

(4) $\rho = \dfrac{RA}{L}$

where L = length between two points in the soil or metal surface.

(5) Four Pin, DC Method for soil resistivity

$$\rho = 2\pi AR$$

A = distance between two points, in cm
R = resistance between two points

(6) Four Pin, AC Method for soil resistivity

$$\rho = 191DR$$

D = pin spacing, feet
R = instrument (potentiometer reading)
ρ = resistivity

(7) Life of Mg anode (years)

$$L_{Mg} = \frac{57.08 \times W}{I}, \quad L_{Zn} = \frac{38.24 \times W}{I}$$

where

L_{Mg} = life of magnesium anode
L_{Zn} = life of zinc anode
W = anode weight (lb)
I = current (mA)

(8) Driving potential of an anode

$$= \text{Solution potential} - \text{Potential of polarized structure} - \text{Polarization potential} (-0.100)$$

(9) Current output of Mg anode and Zinc anode (coated)

$$I_{Mg} = \frac{150\,000\,fY}{\rho}, \quad I_{Zn} = \frac{50\,000\,fY}{\rho}$$

where

I_{Mg} = current output of Mg anode (mA)
I_{Zn} = current output of Zn anode (mA)

ρ = soil resistivity, ohm-cm
f = factor from Table I
Y = factor from Table II

(Condition: The above equations apply if the soil resistivity is above 500 ohm-cm and the distance between the anode and structure is not more than 10 ft.)

(10) E. D. Sundae equation for resistance to earth.

(a) Using multiple vertical anodes

$$R_V = \frac{\rho}{191.5\,NL}\left[2.303\log_{10}\frac{4L}{d} - 1\right.$$
$$\left. + \frac{2L}{s}\times 2.303\log_{10} 656\,N\right]$$

(b) H. E. D'Wight equation for resistance to earth for a single anode (horizontal)

$$R_h = \frac{0.00521}{L}(\rho)\left[2.303\log_{10}\frac{4L}{10}\right.$$
$$\left. + 2.303\log_{10}\frac{L}{h} + \frac{2h}{L} - 2\right]$$

where

R_V = resistance of vertical anode
R_h = resistance of horizontal anode
L = length (ft)
S = anode spacing, ft
h = depth from surface to center

(c) D'Wight equation for single vertical anode

$$R_V = \frac{0.00521}{L}(\rho)\left[2.3\log\frac{8L}{d} - 1\right]$$

(d) Sundae's equation for multiple vertical anodes

$$R = \frac{0.00521}{NL}(\rho)\left[\left(2.3\log\frac{8L}{d} - 1\right)\right.$$
$$\left. + \left(\frac{2L}{S}\,2.3\log 0.656N\right)\right]$$

where N is the number of anodes

(e) R_V (Total resistance of vertical anode) = R_1 (Resistance to vertical anode-to-backfill) + R_2 (Resistance of vertical anode to earth)

(f) Working diameter of magnesium anode

$$\text{Area} = \frac{\pi d^2}{4}$$

(g) Internal resistance of the anode to backfill

$$R_I = R_{V(\text{anode})} - R_{\text{backfill column}}$$

(h) Resistance of a single graphite high silicon cast iron anode installed vertically (no backfill)

$$R_I = \frac{K\rho}{1000}$$

where

K = constant
ρ = soil resistivity

(i) Resistance of graphite of high silicon cast iron anode with and without backfill

$$R = \frac{K\rho}{1000}\times NY$$

where

ρ = soil resistivity
N = number of anodes
Y = spacing factor from the curve
R = resistance in ohms

(11) Current requirement on a coated pipe

$A_s\times$ % uncoated pipe $\times$ current density
A_s = surface area of pipe
% uncoated pipeline current density

(12) Wire resistance

Length of wires + 10% safety factor + resistance of wire ohms/ft

(13) Total volume of backfill needed

= (Number of anodes) (Volume of backfill − Anode volume) + 20%

(14) (a) Induced emf

$$= \frac{\text{Secondary winding} \times \text{Applied voltage}}{\text{Primary winding}}$$

(b) Efficiency of rectifier

$$= \frac{\text{DC volts} \times \text{DC amps} \times \text{Seconds} \times 100}{Kh \times 3600 \times \text{Revolutions}}$$

where

Kh = meter constants

or $\dfrac{\text{DCV} \times \text{DCA}}{\text{Input watts}} \times 100$

Input Watts = $K \times N \times 12$
(N = number of dial revolutions in a 5 min period)

(15) Number of anodes (N)

$$= \frac{\text{Total current requirement}}{\text{Current output of one anode (A)}}$$

(16) Rectifier current rating AC

$$I_{ac} = \frac{E_{dc} \times I_{dc}}{F \times E_{ac}}$$

where

I_{ac} = AC current (A)
E_{dc} = DC volts
I_{dc} = DC current (A)
F = rectifier efficiency
E_{ac} = AC volts

(17) (a) Number of anodes required based on the anode consumption rate

$$N = \left(\frac{Y \times I \times C}{W} \right)$$

where

N = number of impressed current anodes

Y = impressed current system design life, years
I = total current required in A
C = anode consumption rate in kg/A-yr
W = weight of anode in kg

(b) Number of anodes based on weight (lb)
Number of anodes (lb)

$$= \frac{\text{Total weight of anode material}}{\text{Weight of one anode}}$$

(c) Number of anodes based on current output, N

$$N = \frac{\text{Total current output (A)}}{\text{Output of one anode (A)}}$$

(use manufacturer's data)

(d) Current output of one anode (A)

$$= \frac{\text{Driving potential of the anode (V)}}{\text{Resistance of the anode (ohms)}}$$

(18) Structure to electrolyte resistance

$$R_s = \frac{V_{on} - V_{off}}{I_{on}}$$

where

V_{on} = potential (on)
V_{off} = potential (off)
I_{on} = current applied to give V_{on}

(19) Maximum circuit resistance

$$R_{max} = \frac{E_D}{I}$$

where

E_D = driving potential of anode
I = current requirement

(20) Allowable ground-bed resistance

$$R_{agb} = R_{max}(R_s + R_{Lw})$$

where

R_s = structure to electrolyte resistance
R_{Lw} = lead wire resistance

QUESTIONS

A. MULTIPLE CHOICE QUESTIONS

Select one correct answer for the following questions:

1. A bare structure requires more current than a coated structure because

   [ ]   the coated structure corrodes rapidly
   [ ]   the bare structure has a more negative potential than a coated structure
   [ ]   a coated structure is more rapidly polarized than a non-coated structure
   [ ]   the bare structure takes a very long time to polarize than a coated structure

2. A backfill is used *around* an anode

   [ ]   to provide a uniform environment around the anode
   [ ]   to accelerate the rate of consumption of the galvanic anode
   [ ]   to increase the anode-to-earth resistance
   [ ]   to increase the magnitude of the current which is to be provided to the structure

3. Cathodic protection in a metallic structure is achieved by

   [ ]   polarizing the cathode to the open circuit potential of the anode
   [ ]   polarizing the anode to the open circuit potential of cathode
   [ ]   shifting the potential of the structure to less negative values
   [ ]   producing a film of oxide on the surface of the metallic structure to be protected

4. The following is the criteria for cathodic protection:

   [ ]   A shift in the pipe-to-soil potential in the negative direction by 0.40 to 0.50 V from the initial potential for bare structures
   [ ]   To achieve a pipe-to-soil potential of 0.85 V with respect to Ag–AgCl electrode
   [ ]   To achieve a potential of −0.85 V with respect to a copper sulfate electrode
   [ ]   To polarize the whole structure to the cathodic potential of the structure

5. In the *impressed* current *cathodic protection* system

   [ ]   the pulsating direct current goes from the positive (+) terminal of the rectifier to the ground-bed
   [ ]   the pipeline is the positive return or external circuit of the electrolytic cell
   [ ]   AC current is directed to a rectifier where a step-down transformer increases the voltage
   [ ]   it is not necessary to install a magnetic circuit breaker in the AC circuit

6. The following are the disadvantages of the rectifier ground-beds:

   [ ]   Larger driving voltages
   [ ]   Higher current outputs
   [ ]   Protection of larger and more expensive structures
   [ ]   High installation costs

7. The following is the H. B. D'Wight equation for resistance to earth of a single vertical anode:

   [ ] $R_V = \dfrac{\rho}{191.5\,NL}\left[2.303\log_{10}\dfrac{4L}{d} - 1 \right.$
   $\left. + \dfrac{2L}{S}\,2.3\log_{10}0.656N\right]$

   [ ] $R_V = \dfrac{0.00522\,\rho}{L}\left[32.303\log_{10}\dfrac{4L}{d}\right.$
   $\left. + 2.303\log_{10}\dfrac{L}{h} + \dfrac{2h}{L} - 2\right]$

   [ ] $R_V = \dfrac{0.00521}{L}\,\rho\left\{2.3\log\dfrac{8L}{d} - 1\right\}$

   [ ]   None of the above

8. The pipe-to-soil potential of a pipe is -1.35 V and the potential of the polarized structure -0.90 V. The resistance of the anode-to-earth has been found to be 0.150 ohm. The anode output would be

[ ] 5 A
[ ] 4 A
[ ] 3 A
[ ] 1.15 A

9. Cathodic protection systems can be described as

[ ] transferring corrosion from the protected structure to the anodes
[ ] transferring corrosion from the protected structure to the soil
[ ] effective when all current has stopped flowing in the system
[ ] effective when all piping is made sufficiently negative

10. The surface potential survey requires

[ ] the use of a high resistance voltmeter or potentiometer connected to the copper–copper sulfate reference electrode
[ ] the use of a rectifier
[ ] the use of a low resistance voltmeter
[ ] a strip-chart recorder

11. Which one of the following is a good indicator to ensure that the structure is receiving current?

[ ] The achievement of -850 mV$_{Cu-CuSO_4}$ with C. P. applied
[ ] A change of 200 mV from the original potential
[ ] The structure become passive
[ ] Polarized potential of -600 mV$_{Cu-CuSO_4}$

12. While determining the current requirements for a bare structure, it is essential to

[ ] place the electrode over the structure directly

[ ] place the electrode 100 ft from the structure
[ ] place the electrode contacts far enough until there is *no* increase in the negative reading
[ ] place the electrode always 50 ft away from the structure

13. The driving potential is the

[ ] difference between the 'on' value and the 'off' value of potentials
[ ] difference between the 'on potential' and 'static potential'
[ ] difference between the solution potential and polarization voltage
[ ] None of the above

14. The total resistance of the vertical anode to earth is

[ ] resistance of the anode to backfill + resistance of the backfill to the soil
[ ] resistance of the anode to backfill − resistance of the backfill to soil
[ ] resistance of the backfill to soil + resistance of the pipe
[ ] None of the above

15. A low resistance backfill is specified in anode bed design, because it

[ ] increases the area of contact between anode and the soil
[ ] decreases the area of contact between the anode and the soil
[ ] provides a lower anode to earth resistance
[ ] has a higher current output

16. A deep well anode bed design is selected because

[ ] the soil has a higher resistance at a greater depth
[ ] the soil is more aerated at a higher depth compared to a lower depth

[ ] the soil has a high resistance near the surface and a low resistance at a greater depth

[ ] it can be useful in congested areas and will

17. Which of the following is true?

[ ] A remote ground-bed protects a smaller area of pipe than a close ground-bed

[ ] A remote ground-bed protects a larger area of the pipe

[ ] Soil nearest to the anode is most negatively charged

[ ] The further is the soil away from the anode, the more positively charged it becomes

18. Which of the following is true for current originating from a rectifier?

[ ] Return to the rectifier through the pipeline

[ ] Travel to the metallic structure and to get discharged from the metallic structure to the ground

[ ] The current flow reverses 240 times per second

[ ] None of the above

19. When the current enters the remote earth

[ ] the resistance of the soil increases

[ ] the resistance of the soil decreases

[ ] the soil offers no resistance at all

[ ] there is an increase in the potential of the pipe

20. A stray current is indicated if

[ ] a positive potential is indicated at the point of corrosion

[ ] a positive potential is indicated at the point of protection

[ ] the current flows from the electrolyte, like soil, into the metal

[ ] very small magnitude of current flows between the anode and the protected structure

21. The best remedy to minimize stray current corrosion is to

[ ] relocate the metallic structure

[ ] coat the metallic structure

[ ] bury the metallic structure deeper in the ground

[ ] install a metallic bond between the structure and the source of stray current

22. In measurement of soil resistivity by four-pin method

[ ] an alternate current is passed between the outer electrode and the resulting voltage drop is measured between these two electrodes

[ ] the galvanometer is adjusted to read zero by means of a potentiometer which is calibrated to read directly in ohms

[ ] a direct current is passed between the two outer electrodes and the voltage drop is measured between the inner electrode

[ ] the depth of pins is made to be the same as the distance between two electrodes

B. HOW AND WHY QUESTIONS

Explain why (very briefly) the following:

1) A bare structure requires approximately 10 000 times more current than a coated structure.

2) The resistivity of soil decreases with higher salt content.

3) The reference $Cu–CuSO_4$ electrode is placed directly over a coated pipe and away from the pipe if it is not coated.

4) Stray currents cause the uncontrolled corrosion of underground pipes.

5) Well-coated structures polarize more rapidly than bare structures, when cathodic protection is applied.

6) Long slender anodes are preferable over short squat anodes for high resistivity soils.

7) Cathodic current protects the outside of the pipes only.

8) The lowest resistivity soil is often the best location for placement of galvanic anode.
9) High resistance voltmeter must be used to make accurate measurement of pipe-to-soil potential.
10) Over-protection causes damage to the pipeline.

C. PROBLEMS

1. Calculate the current output of a single anode bed and its life. Assume a pipe-to-soil potential of 0.4 ohm. The following data is given:

   Driving potential $= 0.5$ V
   Resistance $= 7.18$ ohms
   Weight of Mg per bed $= 128$ lbs
   Consumption rate $= 0.116$ A-year/lb
   Efficiency $= 0.5$
   Utilization factor $= 0.81$

2. The open circuit potential of a magnesium anode is $-1.55\,V_{Cu-CuSO_4}$. It protects a steel tank polarized to a potential of $-0.950\,V_{Cu-CuSO_4}$. Estimate the driving potential of magnesium anode.
3. The open circuit potential of a single 48 lb Galvomag magnesium alloy anode is -1.75 V. The surrounding backfill has dimensions of $8'' \times 30''$. The anode has polarized the steel in a soil of resistivity 3000 ohms-cm to -0.85 V. Estimate the current output of the anode.
4. Calculate the life of the anode given in Problem (3), if it has an efficiency of 50% and a utilization factor of 0.85.
5. Calculate the expected current output of a 725 lb Galvalum anode, from the data given below:

   Resistivity of seawater $= 30$ ohms-cm
   Anode length $= 96''$
   Anode width $= 10'' \times 10''$
   Anode core $= 4''d$
   Steel polarized to -0.900 V
   Open circuit potential of anode $= -1.15\,V_{Cu-CuSO_4}$

Estimate anode radius from $c = 3.14 \times d$
Hint: $c = \pi d$
Working diameter $=$ Core diameter
$$+ \frac{\text{Anode diameter}}{2}$$

$$R = \frac{0.00521\,\rho}{L}\left[2.3\log\frac{8L}{d} - 1\right]$$

6. From the following data calculate the life of a magnesium anode:

   Weight of magnesium anode $= 50$ lb
   Number of available ampere hour per pound $= 500$
   Current output $= 35$ mA
   (8760 h in one year)

7. Find the number of 17 lb magnesium anodes and the spacing between the anode to protect a 10 000 ft of a bare $4''$ diameter pipeline in a corrosive soil having a resistivity of 800 ohms-cm. Assume a current requirement of $2\,mA/ft^2$ and an anode output of 100 mA per anode.
8. Determine the internal resistance from the anode to the outer edge of the backfill column (R_1) from the following data:

   Resistivity of backfill $= 8000\,\Omega$-cm
   Anode dimension $= 3'' \times 5'$
   Backfill column dimension $= 8'' \times 11'$

9. In Problem (8), R_1 is to be determined. The total resistance of a vertical anode-to-earth $R_v = R_1 + R_2$, where R_1 is the resistance of the backfill column to earth. If the soil resistivity is 2000 ohms-cm and the dimension of the backfill column are $0.50' \times 8'$, determine the total resistance of the anode-to-earth.
10. Calculate the total resistivity of the anode bed from the following data:

   Soil resistivity $= 1000$ ohms-cm
   Backfill resistivity $= 70$ ohms-cm
   Backfill dimensions $= 5'' \times 7'$

Anode weight = 16 lbs

Anode dimensions = 15″ × 5′

Hint: Use D'Wight's equation to determine the resistance of a single anode.

11. Determine the amount of current needed to protect a bare pipe, 3 miles long, 6″ OD. The estimated current density required for protection is 2 mA/ft^2.

12. From the data provided below, determine the coating resistance of a 5 mile, 10¾ OD pipe:

   a) Current output as recorded by ammeter = 0.50 A

   b) The following volts (on) and volts (off) reading were taken at three different locations:

Location	Volts (on)	Volts (off)
A	−1.10	−0.83
B	−1.45	−1.17
C	−1.28	−0.99

13. Calculate the surface area of a 10 mile long, 4–5 ft diameter, 2% uncoated pipeline, if the current density required for protection is 2 mA/ft^2.

14. Calculate the life of an anode bed and its current output from the following data:

   Driving potential of Mg anode = 6 V

   Resistance = 5.18 ohms

   Weight of Mg per anode
   = 5 × 32 = 120 lbs (5 anodes to a bed)

   Consumption rate = 0.116 A-year/lb

   Efficiency of Mg anode = 0.6

   Utilization factor = 0.85

15. In DC four-pin method, a direct current of 221 mA is passed between the two outer electrodes and a voltage drop of 191.8 mV is observed between the two inner electrodes. The pin spacing and anode depth is 7.5 inch. Determine the soil resistivity.

16. If $\rho = 2\,\pi\,AR$ and $A = 30.48\,D$, prove that $\rho = 191\,DE/I$.

   D = distance between electrode, E = volts,
   I = mA

17. An appropriate rectifier is to be selected for designing an impressed current system. The minimum current required has been estimated to be 2.36 A and the total circuit resistance is determined to be 2.54 ohms. Specify the nearest commercial size of the rectifier. (*Hint:* $E = I \times R$)

18. Calculate the minimum potential *vs* Cu–CuSO$_4$ reference electrode to which cadmium must be polarized for complete protection.

$$\left(K_{sp;Cd(OH)_2} = 2 \times 10^{-14}, \right.$$
$$\left. \text{where } K_{sp} \text{ is solubility product}\right)$$

19. Iron corrodes at a rate of 3 mdd in Arabian Gulf water. Calculate the minimum initial current density (A/m^2) necessary for complete cathodic protection.

RECOMMENDED LITERATURE ON CATHODIC PROTECTION

[1] Peabody, A.N. (2001). *Control of Pipeline Corrosion*. Blanchetti, R. L. ed. NACE Int., 2nd ed. Texas: Houston, USA.

[2] Von Beckmann, W., Schwenk, W. and Prinz, W. eds (1998). *Cathodic Corrosion Protection*. Houston: Gulf Publishing Company, Houston, Texas, USA.

[3] Edward Pope, J. (1996). *Rule of Thumb for Mechanical Engineers*. Gulf Publishing Company, Texas: Houston, USA.

[4] Kent Muhlabaner, W. (1966). *Pipeline Risk Management Manual*, 2nd ed. Gulf Publishing Company, Houston, Texas, USA.

[5] Parker, M.E. (1954). *Pipeline Corrosion for Cathodic Protection*. Gulf Publishing Company, Texas: Houston, USA.

[6] West, L.H. and Lewicki, T.F. (1974). *Cathodic protection design*. Civil Engineering Corrosion Control, Vol. 35, 61-73, AFCEC Technical Repot No. 74–76, Tyndall, Florida: AFB, USA.

[7] Uhlig, H.H. (2000). *Corrosion Handbook*. (Revised by Revie, R.W.) New York: John Wiley.

[8] Schultz, M. ed. (2000). *Corrosion and Environmental Degradation*. Vol. 1, Weinheim: Wiley VCH.

KEYWORDS

Anode It is the electrode in a corrosion cell which corrodes by passage of electrical current into the electrolyte.

Anode bed (ground-bed) The specific area where anodes are buried in soil and a backfill is placed around them.

Anode life The number of years taken by an anode to be consumed at a certain current output.

Anodic polarization That portion of polarization which takes place at the anode. The potential becomes more noble as anodic polarization proceeds.

Attenuation curves Curves obtained by plotting driving voltage (ΔE) and polarization potential (ΔV_p) on semi-log paper against distance are called attenuation curves.

Backfill The special soil placed around the anodes to provide uniform resistivity. The material used as backfill in an impressed current system is generally coke breeze whereas in the galvanic system the backfill is composed of a mixture of gypsum, bentonite and clay, the composition being dependent on soil resistivity.

Bond An electrical connection between two metallic structures.

Casing voltage profile A plot of the voltage *vs* the depth of an oil well casing.

Cathode The electrode in a corrosion cell through which conventional (positive) direct current leaves the electrolyte. Reduction takes place at the cathode processes, for example, oxygen reduction and hydrogen reduction takes place at the cathode. Electrons from the cathode are consumed at the cathode surface by $O_2 + H_2O$ forming OH^- ions, or by H^+ ions forming H_2.

Cathodic polarization That portion of polarization which takes place at the cathode. The potential becomes more negative as cathodic polarization proceeds.

Cathodic protection Elimination or reduction of corrosion of a metal surface by making it the cathode (negative), using either a galvanic or an impressed current.

Cell An anode and a cathode is an electrolyte.

Close ground-bed In this arrangement series of anodes are used. The length of the pipeline protected by a single closed anode depends on changing the potential of earth around the pipeline with respect to earth. The earth is made more positive in areas where protection is needed.

Coating resistance The electrical resistance of a coating to the flow of current.

Concentration cell An electrolyte cell in which the emf is the result of a difference in concentration of the electrolyte or active metal at the anode and the cathode.

Copper sulfate half-cell A reference electrode consisting of copper rod in a tube containing a saturated solution of copper sulfate. It is used for measurement of potential of buried structures.

Current density The amount of current required to protect a metallic structure. The magnitude of current varies with the environment.

Depolarization The reduction of a counter emf by removing or diminishing the cause of polarization.

Diode An electrical device with two electrodes, which allows electrons to pass through it in one direction only, hence converting AC to DC.

Drainage Conduction of positive electricity from an underground metallic structure by means of a metallic conductor.

Drainage (forced) Drainage applied to an underground metallic structure by impressed current or by sacrificial anode.

Drain point The point of connection between a cable and protected structure.

Efficiency of rectifier

$$\frac{\text{DC volts} \times \text{DC ampere}}{\text{Input watts}}$$

Electronic current Charge flow by electrons.

Energy content Maximum capability of current output of an anode expressed in either Ah/lb or lbs/A-year. For instance, the energy content of a standard magnesium anode is 1230 Ah/kg or 559 Ah/lb.

Faraday Faraday = 96 400 Coulombs per gram equivalent.

Forward bias The current proceeds in the forward direction and is blocked in the backward direction.

Full-wave rectification Rectification producing both AC waves in the DC output.

Galvanic anode A sacrificial anode that cause a spontaneous current flow.

Galvanic cell A corrosion cell formed by combination of metals differing in potential.

Ground-bed Anodes (impressed current or galvanic) buried in a soil with special backfill and connected to the positive terminal of a current source.

Impressed current anode Electrodes, such as scrap iron, titanium, lead-silver and silicon cast iron which provide current to an underground structure under cathodic protection.

Impressed current system A cathodic protection system which receives the required current for protection from a transformer–rectifier.

Insulation flanges Flanges employed to electrically isolate the over-ground pipeline from the underground pipeline.

Internal resistance of anode Resistance of anode to backfill − Resistance of backfill to earth.

Interrupter-current It is a device which momentarily stops current.

Ion current Charge transfer taking place by ions.

IR drop Voltage drop caused by a current flow (I) through the conductor of resistance R.

Isolating joint A joint or coupling between two lengths of pipe inserted to provide electrical discontinuity (insulation).

Junction boxes Connect electrical cables used in cathodic protection system.

Microampere 0.000001 ampere.

Mote earth A position in earth which offers no resistance to the flow of electrical current.

Overload protection It is a device which protects an electronic component from destruction by excess current.

Over-protection Current in excess of that required for protection.

Pipe-to-soil potential It is the potential of a pipe measured in a soil which acts as an electrolyte. Also, the corrosion potential of a metal in soil.

Polarization A shift in the potential of a metal in an electrolyte by passage of current flow.

Positive current Current flow of cations (e.g. Fe^{++} ions) or hypothetical positron (electron holes).

Potential criteria Attainment of a potential of $-0.85\,V$ vs $Cu-CuSO_4$ for steel structure in soil, such as steel pipe.

Potential decay (attenuation) The drop in the potential of a pipeline with increasing distance. For instance, the pipe-to-soil potential is at maximum near the area of influence where the soil is positive and decreases with the distance.

Potential gradient Potential difference per unit distance.

Potential 'on' Pipe-to-soil potential with switch on. It includes also IR drop.

Potential 'off' Pipe-to-soil potential without any IR drop. Varies with time.

Potential shift criterion Shift of potential required to completely protect a structure. A shift of 200–300 mV negative from the original value of potential brought about by external current to the structure to be protected is considered safe for protection.

Potential survey Survey of the potential of a pipeline with respect to soil over a defined distance. A plot of potential with respect of $Cu-CuSO_4$ electrode vs the distance is called a potential profile.

Rectifier An electrical device which converts AC to DC. A diode.

Resistance bond A metallic connection between the point of drainage of current from a structure to the origin of the current to avoid interferences.

Resistivity (soil) Resistance in ohms of a cm^3 of a material, measured across opposite faces

$$\left[\rho = R\,(l/a)\,,\text{where } l = \text{length}; a = \text{area}\right].$$

The unit is ohms-cm.

Reverse bias diode It has extremely high resistance and blocks all current flows.

Silicon diode Silicon anode is positive with respect to cathode.

Stray current Current flowing in the soil or water environment of a structure and arising mainly from electric power or traction installation. It is the uncontrolled current.

Structure-to-soil potential Potential in a buried structure and a non-polarizable electrode placed in soil.

Surface potential survey Survey of pipe-to-soil potential by two copper sulfate electrodes.

Tap transformer. A connection brought out of a winding at some point between its extremities to permit changing voltage or current ratio.

Test stations Special devices installed above in cathodic protection systems. They are used to measure pipe-to-soil potential, line current and current flow of a bond, to monitor potential measurements and also to measure stray current corrosion.

Transformer It is a device which makes it possible to transfer power from one circuit to another by mutual induction. In a step-up transformer, the energy transferred is from a lower voltage circuit to a higher voltage circuit. In a step-down transformer, the transfer of energy is the reverse of the step-up transformer.

CHAPTER 6

CORROSION CONTROL BY INHIBITION

6.1 INTRODUCTION

Corrosion phenomena are well-known in the petroleum industry and cause a maximum damage to oilfield equipment. Significant corrosion protection efforts have been made by petrochemical industries to prevent corrosion damage. The practice of corrosion prevention by adding substances which can significantly retard corrosion when added in small amounts is called inhibition. Inhibition is used internally with carbon steel pipes and vessels as an economic control alternative to stainless steels and alloys, and to coatings on non-metallic components. One unique advantage is that adding inhibitor can be implemented without disruption of a process. The addition of an inhibitor (any reagent capable of converting an active corrosion process into a passive process) results in significant suppression of corrosion. Corrosion inhibitors are selected on the basis of solubility or dispersibility in the fluids which are to be inhibited. For instance, in a hydrocarbon system, a corrosion inhibitor soluble in hydrocarbon is used. Two-phase systems composed of both hydrocarbons and water, utilize oil soluble water-dispersible inhibitors. Corrosion inhibitors are used in oil and gas exploration and production, petroleum refineries, chemical manufacturing, heavy manufacturing, water treatment and product additive industries. The total consumption of inhibitors in USA alone costs over one billion dollars annually.

6.2 SCOPE OF INHIBITOR

Corrosion control by use of inhibitors is extremely useful in many environments, however, there are certain exceptions, such as:

(a) equipment and components subjected to turbulent flow.
(b) systems operating above the stability limits of inhibitor.
(c) equipment subjected to high velocity, beyond 4 m/s.

However, adding inhibitors can raise the value of the critical flow rate above which erosion–corrosion starts.

6.3 EXAMPLES OF APPLICATION OF INHIBITORS

(1) Corrosion is a serious problem in all cooling water systems. The cooling water may be salt water (35 000 ppm TDS), brackish water (3000–5000 ppm) or fresh water (<300 ppm TDS). Inhibitor treatment is required for heat exchanger and distribution lines.
(2) Corrosion may be caused in the feedwater and boiler sections, if dissolved oxygen and

CO_2 is not removed by water treatment. Scales and deposits may also be formed by dissolved and suspended solids. Excessive alkalinity in boilers can lead to caustic cracking. High alkalinity is caused by high TDS (total dissolved solids) and alkalinity. External treatment includes demineralization and reduction of alkalinity, corrosion inhibition and biological control. Morpholine inhibitor is added as inhibitor for treatment of condensate corrosion.

(3) **Petroleum Industry.** Corrosion phenomena in the petroleum industry occur in a two-phase medium of water and hydrocarbon. It is the presence of a thin layer of water which leads to corrosion, and rigorous elimination of water reduces the corrosion rate to a negligible value. The inhibitors used in petroleum industry, both in production and refining are either oil soluble–water insoluble types or oil soluble–water dispersible compounds. New inhibitors are being developed. For instance, traditional filming amines are being replaced by several others, such as propylenedramine, and they work by adsorption on the surface.

(4) **Sour Gas Systems.** A major problem is encountered in steel pipelines in various sour gas environments. Chemical inhibition is one of the effective methods used to mitigate sulfide induced corrosion. Inhibitors containing alkylammonium ions are found to suppress corrosion effectively.

(5) **Potable Water Systems.** Corrosion is experienced in potable water transportation pipes of steels and cast iron. Inhibitors, such as $Ca(HCO_3)_2$ and polyphosphates are commonly used to combat corrosion.

(6) **Engine Coolants.** Inhibitors, such as $NaCrO_4$ (sodium chromate), borates and nitrites ($NaNO_2$) and mercaptabenzothiazole are widely used for protection of automobile engines. Chromates are a health hazard.

(7) **Packaging Industry.** For transportation of machinery, components and equipment by sea, vapor phase cyclohexylamine and hexamethylamine are used.

(8) **Construction Industry.** Corrosion of rebars in concrete poses a serious threat to building structures. Inhibitors, such as chromates, phosphates, nitrates and sodium metasilicates are used to suppress corrosion. Addition of sodium tetraborate and zinc borate has shown promising results. Commercial inhibitors, such as MCl 2022, MCl 2000 and Rheocrete 222, have shown good promise. Baker Petrolite, Du-Pont and NALCC are some of the known names in inhibitor manufacturing.

The above summary shows the wide range of applications of inhibitors in wide spectrum of environments.

6.4 IMPORTANT CONSIDERATION IN SELECTION OF INHIBITORS

(1) The magnitude of suppression of uniform and localized corrosion.
(2) Long range effectiveness.
(3) Effect on bimetallic coupling to other metals joined to the main system.
(4) Effect of temperature and concentration on the performance of inhibitors.
(5) Effect on the existing condition of the system to be protected. For instance, a metallic structure may be partly corroded; the important point would be to observe the effect of inhibitor on the corroded areas.
(6) Effect of inhibitor on heat transfer characteristics.
(7) Toxicity and pollution problems.
(8) Economically and technically competitive with other considered inhibitors.

6.5 IMPORTANT TERMS RELATED TO INHIBITORS

(1) **Additives.** A typical reagent for treatment of a corrosive fluid contains one main active gradient and one or more additives which assist in achieving the purpose of the reagent.
(2) **Solvent.** Keeps the active reagents in the liquid form and controls their viscosity.
(3) **Solubility.** The active reagent should have the ability to dissolve in the solute.

(4) **Dispersibility.** It is a measure of the reagent's ability to be transported by fluids or gases. The treating reagents must exhibit a high dispersibility.

(5) **Emulsion.** A heterogeneous system consisting of an immiscible liquid dispersed in another liquid in the form of droplets.

(6) **Surfactants.** It is a molecule with two components, each having different chemical properties, one end is polar (hydrophillic), and the other end non-polar (hydrophobic).

6.6 CLASSIFICATION OF INHIBITORS

Inhibitors may be classified as shown in Fig. 6.1. There are two major classes: inorganic and organic. The anodic type of inorganic inhibitors includes chromates, nitrites, molybdates and phosphates, and the cathodic type includes zinc and polyphosphate inhibitors. The film forming class is the major class of organic inhibitor as it includes amines, amine salts and imidazoilnes – sodium benzoate mercaptans, esters, amines and ammonia derivatives.

Inhibitors can also be classified on the basis of their functions. For instance, chromates and nitrates are called passivating inhibitors because of their tendency to passivate the metal surface. Some inhibitors, such as silicates, inhibit both the anodic and cathodic reactions. They also remove undesirable suspended particles from the system, such as iron particles, by precipitation. Certain types of inhibitors make the surrounding environment alkaline to prevent corrosion. Such inhibitors in the gas phase are called '*vapor phase inhibitors*,' and they consist of heterocyclic compounds, such as cyclohexylamine. These inhibitors are used within packing crates during transportation by sea.

6.7 DESCRIPTION OF INHIBITORS

6.7.1 ANODIC PROCESS AND ANODIC INHIBITORS

Consider an anodic dissolution process:

$$(1) \quad Fe \rightarrow Fe^{2+} + 2e \tag{6.1}$$

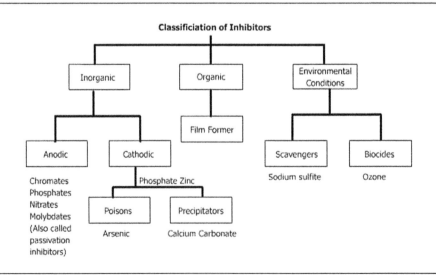

Figure 6.1 Classification of inhibitors

During the dissolution process in an aqueous media, the phenomena of adsorption and disruption of species is very predominant (equation (6.1)). It can be represented in a stepwise manner as:

(2) $Fe + OH^- = FeOH^-$ (ad)   (6.2)

(3) $\underline{FeOH^-} = \underline{FeOH} + e$ (6.3)

(4) $\underline{FeOH} = \underline{FeOH^+} + e^-$ (6.4)

(5) $\underline{FeOH^+} = FeOH^+$ (6.5)

[underline indicates that the species are adsorbed on the surface]

How corrosion slows down would depend on the electron transfer rate limiting step and the environment. Corrosion could be increased or decreased depending on the electrolyte constituents, such as halogen ions and benzoate, for instance, other ions. If, for example, the reaction $Fe + OH^- \rightarrow FeOH^-$ is suppressed by electrolyte constituents, corrosion is decreased, and if the reaction $FeOH^-/ads = FeOH^+$ is promoted, corrosion is accelerated. Species of $FeOH^+$ are at the kink sites at the edge of lattice and they are liable to dislodge easily, or at the dissolving edge of a terrace and are ready for active dissolution. All reactions shown above are subject to interaction with environment and the adsorption or deadsorption would depend on the environmental species and the effect of coordinating groups. Consider once again reaction (6.5) and suppose it is rate controlling or the slowest of all (reactions (6.1–6.4)). On increasing the concentration of the OH^- ion, ferrous hydroxide would be precipitated which would retard the rate of desorption and lead to suppression of corrosion. In anodic control, passivation of the surface is the controlling factor, such as in stainless steels. Inhibitors broaden the range of passivation.

A passive film formed on the metallic surface block will suppress corrosion. They are very thin films (50 to 100 Å). In aluminum, thin films of boehmite ($AlOOH$) and bayerite ($Al(OH)_3$) are formed which affect the corrosion process. Thick films are liable to breakdown and accelerate corrosion. Copper and nickel also have good passive film forming properties. The stifling of anodic sites by corrosion products, such as iron hydroxide, is observed in steel. Both the processes, passive film formation and stifling of anodic sites, lead to suppression of corrosion. However, the kinetics of film formation is different from anodic stifling, and the mechanisms of the process are beyond the scope of this text. Identification of thin film can be conveniently performed by transmission electron microscope. The number of metal ions dissolving is effectively reduced by anodic inhibitors. The effect of adding anodic and cathodic inhibitors on potential difference is shown in Fig. 6.2.

In terms of potential, addition of anodic inhibitors reduces the difference of potential between the anodic and cathodic sites, and consequently reduce the driving force for corrosion reaction to occur. The potential of the anode shifts to the potential of the cathode. The number of metal ions dissolving as a result of anodic reaction is reduced, and the potential shifts in a more noble direction. Figure 6.2 illustrates the effect of anodic and cathodic inhibitors on the potential and the driving voltage.

6.7.2 CATHODIC PROCESSES AND CATHODIC INHIBITORS

As the name suggests, the class of inhibitors which decrease the rate of cathodic reaction in a metal surface are called 'cathodic inhibitors.' To understand the mechanism, consider the two major cathodic reactions:

(1) $2H_2O + O_2 + 4e \rightarrow 4OH^-$ (oxygen reduction)

(2) $2H^+ + 2e \rightleftarrows H_2$ (hydrogen reduction)

It may be pointed out that both the cathodic reactions take place in several steps and the step with the slowest rate of reaction is generally the rate controlling step.

Consider a cathodic site where oxygen is diffusing to the metal/electrolyte interface. If an inhibitor, like zinc and magnesium, is added to the metal/electrolyte system, it would react with the hydroxyl ion and precipitate insoluble compounds which would, in turn, stifle the cathodic sites on the metal. In oxygen-induced corrosion,

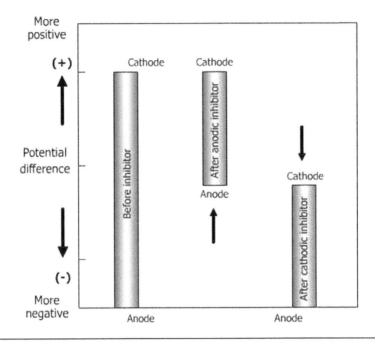

Figure 6.2 Effect of adding anodic and cathodic inhibitors on potential difference

the controlling step is the mass transfer of oxygen to the metal (cathode/electrolyte interface). Consider now that the metal (cathode/electrolyte interface) is in a stagnant condition. The oxygen is rapidly depleted in this condition and the reaction rate is slowed down. On the contrary, in a flowing system, a high rate of reaction would be maintained because of the continuous supply of fresh oxygen to the system. All factors, such as temperature, pressure and salt contents, which affect the solubility of oxygen would also affect the reaction rate (reduction of rate) of oxygen. On arrival of oxygen molecule to the metal (cathode/electrolyte interface), it must be absorbed.

Consider:

$$2\underline{H} \rightleftarrows H_2 \text{ (underline represents adsorption)}$$

A process of reduction follows the adsorption process:

(a) Oxygen molecule is reduced in two one-electron step to hydrogen peroxide which is reduced in a one-step reducing step to produce OH^- ions. The slowest step is rate controlling.

(b) An oxygen-oxygen bond is ruptured producing two chemisorbed oxygen atoms which transform to hydroxyl ion after each atom picks two electrons and a proton.

In either case, (a) or (b), the reaction is completed after desorption of hydroxide ion. The surface of an anode changes continuously to make fresh surface available for the anodic reactions to continue.

Consider now hydrogen evolution on the metal (cathode/electrolyte interface) in acid solution. Similar to oxygen reduction, hydrogen reduction also involves several steps, such as

(1) $H^+ \rightleftarrows H^+$

(2) $H^+ + e \rightarrow \underset{\text{(ads)}}{H}$

$As_2O_3 + 6H_{ads} \rightarrow 2As \downarrow + 3H_2O$

Despite several steps leading to the formation of molecular hydrogen, two steps are common:

(a) Adsorption of hydrogen ions
(b) Reduction of adsorbed hydrogen ion to atomic hydrogen

Hydrogen may then be formed either by

$$2H \rightleftarrows H_2$$

or $\qquad H + H^+ + O_2 \rightleftarrows H_2$

The reaction is completed by desorption of the hydrogen molecule

$$\underset{\text{(Adsorbed)}}{H_2} \quad \rightleftarrows \quad \underset{\text{(De-adsorbed)}}{H_2}$$

It is primarily the adsorption character which determines the rate of cathodic reaction and the degree of inhibition. The evolution of hydrogen is affected by an increase in the over-voltage. Salts of bismuth and antimony are added to obtain a layer of adsorbed hydrogen on the cathode surface. The over-voltage for hydrogen evolution is increased and the cathodic reaction is suppressed. In acids, the formation of hydrogen on the cathodic sites is retarded by the addition of arsenic, bismuth and antimony. For instance, if arsenic trioxide is added, it plates out to form arsenic on the cathodic sites.

Arsenic thus suppresses the cathodic reaction. Elements, like P, As, Sb and Bi, are called poisons and they retard the cathodic reduction of hydrogen. To avoid toxicity, organic inhibitors to achieve the same objectives have been used with success. There is, however, one inherent danger in adding poisons. A good example is H_2S, which poisons (prevents) the recombination of atomic hydrogen. H has an inherent tendency to diffuse into steel surface and cause blistering of the steel structure, if it is not converted to molecular hydrogen (Fig. 6.3). Another example is hydrogen sulfide cracking of steel equipment in oil drilling. It must be remembered that at the cathode, hydrogen ions and oxygen molecules consume electrons and the alkalinity of the electrolyte at the metal (cathode/electrolyte interface) is increased which leads to the precipitation of cathodic inhibitor on the surface and formation of a protective layer. Contrary to the situation shown for anodic inhibitors, the open circuit potential of the cathode shifts to the potential of the anode in the more negative direction (Fig. 6.2).

6.7.3 EFFECT OF INHIBITORS ON POLARIZATION BEHAVIOR

The effect of inhibitor on the metal/electrolyte system can be successfully evaluated from the polarization diagrams discussed in Chapter 2.

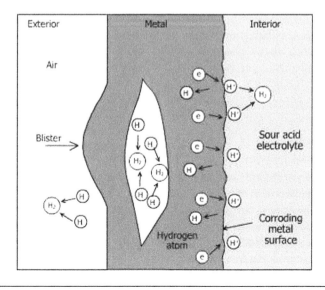

Figure 6.3 An example of hydrogen blistering

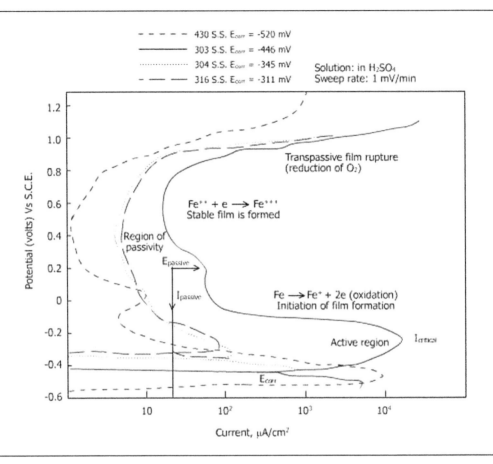

Figure 6.4 Polarization plots of ASTM standard type stainless steels showing the passive region and initiation of film formation

Consider, for example, a polarization diagram shown in Fig. 6.4 which shows the polarization plots of three different types of steels in 1N H_2SO_4. The electrochemical parameters, $I_{critical}$ (critical current density), $I_{passive}$ (passive current density) and E_{pp} (passivation potential), are sensitive to changes in composition of electrolyte. Clearly, Steel 303 is very active as it shows the highest $I_{critical}$, highest $I_{passive}$ and lowest passivation range. The effect of chloride ions and the behavior of the anodic polarization curve is shown in Figs 6.5 and 6.6. The critical current density increases from I_a (no chloride) to I'_a (with Cl^-). The three points, a, b and c, represent an anodic dissolution current corresponding to a reduction current (cathodic current). These points represent corrosion potential. At location a, active dissolution would proceed, whereas at point c,

the passivity is retained and, hence, corrosion would not occur. Now consider Fig. 6.6 which shows a polarization curve in a higher concentration of chloride. Passivation is not observed. The intersection of the cathodic curve with the anodic curve shows at point 'a' a larger corrosion current, and hence a higher corrosion rate. To clarify the effect of electrolyte composition on the anodic polarization behaviors, consider the three cases shown in Fig. 6.7.

Case I. The cathodic curve intersects the anodic polarization curve in the active region. The corrosion potential is in the active region, E_1. A high rate of metal dissolution is expected.

Case II. The cathodic curve intersects the anodic polarization curve at points E_2 and E'_2.

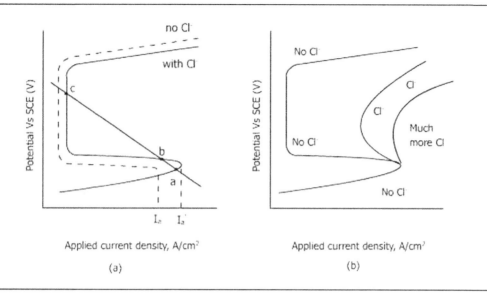

Figure 6.5 Schematic representation of the effect of a small chloride concentration on the polarization curve

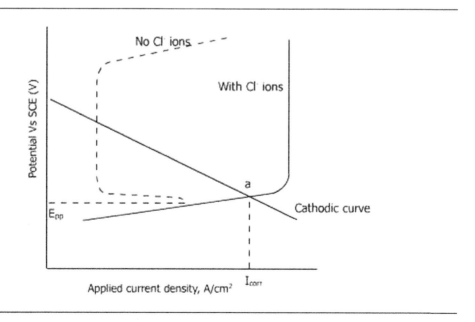

Figure 6.6 Schematic polarization curve in the presence of a large concentration of chlorides

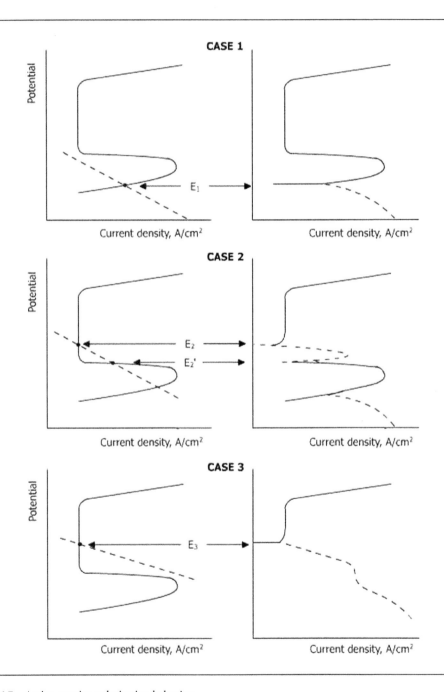

Figure 6.7   Active–passive polarization behavior

This shows an unstable state of potential which oscillates between E_2 and E'_2. It represents a case of serious corrosion.

Case III. The cathodic curve intersects at E_3 in the passivation range. The potential is stable. It represents a case of low corrosion.

The above three cases make it convenient to understand and interpret the effect of inhibitors. Consider inorganic inhibitors, like chromates, which passivate the surface (also called passivators). The effect of such inhibitors can be evaluated from the diagrams shown above. Examine Fig. 6.8, which shows the effect of concentration of a hypothetical inhibitor. For reasons stated earlier, Case I represents a situation where the concentration of inhibitor is sufficient to passivate the metal surface.

The effect of adding anodic and cathodic inhibitors is shown in Figs 6.8a and b. It is to be remembered that:

(a) a cathodic inhibitor shifts the corrosion potential in the negative direction; and

(b) an anodic inhibitor displaces the potential in the positive direction.

Figure 6.8a shows that I_{corr} of the uninhibited electrolyte is higher than the I'_{corr} of the electrolyte to which an anodic inhibitor is added. On adding a cathodic inhibitor (Fig. 6.8b), the I_{corr} is lowered compared to the I'_{corr} of the uninhibited electrolyte ($I_{corr} < I'_{corr}$) and E_{corr} is displaced in a more negative direction.

The effect of addition of mixed inhibitors (anodic and cathodic both) is shown in Fig. 6.9. Mixed inhibitors show the characteristics of both the types of inhibitors seen in Figs 6.8a and b.

As stated earlier, the corrosion potential is displaced in the positive direction on addition of anodic inhibitor and in the negative direction on addition of cathodic inhibitor. The mixed inhibitors protect the metal in three possible ways:

(a) Physical adsorption.

(b) Chemisorption.

(c) Film formation.

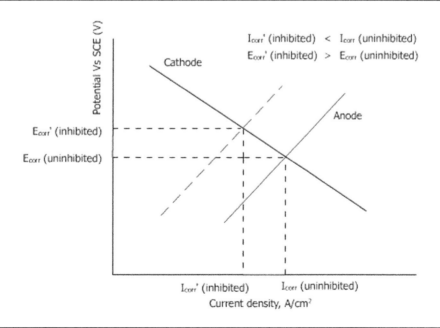

Figure 6.8a Effect of addition of anodic inhibitors

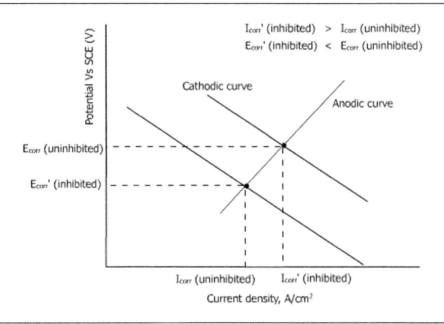

Figure 6.8b Effect of addition of cathodic inhibitors

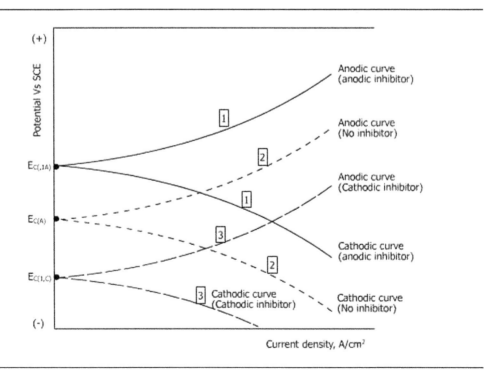

Figure 6.9 Effect of addition of inhibitors on potential

Physical adsorption is caused by electrostatic forces which exist between the inhibitor and the metal surface. The metal surface can be either positively charged or negatively charged. For instance, during cathodic polarization, metals get negatively charged due to the discharge of cations on the metal surface. The reverse happens on anodic polarization. When a metal surface is positively charged, negative charged (anionic) inhibitor is adsorbed on the metal surface (Fig. 6.10a). On the other hand, catonic species would not be adsorbed on a positively charged surface(Fig. 6.10b). Similarly anionic species would not be adsorbed on a negatively charged surface (Fig. 6.10c).

Physically adsorbed species can be removed from the surface by physical force such as increased temperature and increased velocity.

On the other hand, chemisorption results in a strong binding of the inhibitors with the metal surface. Positively charged ions in the presence of negatively charged ions can be adsorbed on a positively charged metal surface in the presence of negative ions, such as Cl^- ions, which act as a bridge between the two. The negative ions are adsorbed on the positively charged metal surface and the positive ions are attached synergistically to the dipole (a molecule with an unique distribution of charges, such as water). The adsorption of positively charged inhibitor on a negatively charged surface in the presence of negatively charged ions

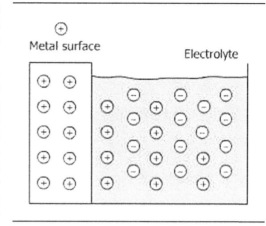

Figure 6.10b No adsorption of positively charged species with a positively charged metal surface

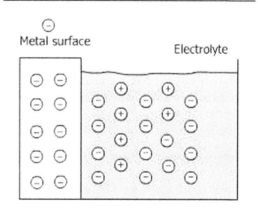

Figure 6.10c No adsorption of negatively charged species with a negatively charged metal surface

is called '*synergistic adsorption*' (Fig. 6.10d). This phenomena is caused by sharing of charges or charge transfer between the inhibitor species and the metal surface. The process of chemisorption is accelerated with time and temperature. While in the process of physical adsorption, de-adsorption may take place under adverse conditions. Chemisorption is not reversible, however, it is a more effective process.

The process of film formation is highly complex and the properties of films are dependent upon its thickness, composition, solubility, temperature and other physical forces. For instance, films of Al_2O_3 produced by anodizing

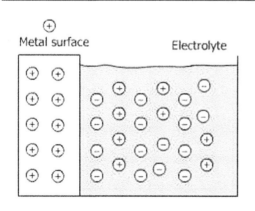

Figure 6.10a Adsorption of negatively charged species on a positively charged metal surface

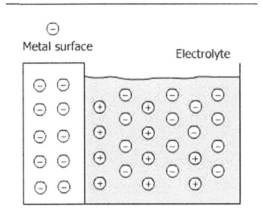

Figure 6.10d Adsorption of positively charged species with a negatively charged metal surface

are highly resistant to corrosion and they produce a very high resistance. Thick films may loose their adhesion due to mechanical damage, like air formed films on steel surface, if the thickness reaches beyond a critical point. Inhibitors of passivating type, like chromates and molybdates, produce passive films and resist corrosion. Passivating inhibitors, like $NaCrO_4$ and $NaNO_2$

do not require oxygen for passivation, whereas non-oxidizing types, like phosphates, tungstates and molybdates, passivate the surface only in the presence of oxygen. The action of anodic, cathodic and mixed inhibitors is summarized in Figs 6.11(a–c).

6.7.4 CHARACTERISTICS OF CATHODIC INHIBITORS

(A) Polyphosphates

The structure of a sodium polyphosphate molecule is

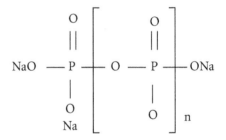

O = orthophosphate $n = 1$
pyrophosphate $n = 2$

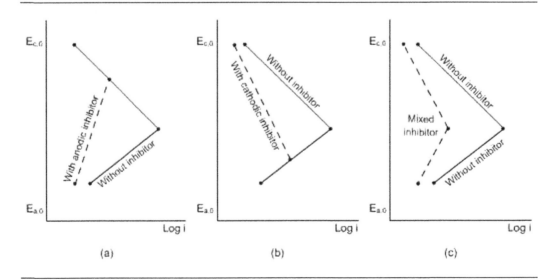

(a)   (b)   (c)

Figure 6.11 Action of corrosion inhibitors: (a) anodic inhibitors examples: chromate, nitrite, molybdate, tungstate, orthophosphate, silicate, benzoate, (b) cathodic inhibitors examples: $Cu(HCO_3)_2$, $ZnSO_4$, $Cr_4(SO_4)_3$, $NiSO_4$, polyphosphate, aminoethylene phosphate and (c) mixed inhibitors examples: organic inhibitors containing nitrogen and/or sulfur, like amines, triazoles, alkythiourea

They are also referred to as condensed or polymer phosphates. The chain length is determined by repetition of the portion of the structure denoted by *n*. The characteristic glassy structure is observed with longer chain lengths of the polymer. One good example of a glassy structure is sodium hexametaphosphate which is extensively used as an inhibitor.

The polyphosphate molecule bonds with divalent calcium and other ions to form positively charged colloidal particles which are attracted to the cathode and form a protective film. As some metal ions, such as iron, may also be adsorbed on the film, polyphosphate also shows a partial anodic behavior although basically they are cathodic inhibitors. The mechanism of corrosion prevention by polyphosphates is shown in Fig. 6.12.

Metallic ions, such as iron and copper affect polyphosphate. If an iron–polyphosphate complex is formed, dissolution rather than inhibition would proceed. If copper ions are present, a galvanic couple would be formed by the passage of copper ions from the polyphosphate film to the iron substrate, and corrosion would progress.

One major disadvantage of using polyphosphate is the hydrolysis of the phosphorus oxygen bond which converts the polyphosphate to orthophosphate, which is a proven weak inhibitor. A pH range of 6.5–7.5 is considered generally suitable to avoid reversion to orthophosphate which allows undesirable algae growth. Polyphosphates are blended with silicates and ferrocyanide to overcome this limitation. In actual operation, two or three inhibitors are blended to maximize the advantage of each other and to minimize their limitations. Frequently anodic and cathodic inhibitors are combined to maximize metal protection. The process is called 'synergistic blending.'

(B) Zinc

Zinc salts are well-known cathodic inhibitors in cooling water systems. They are, however, not used alone as the films formed by them are unstable. They are, however, used very effectively with polyphosphate as a synergistic blend to maximize the effect of inhibition. These synergistic blends

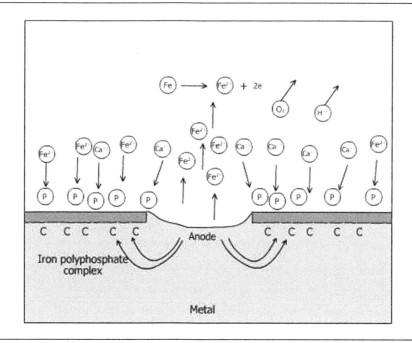

Figure 6.12   Formation of polyphosphates and their reaction with divalent ions in water

minimize the inhibitor concentration. Chromates can also be used in synergistic blends with polyphosphates. Zinc and chromate inhibitors are toxic and not environmentally friendly. They also need a careful pH control as they have a tendency to form scales above a pH of 8.00.

6.8 Inorganic Anodic Inhibitors

The addition of inorganic inhibitors causes suppression of electrochemical reaction at anodic–cathodic areas. Most of the times, inhibitors are used in a blended form. These inhibitors only react at an adequate level of concentration.

(A) Chromate Inhibitors

They are most effective inhibitors, but they are toxic and, hence, their application is restricted and is not advised. In industrial water, the threshold concentration is 120 mg/L. A high concentration is required if the systems to be inhibited contain bi-metal junctions or a high chloride concentration. They are oxidizers and raise the anodic current density above the limiting value needed for passivity. Chromate inhibitors contain either Na_2CrO_4 or $Na_2Cr_2O_7$. The protective passive film which is formed contains iron oxide and chromium oxide which makes the chromate inhibitors very effective (Fig. 6.13). Chromates are reduced to form chromium (III) according to the following reaction:

$$Fe \rightarrow Fe^{2+} + 2e \text{ (Oxidation of iron)}$$

$$CrO_4^- + 8H + 3e^- \rightarrow Cr^{3+} + 4H_2O$$
$$\text{(Formation of } Cr^{3+})$$

A mixed potential is created by the oxidation of iron and reduction of chromium and this potential lies somewhere in the passivation

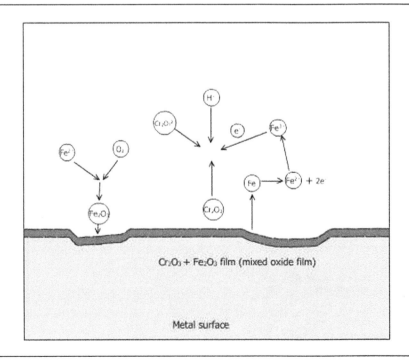

Cr₂O₃ + Fe₂O₃ film (mixed oxide film)

Metal surface

Figure 6.13 Formation of a mixed iron oxide and chromium oxide film

range, which accounts for the passivating effect of chromate inhibitor.

(B) Nitrites

They are effective inhibitors for iron and a number of metals in a wide variety of waters. Like chromates, nitrites are anodic inhibitors and they inhibit the system by forming a passive film with ferric oxide. These are environmentally-friendly inhibitors. Besides steel, nitrites also inhibit the corrosion of copper, tin and nickel alloys at pH levels 9–10. Chromate is an extremely effective inhibitor for corrosion prevention of aluminum alloys. Nitrites should not be used in open systems as they would oxidize to nitrates in the presence of oxygen.

$$2NO_2^- + O_2 = 2NO_3^-$$

Nitrites are not effective inhibitors. The presence of chloride and sulfate ions can damage the protective film formed by nitrites. They are often blended with borax in closed recirculating system.

(C) Nitrates

They protect solder and aluminum. They are not very effective and limited to use only in closes recirculating systems.

(D) Phosphate Inhibitors

Phosphate retards corrosion by promoting the growth of protective iron oxide films and by healing the defects in protective films. The effectiveness of phosphate inhibitor is reduced by chloride ions which damage the protective film formed by phosphate.

(E) Molybdates

Molybdenum is an alloying element which is known to increase passivation of stainless steels. Steels of type 316 contain molybdenum as a minor constituent and promote passivation. Sodium molybdate forms a complex passivation film at the iron anode of ferrous–ferric molybdenum oxide.

The passive film can only be formed in the presence of oxygen. They are very expensive and used with other inhibitors in synergistic blends.

(F) Silicates

They have been used with success for years in potable water systems. The complex silicon ion has a tendency to form negatively charged colloidal particles which migrate to anodic areas and form passive films. Silicates are strong anodic inhibitors and passive films can be formed even on the corroded surface. The monomeric silica does not provide any protection. In waters below pH levels of 6.0, the silicate used is $Na_2O \cdot 2SiO_3$ and with a pH greater than 6.0, it is $Na_2O_3 \cdot 3SiO_3$. Silicate inhibitors are also useful to prevent red water formation in plumbing systems by oxidation of ferrous carbonate in natural water or steel pipes encountering soft water. Red water formation seriously affects plumbing fixtures and the problem appears particularly in galvanized pipes if the temperature exceeds 65°C, due to reversal of polarity. Silicate treatment also prevents dezincification in brass and corrosion of copper. Mixtures of silicates and phosphates have been effectively used as inhibitors.

6.9 ORGANIC INHIBITORS

Organic inhibitors are abundantly used in the oil industry to control oil and gas well corrosion. Most common types are long chain (C_{18}) hydrogen and nitrogen containing compounds. Organic inhibitors are neither anodic nor cathodic, but they inhibit both the anodic and cathodic areas to varying degrees depending on the type of inhibition. The most common types of organic inhibitors are shown below:

(1)   Monoamine:
      Primary amine, RNH_2
      Secondary amine, R_2NH
      Tertiary amine, $R-N(CH_3)_2$
(2)   Diamines
      $R - NHCH_2CH_2CH_2NH_2$
(3)   Amides
      $R - CONH_2$

(4) Polyethoxylated compounds

    (a) Amines

$$R-N \begin{cases} (CH_2 \cdot CH_2O)_x H \\ (CH_2 \cdot CH_2O)_y H \end{cases}$$

    (*x* and *y* vary between 2 and 50)

    (b) Diamines

$$R-NCH_2CH_2N \begin{cases} (CH_2 \cdot CH_2O)_x H \\ (CH_2 \cdot CH_2O)_y H \end{cases}$$
$$|$$
$$(CH_2 \cdot CH_2O)_z H$$

    (*x* + *y* + *z* varies between 3–10)

(5) Quaternaries

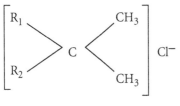

    ($C_{12} - C_{18}$ chains)

    R_1 and R_2 are C_{12} to C_{18} chains.

Organic inhibitors react by adsorption on a metallic surface. Cationic inhibitors (+), like amines, or anionic inhibitors (−), like sulfonates, are preferentially adsorbed depending on the charge of the metal surface (+) or (−). At zero point of charge, there is no particular preference for an anodic or cathodic inhibitor. In such a situation, a combination of an inhibitor which would be strongly adsorbed at more negative potentials along with cathodic protection would provide a greater degree of inhibition than either applying cathodic protection or using inhibitor separately. The formation of a bond between the metal substrate and the organic inhibitor (chemisorption) bonds impedes the anodic and cathodic process and protects the metal surface. Consider an inhibitor molecule, e.g. containing a polar amine nitrogen group at the end of a hydrocarbon chain. The active (-NH$_2$) group contains a pair of unshared electrons which it donates to the metal surface. A chemisorption bond is, therefore, formed which impedes the electrochemical reaction. The polar amine group displaces water molecules from the surface. On adsorption, most of the metal surface is covered by the adsorbed water molecules. The inhibitors react by replacing water molecules by organic inhibitor molecules.

$$Org.molecule(aq) + nH_2O(ads) \rightarrow$$
$$Org.molecule(ads) + nH_2O(soln)$$

Here *n* represents the number of molecules which are replaced to accommodate the organic molecule. The hydrocarbon part of the organic inhibitor is oil soluble, hence, it repels water from the metallic surface. It, therefore, provides a barrier which keeps water away and thus prevents corrosion. The hydrocarbon chain attracts the organic molecules and forms an oily layer which prevents corrosion by acting as a barrier against fluids.

For instance, diethanolamine effectively inhibits corrosion of carbon steel in petroleum/water mixtures. The organic inhibitors are physically adsorbed on the surface.

A scientific description of the mechanism is beyond the scope of this book. Interested readers should look into the suggested literature. To summarize, the following are the main features of corrosion inhibitions by organic inhibitors.

■ The polar nitrogen groups attached to a hydrocarbon chain donate electrons to the metal surface and form a strong chemiadsorbtive bond. The strength of protection is dependent on this bond.
■ The hydrocarbon portion of the inhibitor is oil soluble and it is water repellent.
■ The large hydrocarbon chain orients towards the solution and forms a hydrophobic network (repels water from the metal surface). The water molecules are desorbed and replaced by organic molecules [$Org(soln) + nH_2O(ads) \rightarrow Org(ads) + nH_2O(soln)$]. Water molecules, which are the main source of corrosion, are thus eliminated.

Figure 6.14 shows a simplified mechanism.

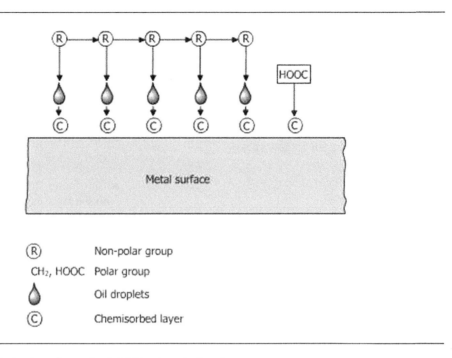

R   Non-polar group
CH₂, HOOC Polar group
        Oil droplets
C   Chemisorbed layer

Figure 6.14   Mechanism of corrosion inhibition by polyphosphates

6.10 SYNERGISTIC INHIBITORS

It is very rare that a single inhibitor is used in systems such as cooling water systems. More often, a combination of inhibitors (anodic and cathodic) is used to obtain better corrosion protection properties. The blends which are produced by mixing of multi-inhibitors are called *synergistic blends*. Examples include chromate–phosphates, polyphosphate–silicate, zinc–tannins, zinc–phosphates. Phosphonates have been used to cathodically protect ferrous materials. Following are the major applications of synergistic blends of inhibitors.

Chromate–polyphosphate	Metal surface cleaning
Chromate–orthophosphate	More effective corrosion control in oilfield
Polyphosphate–silicates	Cooling water system
Polyphosphate–ferrocyanide	Protection of ferrous and non-ferrous constructional materials
Zinc–tannins	Protection of copper and many ferrous materials
Amino–alcohol-sodium nitrite	Combines the precipitation effect of nitrite with the film forming properties of hydroxylklamine

6.11 SCAVENGERS

Oxygen, even in very small amounts, may cause serious corrosion in feedwater lines, stage heaters, economizers, boiler metal, steam operated equipment and condensable piping. It must, therefore, be removed from the closed system. The solubility of oxygen varies with both pressure and temperature. Oxygen is the main cause of corrosion. It reacts by consuming electrons at the cathode causing cathodic depolarization and enhancing the rate of corrosion. Chemicals which eliminate oxygen from the closed systems are called *scavengers*. Ammonium sulfite $(NH_4)_2SO_3$, and hydrazine (N_2H_4) have been successfully used over the years to eliminate oxygen. Oxygen scavengers remove oxygen as shown below:

$$(NH_4)_2SO_3 + \frac{1}{2}O_2 \rightarrow (NH_4)_2SO_4$$

(1) *Org. molecule(aq)* $+ nH_2O(ads) \rightarrow$
    Org. molecule(ads) $+ nH_2O(soln)$

(2) $N_2H_4 + O_2 \rightarrow N_2 \uparrow +2H_2O$

(3) $Na_2SO_3 + \frac{1}{2}O_2 \rightarrow Na_2SO_4$

(4) $NH_4HSO_3 + \frac{1}{2}O_2 \rightarrow NH_4HSO_4$
    (Ammonium hydrogen sulfate)

The oxygen scavenger reaction rate changes by a factor of 2 for every change in temperature by 10°C as shown by reaction (1). Sodium sulfite reacts to form sodium sulfate and increases the total dissolved solid contents of the boiler water. Generally, eight parts by weight of Na_2SO_3 are needed to scavenge one part of O_2, but dosage depends on the purity. Hydrazine can react directly with dissolved oxygen as shown in reaction (1), but the rate of reaction is slow at temperatures below 150°C. The indirect reactions of hydrazine with oxygen, however, proceed rapidly at temperatures as low as 70°C. The indirect reaction proceed as below:

(a) $4Fe_3O_4 + O_2 \rightarrow 6Fe_2O_3$
(b) $6Fe_2O_3 + N_2H_4 \rightarrow 4Fe_3O_4 + 2H_2O + N_2 \uparrow$

For the above reactions, 1ppm of hydrazine is needed for 1ppm of oxygen. It is the most economical and controllable scavenger. Advantages and disadvantages of sodium sulfite and hydrazine are shown in Table 6.1. At higher steam temperatures, excess hydrazine decomposes to form ammonia, nitrogen and sometimes

Table 6.1 Advantages and disadvantages of sodium sulfite and hydrazine

Chemical	Advantages	Disadvantages
Sodium sulfite	■ Rapid reaction	■ Does not reduce ferric oxide to magnetite
	■ Non-toxic	■ May decompose to form corrosive gases
	■ Contributes no solids	
Hydrazine	■ Reduces ferric oxide to magnetite	■ Reacts less rapidly compared to sodium sulfite
	■ Less dosage for scavenging compared to sodium sulfite required	■ More expensive than sodium sulfite
		■ Toxic and flammable

hydrogen. At normal dosages, the quantity of such products is not significant.

At times a combination of both may be required. Hydrazine is considered to be carcinogen by (OSHA). As a result, new scavengers have appeared in the market. The following are new oxygen scavengers which are being promoted in the market:

■ Carbohydrazide
■ Diethyehydroxlamine (DEHA)
■ Ammonium isocascorbate
■ Hydroquinone

Experience with the above inhibitors is presently not sufficient. Factors, such as pH, type of catalyst, temperature, presence of H_2S (such as in oilfields) and biocide must be considered while using oxygen scavengers, as they effect their properties.

6.12 NEUTRALIZERS

Neutralizing inhibitors reduce corrosion by reducing the concentration of H^+ ions in solution. Neutralizers control a small amount of HCl, CO_2, SO_2, H_2S, organic acids and other similar species that react with water to produce H^+ ions. The most commonly used neutralizing inhibitors are morpholine, cyclohexylamine and diethylamine–ethanol. The formation of H^+ by CO_2 is shown below:

$$CO_2 + H_2O \rightarrow H_2CO_3$$

$$\underset{\text{(Carbonic acids)}}{H_2CO_3} \quad \rightarrow \quad \underset{\text{(Hydrogen ions)}}{H^+} \quad + \quad \underset{\text{(Bicarbonate ions)}}{HCO_3^{-}}$$

6.13 SCALE INHIBITORS

Scales are the precipitators which are formed on surfaces in contact with water as a result of physical or chemical changes. Calcium carbonate and calcium sulfate are the major types of scales. The formation of scale creates operating problems, such as blockage of the flow lines, and reduction in the diameter of tubings. Scales are formed when the solubility product of scale forming constituents is exceeded.

Seawater contains dissolved chemicals including carbonates, bicarbonates and hydroxyl compounds. These are identified as soft scale compounds. At ambient temperature, seawater is saturated with $CaCO_3$ and $Mg(OH)_2$.

The mechanism of scale formation of each deposit is different. Calcium carbonate is the most common scale deposit. The solubility of CO_2 decreases with increasing temperature and the equilibrium between CO_2, CO_3^{--} and HCO_3 is upset. Bicarbonate ions break down (Temperature $>70°C$) causing liberation of CO_2 and increase of CO_3^{--} concentration.

$$2HCO_3^- = CO_2 \uparrow + CO_3^{--} + H_2O$$

As CO_2 is evolved, the pH increases and HCO_3^- converts to the less soluble form CO_3^{--}. The CO_3^{--} ion combines with Ca^{++} ions to form relatively insoluble $CaCO_3$ scale as shown below.

$$Ca^{++} + CO_3^{--} = CaCO_3 \text{ (precipitates)}$$

Unfortunately, the scaling tendency of $CaCO_3$ increases as temperature increases, because $CaCO_3$ exhibits an inverse temperature dependence, the solubility decreases with increased temperature. With an extended period of time, the scale ($CaCO_3$) becomes harder and harder. Sodium chloride may be added to decrease the scaling tendency, but it begins to increase when the salt concentration reaches $150\,g/L$. The solubility of CO_2 also increases with increasing pressure. On further heating (beyond $70°C$) the carbonate decomposes.

$$CO_3^{--} + H_2O = CO_2 \uparrow + 2(OH)^-$$

The OH^- ions combine with Mg^{++} ions to form a precipitate of $Mg(OH)_2$ which is

$$Mg^{++} + 2(OH)^- = Mg(OH)_2$$

$Mg(OH)_2$ scaling is predominant above $82°C$ and $CaCO_3$ scaling below $80°C$.

Calcium sulfate scaling is of three types: gypsum ($CaSO_4 \cdot 2H_2O$), anhydrite ($CaSO_4$) and hemihydrate ($CaSO_4 \cdot \frac{1}{2} H_2O$). The gypsum is the more common form. Gypsum exhibits a maximum solubility at $40°C$ followed by a decrease, whereas anhydrite becomes decreasingly soluble on increasing the temperature. Calcium sulfate scale is harder and less porous than $CaCO_3$ scale. Calcium carbonate is soluble in acids, whereas $CaSO_4$ is insoluble in acids. The effect of temperature on the solubility of $CaCO_3$ and gypsum is shown in Fig. 6.15.

6.13.1 SCALE REMOVAL BY INHIBITOR

A large number of compounds are known to inhibit scale formation by removing the scale forming ions and suspended solids from the water. The common inorganic inhibitors used are sodium hexametaphosphate and sodium

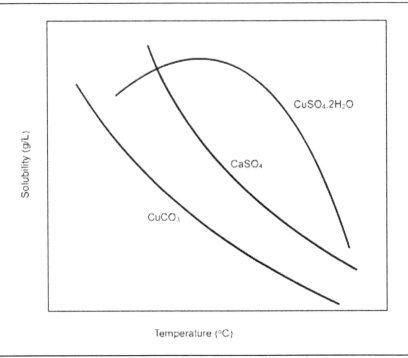

Figure 6.15 Variation of solubility of gypsum and CaCO₃ with temperature

tripolyphosphate. They are effective at low concentration (2–5 ppm for $CaCO_3$ and 10–12 ppm for $CaSO_4$). The hexametaphosphate is, however, subject to reversal to the undesirable orthophosphate form above 140°F (59.5°C). Amongst the organic scale inhibitors, aminotrimethylene phosphoric acid (ATMP) is extensively used. It is stable up to 250°F (120°C) and at all pH values.

In desalination plants $CaSO_4$ scaling can be controlled by allowing normal seawater to concentrate not more than 1.8 times at a time. The maximum temperature attainable in desalination plants without $CaSO_4 \cdot 2H_2O$ formation is 120°C.

Scale removal is an extensive subject and specialized literature must be consulted for further information.

6.13.2 SCALE REMOVAL BY CHEMICAL AND MECHANICAL METHODS

Scales can also be removed by chemicals. Calcium carbonate is soluble in hydrochloric acid, formic acid, acetic acid and sulfamic acid. Hydrochloric acid can be used to remove $CaCO_3$ scale, however, it must contain one of the sequestering agents, like acetic acid, oxalic acid or gluconic acid, to prevent the undesirable precipitation of iron.

Structure of ATMP

Gypsum ($CaSO_4 \cdot 2H_2O$) is insoluble in acids. It should, therefore, be converted to another scale, such as calcium carbonate, which is soluble in acid. Ammonium bicarbonate or sodium carbonate (Na_2CO_3) convert gypsum to soluble $CaCO_3$. Gypsum can also be converted to $Ca(OH)_2$ by KOH. Both the carbonates and hydroxide are soluble in acids. Ethylenediamine tetramine acid (EDTA) dissolves gypsum directly, but it is not economical.

Mechanical softeners, known as '*pigs*,' are sent through the pipe for removal of the hard scale.

A proprietary class of inhibitors, '*threshold inhibitors*,' are also added in small concentration to inhibit crystal growth.

6.14 REBAR INHIBITORS

A serious problem which is faced by building industry is rebar corrosion in concrete. In steel reinforced concrete structure, corrosion attack is prevented by high alkaline environment in concentrate (pH > 12.0) because of the development of a passive layer on steel surface under alkaline environment. The key to success is to restrict the permeability of the concrete. Mitigating corrosion inhibitors can reduce the mobility of corrosive ions and neutralize the corrosion species. The corrosion process in concrete is shown in Fig. 6.16.

The passivation of reinforcement of concrete is provided by the high alkaline environment (pH = 12.8) from lime continuously produced from the hydration reaction of cement. This passive layer may, however, be destroyed due to ingress of chloride and carbon dioxide in certain environments and serious corrosion is caused. Concrete is permeable and allows the ingress of atmosphere. The carbon dioxide reacts with alkalies and forms carbonate and thus reduces the pH values to 9.

$$Ca(OH)_2 + CO_2 + H_2O \rightarrow CaCO_3 + 2H_2O$$

If the carbonated front penetrates deeply into concrete and reaches the reinforcement surface,

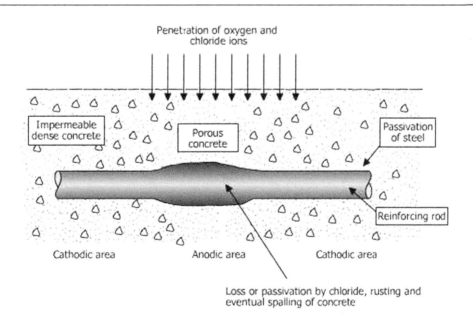

Figure 6.16 Schematic representation of the electrolytic microcell structure in reinforced concrete causing spalling

protection is lost and steel begins to corrode as both moisture and oxygen are available to initiate the corrosion reaction of steel.

$$Fe \rightarrow Fe^{2+} + 2e$$

$$\frac{1}{2}O_2 + H_2O + 2e \rightarrow 2OH^-$$

$$Fe^{2+} + 2OH^- \rightarrow Fe(OH)_2$$

During re-alkalization treatment hydroxyl ions are regenerated by applying a negative voltage and corrosion is brought under control. Inhibitors, such as chromates, phosphates, sodium benzoate and sodium nitrate, have been applied to suppress corrosion. Sodium nitrate is known to preserve passivity on steel rebars and prevents the ingress of chloride. Proprietary blends, such as MCI 2022 (a blend of sulfactants and amine salts in water) and MCI 2000 (alkanolamine), manufactured by CorrTech Corporation, USA, have been used in inhibition of rebars in 3.5% NaCl solution. The observed corrosion rate of rebar, after inhibition, was equal to that of steel 304 and 316 which are highly resistant to corrosion. Experience has shown calcium nitrite to be promising. It competes with chloride to react with Fe^{2+} to maintain passivity by formation of oxide layers as shown below.

$$4Fe^{2+} + 4OH^- + 2NO_2^- \rightarrow 2NOFe_2O_3 + 2H_2$$

Inhibitors, such as sodium tetraborate $(Na_2B_2O_7)$ and zinc borate $(2ZnO\cdot3B_2O_3)$ have proved effective inhibitors.

6.15 BIOCIDES

Bacterial corrosion (MIC) is extremely damaging and causes serious corrosion problems including plugging due to accumulation of solids and slime. Bacteria are unicellular, microscopic organisms found in fresh water and in brine (pH 5–9). The organisms are responsible for plugging and contamination problems in cooling towers and filters. Serious pitting of stainless steels under

biofilms has occurred in many water treatment plants during water pressure testing, even before the plants are put into use. This is due to stagnant water, even for short periods. Once started, it is extremely difficult or impossible to stop such MIC. These include sulfate-reducers, slime formers, iron oxidizing bacteria, sulfur bacteria, algae, yeast and molds. Corrosion attack on steel is caused by one of the following mechanisms:

(a) Generation of iron sulfide cathodic to steel by bacterial action on a steel surface.
(b) Formation of concentration cells by deposition of slime masses on the metal surface (Fig. 6.17).
(c) Corrosion caused by H_2S and carbonic acid generated by aerated alkali and bacteria.

Two types of chemicals are employed to control bacterial growth: bacteriostats or bacteriocides. Bacteriostats put the bacteria in a dormant state without killing bacteria, whereas bactericides kill bacteria. Both types are jointly called biocides. The following is a list of important biocides.

(1) Cocoamine acetate – $R_c \cdot NH_2 \cdot HOOCH$
(2) Dialkyl-benzyl ammonium chloride.

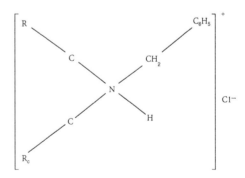

(3) Acrolein, $(CH_2 = CHCHO)$
(4) Chlorine dioxide, (ClO_2)

The objective of the biocide is not to kill the bacteria, but to control their counts and minimize

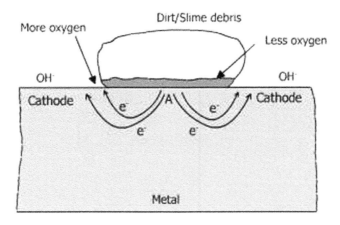

Figure 6.17   Differential oxygen concentration cell formation in slime deposition

their deleterious effect on corrosion and water quality.

6.16 INHIBITOR APPLICATION TECHNIQUES

Basically, there are three well-known inhibitor application techniques:

(1) Continuous injection
(2) Batch treatment
(3) Squeeze treatment

6.16.1 CONTINUOUS INJECTION

As the name suggests, inhibitors are injected in the system to achieve inhibition objectives through the system. Normally the inhibitor is injected into the system by means of an electric or gas driven chemical pump. The inhibitor is added at the point of turbulence to achieve uniform mixing. This method is used for municipal water supplies, cooling towers and oil wells, to minimize scaling and corrosion problems.

In the continuous injection method, a constant supply of chemicals is maintained at a controlled rate. The chemical injection pumps, however, require constant monitoring to check their performance. It is a cost-effective system and widely practiced in oil industry. A diagram of crude oil unit overhead showing NH_3 injection and automated pH control is shown in Fig. 6.18.

6.16.2 BATCH TREATMENT

This is a periodic treatment in which a large quantity of chemicals is used for an extended period of time. It is commonly used to treat flowing oil wells. Batch treatment is also called *slug treatment*. For batch treating, the tube displacement method is employed. Several barrels of inhibitor are pumped into the tubing at the top. The inhibitor is displaced to the bottom of the tubing with the fluids in the oil well. The well is closed for a specific period before operation. The batch is used mainly to treat water with biocides and not to supply inhibitors or scavengers.

It is applied in areas where continuous injection is not practical. The process is not economical as substantial amount of chemicals may be wasted. Also, during the period of shutdown,

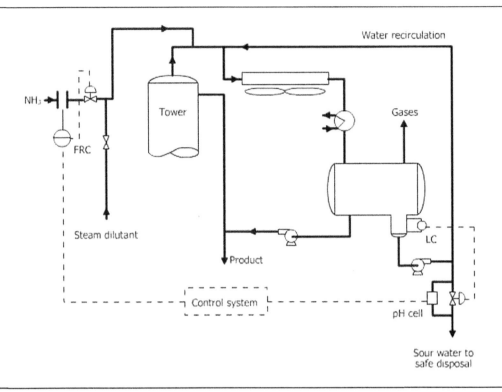

Figure 6.18 Crude oil unit overhead showing NH₃ injection and automated pH control. (From Alley, D.W and Coble, N.D. (2003). *Metal. Perf.*, **44**, May 03)

there is no way that corrosion could be controlled.

6.16.3 SQUEEZE TREATMENT

Continuous treatment of oil wells by inhibitors is achieved by this method. The liquid inhibitor is pumped down through the tubing into the oil producing geological formation under low pressure which acts as a chemical reserve. In oil wells, 1–2 drums of inhibitor is mixed with 10–20 bbl of water (1bbl (British barrel) = 36 gallons), and is pumped into the well followed by pumping in over-flush fluid (50–75 bbl). The inhibitor is absorbed by the formation. It slowly escapes from the formation over a period of time to inhibit the corrosive fluids. A continuous slow release of the inhibitor from this producing formation into the corrosive fluids is the key to

the success of this method. This method allows continuous treatment, however, it can damage the oil producing formation. It is also not cost effective.

6.17 VAPOR PHASE INHIBITORS

By definition vapor phase inhibitors (VPI) are volatile by nature and they are transported to the desired site by volatilization from a given source. They also extend the protection offered by VPI-impregnated wrapping papers to areas out of contact with the wrapping papers so that protection may be achieved by the areas where wrap cannot be applied to complex shapes of the component. Two most popular VPIs are cyclohexylamine carbonate and dicyclohexylamine nitrite (DCHN) with vapor pressures of 0.4 mm and 0.0001 mmHg, respectively. DCHN may be hostile to magnesium, cadmium, zinc and lead. Cyclohexyl carbonate (CHC) may attack copper and its alloys and discolor plastics. The mechanism of inhibition of VPIs is not clearly understood. It is assumed that undissociated molecules of the above inhibitors migrate to the metal surface and undergo hydrolysis by moisture to liberate nitrite, bonzoate or bicarbonate. In the presence of oxygen, they passivate the steel surface. Both DCHN and CHC are impregnated in kraft wrapping papers to effectively protect the steel surface. It is, however, difficult to achieve protection in tropical humid environments. Vapor phase inhibitors are used frequently in storing equipment for a long period of time and for shipping of machinery and components.

6.18 INHIBITOR EFFICIENCY AND INHIBITOR CONCENTRATION

(1) The efficiency of corrosion inhibition can be expressed as

$$E_{inh} = \frac{CR_0 - CR_I}{CR_0}$$

where

E_{inh} = efficiency of a corrosion inhibitor

CR_0 = corrosion rate with zero inhibitor

CR_I = corrosion rate in the presence of

an inhibitor

(2) The quantity of inhibitor required for a fluid to be inhibited can be obtained by the relationship:

$$Q_{inh} = \frac{V_{fluid}}{1.0 \times 10^6} \times C_{inh} \text{ (ppm)}$$

where

Q_{inh} = quantity of inhibitor, kg

V_{fluid} = volume of fluid to be inhibited

C_{inh} = inhibitor concentration, ppm

Example

Calculate the dosage of sodium chromate required to be added to 500 000 liters of water, if the concentration of sodium chromate is 5 ppm.

Solution:

$$Q_{Na_2CrO_4} = \frac{V_{Na_2CrO_4}}{1.0 \times 10^6} \times C_{Na_2CrO_4} \text{ in ppm}$$

$$= \frac{500\,000(\text{kg})}{1.0 \times 10^6} \times 5(\text{ppm})$$

$$= 2.5 \text{ kg}$$

QUESTIONS

A. MULTIPLE CHOICE QUESTIONS

Mark one correct statement in each of the following questions:

1. A corrosion inhibitor protects a metal surface by

   a) eliminating the corrosion from the metal surface
   b) dissolving the corrosion products formed on the metal surface
   c) increasing the thickness of the metal surface
   d) being adsorbed on the metal surface

2. Cathodic inhibitors react by

   a) increasing the rate of cathodic reduction by hydrogen

b) increasing the rate of oxygen reduction at the metal/electrolyte interface

c) shifting the potential of the cathode in a more negative direction

d) shifting the potential of the cathode in a more positive direction

3. Which one of the following is not an important factor in the adsorption of organic inhibitors on the metal surface?

a) Nature of the inhibitor

b) Surface charge of the metal

c) Position of the metal in galvanic series

d) Nature of electrolyte

4. Which one of the following is not characteristic of organic inhibitors?

a) The inhibitors have no effect on surface potential

b) The inhibitor increase metal activity by the process of desorption

c) The inhibitor does not form a complex at the electrode

d) It forms a physical barrier and decreases the diffusion of the reactant

5. Which one of the following inhibitors does not require the presence of oxygen to cause passivation of the surface?

a) Phosphate

b) Tungstate

c) Molybdate

d) Nitrite

6. Which one of the following classes of inhibitors may be termed as dangerous inhibitors?

a) Cathodic inhibitors

b) Anodic inhibitors

c) Organic inhibitor

d) Mixed inhibitor

7. When present in sufficient concentration, organic inhibitors affect

a) only the anodic area

b) only the cathodic area

c) the entire surface

8. The efficiency of certain organic molecules, such as organic amines, improve in the presence of certain halogen ions, such as chloride, bromide or iodides. A combination of amine and chloride ions gives a greater degree of efficiency than either of the two alone. This process is called

a) physical adsorption

b) chemisorption

c) synergism

d) additive effect

9. Vapor phase inhibitors, like dicyclohexylamine, inhibit the corrosion most importantly by

a) liberating carbon dioxide

b) scavenging oxygen

c) making the environment alkaline

d) passivating the surface of the condenser tube

10. Hydrocarbon chains play an important role in the corrosion inhibition process by

a) forming a thin layer of oil and water on the metal surface

b) dispersing the oil layer in the water layer

c) forming an oily layer with a few molecules thick on the surface and creates a barrier to diffuse

d) decreasing the film life of the inhibitor

11. Chromate inhibitors are effective because

a) they form a chromium oxide film that combines with iron oxide fully

b) they remove oxygen from the cooling water system

c) they are not cost effective

d) they are initially applied at low rate

12. Which one of the following kills bacteria?

a) Bacteriostats

b) Bactericides

13. Which one of the following is not an advantage for the continuous treatment process?

a) It supplies chemicals all times

b) It is cost effective

c) The injection rate may be optimized

d) The injection pumps need proper maintenance

14. Which one of the following is not an advantage for a batch treatment process?

a) Allows treatment in an area where continuous treatment is difficult

b) A significant portion of the chemicals in water

c) Corrosion and bacteria are not continuously controlled between treatments

d) Wells are to be closed for a specific period

15. Which one of the following inhibitors is considered effective for inhibition of rebar corrosion?

a) Zinc borate

b) Calcium nitrite

c) Sodium tetraborate

d) MCl 2000

B. How and Why Questions

1) Why are inhibitors needed to be injected in pipelines transporting oil and gas?

2) The bulk of inhibitors in use in oil and gas industries are long chain hydrocarbons, such as aliphatic amines. They are very successfully used as corrosion inhibitor because they significantly suppress corrosion. Explain how the protection is achieved.

3) Why are inhibitors bonded by chemisorption more effective than inhibitors bonded by physical adsorption?

4) As the cathodic reactions consume electrons in the reduction reaction, how is alkalinity affected by these reactions?

5) Why do hydrocarbon molecules keep the water away from the metal surface?

6) How does sodium sulfite scavenge oxygen from the water?

7) What is the major difference between batch treatment and continuous treatment?

8) What is the difference between synergistic adsorption and chemisorption?

9) What is the difference between cathodic protection and corrosion inhibition?

10) What is the effect of increase and decrease of temperature and pressure on the scale forming tendency of calcium carbonate and calcium sulfate?

C. Conceptual Questions

1) State the basic difference between the anodic and cathodic inhibitors in terms of shift of anodic and cathodic potentials.

2) State briefly the mechanism of prevention of corrosion treatment with sodium phosphate and how does it differ from polyphosphate treatment.

3) Describe the essential difference between processes of chemisorption and physical adsorption with examples of inhibitors which react on the metal surface by the above two processes.

4) State the mechanism of cathodic inhibition by addition of poisons of Group VA elements.

5) Summarize the squeeze treatment process in five important steps.

Suggested Reading

[1] Alley, D.W. and Coble, N.D. (2003). Corrosion inhibitors for crude distillation columns. *Materials Performance*, July, **44**, 44–50.

[2] Byars, H.G. (1999). *Corrosion Metal in Petroleum Production*, TCE publication 5, 2nd ed. NACE, Texas: Houston, USA.

[3] Bavarian, B. (2000). Corrosion inhibitors: STE rebar in concrete. *Newsletter*, Cortec Corporation, 1–4, USA.

[4] Bergman, J.I. (1963). *Corrosion Inhibitors*, New York: Macmillan.

[5] Bockris, J.O.M. *et al.* (1967). *J. Electrochem Soc*, **114**, 994, USA.

[6] Bradford, S.A. (2001). *Corrosion Control*. Canada: Casti Publishing Company.

[7] Davenport, P.D. (1986). Protection of metals from corrosion. In *Storage and Transit*. Ellis Horwood, 171–192, UK.

[8] Drew Chemical Corporation (1980). *Principle of Industrial Water Treatment*, 3rd ed.

[9] Dunlop, A.K. (1978). Theory and use of inhibitors, *NWS Corrosion Seminars*. NACE, 147–152, Texas: Houston, USA.

[10] Greene, N.D. (1982). Mechanics and application of oxidizing inhibitors. *Mat. Pref.*, **21**,(3), 20.

[11] Hackerman, N. and Snavely, E.S. (1984). In *Corrosion Basics*, NACE, 127–146, Texas, USA.

[12] Kemmer, F.H. ed. (1988). *The NALCO Water Handbook*, 2nd ed. McGraw Hill.

[13] Munteanu, V.F. and Kinney, F.D. (2000). *Inhibition properties of a complex inhibitor mechanism of inhibition*. CANMET, Canada: Hawa, 255–269.

[14] Mohsen, E. (1978). Overhead corrosion control in crude distillation unit. *Corrosion/78*, Paper No. 132, NACE, Texas: Houston, USA.

[15] Oakes, B.D. (1981). Historical review of inhibitor mechanisms. *Corrosion/81*, Paper No. 248, NACE, Texas: Houston, USA.

[16] Pettus, P.L. and Strickland, L.N. (1974). Water soluble corrosion inhibitors: A different approach to internal pipeline corrosion control. *Corrosion*, **74**, 1–11, Texas: Houston, USA.

[17] Port, R.D. and Herro, H.M. (1991). *The NALCO Guide to Boiler Failure Analysis*. New York: McGraw Hill, USA.

[18] Revie, W. ed. (2000). *Uhlig's Corrosion Handbook*, 2nd ed. New York: John Wiley, USA, 89–1105.

[19] Riggs, O.L. Jr. (1973). In *Corrosion Inhibitors*. Nathan, C. C. ed. Texas: NACE 11.

[20] Simon-Thomas, M.J.J. (2000). Corrosion inhibitors selection: Feedback from the field, Paper No. 56, *Corrosion*, **56**, NACE, Texas: Houston, USA.

[21] Wood, W.G. (Coordinator) (1999). *Metals Handbook*, 9th ed. Vol. 5, ASTM, Ohio: Metals Park, USA.

[22] Wilken, G. (1992). Optimization of corrosion inhibition of sour gas gathering pipelines, Report, Saudi Aramco, AER 5436.

[23] Ash, M. and Ash, I. (Compilers) (2001). *Handbook of Corrosion Inhibitors*, NACE, Texas: Houston, USA.

[24] Davies, M. and Scott, P.J.B. (2000). *Guide to Use of Materials in Water*, NACE, Texas: Houston, USA.

[25] Sastri, V.S. (1998). *Corrosion Inhibitors: Principles and Applications*. New York: Wiley.

[26] Nathan, C.C. (1973). *Corrosion Inhibitors*, NACE, Texas: Houston, USA.

KEYWORDS

Adsorption Attractive interaction of an inhibitor with a metal surface due to ionic charge of the inhibitor and charge on the metal surface. It is a physical process and does not result in bond formation.

Anodic inhibitors They suppress the anodic reaction, displace the corrosion potential in the positive direction, and reduce the current density.

Batch treatment Addition of an inhibitor at one time to provide protection for an extended period of time. Additional quantities may be added later depending on the need.

Cathodic inhibitors Inhibitors which decrease the corrosion rate by suppression of cathodic reduction reaction, such as oxygen and hydrogen reduction.

Cathodic precipitates Inhibitors, like calcium carbonate and magnesium carbonate, which precipitate on cathodic areas.

Chemisorption Formation of a bond between the inhibitor and the metal substrate by donation of a pair of electrons, for example, chemisorption of amines by a metal surface.

Critical concentration The concentration of an inhibitor below which it would not inhibit corrosion.

De-passivation Destruction of a passive film formed on a metal surface. De-passivation increases the rate of corrosion contrary to passivation.

Inhibition Suppression of corrosion by use of inhibitors.

Inhibitor efficiency Degree of inhibition offered by an inhibitor. It is given by $(CR_{(unin)} - CR_{(inh)}/CR_{(unin)}) \times 1000$, where $CR_{(inh)}$ is the corrosion rate in the presence of inhibitor, and $CR_{(unin)}$ the corrosion rate in the absence of inhibitor.

Mixed inhibitors A mixture of anodic and cathodic inhibitors. They inhibit both the anodic and cathodic corrosion reactions.

Neutralizing inhibitors Inhibitors which reduce corrosion by reducing the concentration of hydrogen ions in solution.

Non-oxidizing inhibitors Inhibitors which do not need oxygen to passivate a surface.

Organic inhibitors Long chain nitrogen containing molecules which suppress corrosion by being adsorbed on the metal surface and forming a chemisorptive bond.

Passivating inhibitors Inhibitors which shift the potential of a metal in the noble direction. They stabilize passive films, and also repassivate the metal surface if the passive film formed is damaged.

Poisons Elements from Group VA, like As, Bi, etc. which retard the process of cathodic reduction.

Scaling Formation of scale by precipitation of an ionic material from water. Examples are calcium carbonates as calcium sulfate scale in tubing and flow lines.

Scavengers Chemicals added to reverse oxygen from feed-water, e.g. sodium sulfite and hydrogen.

Squeeze treatment A method of continuous feeding an inhibitor in an oil well by pumping a quantity of inhibitor into the well. The inhibitor is absorbed by the formation from which it slowly escapes to inhibit the produced fluids.

Synergistic inhibitors An increase in the inhibitor efficiency of organic inhibitor in the presence of halogen ions.

CHAPTER 7

COATINGS

7.1 INTRODUCTION

Although coatings have been used for over thousands of years for decorative and identification purposes, the industrial importance of coatings has only been recognized after World War II. The total amount of paint sold annually amounts to billions of gallons. In 2000, USA alone manufactured about 3.5 billion gallons of paint. About one-third of the production of paint is used to protect and decorate metal surfaces. All forms of transport, such as trains, ships, automobiles, aeroplanes, underground buried structures, such as tanks, oil and gas pipelines, offshore structures, iron and structures and all metallic equipment require the use of coatings. The coating industry has, therefore, turned out to be one of the largest in terms of production. The importance of coating can be judged from the fact that coating can hardly be ignored in any corrosion protective scheme.

Corrosion protection of over-ground and underground structures by protective coatings is one of the most proven methods. Other methods include cathodic protection, environmental modification, material selection and design. In contrast to the behavior of rust on steel, the formation of an oxide affords protection against corrosion. If the resistivity of electrolyte is increased and the electron flux is retarded, the rate of corrosion is decreased. By applying coatings of high resistivity, such as epoxies, vinyls, chlorinated rubbers, etc. the flow of electric current to the metal surface is impeded. Also, the higher the thickness of the coating, the higher would be the electrical resistance of the coating. A much higher resistance to the current flow would, therefore, be offered. Thus increasing the electrical resistance of metals by coating offers an excellent method of corrosion prevention. Another method to prevent corrosion is by the use of inhibitors. This can be achieved by using inhibitive pigments, like zinc chromate, red lead and zinc phosphate in coatings. An alternative method is to use a metal more anodic than iron, such as zinc. This is done by using zinc-rich paints. The zinc metal prevents the corrosion of iron by releasing electrons into the iron surface. Thus, coating is an effective method to control corrosion.

Coatings must have the following characteristics for good corrosion resistance:

(a) a high degree of adhesion to the substrate
(b) minimum discontinuity in coating (porosity)
(c) a high resistance to the flow of electrons
(d) a sufficient thickness (the greater the thickness, the more the resistance)
(e) a low diffusion rate for ions such as Cl^- and for H_2O.

Coating and paint technology is adapting to the environmental requirements. The development of water-borne coatings and solvent-free coatings signify new health and safety trends in coating technology. The description of high performance coatings is beyond the scope of this book.

7.2 Objectives

The following are the objectives of coatings:

(1) Protection of equipment and structures from the environment by acting as a barrier between the substrate and the aggressive environment, such as the marine and industrial environments.
(2) Control of solvent losses.
(3) Control of marine fouling; certain constituents in coating control the growth of mildew and marine fouling in seawater.
(4) Reduction in friction; coating reduces friction between two contacting surfaces.
(5) Pleasant appearance; certain types of coatings provide a pleasant appearance and produce attractive surroundings.
(6) Change in light intensity; by selection of appropriate coatings the light intensity in rooms and buildings can be varied as desired.
(7) Visibility; many combinations of colors because of their visibility from large distances are used on TV and radio towers to warn aircraft.
(8) Modification of chemical, mechanical, thermal, electronic and optical properties of materials.
(9) Application of thin coatings on low-cost substrates results in increased efficiency and cost savings.

7.3 Classification of Coatings

Coatings can be classified in the following categories according to corrosion resistance:

(a) Barrier coatings
(b) Conversion coatings
(c) Anodic coatings
(d) Cathodic coatings.

7.3.1 Barrier Coatings

Barrier coatings are of four types – anodic oxides, inorganic coatings, inhibitive coatings and organic coatings.

(1) Anodic Oxides

A layer of Al_2O_3 is produced on aluminum surface by electrolysis. As the oxides are porous, they are sealed by a solution of potassium dichromate. The object of sealing is to minimize porosity. However, chromates have health hazards and are not allowed in some countries.

(2) Inorganic Coatings

These include coatings like ceramics and glass. Glass coatings are virtually impervious to water. Cement coatings are impervious as long as they are not mechanically damaged.

(3) Inhibitive Coatings

In several instances, inhibitors are added to form surface layers which serve as barriers to the environment. Inhibitors, like cinnamic acid, are added to paint coatings to prevent the corrosion of steel in neutral or alkaline media.

(4) Organic Coatings

Epoxy, polyurethane, chlorinated rubber and polyvinyl chloride coatings are extensively used in industry. They serve as a barrier to water, oxygen, and prevent the occurrence of a cathodic reaction beneath the coating. The barrier properties are further increased by addition of an inhibitor, like chromate in the primer.

7.3.2 Conversion Coatings

Phosphate and chromate coatings are examples of conversion coatings. Conversion coatings are so-called because the surface metal is converted into a compound having the desired porosity to act as a good base for a paint. If iron phosphate is used, the following reaction takes place:

$$2Fe + 3NaH_2PO_4 \rightarrow \frac{2FeHPO_4}{Coating} + Na_3PO_4 + \uparrow H_2$$

The corrosion resistance is enhanced by phosphating.

7.3.3 ANODIC COATINGS

By anodic coating, it is meant that a coating which is anodic to the substrate, such as zinc aluminum or cadmium coatings. On steel such coatings are generally called sacrificial coatings. They protect the substrate at the expense of the metallic coating applied. The zinc coatings protect the substrate by acting as a sacrificial anode for the steel which is cathodic to zinc:

$$E^{\circ}_{Zn} = -0.763\,V, \quad E^{\circ}_{Fe} = -0.44\,V$$

Any breaks in the coating cause the anodic oxidation of Zn to occur.

$$Zn \rightarrow Zn^{2+} + 2e$$

The electrons are consumed by the iron substrate which acts as a cathode. The potential is made more negative by electrons and a cathodic reaction is forced to occur on it.

$$2H^+ + 2e \rightarrow H_2 \uparrow$$

A fine film of H_2 is formed on the surface. The steel being cathodic does not corrode. Thus, by acting as a sacrificial material, zinc corrodes while the steel substrate is protected.

7.3.4 CATHODIC COATINGS

In this type of coating, the metals which are deposited are electropositive to the substrate. For instance, for copper coated steel, copper ($E^{\circ} = +0.337\,V$) is positive to steel ($E^{\circ} = -0.440\,V$). The coatings must be pore-free and thick. Electroplated coatings are generally pore-free and discontinuities are not observed. However, if the coating contains a flaw (crater), it acts as the anode with respect to the substrate. Consequently, electrons flow from the crater to the noble coating. At the crater, hydrogen is evolved.

$$2H^+ + 2e \rightleftharpoons H_2 \uparrow$$

Often an intermediate layer is put in between the substrate and the noble coating, such as the nickel–chromium coatings.

Consider nickel coating on a steel substrate. A layer of bright nickel is laid on the dull layer of nickel. Over the bright nickel a layer of chromium is laid. The bright nickel (high sulfur content) is more noble than the steel substrate. Such a coating system is called duplex coating.

7.3.5 MISCELLANEOUS COATINGS

These include glass coatings, porcelain coatings and high-temperature coatings.

7.4 SCOPE OF COATINGS

Sky is the limit for the coating market. The following are the major target areas:

- Aerospace, power plants (turbines) and aircraft.
- Oil and mining industry. Equipment, such as pumps, valves, drilling rigs, slurry pumps, etc.
- Information storage: discs (magnetic coatings), TV display systems.
- Desalination plants: brine heaters, heat exchangers, circulation pumps, valves, etc.
- Automotive industry: gears, valves, pistons, panels, etc.
- Solar energy: photovoltaic cells and solar cells.
- Biotechnology and surgical implantation: artificial hearts, valves and joints.
- Utilities: all household appliances, washing machines, kitchenware and all electrical appliances.
- Pipelines: oil, gas and utilities pipelines.
- Transportation: decks, bridges and railcars.

7.5 PAINTING, COATING AND LINING

(a) A *paint* is a pigmented liquid composition containing drying oils alone or drying oils in combination with resins which combine with oxygen to form a solid protective and adherent film after application as a thin layer.

(b) A *coating* is any material composed essentially of synthetic resins or inorganic silicate polymers which forms a continuous film over a surface after application and is resistant to corrosive environments.

(c) A *lining* is essentially a film of material applied to the inner surface of a vessel or pipeline designed to hold the liquids or slurries.

7.6 Paint Coating System

The coating system comprises:

(a) The primer
(b) The intermediate coat
(c) The top coat

The primer is the most important component of the coating system as the rest of the coating is applied on the primer. In many paint systems, such as those containing a good proportion of natural oils, the pigments may be inhibitive. However, some pigments, such as red lead in linseed oil (red lead primer) react with the oil to produce soap and protect the steel surface although they do not act inhibitively.

The following are the functions of primers:

(1) It must be strongly bonded to the substrate.
(2) It must be resistant to corrosive environments and suppress corrosion.
(3) It must provide good adhesion to the intermediate coat or the top coat.

Primers are allowed to stand for a sufficiently long length of time before any coating is applied. Primers can be inhibitive, impervious or cathodic, as described below, and their applications are dependent on the environment encountered by the metallic structure.

(1) Inhibitive Primers

The pigment contained in the primer reacts with absorbed moisture in the coating and form a passive surface on the substrate, such as steel. Pigments, such as chromate salts and red lead are the examples of inhibitive pigments, but have health hazards.

(2) Impervious Primer

This primer used in impervious coating systems makes the coating much more impervious to the passage of CO_2, oxygen, air, ions and electrons. Conventional thin film coatings cannot prevent oxygen and water permeation. The primer in an impervious system is used with a thick coal tar enamel to form a highly impervious coating. The pigments are generally metal salts, chromates of zinc, lead and strontium, but Cr and Pb contents have health hazards.

(3) Cathodically Protective Primer

One good example is a primer containing zinc. As zinc is anodic to steel, zinc corrodes in preference to steel and protects the steel from corrosion. Experience has shown that a zinc-rich primer can double the life of a chlorinated rubber or an epoxy top coat.

The three types of primers described above form the basis of three important coating systems:

(a) impervious coating system for equipment requiring immersion,
(b) inhibitive system for application in marine atmosphere and
(c) cathodically protective system for severe corrosive environments.

7.7 Paint Coating Components (Table 7.1)

(a) **Vehicle**. This is the liquid portion of the paint in which the pigment is dispersed. It is composed of *binder* and the *solvent* or *thinner* or both.

(b) **Binder**. This binds the pigments in the coating in a homogeneous film. The binder also binds the total coating to the substrate.

Table 7.1 The composition of paints

Component	Function
Resin (binder)	Protects the surface from corrosion by providing a homogeneous film. It controls the desired film properties.
Solvent	Dissolves the binder. It provides the means of applying the painting. The solid resins are transformed to a liquid state.
Additives	Assist in improving the quality of the paint.
Pigments	Provides opacity, color and also resistance to corrosion.
Extender (inorganic matter)	Used for a wide range of purposes in conjunction with pigments to extend the properties of the binders. Examples: talc, china clay, silica and barytes.

The binders provide the basis for the generic terminology of paint. The physical and chemical properties of paints are determined by the binders.

(c) **Pigment.** A pigment not only provides a pleasing color but protects the binder from the adverse effect of ultraviolet radiation on the coating.

(d) **Solvent.** The purpose of the solvent is to provide the surface with a coating material in a form in which it can be physically applied on the surface.

(e) **Additives.** These are used to modify the properties of the coating, such as reducing drying time and enhancing the desired properties. One example is a *plasticizer* which makes the film flexible. Similarly a *drier* may be added. It is a substance, such a compound of lead, manganese, which when introduced in drying oils reduce their time. A drying oil,

such as tung oil, forms a tough solid film in air.

(f) **Extenders.** They are added to improve the application properties of the paint.

Table 7.1 summarizes the major components of paints.

Paints and coatings can be divided into two categories: convertible type which need a chemical reaction, such as oxidation or polymerization, and the non-convertible type which are formed by evaporation of the solvents. The former category includes alkyds, epoxy, esters, polyesters, urethenes, silicon and other resins.

7.8 COMPOSITION AND FUNCTIONS OF PAINT COATING COMPONENTS

There are several ways of classifying binders (resins). One such classification is given below:

(a) Drying oil types, such as alkyd
(b) Epoxy, polyurethane and coal tar (two-pack chemical resistance)
(c) Vinyl and chlorinated rubber (one-pack chemical resistance)
(d) Bituminous coatings
(e) Lacquers.

7.8.1 DRYING OILS

These are natural binders and used as they are or with a certain amount of processing. This group contains natural drying oils, such as linseed oil, soya bean oil and coconut oil. They are modified with synthetic resins to obtain useful properties. The composition of major oils used in coatings is given in Table 7.2. Oils are classified as drying, semi-drying and non-drying on the basis of concentration of fatty acids. Fatty acids are called saturated if no double bond is present, unsaturated if one double bond is present and polyunsaturated if more than two double bonds are present.

Table 7.2 Composition of major oils used in surface coatings

	Saturated acid	Oleic acid	9, 12 Linoleic acid	9, 12, 15 Linolenic acid	Conjugated acid
Tung	6	7	4	3	80***
Linseed	10	20–24	14–19	48–54	0
Soya bean	14	22–28	52–55	5–9	0
Castor oil	2–4	90–92*	3–6	0	0
Dehydrated castor oil	2–4	6–8	48–50	0	40–42**
Tall	3	30–35	35–40	2–5	10–15***
Coconut	89–94	6–8	0–2	0	0

*Principally ricinoleic acid, not oleic
**Conjugated 9, 11 linoleic acid
***Conjugated linoleic acid, isomerized linolenic acid and 9, 11, 13 eleostearic acid, proportions dependent on source and degree of refinement

A drying oil must contain at least 50% polyunsaturated fatty acid. For instance, linolenic acid is a polyunsaturated fatty acid $[CH_3 \cdot CH_2CH = CH \cong CH_2CH = CH_2CH_2CH = CH(CH_2)_7 COOH]$. Linseed oil contains over 50% of 9, 12, 15 linolenic acid, hence it is a good drying oil. Structures of some fatty acids are shown in Table 7.2.

Common to all oils is that they are treated with various chemicals and heated to manufacture alkyd resins.

Oleoresinous binders are those which are manufactured by heating oils and either natural or pre-formed synthetic resins together. The oleoresinous vehicles have largely been replaced by alkyds and other synthetic resins.

The following are typical structures:

(a) *Oil*

(Fatty acids)

CH_2-O-R
|
$CH\ -O-R$
|
CH_2-O-R

(b) *Glycerol*

CH_2OH
|
$CHOH$
|
CH_2OH

(c) *Phthalic anhydride*

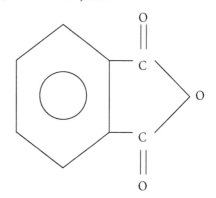

7.8.2 ALKYD RESINS

The alkyd resin is made through reacting an oil with an acid and alcohol. Alkyd resins are the most extensively used synthetic polymers in the coating industry. The principal raw materials are oils, fatty acids, polyhydric alcohols and dibasic acids. Examples of the above are:

(1) Oils – castor oil, soya bean oil, etc.
(2) Polyhydric alcohol (glycerol).
(3) Dibasic acid (phthalic acid).

An oil derived fatty acid is chemically combined into a polymer structure. All natural fats are triglycerides [1 mole of glycerine + 2 molecules of fatty acid (saturated)]. Oils are formed from unsaturated fatty acids. All saturated fatty acids contain double bonds.

To form an alkyd resin, a drying oil fatty acid, such as oleic acid, or linoleic acid is added to a polyhydric alcohol and a dibasic acid.

$$CH_2-OH$$
$$|$$
$$CH-OH$$
$$|$$
$$CH_2-OH$$

Glycerol is a traditional alcohol used in alkyd production.

Phthalic acid is a typical dibasic acid.

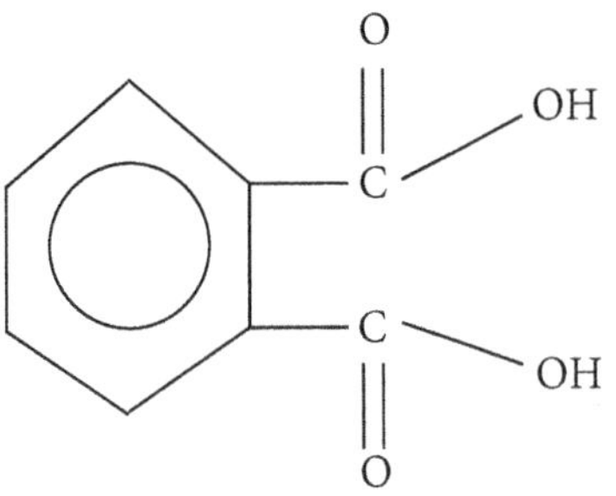

[Phthalic Anhydride + Glycerol + Drying Oil + Fatty Acid $\longrightarrow$ Alkyd]

A simplified structure of oil modified alkyd resin is shown below:

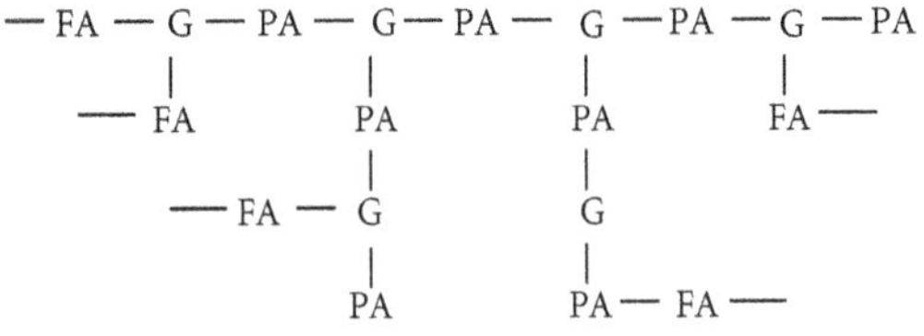

FA = Fatty Acid
G = Glycerol
PA = Phthallic Anhydride

The alkyd has important properties over the original drying oil. Alkyds are often mixed with other generic types of binders to enhance properties.

The most common binders used are chlorinated rubber, vinyls, silicones and urethanes. They are widely used paints, successfully applied under a wide range of conditions. They also belong to the convertible type of coatings.

7.8.3 EPOXY ESTERS

They also belong to the class of convertible type of coatings. A number of epoxy-based coatings are available, such as air drying (epoxy esters), catalyzed (amine or polyamide), epoxy co-polymers (coat tar, phenolic, amine or polyamide), heat-cured (epoxy-phenolic) and high-build epoxy-based coatings. The group of synthetic resins produce some of the strongest adhesives in current use. One component epoxy, is epoxy ester paints which are cured by oxygen.

A special class of epoxies which has come out in recent years is epoxy mastics which are solventless materials. They need less surface preparation and can be cured down to $-10°C$. They form very homogeneous coatings. Another class is termed coal-tar epoxy, in which some epoxy is replaced by tar. They are used for tank lining. Epoxy resins are combined chemically with drying oils to form epoxy esters. Epoxy resins are formed by the reaction between diphenylolpropane (bisphenol A) and epichlorohydrin (DPP + ECH). The epoxy group $H_2C-O-CH_2$ is found at both the ends of the chain. Amines or polyamides are used as catalysts for epoxy resins to cause cross-linking of molecules. These catalyzed epoxy coatings have a good resistance to alkalies solvents and acids. Both the terminal epoxide group and the secondary hydroxy groups of solid epoxy resins can be reacted with fatty acids to produce the epoxy ester.

The epoxide group is terminal in the molecule and the resistance is due to the C–C bond and ether linkage in the polymer. Adhesive properties are because of the polar nature of the polymer. Their overall properties are superior to alkyds. They have improved water and hydrocarbon resistance. They have, however a tendency to 'chalk.'

The reaction between DPP and epichlorohydrin in presence of an alkali produces epoxy resin as shown in Section 7.8.3.

7.8.4 URETHANE OILS AND URETHANE ALKYDS

The term polyurethanes denotes a class of polymers containing the group:

$$-NH-C-O$$
$$\overset{\|}{O}$$

The urethane oils and urethane alkyds constitute about 50% of the total products of alkyds. They correspond in composition to the conventional air drying alkyds discussed earlier, the only difference being that the dibasic acid (phthalic acid) used in the case of air drying alkyds is replaced by an isocyanates ($R-N=C=O$). The most common isocyanate used is toluene diisocyanate (abbreviated as T.D.I.). The T.D.I. reacts with an active hydrogen atom obtained from hydroxy containing compounds, such as polyethers and vegetable oils. The performance depends upon the characteristics of isocyanate and hydroxy compound used. They do not react with moisture.

The following is the sequence of preparation of urethane oils:

Drying oil   Glycerol   Mixed mono- & diglycerides

Urethane oil

The alkyd molecule retains the fatty acids groups. The oil undergoes oxidation, hence the curing of this alkyd takes place by oxidation of oils. These alkyds have a high resistance to abrasion. The polyurethane resins have a good resistance to water, solvents, organic acid and alkaline. They have a high degree of abrasive resistance.

7.8.5 SILICONE ALKYDS

There are numerous types of paints based on silicone which are combined either with alkyd or acrylic resin. Pure silicone resists temperature up to 600°C and has excellent weathering properties. The copolymer is obtained by reaction between the hydroxyl groups in the alkyd resins with the hydroxyl groups in the silicon intermediates.

The silicon intermediate has a low molecular weight. Silicone resin contains a relatively high percentage of reactive hydroxyl ($Si-OH$) or aloxyl ($Si-OR$) groups attached to silicon. Thus a copolymer structure is obtained which combines alkyd with silicone.

$$\equiv Si-OH + HOC-ALKYD \rightarrow$$
$$\equiv Si-O-C-ALKYD + H_2O$$

The silicone intermediates commercially available for polymer modifications are

This combines the properties of the alkyds with the properties of silicones. Silicone alkyds are highly resistant to weather and heat because of silicon contents.

Paints made from silicone modified resins have excellent weatherability, with good chemical resistance. Silicone copolymers can withstand most severe weather conditions. They are extensively used for coating of steel coils.

7.8.6 LACQUERS

These are solutions of organic film forming materials in organic solvents from which the solvent evaporates after application to a substrate. The solid film is formed after evaporation. The properties of lacquers depend on the type of resin used. The lacquers dry very rapidly. The following are the major types of lacquers:

- Cellulose derivatives
- Polyvinyl chloride
- Chlorinated rubber
- Polyurethane elastomers

(1) Cellulose Nitrate

The structure of cellulose nitrate is shown below:

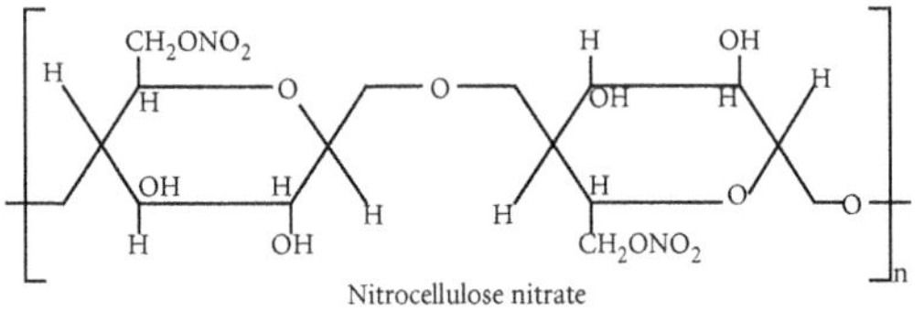

Nitrocellulose nitrate

The raw material is the wood pulp or cotton linters which are the sources of cellulose. The cellulose is nitrated and heated in water until the desired characteristics re-obtained. It is used as a film former for lacquers.

(2) Polyvinyl Chloride

Vinyl chloride is a gas produced by reacting ethylene or acetylene with hydrochloric acid. The reaction replaces one hydrogen atom in ethylene with a chlorine atom which makes it non-burning. Polymerization produces polyvinyl chloride. Polyvinyl chloride is a linear chain polymer with bulky chlorine side groups. It is conveniently represented as below:

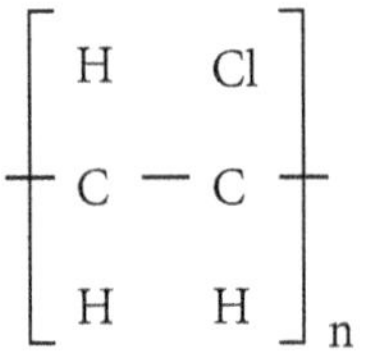

The vinyl chloride resin coatings are well-known for their durability. The vinyl copolymer contains 86% vinyl chloride and 14% vinyl acetate. The material containing the vinyl group when combined with catalysts react to form long chain polymers.

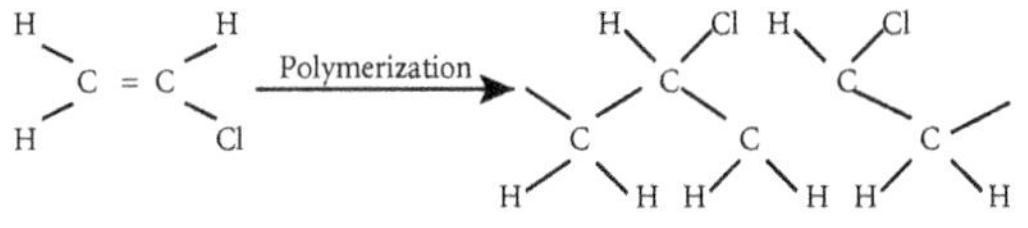

PVC–PVA Copolymer

The above is a double bond polymerization process. One vinyl chloride molecule is made to react with another molecule to form the PVC polymers.

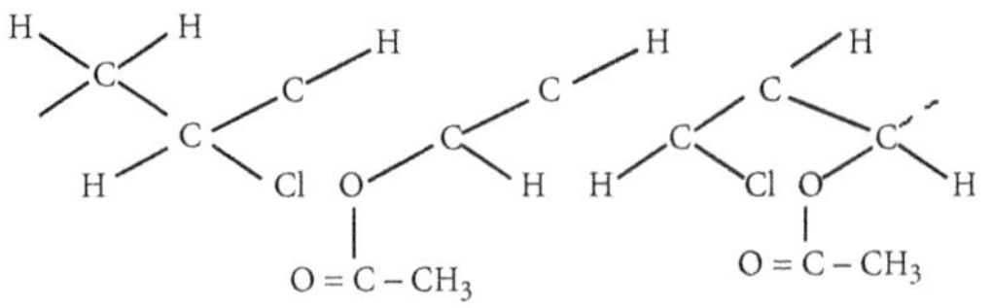

The vinyl acetate and vinyl chloride monomers react together to form a polyvinyl polychloride acetate resin. A typical resin is bakelite which contains 14% vinyl acetate and 86% vinyl chloride.

Vinyl coatings are inert to alkalies, water, oils and greases. Vinyl solutions containing PVC are used for pipelines, dams and several important industrial applications. They have an excellent resistance to water and weather.

(3) Chlorinated Rubber

It is derived from natural rubber which is a polymer of isoprene. Tetrachloroisoprene is produced by chlorination of natural rubber. A chlorinated isoprene molecule and its conversion to chlorinated isopropane molecule is shown below.

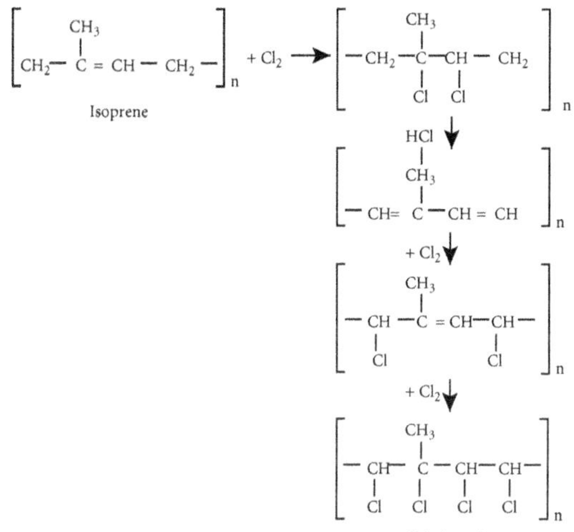

Isoprene

Chlorinated
Isoprene Molecule

methacrylic acid or copolymers with ethyl, –butyl or methacrylate. Thermoplastic acrylics are prepared by the co-polymerization of a mixture of acrylic and methacrylic monomers.

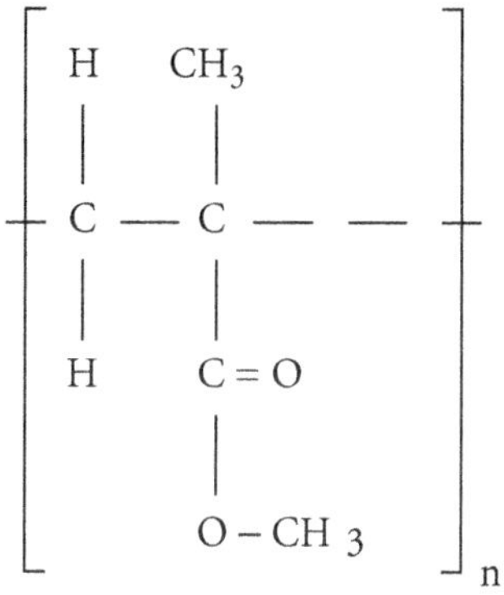

Methyl Methacrylate

This provides strong resistance to inorganic acids, water and general chemicals and is heat resistant and possesses good durability. They dry at high speed and show a good versatility for blending with other coatings. Chlorinated rubbers are thermoplastic which means that when they are subjected to high temperature, they become soft. Above 60°C, chlorine may be released. Resistance to hydrocarbons is poor.

(4) Acrylics

The acrylics have been, to a great extent, replaced by chlorinated rubber and vinyl paints for use both in land-based and marine industry. They have a general chemical structure shown below, where R and R' are either hydrogen or alkyl group.

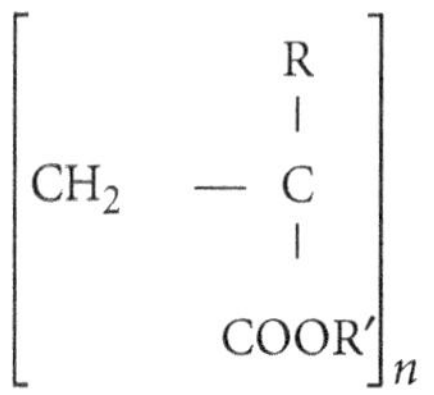

Acrylic polymers are mainly used for their excellent strength, chemical and weather resistance. They are made by polymerization of esters of acrylic and methacrylic acid. The main resins are polymers of methyl and esters of acrylic and

They have a water white color and show a high resistance to change in color with time. Generally acrylics when co-reacted with other resins, such as epoxy and isocyanide, show a greater durability. They have found a wide use in trains, aeroplanes and buses.

7.8.7 BITUMINOUS MATERIALS

Asphalt is a component of crude oil, and coal tar is a residue of coal tar distillation. The above two are generally combined with a solvent, such as aliphatic or aromatic hydrocarbons, to form a lacquer. They dry by solvent evaporation.

Water emulsions are prepared by suspending minute particles of bitumen in water and emulsifying with fillers, such as manganese and aluminum silicates, coal dust and limestone dust. The bitumen is heated to a temperature 175–245°C for hot metal applications.

They provide the best moisture and chemical resistance. They are economical and used for foundries, underground storage tanks, truck chassis, radiators and water proofing. They have a good resistance to non-oxidizing acid, hence, they are attacked by non-polar solvents.

When coal is heated to 1095°C in the absence of air, it decomposes into gas and coke. Coal tar is formed and gas is condensed. All coal tar coatings are subjected to degradation by ultraviolet radiation and cracking.

Asphalt is a hard natural brittle resin. Asphaltic compounds are composed of complex polymeric aliphatic hydrocarbons. They have a good water and ultraviolet resistance. Coatings based on bituminous resins fall into four major groups:

(1) **Solid asphalt.** Solid blocks of asphalts are melted and applied over the substrate.
(2) **Solvent cutbacks.** Bitumen is dissolved by a solvent to produce a non-convertible coating by solvent evaporation. It is called as cutback asphalt.
(3) **Emulsion.** The bitumens are emulsified with water and applied as top coat.
(4) **Varnishes.** The bitumen replaces some or all parts of a resin in oleoresinous paints (oils combined with resin to obtain paints, such as alkyd, phenolics, etc.).

The bitumens and tars both have been used extensively in blast tanks.

7.8.8 HEAT CONVERSION BINDERS

Examples of heat conversion binders include asphalt or coal tar which are melted and applied on the surface. They also include the resins which do not solvate until the heat is applied. They also include high molecular weight thermoplastics or semi-thermoset resins, such as epoxies. Examples:

Fusion Bonded Epoxy (F.B.E.). The steel is decontaminated, abrasive blast cleaned and heated by electrical induction heaters to the prescribed temperatures, before uniformly applying epoxy powder. Another example is the use of hot applied coal tar enamels used mainly as pipeline coatings.

Organosols and Plastisols. Vinyl coatings of high film thickness can be applied by dispersing polymers of vinyl chloride in a plasticizer (plastisol) or by dispersing in a plasticizer and an organic diluent (organosol). They provide flexible films of thickness up to 5 mm. Organosols are used mainly for spray applications on furniture and tool handles, etc. Plastisols are used in coil industry and containers.

7.8.9 ADVANTAGES AND DISADVANTAGES OF ORGANIC COATINGS

Table 7.3 shows the advantages and disadvantages of various organic coatings.

7.8.10 INORGANIC SILICATES

The inorganic silicates form a class of binders which are either dissolved in water or solvent and which react with water to give an inorganic film. They are prepared by fusing a mixture of sand or silicon oxide and typically a mixture of potassium or sodium carbonate. The main use of the silicate binders is in conjunction with a high loading zinc dust as a corrosion control primer. They can be classified in three groups:

(a) Post-cured water-based alkali metal silicates
(b) Self-cured water-based alkali metal silicates
(c) Self-cured solvent-based alkali silicates

(1) Post-cured Water-based Alkali Metal Silicates

They come as a three-package system consisting of zinc dust, silicate binder and the curing solution. The pigment portion is metallic zinc or metallic zinc modified with up to 20% of a coloring pigment. The coating has three components: zinc dust, silicate solution and curing solution.

(2) Self-cured Water-based Alkali Metal Silicates

Silicates, such as sodium and potassium, are combined with colloidal silicon to accelerate the speed of curing. They are based on high ratio of sodium silicates ($Na_2O \cdot SiO_2$) potassium silicates. Lithium hydroxide colloidal silica and quaternary ammonium silicates are also included, but their higher cost tends to restrict their use. The corrosion resistance of solvent-based coatings is a little inferior to the post-cured coating.

Table 7.3 Advantages and disadvantages of various organic coatings

Type of coating	Advantages	Disadvantages
(1) Drying oil	Good wetting, good flexibility, low cost	Slow drying, poor chemical and solvent resistance
(2) Alkyds	Good durability, good gloss, low cost	Poor chemical resistance
(3) Epoxy Esters	Good resistance to moisture, moderate cost	Poor gloss retention, tendency to chalk
(4) Chlorinated rubber (cured by evaporation)	Excellent resistance to water, excellent durability	Poor heat resistance
(5) Bitumens (cure by evaporation)	Good performance, good resistance to water	Poor resistance to hydrocarbons
(6) Polyurethane	Excellent chemical resistance, excellent adhesion, good gloss retention	Relative high cost
(7) Silicones	Excellent heat resistance	Low resistance to chemicals, high cost

(3) Self-cured Solvent-based Alkali Silicates

These are solvent born alkyl silicates coatings in which zinc is incorporated. The binders are mostly modified alkyl silicates consisting of partially hydrolyzed silicates (generally ethyl-silicate type).

The following are the ingredients of a typical ethyl-silicate zinc-rich coating:

(a) Pigments: zinc dust (38% wet volume; 75% dry volume)
(b) Solvent: ethyl alcohol (C_2H_3OH)
(c) Binder: ethyl-silicate (partially polymerized)

Reaction Chemistry

Firstly the silicate film is concentrated as the solvent is evaporated. The binder (tetraethyl-silicate) hydrolyzes to form silicic acid which reacts with zinc to form silica-oxygen zinc polymer. Zinc reacts and combines with the matrix. The unreacted zinc surrounds the large molecules. A film of silica matrix will surround zinc polymer.

The binder becomes extremely hard and corrosion resistant. The inorganic zinc silicates have an excellent resistance to chemicals, wears and weathering. They also afford cathodic protection to metal substrate. The coatings can be applied at higher humidities and lower temperatures.

7.9 CLASSIFICATION OF BINDERS

Binders are classified according to their chemical reactions. The binders belong to one of the following types:

(a) **Oxygen reactive binders.** The resins belonging to this category produce coatings by intermolecular reaction with oxygen. Urethane alkyds, epoxy esters and silicone alkyds are examples.
(b) **Lacquers.** The coatings are converted from a liquid material to a solid film by evaporation, e.g. PVC copolymers, chlorinated rubbers and bituminous material.
(c) **Heat conversion binders.** Materials, such as sulfate or coal tar, are applied in a melted condition. Other examples are organisols and plastisols.
(d) **Co-reactive binders.** Binders, such as epoxies and polyurethanes, are formed by combination of two low molecular weight resins. A solid film is formed after they react upon application on the solid surface.
(e) **Condensation binders.** Binders, such as phenolic resins, release water during polymerization. Phenolic resins are the product of phenol and formaldehyde.
(f) **Coalescent binders.** These include emulsion resins (insoluble) particles in water. These are two-phase systems of two immiscible liquids, where small droplets of one form the dispersoid phase in the other (continuous phase). The particle size is generally 0.1–0.5 µm.
(g) **Inorganic binders.** These are generally inorganic silicate polymers used with zinc dust to provide high quality coating. The inorganic silicates are prepared by fusing a mixture of sand or silica and a mixture of sodium or potassium carbonate or zinc oxide.

Table 7.4 shows composition and application properties of binders and lacquers.

Table 7.4 Composition and application properties of binders and lacquers

Type	Composition	Advantages	Disadvantages
Oils	Linseed oil, soya bean oil, castor oil, tall oil, coconut oil	Economical and easy to apply	Poor weather and poor corrosion resistance
Oleoresinous	Oils + natural or synthetic resins	Economical and easy to apply	Inadequate corrosion resistance
Alkyds	Drying oil modified glycerol phthalate (polybasic acid + polyhydric alcohol + oil (oil does not react directly)) Glycerine + Phthalic acid + Fatty acids → Alkyd Ester	Good resistance to weather	Inadequate corrosion resistance
Epoxy	(i) Fatty acids of vegetable oils + epoxy resin → Epoxy ester resin (ii) Principal type of epoxy resin is formed by reacting diphenyl propane (bisphenol) with epichlorohydrin	High corrosion resistance of 2-pack epoxy coal tar epoxy resistant to marine conditions	High cost and poor resistance to weathering

(Contd)

Table 7.4 (*Contd*)

Type	Composition	Advantages	Disadvantages
	(iii) Coal tar epoxy are combination of basic epoxy resin with coal tar		
Polyurethane	(i) Toluene di-isocyanate-linseed oil Tdi + polyester pre-polymer (isocyanate group) + hydroxy compound in oils polyesters, or polyhydric alcohols. (ii) Typical example is the reaction between TDI and polypropylene glycol to form an isocyanide terminated pre-polymer	High resistance to corrosion	Adequate tendency for discoloration
Silicone	Silicone resin silicone/alkyd resin	Good weather resistance good chemical resistance	
Vinyl	Vinyl chloride vinyl acetate co-polymer	Resistance to acids, alkalies and salts	Poor resistance to solvents
Acrylic	Methyl methacrylate copolymer	Good weather resistance	Poor chemical resistance
Phenolic	Phenol formaldehyde resin	Good chemical resistance	Tendency for discoloration
Chlorinated rubber	Chlorinated rubber + a plasticizer (made by reaction of rubber with chlorine)	Good resistance to mineral oils, acids and alkalies	Upper temperature limit is 80°C. Limited resistance in ultraviolet radiation
Bituminous materials	Asphalts, Coal tars	Resistance to water. Good adherence. Adequate resistance to rusting. Economical	Upper limit 65°C
Nitrocellulose	Nitrocellulose + alkyd	Dries fast	Limited resistance to solvent
Grease	Grease, oils	Economical	Only used for temporary protection

7.10 PIGMENTS

Pigment is a substance which is dispersed in the binder to give specific physical and chemical properties. The following are the functions of pigments:

(1) Aesthetic appeal: color or special effects.
(2) Surface protection: provides resistance to corrosion and weathering, e.g. iron oxide and chrome oxide.
(3) Corrosion inhibition: many pigments, such as chromates, act as inhibitor by helping in forming a barrier film.
(4) Film reinforcement: many particles of pigments provide toughness and strength to the paint film, e.g. silicates and oxides.
(5) Increased coverage: many pigments increase the volume of the solids present without decreasing the effectiveness of the pigment.
(6) Adhesion: many pigments, such as aluminum flakes, increase the adhesion characteristics.

7.10.1 SELECTION OF PIGMENTS

(a) The pigment used must be compatible with the type of resin used, as the pigment selected must not be affected by the solvent, diluent or any ingredient of the paint.
(b) Performance of the paint. A pigment must be compatible with the application of the paint. For instance, titanium dioxide is excellent for industrial plants whereas carbon black is not.
(c) Opacity or transparency. The selection of pigment depends on the requirement of opacity. If opacity is required, black, brown and green pigments can be used. Transparency can be achieved with yellow and red iron oxide pigments.
(d) Pigment blending capability. Several organic and inorganic colored pigments can be blended to produce different blends of colors. For instance, a mixture of yellow and blue color produces a green color and a red and white combination produces a pink pigment.

(e) Durability. This is the degree to which the pigments can withstand the effect of conditions without loss in color retention time.
(f) Gloss. The pigments should provide a good glossy surface. Gloss is the degree to which a surface simulates a perfect mirror image in its capacity to reflect incident light.

7.10.2 TYPES OF PIGMENTS

(A) White Pigments

Titanium dioxide. It is manufactured in two modifications: (a) anatase and (b) rutile (Table 7.5). It is a non-inhibitive inert white pigment. It gives good opacity, tinting strength and gloss. It is unaffected by mineral acids and organic acids at high temperature.
White lead [$2PbCO_3 \cdot Pb(OH)_2$]. It provides a tough, elastic and durable paint film. It, however, reacts with acidic binders. It is a health hazard.
Zinc oxide. It is a good absorber for ultraviolet radiation and protects the resin from breakdown. It has a good fungicidal and mildew resistance.
Lithophone [ZnS 30%/$BaSO_4$ 70%]. It has excellent dispensing properties in the resin. It is a very strong pigment. It has good opacity, good tinting strength, but poor resistance to chalking.

(B) Black Pigments

Carbon black. Carbon blacks are prepared by burning carbonaceous substances and

Table 7.5 Properties of titanium dioxide

Property	Rutile	Anatase
Tinting strength	1000	1200
Density	4.2 kg/dm^2	3.9 kg/dm^2
Color	Slight cream	Pure white
Resistance to chalking	Very good	Poor

allowing flue gases to impinge on acid surface. They are classified as high, medium and low color blacks depending on the intensity of the blackness. The fine particle size (0.010–0.25 μm) is used for the painting of motor cars. The coarser particle size is used for decorative paints. They disperse in paints easily.

Graphite. It has a low tinting strength and gives low intensity blacks.

Black iron oxide [ferroso–ferric oxide (FeO·Fe₂O₃)]. Black iron oxide gives a coarse particle size pigment. It is a cheap and inert pigment. It is resistant to ultraviolet radiation.

(C) Brown Pigments

Iron oxides [Fe_2O_3]. These are transparent and have a low tinting strength.

Calcium plumbate. It is made by heating lime with lead oxide in a stream of air. It is resistant to corrosion. It provides an excellent adhesion and toughness.

Monoazo-brown and diazo-brown. Both the pigments have a good resistance to corrosion. They are economical and stable against light.

(D) Blue Pigments

Ultramarine blue [$Na_7Al_6Si_6O_{24}S_2$]. It is resistant to heat. It has a low tinting strength and low opacity. It has a poor resistance to acid and a good resistance to alkali.

Pthalocyanine [copper–phthalocyanine]. These have to flocculate in mixture with other pigments. They have high tinting strength, excellent resistance to chemicals, ultraviolet radiation and weathering.

(E) Green Pigments – Brunswick Greens

Chrome greens. They are economical and have a high opacity. They are toxic. They may become blue on exposure. They offer a good heat resistance.

(F) Yellow Pigments

Lead chromate [pigment yellow 34 – $PbCrO_4$, $PbSO_4$]. The following are the important chromate pigments:

- Primrose chrome (greenish-yellow lead sulfochromate)
- Lemon chrome (yellow lead sulfochromate)
- Middle chromes (reddish-yellow chromates)

They have good opacity and tinting strength. Lead chromates with lead sulfate are used to provide rays of color from pale primrose to scarlet. They are economical and have inhibitive properties largely used in automotive coatings.

Yellow iron oxide [hydrated ferric oxide – $Fe_2O_3 \cdot H_2O$]. These have excellent resistance to weathering and chemical attack.

Monoazo yellow (2-nitro-p-toluidine coupled with a acetoacetanilide) or (hansa) yellow. They have high tinting strength and a good chemical resistance.

(G) Orange Pigments

Basic lead chromate [$xPbCrO_4$, $yPbO$ with some lead sulfate $PbSO_4$]. They are toxic like other chrome pigments. They darken on exposure to sunlight. They are resistant to alkalies.

(H) Red Pigments

These consist of pigments of mineral origin with the exception of red lead and also contain Fe_2O_3 which gives the red color.

Red iron oxides [natural and synthetic ferric oxide (Fe_2O_3)] [Al_2O_2, SiO_2, $2H_2O$ Fe_2O_3 and Fe_2O_3]. Haematite is the red iron oxide available in earth's crust. The synthetic type has an excellent resistance to chemicals. It has excellent opacity and light fastness.

Diazo-red pigments [pyrazolone]. It has an excellent resistance to solvents and light fastness.

(I) Special Effect Pigments

Aluminum flakes [leafy and non-leafy types]. These metallic pigments provide an excellent protection to metal substrate subjected to marine environment. The difference between these two types is shown in Table 7.6.

Fluorescent pigments [rhodamine]. Fluorescent dyes are made by incorporating fluorescent dyes in resins. They fluoresce by absorbing radiation in the ultraviolet range at a particular frequency and retransmitting the energy at lower frequencies. They are highly expensive and have a poor opacity. They have a poor resistance to weathering.

(J) Corrosion Inhibiting Pigments

They prevent corrosion by providing a barrier to the environments. The following is the example.

Table 7.6 Difference between leafy and non-leafy aluminum flakes

Leafy	Non-leafy
(1) The leafy pigments flow to the surface of the coating film	(1) They remain uniformly distributed throughout the film
(2) They give the appearance of a bright aluminum surface	(2) They provide a gray color without luster
(3) They are highly resistant to heating up to 1000°F	(3) They prevent the surface from the effect of actinic rays of sun
(4) They have excellent durability and reduce the transfer rate of moisture	

Zinc phosphate $[Zn(PO_4)_2 \cdot 4H_2O]$. It is highly corrosion inhibiting. It is non-toxic and is highly resistant to weathering. It provides an excellent adhesion to the intermediate coating.

(K) Red Lead $[Pb_3O_4]$

It is used extensively as an inhibiting primer for steel. It is economical and resistant to atmosphere. It is a health hazard.

A summary of important pigments is shown in Table 7.7.

7.11 SOLVENT

It is defined as a substance by means of which a solid is brought to a fluid state. Solvents are organic compounds of low molecular weight. A solvent dissolves the binder in the paint. After application a solvent is no longer required and should be evaporated from the paint film, however, before it evaporates, it must dissolve the binder thus giving the paint suitable application properties.

Solvents are two types – hydrocarbons, such as aliphatic and aromatic compounds and chemical solvents, such as alcohols, ethers, ketones, esters, etc. Chemical solvents contain C, H and O molecules.

7.11.1 HYDROCARBON SOLVENTS

Normal paraffin, –hexane, boiling point 69°C.

$$
\begin{array}{ccccccc}
& H & H & H & H & H & H \\
& | & | & | & | & | & | \\
H\!-\! & C & -\!C & -\!C & -\!C & -\!C & -\!C & \!-\!H \\
& | & | & | & | & | & | \\
& H & H & H & H & H & H
\end{array}
$$

Table 7.7 Summary of important pigments

Color	Pigment	Formula
Black	Iron oxide	FeO, Fe_2O_3
	Carbon black	C
	Copper chromite	$CuCr_2O_4$
	Manganese dioxide	MnO_2
	Aniline black	Pyrazine ring
Yellow	Lead chromate	$PbCrO_4 \cdot PbSO_4$
	Cadmium sulfide	CdS
	Yellow iron oxide	$Fe_2O_3 \cdot H_2O$
	Nickel azo yellow	Nickel complexes of p-chlor aniline coupled with 2-4 dihydroxyquinoline
Blue	Ultramarine blue	$Na_7Al_6Si_6O_{24}S_2$
	Prussian blue	$KFe \cdot Fe(CN)_6$
	Cobalt blue	Cobalt chromium aluminate
	Phthalocyanine blue	Copper phthalocyanine
	Indanthrone	—
Green	Chromium oxide green	Chromium sequioxide (Cr_2O_3)
Orange	Chrome orange	($xPbCrO_4$, $yPbO$)
Red	Red iron oxides (pigment red 102)	Aluminum Silicate/Ferric Oxide $Al_2O_3 \cdot 2SiO_2 \cdot 2H_2O – Fe_2O_3$
	Cadmium red	Synthetic ferric oxide (FeO_3)
	Toluidine red	Toluidine in red

Toluidine Red

White		
(A) Opaque	Titanium dioxide	TiO_2 (Opaque)
	Zinc oxide	ZnO
	Antimony oxide	Sb_2O_3
	Aluminum hydrate	$Al(OH)_3$
	Gypsum	$CaSO_4 \cdot 2H_2O$
(B) Transparent	Mica	Hydrous aluminum potassium silicate
	Chalk	$CaCO_3$
	Barium carbonate	$BaCO_3$

Iso-paraffin, 2-methyl-pentane, boiling point 60.3°C.

$$H - \overset{\underset{\displaystyle H}{|}}{\underset{\displaystyle H}{\overset{\displaystyle H}{C}}} - \overset{\underset{\displaystyle |}{|}}{\underset{\displaystyle H-C-H}{\overset{\displaystyle H}{C}}} - \overset{\underset{\displaystyle H}{|}}{\overset{\displaystyle H}{C}} - \overset{\underset{\displaystyle H}{|}}{\overset{\displaystyle H}{C}} - \overset{\underset{\displaystyle H}{|}}{\overset{\displaystyle H}{C}} - H$$

Cyclo-paraffin (naphthene), cyclohexane, boiling point 81°C.

$$
\begin{array}{c}
H_2 \\
C \\
H_2C \qquad CH_2 \\
H_2C \qquad CH_2 \\
C \\
H_2
\end{array}
$$

7.11.2 CHEMICAL SOLVENTS

The following are the major groups:
Methyl iso-butyl carbinol (MIBC), boiling point 130 to 133°C.

$$CH_3 - \underset{\underset{\displaystyle CH_3}{|}}{CH} - CH_2 - \underset{\underset{\displaystyle CH_3}{|}}{\overset{\overset{\displaystyle OH}{|}}{CH}}$$

Diacetone alcohol (keto-alcohol): DAA. Boiling point 166°C with decomposition to acetone.

$$CH_3 - \underset{\underset{\displaystyle CH_3}{|}}{\overset{\overset{\displaystyle OH}{|}}{C}} - CH_2 - \overset{\overset{\displaystyle O}{\parallel}}{C} - CH_3$$

Ethanol, boiling point 78.3°C.

$$CH_3 - CH_2 - OH$$

Hexylene glycol (HG), boiling point 192 to 200°C.

$$CH_3 - \underset{\underset{\displaystyle CH_3}{|}}{\overset{\overset{\displaystyle OH}{|}}{C}} - CH_2 - \underset{\underset{\displaystyle CH_3}{|}}{\overset{\overset{\displaystyle OH}{|}}{CH}}$$

7.11.3 PROPERTIES OF SOLVENTS

The following are the main properties of solvents:

(A) Solvency

The more complex the polymer, the stronger is the solvent required to dissolve it. White spirit is suitable for oil and alkyd resins, and toluene is a good solvent for chlorinated rubber as it has higher solvent power.

(B) Viscosity and Solid Content

The higher the solvent power the less solvent is needed to achieve a particular viscosity so that solid contents remain high.

(C) Ease of Application

Solvents with a high rate of evaporation have a tendency to form a dry mist during spray application. With brush application the evaporation must be slow.

(D) Drying

Physical drying of the paint is slowed down by a slower evaporation rate. Rapid evaporation may lead to several defects.

(E) Boiling Points and Evaporative Rates

The boiling point is related to the rate of evaporation. As a rule, the lower the boiling point, the higher the rate of evaporation.

(F) Flash Point

Flash point indicates a fire risk. Paints with flash points below 23°C are considered highly flammable.

(G) Maximum Acceptable Concentration (MAC) Value

It indicates the danger limit which the body can accept without any hazard to the body. The concentration is given in per cubic meter of air. A lower value indicates a higher toxicity than a higher value. Following is a summary of MAC value of important solvents (Table 7.8).

(H) Solvent Retention

If any amount of solvent is retained in the coating it would cause poor adhesion and retard the chemical resistance of the coating.

(I) Compatibility

Prior to 1960s the industrial coatings were solvent-borne coatings. The resins were primarily dissolved in acids diluted with organic solvents. These solvents were subsequently evaporated.

Their evaporation, however, added to environmental pollution. Hence, great emphasis has been laid by environmentalists to lower the volatile organic contents (VOC) of the solvents to minimize their hazardous influence on the environment. In recent years, coating formulations requiring lesser solvents or no solvents have been commercially developed. The VOC regulations imposed by environmental protection agency (EPA) have seriously restricted the choice for formulation of coatings.

The solvent used must be compatible with the formulation of the coating to obtain a good quality and highly resistant film. Because several resins may be incorporated in the formulation, a proper solvent combination must be selected to improve the compatibility of the resins used in the formulation.

Although the solvents evaporate, they may affect the coating by creating porosity, discoloration, poor gloss and poor adhesion. Hence, selection of solvents is of a prime importance.

7.12 THINNERS (DILUENTS)

They are added before the application to achieve the right viscosity for the paints. These thinners must have the same solvent power as the most volatile element of the particular paint. The rate of evaporation must be compatible with the application method. Table 7.9 summarizes the boiling points and flash points for important solvents and diluents.

7.13 ADDITIVES

These are anti-corrosive pigments enhancers, anti-foams, anti-settling agent, anti-skidding agents, driers and preservatives, which are added in small amount to improve the properties of the paints. Examples of additives are calcite ($CaCO_3$), quartz (SiO_2) and kaolin clay ($Al_4(OH)_8 Si_4O_{10}$). A comparison of different types of coatings is shown in Table 7.10.

Table 7.8 MAC value of important solvents

Solvent	Flash point (°C)	MAC (ppm)
Toluene	6	100
Ethanol	12	1000
Butyl acetate	25	200
Xylene	27	100

Table 7.9 Boiling points and flash points of solvents and diluents (thinners)

Class	Main types	Boiling point or boiling range (°C)	Flash point (closed cup) (°C)
Aliphatic hydrocarbons	(1) Special boiling points spirits, free from aromatics	80/100 to 140/160	−20 to 27
	(2) Aromatic-free mineral spirits (odorless white spirit)	187–212	58
	(3) Mineral spirits (white spirit, contains about 15% aromatics)	160–200	39
	(4) Aromatic white spirit: contains about 40% aromatics	160–200	41
Aromatic hydrocarbons	(1) Toluene	111	6
	(2) Xylene	144	27
	(3) High boiling aromatics (solvent napthas), such as Aromasol H and Solvesso 150	170–200 to 195–215	42 and 66
Alcohols	(1) Ethanol	78	12
	(2) Isopropanol (isopropyl alcohol. IPA)	82	12
	(3) Isobutanol (isobutyl alcohol)	108	27
	(4) Butanol (butyl alcohol)	118	35
Esters	(1) Ethyl acetate	77	−4
	(2) Butyl acetate	127	25
	(3) Isobutyl acetate	118	18
Ketones	(1) Methyl ethyl ketone (MEK)	80	−4
	(2) Methyl isobutyl ketone (MIBK)	117	15
Glycol derivatives	(1) Ethyl glycol (glycol monoethyl ether, cellosolve oxitol)	135	40
	(2) Butyl glycol (glycol monobutyl ether, butyl cellosolve, butyl oxitol)	171	61
	(3) Ethyl glycol acetate (cellosolve acetate, oxitol acetate)	156	51

7.13.1 ADVANCES IN COATINGS, WATERBORNE, SOLVENT-FREE AND SOLVENT-LESS COATINGS

By the rapid advancement in coating technology, a considerable confusion has been caused by terminologies related to the above coatings. A brief description of each type is given below to clear the confusion.

(A) Waterborne Coatings

The waterborne coatings gather popularity in recent years because of their avoidance of

Table 7.10 Summary of selected protective coatings

Coating type (curing mechanism)	Advantages	Disadvantages
Drying oils (oxidation)	Very good application properties Very good exterior durability Outstanding wetting and penetration qualities Excellent flexibility Good film build per coat Relatively inexpensive	Slow drying Soft films – low abrasion resistance Poor water resistance Fair exterior gloss retention Poor chemical and solvent resistance
Alkyd (oxidation)	Excellent exterior durability Low cost Excellent flexibility Excellent adhesion to most surfaces Easy to apply Very good gloss retention	Poor chemical and solvent resistance Fair water resistance Poor heat resistance
Epoxy ester (oxidation)	One-package coating – unlimited pot life Hard, durable film Very good chemical resistance Moderate cost	Poor gloss retention Film yellows and chalks on aging Fair heat resistance
Acrylic (evaporation)	Rapid drying Excellent durability, gloss and color retention Very good heat resistance Hard film	Poor solvent resistance High cost Low film build per coat Blasted surface desirable
Chlorinated rubber (evaporation)	Rapid drying and recoating Excellent chemical resistance Excellent water resistance Good gloss retention	Poor solvent resistance Poor heat resistance Low film build per coat Blasted surface desirable
Epoxy (polymerization)	Excellent chemical and solvent resistance Excellent water resistance Very good exterior durability Hard, slick film Excellent adhesion Excellent abrasion resistance	Two-package coating – limited pot life Curing temperature must be above 50°F Poor gloss retention Film chalks and yellows on aging Sandblasted surface desirable Fair acid resistance

(Contd)

Table 7.10 *(Contd)*

Coating type (curing mechanism)	Advantages	Disadvantages
Epoxy emulsion (evaporation, polymerization)	Very good chemical and solvent resistance (fumes and spills) Hard, abrasion-resistant film High film build per coat Excellent adhesion, particularly to aged, intact coatings Easily topcoated after extended periods of of time with a variety of coating types May be applied directly to clean, dry concrete surfaces	Two-package coating – limited pot life Curing temperature must be above 50°F Not recommended for immersion service (chemicals, solvents, water) Develops surface chalk on aging
Silicone (polymerization – heat required)	Excellent heat resistance Good water resistance and water repellency	Must be heat-cured Very high cost Poor solvent resistance Requires blasted surface Limited chemical resistance

health-hazardous solvents, environmental friendliness, ease of cleaning and lack of repulsing odor. Waterborne epoxy coatings and waterborne polyurethane are being used widely in industry. Latex is a very popular waterborne coating. It is a stable suspension of polymeric particles in water. The water forms a continuous phase. Once applied, the water evaporates and polymer particles are forced together overcoming the streak and ionic forces. The particles form a continuous film by coalescence. The particle size in latex has a very strong influence on the coating properties. Suspension polymerization is a technique for preparing polymer dispersed in water. The monomer is dispersed in water in small droplets in the media and is maintained by rigorous stirring. Polymerization occurs in each monomer droplet. Some of the inter-relative terms for polymer in liquid media are defined below.

Dispersion. This is a stable two-phase mixture, comprising at least two compounds, a continuous phase and a dispersed phase (discontinuous phase).

Emulsion. This is a two-phase liquid system in which small droplets of one liquid are immiscible and dispersed in a second continuous liquid phase.

Suspension. Dispersion in which the dispersed phase is a solid and the continuous phase is a liquid.

Foam. Dispersion of a gas in a liquid or a solid.

Aerosol. Dispersion of solid or liquid in gaseous media (air).

(B) Solvent-less and Solvent-free Coatings

Both types are based on low molecular epoxy resin. Solvent-free coatings, for instance epoxy and polyurethane, do not contain an active solvent unlike the conventional solvent-borne coatings (coatings in which a solvent is used to dissolve the binder). Solvent-free coatings offer the following advantages:

- Outstanding impact and wear resistance
- High impermeability to water
- Excellent resistance to immersion
- Excellent abrasion resistance
- Environmental friendly

They have been successfully used in ships and oil tankers.

The solvent-less coatings contain a very low solvent content (5.2 volume%). The coatings are based on low molecular epoxy resins. Health regulations now often specify their use, to avoid solvent vapors.

7.13.2 SHOP PRIMERS AND WASH PRIMERS

(A) Shop Primers (Non-designated Prefabrication Primers)

Steel plates, profiles and steel products are passed through a prefabrication priming plant where steel is coated by a special primer after being cleaned in a centrifugal blast machine. The binders used in the prefabrication primer are alkyd, epoxy, zinc silicate. Iron oxide and zinc are used as pigments. Zinc is used both in organic and inorganic prefabrication primers. Iron oxide is used only in organic binders.

(B) Wash Primers

They are used as primers in degreased aluminum wherein the acid part would etch the surface of aluminum and increase adhesion. They are also applied as primers on a metallized surface prior to coating. They are widely used in automobile coatings.

7.14 METALLIC COATINGS

The use of metallic coatings is only justified by longer life as they are substantially more expensive than the organic coatings. Metallic coatings protect the substrate either by acting as a barrier between the environment and the substrate or by corroding preferentially with reference to the substrate. The following are the major methods used for applying metallic coatings:

(a) Electroplating or electrodeposition
(b) Diffusion
(c) Spraying
(d) Hot dipping
(e) Chemical vapor deposition
(f) Physical vapor deposition
(g) Welding, cladding, bonding
(h) Miscellaneous methods.

7.14.1 ELECTROPLATING

Metallic coatings are obtained in a conducting substrate by electro-deposition. The metal which is coated is exposed to a solution container containing a salt of the coating metal in a specially designed tank. The metal which is to be electroplated (workpiece) is made the cathode whereas the anode consists of a rod or sheet of the coating metal. The cathode is connected to the negative terminal of a DC power source and anode to the positive terminal. A specified voltage is given to the system depending on what metal is to be electro-deposited. The anode contains the rod of the metal which is to be electro-deposited, for instance, a copper rod is made the anode if copper is to be electro-deposited. If an anode of inert metal is to be used, a suitable metal salt must be added to the electrolyte. Pure metals, alloys and mixed metals can be electro-deposited (Fig. 7.1). Three methods are used – vat, selective plating and electroless plating.

(A) Vat Plating

The electro-deposition process is performed in tanks or vats having capacities up to thousands of liters. The workpiece is fixed in a jig and suspended in the electrolyte. Inert anodes which do not dissolve in the electrolyte are suspended a few centimeters from the workpiece. The DC current is supplied by a transformer–rectifier (4–8 V). Thin coatings obtained are mainly used for

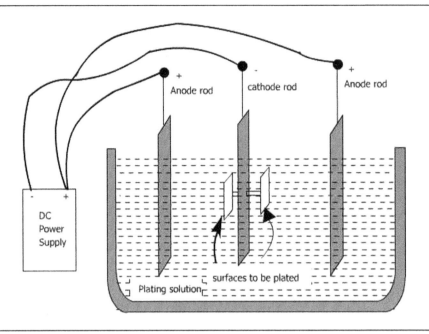

Figure 7.1 Plating bath

decorative purposes. Dense coatings are produced by vat plating. The thickness of the coating is proportional to the current density.

(B) Selective Plating

In this process, electro-deposition can be made on the desired localized areas without the need for masking and without immersion of components. The anode is mounted in an insulated handle and covered by an absorbent pad soaked in the electrolyte. The work is connected to the negative side of a DC power source and circuit is completed by the contact of absorbent pad with the workpiece. The process makes masking unnecessary. The deposition rate is higher than vat plating and this process enjoys the benefit of portability. The process is illustrated in Fig. 7.2.

(C) Electroless Plating

The process involves basically the reduction of metal ions to produce metal atoms which are deposited on the cathode (workpiece). Pure metals, such as copper, nickel, cobalt, gold, silver, etc. can be deposited from their salts by the reduction process. For instance, the following is the reduction chemistry for deposition of nickel:

$$Ni^{++} + H_2PO_2^- + H_2O \rightarrow Catalyst$$

$$\rightarrow Ni(metal) + 2H^+ + H_2PO_3$$

The process is conducted in PTFE-lined stainless steel tank at 90°C. Uniform deposits are obtained.

7.14.2 DIFFUSION COATINGS

In this method, the surface of the metal to be coated is modified by diffusing into it at a high temperature into a metal or an element, which would provide the required resistance when combined with the parent metal. Such coatings are called '*diffusion coatings*.' Diffusion coatings can be applied to a range of metals and alloys, such as nickel, titanium and molybdenum, but the

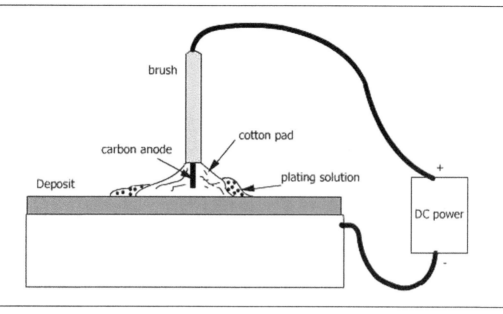

Figure 7.2 Brush or selective plating

widest use is on ferrous metals. Examples are zinc diffusion coatings and aluminum diffusion coatings.

7.14.3 FLAME SPRAYING

The general procedure is to melt the coating material and blow it on to the surface to be coated. The coating material is in the form of very small molten particles or droplets. Four methods based on the form of the coating material are generally used:

(1) The coating material is in the form of rod which is melted by an oxyacetylene flame and blown onto the surface to be coated.

(2) The coating material is in the form of a powder which is heated by an oxyacetylene frame, atomized and blown onto the surface. A cross section of a typical powder flame gun is shown in Fig. 7.3.

(3) The coating is in the form of a wire. The wire is fed into the central orifice of a nozzle. It passes through an oxyacetylene flame and sprayed to the metal surface.

(4) The coating material in the form of a wire is heated by passing it through the plasma of

an electric arc. The resulting molten metal is blown out of the arc by an auxiliary gas stream, as droplets.

The process is versatile with low capital investment. In the electrostatic spraying process, the particles released from the spray gun are electrostatically charged and propelled at low velocity by air or revolving spray head. Too much air pressure is to be avoided. This procedure produces a good wrap around without the need to positioning the workpiece. The process is suitable for tubular articles.

7.14.4 HOT DIP GALVANIZING

This is a process by which iron and steel can be treated to prevent rusting. It involves dipping an article into a molten bath of zinc which reacts with the iron and forms a coating.

Firstly, heavy greases, paints and lacquers are removed by hot alkali and cold alkali solutions. Surface contamination is removed by grit blasting. Scales and oxides are removed by pickling solution of H_2SO_4 or HCl.

The object to be galvanized is immersed in a molten bath of zinc at 450°C. When immersed

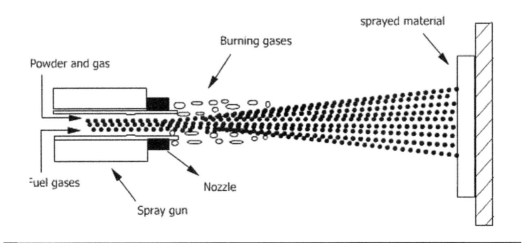

Figure 7.3 Cross section of a typical powder flame gun

in galvanized bath, both iron and steel are immediately wetted by the zinc and react to form an iron zinc layer. At the normal galvanizing temperature, 445–465°C, the initial rate of reaction is very rapid. The main thickness is formed during this period. The normal immersion period is 1–2 min.

A hot dipped galvanized coating consists mainly of two parts: an inner layer of zinc and iron in contact with the base metal, and an outer lay of unalloyed zinc. Alloying additions are made to reduce alloy layer formation and improve ductility. When the coating is entirely made of alloy layer, it is called 'gray coating.' Firstly, heavy greases, paints and lacquers are removed by hot alkali and cold alkali solutions. Grit blasting is done to remove surface contamination. The oxides are removed by pickling as shown earlier.

Various metals, such as aluminum and zinc, are added to the zinc bath to modify the coating. Addition of aluminum produces an even coating and addition of tin improves adhesion. Fluxing is necessary to absorb the impurities on the surface and to ensure that only clean steel comes in contact with molten zinc. In the dry process, the workpiece is first degreased in a hot alkaline solution, and rinsed with hot and cold water, respectively. After rinsing it is pickled in HCl or

H_2SO_4 to remove the mill scale, oxides and rust. It is rinsed again in water to remove iron salts. After degreasing, the workpiece is immersed in a tank of flux and dried. Drying can be done on hot plates or in ovens at 120°C. The dried workpiece is then immersed in the molten bath of zinc as described above.

In the wet galvanizing process, the workpiece is not fluxed and dried prior to immersion in the molten zinc bath. It is directly immersed in a molten bath of zinc containing a blanket of flux on the surface of zinc. The flux cover used in the wet process is composed of zinc ammonium chloride. The workpiece is withdrawn through a flux blanket.

The surface of the workpiece is wetted and cleaned as it passes through the flux into the molten bath. The dry process produces a lower quality of dross and a cleaner environment can be maintained in the plant. Less space is required for the wet process and a better surface for galvanizing is produced as the workpiece passes through the flux blanket.

A microphotograph of a section of hot dipped galvanized coating in mild steel is shown in Fig. 7.4.

Small quantities of 0.005% iron and aluminum reduce the rate of oxidation and improve the brightness.

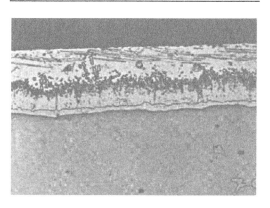

Figure 7.4 Photomicrograph of a section through a hot dip galvanized coating on mild steel

Important Considerations

- Bath temperature = 445–465°C (830–870°F)
- Period of immersion (generally 1–5 min)
- Withdrawal rate (optimum) = 1.5 m/min (5 ft/min)
- Cooling: free circulation of air
- Coating thickness (ounces per square foot)
- Coating weight requirements: given in ASTM A-123 and ASTM A-153.

(A) Electrogalvanizing

In electrogalvanizing, the furnaces, galvanizing pots and cooling tower of hot dipped processes are replaced by a series of electrolytic cells through which steel strip passes. In each of the plating cells, electrical current flows through a zinc solution from anode to cathode (workpiece). Generally soluble anodes made of zinc slabs are used. These anodes are dissolved to provide zinc which is deposited on the cathode (workpiece). Zinc is bonded to steel electrochemically and not chemically as in the hot dipping process. There is no alloy formation in the electrogalvanizing process. Continuous hot dipped galvanizing and electrogalvanizing lines are shown in Fig. 7.5.

7.14.5 MECHANISM OF PROTECTION

Zinc protects the steel substrate from corrosion by (a) physically protecting the steel from the atmosphere and (b) by sacrificial corrosion of zinc in the environment. Zinc has a potential of $-0.763\,V$ and iron a potential $-0.44\,V$. Zinc is clearly anodic to steel.

Zinc is oxidized to ZnO in dry air. In the presence of moisture, ZnO reacts further to form $Zn(OH)_2$.

$$Zn + 2H_2O \rightarrow Zn(OH)_2 + H_2$$

$Zn(OH)_2$ may react further to form a protective film of $ZnCO_3$ by reacting further with CO_2.

The corrosion rate of Zn is, therefore, retarded. Zinc coating tends to dissolve in chloride and sulfur dioxide containing environment. The formation of $ZnCO_3$ retards the rate of corrosion.

When the protective film, however, dissolves zinc undergoes oxidation,

$$Zn \rightarrow Zn^{2+} + 2e$$
$$Zn + 2H_2O \rightarrow Zn(OH)_2 + H_2$$
$$2AlCl_3 + 3Fe \rightarrow 3FeCl_2 + 2Al$$
$$Sn \rightarrow Sn^{2+} + 2e$$
$$2H_2O + O_2 + 4e \rightarrow 4OH^-$$

which releases two electrons. These electrons are accepted by iron which consumes the electrons (Fig. 7.6). Zinc thus becomes the anode and iron (the substrate) becomes the cathode. Hence, while zinc corrodes it protects the steel substrate which consumes the electrons and becomes the cathode. The potential of steel becomes more negative by consuming electrons and a cathodic reaction is formed on its surface. Hence, steel is cathodically protected. Thus, steel is protected by zinc mainly by (a) formation of a partially protective film of Zn on steel surface and (b) cathodically protecting the steel substance from corrosion by sacrificial protection when the zinc protective film breaks down in an aggressive environment.

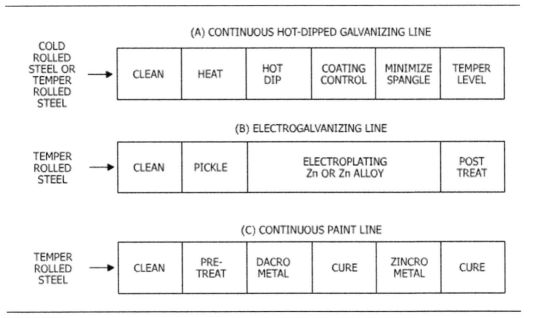

Figure 7.5   Finishing and coating flow line

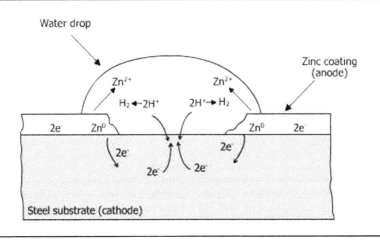

Figure 7.6   Corrosion mechanism under a water droplet. The Zn coating prevents corrosion

7.14.6 RESISTANCE OF GALVANIZED COATINGS TO CORROSION

(a) **Environment**. Zinc in dry air forms ZnO and in moisture reacts further to form zinc hydroxide ($Zn(OH)_2$). It can further react with CO_2 to form $ZnCO_3$ which is protective.

(b) **Natural water**. In hard water the calcium and magnesium salts act as cathodic inhibitors and insoluble hydroxides of Ca^{++} and Mg^{++} are precipitated. The rate of corrosion in soft water is ten times higher than in hard water.

Corrosion rate is reported to be lowest in a pH range of 6 to 12.5.

(c) **Seawater.** The rate of corrosion of glavanized steel is in the range of 5–10 mg/dm^2/d in seawater.

(d) **Soils.** In soils of poor drainage and low resistivity, the life of a galvanized coating becomes shorter compared to that with good drainage and high resistivity soils.

(e) **Temperature.** Reversal of polarity order of iron and zinc in many hot aerated waters at 60°C or above has been observed in many instances. This causes the corrosion of iron to take place in place of zinc in this situation which is very undesirable. Also, water which is high in carbonates and nitrates may bring about the reversal of polarity whereas chloride and sulfate contents generally decreases the tendency of reversal.

7.15 ALUMINUM COATINGS

The following methods are available for coating metals with aluminum:

(1) Spray aluminizing.
(2) Hot dipping.
(3) Calorizing.
(4) Vacuum deposition.
(5) Electroplating.
(6) Vapor deposition.
(7) Cladding.

Two types of aluminum coatings are generally employed: (a) aluminum alloy containing 5–11% Si and (b) pure aluminum.

7.15.1 SPRAYED COATINGS

Either pure aluminum or an aluminum alloy is applied to the surface to be coated by a pistol fed by either aluminum wire or aluminum powder. The coating applied is generally 0.1–0.2 mm thick.

7.15.2 HOT DIPPED

Molten baths of aluminum for hot dipping usually contain silicon (7–11%) to retard the growth of brittle aluminum intermetallic layer. The steel strip is heated in the high temperature furnace in an oxidizing atmosphere to remove organic oils followed by heating in a reducing furnace to reduce the oxide layer. The process consists of

(a) surface preparation,
(b) heat treatment and
(c) immersion coating with aluminum.

The bath temperature is maintained between 620–710°C. The coating thickness is generally 0.025–0.75 mm. The batch method is shown in Fig. 7.7.

7.15.3 CALORIZED COATINGS

Articles are heated in a hydrogen atmosphere in contact with a mixture of aluminum oxide and 3% aluminum chloride. After removal from the mixture the articles are heated in the range 620–710°C for 48 h. The coating thickness is generally 0.025–0.75 mm. The calorized coating must not be deformed.

7.15.4 VACUUM DEPOSITION

In vacuum metallizing aluminum is evaporated at high temperature by passing a heavy current through tungsten filaments around which aluminum is wound. The operation chamber is maintained in a high vacuum. The surface is at room temperature or relatively lower temperature. The vaporized aluminum is condensed on the surface of the workpiece. Coatings of the order of 0.021–0.075 mm are generally applied. Sputter application is similar, however, it is carried out at a vacuum of 13 N/m^2 at a potential of 1 kV.

7.15.5 ELECTROPLATED COATINGS

Aluminum coatings can be deposited from organic solution if a sufficient control is

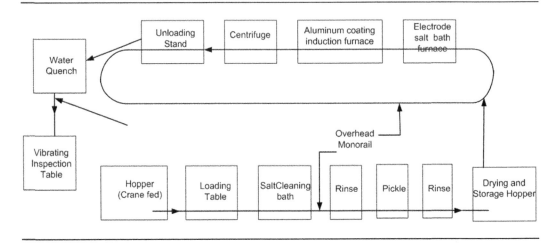

Figure 7.7 Automatic line for high-production aluminum coating of small parts by bath hot dip method. (By kind permission of ASM, Metals Park, Ohio, USA)

maintained. Solutions of aluminum chloride, benzoyl chloride, nitrobenzoane and formamide are used. A current density of 3.2–3.5 kA/m^2 is required. Aluminum coating can also be obtained from aluminum chloride. The bath is operated at 50°C and a current density of 30 A/dm^2 is required.

7.15.6 VAPOR DEPOSITION

In this process, vapors of a metal bearing compound are brought into contact with a heated substrate and a metal compound is deposited on the surface. Steel is exposed to dry aluminum chloride in a reducing atmosphere at 1000°C and aluminum is deposited on the surface.

$$2AlCl_3 + 3Fe \rightarrow 3FeCl_2 + 2Al$$

7.15.7 CLADDING

Bonding of steel to aluminum is obtained by rolling, extrusion or drawing. The bonding of aluminum to steel is obtained by rolling at 540°C. Aluminum alloy-base metal coated on both sides by pure aluminum is called '*Alclad*.' The method of application has a significant effect on the corrosion resistance of aluminum coating.

7.15.8 CORROSION RESISTANCE OF ALUMINUM COATING

(A) Atmosphere

Aluminum coatings are more resistant to corrosion in the atmosphere than zinc coatings. The formation of an Al_2O_3 layer affords protection to the base metals. A film of hydrated aluminum sulfate is formed which is protective.

(B) Natural Water

The air formed oxide/hydroxide film is destroyed as soon as it is immersed in water, unless the rate of reformation of film from oxygen in water is more than the rate of dissolution of film by Cl, NO_3 and SO_4^{2-}. Soft waters are least aggressive to aluminum. The coatings become passive in a pH range between 4 and 9 and corrode rapidly in acid and alkaline solution.

(C) Seawater

Aluminum coated steel is subject to pitting in seawater. The presence of copper in small amounts causes severe microgalvanic corrosion.

Pitting may be initiated by the breakdown of the oxide film at weak points in seawater or brackish water containing a high chloride content.

7.16 TIN COATINGS

Tin coatings are widely used on non-ferrous metals, such as copper, lead, cadmium and nickel. However, tin coatings on low carbon steel are extensively used in the food industry in tin cans.

The base metal for tin-plating is low carbon steel. The term 'tin-plate' is reserved for a low carbon steel strip coated on both sides with a thin layer of tin. In recent years, the hot dipping process for tin plating has been replaced by electroplating because of the improved film properties obtained by the latter process.

The metal to be plated is in a strip form (low carbon steel in this case) of 0.15–0.50 mm thickness. To improve the mechanical properties, the strip is a tempered roll. Medium tinning process plant operates on a continuous strip basis, each coil of steel sheet being welded to the next. The steel sheets travel at a speed of 8 m/s. The steep strip passes through tanks of acid cleaning and alkaline cleaning solutions and pickling, if required. The following is the typical composition of a typical electrolyte of an alkaline tinning both. The bath is operated at 75°C.

Sodium stannate 90 g/l
Sodium acetate 15 g/l
Sodium hydroxide 8 g/l
Hydrogen peroxide 0.1 g/l

This may be deposited from alkaline or acid solution. Acid tinning baths contain 70 g/l of stannous sulfate together with substances, such as phenolsulfonic acid gelatins and phenolic compounds. The electrolyte temperature is 50°C.

Four layers are observed on a substrate of tin coated steel.

- Steel sheet (200–300 μm)
- A thin layer of Fe–Sn intermetallic compound on both sides (0.08 mm)
- Free tin layer (0.3 μm)

- Passivation film created by immersion in H_2CrO_4
- Oil film of a lubricant applied electrostatically (0.002 μm).

Tin plating on food cans are extremely thin (0.5 μm). Additional protection is given by application of lacquers. Can exteriors are painted for decoration and corrosion prevention. Coatings based on acrylic resins or polyester is used. Interior coatings are based on organosalt, to prevent unacceptably high Sn dissolution with time, for some types of foodstuffs.

7.16.1 MECHANISM OF CORROSION AND CORROSION PROTECTION OF TIN CANS

Tin is cathodic to low carbon steel $\left(E^\circ_{Sn|Sn^{++}} = 0.136\,V, E^\circ_{Fe} = -0.440\,V\right)$. Hence, steel corrodes in preference to tin. This may be true on the outside of the tin can but not inside the tin can.

Inside the tin can Sn^{++} forms complex ions with the organic liquids food. This causes reversal of polarity and tin becomes active and corrodes as Sn^{++}. The corrosion of the base metal (carbon steel) is thus prevented by the sacrificial action of tin coating. The tin ions also inhibit the corrosion of base carbon steel by plating on the iron. Tin is plated on steel as Sn–Fe alloy which has a more noble potential than steel. Thus, within the food container the corrosion of steel is prevented.

Tin corrodes in oxygenated solutions. Some oxygen is trapped within the tin can. However, as soon as the amount of oxygen is consumed by cathodic reduction, the dissolution of tin stops.

$$\text{Anodic} \quad Sn \rightarrow Sn^{2+} + 2e$$

$$\text{Cathodic} \quad 2H_2O + O_2 + 4e \rightarrow 4OH^-$$

If, however, the amount of oxygen is very large, the magnitude of corrosion of tin would be high and hydrogen would also evolve. The amount of oxygen depends upon the volume between the top and bottom contents of the tin can. If hydrogen is liberated, the tin can swells up and indicates can failure. A consumer should not buy a bulged can. The presence of sulfur, copper, phosphorus and silicon have a degrading effect on steel. Several can lacquers are applied to protect the tin can from corrosion failure. The selection of lacquer would, however, depend upon the food content in the tin can. Thick coatings of tin are applied to prevent external corrosion.

7.17 NICKEL COATINGS

Nickel is one of the most important coating materials because it has the same strength as steel, but it is extremely resistant to corrosion. Nickel is most commonly applied electrolytically, but it can also be applied by chemical and cladding techniques.

Several plating baths are used. Three general purpose baths are commonly used: Watts, Sulfamate and Fluoborate. The Watts bath contains the following:

- Nickel sulfate, 225–410 g/l
- Nickel chloride, 30–60 g/l
- Boric acid, 30–45 g/l
- Temperature, 46–71°C
- Current density, 1–10 A/cm^2

As the surface of the nickel coating is dull auxiliary brighteners, such as sodium allylsulfonate, diphenyl sulfonate, as well as salts of Zn, Co and Cd are added into the electrolyte.

7.17.1 CHEMICAL DEPOSITION OF NICKEL

Steel articles are plated by immersing them in a solution of nickel sulfate or nickel chloride at 70°C and at a pH of 3.5. The workpiece is immersed for five minutes, washed with a solution of sodium carbonate, and heated to 750°C. Metals, such as cobalt, palladium and aluminum are added to catalyze the reaction. A coating thickness of 0.03 mm is produced. The chemically deposited coatings have a higher resistance than electroplated coatings.

7.17.2 ELECTROLESS NICKEL PLATING

In contrast to electrolytic nickel coating, electroless nickel coating, is deposited without any current as the name indicates. The metal is formed by the reduction of nickel ions in solution by a reducing agent. Sodium hypophosphite is used as a reducing agent. The following is the mechanism of reduction.

(a) Oxidation of hypophosphite to orthophosphate in the presence of a catalyst.

$$(H_2PO_2)^- + H_2O \xrightarrow[\text{Heat}]{\text{Catalyst}} H^+ + (HPO_3)^=$$
$$+ 2H_{(abs)} \uparrow$$

[H_2 is absorbed on the catalyst surface and $(H_2PO_2)^-$ is transformed to $(HPO_3)^=$.]

(b) Reduction of Ni^{++} by adsorbed hydrogen.

$$Ni^{2+} + 2H_{absorbed} \rightarrow Ni + 2H^+$$

(c) Some hydrogen is used to reduce the amount of $(H_2PO_2)^-$ to form H_2O, OH^- and P.

$$(H_2PO_2)^- + H_{(abs)} \rightarrow H_2OOH^- + P$$

(d) Oxidation of hypophosphite to $(HPO_3)^=$ and release of gaseous hydrogen.

$$(H_2PO_2)^- + H_2O \rightarrow H^+HPO_3^= + H_2 \uparrow$$

Complexing agents are added to prevent the oxidation of reduced nickel and to control the pH. Inhibitors are also added to prevent the decomposition of the solutions in the bath. Typical compositions of alkaline and acid baths are given in Table 7.11.

7.17.3 PROPERTIES

The corrosion resistance of electroless nickel is superior to that of electrodeposited nickel. The coating may contain 6–12% phosphorus which increases the resistance to corrosion. A uniform coating thickness is obtained by electroless nickel.

Table 7.11   Typical compositions of alkaline and acid baths

Alkaline bath		Acid bath	
Contents	Concentration (g/l)	Contents	Concentration (g/l)
Nickel chloride	45	Nickel sulfate	21
Sodium hypophosphite	11	Sodium hypophosphite	24
Ammonium chloride	50	Lactic acid	28
Sodium citrate	100	Propionic acid	2.2
pH	8.5–10	Lead	1
Temperature	90–95°C	pH	4.3–4.6
		Temperature	190–205°C

The adhesion of electroless nickel coating to the substrate is excellent. The coatings have high strength, limited ductility and a high modulus of elasticity. Nickel coating is not resistant to caustic solutions.

7.17.4 CORROSION RESISTANCE OF NICKEL COATINGS

Nickel coatings are resistant to dry gases, such as carbon dioxide, hydrogen, ammonia. They are also resistant to carbon tetrachloride, oil, soaps and petrol. Nickel coating increases fatigue strength. Nickel coatings also minimize corrosion fatigue. Nickel coatings are not resistant to nitric acid and environments containing chloride.

The life of nickel coatings may be further increased by a thin overlay of microcracked chromium because corrosion would not penetrate the nickel directly but spread laterally.

Nickel is generally plated as a part of multilayer coating system. The outermost layer is generally 120 mV more active than the substrate. Hence, if any corrosion were to occur, it would be confined to the top layer as it would be anodic to the substrate. The coating system comprising the dull nickel, bright nickel and chromium layer is called the composite nickel coating.

7.17.5 APPLICATIONS

Nickel coatings find wide applications in automotive industry, such as bumpers and exhaust trims,

household appliances and tools. They are widely used in pump bodies, heat exchangers, plates, evaporator tubes, alkaline battery cases and food handling equipment.

Nickel tarnishes rapidly in a corrosive environment. Nickel coatings are very useful as long as a decorative appearance is not required. However, if the nickel coatings are to maintain the decorative appearance, they are given a bright chromium overlay. Nickel offers a high degree of resistance to corrosion and oxidation in seawater.

7.17.6 CHROMIUM COATINGS ($E° = -0.74$ V)

As mentioned earlier, a thin overlay of chromium is applied on nickel coating to prevent the corrosion of nickel coatings. Chromium is generally applied in the form of electrodeposit. As chromium coating is resistant to corrosion and has a bright luster, it is used as a decorating coating. Chromium is more active than iron ($E° = -0.76$ V) and has a tendency to become passive.

The electrolyte for chromium plating has the following nominal composition:

- Chromic acid, 300 g/l
- Sulfuric acid, 2.6 g/l
- Sodium fluorides, 0.4 g/l
- Sodium fluosilicate, 0.2 g/l
- Temperature, 90°C
- Current density, 30 A/dm^2

The chromium deposits are quite porous, hence, corrodant can find easy entry. If the layer is more than 0.003 mm thick, microcracks appear in the layer. In modern practice, a microcracked layer (more than 0.003 mm) is laid over a microcrack free layer, which is laid on a nickel coating.

7.17.7 CORROSION RESISTANCE

Microcracked chromium coatings are resistant to corrosion in the atmosphere. The coatings are not suitable for use in strongly acidic environments. The chromium coatings may not be impressive on their own merit but their contribution in increasing the life of nickel, copper and other coatings is significant.

7.17.8 APPLICATIONS

Chromium coatings are mainly used in automotive applications. The other important application is in food industry. Tin coating of food cans have been replaced in several instances by chromium coatings. The steel is coated with chromium $(0.008–0.01\,\mu m)$ thin and an organic top coat is laid over the chromium coating. This practice increases the resistance of the outside of the containers to corrosion and increased adhesion to steel.

7.18 CADMIUM COATING

Cadmium is anodic to iron $\left(E_{Cd}^{\circ} = -0.403\,V\right.$ and $E_{Fe}^{\circ} = -0.440\,V\right)$. Cadmium can resist a humid atmosphere better than zinc, but it has a far less protecting power than zinc. The most common method of applying cadmium coating is by electroplating. Cadmium plating is done in cyanide baths containing a mixture of cadmium oxide and sodium cyanide to produce $Na_2Cd(CN)_4$. The following is the formation:

- Cadmium oxide, 24 g
- Sodium cyanide, 75 g
- Sodium hydroxide, 15 g
- Triethoanolamine, 20 ml
- Nickel oxide, 0.2 g
- Current density, 5 A/dm^2

Alternate (if cyanide is to be avoided):

- Hydrated cadmium sulfate, 50 g
- Sulfuric acid, 50 g
- Gelatine, 10 g
- Sulfonated naphthalene, 5 g
- Current density, 1.8 A/dm^2

7.18.1 CORROSION RESISTANCE

In an industrial atmosphere, a $25\,\mu m$ thick coating may last for one year but in a marine environment, the life is significantly increased. In the marine environment, the chloride and insoluble carbonates are produced which are not washed out from the surface. They provide good protection to steel in stagnant condition, soft waters and acid, or alkaline conditions. But cadmium is a health hazard.

7.19 CONVERSION COATINGS

Conversion coatings refer to the types of coating which on application convert the substrate into a compound with desirable properties. The surface so prepared provides a high degree of adhesion and corrosion resistance. Some important conversion coatings are described below.

7.19.1 PHOSPHATE COATINGS

Phosphate conversion coatings are well-known and widely used for applications on steel, zinc and aluminum.

7.19.2 TYPES OF PHOSPHATE COATINGS

Phosphate coatings are of three types: (a) zinc phosphate, (b) iron phosphates and (c) manganese chromium phosphates.

7.19.3 PROCESS OF PHOSPHATING

Phosphate coatings are generally applied by immersion or spraying (Fig. 7.8). The bath contains generally zinc phosphate in phosphoric acid and an oxidizing agent, such as nitrate. The metallic workpiece passes through the following stages. A spray zinc phosphate process for steel and zinc is shown in Fig. 7.9. The following is the process cycle:

(1) Alkaline cleaning with an alkali (3–8 g/l) at 82°C.
(2) Hot rinsing with water at 77°C.
(3) Cold rinsing with water at 25°C.
(4) Acid pickling by H_2SO_4 (15 wt%) at 60°C.
(5) Zinc phosphate bath at 25°C.
(6) Neutralization by $NaNO_2$.
(7) Lubrication with soap at 66°C.

7.19.4 MECHANISM OF ZINC PHOSPHATING

(1) The metal reacts with phosphoric acid producing iron phosphate.

$$3Fe + 2H_3PO_4 \rightarrow Fe_3(PO_4)_2 + 3H_2 \uparrow$$

(2) Formation of an insoluble tertiary zinc phosphate from soluble primary zinc phosphate.

$$3Zn(H_2PO_4)_2 \rightarrow Zn_3(PO_4)_2 + 4H_3PO_4$$

(3) Formation of sludge (ferrous phosphate) by addition of $NaNO_2$.

$$H_3PO_4 + Fe_3(PO_4)_2 + 3HNO_2 \rightarrow 3FePO_4 \downarrow$$
$$+ 3NO + 3H_2O$$

Sludge is the precipitation suspended in the bath. It settles down to the bottom.

7.19.5 IRON PHOSPHATE COATINGS

The basic composition is modified by addition of chemicals, like fluoride. They do not offer as good resistance as zinc phosphate coatings. The following are the reactions:

(a) $2Fe + 3NaH_2PO_4 \rightarrow 2FeHPO_4 + Na_3PO_4$
$\qquad + 2H_2 \uparrow$

(b) $4FeHPO_4 + O_2 \rightarrow 4FePO_4 \downarrow + 2H_2O$

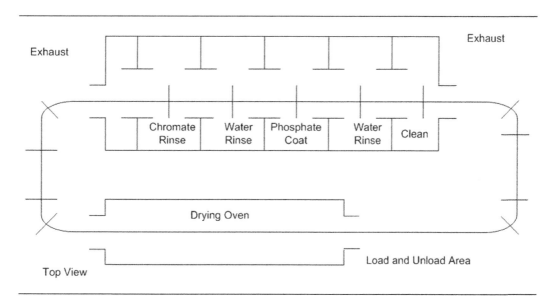

Figure 7.8 Continuous conveyorized spray line for phosphating. (By permission of ASM, ASM Metals Handbook, Vol. 5, p. 446. By kind permission of ASM, Metals Park, Ohio, USA)

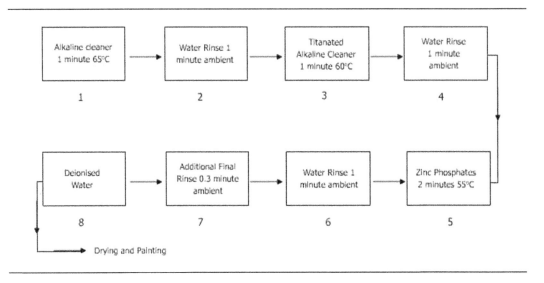

Figure 7.9 Spray zinc phosphating for steel

7.19.6 CHROMIUM PHOSPHATE COATINGS

The coating bath consists of an acidic mixture of phosphate, chromate and fluoride. Fluoride acts as an accelerator. It is applied to aluminum. The mechanism is given below:

(a) $Al + 3HF \rightarrow AlF_3 + 3H^+$
(b) $CrO_3 + 3H^+ + H_3PO_4 \rightarrow CrPO_4 + 3H_2O$

The coating is characterized by the formation of $Al_2O_3 \cdot 2CrO_4 \cdot 8H_2O$.

7.19.7 ADVANTAGES OF PHOSPHATING

Phosphate coatings have a good resistance to corrosion. They are porous and so make an attractive base for application of paints and organic coatings. The rate of corrosion is substantially decreased when paints are applied on phosphate coatings. They are also used as protection against corrosion in combination with sealing film of oil or grease instead of paint. Phosphate coatings find application in military equipment, cars, refrigerators, washing machines, toys and steel strips for painting.

7.19.8 CHROMATE CONVERSION COATINGS

Chromate coatings are generally used on aluminum as well as on zinc and certain other metals. Chromium coating is formed as a consequence of chemical attack on a metal when it is brought in contact with an aqueous solution of chlomic acid, chromium salts, hydrofluoric acid, salts, phosphoric acid or mineral acids. Chromate ions are known for their inhibition properties. If a chromate coating is to be deposited, the passivity on the metal surface in solution must be broken down so that the coating is deposited on the metal surface. This is done by adding ions like chlorides and sulfates which break down the passivity. When the passive film is broken down by chloride ions, hydrogen is released which reduces the chromate ion to $Cr_2O_3 \cdot CrO_3 \cdot xH_2O$, which is sparingly soluble. It is deposited on the metal surface. The chromate treatment process for aluminum is shown in Fig. 7.10. However, chromium is a health hazard.

Corrosion resistance. Chromate coatings are mostly applied on aluminum components, particularly on aircraft components and on aluminum objects which are to be coated with paints. They are beneficial in marine environments. Chromium coatings provide an non-porous bond for all paints.

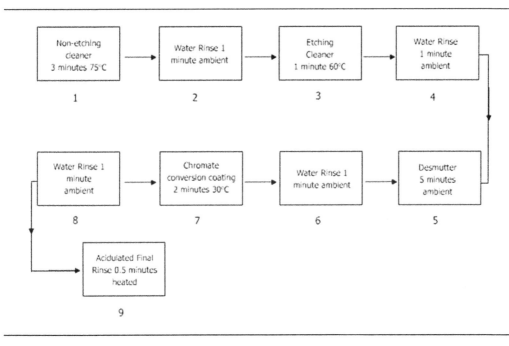

Figure 7.10 Immersion chromate treatment for aluminum. (By permission of oil and colour chemists association, Australia, Paints and their application, Vol. 2, Tafe Educational Books, Kensington, NSW, Australia, 2003)

7.20 ANODIZED COATINGS

Anodizing is the formation of a relatively thick coating and a coherent oxide film on the surface of a metal by making the metal the anode in an electrolytic cell. The most common electrolytes used are chromic acid, sulfuric acid, phosphoric acid and mixtures of theirs. The following are the typical compositions of the bath for anodizing of aluminum.

(a) Sulfuric Acid Bath

Sulfuric acid, 10 wt%
Water, 90 wt%
Temperature, 18°C
Duration, 15–30 min
Current density, 1.2 A/dm^2

(b) Alumilite Bath

Sulfuric acid, 15 wt%
Water, 85 wt%
Current density, 1.3 A/dm^2

A number of trade names have been associated with anodized coatings. Often dyes are applied on anodized coatings before the sealing process. The drying materials are generally hydroxide of copper, cobalt and aluminum.

Bright anodizing is a process in which aluminum is polished followed by the protection of the shining aluminum surface with the deposition of a thin film of oxide. Bright anodizing is carried out on aluminum alloys containing some magnesium. No impurity of any kind should be present in the alloy. For bright anodizing a mixture of phosphoric acid and nitric acid containing some acetic acid is used.

$$\text{Reaction}: \begin{bmatrix} \text{Al} \rightleftarrows \text{Al}^{3+} + 3e\,(\text{Anodic}) \\ \text{H}_2 \rightleftarrows 2\text{H}^+ + 2e\,(\text{Cathodic}) \end{bmatrix}$$

7.20.1 CORROSION RESISTANCE OF ANODIZED ALUMINUM

Aluminum anodized coatings are resistant to corrosion between pH 4 and 8.5. The coatings are suitable for outdoor applications. An optimum resistance to corrosion is obtained if the coating is in the thickness range of 18–30 μm. They show a good resistance to deicing salts. Anodized aluminum is subject to corrosion by magnesium hydroxide, calcium hydroxide, mortar, plasters and cement.

Aluminum anodized coatings find a wide use in food packaging and processing industry as well as packaging of pharmaceutical products because of the high resistance of these coatings to the food and pharmaceutical products. A large amount of anodized aluminum is used in food cans. Within a pH range 4–9, anodized coatings resist almost all inorganic chemicals, however, they are subjected to pitting in aerated chloride solutions. Anodized coatings are resistant to halogenated organic compounds, however, in the presence of water these chemicals hydrolyze to produce mineral acids which may destroy the oxide film on the surface. The corrosion resistance of anodized coatings can be further enhanced by sealing the pores of the coating and incorporating inhibitors, such as dichromate in the sealing solution. Several formulations are available for sealings. Chromium compounds should not be used for food and pharmaceutical products.

7.21 GLASS FLAKE COATINGS

Flakes of chemically resistant glass 3–4 microns in thickness are dispersed in alkyl resins. TiO_2 is used as pigment. A glass flake layer of 5 microns is obtained. The use of glass flakes increases the impermeability and erosion resistance.

7.21.1 PORCELAIN ENAMELS

In the wet process, powdered frit (ground glass) is suspended in water together with clay, and the mix which contains 15% solids. The mix is formed by melting mixtures of B_2O_3, SiO_2, Al_2O_3, ZrO_2, Na_2CO_3 and PbO. For acid resistance, the proportion of SiO_2 in increased. The mix is sprayed onto the metal surface to be coated. The workpiece is heated to dry the coating followed by firing

to 850°C for 5 min. This process is repeated until the required thickness coating is obtained.

In the dry process, the object to be coated is heated to fusion temperature of the frit and the powdered frit is spread on the workpiece. The firing temperature is 700°C for iron and steel coating. Steels coated with porcelain enamels can be used up to 1000°C.

7.21.2 CORROSION RESISTANCE

Porcelain enamels have excellent resistance to most acids and salts but are attacked by alkalies. Resistance to atmosphere is excellent.

Recommended coatings for industrial plants are given in Table 7.12.

7.22 FAILURE OF PAINTS AND COATINGS

A mechanistic understanding of failures of paints and coatings is important for engineers. The failures can be divided in four categories:

(a) Formulation-related failures
(b) Adhesion-related failures
(c) Application-related failures
(d) Design-related failures

An account of important coating failures is given below. Only selected failure categories are described here.

7.22.1 FORMULATION RELATED FAILURES

In this category, failures of coating due to improper formulation are described.

(A) Chalking

(a) **Definition.** It is the failure of a coating by the formation of a powdery layer on a coating surface. As the powder is white, the failure is referred to as chalking (Fig. 7.11).

Table 7.12 Recommended coatings for industrial plants

Service	Mild corrosive conditions					Severe corrosive conditions				
	Surface prep.	Primer 1st	Primer 2nd	Finish coats 1st	Finish coats 2nd	Surface prep.	Primer 1st	Primer 2nd	Finish coats 1st	Finish coats 2nd
Bare steel surfaces operating below 20°C (sweating) and equipment adjacent to cooler towers	Blast Sa 2½	Inorganic zinc 75 microns	—	Phenolic or chlorinated rubber 40 microns	—	Blast Sa 2½	Inorganic zinc 75 microns	Epoxy converted 40 microns	Epoxy converted 50 microns or polyurethane 50 microns	—
ALTERNATE	Blast Sa 2	Epoxy 25 microns	—	Coal tar epoxy 125 microns	Coal tar epoxy 125 microns	Blast Sa 2½	Epoxy 25 microns	—	Coal tar epoxy 150 microns	Coal tar epoxy 150 microns
(a) Metal surfaces operating up to 93°C	Blast Sa 2½	Inorganic zinc 75 microns	—	—	—	Blast Sa 2½	Inorganic zinc 75 microns	—	Epoxy converted (mastic) 150 microns	—
ALTERNATE	Blast Sa 2	Epoxy ester 40 microns	—	Epoxy ester 40 microns	Epoxy ester 40 microns	Blast Sa 2½	Epoxy converted 40 microns	—	Epoxy converted (mastic) 125 microns	Epoxy converted 50 microns
(b) Machinery and electrical equipment	Blast Sa 2 or St 2/3	Epoxy ester 40 microns	—	Epoxy ester 40 microns	Epoxy Ester 40 microns	Blast Sa 2½	Vinyl 30 microns	—	Polyurethane 40 microns	Polyurethane 40 microns

(Contd)

Table 7.12 (*Contd*)

Service	Mild corrosive conditions					Severe corrosive conditions				
	Surface prep.	Primer		Finish coats		Surface prep.	Primer		Finish coats	
		1st	2nd	1st	2nd		1st	2nd	1st	2nd
ALTERNATE						Blast Sa 2	Epoxy ester 40 microns	Epoxy ester 40 microns	Epoxy ester 40 microns	Epoxy ester 40 microns
(c) Galvanized structure and equipment	Solvent clean	Etch primer 8–10 microns	—	Chlorinated rubber 100 microns	Chlorinated rubber 30 microns	Solvent clean	Epoxy con-verted 40 microns	—	—	Epoxy converted or Polyurethane 50 microns
Metal surfaces operating above 90°C. Equipment, manheads, exchanger heads, piping pumps, machinery, furnace, stacks, etc.										
(a) Not over 260°C	Blast Sa 2	Oil/graphite/ zinc 40 microns	—	Oil/graphite/zinc— 40 microns	—	Blast Sa 2½	Inorganic zinc 75 microns	—	Silicone acrylic 35 microns	—
ALTERNATE	—	—	—	—	—	Blast Sa 2½	Silicone acrylic 25 microns	—	Silicone acrylic 35 microns	Silicone acrylic 35 microns

Surface / Condition									
(b) Over 260°C	Blast Sa 2½	Inorganic zinc 75 microns	—	Silicone 30 microns	—	Blast Sa 3	Inorganic zinc 75 microns	Silicone 40 microns	—
(c) Not over 482°C	Blast Sa 2	Silicone 40 microns	—	Silicone 40 microns	—	Blast Sa 3	Silicone 40 microns	Silicone 40 microns	—
Externally insulated metal surfaces: Metal temperature									
(a) −10 to 93°C	Blast Sa 2	Phenolic 100% 35 microns	—	Phenolic 100% 35 microns	—	Blast Sa 2½	Phelolic 100% 35 microns	Phenolic 100% 35 microns	—
(b) −10 to −50°C	Blast Sa 2	Epoxy converted 40 microns	—	—	—	Blast Sa 2½	Epoxy converted 40 microns	—	—
Storage vessels, external surfaces Heavy gravity service (reid vapor pressure 2.0 pounds and under) Shell and roof	Blast Sa 2	Chlorinated rubber 40 microns	—	Chlorinated rubber 40 microns	Chlorinated rubber 40 microns	Blast	Inorganic zinc 75 microns	—	Epoxy ester 35 or chlorinated rubber 35 microns

All microns thicknesses are dry film

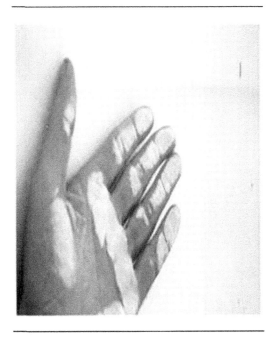

Figure 7.11 Chalking (Courtesy: Sigma Coatings)

(b) **Mechanism**. The binder of the coating disintegrates due to severe sunlight and leaves the pigments which were held by it. The binder continues to disintegrate until corrosion starts to develop on the substrate and the coating surface is worn out. A pigment may interact with ultraviolet radiation and breakdown. A change in the color of a pigment indicates chalking. Previous coatings may not have satisfied the porosity of the substrate.

(c) **Prevention**

(1) Select an appropriate combination of binder and pigment on the basis of experience.
(2) Select pigments with a least tendency for chalking.
(3) Remove loose chalk deposits and apply a new coating system.

(B) Erosion

(a) **Definition**. It is the failure of a coating due to erosion by high winds and sand particles, mostly encountered in deserts.

(b) **Mechanism**. In erosion, it is the high winds or sand particles which hit the coating and disintegrate it. Once the coating is disintegrated and the substrate is exposed, corrosion proceeds rapidly on the substrate, particularly in a humid environment.

(c) **Prevention**. Same as for the chalking.

(C) Checking, Alligatoring and Cracking

Although different in appearance, all the three failures mentioned above are related to the deterioration of the coating with time.

Definitions

(a) **Checking**. Small check-board patterns are formed on the surface of a coating as it ages. The coating becomes harder and the brittleness increases as it ages (Fig. 7.12).

Figure 7.12 Checking (Courtesy: Sigma Coatings)

(b) **Alligatoring**. It is a film rupture caused by the application of a hard dry brittle film over a more softer and extensible film. The failed surface of the coating resembles the hide of an alligator, hence the name alligatoring (Fig. 7.13). It is also called crazing.

(c) **Cracking**. The surface of a coating cracks due to weathering and aging are similar to alligatoring, however, in this case the cracks reach the substrate and cause a more serious defect than either checking or alligatoring (Fig. 7.14). Cracking may also be caused by a fast curing rate.

Mechanism

The basic reason in the above failures is the introduction of stresses in the coating. In the case of checking, stresses are caused by shrinkage of the coating due to weathering, whereas in the case of alligatoring, the stress are caused by the shrinkage of the surface of the coating at a higher rate than the rate at which the body of the coating shrinks. In the case of cracking, stresses are set by continuous weather of the coating. The end result is the rupture of the coating by the stress.

Prevention

Alligatoring can be prevented by selecting proper formulation and allowing specified drying time between coats. The ingredients selected must

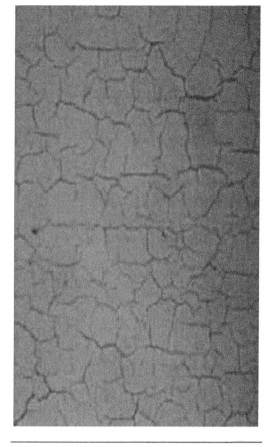

Figure 7.14   Crakling (Courtesy: Sigma Coatings)

show a minimum difference in the rates of expansion and contraction so that stresses created by the above processes are minimized.

Cracking can be minimized by selecting a coating which is resistant to weathering and oxidation. Remove all dust and contaminations and prime, fill and then reapply full paint system. The coating selected must also be flexible. Similar prevention is required for checking. The pigments selected for the latter must be inert and reinforcing.

(D) Wrinkling

Definition. It is the appearance of wrinkles on the surface of the coating (Fig. 7.15).

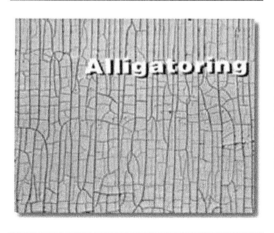

Figure 7.13   Alligatoring (Courtesy: Sigma Coatings)

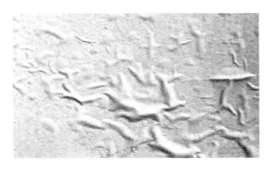

Figure 7.15 Wrinkling (Courtesy: Sigma Coatings)

Mechanism. During curing, the surface is cured more rapidly than the body of the coating. Because of the difference in the rate of expansion of the surface of the coating and body of the coating, serious stresses are created in the coating. These stresses cause wrinkles. Such types of failures occur generally on oil-based coatings.

Prevention. Use of slow drying solvents and mixtures of silicates to control the drying rate. Thin coating should be applied to avoid excessive thickness.

7.22.2 ADHESION-RELATED FAILURES

The failures described under this heading are due to the lack of adhesion of the coatings to the metal. The following may be the reasons for lack of adhesion.

- **Poor surface preparation**. A good surface preparation is a primary prerequisite for coating. Any dirt, grease, rust or scale, if present, must be removed. Before the application of new coating, the old coatings must be removed completely.

    This mill scale must be completely removed from the surface. If it is not removed, the oxygen and moisture may penetrate under the scale and cause the surface to corrode. Once the surface corrodes, the substrate and the coating is destroyed. Loss of adhesion may not occur immediately. However, it is dependent on how severe the environment is.

- **Poor application**. Porosity can result from improper application.
- **Film thickness**. If the film thickness is very high, stresses are introduced by shrinkage and the coating peels off.

Examples

(A) Blistering

(a) **Definition**. This is the failure of a coating by the development of large or small round hemispherical pimples from the surface either dry or filled with a liquid (Fig. 7.16).

(b) **Mechanism**. The blister may develop by one of the following mechanisms:

(1) **Absorption of moisture by the soluble pigments**. When the soluble pigments absorb moisture, the solution becomes concentrated and water is pulled into all the areas where the pigment is dissolved, by the process of osmosis. Osmosis is the transfer of the liquid through the coating from a lesser concentration to a higher concentration. Blisters filled with liquid are, therefore, formed. If some soluble salts are present between the inter-coats or in the substrate, absorption by osmosis would occur and blisters would be formed.

Figure 7.16 Blistering (Courtesy: Sigma Coatings)

(2) **Inadequate solvent release**. If the coating is a rapid drying type, some of the solvent may not be released and is trapped in the coating. When the temperature of the coated surface is raised, the solvent may create a significant vapor pressure and cause the formation of blisters. A similar condition may arise when a top coat is applied over a porous undercoat.

(3) **Poor coating adhesion**. In the areas of poor adhesion, adhesion of coating to substrate or adhesion of inter-coats, the liquids and gases may be trapped and this can lead to the development of blisters.

Salt contamination on a surface may also cause blistering. They may occur in extremely wet environment.

(4) **Improper film thickness**. The thickness of the coating must be strictly in accordance with the specifications for a particular coating.

(5) **Proper choice of primer**. As primer is the foundation of a coating, it is important to select a proper primer according to the intended application of the coating. For instance, if the coating is to be applied in a marine environment, an inhibitive primer must be selected. For immersion service, an impermeable primer must be selected.

(6) **Weather**. High humidity, fluctuating temperatures and wind speeds can lead to failure of coatings by either creating conditions inappropriate for curing or by introducing contamination during coating application. Inhibitive primers can be usefully applied to overcome the effect of moisture adsorption.

(7) **Cathodic protection**. If a structure is cathodically protected, hydrogen gas may be formed in sufficient volume. It may also be formed by stray currents. Hydrogen vapor pressure pushes the coatings up and blisters are formed. The size of the blisters depends on the degree of adhesion of the coating, the salt content in the blisters and the internal pressure of either the liquids or gas inside the blisters. Generally, blistering leads to rust formation. The blister may expand to a sufficient degree to cause lifting of the coating.

(c) **Prevention**. Blistering is a serious type of failure. The only practical remedy is to remove the coating and repair it, if blistering is localized. If blistering occurs all over the surface, completely change the coating which is more expensive but provide a better adhesion. Water soluble pigments in humid environments must not be avoided. Ensure that prior to application of coating the substrate is completely dry and free from any contaminations. Sometimes blistering may be harmless, e.g. blistering of sub-sea coal tar epoxy paints on oil rigs can be caused by cathodic protection but can be a harmless effect.

(B) Peeling

(a) **Definition**. Peeling is the loss of adhesion resulting in detachment and curling out of the paint film from the substrate or between inter-coats. It may occur by a number of different causes given below (Fig. 7.17).

A combination of one or more of the following factors may lead to peeling:

(1) If the tensile strength of a coating exceeds the bond strength between coating and substrate.
(2) If the solvent used for one coating softens the other coating.
(3) If corrosion proceeds under the coat.
(4) Painting of a surface when it is wet with rain or dew.
(5) Inadequate removal of the old coat before the application of a new coat.
(6) Ingress of moisture to the substrate.

(b) **Mechanism**. One or more than one of the factors given above may cause a loss in adhesion of the coating. The paint film is curled from the surface.

(c) **Prevention**. Remove the old coat completely before applying the new coat. Never coat the surface under wet conditions. Do not allow moisture to penetrate. Minimize or eliminate

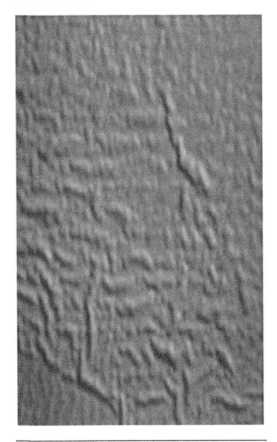

Figure 7.17   Peeling

Figure 7.18   Flaking (Courtesy: Sigma Coatings)

the conditions from (1) to (6) above, which cause peeling.

(C) Flaking

(a) **Definition**. Flaking is the pulling away of the coating from the substrate because of the hard and brittle nature of the coating (Fig. 7.18).

(b) **Mechanism**. It is similar to the mechanism of peeling. The coating which flakes is generally hard and brittle.

(c) **Prevention**. Avoid contamination between coatings and do not apply very thick coatings. Never paint a wet surface and use coatings which are compatible.

(D) Inter-coat Delamination

(a) **Definition**. As the name suggests, this is the loss of adhesion between two coats (Fig. 7.19).

(b) **Causes of delamination**

(1) Not using a compatible coating with the old coating during repairs.
(2) Lack of removal of contamination from the substrate before applying a new coat.
(3) Application of very thick coatings.
(4) An increase in the rate of the curing by sunlight, such as in case of coal tar epoxy coatings.

(c) **Mechanism**. One or more of the above factors may come into play and cause delamination.

(d) **Prevention**. If the delamination is localized, the coating may be repaired by sanding off and recoating with a compatible paint. However, when the delamination spreads over a

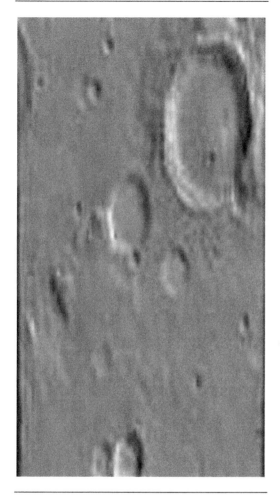

Figure 7.19 Inter-coat delamination (Courtesy: Sigma Coatings)

Figure 7.20 Holidays/Misses (Courtesy: Sigma Coatings)

very large area, the best remedy is to reapply the coating all over again, after removing the old coating.

7.22.3 FAILURE DUE TO IMPROPER APPLICATIONS

(A) Holiday

This is a particular area in which coating has been left uncoated while the rest of the surface has been coated (Fig. 7.20). 'Holidays' includes pinholes.

Holidays can also be caused by dust particles being incorporated during painting.

Remedy. None of the areas must remain uncoated.

(B) Pinholes

These are minute holes which are formed during the application of a paint, which exposes the underlying substrate. The flow of applied paint is not continuous due either to poor mixing of two pack materials or insufficient agitation leading to insufficient wetting of substrate.

Remedy. Apply coating evenly and pay proper attention to the surface. Apply coating by brush evenly, and allow to dry. Apply another coat of paint.

(C) Spattering

This is the defect caused when only droplets of the paint reach the substrate and a gap of bare metal exists between the droplets. The bare metal serves as the '*holiday area.*'

Remedy. Apply paint by spray gun adjusted properly.

(D) Cratering

It is the formation of small bowl shaped depressions in the paint or varnish film. It is caused by the falling of impurities in the wet paint.

Remedy. It can be removed by addition of antifoamers and bubble releasing agents (for solvent-based paints).

7.22.4 FAILURE RELATED TO WELDING AND DESIGN

Weld areas can cause failure of coatings by several causes. Some of the causes are given below:

(a) Interference of the weld flux with the coating creates a lack of proper adhesion.
(b) Splatters (droplets) of welds create cavities which expose the bare metal surface.
(c) Underfilm rusting may proceed by creation of cavities.

Prevention
The first cause (flux interference) can be removed by adopting a proper cleaning process. Welds are the potential holidays in underground pipelines. Weld areas are generally the first regions to show coating failures in a non-aqueous environment.

(A) Edge Failure

The failure of coating on the edge is a design related failure because coating is thinner at the edge compared to the rest of the surface. An edge provides a break in the coating. The coat fails by the development of rust in the coating which penetrates to the substrate.

Remedy. Apply an extra coat on the edges.

7.23 SELECTION OF COATINGS

Selection of an appropriate coating system is a prerequisite for the durability of coating. What are the major factors that would influence the choice? There may be several factors, some of them are more important than others. Below is a list of such factors which deserve consideration:

(1) Cost and life to the first maintenance.
(2) Compatibility with the environment.
(3) Resistance to atmospheric pollution.
(4) Resistance to extreme climates.
(5) Ease of maintenance.
(6) Safety problem.

Recommended coatings for industrial plants are given in Table 7.9.

7.23.1 COST

It is the most important factor for the selection of coating. The cost is related to the importance of the structure to be coated and life of the coating system. The life of the system would, in turn, depend on several factors, such as surface preparation, quality of the coating, proper application, proper curing, maintenance and monitoring. The cost factor is important but not a deciding factor in areas of civil and military strategic importance. Frequency of maintenance is an important factor.

7.23.2 COMPATIBILITY WITH THE ENVIRONMENT

Environment consideration is critical to the proper selection of coating system. For instance, for structures close to seawater, an inhibitive primer of proven quality needs to be considered and preferred over a vinyl primer. The coating components should be compatible with the seawater environment. For a desert environment, the system needs to withstand high wind and fine dust particles to avoid erosion and chalking of the coating.

7.23.3 RESISTANCE TO ATMOSPHERIC POLLUTION

The pollution problem may be caused by a large emission of C and CO_2 from automobiles, in certain cases SO_2 from refineries. In agricultural areas, ammonia may contaminate the atmosphere. These factors deserve consideration before the selection of coating system.

7.23.4 RESISTANCE TO EXTREME CLIMATES

Lack of consideration of climatic factors may lead to premature failure of the coating. For instance, consider three towers, one located in Malaysia (warm and humid with frequent rains), one located in Alaska (with extremely cold climate) and the last one located on the Eastern Coast of Saudi Arabia (with very hot and humid climate with suspended salts and sand particles in the air). In the first situation (warm and humid with frequent rains), the coating system needs to be highly resistant to humidity and wet conditions, and coatings based on epoxy esters binders may offer an ideal choice as these resins are highly resistant to moisture and wet conditions. In the second case, any modest coating system would help, the important point would be to ensure that the coating would not shrink and would maintain its strength in extreme temperatures below freezing. In the third instance, the coating would need to be resistant to humidity, dust particles, heat and corrosive environment. Coatings, such as epoxy esters would be suitable.

7.23.5 EASE OF MAINTENANCE

A system which requires least maintenance is a better choice even at a high initial capital cost and compromise is to be made with the original life of the system.

7.23.6 SAFETY PROBLEMS

Safety is related to the integrity of a coated structure. The coating system, therefore, needs to offer an excellent corrosion protection. Any defects, such as cracks and leakages caused by corrosion, may lead to explosion and fires and result in an economical and technical catastrophe. Hence, safety related issues deserve serious considerations.

QUESTIONS

FUNDAMENTALS OF COATING

Multiple Choice Questions

1. Mark the statement which best defines '*paint.*'

   a) A product in a liquid form when applied to a surface forms a dry film having protective, decorative or specific technical properties
   b) A layer of dried paint film resulting from the application of a paint
   c) The formation of a tenacious, homogeneous and adherent film on a substrate
   d) A barrier which prevents the substrate from the environment.

2. In an inhibitive type of coating

   a) the pigment contained in the primer reacts with the moisture contained in the air and passivates the steel surface
   b) the primer forms a barrier between the steel surface and the environment
   c) the moisture does not penetrate to the inhibitive primer
   d) the water absorbed by the coating does not play any role in the inhibition

3. A zinc-rich primer protects the steel surface by

   a) forming a strong bond with the steel surface through alloy formation between zinc and steel
   b) acting as an impervious barrier between the steel surface and the environment
   c) cathodic action of zinc-rich primer with the steel surface
   d) forming a physical bond with the steel surface

4. An impervious coating protects the steel surface by

   a) acting as an inert barrier against air, oxygen and carbon dioxide
   b) forming a chemical bond with the metal surface
   c) acting as a cathode to the steel surface
   d) passivating the steel surface

5. Which one of the following is not a function of the primer?

   a) Binding to metal surface
   b) Adhesion of topcoats
   c) Chemical resistance
   d) Providing a seal for the coating system

6. The primary function of a binder is to

   a) blend the pigments together into a homogeneous film
   b) provide a pleasant appearance to the coating
   c) improve the flexibility of coating
   d) control the thickness of the film

7. Chlorinated rubber is an example of

   a) lacquers
   b) oxygen-reactive binders
   c) co-reactive binders
   d) condensation binder

8. A pigment may reinforce a film by

   a) forming a barrier to ultraviolet radiation
   b) increasing the thermal coefficient of expansion
   c) dispersing of pigment particles
   d) transferring heat to the localized heat site

9. In general, solvents

   a) convert binders into workable fluids
   b) distribute the pigment in the binder evenly
   c) are universal
   d) are not compatible with binders

10. Which of the following is not affected by the choice of solvents?

   a) Viscosity
   b) Drying speed
   c) Gloss
   d) Hiding power

Conceptual Questions

1. What is the difference between 'paint' and 'coating?'
2. From a purely functional point of view, state the important components of a coating system?
3. State two materials from each category used for castings:

   a) metals
   b) inorganic materials
   c) organic materials

4. State the chronology of organic coatings starting from the raw materials to solid film formation.
5. State three characteristics of PVC coatings.
6. What improvements in the properties of paints are expected by addition of copolymer binders?
7. What is the main ingredient which determines the properties of the oils?
8. What determines the drying properties of the oils?
9. State three factors which influence the film properties.
10. Differentiate between active solvents and diluents.

METAL COATINGS

Multiple Choice Questions

1. In calorizing

   a) the steel strip is heated in a high temperature furnace in an oxidizing atmosphere
   b) either pure aluminum or an aluminum alloy is applied to the surface

c) articles are heated in a hydrogen atmosphere in contact with a mixture of aluminum oxide and 30% aluminum chloride and after removal from the mixtures, the articles are heated in the range of 620–710°C

d) articles in an oxidizing atmosphere are heated in a mixture of 30% aluminum chloride and 3% aluminum oxide

2. The process represented by the equation, $2AlCl_3 + 3Fe \rightarrow 3FeCl_2 + 2Al$, is called

a) vapor deposition process
b) vacuum deposition process
c) electroplating
d) hot dipping

3. Aluminum coatings are

a) more resistance to atmosphere than zinc coatings
b) not subjected to pitting in seawater
c) not satisfactory for application in soils
d) not suitable for application in a marine environment

4. Zinc coatings are generally applied by

a) hot dipping
b) hot spraying
c) airless spraying
d) electroplating

5. Tin corrodes on the inside of the tin cans because

a) tin becomes noble by reacting with the organic food liquids
b) tin becomes cathodic to iron
c) organic substance at the cathode oxidize
d) food is contaminated with bacteria

6. Tin can corrode if

a) oxygen is trapped within the tin can
b) tin is reduced because of high over-voltage of Sn^{++} ions
c) temperature inside food cans exceeds 50°C

d) moisture enters the tin cans

7. Corrosion of nickel coating in industrial atmosphere is minimized by

a) bright nickel coating which acts as sacrificial coating
b) applying a coating thickness of 0.8–1.5 mil
c) electrodepositing a very thin layer (0.01–0.03 mil) of chromium coating on top of nickel coating
d) applying a coating thickness of 1–10 mil (0.025–0.15 mm)

8. Chromium coatings are most extensively used in

a) food industry
b) ship industry
c) aerospace industry
d) steel industry

9. Zinc coatings protect the steel surface mainly by

a) acting as anode to the substrate
b) forming an intermetallic compound with steel
c) acting as a noble coating inert barrier to environment

10. The main objective of phosphate coating is to

a) provide an excellent adherence of paint to steel
b) protect steel from corrosion
c) remove any unevenness from the surface
d) provide an inert barrier for the atmosphere

Conceptual Questions

1. Which types of coatings is signified by the following coating processes?

a) Calorizing
b) Chromizing
c) Aluminizing

State one method for chromizing.

2. What is the main difference between the dry and wet galvanizing process?
3. What is the effect of the following factors on galvanizing?

   a) Bath temperature
   b) Immersion time
   c) Cooling after galvanizing
   d) Fluxing

4. State three advantages and three disadvantages of hot dip aluminum coating.
5. What is the mechanism of corrosion of tin inside the food cans?

How and Why Questions

1. Why calorizing, siliconizing and chromizing are considered diffusion coatings? In which industry aluminized coatings are used?
2. State why galvanized pipes are no longer protective if water heated above 85°C is passed through the pipes.
3. How a microcracked chromium coating protects a nickel coating from corrosion?
4. Why phosphating is used as a base for organic paints?
5. What are the advantages of using chromium coatings over tin coatings in food cans?

FAILURES OF COATINGS

Multiple Choice Questions

1. '*Chalking*' is

   a) an application-related failure
   b) formulation-related failure
   c) design related to failure
   d) adhesion-related failure

2. Checking

   a) is surface phenomenon
   b) penetrates through the coating
   c) is visible to the eyes
   d) is an adhesion-related failure

3. Cracking

   a) is a surface problem
   b) is less serious than either checking or alligatoring
   c) occurs as the coating ages with time
   d) is adhesion-related

4. Cratering

   a) is also called spatter coat
   b) is used by development of pin holes
   c) is due to the falling of dust particles in the paint when it is dry
   d) is also called crawling

5. Retained solvents create blistering because

   a) the solvents make the coating solver and flexible
   b) the vapor pressure of solvent is reduced
   c) combining of the moisture vapor with the solvent or dissolving into the solvent
   d) rapid evaporation of the solvent

6. Peeling is

   a) a design-related failure
   b) an application failure
   c) occurs because the coating becomes very hard and brittle as it ages
   d) takes place if the topcoat is not compatible with the undercoat

7. Chalking may be controlled by

   a) using white pigments, such as titanium dioxide
   b) using blue pigments
   c) selection of a proper solvent
   d) controlling reduction in the thickness of the coating

8. Mud cracking may be prevented by

   a) using a proper pigment to vehicle ratio
   b) using blue pigments
   c) using highly filled water-base coating
   d) apply coatings during wet conditions that are moderate

Suggested Literature for Reading

[1] ASM, *Metals Handbook: Surface Cleaning, Finishing, and Coating*. 10th ed., **5**, ASM, 2000, Ohio: Metals Park, USA.

[2] *Surface Coatings*, **1 and 2**, Oil and Color Chemists Association, Tafe Educational Books, 1974, NSN: Kensington, Australia.

[3] American Society of Metals, *Metals Handbook: Corrosion*, 10th ed., **13**, ASM, 2000, Ohio: Metals Park, USA.

[4] *Galvanizing Guide*, 3rd ed., Zinc Development Association, London, 1971, Australia: Melbourne.

[5] *Zinc, Its Corrosion Resistance*, Australian Zinc Development Association, 1975.

[6] Munger, C.G. (1999). *Corrosion Prevention by Protective Coatings*, 2nd revised ed. Vincent, L.D. ed. NACE, Texas: Houston, USA.

[7] Stoye, D. and Freitag, W. eds (1998). *Paints, Coatings, and Solvents*, 2nd revised ed. New York: Wiley-VCH.

[8] Gainger, S. and Blunt, J. eds (1998). *Engineering Coating Design and Application*, 2nd ed. Cambridge: Abington Publishing, England.

[9] Fedrizzi, L. and Bonora, P.L. (1997). Organic and inorganic coatings for corrosion prevention, Paper from Eurocorr 96, Institute of Material, London, England.

[10] Lamboune, R. (1987). *Paint and Surface Coatings: Theory and Practice*. England: Ellis Horwood.

[11] Galvanizers Association, *General Galvanizing Practice*, London, 2000, England.

[12] Zinc Development Association, *Galvanizing Guide*, London, 1988.

[13] Satas, D. and Tracton, A.A. (2001). *Coatings Technology Handbook*. New York: Marcel Dekker.

[14] Bierwagen, C. P. (1987). The science of durability of organic coating. *Progress Org. Coat*, **15**, 179–185.

[15] Simpson, T.C. (1993). Electrochemical methods to monitor corrosion degradation of metallic coatings. *Proc. 12th Int. Corrosion Congress*, **1**, Texas: Houston.

[16] American Society of Metals, *Surface Engineering*, ASM Handbook, **6**, 2000, Ohio: Metals Park, USA.

[17] Vincent, L.D. (2004). *The Protective Coaters Handbook*. NACE, Texas: Houston, USA.

[18] Grainger, S. and Blunt, J. (1998). *Engineering Coatings: Design and Applications*, 2nd ed. NACE.

[19] Marshall, A. (1996). *Corrosion Inspectors Handbook*, 3rd ed. NACE, Texas: Houston, USA.

[20] Lesoto, S. *et al.* (1978). *Paint/Coating Dictionary*, Fed. Of Soc. For Coating Technology, Philadelphia, Penn.

[21] NACE, Coating Collections on CD Rom, Network Version, Texas: Houston, USA.

[22] Schweitzer, P.A. (1996). *Corrosion Engineering Handbook*, New York: Marcel Dekker, USA.

[23] NACE, *Coating Inspector's Program*, Module 1–51, Coating Inspector's Training Course, NACE, Texas: Houston, USA.

Keywords

Fundamentals of Coating

Acrylic resin A synthetic resin formed from either the polymerization or co-polymerization of acrylic monomer.

Alkyd resin A synthetic resin made by condensation reaction (release of water) between a polyhydric alcohol (glycerol, etc.) and dibasic acid (or phthalic anhydride).

Binder The non-volatile portion of the vehicle of a paint. After drying, it binds the pigment particles together with the paint film as a whole.

Body It is used to indicate the consistency of the paint.

Brush mark Lines of unevenness left after the paint has been applied by a brush.

Coverage The rate at which the paint spreads. The rate of spreading is expressed in square meters per liter.

Cross-linking Establishment of chemical links between the molecular chains to form a three-dimensional network of polymers. Cross-linking toughens the coating.

Diluent A liquid which is used in conjunction with a solvent to improve the film properties.

Epoxide resin A resin formed by interaction between epichlorohydrin and bisphenol.

Filler A compound for filling fine cracks and indentation.

Gloss The property by which light is scattered specularly.

Lacquer A fast drying clear or pigmented coating which dries slowly by evaporation of the solvent, e.g. chlorinated rubber.

Leafing A particular orientation of flaky pigments to form a continuous sheet at the surface of the film.

Oil length The percentage of oil in the binder.

Paint A product in a liquid form which, when applied to the surface, forms a continuous film having decorative, protective or other specified properties.

Phenolic resins A class of synthetic resin produced by the condensation of a phenol with an aliphatic aldehyde, such as formaldehyde.

Pigment The insoluble solid particles dispersed in a paint which gives the desired color and the properties required, such as color, opacity and durability.

Primer The first coat of a painting system which helps to bind the coats applied subsequently to the substrate. Primer inhibits corrosion. Red lead is an example.

Resin A synthetic or natural material which helps to bind the pigments. Polyester is an example of resin.
Sagging It is the excess flow of paints usually downwards due to gravity which causes unevenness.
Sealer A compound which is used to seal the substrate by restricting the absorption of moisture to the substrate.
Solvent A liquid used to dissolve or disperse the film forming constituents to regulate the desired properties.
Substrate The surface in which a coating is applied.
Wash primer 91% polyvinyl butyl and 9% zinc tetraoxychromate by weight in a mixture of isopropanol and butanol which is mixed with 18% PO_4 in isopropanol and water just before use.

Metal Coatings

Aluminizing Coating a substrate with aluminum.
Anode Electrode attached to the positive terminal of a DC power source. The main reaction at the anode is oxidation.
Batch galvanizing Multiple units of articles are galvanized in a specific sequence.
Calorizing Coatings produced by tumbling the work in a mixture of aluminum powder Al_2O_3, and a small amount of NH_4Cl as flux in a hydrogen atmosphere at about 1000°C.
Cementation See calorizing.
Chemical vapor deposition Formation of coatings by the chemical reaction of vapors and gases at a heated surface and not with the heated surface as in diffusion coatings. A chemical compound is volatilized in a separate unit, mixing it with other gas or gases and metering and passing it into a chemical containing heated objects. At the heated surface a chemical reaction takes place.
Chromizing Heating a mixture of 55% chromium powder with 45% Al_2O_3 at a temperature up to 1350°C for 4 h in an atmosphere of hydrogen.
Cladding A physical process in which a thin layer of one metal (cladding) is brought in contact with a heavy layer of a base metal and binding by a combination of heat and pressure.
Decorative coatings These are mostly architectural coatings for aesthetic appeal. Bulk of coatings are applied to molded plastics.
Electroplating Application of a coating by use of electric current. An anode (+) and cathode (−) is immersed in an aqueous medium (electrolyte) through which current is passed. For instance, in chromium plating the electrolyte contains chromic acid, sulfuric acid, sodium fluoride and sodium fluosilicate. The bath is operated at 90°C at current densities up to 30 A/cm^2.
Fluxing Is the process to remove residual impurities that remain on the surface of work after degreasing and pickling and to clear the oxides from the surface of the coating bath at the point where steel enters. The flux cover used in wet galvanizing is zinc ammonium chloride with some amount of tallow or glycerin.
Fogging Loss of luster in a nickel coating.
Galvanizing Zinc coating on a steel substrate.
Hot dip galvanizing Applying zinc coating by dipping a steel object in a molten bath of zinc maintained between 810–870°F.
Ion plating The metal to be applied is heated and connected to a high positive potential charge. The substrate is unheated and negative. The metal vapor becomes ionized and the ions are attracted to the substrate.
Metallizing The process of applying metal coating on a substrate.
Phosphating A steel object is immersed in a bath containing zinc phosphate and phosphoric acid to precipitate zinc phosphate on the surface.

Failures of Coatings

Aging Deterioration of coating occurring with the passage of time.
Alligatoring This is caused by hardening and shrinking of the surface of the coating at a higher rate than the body of the coating. Coal tar coating tends to undergo alligatoring when exposed to sunlight.
Blistering Isolated convex deformation of a paint film in the form of blisters arising from the detachment of one or more of the coats. It may be caused by solubility of pigment due to absorption of water, incompatibility of coatings, inadequate solvent release and cathodic overprotection.
Bleeding Discoloration in the coatings caused by migration of components from the underlying films.
Chalking It is generally the disintegration of the binder by the action of the rays of the sun leaving the pigments in the form of loosely adherent powder. This defect does not reach the substrate.
Checking Formation of small breaks in the coating as the coating becomes more brittle and harder. It does not extend to the full depth of coating.
Cracking Formation of breaks in the paint film that expose the underlying surface. It is a serious type of failure which is caused by high stresses in film.
Cratering Formation of small bowl-shaped depression in a paint.
Curtaining Excessive flow of paint on vertical surfaces causing imperfection with thick layer edges in the faint film. It may be caused by poor application techniques (see also sagging).
Discoloration Any change in the color of the film exposure in the environment.
Erosion Deterioration of coating when exposed to natural weathering, such as wind, rainfall and hailstorm.
Fading Fading away of color because of poor selection of pigments.
Flaking Peeling away of a hard and brittle surface upon cracking. The edges curl away from the surface and create a flaking tendency.

Flow The ability of a paint to spread.

Foaming The formation of stable gas in liquid dispersion in which the bubbles do not coalesce with one another.

Hiding power The ability of a paint to obliterate the color difference of a substrate.

Holidays Defects characterized by a film having areas from thin coatings to areas of no coating. Includes pinholes.

Hopping A small bubble-like defect in a paint film.

Inter-coat delamination Partial or complete loss of adhesion between the different coats which may be caused by contamination, chalking and over-curing of the surfaces.

Peeling Loss of adhesion resulting in the detachment and curling out of the paint film.

Pinholes Small holes in a dry film which form during application and drying of paint.

Sagging See curtaining.

Saponification Defect arising out of the degradation of a binder by alkali.

Wrinkling Development of wrinkles in a coating after it has dried.

CHAPTER 8

CORROSION PREVENTION BY DESIGN

8.1 INTRODUCTION

As an old adage says, corrosion prevention must start at the blackboard, at the design stage. A good design at the blackboard is no more costly than a bad design, a bad design is always more expensive than a good design in reality. Technical design includes the aspects of design that directly bear on the proper technical functioning of the product attributes that describe how it works and how it is made. Design configuration has a critical role to play in the service life of components. The important point is that the designers must have an understanding and awareness of corrosion problems. Corrosion is, however, only one of the several parameters with which the designer is concerned and it may not be, however, important to a designer to give consideration to corrosion unless dictated by a requirement. In many instances, corrosion is incorporated in design of an equipment only after its premature failure. More often, more attention is paid to the selection of corrosion resistant materials for a specific environment, and a minimal consideration is given to design, which leads to equipment failure. For instance, even a material, like 90-10 copper–nickel may fail prematurely as a condenser tube material, if the flow velocity of salt water or seawater is not given a due consideration for a smooth flow in the tube design. This has been a common observation in desalination plants in the Gulf region. This chapter would highlight how corrosion could be prevented by adopting good design practices.

8.2 SERVICE LIFE OF EQUIPMENT

Selection of a corrosion resistant material for the environment is a prerequisite to a good design. Materials and design are complimentary to each other and neither of the two can be ignored. The following factors influence the service life of equipment (Fig. 8.1):

1) Environments and geographic location
2) Selection of materials
3) Maintenance
4) Corrosive environment and velocity of flow
5) Design
6) Feature promoting corrosion
7) Bimetallic connection.

Environmental factors affecting the service life of equipment are shown in Fig. 8.2.

8.3 CAUSES OF FAILURES IN THE CONTEXT OF DESIGN

A good engineering design should provide a maintenance-free service, satisfy the end user, and

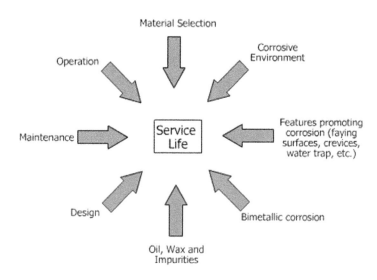

Figure 8.1   Factors influencing the service life of equipment

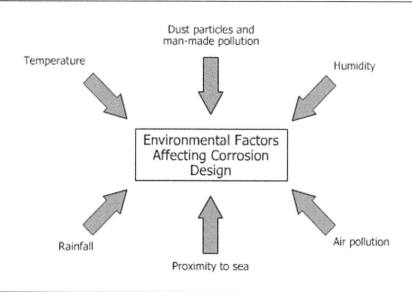

Figure 8.2   Environmental factors affecting service life of equipment

provide a maximum return on capital in a shortest return period. However, there are several areas related to failure as show below.

1) **Breakdown of protective system**. Many protective surface treatments, such as coating and welding, may not be very effective because of the presence of surface irregularities, voids, surface porosity, undercuts, and general surface roughness. The surface heterogeneities act as moisture traps and cause the damage.

2) **Poor fabrication**. Factors, such as improper welding, excessive cold working and excess machining lead to failure.

3) **Lack of accessibility**. In complex systems, machinery, and components, there might be inaccessible areas due to lack of design insight where it may not be possible to carry out the corrosion protection measures. Interiors of car doors are examples which are subjected to intensive localized corrosion. Figure 8.3 shows a design which provides adequate air circulation and spraying accessibility.

4) **Structural heterogeneity in materials**. Joining similar materials with structural differences, such as differences in thermo-mechanical processing, grain size, number of impurity elements, grain boundary segregates, may cause deviation from the performance expected.

5) **Operating conditions**. Factors, such as temperature, pressure, and velocity, influence the service life if allowed to exceed the prescribed limits.

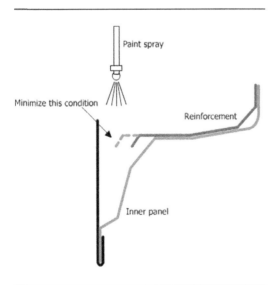

Figure 8.3 Preferred design of sheet metal such as above allows for adequate air circulation and paint spraying accessibility. (Reproduced by kind permission of SAE, USA)

8.4 CORROSIVE ENVIRONMENT

The following are the major ingredients of an atmospheric corrosive environment:

(1) Temperature
(2) Humidity
(3) Rainfall
(4) Air pollution
(5) Proximity to sea
(6) Dust storms and dust particles.

8.4.1 TYPE OF METALS OR ALLOYS

The metal or alloy must have a proven compatibility to the corrosive environment. For instance, stainless steel (SS) 316 with 2% Mo is a better material for seawater service than SS 304 without molybdenum. Brass, bronze and copper based alloys are highly desirable for salt water transportation, however, they are vulnerable for an environment containing ammonia frequently encountered in agriculture. A good design to prevent corrosion must be compatible with the corrosive environment. Following is a summary of the effect of major contributors to corrosive environments.

1) **Temperature**. Temperatures slightly in excess of 50°C are observed in several countries, like Kuwait, Saudi Arabia and the United Arab Emirates. High temperatures in combination with high humidity produce an accelerating effect on corrosion. In a survey conducted in one of the towns in Saudi Arabia, the corrosion-free life of an automobile is only six months. A rapid fall in temperature can cause condensation.

2) **Humidity**. Corrosion progresses fast when the relative humidity exceeds 75%. Humidity in Europe and the British Isles often exceed 75%. In certain areas of Ghana, Nigeria, Congo Basin, South America, South-east Asia and Gulf region, humidity may approach 100% and cause condensation.

3) **Rainfall**. Rain can be beneficial or harmful. Excess rainfall washes corrosive materials and removes dirt, debris and other deposits which may initiate corrosion, whereas scanty rainfall may leave water droplets on the surface and lead to corrosion as salt is present in the air. The frequency of rainfall contributes to humidity.

4) **Pollution**. In addition to sodium chloride particles in coastal areas, the atmosphere may contain sulfur dioxide, sulfurous acid and sulfuric acid which are considered as the worst offenders as far as corrosion is concerned. They originate from power stations, refineries, chemical and steel manufacturing plants. The environment is abundantly populated by them in oil-producing countries in the Gulf region.

5) **Man-made pollution**. To the above factors, must be added the cumulative effect of man-made pollutants, such as the presence of sodium chloride which is extensively used in deicing of roads in North American and European countries. Use of small amounts can induce high levels of corrosion in road vehicles. In desert regions, the abundance of sand particles accelerates corrosion because of the hygroscopic nature of some constituents of sand particles. The atmosphere may also contain other pollutants, such as carbon monoxide, nitrogen oxides, non-methane hydrocarbons and methane. Closeness to sea in many tropical areas creates a condition highly conducive to the onset of corrosion.

6) **Proximity to Sea**. Seawater is considered to be equivalent to a 3.5% solution of sodium chloride. The salinity of most oceans is 35 grams per thousand and the conductivity of seawater at 15°C is 0.042 ohm/cm. There is abundance of chloride in the marine environment and in industrial zones located in marine environment. A cumulative corrosive effect is caused by both chloride and sulfur dioxide. Chlorides can absorb moisture at low relative humidities. Saturated NaCl solution is in equilibrium with a relative humidity of 78%, but saturated $ZnCl_2$ solution is in equilibrium with only 10%.

8.5 STAGES IN THE DESIGN PROCESS

The following are the four important stages in the design process:

1) Clarifying objectives – To clarify design objectives
2) Establishing functions – To define the function requirements
3) Setting requirements – To specify the performance
4) Improving details – To increase the value of a product and making it cost effective

8.6 CONSIDERATION OF AREAS REQUIRING ATTENTION AT DESIGN STAGE

The following are the areas which require attention to minimize corrosion:

- Bimetallic contacts
- Faying surface
- Crevices
- Moisture traps
- Water traps
- Metals in contact with moisture absorbent materials
- Inaccessibility
- Areas of condensation
- Features which reduce the paint thickness
- Welds
- Oil, grease and rust patches
- Fluid movements
- Joints (threaded, riveted and screwed)
- Closed sections and entrapment areas
- Mechanical factors
- Corrosion awareness

Effects of some factors stated above on design are briefly described below.

8.6.1 BIMETALLIC CONTACT

Bimetallic corrosion is serious and it occurs when two materials differing in electrochemical potential are joined together. The galvanic series is a practical guide for engineers. The position of two metals or alloys in galvanic series dictates the extent of bimetallic corrosion. Consider, for instance, aluminum and copper. It can be observed from the table that aluminum is active to copper in the galvanic series and hence it would act as anode to copper which become the cathode due to its relatively more noble position. Joining the two would, therefore, give rise to bimetallic corrosion. The joining of two metals and alloys close to each other in the galvanic series would not cause bimetallic corrosion. The closer together are the materials in the galvanic series, the less the potential difference (the driving force for corrosion) and further apart the materials in the galvanic series the greater would be the potential difference and a greater driving force for

corrosion. The following are the factors affecting bimetallic corrosion:

a) Difference of potential.
b) A small anodic area to a large cathodic area.
c) Conduction of electrolyte.
d) Deposition of impurities and deposits, such as hygroscopic particles, sand or salt particles.
e) Contact with insulation materials.

a) Potential Difference

The greater the difference of potential between the two metals, the greater is the magnitude of bimetallic corrosion. Figure 8.4 shows a valve from a condensate pipe. The cast iron valve was incorporated in AISI 304 stainless steel condensate pipe of a copper heat exchanger. The difference of potential between copper, steel and cast iron caused bimetallic corrosion.

Figure 8.4 Galvanic corrosion in a cast iron valve which was attached to a stainless steel condensate pipe

Figure 8.5a   Steel rivets in copper sheets

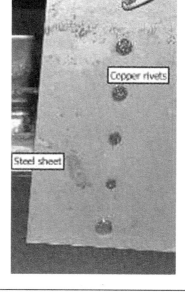

Figure 8.5b   Copper rivets in steel sheets

b) Anode to Cathode Ratio

A small anode area to a large cathode area causes serious bimetallic corrosion because of a large current density on a smaller anodic area.

Figure 8.5a shows a steel rivet in a copper sheet and Fig. 8.5b shows copper rivets in a steel sheet. More severe corrosion in the first case is observed and is concentrated on small anodes (steel rivets).

c) Deposition of Impurities

The deposition of impurities must not be allowed. The deposition of impurities, as debris cause the formation of differential aeration cells, and allows the absorption of moisture from the air which leads to corrosion (Figs 8.6(a) and (b)). The passive surface of steel may be destroyed by such deposits. The sites under the deposits become the anodes and the lead to pitting. A significant difference between the anodes and the cathodes may be observed. Hence, a good design should not allow the built-up of impurities on the surface. In

Figs 8.7(a) and (b) the dirt and impurities cannot be easily removed, hence, the two designs are poor designs. Impurities or deposits can easily be removed in the designs shown by Figs 8.7(c) and (d). Figure 8.8 shows good and poor designs against rain.

In automobiles, structural members must be designed to minimize the retention of water on the surface. Some examples of preferred orientation and design are shown in Figs 8.9(a) and (b).

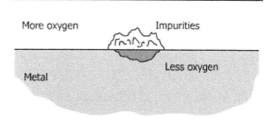

Figure 8.6a   Formation of differential aeration cells under impurities deposited on metal surface

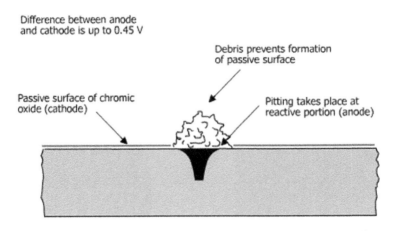

Difference between anode
and cathode is up to 0.45 V

Debris prevents formation
of passive surface

Passive surface of chromic
oxide (cathode)

Pitting takes place at
reactive portion (anode)

Figure 8.6b Mechanism of corrosion on stainless steel where portions are covered by debris

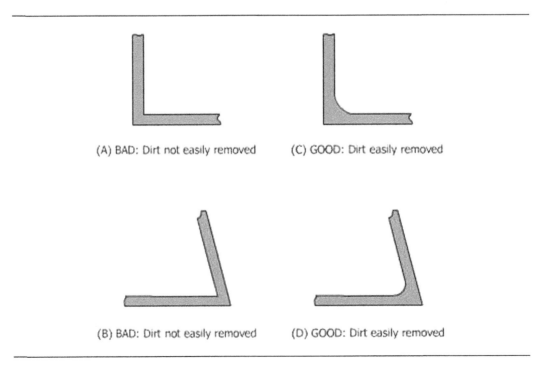

(A) BAD: Dirt not easily removed

(C) GOOD: Dirt easily removed

(B) BAD: Dirt not easily removed

(D) GOOD: Dirt easily removed

Figure 8.7 Good and bad designs for removal of dirt

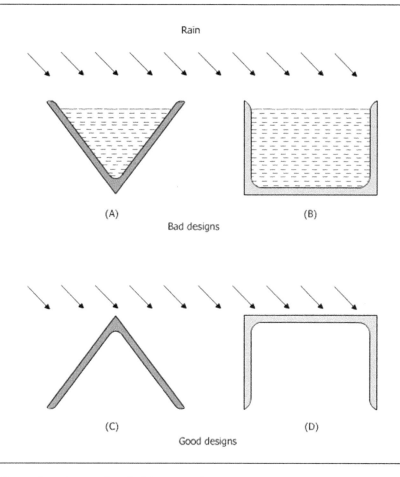

Figure 8.8   Preventing rainwater from lodging on steel structures

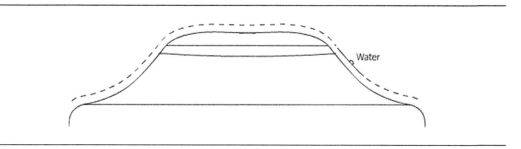

Figure 8.9a   A structural member of automobile

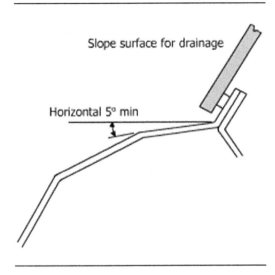

Figure 8.9b Proper design of panels

d) Wet Atmosphere

If the wet surface is maintained for a longer duration, in particular if salt particles are present, bimetallic corrosion is aggravated. The design of components must not allow the retention of a layer of water on the surface. A smooth polished surface would not allow water retention. A smooth design of polished surface sloping downwards would not allow the retention of water layer for longer periods of time. More examples of good and bad channel designs are shown in Fig. 8.10.

e) Contact with Wet Insulation Materials

Insulation materials, such as glass wool, glass fibers, polyurethane foam, do not cause the corrosion of steel in contact with insulation

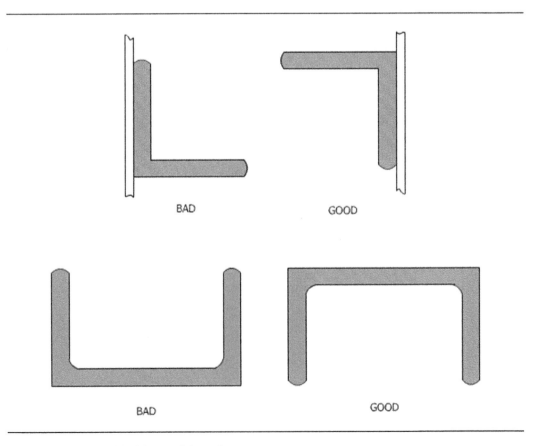

Figure 8.10 Good and bad design of channels

as long as the insulation material remains dry. Corrosion underneath insulation, however, assumes significant proportions if the insulation becomes wet during storage, operation or field erection. Following are the factors leading to insulation-induced corrosion:

1) Ingress of moisture in the insulation resulting in the leaching of soluble salts of low pH (about 2–3).
2) Release of chloride ions.
3) Destruction of passivity of steel by chloride ions and initiation of pits.
4) Weather cycles, such as wet–dry, hot–dry, and damp–warm cycles, which induce high corrosion rates.
5) Inadequate moisture barriers, because of inadequate spacing of insulation.

The effect of moisture ingress is shown in Figs 8.11, 8.12 and 8.13 shows the corrosion caused by ingressing of moisture in insulation in a water heater after six months of service.

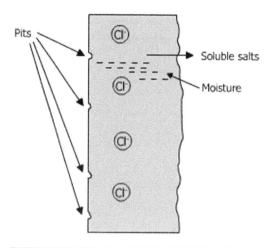

Figure 8.11   Effect of moisture ingress in insulating materials

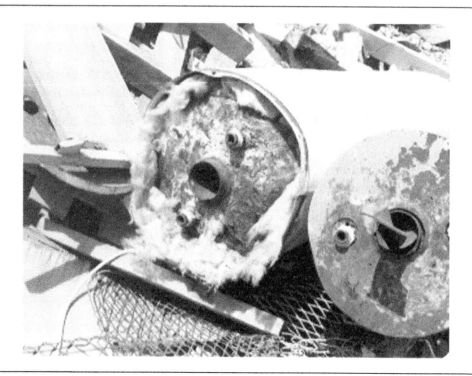

Figure 8.12   Insulation corrosion in water heaters

Figure 8.13 Initiation of corrosion of water heaters by ingress of moisture in insulation

Corrosion induced by insulation can be prevented by the following measures:

1) Eliminating flat horizontal surfaces.
2) Structural designs which trap water (example, H-beams and channels) (Fig. 8.14).
3) Strict compliance of insulation thickness.
4) Providing an adequate moisture barrier and waterproofing.
5) Addition of sodium silicate as an inhibitor.

An idealized design to prevent insulation-induced corrosion is shown in Fig. 8.15.

8.6.2 JOINTS AND FAYING SURFACES

In joining of two surfaces, crevice formation and galvanic effects are two major considerations.

The formation of crevices must be avoided by sealing, or otherwise shielding the crevice from the environment. To minimize galvanic corrosion contact between the two metals must be insulated.

In good design practice all joints should be designed to be permanent and watertight and the direct contact of the two metals must be avoided by means of insulating materials or by applying protective coatings. All components must be effectively insulated. Consider Fig. 8.16(a) where two steel plates are joined by an aluminum bolt. Aluminum is anodic to steel (-1.23 *vs* -0.443 V) and so will undergo galvanic corrosion and it has a small anodic area compared to the large cathodic area of steel plate. Because of the unfavorable area ratio, the aluminum rivet would corrode at a greatly enhanced rate. The opposite situation is observed in Fig. 8.16(b), where aluminum still corrode in preference to steel. Figure 8.17 illustrates how two metals differing in potential, e.g. copper ($E_o = 0.334$ V) could be joined with Al ($E_o = -0.162$ V) without the risk of galvanic corrosion by inserting insulation sleeves and washers between copper and aluminum. Figure 8.18 illustrates how two plates of aluminum and steel could be joined without the risk of galvanic corrosion by insulation with either jointing compounds, or insulating inserts and insulating washers. The galvanic corrosion can be minimized by proper insulation as illustrated in Fig. 8.19. When two dissimilar metals are to be joined, either paint both the metals or only the more noble metal.

8.6.3 WELDING AND ITS IMPACT ON CORROSION AND DESIGNS TO PREVENT CORROSION

Welding is a most common method of joining of metals which has a significant influence on corrosion resistance and design to prevent corrosion. The metallurgical properties of metals are significantly affected by the metallurgical changes brought about by welding which include melting, freezing, thermal strains and solid state transformation. The formation of different phases

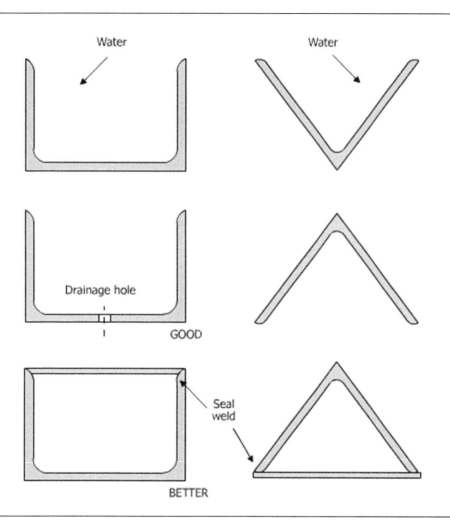

Figure 8.14   Good and bad designs for water trapping

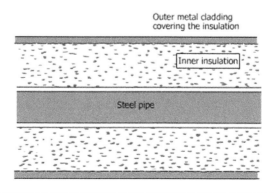

Figure 8.15   Suggested design to minimize corrosion induced by thermal insulation

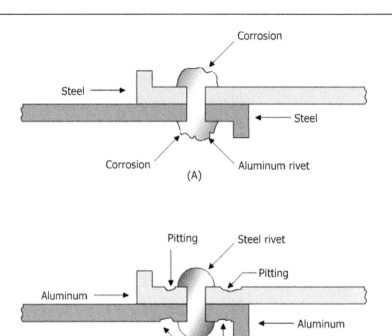

Figure 8.16   A bad design approach (no insulation)

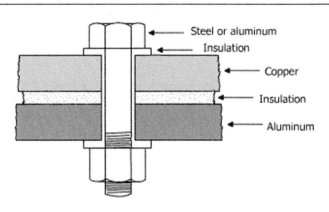

Figure 8.17   Illustration of a good bolted joint

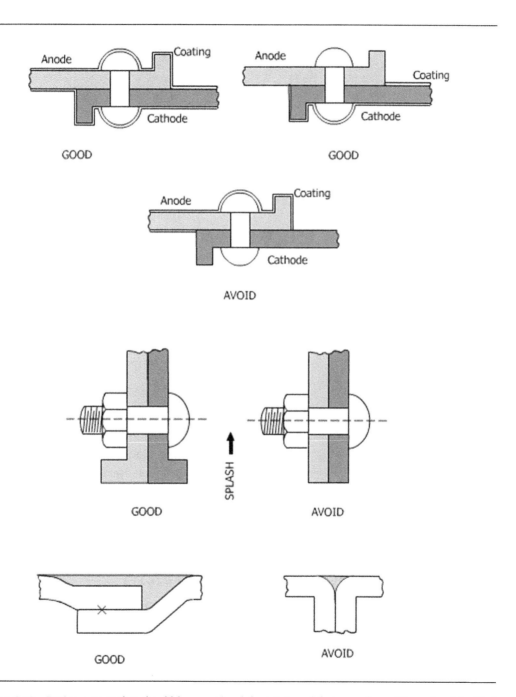

Figure 8.18 Design aspects that should be considered for joints and faying surfaces. (Reproduced by kind permission of SAE, USA)

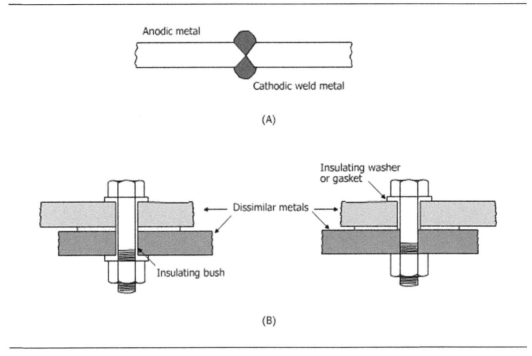

Figure 8.19   Preferred design features for joints to avoid galvanic corrosion

and intermetallic compounds not only affects the mechanical properties but also the resistance of materials to corrosion. For instance, the formation of $Cr_{23}C_6$ intermetallic in the temperature range 510–680°C makes 18-8 stainless steel highly sensitive to intergranular corrosion. In a similar manner, the transformation to austenite to ferrite and martensite leads to changes in the corrosion behavior because each phase formed has a different corrosion resistance. Similarly, in aluminum alloys, such as Al-Mg-Si alloys, the formation of Mg_2Si intermetallic leads to an increase in corrosion susceptibility. The segregation of impurity elements in steels, like phosphorus and sulfur, the grain boundary makes them susceptible to corrosion. The susceptibility of austenitic steel to corrosion is, for example, reduced by restricting impurity elements segregated at the grain boundaries. Important factors affecting weldability are:

1) Chemical composition
2) Microstructure
3) History
4) Welding processes

The major problem caused by welds is the creation of an inhomogeneous and discontinuous surface which creates potential sites for trapping of dirt, moisture and impurities.

8.6.4 CARBON AND WELDABILITY

The carbon content in plain carbon and low steels is restricted to 0.30 and 0.15%, respectively. The cooling rates and carbon contents are controlled during the welding of carbon steels to maximize the formation of soft α-ferrite and minimize the formation of pearlite and cementite which are hard components. Similarly in low alloy chromium steels, the formation of brittle martensite is minimized. Martensite is sensitive to hydrogen-induced cracking and ferrite is more resistant to corrosion than either pearlite or cementite (Fe_3C).

8.6.5 EFFECT OF ALLOYING ELEMENTS

The effect of other elements can be determined by equating them to an equivalent amount of carbon using the equation below:

$$CE \text{ (carbon equivalent)} = \frac{C + Mn}{6} + \frac{Ni}{15} + \frac{Cu}{15} + \frac{Cr}{5} + \frac{Mo}{5} + \frac{V}{5}$$

The lower value below 0.40 exhibits excellent weldability.

8.6.6 SENSITIZATION OF STEELS

Austenitic steels are susceptible to sensitization when heated in the range of 399–870°C because of the precipitation of chromium carbide in the above temperature range. This problem is overcome either by use of low carbon steels or by using a stabilized grade of steels containing titanium and niobium which prevents the formation of chromium carbide.

8.6.7 FILLER METAL CONTROL

The control of filler is important to ensure that the welded joint has the desired mechanical composition and the chemical strength.

8.6.8 WELD METAL OVERLAYS

Weld metal overlays, Gas Metal Arc Welding (GMAW) or Submerged Arc Welding (SAW), followed by machining or grinding are standard measures adopted to repair corroded surface and to restore the designed thickness. The weld metal overlays are applied by depositing corrosion-resistant weld metal on to the surface.

Monel, Inconel and austenitic steels are frequently used as overlay materials. Overlays are used generally by Shielded Metal Arc Welding (SMAW) and Gas-Tungsten Arc Welding (GTAW).

8.6.9 OVERLAY DEFECTS

All types of welding defects must be minimized as they directly affect the properties. The most important defects are micro-fissures, cracking caused by slag inclusion, porosity, oxide tints and disbonding from the substrate. Material fissures caused by non-metallic elements, like sulfur and phosphorus, increase tensile stress between the grains. The stresses of the grain fissures are eliminated by weld deposits containing 12% ferrite. Porosity, disbanding and cracking can be controlled by adoption of proper welding technique. Cracking can be minimized by minimizing the formation of intermetallic compounds, like chromium carbide in 18-8 stainless steel. In principle, the formation of any brittle compound in the overlay zone must be minimized to control cracking defects.

8.6.10 EXAMPLES

Figure 8.20 shows that a continuous welding offers better corrosion protection and strength than intermittent welding. Continuous welding should be used to close the crevices. Welded butt joints should be preferred over bolted joints particularly in new structures (Figs 8.21(a) and (b)). The welds should be cleaned and removed from the parent plate. It is recommended that the smaller side of the weld be placed in contact with the corrosive medium to provide minimum exposure to the weld (Fig. 8.22).

8.7 SOLDERING AND THREADING

Threaded joints are susceptible to the formation of differential aeration cells (Fig. 8.23). The low oxygen areas form the anode and hence they are liable to corrosion. Soldering is

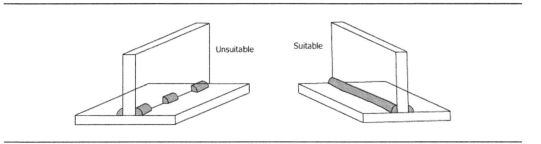

Figure 8.20 Continuous and intermittent welds. (From Costa, W. (1985). *An Introduction to Corrosion and Protection of Metals*. Chapman and Hall, London. By kind permission of Chapman and Hall, London)

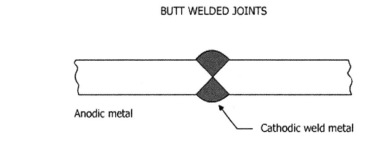

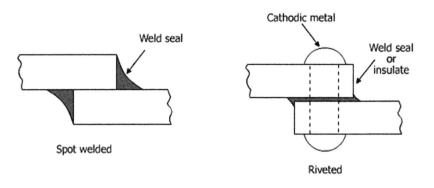

Figure 8.21a Comparison of lap and butt joints. (Reproduced by kind permission of SAE, Warrandale, PA, USA)

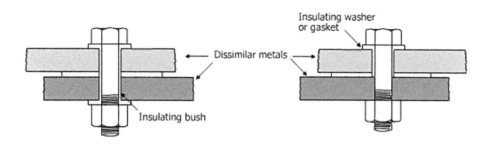

Figure 8.21b   Preferred design features for joints. (By permission of SAE, Warrandale, PA, USA)

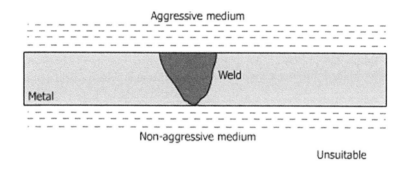

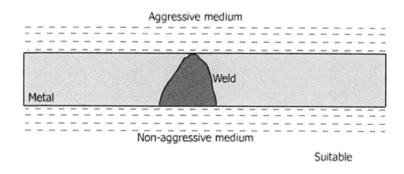

Figure 8.22   The smaller side of the weld should be turned towards the aggressive medium

preferable to threaded joints. A metal with a more noble potential than the parent material must be used for soldering. Bolted joints and threaded joints cannot be always avoided in non-permanent structures for reasons of assembling and disassembling. However, whenever they are used, they should be treated for corrosion prevention. Threaded joints can be treated with an inhibiting primer. For the same reasons, the nuts and bolts should be galvanized.

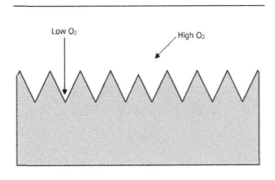

Figure 8.23 Low and high oxygen areas in a screw threaded zone

Figure 8.24 shows the suitability of soldering over threaded joints. Figure 8.25 shows comparison of spot welded joints and riveted. The figure also illustrates why welded joints are preferable to riveted.

8.8 CREVICES

A crevice may lead to pitting of the metal surface by forming differential oxygen cells. The crevice becomes oxygen starved, compared to outside, the crevice which functions as cathode and causes reduction of oxygen. The crevice becomes the anode and oxidizes the surface, such as $M \longrightarrow M^+ + 2e$. The positive ions react with the Cl^- ions in the liquid in contact and form MCl. Upon hydrolysis, MCl transfers to HCl (MCl + HOH $\longrightarrow$ MOH + HCL) (refer to Chapter 4 - Mechanism of Crevice Corrosion). The process is autocatalytic and very similar to pitting. It is, therefore, extremely important for engineers to design against crevice corrosion. Figure 8.26 illustrates a few designs to minimize crevice corrosion.

To minimize crevice corrosion:

1) Use welded joints in preference to bolted or riveted joints.
2) Minimize contacts between metals and non-metals which might cause a crevice.
3) Avoid sharp corners, edges and packets.
4) Use fillers and mastics to fill any crevice gaps.

Figure 8.27 illustrates the effect of spot welding and continuous welding. Spot welding creates micro-crevices, whereas continuous welding eliminates crevices.

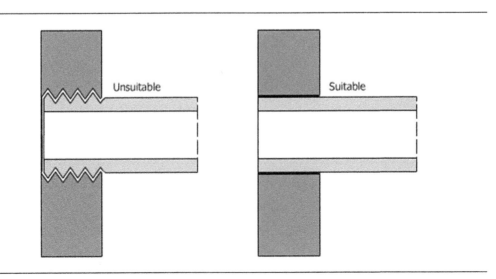

Figure 8.24 Soldering is to be preferred to threading. (From Costa, W. (1985). *An Introduction to Corrosion and Protection of Metals*. Chapman and Hall, London. Reproduced by kind permission of Chapman and Hall)

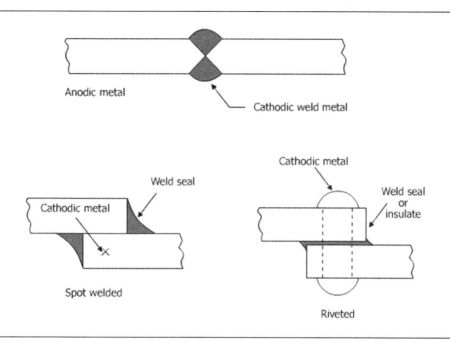

Figure 8.25   Comparison of welded joints and riveted joints. (From Rowe, L.C. (1977). GM Research Lab, Warren, MI, RPT24, Pub MR 2294 (GMR-3101-PCP (III) updated 1979). Reproduced by kind permission of GM, USA)

8.9 DESIGN TO CONTROL CORROSION IN WATER DISTRIBUTION, OIL AND GAS PIPELINES

Typical problems arising due to corrosion in water distribution systems are the following (Table 8.1).

Major problems encountered are galvanic corrosion, pitting, crevice, erosion and cavitation which need to be addressed.

8.9.1 FACTORS CONTRIBUTING TO POOR DESIGN

1) Ignoring Specifications

There is a general trend to use PVC pipes in gas and water distribution systems. Concrete pipes have been used widely for water mains. There are specifications on soil compaction, the pressure the pipes can withstand, and the composition of soils. Non-adherence of these specifications lead to serious failures. Copper pipes are joined in several instances with steel pipes without proper insulation and coatings which leads to service problems of galvanic corrosion.

2) Putting Dissimilar Pipes in the Same Trench

Pipes of different materials, such as copper, steel mild steel and galvanized iron are often buried very close to each other in the same trench without any concern for galvanic corrosion. Figure 8.28 shows a method to prevent galvanic corrosion of mild steel pipe which is put in the same trench as close to the copper pipe. The copper pipe is coated and insulated to minimize galvanic corrosion. The mild steel pipe may be protected by a galvanic anode but this is not cost effective.

Note: Normally galvanic corrosion will not occur unless there is metal-to-metal-contact. If this were not so, galvanic corrosion could not

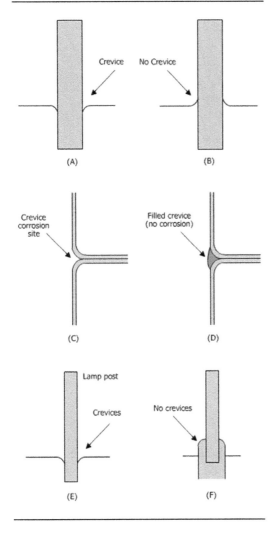

Figure 8.26 A few designs to minimize crevice corrosion

be prevented by separating dissimilar metals with an insulator. If two pipes are buried in a trench it is likely that they are bonded together with a metal strap somewhere, which will thus give a path for electrons and cause galvanic corrosion. Figure 8.28 shows this may then be prevented.

3) Insulation

Non-metallic couplings, unions and flange insulation are widely used for insulating against bimetallic corrosion. However, they are not always wisely used. All underground pipes must be insulated from the above-ground pipes.

Insulators must be installed on distribution mains when connecting new steel pipes to old steel pipes, when connecting steel to cast iron and when installing a newly coated pipeline at every 2000 ft. When connecting a copper service pipe to a steel main, insulation is needed where the copper pipe connects with steel main pipe (Fig. 8.29). No insulation is necessary when joining a plastic pipe to steel mains (Fig. 8.30). The meter outlet is to be insulated. The insulation device used must be made of non-metallic couplings, unions and flange gaskets, to minimize the risk of bimetallic crevices. Figures 8.31 and 8.32 show the construction details of an insulated coupling and an insulated union, respectively.

8.10 FLOWING WATER SYSTEMS

The majority of corrosion problems in flowing water systems are caused by obstruction to smooth flow. Turbulence and impingement adversely affect smooth flow conditions and lead to erosion–corrosion attack. One reason for failure are imperfections created at the manufacturing stage. If operations, such as heat treating, drawing and straightening, are not performed properly they may induce high residual stresses and bending stresses. Proper alignment of tube steels and plates minimize the stresses induced during fabrication. The tubes hardly fail within the tube sheet, they fail behind the tube sheets. Over-rolling of tubes is a frequent cause of such failures and it must be minimized to minimize the stresses. A corrosion problem is commonly observed at the inlet of condenser tubes and extending the pipe to the tube plate has proved effective in minimizing corrosion. Use of plastic inserts at tube ends is also effective in preventing turbulence (which occurs at tube inlets) from damaging the metal tube. The problems in flowing systems are often caused by refitting of the components, like gaskets and washers, during the service period. Changes in smooth flowing patterns may be caused by such operation if specifications and tolerances are overlooked.

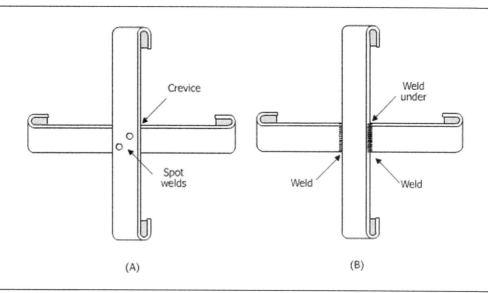

Figure 8.27 Spot welding and continuous welding. (From Chandler, K. A. 1985. Marine and Offshore Corrosion. Butterworth. Reproduced by kind permission of Butterworth, UK)

Table 8.1 Typical problems arising due to corrosion in water distribution systems

Indication	Causes
Red water	Corrosion of mainly galvanic iron pipes with operational temperature in excess of 60°C
Bluish stains	Corrosion of copper pipes
Black water	Sulfide corrosion of copper and iron
Foul taste	Microbial activity
Loss of pressure	Excess scaling
Short life of pipes	Leaking due to pitting

The best way to minimize corrosion in flowing water system is at the fabrication stage as discussed above. Other methods to minimize flow-induced corrosion are summarized below.

- Design replaceable parts for the system areas which are most likely to corrode.
- Select materials which are compatible and do not offer any risk of bimetallic corrosion.

- The pipe should be designed for a smooth flow and all valves, flanges and other fittings should be installed in accordance with the design specifications to allow a minimum disturbance to a smooth flow. The fittings, such as gaskets and flanges, used should have an equal inside diameter or a tapered join. The turbulence at screwed joints is shown in Fig. 8.33. Use of straight-through type of valves, such as gate, butterfly and plug valves, offers a lesser resistance to flow.
- Impingement attack generally occurs in condenser tubes handling seawater which circulates at high velocities with turbulent flow. The problem can be overcome by decreasing the velocity and streamlining the design of pipeline, water boxes and injector nozzles (Fig. 8.34(a)). Abrupt changes in flow direction must not be allowed. The use of sacrificial baffle plates is effective in minimizing damage by unavoidable impingement corrosion. A good impeller design in a water pump is shown in Fig. 8.34(b).
- Increasing the pipe diameter is another way of reducing velocity and minimizing corrosion in a flowing water system.
- Regular maintenance and cleaning of the pipe is important as accumulation of impurities

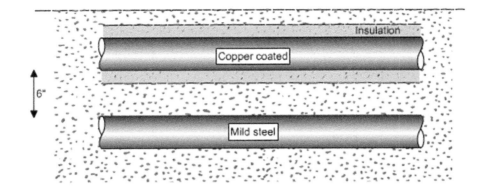

Figure 8.28 Copper coated pipe at a distance of 6″ from the mild steel pipe

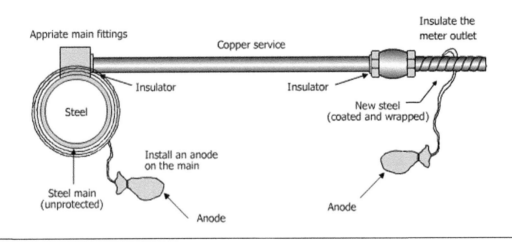

Figure 8.29 Copper service connected to a steel main. (From Right, J.E. (1981). Practical Corrosion Control Methods in Gas Utility Piping, NACE. Reproduced by kind permission of NACE, USA)

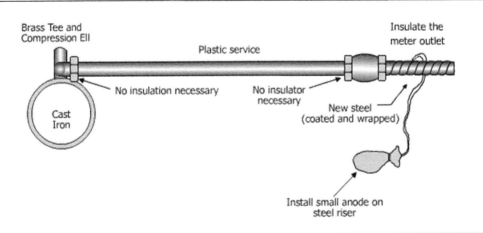

Figure 8.30 Piping practices (new services) plastic service connected to a cast iron main (From Right, J.E. (1981). Practical Corrosion Control Methods in Gas Utility Piping, NACE). (Reproduced by permission of NACE, USA)

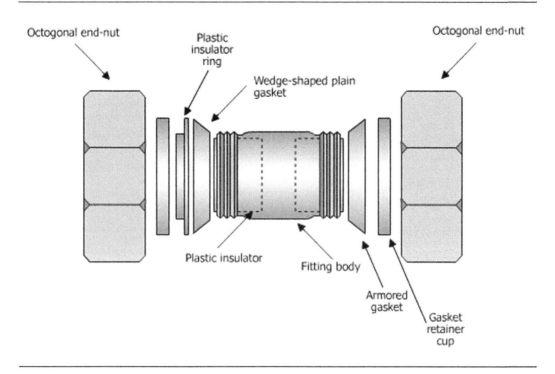

Figure 8.31   Construction details of an insulated coupling

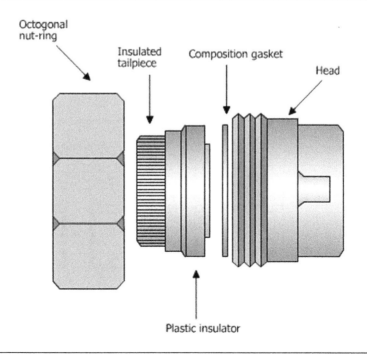

Figure 8.32   Construction details of an insulated union. (From Right, J.E. (1981). *Practical Corrosion Control Methods in Gas Utility Piping*, NACE. Reproduced by permission of NACE, USA)

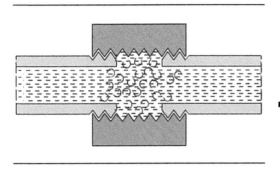

Figure 8.33   Turbulence in flow at screwed joints

may lead to changes in flow pattern. Figures 8.35 (a–c) show erosion–corrosion and cracking of water distribution pipe caused by sudden changes of flow conditions. It is a good idea to replace steel elbows by C-PVC or V-PVC.

■ Avoid placing pipes in direct contact with sand to avoid corrosion at the bottom and minimize vibrations by utilizing a good supporting system, such as a concrete foundation wherever economic justifies (Fig. 8.36).

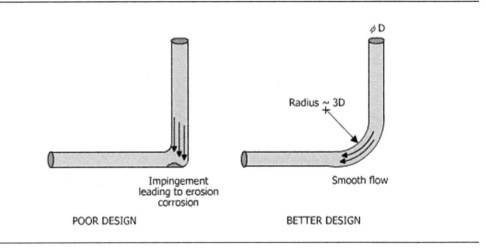

Figure 8.34a   Larger radius is preferred for higher velocities in order to avoid erosion–corrosion

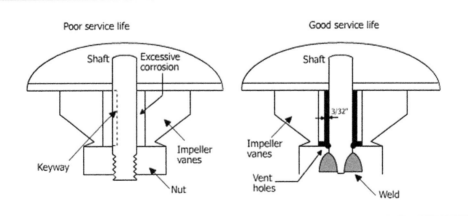

Figure 8.34b   Impeller design

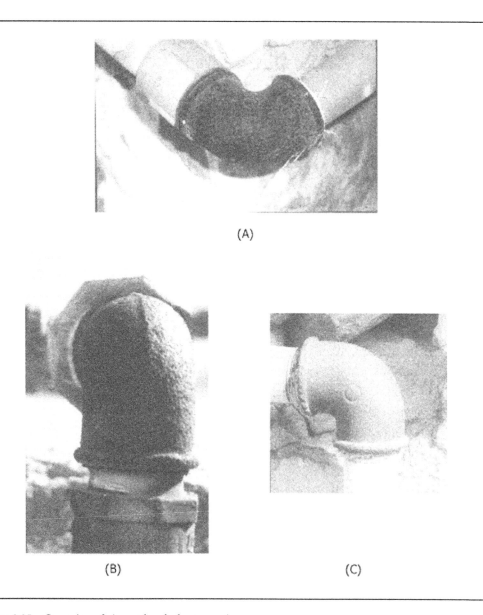

(A)

(B)   (C)

Figure 8.35   Corrosion of pipes at bends due to erosion

8.11 ACCESSIBILITY FOR MAINTENANCE

The design must be able to allow easy access to the areas requiring repair or maintenance. Appropriate long-life paints should be applied in areas which may not be accessible for a sufficient length of time. Figure 8.37 shows access to areas suitable and unsuitable for painting.

A good design should allow uniform painting to be applied on the surface. Areas of uneven coating thickness are potential sites for initiation of corrosion. Figures 8.38a and b illustrate the point. For uniform coating application grind all sharp edges and apply an extra coat of paint. Keep sharp edges to a minimum.

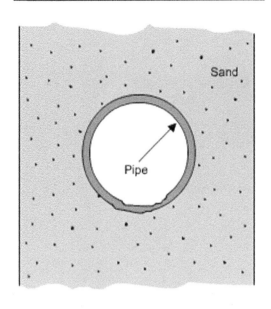

Figure 8.36 Corrosion takes place at the bottom, if the pipe is placed in direct contact with the sand

8.12 DESIGNS FOR LIQUID CONTAINERS

A good design for liquid container must offer the following:

a) Freedom from sharp corners and edges.
b) Smooth flow of liquid from the container.
c) Freedom from the buildup of water traps around the corners.

d) Complete drainage from the corners without any water traps. The elimination of water traps is essential to minimize the formation of differential oxygen cells which lead to corrosion. As an operational matter, it is essential to remove water and dry out stainless steel tubing without delay as soon as leak testing of new water treatment plant is completed; there are many examples of microbial corrosion causing severe pitting of new plant soon after leak testing.

e) Minimizing of bimetallic corrosion by joining compatible materials without the risk of galvanic corrosion.

f) Complete internal and external coating of the containers, if cost effective.

Some of the above measures to prevent corrosion in liquid containers are shown in Fig. 8.39. Figure 8.39(a) shows the best design because of the capability of the liquid containers for complete drainage and absence of water traps. Figures 8.39 (b) and (c) are examples of bad design because of the incapability for complete drainage and presence of water and moisture traps around the corners. Better designs are shown in Figures 8.39(d) and (e). Figure 8.40(a) shows a bad design because of the joining of a copper pipe with the galvanized steel tank. The copper ions may be plated on the surface of galvanized steel and lead to pitting. An aluminum inlet pipe joined to an aluminum tank would not cause galvanic corrosion (Fig. 8.40(b)). The design also offers a good drainage of the liquid. The design could be further improved by further smoothing the corners.

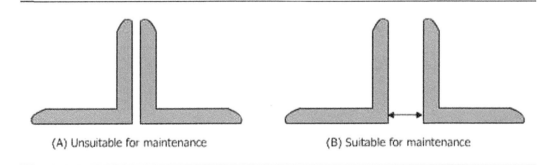

(A) Unsuitable for maintenance   (B) Suitable for maintenance

Figure 8.37 Design aspects with respect to accessibility for maintenance

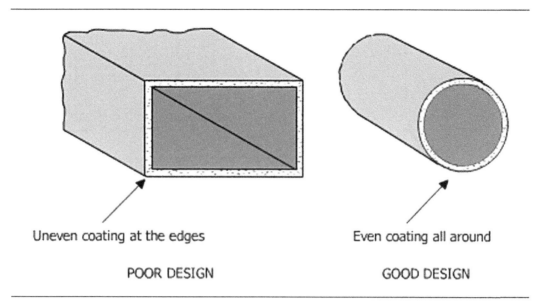

Uneven coating at the edges

POOR DESIGN

Even coating all around

GOOD DESIGN

Figure 8.38a   Design for uniform coating protection

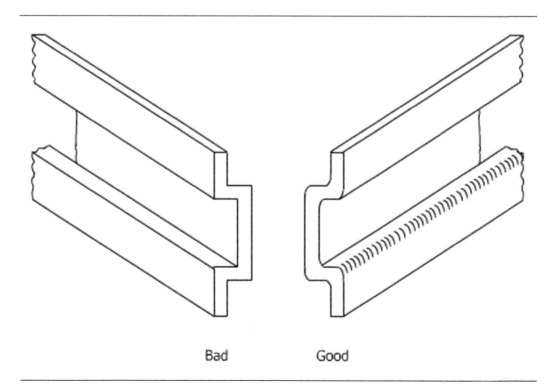

Bad   Good

Figure 8.38b   Suitability of designs for coating

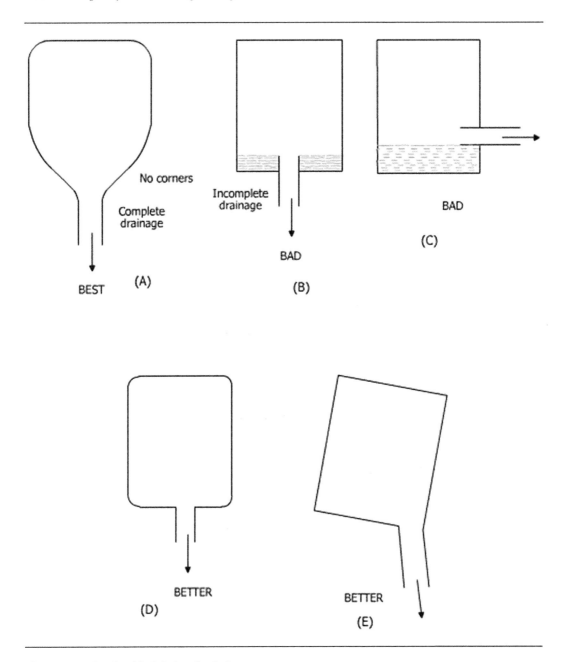

Figure 8.39 Good and bad designs for drainage

8.13 DESIGN IN PACKAGING

Temporary corrosion protection is required for storage and transit of equipment and machinery. Such corrosion protection is offered by volatile corrosion inhibitors (VCI), also called vapor phase inhibitors (VPI). Volatile corrosion inhibitors extend the protection offered by impregnated wrapping papers to areas out of contact with paper so that the protection may reach those complex areas where contact

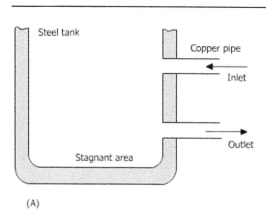

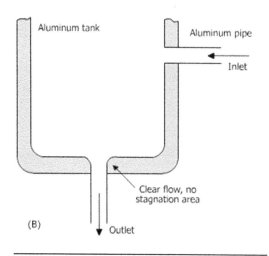

Figure 8.40 Design for liquid containers

with paper is not possible. Two well-known VCIs are cyclohexylamine carbonate (CHC) and dicyclohexylamine nitrite (DCHN). Both are effective on steel.

The following are the important factors affecting packaging design:

a) Length of protection needed.
b) Domestic or export.
c) Climatic conditions.
d) Size and weight of product.
e) Processing and cleaning methods.

Packaging Design Consideration is $1\,ft^2$ of paper for every $3\,ft^2$ of surface area. The VCI

protection depends upon

- Metals and alloys to be protected.
- Quantity of VCI in the paper or in the film.
- Effectiveness of inhibitor in the presence of humidity and moisture.
- Packaging design and the conditions to be encountered during storing and shipping of the equipment and machinery.
- Type of carrier paper or film for the vapour phase inhibitors.

Example:
Figure 8.41 shows important packaging and design consideration for ten years engine lay-up and storage and correct packaging of pumps and valves. The VCI offers protection to the engine block and steel components. Uniwarp-A (proprietary product) provides protection to non-metallic parts, while VCI-2000 fly wheel protects gears from atmospheric oxidation. In Fig. 8.42, the water soluble VCI provide protection against condensation. The VCI polyethylene acts as a barrier against moisture. For overseas shipping by ocean, wood crates lined internally with polythene (to prevent wood acids vapours) and containing VCI are used.

8.14 COATING AND DESIGN

Coating is an essential tool for corrosion prevention and repair of damage caused by corrosion to structure and equipment. The application of coatings covers all sectors of industry and is a process technology for corrosion prevention, corrosion maintenance and corrosion repairs. However, there are certain factors which affect the life of a paint system. These factors are summarized in Table 8.2.

The major components affecting the life of paints is the application error and one controlling factor is the inappropriate design or the complexity of the equipment which prevents the proper application of paint.

8.14.1 EDGE FAILURES

Coating failures are common on critical areas of structure, such as the edges because of the

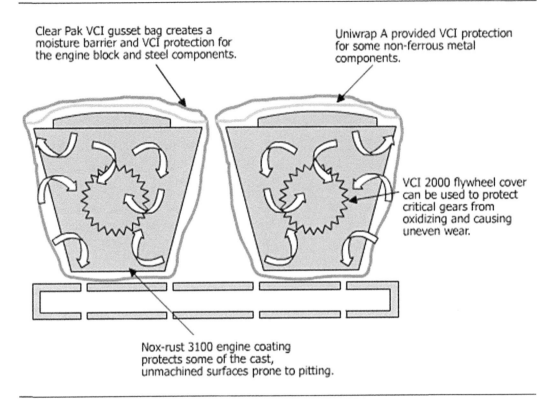

Clear Pak VCI gusset bag creates a moisture barrier and VCI protection for the engine block and steel components.

Uniwrap A provided VCI protection for some non-ferrous metal components.

VCI 2000 flywheel cover can be used to protect critical gears from oxidizing and causing uneven wear.

Nox-rust 3100 engine coating protects some of the cast, unmachined surfaces prone to pitting.

Figure 8.41   Ten years engine lay-up and storage (Courtesy: Daubert, VCI Inc.)

difficulty of achieving a uniform thickness. Undercutting of coatings can be observed on coatings of uneven thickness applied on the edges of structural components in a marine environment.

Figure 8.43 shows undercutting caused by abrasives. Corrosion at exterior corners is caused for the same reason as on edges. In the interior corners, corrosion is initiated by dirt and debris which is not properly removed prior to coating. Blistering is also caused by shrinkage of coating from the interior corners. Figure 8.44 shows a blistered surface after the removal of coating. In such an instance the entire coating should be replaced. The mechanism of corrosion caused by dirt and debris inside at the bottom of an oil tank is shown in Fig. 8.45.

The best measure to prevent corrosion of an oil tank bottom is to apply a thin film of epoxy copolymer and coal tar epoxy coating. A drainage facility may be incorporated for minimizing the buildup of water at the tank bottom. At the exterior of the tank bottom, corrosion penetrating from inside may be prevented by a glass fiber reinforced plastic (GRP) to strengthen the bottom. An additional measure would be the application of cathodic protection. One of the best examples of beneficial effects of design on coating is illustrated by automobiles. Figure 8.46 shows the design of different components prior to galvanizing and assembling the body.

The corrosion of buildings and concrete structures is a major area of concern to engineers and builders. Repairs to concrete structures are essential to maintain their integrity. The design of the structure should have sufficient accessibility for repairs as shown in Fig. 8.47. During repair, spalled concrete is taken out ensuring that the salts have been sufficiently removed and the steel is cleaned. The concrete is replaced by a suitable mortar or concrete with proprietary additives. It has been observed in recent years that epoxy coated reinforcement provides an excellent protection against reinforced concrete corrosion.

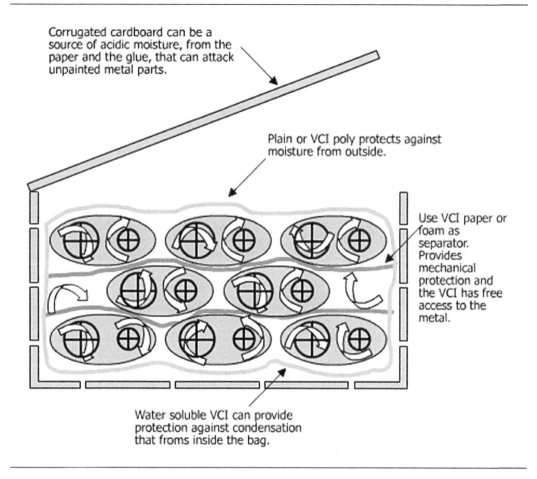

Corrugated cardboard can be a source of acidic moisture, from the paper and the glue, that can attack unpainted metal parts.

Plain or VCI poly protects against moisture from outside.

Use VCI paper or foam as separator. Provides mechanical protection and the VCI has free access to the metal.

Water soluble VCI can provide protection against condensation that froms inside the bag.

Figure 8.42   Example of correct packaging (Courtesy: Daubert, VCI Inc.)

8.14.2 DESIGN DEMANDS FOR FABRICATION

Proper coating applications demand special fabrication and surface preparation procedures for equipment, like tanks and pipes, to be properly coated for corrosion control. Following are the major fabrication requirements:

1) Vessel should be completely back-welded from inside.
2) Fittings and attachments should be flanged types.
3) Manholes should be large.
4) Flanges, trays, nozzles, valves and baffles should be seal welded with continuous welds or designed to be replaceable.
5) Plate thickness for coating should be such that edges could be radiused to one quarter of an inch (Fig. 8.48). Inaccessible areas should be minimized (Fig. 8.49).
6) Avoid shapes that cannot be properly coated. Some difficult-to-coat shapes are shown in Fig. 8.50.

Table 8.2   Factors affecting the life of a paint system

Affecting factors	Percentage
Application error	46%
Incorrect specifications	41%
Change in environment	11%
Faulty paint	2%

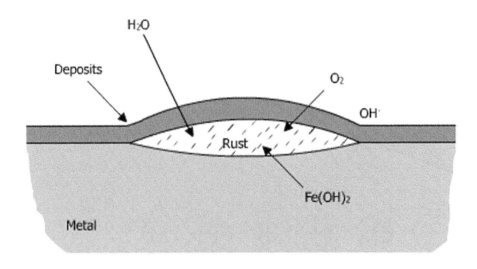

Figure 8.43 Mechanism of corrosion caused by undercutting

7) Manifold outlets should be welded (Fig. 8.51).
8) All welds should be smooth without defects. Tints (thin oxides) must be removed from stainless steel welds.
9) All sharp edges and corners should be rounded by grinding to 1/8″ radius, as they give rise to oxidation.
10) All inaccessible areas should be sealed by welding.
11) Sufficient clearance should be available for cleaning and painting (Fig. 8.52).

8.14.3 FOUNDATION CORROSION

Storage tanks and pipelines should never be placed directly on concrete supporting to minimize corrosion, as the narrow gap between the pipe and concrete support would lead to severe corrosion of the bottom side of the tank. To prolong the service life of the tank, a metal pad made from the same metal as the tank must be welded around the tank and placed on metallic saddle supported on a concrete structure. The tank after welding with the metal pad may be placed directly on the concrete support, depending on the period of service.

It is also possible to place insulation material, such as glass-fiber mat, between the concrete support and the tank after coating the concrete support with bitumen coating. A design recommended for pipeline supports is shown in Fig. 8.52.

Design for support must provide sufficient clearance for cleaning and painting (Fig. 8.53).

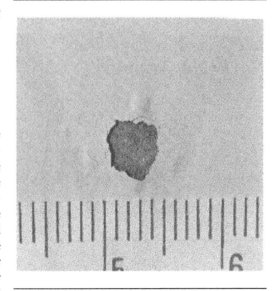

Figure 8.44 Blister burst with rust visible

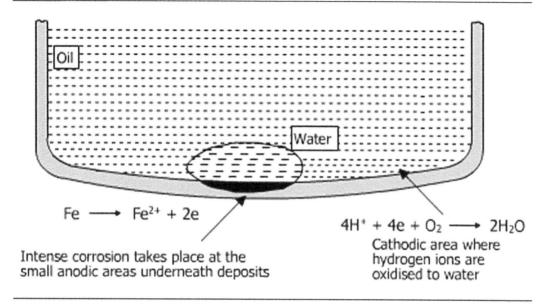

$$Fe \longrightarrow Fe^{2+} + 2e$$

Intense corrosion takes place at the
small anodic areas underneath deposits

$$4H^+ + 4e + O_2 \longrightarrow 2H_2O$$

Cathodic area where
hydrogen ions are
oxidised to water

Figure 8.45   Corrosion at the bottom of oil tanks

8.15 STORAGE OF COMBAT VEHICLES

The readiness rates for track wheel vehicles, aircraft and weapons stored outside without protection, deteriorates at an unacceptable rate of corrosion. It is known that relative humidity below 50% provides the most efficient environment for protection of equipment and material. It prevents rusting of iron and steel without the application of preservatives. Controlled humidity storage is a mandatory requirement for maintenance of expensive and sophisticated army vehicles. There are three alternatives:

a) Open storage
b) Controlled humidity warehouse storage (CHW)
c) Enclosed dry air method (EDAM)

8.15.1 ADVANTAGES AND DISADVANTAGES

Table 8.3 shows the advantages and disadvantages of the three alternatives.

Corrosion remains a significant problem until relative humidity (RH) is reduced to less than 45%. It should be remembered that even less than 25% humidity can cause problems, such as cracking of steel.

8.15.2 PREVENTION

Application of dynamic dehumidification preservation technology (DP) has been successfully applied to preserve weapon systems. DP technology has been applied to ground combat vehicles, helicopters, combat aircraft and air warning and control systems. Currently employed moisture prevention technologies include changes in material design and use of physical barriers to exclude moisture from the air.

The other approach is to extract moisture from the air and control the humidity to an acceptable level. Two systems, static or dynamic dehumidification, can be employed. In static dehumidification desiccant bags are placed or scattered throughout the area for absorption of moisture. This practice has limitations for larger areas and efficient handling as the bags need

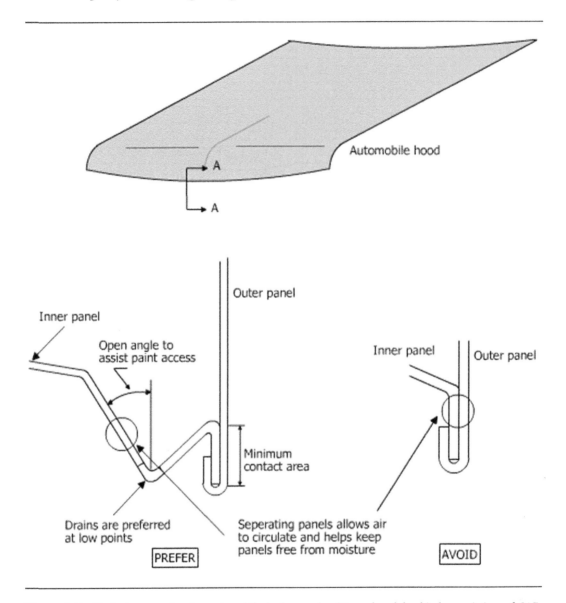

Figure 8.46 Designing practice for automobile body panels. (Reproduced by kind permission of SAE, Warrandale, PA, USA)

to be reactivated after they absorb moisture. In dynamic dehumidification, mechanical dehumidifiers convert moisture laden air into the air containing a specified level of humidity and this air is circulated around the equipment to be preserved.

In the technique called EDAM (Enclosed Dry Air Method), a desiccant impregnated honeycomb construction wheel is impregnated with desiccant. A dry air system is shown in Fig. 8.54. Air is humidified in one section of a wheel and the desiccant is dried and activated in another section. This method utilizes flexible plastic covers as well the humidifiers described above.

Special storage areas can be constructed in existing buildings to utilize dynamic dehumidification technology.

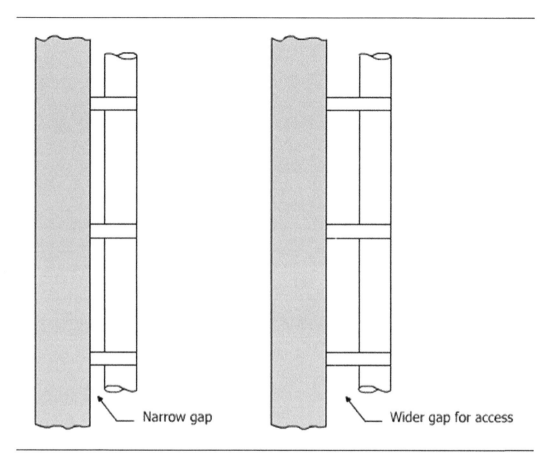

Figure 8.47   Access for maintenance of pipes. (Chaldler, K. A. 1985. *Marine and offshore Corrosion*. Butterworth. By permission of SAE, Warrandale, PA, USA)

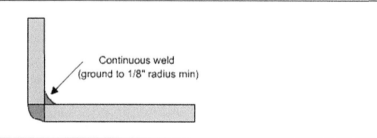

Figure 8.48   Property rounded continuous welds

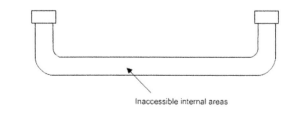

Figure 8.49   Pipe showing areas inaccessible to coating

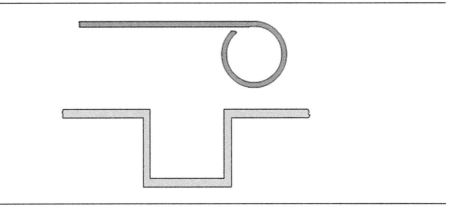

Figure 8.50   Difficult to coat shapes

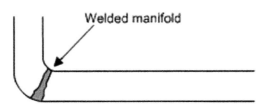

Figure 8.51   Welded manifold outlets

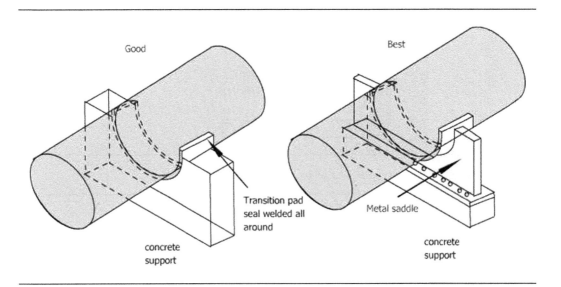

Figure 8.52   Support of horizontal tanks

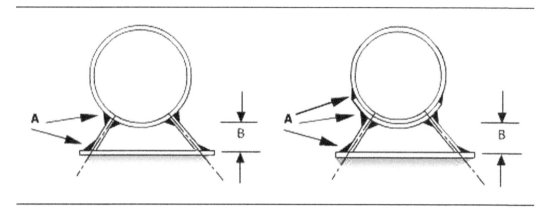

Figure 8.53 Methods of supporting pipelines

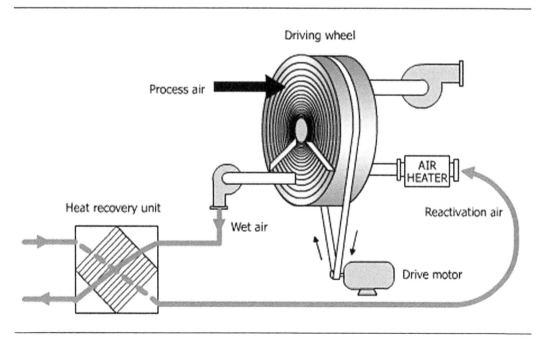

Figure 8.54 Dry air system for dehumidification. (From Laurent, C.S. Munters Cargocaire, 79, Monroe St. PO Box 640, Amesbury, MA. By permission of Laurent, C.S. Amesbury, MA, USA)

Table 8.3 Advantages and disadvantages of three alternatives of storage

Alternatives	Advantages	Disadvantages
Open storage	■ Low cost ■ Less space ■ Low investment cost	■ Adverse effects of environment on vehicles ■ High moisture ingress ■ Low protection against chemicals
Controlled humidity warehouse storage	■ Accessibility of vehicles ■ Good chemical protection ■ Reduced corrosion ■ Good security control	■ High initial cost ■ High electrical operating cost-energy inefficient
Enclosed Dry Air Method (EDAM)	■ Best degree of protection ■ Smallest volume of dry air required ■ Maximum degree of security	■ Requires installation of power lines and transformers ■ Expensive

QUESTIONS

A. MULTIPLE CHOICE QUESTIONS

In the following questions given below, mark one correct answer:

1. Which one of the following features promotes least corrosion?

   a) Entrapment of moisture
   b) Crevices
   c) Shape
   d) Thickness of the metal or alloy used

2. To eliminate entrapment of moisture, it is advised to

   a) make corners smooth
   b) coat the containers
   c) use corrosion resistant alloy, like cupro-nickel
   d) provide a drainage hole

3. Steel columns rest on ground levels. In certain situations water and debris collects at the junction of the ground and the columns.

The most practical way to resolve the problem is to

   a) keep the junction clean by physical means
   b) eliminate the source of water and debris
   c) use coated steel columns
   d) use concrete plinths with provision for water to runoff

4. It is not always practical to avoid crevices, however, the following method of control may prove effective:

   a) use fillers to fill the crevice in all instances
   b) use bolted joints
   c) use spot weld
   d) always use continuous welding to fill the crevices

5. Storage containers should be designed in such a way that

   a) no crevices are formed
   b) no obstacle to drainage occurs
   c) the design should ensure that the containers are drained in the minimum time

d) the containers should be constructed from plastics rather than the metals or alloys

6. In storage tanks, design plays an important role. Which one of the following should be a major consideration?

   a) The lining of the tank must be uniform
   b) The tank material should be highly resistant to corrosive fluids
   c) The exterior of the tank should be protected
   d) All outlet materials, such as pipes, should be galvanically compatible with the tank material

7. If moisture and dirt entrapment is a major problem, it would be a good practice to

   a) spot weld
   b) skip weld
   c) stitch weld
   d) butt weld

8. The corrosion resistance of metals, such as stainless steels and copper alloys, depends upon the buildup of uniform surface films. Which one of the following would require a maximum attention of the designer?

   a) Accessibility of oxygen in the operating medium
   b) Coating of the metal
   c) The drainage capability of the vessel
   d) The period of storage of the liquid in a container

9. To maintain a smooth flow in pipes, it is usual to avoid throttles, valves and orifices unless absolutely necessary. This is usually done to

   a) maintain smooth flow
   b) avoid impingement of fluids
   c) eliminate surging of pressure
   d) change the fluid directions

10. Pipes of different materials, such as copper and steels, should not be embedded in a trench in close proximity to avoid

    a) deposition of copper on steel pipe
    b) depassivation of steel
    c) corrosion of copper pipes
    d) galvanic corrosion, in general

B. How and Why Questions

1. What is the major impact of environment on corrosion designing to prevent corrosion?
2. Which one of the following contributes most to corrosion damage and why?

   a) Rainfall
   b) Humidity
   c) Proximity to sea
   d) Dust storms

3. How does deposition of impurities on a metallic surface lead to corrosion?
4. Describe very briefly the mechanism of initiation of corrosion in threaded joints. What can be done to improve the corrosion resistance of threaded joints?
5. What is the major cause of leakage of water mains? What can be done to temporarily stop the leakage?
6. How could incomplete drainage contribute to the corrosion of liquid containers? Why is complete drainage necessary?
7. What is the objective of circulation of hot air, in and around aircraft?
8. What is the basic cause of corrosion of steel columns in the ground, and what measures can be taken to prevent them?
9. Why are butt weld joints preferred over lap joints?
10. How can vibration be minimized in water transport pipes? Sketch a design.

C. Conceptual Questions

1. Describe an ideal design for prevention of corrosion under thermal insulation.

2. State four features which promote corrosion and must be taken into consideration for a successful design.
3. It is very difficult to avoid crevices in design. State the techniques which can be used to control crevice corrosion in design.
4. State one simple method to reduce erosion–corrosion effect in pipes.
5. State one simple technique of minimizing impingement attacks.

SUGGESTED READING

[1] Ashby, M.T. (1999). *Materials Selection in Mechanical Design.* London: Butterworth.
[2] *British Standards: Codes of Practice.* C. P. 2008. The Council of Codes of Practice, London.
[3] Chandler, K.A. (1984). *Marine and Offshore Corrosion.* London: Butterworth.
[4] Charles, J.A. and Crane, F.A.A. (1989). *Selection and Use of Engineering Materials.* 2nd ed. London: Butterworth. pp. 530–540.
[5] Costa, W. (1985). *An Introduction to Corrosion and Protection of Metals.* London: Chapman & Hall.
[6] Donavan, P.D. (1986). *Protection of Metals from Corrosion in Storage and Transit.* Ellis Horwood.
[7] Jones, D.A. (1991). *Principles and Prevention of Corrosion.* 2nd ed., upper saddle river, NJ: Prentice Hall, USA.
[8] Nigel, C. (1994). *Engineering Design Methods.* London: John Wiley.
[9] Elliott, P. (2002). In *Corrosion*, Vol. 13, *ASM Metals Handbook*, ASM International, Texas: Houston, USA.
[10] Pludek, V.R. (1977). *Design and Corrosion Control.* London: Macmillan Press.
[11] Shigley, J. and Mischke, C. (2001). *Mechanical Engineering Design.* 6th ed., Inc. New York: McGraw-Hill.
[12] Davis, J.R. (2000). *Understanding the Basics*, ASM Int., Ohio: Metals Park, USA.

COMMERCIAL TECHNICAL INFORMATION SOURCES:

[13] Minox Technology (Siste NYTT FRA Industriei-Fax 4735017465 - Daubert VIC/NC (India), 2003.
[14] Controlling Corrosion Case Studies, US Department of Industry, Committee on Corrosion, Pamphlet # 5, 1975.

INTERNET SOURCES

[15] www.inter-corr.com
[16] www.corrosion-doctors.com
[17] http://web1.desbook.usd.mil/html/files/dby
[18] www.soe.org/about/index.htm

KEYWORDS

Accessibility Availability of sufficient space to carry out maintenance or paint work.

Bimetallic contact A metallic contact between two metals differing in potential as exemplified by galvanic series.

Brazing Fusion with a filler metal that has a liquidus temperature above 1000°C but lower than the solidus temperature of the parent metal to be protected.

CHW Abbreviation for controlled humidity warehouse.

Combat vehicles Army equipment in battlefield, such as tanks, vehicles, etc.

Crevice corrosion Corrosion caused in an annular space or cavity by formation of differentiation aeration cells.

EDAM Enclosed dry air method (method of storage of military equipment using flexible covers and dehumidifiers).

Insulation A procedure to conserve energy by keeping cold processes cold and hot processes hot. Glass wood is one example of insulation material.

Relative humidity Ratio of vapour pressure of water in a given environment to the vapour pressure at saturation at the same temperature.

Service life The number of years that equipment/component serves in a specified environment.

Soldering A fusion process that uses a filler metal with a liquidus temperature less than 538°C.

Undercutting In welding, a defect that involves the melting away of the base metal at the side of the weld without filling one of the welding grooves.

CHAPTER 9

SELECTION OF MATERIALS FOR CORROSIVE ENVIRONMENT

9.1 INTRODUCTION

The world of materials comprises of polymers, metals, ceramics, glasses, natural materials and composites. Revolutionary developments have taken place in recent years because of the highly competitive materials market and emergence of new materials and new processing techniques.

Despite the global oversupply, job losses and plunging prices, the steel industry is on the road to a '*new era*' and the positive change is not very far off [1]. The consumption of carbon steel at 800 million tons per annum, dwarfs that of aluminum at 30 million tons, and copper at 15 million tons per annum. It is expected that the steel industry would continue to grow at a rate of 2% driven by developing economics. High strength steel is put to new use in construction of car bodies with the advantage of lower body weight and more efficient fuel consumption without sacrificing the safety standards [2]. New steel sheet forming methods have been developed to improve formability while maintaining productivity. The polymeric materials showed the fastest growth rate (9%) until they began to be substituted by metals, glass and papers, in packaging and construction. Ceramic materials have maintained a steady growth rate and they are finding increased applications in aerospace and the automobile industry.

Composite materials have shown a steady growth rate of about 3% per year. Advanced composite materials, such as fiber glass epoxy and graphite epoxy are finding increased popularity because of their high performance and critical structural applications. For instance, a C-17 transport plane with a wing span of 165 ft uses 15 000 lb of composite materials. Innovative methods of extraction of pure metals are being worked out to cut down the cost of expensive materials, like copper, cobalt, nickel, tantalum, chromium and silicon, from their oxides [3].

The accelerated pace of growth of new materials and the improvements brought about in conventional materials have made the task of selection of materials more challenging. New methods of material selection, such as composite and material selection, have been developed to facilitate the selection of materials to perform a specific function in a specific environment [4].

9.1.1 MATERIALS EVALUATION AND SELECTION

Material selection is critical to engineering design. Corrosion may be minimized by employing an appropriate design as discussed in Chapter 7. The selection of appropriate materials in a given environment is a key factor for corrosion control strategy. The material selected has to meet the criteria for mechanical strength, corrosion and erosion resistance for specific service conditions. For instance, high performance corrosion resistant alloys are specified for valves and piping systems in aggressive environments encountered in refineries and chemical processing plants.

9.2 FACTORS AFFECTING THE PERFORMANCE OF MATERIALS

The following is a list of factors affecting the performance of materials:

(1) Expected performance and functions of the product.
(2) Physical characteristics.
(3) Strength and mechanical characteristics.
(4) Corrosion and wear characteristics.
(5) Fabrication parameters.
(6) Recycling possibilities.

Following is a brief description of each of the above factors.

9.2.1 EXPECTED PERFORMANCE AND FUNCTIONS OF THE PRODUCT

For example, if a seawater storage tank is to be fabricated, fiber glass would be an ideal material because of its advantage of density and resistance to corrosion over the conventional steels, such as AISI 304. For industrial water supply, discharge of used water and irrigation water, PVC (polyvinyl chloride) would be ideal because of the advantage of increased durability, non-corrosivity, ease of maintenance and quality-to-price ratio, they offer. However, knowledge of the expected performance of a material in a given environment is a primary step for the right selection of a material. For aerospace applications, aluminum alloy 6013 (Al–Mg–Si–Cu alloy) is very promising and finding increased application because of its combination of mechanical properties, corrosion resistance, weldability and high strength-to-weight ratio [5].

9.2.2 PHYSICAL CHARACTERISTICS

Physical characteristics, such as electrical conductivity, thermal coefficient of expansion, and thermal conductivity, are important considerations for selection of materials. For instance, cupronickel (90-10 Cu–Ni) are excellent materials for heat exchanger tubes in thermal desalination plants employing raw seawater, because of their excellent conductivity and corrosion resistance. The choice of material is narrowed down by eliminating low conductivity and low corrosion resistant materials.

9.2.3 STRENGTH AND MECHANICAL CHARACTERISTICS

Factors, such as yield strength, tensile strength, fatigue strength and creep strength are crucial to design as they dictate the ability of a product or component to withstand different types of stresses imposed upon them while in service. Mechanical properties are also affected by environment. For instance, the yield strength of steels may be significantly reduced by saline water over a period of time. The level of strength in engineering alloys is to be reviewed in relative terms. For instance, a tensile strength of 1400 MPa for steels is a high strength both in absolute terms, however, an aluminum alloy, such as 6013 (Al–Mg–Si–Cu alloy) having a strength of 475 MPa would also be designated as a high strength alloy in relative terms. For instance, aircraft flying at lower altitudes in sea coastal areas beset with high humidity, salt particles and temperatures exceeding 50°C, suffer loss of performance. The maximum takeoff weight is considerably reduced when the temperature exceeds 40°C [6].

The reduction in aircraft performance, reduces the efficiency of the operation which results in additional flights, and hence an indirect increase in the operational costs.

9.2.4 CORROSION AND WEAR CHARACTERISTICS

Improper selection of materials without consideration of their corrosion behavior in aggressive service environment can lead to premature failure of components and plant shutdown. To avoid failures caused by corrosion:

(a) Materials selected should be compatible with the environment. They must possess sufficient resistance to corrosion for the designed life.
(b) Appropriate preventive maintenance practice must be adopted to maintain the integrity of the equipment/component.

The selection of materials must be based on an extensive knowledge of the service environment. It is to be realized that the behavior of a material may dramatically change when exposed to a corrosive environment. The behavior of materials largely depends upon the following:

(a) Corrosive medium parameter.
(b) Design parameters.
(c) Materials parameters.

9.2.4.1 Corrosive Medium Parameters

Corrosive medium parameters include (Fig. 9.1):

(a) Composition of the corrosive medium. For instance, if a metal is exposed in seawater, a quantitative analysis of all constituents must be made to determine the composition.
(b) Physical and external factors affecting the medium, such as pH, conductivity, temperature and velocity affect the magnitude of corrosion induced by the medium.
(c) Presence of dissolved gases, such as oxygen, carbon dioxide, hydrogen sulfide in fluids promotes the corrosivity of the medium.

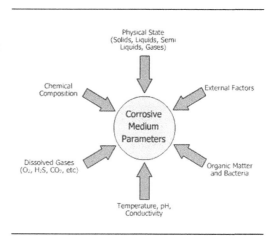

Figure 9.1 Parameters of corrosive fluids contributing to corrosion

(d) Presence of organic matter and bacteria promote the corrosivity of the environment. For instance, sulfide reducing bacteria, desulfovibrio and clostridium, induce corrosion by producing hydrogen sulfide, a serious corrodant for steel. Algae, yeasts and molds also contribute to corrosion.
(e) Physical state of the corrosive medium has a pronounced effect on the corrosivity of the medium. For instance, a dry soil would be less aggressive to corrosion than a wet soil in which salts are dissolved. Similarly, a frozen soil with a very high resistivity exceeding 10 000 ohm-cm would not corrode a pipe, whereas, a wet soil with a low resistivity, such as 500 ohm-cm would cause severe corrosion to structure buried in the soil. Liquids in intimate contact with metals, such as seawater, acids alkides and alkalies, are serious corroders for a large variety of metals and alloys. The presence of gases leads to adsorption on the surface and the formation of an oxide, sulfide, and chlorides. Dry oxidation refers to an electron producing reaction in a dry gaseous environment. It is also known as '*hot corrosion*' if sulfur is involved. Accelerated oxidation (catastrophic oxidation) occurs at high temperatures and leads to accelerated failure of metals.

9.2.4.2 Design Parameters

Design parameters which affect the rate of corrosion include the following:

(1) Stresses acting on the materials in service.
(2) Relative velocity of the medium and obstacles to flow.
(3) Bimetallic contacts.
(4) Crevices.
(5) Riveted joints.
(6) Spacing for maintenance.
(7) Drainage and directional orientation of loop.
(8) Joints to avoid entrapment.
(9) Sharp corners.
(10) Non-homogeneous surface.

All the above parameters have been fully discussed in Chapter 7. Corrosion can be prevented by intelligent designing as discussed in the chapter.

9.2.4.3 Materials Parameters

The following are the material parameters which may affect corrosion resistance:

(1) Impurity segregation on grain boundaries leads to weakening of grain boundaries and accelerates corrosion attack. In 18-8 steels, the depletion of chromium due to the formation of chromium carbide promotes intergranular attack. The formation of Mg_2Si at the grain boundaries in Al–Mg alloys leads to weakening of grain boundaries and promotes corrosion of Al–Mg alloys.
(2) Microstructural constituents. A heterogeneous microstructure forms anodic and cathodic sites which promotes corrosion. The intermetallic precipitate serves as anodic and cathodic sites and they may be anodic or cathodic to the matrix. For instance, a $CuAl_2$ precipitate is cathodic to the aluminum matrix and Mg_2Si is anodic to the matrix. Microstructural variation plays a leading role in metal matrix composites. For instance, Al 6013-SiC composite corrodes at the Al matrix/SiC interface because

of the preponderance of intermetallic secondary phases [7]. Also a non-homogeneous distribution of SiC contributes to accelerated corrosion [8].

(3) Surface treatment, such as galvanizing, phosphating and painting increase the resistance of materials to corrosion. Galvanizing improves the resistance of steels to corrosion and is widely used in automobile industry. Galvanized pipes are widely used for transport of hot water in domestic plumbing systems. If the temperature, however, exceeds 65°C, reversal of polarity (steel becomes anodic) can cause corrosion of the galvanized pipes.

(4) Alloying elements and film formation. Alloying elements in steels, such as chromium, nickel or molybdenum contribute to the production of a protective oxide layer which makes steel passive. Addition of copper to aluminum alloy increases the strength but decreases the corrosion resistance. Scandium addition to Al–Mg–Si alloys significantly increases their strength without effecting a decrease in their corrosion resistance [9]. Coatings of epoxy and polyurethanes provide a longer life to structures exposed to corrosive environments. Amorphous thermally sprayed coatings using physical vapor deposition (PVD), chemical vapor deposition (CVP), laser or electron surface remelting and thermal spraying techniques offer excellent resistance to corrosion [10,11]. Recently electron beam physical vapor deposited coatings (EB-PVD) has been found to increase the component life significantly [12].

9.2.5 FABRICATION PARAMETERS

Following are the fabrication parameters required for analyzing material selection:

(1) Weldability. Welding procedures, such as electric arc welding, friction welding, spot welding, need to be carefully selected to minimize the effect of corrosion. For instance, gas

welding in the sensitive temperature range may cause intergranular cracking.

(2) Machinability. Machining operations, such as drilling, milling, shearing, turning, may lead to enhancement of corrosion if they are not properly controlled. Drilling fluids are highly corrosive and need to be handled with care.

(3) Surface modification procedures, such as cladding, galvanizing and metallizing (metallic coatings) increase the resistance of the materials to corrosion. The success of the coating depends on the bonding between the coating and the substrate and surface preparation before the application of coatings according to the international standards. Surface preparation techniques include hand, power tool cleaning and abrasive blast cleaning according to the American Standards for Surface Cleaning.

9.2.6 SALVAGE OR RECYCLING

Often it might be cost-effective to use an expensive material if it can be recycled with a high salvage value on completion of the useful life of the equipment/product. Environmental considerations also demand the recycling of metallic materials. The Association of European Producers of Steel for Packaging (AEPS) has stated that 50% of all steel packaging was recycled, totaling 1 670 000 tons. The leading country in European Union is Luxembourg with 93% followed by Belgium and The Netherlands [13].

9.3 THE MATERIAL SELECTION PROCESS

With the phenomenal growth in the development of new materials and availability of over 40 000 metal alloys and a larger number of nonmetallic materials, the process of selection of materials has become very complex and challenging. The selection process should start at the drawing board stage where all important factors affecting the products, such as the choice of engineering design and processing methods should be considered.

9.3.1 FACTORS IN MATERIALS SELECTION

9.3.1.1 Physical and Mechanical Factors

Amongst the physical factors the size, shape and the weight of material are important factors. The weight of a material is crucial for application, such as in aerospace industry where light-weight and high performance alloys are needed to save energy and increase operational efficiency. The size and shape dictate the heat treating procedures.

The mechanical factors relate the ability of material to withstand stresses imposed on them. The mechanical properties are used as a failure criteria for design. The mechanical properties commonly used for design are the following:

(1) Density (ρ)
(2) Modulus of elasticity (E)
(3) Strength (σ)
(4) Ductility
(5) Fracture toughness
(6) Fatigue
(7) Corrosion fatigue
(8) Creep
(9) Impact
(10) Hardness.

(A) **Functional requirements**. Selection of materials is normally based on optimizing materials performance according to the functional requirement, geometry and properties of materials. Functional requirement is directly related to the expected characteristics of the product. For example, if a product carries a uniaxial load, the load carrying capacity can be related to the yield strength of the material. For heat exchanger tubes, the material of the tubing must conduct heat, have a maximum operating temperature above the operating temperature of the design and should have a high resistance to corrosion. Similarly, the resistance to stress corrosion cracking can be related to $K_{i,SCC}$ (limiting value of stress intensity, K_i) below which stress corrosion cracking does not occur. The threshold, Rockwell hardness, for low alloy steels in a sulfide environment is 20–23 below which SCC would not occur [14]. Phosphorus control affects $K_{i,SCC}$ for SCC of 304 stainless steels in 45% $MgCl_2$ solution at 154°C. Steels containing low phosphorus content show a large $K_{i,SCC}$, whereas, steels containing larger phosphorus contents show a smaller $K_{i,SCC}$ indicating that cracks arise at smaller stress level [15]. The functional requirement is the first basic step in the selection process for materials.

(B) **Processing requirements**. The type of materials to be used affects the choice of processing method. The processability of a material reflects its ability to be hot worked or cold worked and shaped into the desired finished product. Processibility includes castability, machinability, weldability and deformation process. Complicated shapes are best produced as castings and small shapes are usually investment casted. Thermoplastic and ductile metals are shaped by deformation processing. Fabricability includes welding, brazing, soldering and machining process. Thermal and heat treatment processes are implemented before the finishing process, such as surface coating and polishing. In all operations mentioned above, materials properties would be affected, hence the processing is very closely related to functional requirements.

(C) **Life of components**. The following properties affect the service life of materials in the environment:

(1) Corrosion and wear
(2) Creep
(3) Fatigue and corrosion fatigue.

The properties of materials affected by the factors mentioned above cannot be accurately predicted partly because of the variation in the composition of the environment and lack of technique to accurately predict the life span.

(D) **Cost**. The selection of materials is strongly influenced by cost and availability. Titanium is an excellent material for heat exchanger service in brine and seawater and other aqueous environments, however, its high cost compared to 70:30 cupronickel can be justified by using thin walled titanium tubes (19 mm OD × 0.5 mm) compared to 19 mm OD × 1.20 mm 70:30 CuNi tubes.

(E) **Reliability**. The material selected should perform the intended function for a specified period without a failure. Premature material failure can lead to catastrophic failures and expensive plant shutdowns. Corrosion of steel structures, decks and bridges, can lead to serious failures and loss of human lives. Failure of horizontal stabilizers in aircraft and landing gears can lead to fatal accidents.

(F) **Resistance to service conditions**. Corrosive environments seriously shorten the performance of a material. The selection process requires materials compatibility with the environment. For instance, austenitic steels (304, 304L, 316, 316L) withstand the corrosive attack of polluted seawater much better than the copper alloys.

(G) **Codes and other factors**. Codes are sets of requirements imposed by customers or by technical organizations, such as ASTM (American Society for Testing of Materials) and SAE (Society of Automobile Engineers). Statutory factors comprise regulations regarding health, safety, disposal, environmental requirements and recycling which need consideration.

9.4 MATERIALS CLASSIFICATION

The world of materials basically comprises families of polymers, metals, ceramics, glasses, natural materials and composites, which can be synthesized by combination of the above materials. Table 9.1 shows the subdivisions of the material kingdom in families, classes, subclasses and

Table 9.1 Materials classification and their attributes [4]

Kingdom	Family	Class	Subclass	Members	Attributes
Materials	Polymers Ceramics Glasses Metals Natural Composites	Steels Cu alloys Ti alloys Al alloys Ni alloys Mg alloys Zn alloys	1xxx (Pure Al) 3xxx (Al–Mn) 4xxx (Al–Si) 5xxx (Al–Mg) 6xxx (Al–Mg–Si) 7xxx (Al–Zn) 2xxx (Al–Cu)	6061-O 6061-T1 6064-T4 6063-O 6066-T4 6162-T5 6162-T6 6463-T1	Density, ρ Modulus of elasticity, E Strength, σ Fracture toughness, K_{ic} Thermal conductivity, λ(W/m.K) Coefficient of expansion α Corrosion, mm-year^{-1} Cost (relative cost per unit) Fatigue Corrosion fatigue

attributes for each member of the subclass. For example, Table 9.1 shows Al as a class of the metal family, and the subclass shows the members belonging to the subclass i.e. Al 1000, Al 2000, represented by 1xxx, 2xxx, etc. Attributes comprise the mechanical, thermal, and electrical properties, the processing characteristics, cost and the environmental impact. The set of attributes is called the property profile.

9.4.1 FUNCTIONS, OBJECTIVES AND CONSTRAINTS

The design requirements need to be translated in terms of the suitability of the materials with the design requirements. However, the functions required, the objectives to be achieved, and the negotiable and non-negotiable constraints must be met (Table 9.2). The function of a product or component, such as a beam carrying bending and ties carrying tension, need to be clearly specified. The objective should specify the variables, such as cost, mass, stored energy, etc. which has to be maximized and minimized. The essential

Table 9.2 Analysis of functions

Function	Basic function expected of a component
Objectives	Variables to be minimized or maximized
Constraints	Non-negotiable and desirable conditions which must be met
Free variables	Variables not specified by design and a freedom to choose value to satisfy the design constraints

constraint conditions, such as functioning of a component in a given range of temperature and environment without failure, carrying of specified load and minimum deflection allowed, must be met to achieve the objectives. Free design variables have usually a geometric dimension, such as the wall thickness of a tube or the area of a tie. From the design concepts mentioned above, we obtain a screening criteria, expressed as numerical limits on material property values and material indices which can be used to characterize their performance.

9.4.2 STRATEGY OF MATERIALS SELECTION

The following steps lead to the screening and ranking of candidate materials:

(a) The material attributes must meet the required design requirements. For example, if the service temperature for a commercial steel component in an oxidizing environment is prescribed at 1100°C, 25 Cr–20 Ni steel would be a suitable choice and other steels, such as 13 Cr Steel, 18 Cr steel would be screened out because their limiting service temperature are below 750°C. The screening is thus dictated by property limits.
(b) Find candidate materials which can do the job. How good a job can be done by other materials can be known by determining their material indices. A combination of mechanical properties which characterize the performance of material gives maximized or minimized material attributes.
(c) Supporting information: Screening helps to short list the candidate materials and considerably narrows down the range of selection. To further narrow the choice of candidates, support information on material is explored which comprise the following:

    (1) Handbooks, such as ASM Handbook, CRC Handbook of Materials, etc.
    (2) Data sheets from suppliers.
    (3) Data sheet from professional organizations, such as Alcoa, USA; Nickel Development Association (NDA); Copper Development Association (CDA); American Iron and Steel Institute (AISI), etc.
    (4) Websites of suppliers.
    (5) Personal contacts.

(d) Local environment: The suitability of a product or equipment in a local environment is essential to service life. Helicopter rotor blades have a design life of fifteen years in normal operating environment, however, the life span was reduced to one year in a desert environment in the Gulf, as witnessed in the Gulf War.
(e) The final material choice will evolve the completion of a systematic screening process.

9.4.3 PERFORMANCE INDICES

The most significant steps in the new material selection procedure described by Cebon and Ashby [16] is the determination of performance indices of properties which maximize material properties. The design of a mechanical component is specified by the following three factors:

(1) Functional requirements (F)
(2) Geometric parameters (G)
(3) Material properties (M)

The performance of a component can be best described by:

$$P = \left[\left(\begin{array}{c} \text{Specified functional} \\ \text{requirements, } F \end{array} \right), \right.$$
$$\left. \left(\begin{array}{c} \text{Geometric} \\ \text{parameters, } G \end{array} \right) \cdot \left(\begin{array}{c} \text{Material} \\ \text{properties, } M \end{array} \right) \right]$$
$$(9.1)$$

where P may refer to mass, volume, life or cost. The above three groups are separable when equation (9.1) is written as:

$$P = f_1(F) \cdot f_2(G) \cdot f_3(M) \qquad (9.2)$$

where f_1, f_2, and f_3 refer to functions.

Separation of the groups allows significant simplification because the optimum choice of material becomes independent of all geometries (G) and functional requirements (F) associated with design. On maximizing $M(f_3)$, the performance of $F(f_1)$ and $G(f_2)$ is maximized.

9.4.4 PROCEDURE [16]

(1) Identify the functions, objectives, constraints and free variables of the design.
(2) Write an expression for the objective functions.

(3) Derive an expression for free variable using the constraint equation.
(4) Substitute free variable into the objective function and rearrange the expression to find the material index D.

Example:
Find the material index for a strong light tie rod (Fig. 9.2 as an example [16])

Solution:
Specify function, objective, constraints and free variables.

Function in this case: Tie carries a tensile force.
Objective: Minimizing of mass.
Constraints: A specified load is to be carried.
Free variables: Cross-sectional area of the tie, they are generally geometric dimensions.
The object function is described by

$$m = A l \rho \quad \text{(equation for the objective)}$$

where A is the cross-sectional area, l is length and ρ is density.
The length (l) is specified and must not fail under load F (constraint). The cross-sectional area (A) is free (free variable). To summarize, function: tie rod. Objective: minimize mass constraint. Length is specified, must not fail under load F (adequate fracture toughness).

$$F/A < \sigma_Y \quad \text{(equation for the constraint)}$$

The formula for the mass of tie is: density (ρ) × area (A) × length (l)

$$\frac{\text{Force}(F)}{\text{Area }(A)} \quad (9.3)$$

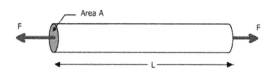

Figure 9.2 Design of tie for minimum mass

Hence, the area (free variable) is:

$$A = \frac{F(\text{Force})}{\sigma_Y(\text{yield strength})} \quad (9.4)$$

Stronger materials, like stainless steel require a small cross-sectional area, whereas weaker materials, like polymers, require large cross-sectional areas, to carry the same load. In this design both can be used, small and large cross-sectional areas, as they are no constraint. The important question is which material would give the smallest mass for a light and stiff beam. This can be determined as: generate m in terms of F, G, M. Define m in terms of any combination of F, G, M. It is shown that $m = A \times l \times \rho$. Missing parameter is F. Define F in the above equation, $F/A = \sigma_Y$, eliminating area.

$$m = (F)(l) \left(\frac{\rho}{\sigma_Y} \right) \quad (9.5)$$

Hence, the performance index for a light strong tie is, therefore,

$$\left(\frac{\rho}{\sigma_Y} \right) \quad (9.6)$$

Choose material with smaller l.
The first and second brackets contain the specified load and the specified length. The material property represented by (ρ/σ_Y) in the last bracket suggests that the lightest tie rod which would carry the specified load F is that of a material with the smallest value of ρ/σ_Y. In order to maximize the performance, the material properties term (ρ/σ_Y) is inverted. The materials property are defined by an index M. In the above case $M = \sigma_Y/\rho$. The lightest tie rod would carry the load F would be with the largest value of M.
Similarly for a light strong beam, the material index, M is

$$M = \frac{\sigma_f^{3/2}}{\rho}$$

and the mass is minimized by selecting material with the largest value of the index. In the above example, l and f were the constraints.

Table 9.3   Material indices [16]

Functions, objectives and constraints	Performance index to be maximized
1. Ties – minimum weight, stiffness prescribed	$\frac{E}{\rho}$
2. Beam – minimum weight, strength prescribed	$\frac{\sigma_y^{2/3}}{\rho}$
3. Beam – minimum cost, stiffness prescribed	$\frac{\sigma_y^{2/3}}{C_m \rho}$
4. Thermal insulation – minimum cost, heat flux prescribed	$\frac{1}{\lambda C_m \rho}$

Abbreviations: C_m = cost/kg, λ = thermal conductivity, σ_y = elastic limit, ρ = density.

Typical examples of material indices are given in Table 9.3.

The performance depends on two or more material properties which are presented by plotting one mechanical property or a mathematical combination of properties on each axis of materials chart. One way to consider materials for a particular property or material index is to use material selection chart. A material selection chart (ρ vs E) is shown in Fig. 9.3. It shows the modulus of materials stretching four decades from 0.1 GPa to 1000 GPa. The log scales allow large information to be displayed in a small space. The data for a number of a particular class of materials is clustered and enclosed by an envelope (heavy line). The bubbles in the envelope represent a subset of materials. For example, in Fig. 9.3, an envelope of engineering composites contains KFRP (Kelver Fiber Reinforced Polymer), GFRP (Graphite Fiber Reinforced Polymer) and CFRP (Carbon Fiber Reinforced Polymer), as subsets of materials in a bubble. The envelope encloses all members of the class, and the bubbles members of sub classes. A range of materials are included in the chart. These charts display data for important properties which are of primary interest in characterization of materials and engineering design. These properties include density, modulus, strength, toughness, thermal conductivity, diffusivity and expansion.

From the information provided in the charts, the subset of materials which maximize the performance of components after having fixed the primary constraints are found. Resistance to corrosion, modulus, strength and density are all primary constraints which must be met. The constraints may eliminate several classes of materials. The constraints are given by

$$P > P_{critical}$$

where P is the property, and $P_{critical}$ is the critical value which must not be exceeded.

For instance, if corrosion is involved, the corrosion rate should not exceed the prescribed critical value. Primary constraints appeared as horizontal or vertical lines on the material selection chart. For instance, if a constraint is $E > 10$ GPa, and $\rho = 3$ mg/m^3, the intersect of the lines would yield a search region for selection of appropriate materials.

From the information provided by the materials property chart, the materials which maximize performance are selected from the subsets which meet the property limits using material indices.

For instance, select materials which have an elastic modulus of as low as 10 GPa, a line is drawn at that value (Fig. 9.4). All the materials above the line would show a set of candidate materials. If the density requirement is less than 2.0 mg/m^3, a line is drawn on the specified density value. The subset of materials which meet both the criteria would be found on the upper left quadrant.

The material indices $\frac{E}{\rho}$, $\frac{E^{1/2}}{\rho}$, $\frac{E^{1/3}}{\rho}$, for light and stiff components can be plotted on to the chart (Fig. 9.5).

Consider, for example, the maximizing factor for a light, stiff beam and give it a value of

$$\frac{E}{\rho} = C \quad \text{or} \quad E = C\rho$$

Taking logs on both sides, $\log E = \log \rho + \log C$, which is a straight line equation with slope 1. The value of C is obtained when $\rho = 1$ and equals E, and dictates the value of the guidelines. Materials intersected by the guideline have the same value for E/ρ (specific modulus).

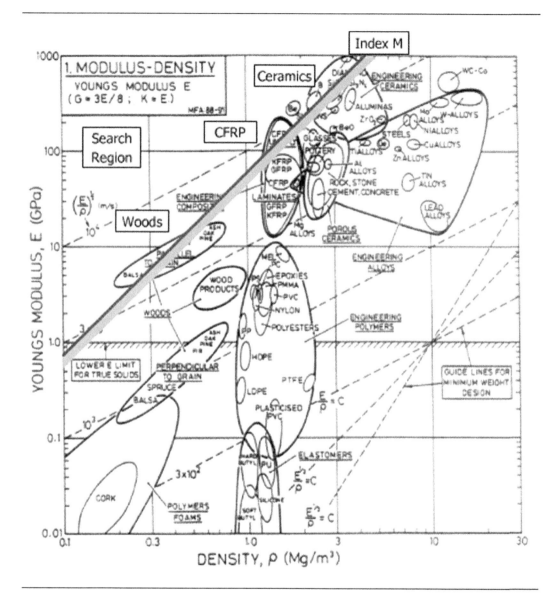

Figure 9.3 Materials for oars. CFRP is better than wood because the structure can be controlled. (From Ashby, M.F. (1999). *Materials Selection in Mechanical Design*, Butterworth. By kind permission of Butterworth Publication)

For the condition:

$$C = \frac{E^{1/2}}{\rho}$$

$$\log C = \frac{1}{2} \log E - \log \rho$$

$$\log E = 2 \log \rho + 2 \log C$$

We thus obtain a family of straight line (Fig. 9.5a). Other lines in the family would be parallel lines with a gradient 2.

Similarly, for the condition:

$$\frac{E^{1/3}}{\rho} = C$$

we would obtain another set of lines with slope 3. Fig. 9.5b shows a guideline and a

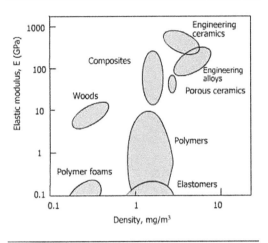

Figure 9.4 Property chart

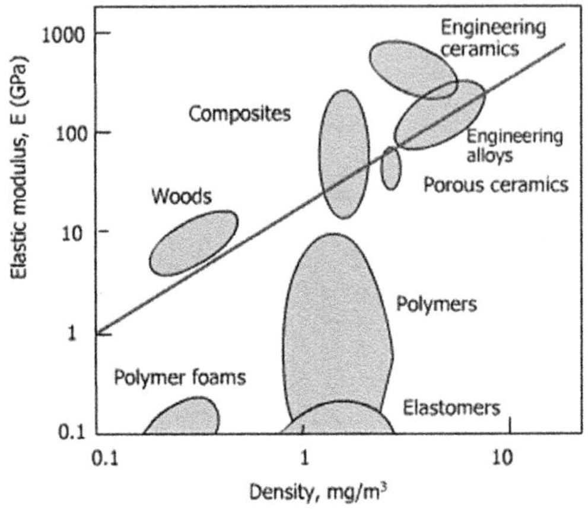

Figure 9.5b Line for $E/\rho = 1000$

subset of material. The guidelines show the slopes of a family of parallel lines for a particular performance index.

Consider selecting material with values of E/ρ greater than, approximately, 1000.

$$\log E - \log \rho = \log 1000$$

$$\log E = \log \rho + 3 \; [\log E - \log \rho = \log C]$$

This would give a line with a slope of one and intercept of 3 on the log E axis. The materials intersected by the guideline have the same E/ρ value (specific modulus). Materials above this line have values higher than C and those below have lower values and not suitable. A schematic $E-\rho$ chart showing guidelines for three material indices for stiff, light-weight design is shown in Fig. 9.6.

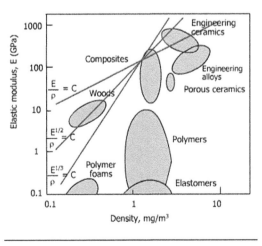

Figure 9.6 A schematic $E - \rho$ chart showing guidelines for the three material indices for stiff, lightweight design. (From Ashby, M.F. (1999). *Materials Selection in Mechanical Design*, Butterworth)

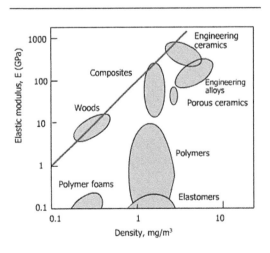

Figure 9.5a Guideline for $E^{1/2}/\rho$ maximized

9.4.5 COST FACTOR IN MATERIALS SELECTION

Cost puts a constraint on material selection. In simple cases, the cost per unit property can be used as a criterion for selecting the optimum material. Because of the changing cost of materials in the market, inflation, supply and scarcity, the cost of materials is measured as a relative cost (C_R) where the cost of material is normalized to the cost of mild steel.

$$C_R = \frac{\text{Cost per kg of the material}}{\text{Cost per kg of mild steel rod}}$$

C_R (the relative cost per unit volume) is plotted vs the modulus of elasticity (Fig. 9.7). The most cost-effective materials are located in the upper left-hand corner. Chip and stiff materials are located at the bottom of the chart. The figure shows that stone, bricks, concrete, in the porous ceramic group and mild steel and cast iron in the alloys group are cost-effective materials because of their increased consumption by the building industry.

9.4.6 LIMITATIONS OF THE MATERIALS PROPERTY CHART

With the unprecedented growth in the number and variety of materials, the selection process is becoming more demanding and it has to meet stringent technical, economical, environmental and aesthetic requirements. The use of material indices allows the optimum selection of material. A wealth of data is stored in the data base and each material has a unique identifier. The CES educational software is a unique tool for selection of materials from beginner to advanced levels [16]. To work out material indices, a good knowledge of engineering mechanics is required. Basic knowledge in structural engineering and CAD is required for a better understanding. A wealth of information on material properties is given in the software package which can be used for materials selection.

Important steps in selection of materials are:

(1) From a wide range of class of materials, such as metals, alloys, plastics, elastomers, ceramics, composites, etc., screen out the materials not meeting the functional requirements.
(2) Identify constraints. There are some conditions which must be met, for instance, resistance to corrosion can put an important constraint for turbine heat exchanger materials. Strength and stiffness may impose limitations in the selection process. Materials not meeting the minimum constraint requirements would be eliminated. In the case of a tie rod, the specified length and the tensile load to be supported would be the constraint.
(3) Decide what is to be maximized or minimized. For instance, we minimize the mass of a light tie rod.
(4) Identify the variables. For instance, the cross-sectional area A in the expression $m = A \cdot l \cdot P$ is variable, the mass m can be reduced by reducing the cross-sectional area (A), however, it should carry the specified tensile load (P) and maintain a prescribed length (l), both being constraints.
(5) Develop an equation for constraint in terms of G (general) and M (material properties).
(6) Express free variable in terms of constraints, such as $A = E/\sigma_f$.
(7) Read the performance to be maximized. Several indices can be found as cited in reference [16].

9.5 MATERIALS AND FLUID CORROSIVITY

The corrosion resistance of materials depends on the nature of the corrosive environment. A material may be very corrosion-resistant in an environment and yet fail in another environment. A knowledge of corrosion resistance of a material in a given environment is, therefore, of a fundamental importance to their successful application. A knowledge of the corrosivity of the environment, is therefore, very

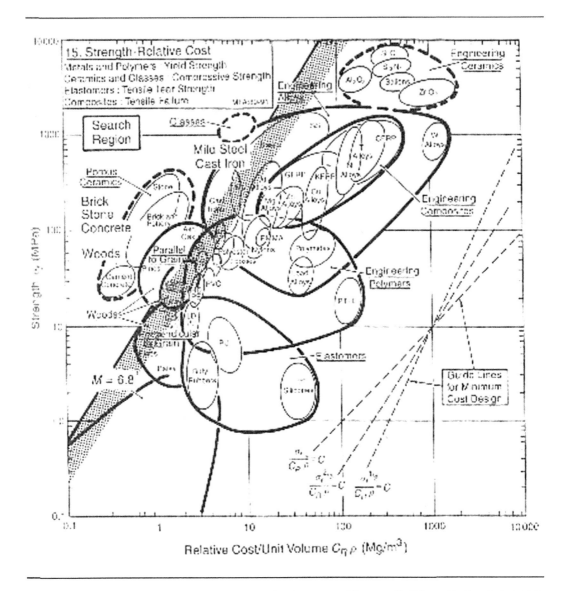

Figure 9.7   The selection of cheap, strong materials for the structural frames of buildings. (Ashby, M.F. (1999). *Materials Selection in Mechanical Design*, Butterworth. By kind permission of Butterworth Publication)

useful for selection of materials. Water plays a predominant role in a corrosion reaction since it is an electrolyte, an essential component of a corrosion cell. Pure water is a poor conductor and not significantly corrosive up to 170°C. In the presence of salts, acids and dissolved gases, like hydrogen sulfide, carbon dioxide (CO_2) and oxygen (O_2), the degree of corrosivity of water

is significantly increased. The following are the common aggressive environments encountered by materials:

(1) Marine environment.
(2) Industrial environment.
(3) Oilfield environment.
(4) Pollution in environment.

9.5.1 MARINE ENVIRONMENT

Seawater is the major contributor to the marine environment. It is almost a universal fluid to evaluate the corrosion resistance of metals and alloys. Any material showing a good resistance to corrosion in seawater is considered as a suitable material for application in a wide variety of applications encountering water. The following are the major seawater environments:

Classification	Corrosivity
■ Atmosphere	Medium
■ Splash	High
■ Tidal	Medium-High
■ Shallow water	Medium
■ Deep ocean	Low
■ Mud	High

Table 9.4 Typical seawater composition

Component	Concentration (mg/l)	% Total salt
Cl^-	18 980	55.04
Br^-	65	0.19
SO_4^{--}	2649	7.68
HCO_3^-	140	0.41
F^-	1	0.00
H_3BO_3	26	0.07
Mg^+	1272	3.69
Ca^{++}	400	1.16
Sr^{++}	13	0.04
K^+	380	1.10
Na^+	10 556	30.61
	34 482	99.99

9.5.2 FACTORS AFFECTING THE CORROSIVITY OF SEAWATER

A. Movement

Moving seawater is less corrosive than stagnant seawater. Under stagnant conditions, oxygen concentration cells are setup on the surface of materials and corrosion is accelerated. Salt water must be rinsed without delay from stainless steel equipment before plant shutdown, otherwise heavy damage by pitting may be caused to the equipment necessitating replacement.

B. Corrosive agents

There is a large variation in the total dissolved solid contents (TDS) from one ocean to another. The total dissolved solid content in the Baltic Sea is represented to be on an average of 8000 ppm as against the TDS content of 44 000–60 000 ppm in Arabian Gulf water. The degree of corrosivity of seawater varies with the TDS content. The chloride ion is the most aggressive (corrosive) constituent in seawater, as it has a tendency to penetrate the passive films on the metal surface and destroy them. A typical seawater composition is shown in Table 9.4.

The corrosion resistance of materials in seawater is generally accepted as a standard criterion for the service performance of materials. Where seawater is not available, the performance of material is evaluated in 3.5 wt% NaCl.

9.5.3 INDUSTRIAL ENVIRONMENT

The industries contributing to the industrial environment include chemical, petrochemical, fertilizer, pulp and paper, and all other process industries. The following are the major corrosive fluids encountered in a wide spectrum of industry:

(a) Acids (Inorganic). Acids, such as hydrochloric, hydrofluoric, sulfuric and sulfurous, are encountered in highly corrosive environments in refineries.
   Acids (Organic). Acids, such as formic, acetic, propionic, dicarboxylic and napthenic, are organic acids present in crude oil.
(b) Strong alkalies.
(c) Salt water.
(d) Dissolved gases, such as hydrogen sulfide (H_2S), carbon dioxide (CO_2) and oxygen (O_2).

(e) Pollutants, such as particulate matter, sulfur oxides, carbon monoxides, nitrogen dioxides, ozone and lead.

(f) Soil pollution, such as bacteria, oil spills, natural gas contaminants sewage contaminants and pesticide degradation products.

The success of service performance of materials would depend on their ability to offer sufficient resistance to corrosion in industrial environments.

9.5.4 OILFIELD ENVIRONMENT

The corrosion problems caused by environment can be classified in to two categories: (a) downhole environment and (b) surface environment. Whereas the surface environment is mainly the corrosion related to the absorption and atmospheric contaminants, the downhole environment is highly aggressive as it may contain brine, carbon dioxide, acids and hydrogen sulfide, all of which are highly corrosive. The downhole environment is also subjected to elevated temperatures. Corrosion protection of drilling equipment is a challenging job in the oil industry.

9.5.5 POLLUTION

Atmospheric contaminants, mainly salts and SO_2, in combination with humidity contribute to the corrosion of most metals. Chloride ions breakdown the passive layer of oxides on steels and initiate pitting. Compounds of nitrogen, such as ammonia, cause serious problem on copper and brass structures by formation of $Cu(NH_3)$ complexes. Copper piping networks can be seriously damaged by corrosion in agricultural areas where ammonia may be present. Aluminum alloys are sensitive to corrosion in an environment with a pH range less than 4 and greater than 10, the former representing the acid regimes and the later the alkaline regimes corrosive to aluminum.

9.6 RANKING OF PERFORMANCE OF MATERIALS

The ranking of materials is based on the basis of the degree of corrosion resistance they offer in a given environment. The best way to evaluate corrosion resistance is to determine their rate of corrosion according to ASTM (American Society for Testing of Materials) and NACE (National Association of Corrosion Engineers) standards. Important tests to determine the rate of corrosion are given below:

(1*) ASTM Designation G5. Annual Book of Standards, Vol. 03, 02. Test Method for Making Potentiostatic and Potentiodynamic Anodic Polarization Measurement. (American Society for Testing of Materials, 2001)

(2*) ASTM Designation G31. Annual Book of Standards, Vol. 03, 02. Laboratory Immersion Corrosion Testing of Metals. (American Society for Testing of Materials, 2001)

(3*) ASTM Designation G59. Annual Book of Standards, Vol. 03, 02. Standard Practice for Conducting Potentiodynamic Polarization Resistance Measurements. (American Society for Testing of Materials, 2001)

(4*) ANSI/ASTM Designation G51. Annual Book of Standards, Vol. 02, 03. Conducting Cyclic Potentiodynamic Polarization Measurement for Localized Corrosion. (American Society for Testing of Materials, 2002)

(5*) ASTM Designation G4. Recommended Practice for Conducting Plant Corrosion Tests. (American Society for Testing of Materials, 2003)

(6*) NACE Standard TMO169-76. Laboratory Corrosion Testing of Metals for the Process Industry. (NACE, 1969)

(7*) Z. Ahmad and B. J. Abdul Aleem, "Effect of Velocity and Elevated Temperatures on Pitting of Al 6013-20 SiC(P) Composite in 3.5 wt% NaCl," *Br. Corros. Journal*, **33**, (3), (1995). The above paper describes

Table 9.5 Ranking of materials

Ranking	Range (mmy^{-1})	Example
Excellent	Up to .01 mmy^{-1}	Ti in high velocity seawater at 41 m/s
Good	Greater than .01 mmy^{-1} but not higher than 0.5 mmy^{-1}	6% Al–Brass in polluted seawater 70-30 Cu–Ni in seawater at 41 m/s
Satisfactory	Greater than 0.5 mmy^{-1} but not higher than 1.0 mmy^{-1}	Al–Bronze in pump impellers in seawater
Unsatisfactory	Higher than 1.0 mmy^{-1}	Carbon steel in high velocity seawater

the experimental techniques to conduct polarization measurements at elevated temperatures.

(8) Z. Ahmad and B. J. Abdul Aleem, "Corrosion Resistance of a New Al 6013-20 SiC(P) in Salt Spray Chamber," *Jr. Mat. Eng. and Performance*, **9**, (3), p.338 (2000). The technique of conducting corrosion studies in a salt spray chamber are described in the experimental section of this paper.

(9) Z. Ahmad and B. J. Abdul Aleem, "Effect of Suspended Solids in the Flow Induced Corrosion of Modified Al-2.5 Mg Alloy in Arabian Gulf water," *Jr. Mat. Sc. and Eng.*, **1**, p.61, 1992. The technique of studying the effect of velocity on the corrosion resistance of materials is described in this paper.

(10) Z. Ahmad and B. J. Abdul Aleem, "The Role of Al/SiC Interface on the Corrosion Behaviour of Al 6013-20 SiC Composites," *Corrosion*, NACE, **60**, (10), (2004), 954. The technique of studying elevated temperatures and elevated pressure corrosion studies using an autoclave is described in this paper.

(11) ASTM STP-665, Stress Corrosion Cracking, The Slow Strain Rate Technique, STP-665, 1977.
[*The ASTM Standards are subject to revision every year and the latest revision must be consulted. ASTM Annual Standards are published each year by the American Society for Testing of Materials, located in Philadelphia, USA. ASTM Annual Book of Standards (vOL.03, 02) is a highly source of information on experimental procedures for corrosion investigation. Procedures for determining the resistance of materials to localized corrosion, such as pitting, intergranular corrosion and stress corrosion cracking, and others are also described in this volume.]

The ranking of the materials is based on the proven record of performance of materials and the corrosion rates exhibited by the materials. The ranking only serves as a guideline for selection of materials resistant to corrosion and the choice of corrosion rates is purely arbitrary (Table 9.5).

9.6.1 OTHER PARAMETERS FOR RANKING OF MATERIALS SUSCEPTIBLE TO PITTING RESISTANCE OF STEELS

The resistance of materials to pitting can be determined by the measurement of the depth of the deepest pit. The following guidelines can be used to estimate the pitting resistance [17]:

Excellent	25–100 μm	Steel 316 in seawater Titanium in seawater
Good	Above 100 μm, but below 150 μm	Austenitic chromium steel in seawater
Satisfactory	Above 150 μm but below 200 μm	Steel 316 in seawater at a velocity of 1.5 m/s

9.6.2 PITTING RESISTANCE EQUIVALENT OF STEEL (PRE) [17]

The alloying element that increase the resistance to pitting and crevice corrosion are chromium and molybdenum which is expressed as pitting resistance equivalent (PRE).

The PRE number is calculated using an empirical formula.

For austenitic stainless steels with <3% Mo and for X3GNiMoN27-5-2:

$$PRE = \%Cr + 3.3 \times \%Mo$$

where Cr and Mo represent the percentage by weight,

and for austenitic stainless steel with = or >3 % Mo

$$PRE = \%Cr + 3.3 \times \%Mo + 30 \times \%N$$

In the relationship above nitrogen is quantified by a multiplying factor of 30.

For austenitic–ferritic stainless steels with = or > 3% Mo:

$$PRE = \%Cr + 3.3 \times \%Mo + 16 \times \%N$$

A good resistance to pitting corrosion is indicated if PRE is equal to or greater than 40. The PRE value does not always correspond with pitting potentials (the minimum potential at which a sudden surge of current in an anodic polarization occurs because of the breakdown of the oxide film) and several anomalies are observed for steels.

A substantial interest has developed between PRE numbers and the measuring of resistance to localized corrosion, such as CPT (critical pitting temperature) and CCT (critical crevice temperature) and the relationship to PRE numbers. The best known test to determine the resistance of steels to pitting is the ASTM Standard G48 [18]. The steel is immersed in a 6% $FeCl_3$ solution for a reasonable length of time to assess the degree of corrosion. In CPT (critical pitting temperature test), the minimum temperature at which pitting initiates in 1.0% $FeCl_3$ solution after twenty-four hours is determined.

Table 9.6 CCT, CPT and PRE numbers for some stainless steels and nickel alloys [19]

Alloy name (UNS)	CCT°C	CPT°C	PRE
304L (S30403)	—	<2.5–4	19
316L (S31603)	—	<5–15	26
2507 (S32750)	25	56	35
2304 (S32304)	4–5	55–80	43

A close relationship between PRE and CPT was observed. The PRE value is, however, a rough estimate of resistance to pitting as the formula does not include certain elements contributing to pitting. It also excludes the effect of chromium and Mo depletion which contributes to the pitting process. The effect of surface finishing and the type of welding is also not taken into account. The PRE technique, however, provides a fairly good estimate of ranking steel alloys based on composition. Generally speaking, the higher the PRE, the higher would be the CPT and the better would be the corrosion resistance of stainless steels.

Table 9.6 gives some CCT, CPT and PRE numbers for some stainless steels and nickel alloys.

Although the PRE, CPT and CCT numbers indicate the tendency of stainless steels and nickel alloys to resist localized corrosion, they cannot be used alone to predict the localized corrosion behavior because of other external factors which contribute to the corrosion behavior.

9.7 ELECTROLYTE FACTORS

Water in any form plays an important role in corrosion reaction and affects the performance of materials. For instance, the corrosion rates of Cu–Ni alloys are almost tripled in polluted seawater compared to clean seawater. Mine water is highly corrosive compared to potable water and highly corrosion–resistant material would be required for the construction of equipment to handling of mining water. The corrosive nature of the electrolyte is, therefore, an important parameter for selection of materials in a corrosive environment. The following are the major factors related to the corrosivity of the electrolyte which influence the material selection process.

9.8 CORROSION INDICES

Several attempts have been made to develop an index that would predict whether a water is corrosive or not, however, these attempts have not been successful. Several indices have been developed over the years. Some of these indices can be used in corrosion control programs. The two most common indices used are Langelier Saturation Index (LSI) and Aggressive Index (AI), which are summarized below.

9.8.1 LANGELIER SATURATION INDEX (LSI) [20]

This index is used to predict the scaling tendency of water. Scaling occurs by deposition of $CaCO_3$.

$$Ca^{++} + HCO_3^- \rightarrow CaCO_3 + H^+$$

If the reaction proceeds from L $\rightarrow$ R, $CaCO_3$ scales would deposit and if it proceeds from R $\rightarrow$ L, $CaCO_3$ scales would dissolve. The Langelier index is defined as

$$LSI = pH - PH_S$$

where PH_S is the pH at which water is saturated with $CaCO_3$. The results can be interpreted as below:

LSI > 0 – water is supersaturated and tends to precipitate $CaCO_3$.
LSI = 0 – Water is in equilibrium with $CaCO_3$, a scale layer of $CaCO_3$ is neither precipitated nor dissolved.
LSI < 0 – water is under-saturated and tends to dissolve $CaCO_3$.
The following data is needed to calculate LSI:

- Total alkalinity ($CaCO_3$) in mg/l
- Calcium as $CaCO_3$

- Total dissolved solids
- pH
- Temperature
- pH at saturated which is calculated as:

$$pH = A + B - log\ [Ca^{++}] - log\ total\ alkalinity$$

A and B are constants (Tables 9.7a and 9.7b) related to temperatures and dissolved solids of the water. The log Ca^{++} and alkalinity are given in Table 9.7c.

Example:
Determine the LSI for a water having the following analysis:
Calcium = 88.0 mg/l
Total alkalinity = 110.0 mg/l
Total dissolved solids = 170.0 mg/l
pH = 8.20
Temperature = 25°C

Table 9.7 a Constant 'A' as a function of water temperature

Water temperature		Constant*
°F	°C	
32	0	2.60
39.2	4	2.50
46.4	8	2.40
53.6	12	2.30
60.8	16	2.20
68	20	2.10
77	25	2.00
86	30	1.90
104	40	1.70
122	50	1.55
140	60	1.40
158	70	1.25
176	80	1.15

*Calculated from K_2 as reported by Harned and Scholes and K_2 as reported by Larson and Buswell. Values above 40°C have been extrapolated. (*Source:* Federal Register, 1980) (J. E. Singley, B. A. Beaudet, P. H. Markey, *et al. Corrosion Prevention and Control in Water Treatment Supply System*, Noyes Publications, N. J. (1985).

Table 9.7 b Constant 'B' as a function of total filterable residue

Total dissolved solids (mg/l)	Constant
0	9.70
100	9.77
200	9.83
400	9.86
800	9.89
1000	9.90

Source: Federal Register, 1980.

Table 9.7 c Logarithms of calcium and alkalinity concentration

Ca^{+2} or Alkalinity (mg/l $CaCO_3$)	Log
10	1.00
20	1.30
30	1.48
40	1.60
50	1.70
60	1.78
70	1.84
80	1.90
100	2.00
200	2.30
300	2.48
400	2.60
500	2.70
600	2.78
700	2.84
800	2.90
900	2.95
1000	3.00

Source: Federal Register, 1980.

Solution:
PH$_S$ = A + B log[Ca^{++}]−log alkalinity
A = 2.0 from Table 9.7a.
B = 9.81 from Table 9.7b on extrapolation
log Ca^{++} = 1.94
log Alkalinity = 2.04 (Table 9.7c)

Applying the above terms in equation for LSI,

$$PH_S = 2.00 + 9.81 - 1.94 - 2.04$$

$$= 7.83$$

$$LSI = pH - PH_S$$

$$= 8.20 - 7.83$$

$$= 0.37$$

Results: The water is supersaturated with $CaCO_3$ and forms scale.

9.8.2 AGGRESSIVE INDEX (AI) (STANDARD AWWA C-400)

$$AI = pH + \log[(A)(H)]$$

where
A = total alkalinity (mg/l of $CaCO_3$)
H = calcium hardness
Interpretation of values:
AI < 10 – Highly aggressive
AI = 10–12 – Mildly aggressive
AI > 12 – Not aggressive.

9.8.3 RYZNER INDEX [21]

Another index was developed by Ryzner:

$$RSI = 2PH_S - pH$$

He also developed a curve showing corrosion of steel mains as a function of RSI (Fig. 9.8).

9.8.4 RIDDICK CORROSION INDEX [22]

The formula weighs several corrosion influencing factors including dissolved oxygen, chloride ion concentration, non-carbonate hardness and silica.

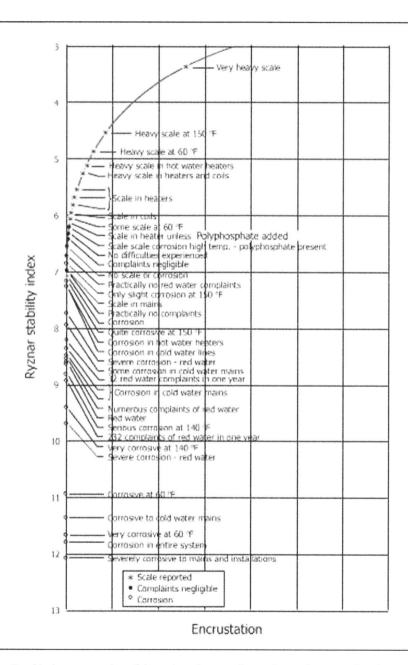

Figure 9.8   Graphical representation of the various degrees of corrosion and encrustation. (From Singley, J. E., Beaudet, B. A., Markey, D. H. *et al., Corrosion prevention and control in water treatment and supply system*, Noyes Publication)

The index is given by the following expression:

$$RCI + \frac{75}{Alk} \left[CO_2 + \frac{1}{2}(Hardness - Alk) + Cl^- \right.$$
$$\left. + 2N \right] \left(\frac{10}{SiO_2} \right) \left(\frac{D.O. + 2}{Sat.D.O.} \right)$$

where

N = nitrate ion concentration, CO_2 (mg/l), D.O. (mg/l).

Ranking of corrosivity:

0–5	Extremely non-corrosive
5–25	Non-corrosive
26–50	Moderately corrosive
51–75	Corrosive
76–100	Very corrosive
>100	Extremely corrosive.

9.8.5 STABILITY INDEX (STIFF–DAVIS MODIFICATION OF LANGELIER EQUATION)

The solubility product of salt changes with temperature and the concentration of the ions present in solution. In oilfield brines, if the surface solubility produced is smaller than their solubility product under down-hole formation, scale formation would occur. Under such conditions, the water is called 'unstable water.' The Stability Index (*SI*) is used to predict the tendency of ions to form scales from oilfield brines which are highly corrosive and encountered during drilling process. The Stiff–Davis Modification of the Langelier equation is used to predict calcium carbonate precipitation from the brines and is given by

$$SI_{(T)} = pH - P_{Ca} - P_{Alk} - K_T$$

where

$SI_{(T)}$ = stability index
pH = $-\log [H^+]$
P_{Ca} = $-\log [Ca^{++}]$
P_{Alk} = $-\log [alk]$
K_T = total ionic strength of brine at a temperature in °C.

Total ionic strength is:

$$\mu = \frac{1}{2} \left(C_1 Z_1^2 + C_2 Z_2^2 + \cdots + C_n Z_n^2 \right)$$

where

μ = total ionic strength.
C_i = molal concentration of ion, i.
Z_i = electronic charge of ion, i.

Figs 9.9 and 9.10 can be used to obtain K, P_{Ca} and P_{Alk} for the stability index. Similar indices may be used to predict barium sulfate or strontium sulfate or calcium sulfate deposition. If the product of concentration of Ca^{++} ions and SO_4^{-2} ions exceeds the solubility products (K_{SP}) of $CaSO_4$, calcium sulfate would be precipitated. The solubility of calcium sulfate decreases with an increase in the temperature. $CaCO_3$ and $CaSO_4$ scaling cause major problems in plants and equipment encountering salt water.

All the indices described above are qualitative in nature and they must be used with caution under appropriate conditions.

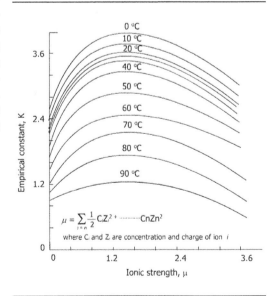

Figure 9.9 Values of K as a function of ionic strength

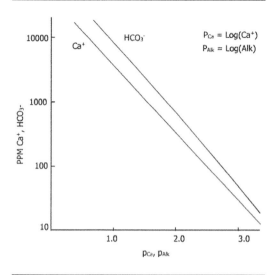

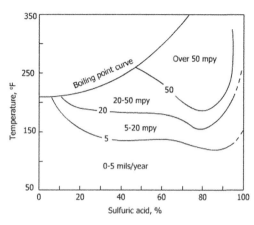

Figure 9.10 Determination of P_{Ca} and P_{alk} from the ionic concentration

Figure 9.11 a Corrosion of Durimet 20 by sulfuric acid as a function of concentration and temperature. (From Fontana, M.G. in *Home Study and Extension Courses*, Metals Engineering Institute, Course 14, Lesson 10, ASM, Int., Metals Park, Ohio, USA. By kind permission of ASM, Metals, Park, Ohio, USA)

9.9 Iso-Corrosion Charts [23]

Iso-corrosion charts provide a convenient method of selecting materials for the process industry. Sulfuric acid is a widely used acid in fertilizer manufacturing, petroleum refinery and product of a wide range of chemicals. Figure 9.11a shows the corrosion rate of durimet as a function of temperature and concentration of sulfuric acid. The curves in the figure represent corrosion rates of 5, 20, 50 and 200 mils per year. These curves are constant corrosion rate lines (iso-corrosion curves). The corrosion rates shown in the diagram are obtained from simple immersion tests. It can be observed from the diagram that steel is a suitable material for handling H_2SO_4 at concentration higher than 70%. Steel shows a good resistance to corrosion at strengths exceeding 100% at moderate temperature. Dilute sulfuric acid is highly corrosive to steel. A sharp dip in the curves at all temperatures is observed around 100% concentration. The iso-corrosion charts present extensive experimental data in a highly condensed form which can be used to predict the corrosion resistance of metals and alloys at a glance.

With a simple experimental technique, iso-corrosion charts for important alloys in specific environments can be prepared. The iso-corrosion chart for titanium in H_2SO_4 is shown in Figure 9.11b.

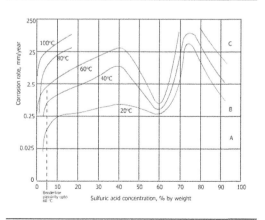

Figure 9.11 b Corrosion rate of titanium in sulfuric acid solutions (natural aeration). (From Corrosion Resistance of Titanium, I.M.I., Witton Birmingham, UK)

9.10 Cost Effectiveness

A. **Design life**

This is based on a specified amount of time required to recover investment and operating costs, plus a reasonable profit margin.

B. **Quality level**

A quality level assigned to an item of fabricated equipment indicates the required level of testing and inspection. Quality can be defined as 'fitness for the purpose.'

C. **Corrosion allowance**

Corrosion allowance is related to the corrosion rate and the designed life of the equipment. Carbon steel is a cost-effective material because it remains economical even after the adding cost of extra wall thickness required to maintain the corrosion allowance.

D. **National codes**

National codes and standards can be used as guidelines for selection of material and fabrication of equipment.

E. **Economic consideration**

Initial cost, operating costs and maintenance costs must be given full consideration in the process of materials selection.

F. **Availability**

This is an important consideration and an alternate material must also be specified in case the recommended material is not available.

9.11 Selection of Materials for Corrosive Environments

Corrosive environments have been described in Section 9.5 (metals and fluid corrosivity). It is not within the scope of this chapter to discuss the selection of materials for a wide spectrum of corrosive environments because of their very wide range and complexity. The discussion is, therefore, limited mainly to aqueous systems and important industrial environments. Resistance of materials to seawater corrosion has been accepted as a sufficient criterion for the corrosion resistance of metallic materials. The performance of important classes of steels is judged by their performance in natural seawater and their applications. The resistance of metallic materials is a major consideration in desalination plants and process industries. Resistance to seawater corrosion is taken as an index for selection of materials in major industrial applications and water transportation systems. The description of selection of materials in a corrosive environment is, therefore, restricted mainly to seawater and in certain cases to other environments. The classes of materials considered are iron and steel, copper, aluminum, titanium and nickel alloys, because of their wide range of applications.

9.11.1 Carbon Steels

Carbon steels are a series of alloys of carbon and iron containing up to about 1% carbon and up to 1.65% Mn, with elements added in specific quantities for deoxidization and residual quantities of other elements. In these steels carbon is the major strengthening element. The identification systems adopted by the American Iron and Steel Institute (AISI) and the Society of Automotive Engineers (SAE) are the most accepted. For instance, digit one of the four digit indicates that carbon is the major alloying element (1xxx). The second digit in some cases suggests the percentage of primary alloying elements. A 23xx steel has 3% Ni, a 25xx steel suggests 5% Ni. The last two or three digits suggest the amount of carbon, such as the amount of carbon in 1040 steel is 0.40%.

Alloy steels are generally designated by a four digit AISI–SAE system. The first two digits indicate the principal alloying element and the last two digits indicate the hundredths of percent of carbon in steel as illustrated below:

AISI series	Constituents
13xx	Manganese 1.75%
31xx	Ni (1.25%), chromium (0.65–0.80%)
40xx	Molybdenum 0.25%
50xx	Chromium (0.3–0.6%)

9.11.2 LOW, MEDIUM AND HIGH CARBON STEELS

(a) **Carbon steels**. Steels containing less than 1% carbon and are soft, ductile and not heat treatable. They are used for high cold formability.

(b) **Mild steels**. Steels containing <0.25% carbon are general purpose steels, commonly used for food and beverage cans. They have a tensile strength of about 430 N/m^2 and yield strength of 230 N/m^2.

(c) **Medium steels**. They contain carbon in the range of 0.25–0.50% carbon. They are used where strength and toughness is the demanding factors.

(d) **High carbon steels**. They contain more than 0.5% carbon. They are used for hammers, punches, drills, razors, etc.

9.11.3 ALLOYING ELEMENTS

These steels which contain one or more of the alloying elements are called alloy steels. The effects of major alloying elements are given in Table 9.8.

9.11.4 HIGH STRENGTH LOW ALLOY STEELS

These steels have strength in the range of 29–48 MPa, and tensile strength in the range of 41–62 MPa. They are mainly used as structural steels. One form of the above steel has 0.2% C, 1.5% Mn, between 0.003 and 0.1% Nb and between 0.003 and 0.1% vanadium, with a yield strength of 450 MPa. They are weldable. The carbon content is usually above 0.3%. To maximize strength, alloying elements, such as Ni, Mn and titanium, are adjusted to produce the desired strength. A carbon steel with 0.2% C and 1.5% Mn gives a strength of 250 MPa. Typical properties are given in Table 9.9.

Table 9.9 Typical properties of structural steels

Grade*	Tensile strength (MPa)	Minimum yield stress (MPa)
S235JR	340–470	235
S275JR	410–560	275
S355K2G3	490–630	355

*European Specifications: BSEN10025:1993.

Table 9.8 Effect of alloying elements on steel

Elements	Concentration of alloying elements, %	Principal effects
Nickel	0.3–0.5	Austenitic stabilized, increases toughness, increases corrosion resistance
Aluminum	<2	Promotes nitriding
Chromium	0.3–0.4	Increases corrosion resistance, increases oxidation resistance
Molybdenum	0.1–0.5	Improves corrosion abrasion resistance, and hardenability
Copper	0.2–0.5	Promotes resistance to atmospheric corrosion
Silicon	0.2–2.5	Deoxidizes liquid steels, improves toughness and hardenability. Decreases resistance to corrosion
Sulfur	<0.5	Reduces ductility, improves machinability. Not beneficial for corrosion resistance
Vanadium	0.1–0.3	Increases hardenability

9.11.5 HIGH STRENGTH HIGH ALLOYED STEELS

Stainless steels containing chromium (18%) and nickel (8%) are well-known in this category. Alloying elements exceed 12% in this category. Chromium nickel steels, 3000 series are well known for their excellent corrosion resistance. Steels of the type 304, 304L, 347, 316, 316L, are used in corrosive aqueous solutions of acids and in marine environment. The composition and properties of selected chromium nickel austenitic steels is shown in Table 9.10.

9.11.6 ULTRA-HIGH STRENGTH STEELS

These steels are characterized by discriminating yield strengths and measurable toughness.

Maraging steels are the best examples of ultra-high strength steels. They have a high nickel content (18–22%) and a carbon content less than 0.3%, other elements being cobalt, titanium, molybdenum, which form intermetallic components with nickel. Properties of two maraging steels are shown in Table 9.11.

9.11.7 CAST IRONS

Cast irons have carbon content between 2 and 4%, and often significant amounts of carbon and smaller amounts of other elements. The following factors determine the properties of cast iron:

- Chemical composition of cast iron
- Rate of cooling of the casting in the mold
- The type of graphite formed.

Table 9.10 Composition of selected chromium nickel austenitic steels

AISI Type number	Chemical composition (%)					
	C (Max)	Mn (Max)	S (Max)	Cr	Ni	Others
200 Series – Economy steels with Ni partly replaced by Mn						
303	0.15	2.0	1.0	17–19	8–10	Mo: 0.6
303Se	0.15	2.0	1.0	17–19	8–10	Se: 0.15
304	0.08	2.0	1.0	18–20	8–12	
304L	0.03	2.0	1.0	18–20	8–12	
305	0.12	2.0	1.0	17–19	10–13	
308	0.08	2.0	1.0	19–21	10–12	
309	0.20	2.0	1.0	22–24	12–15	
309S	0.08	2.0	1.0	22–24	12–15	
310	0.25	2.0	1.5	24–26	19–22	
310S	0.08	2.0	1.5	24–26	19–22	
314	0.25	2.0	1.5–3.0	23–26	19–22	
316	0.08	2.0	1.0	16–18	10–14	Mo: 2–3
316L	0.03	2.0	1.0	16–18	10–14	Mo: 2–3
317	0.08	2.0	1.0	18–20	11–15	Mo: 3–4
321	0.08	2.0	1.0	17–19	9–12	5 C min Ti
347	0.08	2.0	1.0	17–19	9–13	10 × C min Nb
348	0.08	2.0	1.0	17–19	9–13	0.2 CO, 10 × %C min Nb, 0.10 Ta

Table 9.11 Composition and properties of maraging steels

Grade	Nominal composition	Tensile strength	0.2% proof stress	% Elongation
18–Ni 1400	1.7–1.9 Ni 8–9 CO 3.0–3.5 Mo 0.15–0.25 Ti 0.05–0.15 Al 0.03 C (max)	1565	1400	9
18–Ni 2400	17–18 Ni 12–13 CO 3.5–4.0 Mo 1.6–2.0 Ti 0.1–0.2 Al 0.001 C (max)	2460	2400	8

Following is a brief description of different types of cast iron and their properties.

9.11.7.1 Gray Cast Iron

Consider cooling of iron containing between 2.0 and 4.3% carbon from the liquid state. When solidification starts, austenite is formed from the liquid. At 1147°C, the solidification of austenite is complete as observed in an iron–carbon diagram. However, if the cooling is slow, graphite flakes are formed. At 7236°C, the remaining austenite changes to α-ferrite and at room temperature, a structure comprising of pearlite and graphite (flakes) is formed. Because of the gray appearance, this type of cast iron is called 'gray cast iron.' It is the most commonly used cast iron and easiest to machine. The chemical composition of gray iron ranges from 2 to 4% of total carbon with at least 1% silicon.

9.11.7.2 White Cast Iron

A fast cooling rate gives a structure which solidifies at 1147°C, which gives a structure of cementite and pearlite at 723°C. More iron carbide (Fe_3C) is formed instead of graphite upon solidification. The white cast iron fracture to produce a white surface, hence they are called 'white cast iron.' They serve as raw materials for malleable irons.

9.11.7.3 Malleable Cast Irons

Cold white iron castings are heated in a malleabilizing furnace to decompose the iron carbide of white iron to graphite and iron. The graphite is in the form of irregular nodular aggregates, called 'temper carbon.' Malleable irons are produced from white cast iron in two stages. In the first stage, the white cast irons are heated above 940°C and held for 3–20 h. In this stage iron carbide is transformed to carbon (graphite) and austenite (γ). In the second stage, the austenite can be transformed to either ferrite, pearlite or martensite matrices, depending on heat treatment as below:

(a) Ferrite malleable iron. The casting is cooled rapidly to 740–760°C followed by slow cooling at a rate of 3–11 Celsius degrees per hour. During cooling austenite is transformed to ferrite or graphite. The graphite deposits on particles of temper carbon (nodular aggregates). This malleable cast iron is called 'blackheart malleable iron.' In the white heat process, the castings are packed in canisters with haematite iron ions. On slow cooling a duplex structure (ferrite near the surface and pearlite near the center) of large section is produced.

(b) Pearlitic malleable iron. White iron castings are heated in a oxidizing atmosphere to 900°C, soaked and rapidly cooled. The rapid cooling transforms the austenite to a pearlite and forms pearlitic iron. The structure consists of temper carbon nodular in a matrix of pearlite.

(c) Tempered martensite malleable iron. It is produced by cooling castings to 845–870°C, holding for 15–30 min and quenching in oil to develop a martensite matrix. It is tempered at 590–725°C to obtain the desired mechanical properties. The structure comprises carbon nodular in a tempered martensite matrix.

9.11.8 SPHEROIDAL GRAPHITE (S.G.)

It is also known as ductile iron in USA and is produced by treating a high carbon equivalent liquid iron with manganese or cerium to prevent the formation of graphite flakes. The structure at room temperature comprises a graphite spheres in a matrix of ferrite. Ductility is increased by heating to 900°C followed by a slow quench. Typical properties of cast irons are given in Table 9.12.

9.12 CORROSION CHARACTERISTICS OF CARBON AND LOW ALLOY STEELS

Carbon steel is the most universally used material for service in seawater. It is used for a myriad of applications in ships and shipping industry.

Table 9.12 Summary of properties of typical gray cast, blackheart, white heart, pearlite and ductile cast iron

Material	Grade*	Tensile strength (MPa)	% Elongation
Gray cast iron	150	150	0.6
Gray cast iron	350	350	0.5
Blackheart	B 30–06	300	6
Blackheart	B 35–12	350	12
White heart	W 3504	360	3
White heart	W 45–07	480	4
Pearlitic	P 45–06	270	6
Pearlitic	P 70–02	700	2
Spheroidal cast iron	350/22 (ferrite matrix) (*A*)	350	22
Spheroidal graphite cast iron	500/7 (ferrite + pearlite) (*A*) Matrix	420	12
Spheroidal graphite cast iron	800/2 (tempered martensite) (*Q + T*)	800	2

Note: A = annealed
$Q+T$ = quenched and tempered
*BS Grade.

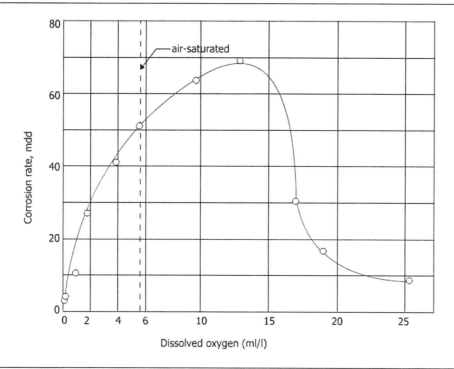

Figure 9.12 Effects of dissolved oxygen level of corrosion of mild steel in distilled water (25°C, 48 h). (From Uhilg, H.H., Triadis, D. and Stern, M. (1995). *J. Electrochem. Soc.,* **102**, 59)

It offers the advantages of easy availability, low price, a satisfactory rate of corrosion and ease of fabrication. On exposing to water, iron corrosion products would be formed and may create operational problems of equipment and plants. The following factors affect the corrosion of carbon and low alloy steels.

a. **Dissolved oxygen**
The effect of dissolved oxygen is shown in Fig. 9.12 [24]. The corrosion rate increases initially up to a certain level, however, above a threshold, it decreases due to passivation of the surface.

b. **pH**
In the pH range, 4–10, corrosion rate is independent of pH and it is controlled by oxygen diffusion. At pH <4.0, hydrogen evolution is the rate controlling factor. Above pH >10.0, the corrosion is decreased because of the passivation of the surface caused by oxygen and alkalies. Effect of pH on corrosion of iron in

aerated water at room temperature is shown in Fig. 9.13 [25].

c. **Temperature**
The rate of corrosion increases monotonically with increased temperature, however, above 80°C, the corrosion rate tends to drop because of a fall in the level of dissolved oxygen.

d. **Material factors**
Alloying elements of Mn, Al, Si, P, minimize flow induced corrosion of steels (annealed) in 3.0% NaCl, whereas C promotes corrosion. The effect of alloying elements on flow induced corrosion are shown in Fig. 9.14 [26].

e. **Alkalies**
The service performance of a heat transfer surface is adversely affected by building up of elevated concentrations of alkalies. For instance, a high concentration of NaOH destroys the passive film on the heat transfer surface and exposes iron which reacts directly with sodium hydroxide to form the destructive

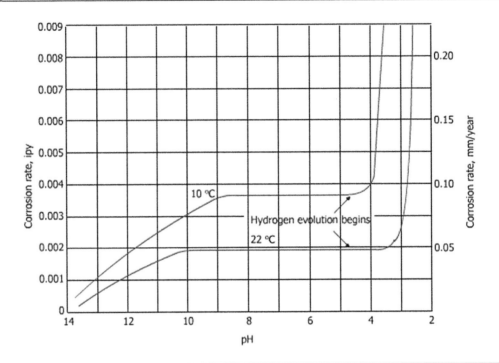

Figure 9.13 Effect of pH on corrosion of iron in aerated water at room temperature. (From Whitman, G.W., Russel, R.P. and Altieri, V.J. (1924). *Ind. Eng. chem.*, **16**, (71), 665)

Na_2FeO_2 compound.

$$4NaOH + Fe_3O_4 \rightarrow Na_2FeO_2 + 2NaFeO_2$$
$$+ 2H_2O$$

$$2NaOH + Fe \rightarrow Na_2FeO_2 + 2H$$

Magnetite (Fe_3O_4) is not sufficiently protective at higher temperatures.

f. **Hydrogen attack**
Cases of hydrogen attack in boilers have been widely reported in the USA. Water-wall tube scales may cause hydrogen attacks producing hydrogen.

$$3Fe + 4H_2O = Fe_3O_4 + 4H_2 \uparrow$$

The hydrogen penetrates the steel surface and reacts with the cementite (Fe_3C) to form methane:

$$Fe_3C + 4H = 3Fe + CH_4$$

This leaves a decarburized structure. The attack may be caused by anodic water or decrease of pH of alkaline solution. Alkali corrosion occurs on the flame side, whereas hydrogen attack occurs on the water side in boiler tubes.

g. **Attack by geothermal fluids**
Geothermal fluids are usually deficient in oxygen, but oxygen is not a cause of concern. The H_2S levels in geothermal fluids are not as high as sour gas wells, but they cannot be neglected. Carbon steel is not a suitable material to be used against H_2S. High nickel alloys resist sulfide corrosion in sour gas well environments [27].

h. **Effect of CO_2 corrosion in oil and gas fields**
Several oil and gas fields are very active in CO_2 corrosion (sweet corrosion). In Holland, the bottom temperature of gas wells was reported to be 97–124°C and the pressure was 0.9–3 bar. The corrosion rate of carbon steel was observed to be maximum at −90°C followed by a gradual decrease above it [28]. The corrosion rate tends to decrease with increased chromium contents.

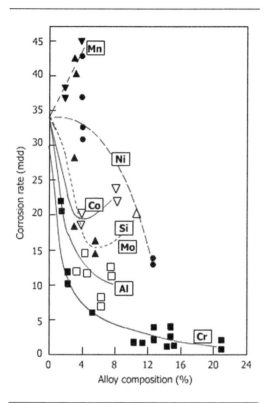

Figure 9.14 Effects of alloying elements on corrosion of Fe base alloys in flowing synthetic seawater (121°C, DO < 50 ppb, flowrate of 0.3 ft/s). (From OSW (1968). R and D Report, No. 3, 94)

The corrosion rate of plain carbon steel increases linearly with CO_2 pressure. In the next stage, the rate decreases as CO_2 pressure increases due to reduction of cathodic reaction and becomes dependent on temperature. In the third stage, it increases linearly with pressure as in the first stage because of the formation of iron carbonate. Corrosion by CO_2 is caused by the formation of $FeCO_3$

$$H_2O + CO_2 \rightleftharpoons H^+ + HCO_3^- \rightleftharpoons 2H^+ + CO_3^{--}$$

$$CO_3^{--} + Fe^{++} \rightarrow FeCO_3 \downarrow$$

Chloride ion is not as corrosive to carbon steel as to stainless steel. The corrosion rate of carbon steel is unaffected by NaCl concentration in the range of 0–10% at 150–240°C, but the rate increases with

NaCl concentration above 10% in the above temperature range [29]. At elevated concentration of chloride, the reduced solubility of CO_2 in chloride accounts for the reduced rate of corrosion of steel in high chloride concentration. Generally speaking, chloride ions in the presence of CO_2 do not have a significant bearing directly on the corrosion rate of carbon steel.

9.13 CORROSION RATE OF CARBON STEEL IN SEAWATER

Table 9.13 shows the corrosion rate of carbon steel in seawater under conditions [30].

9.14 CAST IRONS

9.14.1 PITTING AND CREVICE CORROSION

Cast irons behave similarly to carbon steels. They show susceptibility to pitting and crevice corrosion in seawater and similar environments. Addition of nickel–chromium and molybdenum increase their resistance to pitting.

9.14.2 STRESS CORROSION CRACKING

The following environments cause stress corrosion cracking of cast irons:

(1) Acid chloride   (5) Seawater
(2) Ammonium nitrate (6) Calcium nitrate
(3) Hydrogen sulfide   (7) Sodium hydroxide
(4) Sodium hydroxide (8) Sodium nitrate

9.14.3 GRAPHITE CORROSION

Cast iron is prone to selective leaching in soft water, acidic water, brackish water and water containing low levels of hydrogen sulfide.

Table 9.13 Corrosion rate of carbon steel in seawater

Condition	Rate of corrosion of carbon steel (μm/year)	Rate of corrosion of low alloy steel (μm/year)
Synthetic seawater:		
27°C	155	–
60–63°C	225	–
Seawater aerated, 1.8 m/s	>1250	>1500
Seawater deaerated, 82°C	600	550
Deaerated water, 90.5°C	475	350
Deaerated water, 113°C	1020	875

Graphite flakes are cathodic to iron, and corrosion is localized to iron which starts leaching (becomes a porous mass) and leaves a rich residue of graphite flakes. Graphitization is observed on gray cast iron water pipes.

9.14.4 FLOW INDUCED CORROSION

Addition of chromium, nickel and copper in small concentrations are known to reduce the rate of corrosion in natural, potable, brackish and seawaters. The rate of corrosion is strongly affected by the tendency of cast iron to form scales. The scale forming tendency can be predicted using Langelier's index, discussed earlier in this chapter.

The rate of corrosion of cast iron is directly linked to oxygen content. Carbon dioxide accelerates the formation of scales in fresh water. Cast iron and carbon steels are sensitive to attack by hydrogen sulfide even in the absence of oxygen. Corrosion rates of low carbon steels and cast iron accelerate with velocity, if the water is not treated with inhibitors.

The effect of velocity on the corrosion of carbon steels in fresh water is shown in Table 9.14 [31].

9.14.5 CORROSION OF CAST IRON IN SEAWATER

Cast iron may suffer from graphitic corrosion in seawater. Graphitization is a form of attack

Table 9.14 Effect of velocity on the corrosion of carbon steel in fresh water

Velocity (m/s)	Effect on corrosion
<0.3	Pitting may not occur
0.3	Corrosion rate may accelerate to 1 mm/year
0.3–3.0	Corrosion rate may fall to 0.25–0.76 mm/year because of passivation
>4.6	Acceleration of corrosion rate
≫4.6	Corrosion rate may increase to 5 mm/year at 12 m/s

characteristic of gray iron which considerably weakens the iron. Cast iron with 1–3% Ni are used for pump houses, water boxes and other component for service. Gray cast iron has a greater susceptibility to pitting than malleable cast iron. The pitting factor of cast iron is shown in Table 9.15 [32]. Pitting factor is represented as: Deepest pit depth/Average corrosion rate

Corrosion resistance of malleable irons depends upon whether they are white heart, blackheart, or pearlitic. White heart is more corrosion-resistant than blackheart (2.9 to 24 s/m^2 *vs* 29–240 s/m^2) in seawater. Mechanical properties are weakened as shown in Table 9.16 [33].

Gray cast iron and malleable iron are resistant to atmospheric corrosion. They show a higher resistance than low alloy high strength steels. They are, however, sensitive to attack by H_2S released by sulfate-reducing bacteria.

Table 9.15 Pitting factor of cast iron

Type	Corrosion rate (mm/year)	Pitting factor
Gray cast iron	0.02	10.9
Ferritic spheroidal gray cast iron	0.025	14.7
Malleable cast iron	0.045	4.9
Pearlitic gray cast iron	0.003	9.3
Spheroidal graphite Ni cast iron	0.07	6.7
White cast iron	0.045	10.00

Table 9.17 Selected properties of gray and ductile iron [34]

Pipe material	Nominal internal diameter (mm)	Minimum tensile strength (N/mm^2)	Elongation % (min)
Gray cast iron	<300 300–600 >600	200 180	Negligible elongation
Ductile iron	80–60	420	10%

Typical failures of cast iron pipes are characterized by leaks in pipe wall along the graphite flake/ferrite interface. The ratio of mechanical failure between the cast iron and ductile iron pipes is 2.5 to 1.0. Following is a comparison of pitting depths of ductile and gray cast iron pipes in aggressive soils (Table 9.18)[34]:

9.15 DUCTILE AND CAST IRON PIPES

Cast iron and ductile iron pipes have been extensively used in the USA, Canada and European countries in water distribution systems. Cast iron pipes have been gradually replaced in recent years by ductile irons and PVC pipes. The ductile iron pipes enjoy a mechanical superiority over the traditional cast iron pipes as shown in Table 9.17.

Ductile iron pipes and cast iron pipes, however, corrode at about the same rate in soil.

9.15.1 LOW ALLOY STEELS

Alloying additions of less than 5% have little effect on seawater corrosion. The susceptibility to pitting is least for 3% Cr–steel than plain carbon steel. Table 9.19 shows the corrosion rates of several low alloy steels containing small quantities of alloying elements (Cu, Ni, Cr) in various combinations.

Table 9.16 Effect of corrosion on the mechanical strength of cast iron

Material	Decrease in mechanical strength in synthetic seawater		Decrease in mechanical strength in natural seawater	
	UTS	Elongation	UTS	Elongation
(1*) White heart	2	18	3	22
(2*) Blackheart	12	25	18	32
(3*) Gray	6	–	12	–

*Details of materials:

	C	Si	Mn	P	S	Cu%
(1*)	2.65	0.65	0.38	0.11	0.09	0.09
(2*)	2.45	0.95	0.12	0.09	0.09	0.10
(3*)	3.10	2.10	0.91	0.28	0.11	0.03

Table 9.18 Average maximum pit depth for ductile and gray cast iron pipes in aggressive soils [34]

Type of soil	Exposure (in years)	Maximum pit depth	
		Ductile	Gray
Cinder	3.7	3.3	3.3
	9.4	4.3	5.3
	13.5	3.8	6.8
Alkalies	6.0	2.0	1.5
	8.0	2.0	1.5
	12.0	2.3	3.0
	14.0	3.0	4.5

Table 9.19 Pitting depths of selected low alloy steel [35]

Pit depth (μm)	Composition				
	A	B	C	D	E
Average of deepest pit	165	205	200	140	243
Deepest pit	215	380	430	340	–

A C-steel
B Low alloy steel (Ni, Cu) – 0.08 C, 0.47 Mn, 0.63 Cr, 1.54 Ni, 0.87 Cu
C Low alloy steel (Cr, Ni, Cu) – 0.15 C, 0.45 Mn, 0.68 Cr, 0.49 Ni, 0.42 Cu
D Low alloy steel (Ni, Cu, Mn, Mo) – 0.078 C, 0.75 Mn, 0.058 P, 0.022 S, 0.72 Ni, 0.61 Cu, 0.13 Mo
E Low alloy steel (Ni, Cr, Mn) – 0.13 C, 0.60 Mn, 0.55 C, 0.30 Ni, 0.61 Cu, 0.06 Mo

Selected mechanical properties of chromium and nickel–chromium–molybdenum steels are given in Table 9.20. Selected applications are shown in Table 9.21.

The corrosion resistance of chromium steels (12–14%) or about 17% Cr, is superior to C as low alloy steels but inferior to Cr–Ni steels. They are not susceptible to stress corrosion cracking, but they develop pitting in quite seawater. A 17% Cr steel is fairly resistant to salt water. An exceptional resistance to corrosion is shown on addition of more than 12% chromium to steels. Characteristics of steel containing high chromium contents are discussed under stainless steel. The pitting corrosion of steel containing 13–18% chromium is in the order of 1.73–1.75 mm.year^{-1}. These materials are not suitable for elevated temperatures and for heat exchanger application.

9.16 STAINLESS STEELS

A complete description of stainless steels and major corrosion problems encountered by steels have been described in Chapter 4. Here, a brief description of stainless steels with reference to their corrosion resisting properties in certain selected environments is presented.

9.16.1 BRIEF REVIEW OF CLASSES OF STAINLESS STEELS

These are alloys of iron, chromium and other alloying elements that resist corrosion in several environments. A steel cannot be called 'stainless' unless it has a minimum of 12% of chromium as chromium is the major element responsible for corrosion resistance. There are several classes of stainless steels categorized by their microstructures:

(1) Ferrite. Iron–chromium + low carbon (between 12 and 25% chromium and less than 0.1% C).
(2) Martensitic. Iron–chromium + higher amount of carbon (12–18% Cr and between 0.1 and 1.2% carbon).
(3) Austenitic. FCC iron (16–26% Cr, more than 6% Ni and carbon as low as 0.1% or even less).
(4) PH (Precipitation hardening). These are iron–chromium–nickel alloys containing precipitating element, such as Al. Strength is achieved by precipitation hardening heat treatment. One example is (17–7 PH)

Table 9.20 Mechanical properties of selected nickel–chromium and nickel–chromium–molybdenum steels

Grade (BS)	Nominal composition (wt%)	Tensile strength (MPa)	Yield strength (MPa)	% Elongation
Chromium steel:				
(a) 530M40	0.4 C, 0.6–0.9 Mn, 0.9–1.2 Cr *(OQ 850–880°C, T 550–700°C)	700–850	525	17
Nickel–chromium–molybdenum steels:				
(b) 817M40	0.4 C, 0.45–0.7 Mn, 1.0–1.4 Cr, 0.2–0.35 Mo, 1.3–1.7 Ni *(OQ 880–850°C, T 660°C)	850–1000	650	13
(c) 853M30	0.3 C, 0.45–0.7 Mn, 1.4–1.8 Cr, 0.2–0.35 Mo, 3.9–4.9 Ni *(OQ 810–840°C, T 200–280°C)	1550 max	1235	7
Manganese steel:				
(d) 150M36	0.32–0.4 C, 1.3–1.7 Mn *(OQ 840–870°C, T 550–600°C)	625–775	400	18

*OQ – Oil Quenched; T – Temper

steel containing 0.09 C–17 Cr–7 Ni–1.0 Al–1.0 Mn.

9.16.2 COMPOSITION AND MECHANICAL PROPERTIES OF STAINLESS STEELS

The nominal chemical composition of the four classes of stainless steels and their selected properties are given in Table 9.21.

Applications of the most commonly used type of stainless steels in each class are shown in Table 9.22.

9.16.3 HIGH TEMPERATURE RESISTANCE OF STAINLESS STEELS

High temperature oxidation corrosion effects include oxidation, carburization, nitriding, sulfidizing and vanadium attacks. V_2O_5 is formed when heavy fuel oil containing vanadium burned. V_2O_5 melts at relatively low temperature, it forms a molten layer on metals and causes accelerated corrosion. Na, S and Cl may accompany vanadium which can reduce the melting point further and accelerate corrosion.

Following is a brief summary of fuel ash corrosion to which components in process heaters, boilers and other equipment where high metal content fuel are fired:

$$Na_2O \cdot V_2O_4 \cdot 5V_2O_5 + \frac{1}{2}O_2 = Na_2O \cdot 6V_2O_5$$

$$Na_2O \cdot V_2O_4 \cdot 5V_2O_5 + SO_3 = Na_2O \cdot 6V_2O_5 + SO_2$$

$$Na_2O \cdot V_2O_5 + Fe = Na_2O \cdot V_2O_4 \cdot 5V_2O_5 + FeO$$

The FeO is dissolved in $Na_2O \cdot 6V_2O_5$, hence, corrosion cannot be retarded (protective layer) here by the corrosion products formed. Also $V_2O_5 \cdot Na_2SO_4$ in the flue gas promotes corrosion. Vanadium attack may be accelerated by the presence of NaCl, which reacts with S to

Table 9.21 Typical mechanical properties of stainless steels

Alloy number (AISI)	Composition (%)	Condition (MPa)	Tensile strength (MPa)	Yield strength (MPa)	% Elongation
Ferritic Steels					
403	0.08 C, 1.0 Mn, 12–14	Air-cooled from 700–780°C	420	280	20
430	0.08 C, 1.0 Mn, 16–18 Cr	Air-cooled from 750–870°C	430	280	20
Martensitic Steels					
410	0.09–0.15 C, 1.0 Mn, 11.5–13.5 Cr	Air-cooled from 950–1020°C	500–700	380	18
420	0.14–0.20 C, 1.0 Mn, 11.5–13.5 Cr	Air-cooled from 950–1020°C	700–850	525	15
Austenitic Steels					
304	0.08 C, 2.0 Mn, 17–19 Cr, 8–10 Ni	Softened 1000–1100°C	510	190	40
321	0.08 C, 2 Mn, 17–19 Cr, 9–12 Ni, Ti 5x Carbon	Softened 1000–1100°C	510	200	35
316(L)	0.03 C, 2 Mn, 16.4 Cr, 11.0–14.5 Ni, 2.5–3.0 Mo	Softened 1000–1100°C	490	190	40
17–7 PH	0.09 C, 17 Cr, 7 Ni, 1.0 Al, 1.0 Mn	Precipitation hardened	1450	1310	1–6

form Na_2SO_4. NaCl may be derived from the seawater.

$$2NaCl + SO_2 + \frac{1}{2}O_2 + H_2O \rightarrow Na_2SO_4 + 2HCl$$

$$2NaCL + SO_3 + H_2O \rightarrow Na_2SO_4 + 3HCl$$

The corrosivity of Na_2SO_4 is increased by NaCl. NaCl reacts with vanadium to produce $NaVO_3$.

$$2NaCl + H_2O + V_2O_5 \rightarrow 2NaVO_3 + 2HCl$$

Vanadium attack may be reduced by Ca and Mg. The vanadium attack has not been established but various mechanisms have been proposed. The melting point of V_2O_5 is very low so that it sticks to the metal and promotes oxidation.

The following elements affect vanadium attack of 20 Cr, 10 Ni, 0.3 C, 0.4 Si, 1.0 Mn steel in 90% V_2O_5 + 10% Na_2SO_4 at 2% synthetic air [36].

Element	Effect
Mo	Harmful
Si	Effective
Cr	Effective
Ni, Al, Ti	Effective

Prevention:

■ Use compounds that melt at high temperatures.
■ Select high alloy steels, such as 20 Cr, 10–30 Ni.
■ Apply nickel and chromium plating.

Table 9.22 Commonly used stainless steels

Class	Type	Application
Ferritic	430	Elevated temperatures, turbine parts
Martensitic	416 420 431	Used as cutting tool and for structural components
Austenitic	303 304 316	Chemical service, corrosion-resistant
	304L 316L 347	Tanks and pipings
Precipitation hardening	17.7	Not highly resistant to corrosion, pressure vessels, knives

9.16.4 RESISTANCE OF STEELS TO REFINERY ENVIRONMENT

(a) HCl may be encountered in crude units from hydrolysis of chloride salts. The high nickel alloys, such as Monel, are recommended. Hastelloys are also suitable.

(b) Hydrofluoric acid. HF is corrosive to all steels and Alloys 70–30 Cu–Ni and Monel are recommended. The corrosion rates of 304 and 316 stainless steels exceed the corrosion rates of nickel–copper and high nickel alloys.

(c) Napthenic acid. This is a collective name for organic acids, which are present in crude oils. Stainless type 316 (18-10-2 Mo) is highly resistant to napthenic acid.

(d) Sulfuric and Sulfurous acids. Sulfuric acid is a problem in boilers and process heaters. It leads to 'dew point' corrosion, caused by a mixture of H_2SO_3 and H_2SO_4. The performance of stainless steels is not inferior compared to the performance of high alloy steels. CRIA steel [C < 0.13, Si (0.20–0.82), Mn < 1.40, P < 0.021, S (0.013–0.030),

Cu (0.35–0.35), Cr (1.0–1.12)] are highly resistant to dew point corrosion.

9.16.5 CORROSION OF STAINLESS STEELS IN ACIDS

Stainless steels are subjected to serious corrosion problems in oil refineries as they may encounter very corrosive fluids, such as hydrochloric acid and napthenic acids in the process streams.

Hydrofluoric acid, hydrogen fluoride and fluorine are less corrosive to many metals and alloys than their own halide counterpart. The nickel–copper alloys, typified by Monel alloy 400 have excellent resistance to hydrofluoric acid corrosion. Stainless steels, such as 316, suffered severe transgranular corrosion. Table 9.23 summarizes the corrosion resistance of nickel alloys and stainless steels in anhydrous hydrogen fluoride [37]. The weakness of stainless steel to anhydrous hydrogen fluoride corrosion is shown in Table 9.23.

The corrosion resistance of Cr–Ni–Mo–Fe alloys in aqueous HF is shown in Table 9.24.

Stainless steels, such as AISI 304 and AISI 316, perform well in flowing seawater up to 2.0 $m \cdot s^{-1}$. The corrosion rate of 304 and 316 steels in flowing seawater are compared below (Table 9.25).

Table 9.23 Corrosion of nickel alloys and stainless steels in anhydrous hydrogen fluoride [37]

Material	Corrosion rate, mm/year (mils/year)		
	500°C	550°C	600°C
Inconel 600	1.2 (60)	1.2 (48)	1.8 (72)
SS 347	180 (7200)	460 (18 000)	180 (7200)
SS 310	12 (480)	100 (4000)	300 (12 000)

From "*Corrosion resistance of nickel containing alloys in hydroflouric acid, hydrogen fluoride, and fluorine*," Inco, Report 4156.

Table 9.24 Corrosion resistance of SS 304 in aqueous HF [37]

Concentration	Temperature (°C)	Rate of corrosion μm/year (mils/year)
0.5	60	12 (300)
0.1	60	25 (640)
0.15	60	47 (1200)
10	16	0.4 (10)
90	4	35 (890)
90	21	30 (760)

At velocity 0.4–1.45 ft/s(120 to 440 mm/s).

Table 9.26 TDS content of different oceans

Name	TDS (ppm)
Arabian Gulf	45 000–65 000
Red Sea	42 000
Mediterranean Sea	41 000
Carribean Sea	38 000
Atlantic Ocean	37 000
Black Sea	22 000
Caspian Sea	13 000
Baltic Sea	8 000

It has been established by extensive studies that AISI 316 is a better material for seawater service compared to AISI 304. In the Arabian Gulf, AISI 316 is the recommended steel for desalination plants.

9.16.6 FACTORS AFFECTING SEAWATER CORROSION

Seawater covering 70% of the earth's surface has been offering unlimited material benefits to mankind ever since the dawn of history. The ocean floor is a full hydrospace for minerals, fuel, food and energy. Seawater is, however, known to be quite corrosive and increased exploitation and exploring activities require materials resistant to seawater corrosion. Due to its corrosivity, corrosion resistance in seawater is taken as an index of corrosion tendency of materials. A material exhibiting a sufficient corrosion resistance in seawater is considered satisfactory for major aqueous applications.

a) Total Dissolved Solids

The TDS contents of different oceans around the world are given in Table 9.26.

The higher the TDS content, the more is the aggressivity of the seawater.

b) Chloride Ions

It is observed from Table 9.27 that chloride ions account for almost 55% of the total dissolved solids. The chloride is the most aggressive of all ions in seawater because of its ability to destroy the passive surface of steels and accelerate their corrosion. The sulfate ion is less corrosive than chloride ion.

Table 9.25 Corrosion rate of 304 and 316 steels in flowing seawater

Steel type	Condition	Velocity	Corrosion rate (mm/year)
AISI 304	160°C desalination plant, pH = 7.0 (deaerated)	1.5 ms^{-1}	.0025 mmyear^{-1}
	82°C with trace oxygen pH = 7.0	1.5 ms^{-1}	.225 mmyear^{-1}
AISI 316	Deaerated seawater 113°C, pH = 7.0	1.5 ms^{-1}	.003 mmyear^{-1}
	Deaerated seawater 90.5°C, pH = 7.0	1.5 ms^{-1}	.003 mmyear^{-1}

Table 9.27 Effect of seawater condition on the corrosion rate of selected stainless steels

Type	Duration	Condition	Depth of attack(mm)
AISI 304	7 years	Totally immersed	3.15 (perforated)
	1 year	Tidal zone	0.86
AISI 316	3 years	Strong tidal zone	1.51
	7 years	Fouling	3.15 (perforated)

c) Oxygen

The quantity of oxygen taken from the atmosphere decreases with increasing salt contents and increasing temperature. For instance, the amount of oxygen at 25°C in the presence of 36.11 g/kg of salt content is 6.53 ppm compared to 7.23 ppm at 20°C with the same salt concentration. Increasing temperature is accompanied by accelerated attack. The corrosion rate of steel in seawater is higher in summer than in winter.

d) Depth of Immersion

Materials partially immersed undergo a heavier corrosion attack than materials fully immersed. Metal samples are most severely attacked in tidal zone. (Table 9.27)

9.16.7 CORROSION RESISTANCE OF STEELS IN SEAWATER

The depth of corrosion of selected stainless steels in seawater are given in Table 9.28.

Most stainless steels are sensitive to attack by HCl. The corrosion rates of steels in H_2SO_4 show low corrosion rates at low and high concentrations (> 95%). They are resistant to dilute solution only at low temperature. See the Iso-Corrosion Chart described earlier in this chapter. Steels 320 are resistant to corrosion in nitric acid at all temperatures. Type 430 can be used only up to 80% concentration. Type 316 has a resistance similar to SS 321 and SS 341.

9.17 CORROSION BEHAVIOR OF COPPER AND COPPER ALLOYS

Copper and several of its alloys are known since time immemorial. Copper became the first engineering material to be exploited because of its existence in the metallic form and the ease it offered for extraction. Because of the multitude of benefits they offer, copper alloys have become an unique class of material and enjoy sustained development ever since their first exploitation by mankind.

Due to the very large number of copper alloys existing over a long period of history, the classification and vocabulary associated with these alloys is confusing. The classification of copper alloys is based on a system administered by Copper Development Association (CDA). Alloy numbering C1011 to C7900 are designated as wrought-alloys and alloys numbering from C80000 to C99900 are designated as cast alloys. The classification is shown in Table 9.29.

9.17.1 COPPER BRASSES AND BRONZES

Copper exhibits a high electrical and thermal conductivity, excellent formability and a good resistance to corrosion. As a weak material it is alloyed to improve its mechanical properties and corrosion resistance. The mechanical properties of various coppers are shown in Table 9.30.

9.17.2 MECHANICAL PROPERTIES OF COPPER ALLOYS

a) Brasses

These are copper alloys in which zinc is the dominant alloying element. The strength and

Table 9.28 Depth of attack on stainless steels in seawater (mm/year). (From cobot corporation, stellite division, Kokmo, Indiana, USA)

Alloy	Temperature			Condition
	20°C	50°C	90°C	
AISI 304	0.00254	0.00101	0.0254	96 h exposure in synthetic seawater
AISI 316	0.00254	0.00127	<0.00254	96 h exposure in synthetic seawater
AISI 302	0.0096	–	–	643 days in natural seawater
AISI 316	0.0057	–	–	365 days in natural seawater
Ferralium*	0.000762	0.00348	0.00588	96 h exposure in synthetic seawater
Haynes Alloy 600$^\gamma$	0.00254	<0.00254	<0.00254	96 h exposure in synthetic seawater

*Composition of ferralium – Ni 3.0, Cu 1.7, Cr 25.51, Mo 3.5, C 0.04, Fe balance (Ferralium Langley alloy).
$^\gamma$ Fe 8.0, Cr 16.0, C 0.04, Ni balance (Haynes Alloy 600)

Table 9.29 Classification of copper alloys

Series	Classification
A. Wrought Alloys	
C1xxxx	Copper and high copper alloys
C2xxxx	Copper–zinc alloys (brasses)
C3xxxx	Copper–zinc–tin alloys (tin brasses)
C4xxxx	Copper–tin alloys (tin brass)
C5xxxx	Copper–tin alloys (phosphorus bronze)
C6xxxx	Copper–aluminum alloys, copper–silicon alloys (silicon bronzes), and other copper–zinc alloys
C7xxxx	Copper–nickel and copper–nickel–zinc alloys
B. Cast Alloys	
C8xxxx	Cast copper, cast brasses, cast manganese–bronze, cast copper–zinc–silicon alloys
C9xxxx	Cast copper–tin alloys, cast high copper alloys, cast brasses, cast manganese–bronze, cast copper, nickel and zinc alloys

ductility of brass are maintained over a range of 300–180°C. Brasses are ideal for casting and machining. Typical properties of brasses are given in Table 9.31.

b) Bronzes

The term bronze means a copper alloy with tin as a major alloying element. However, tin is not an essential constituent of bronzes. For instances, tin is an alloying element in tin–bronze, whereas aluminum is a major alloying element in aluminum–bronze. Following are the major types of bronzes:

(1) Tin bronzes

Contains up to 10% tin. Small amounts of phosphorus (0.02–0.04%) are added to deoxidize the metal. Such bronzes are called 'phosphor bronze.' Tin bronzes are stronger than brasses.

Table 9.30 Mechanical properties of various coppers

CDA designation	Copper	Alloying element (wt%)	Tensile strength	Electrical conductivity (IACS%)
C101	Electrolytic tough pitch *hc** copper	99.0 Cu (min) 0.05 O	220(A) 400(H)	101.5–100
C103	Oxygen-free *hc* copper	0.05 O	220(A) 400(H)	101.5–100
C104	Fire refined tough pitch *hc* copper	99.85 Cu (min) 0.05 O	220(A) 400(H)	95–89
C105	Arsenical copper	0.3–0.5 As 0.05 O	220(A) 400(H)	90–70
C106	Deoxidized copper	0.013–0.05 P	220(A) 400(H)	90–70
C107	Deoxidized arsenical copper	0.3–0.5 As 0.013–0.5 P 99.20 Cu (min)	220(A) 400(H)	50–35
C108	Cadmium copper	1 Cd 99.00 Cu (min)	220(A) 700(H)	75–92

*hc = high conductivity
(A) Annealed, (H) Hardened

Table 9.31 Typical properties of brasses (copper–zinc alloys)

CDA number (*)	Designation	Composition	Condition	Tensile strength	% Elongation	Electrical conductivity (IACS%)
CZ108	Basic brass	63 Cu, 32 Zn	(A) (H)	340 530	56 5	23
CZ109	Muntz metal	60 Cu, 40 Zn	(A)	380	40	28
CZ112	Naval brass	60 Cu, 37 Zn, 1 Sn	(A)	370	45	26

(A) Annealed, (H) Hardened
*The British Standards Association coding for wrought-alloys consisting of copper and copper alloys, consists of two letters followed by three digits, indicate the alloy group and the digits the alloy within the group.

Casting bronzes containing zinc are called 'gun metals.'

(2) Aluminum bronzes

Copper–aluminum alloys containing up to 9% aluminum are called alpha aluminum bronzes. If more than 9% aluminum is added, the beta phase is formed, and the bronzes are termed as beta bronzes or duplex alloys. The tensile strength increase with aluminum contents.

(3) Silicon bronzes

Copper–silicon alloys containing up to 3% silicon are called silicon bronzes. They are used in fabrication of explosive-proof equipment. They have an excellent resistance to corrosion.

(4) Cupronickels

Alloys of copper and nickel are soluble in each other in both solid as well as liquid forms and form solid solutions. They are completely alpha phase alloys. In Britain, some coins contain 75% copper and 25% nickel. Cupronickels are available in several compositions as they offer an ideal combination of mechanical strength and corrosion resistance. They are widely used as heat exchanger materials in salt water systems.

Cupronickels (90:10) and (70:30) are the most reliable materials for application in a salt water environment as shown by the proven record of these alloys in desalination service. These alloys can be used with a minimum risk of corrosion in important components of the plants, such as water boxes, condenser tubes, heat exchanger tubes, and tube plates. They exhibit a high resistance to impingement attacks, alloy 70:30 exhibiting a higher resistance than alloy 90:10. Saudi Arabia has the largest density of desalination plants in the world and these alloys are considered as standard materials for these plants. The mechanical properties of selected bronzes are given in Table 9.32.

9.18 CORROSION RESISTANCE OF COPPER-BASE ALLOYS IN SEAWATER

Copper exhibits a good resistance to corrosion both in clean and polluted seawater. Copper tubes (C106) have enjoyed a good reputation in water distribution service. With a clean copper surface in water-containing chloride, copper ions are produced by anodic dissolution. They combine with chloride ions to form cuprous chloride.

$$Cu^+ + Cl^- = CuCl \qquad (1)$$

Cuprous chloride hydrolysis in the neutral pH range to form cuprous oxide which is precipitated

$$2CuCl + H_2O = Cu_2O + 2HCl \qquad (2)$$

Table 9.32 Mechanical properties of selected bronzes (tin bronzes and aluminum bronzes)

Alloy designation	Composition (wt%)	Condition	Tensile strength	% Elongation
7% Phosphor bronze (PB103)	95 Cu, 7 Sn, 0.01–0.4 P	(A) (H)	370 650	65 14
Cast admiralty gun metal (G1)	88 Cu, 10 Sn, 2 Zn	Chill cast	250–310	3–8
5% Aluminum bronze (CA101)	95 Cu, 5 Al	(A) (H)	370 650	65 15
9% Aluminum bronze (CA103)	88 Cu, 9 Al, 3 Fe	(A) (H)	570 650	30 22
Al–bronze (Cast) (AB1)	88 Cu, 9.5 Al, 2.5 Fe	Die cast	540–620	8–12
Cupronickel(70/30) (CN107)	68 Cu, 30 Ni, 1.5 Mn, 0.5 Fe	(A)	390	45
Cupronickel(90/10) (CN102)	87.5 Cu, 10 Ni, 1.5 Fe, 1 Mn	(A) (H)	320 360–400	42 30–12

The supporting cathodic reaction is

$$O_2 + 2H_2O + 4e = 4OH^- \quad (3)$$

When corrosion proceeds, the hydroxyl ions are removed by bicarbonate ions present in the water

$$OH^- + HCO_3^- = CO_3^{2-} + H_2O \quad (4)$$

resulting in precipitation of a mixed calcium carbonate and basic carbonate scales. In the presence of carbon contamination, potential may reach a critical value for pitting to occur. Pitting breaks down the protective Cu_2O layer formed on the alloy surface. Alloying elements reduce the electrical conductivity of Cu_2O, and thereby improve the corrosion resistance in addition to substantially increasing the mechanical properties. Copper-based alloys containing less than 15% zinc include silicon bronze, phosphorus bronze, manganese bronze and aluminum bronze. They are protected by the formation of protective film of oxide on the surface. The composition of the film depends on the chemical composition of alloys. Generally speaking, they offer a better resistance to high velocity than austenitic stainless steels. On the other hand, alloys containing more than 15% zinc, such as naval brass, manganese brass and muntz metal, are subjected to dezincification (leaching) and, therefore, not very suitable for seawater service.

Red brasses (up to 85% Cu), alpha brasses (63 to 85% Cu), and alpha/beta brasses (63 to 40% Cu) behave differently in seawater. The alpha and red brasses are not very corrosion-resistant, however, their resistance increases substantially on increasing the zinc content in the

Table 9.33 Effect of arsenic addition on brass in seawater

Brass type	Corrosion rate (μm/year)	Depth of dezincification pits/m
Al–brass	60	11.5
Al–brass + As	60	Trace
Admiralty brass	75	3.75
Admiralty brass + As	75	Trace

range of 20–35%. Addition of Sn or Sb makes them resistant to dezincification. The effect of As addition is shown in Table 9.33.

Admiralty brass is not particularly resistant to erosion–corrosion. It should not be used in water flowing at velocities exceeding 2 m/s. They give good service as condenser tube materials in ships condensers.

Aluminum brass is resistant to erosion–corrosion and dezincification. It is used as condenser heat exchanger tube material for marine applications. A good example is the use of Al–Brass (76–79 Cu, 1.8–2.5 Al, 0.005 As, 0.007 Pb) as in marine service.

Following is a summary of corrosion rate of selected brasses in seawater (Table 9.34).

The term aluminum brass and aluminum bronze are sometimes confusing. Bronzes are copper alloys in which the major alloying element is either zinc or nickel, whereas brasses are copper alloys in which zinc is the dominant element. The brasses are named after major alloying element added in them, such as silicon, lead,

Table 9.34 Corrosion rate of selected brasses in seawater [38]

Type	Condition	Corrosion rate (μm/year)
Red brass (90 Cu, 10 Zn)	Cold rolled	0.045
Alpha brass (70 Cu, 28 Zn)	Cold rolled	0.001
Admiralty brass (72 Cu, 26.8 Zn, 1.1 Sn, 0.03 Pb, 0.014 P)	Cold rolled	0.018
Al–brass (76 Cu, 21.4 Zn, 2 Al, 0.005 As, 0.007 Pb)	Cold rolled	0.008
Naval brass (62 Cu, 36.7 Zn, 1.01 Sn, 0.23 Pb)	Cold rolled	1.08

aluminum brass. Similarly, there are several categories of bronzes, such as aluminum bronze and nickel–tin bronze, depending on the alloys added.

9.18.1 CORROSION OF BRONZES

Bronzes with tin content in the range of 5–14 are resistant to seawater in both wrought and cast form when water velocities are encountered. Alloys with higher tin contents are preferred. The low Sn-bronzes are resistant to impingement corrosion and used in pump bodies, rotors, valves, liners and water boxes.

Bronze containing 2.4% tin showed a corrosion rate of .05 mm/year at a velocity of 0.6 m·s^{-1}. The corrosion rate of important brasses, bronzes and copper in seawater are shown in Table 9.35.

Copper–tin alloys (76 Cu–22 Zn–2 Al), also known as tin bronze, show increased corrosion rates with increased temperature compared to Cu–Al alloys. The corrosion resistance of copper–aluminum alloys increases with increasing aluminum content. Alpha alloys (α-aluminum bronze) consists of single (alpha) phase up to 7% aluminum and two phase ($\alpha + \beta$ or $\alpha + \gamma$) above 7%. Aluminum bronzes exhibit an outstanding resistance to corrosion in seawater and offer a good resistance to impingement corrosion.

Aluminum bronzes containing 7% Al, 2% Ni, show an outstanding resistance to de-alloying and cavitation corrosion in most fluids and seawater, because of nickel addition which is highly resistant to corrosion. Aluminum bronze, such as 76 Cu–22 Zn–2 Al, are used for marine heat exchangers and condenser because of its excellent corrosion resistance. Aluminum is responsible for increased corrosion resistance. But the velocity must not exceed a safe threshold to avoid erosion–corrosion.

9.18.2 CORROSION PROBLEMS IN CONDENSERS AND HEAT EXCHANGERS

A variety of problems may occur in condensers and heat exchangers using seawater as a coolant. These include impingement corrosion, sand erosion and pitting. External corrosion may also occur if gases, such as carbon dioxide, are dissolved in the condensate. The following are the causes of major tube failures:

(a) Sand erosion.
(b) Polluted cooling water (H_2S).
(c) Steam impingement.
(d) Corrosive condensate.
(e) Build-up of local high temperatures.
(f) Deposition of debris.

Table 9.35 Corrosion resistance of selected bronzes in seawater [39]

Condition	Exposure time	Alloy designation						
		Corrosion rate, mm-year^{-1}						
		Yellow brass (62–38)	Admiralty brass (70–29–1)	Red brass (76–22–2)	Aluminum brass	Al–bronze (7 Al–8 Ni)	Sn–bronze (12.5 Sn)	Cu–Ni (70–30)
Mediterranean sea	81 days	.0089	.0086	–	.0018	.0027	–	.0017
Dead sea	81 days	.0198	.0215	.0054	.0097	.0016	.0241	.0138

Source: White, J.H., Yanii, A.L. and Schide, H. (1966). *Corrosion Sc.*, **6**, 453.

(a) *Sand erosion.* Sand may induce erosion–corrosion in the tubes.

(b) *Polluted water.* Waters containing hydrogen sulfide are highly aggressive. Sulfate bacteria reduce sulfates to hydrogen sulfide.

(c) *Impingement.* High velocity of droplets of water in steam may lead to erosion–corrosion of the metals.

(d) *Condensate.* High quantities of dissolved oxygen and carbon dioxide make the condensate extremely corrosive.

(e) *Build-up of local temperatures.* Severe pitting may occur on the cooling water side at points corresponding to the build-up of high temperatures on the outside of the tubes.

(f) *Deposition of Debris.* Corrosion under deposits due to the differential aeration cell may cause serious problems. The deposits which settle are mostly weed, stick, stones, barnacle, mussels, etc.

Table 9.36 gives some widely known applications of copper alloys in marine service.

Preventive measures:

(1) Select appropriate tube materials.
(2) Allow minimum stresses during fitting.
(3) Remove all debris.
(4) Minimize throttling of water flow.
(5) Apply cathodic protection in water boxes and other components.
(6) Protect tubes from high velocity water droplets by fitting baffle plates.

(7) Remove incondensible gases.
(8) Do not exceed specified velocity.

9.19 ALUMINUM AND ITS ALLOYS

9.19.1 INTRODUCTION

Aluminum is considered as an extremely corrosion-resistant material due to a thin, protective, highly adherent oxide film which it forms on the surface. The history of aluminum development has been governed largely by the quest for increased mechanical properties which culminated in the application of high strength alloys in aerospace industry. Considerable efforts have been directed to develop high performance alloys with high resistance to corrosion. A substantial interest was generated in the development of alloys for seawater service and particularly in the desalination plants in the seventies. Continued efforts are being made in the development of new alloys for aerospace engineering transportation and space vehicles. Apart from the high strength achieved in recent years in aluminum alloys, sustained efforts have been made to enhance their resistance to corrosion in a wide variety of corrosive environments. A brief review of important aluminum alloys and their corrosion resistance is given below.

Table 9.36 Some widely known applications of copper alloys in marine service

Alloy	ASTM designation	Use
70/30 Cu–Ni (Nickel 30, Iron 0.6, Mn 0.8)	715	Condenser and heat exchanger tubes, pipelines, tube plates
90–10 Cu–Ni (Ni 10, Iron 1.6, Mn 0.8)	706	Condenser and heat exchanger tubes, water boxes
Aluminum brass (Zn 22, Al 2, As 0.04)	687	Condenser and heat exchanger tubes, water boxes
Aluminum bronze (Al 9.5, Ni 5, Iron 2.5, Mn 1)	628	Tube plates
Aluminum bronze (Al 6–7)	608	Condenser and heat exchanger tubes
Naval brass (Zn 38, Sn 1)	464	Tube plates

9.19.2 ALLOY CATEGORIES

These alloys can be divided in two categories: wrought alloys and cast alloys. There is a further differentiation on the basis of their response to heat treatment. If they respond to heat treatment, they are called 'heat treatable.' The term applies to both cast and wrought alloys. Some alloys are not heat treatable and used only in the cast condition.

Wrought alloys are formed by rolling, forging and extrusion into useful products, whereas cast alloys are directly cast into finished products.

9.19.3 WROUGHT ALUMINUM ALLOYS DESIGNATIONS

The commonly designated system is that of the Aluminum Association. It is based on four digits. The first digit define the major alloying class of the series. The second digit define variation in the original basic alloy. That digit is zero for the original composition, a one (1) for first variation, a two (2) for second variation and so forth. Variations are defined by one or more alloying elements of 0.15–0.50% or more. The third and fourth digit designate the specific alloys within the series (Table 9.37).

9.19.4 CAST ALLOYS DESIGNATION SYSTEM

The first digit indicates the alloy group. The second and third digits identify the specific aluminum alloy. The fourth digit indicates the product form:

xxx.0   indicates casting
xxx.1   indicates ingot

Table 9.38 shows the designation system for cast alloys.

The following list shows the temper designations:

xxxx F   as fabricated
xxxx W   solution heat treated
xxxx O   annealed (wrought alloys)
xxxx H   strain hardened
xxxx T   thermally treated

9.19.5 COMPOSITIONS OF ALUMINUM ALLOYS

Casting compositions are described by three digit system followed by a decimal value. The decimal 0.0 in all cases pertains to casting alloy limit. Decimals 0.1 and 0.2 pertain to ingot composition. The following are the families:

Table 9.37 Major alloying elements in wrought alloys designation system

Series	Contents
1000	Pure aluminum
2000	Copper
3000	Magnesium
4000	Silicon
5000	Magnesium
6000	Magnesium and silicon
7000	Zinc
8000	Other elements
9000	Unassigned

Table 9.38 Major alloying elements in cast alloys designation system

Series	Contents
1xxx	Pure aluminum
2xxx	Copper
3xxx	Silicon with added copper
4xxx	Silicon
5xxx	Magnesium
6xxx	Not used
7xxx	Zinc
8xxx	Tin
9xxx	Other alloying elements

1xxx   Pure composites
2xxx   Copper as principal alloying element
3xxx   Silicon is the principal alloying element. Magnesium and copper are specified
4xxx   Silicon is the principal alloying element
5xxx   Magnesium is the principal alloying element
6xxx   Unused
7xxx   Zinc is the principal alloying element. Copper and magnesium are specified
8xxx   Tin is the principal alloying element.

The nominal composition and application of cast and wrought alloys are given in Table 9.39.

The nominal composition and applications of casting aluminum alloys are given in Table 9.40.

9.19.6 MECHANICAL PROPERTIES OF WROUGHT ALUMINUM ALLOYS

Typical mechanical properties of cast and wrought aluminum alloys are given in Table 9.41.

9.19.7 CHARACTERISTICS OF WROUGHT ALUMINUM ALLOYS

a)   1xxx Pure Aluminum

This represents commercially pure aluminum ranging from 99.00% minimum to

Table 9.39   Nominal composition of common wrought aluminum alloys and their applications

No	Al	Si	Cu	Mn	Mg	Cr	Zn	Others	Common applications
1100	99.99	–	0.12	–	–	–	–	–	High purity work, chemical process equipment
1350	99.50	–	–	–	–	–	–	–	Conductors
2011	93.7	–	5.5	–	–	–	–	–	Fasteners, machine parts
2014	93.5	0.8	4.4	0.8	0.5	–	–	–	Aircraft and automotive structures
2017	97.8	0.5	4.0	0.7	0.6	–	–	–	Fasteners, machine parts
2024	93.5	–	4.4	0.6	1.5	–	–	–	Gears, cylinders, pistons, naval aircrafts
2219	93.0	–	6.3	0.3	–	–	–	0.06 Ti	High temperature applications
3003	98.6	–	0.12	1.2	–	–	–	–	Fuel tanks, trucks, chemical equipments
3004	97.8	–	–	12.0	1.0	–	–	–	Agriculture and fuel tanks applications
3105	99.0	–	–	0.55	0.50	–	–	–	Kitchen equipment
5052	97.2	–	–	–	2.5	0.25	–	–	Chemical equipment, marine application
5056	95.0	–	–	0.12	5.0	0.12	–	–	Fasteners, automotives
5454	96.3	–	–	0.8	2.7	0.12	–	–	Pressure vessels
5456	93.9	–	–	0.8	5.1	0.12	–	–	Welded assemblies, pressure vessels
6061	97.9	0.6	0.26	–	1.0	0.2	–	–	Automotive parts, aerospace and marine applications
6063	98.9	0.4	–	–	0.7	–	–	–	Marine and saltwater applications
7050	89.0	–	2.3	2.3	–	6.2	–	0.12 Zr	Aircraft structures
7075	90.0	–	1.6	–	2.5	0.23	5.6	–	Aircrafts and machine parts

Table 9.40 Nominal composition and applications of casting aluminum alloys

Alloy	Cu	Mn	Mg	Si	Others	Applications
201.0	4.6	0.35	0.35	–	–	Gears, pumps
208.0	4.6	–	–	3.0	–	Valve bodies, general purpose castings
242.0	4.0	–	1.5	–	2.0 Ni	Pistons, cylinder heads
308.0	4.5	–	–	5.5	–	Permanent mold castings
336.0	1.0	–	1.0	12	2.5 Ni	Automotive wheels, complex castings
356.0	0.25	0.35	0.32	7.0	0.6 Fe, 0.35 Zn	Pump bodies, aircraft wheels
357.0	–	–	0.50	7.0	–	Corrosion-resistant applications
390.0	4.5	–	0.6	17.0	–	Pistons of internal combustion engines, cylinder heads
413.0	–	–	–	12.0	–	Marine equipment, architectural applications
443.0	–	–	–	5.2	–	Marine applications
B443.0	0.15	–	–	5.2	–	Leak–proof, thin walled castings
514.0	–	–	4.0		–	Components exposed to seawater
535.0	–	0.18	6.8		0.18 Ti	Components for corrosive environments
712.0	–	–	0.6	–	5.8 Zn, 0.5 Cr, 0.20 Ti	Office machines, farming equipment
850.0	1.00	–		–	1.0 Ni	Bushing and journal bearings

Table 9.41 Mechanical properties of wrought aluminum alloys

Alloy	Temper	UTS (MPa)	YS (MPa)	% Elongation (50 mm)
1060	0	70	30	43
1060	H-16	115	105	8
1060	H-18	130	125	6
2010	T4, T41	405	310	5
2014	T-6, T-651	485	415	–
2024	T-4, T-351	472	325	20
3004	H-38	285	250	5
5052	H-36	275	240	8
5454	H-34	305	240	10
6061	T-4	240	145	22
6061	T-6	310	275	12
7050	T-6	570	505	9

99.50% minimum. They offer a good combination of excellent formability and high corrosion resistance. They are also used in electrical applications.

b) 2xxx Series

They represent a series of alloys in which copper is the principal alloying element. They possess a good combination of weldability, toughness and high strength. They do not offer a good resistance to aqueous and atmospheric corrosion, however, they are clad with pure aluminum for application requiring corrosion resistance. The corrosion rate decreases linearly in hot seawater with increasing pH. Alloy 2024 is used in naval and military aircraft and truck bodies. Al 2195 (Al–Li alloy) is a very high strength alloy used in aerospace applications.

c) 3xxx Series

These alloys are easily soldered, brazed and welded. They show an excellent resistance to corrosion in a marine environment. They have shown excellent potential as condenser tube material for hot seawater. For example alloy 3003 shows a decreased rate of corrosion with increasing pH in deaerated hot seawater. The above alloy is also used in equipment handling food and chemicals. Alloy 3104 is widely used in beverage cans.

d) 4xxx Series

Silicon is the major alloying element in this series and it can be added up to 12%. They are easily joined by soldering and brazing. Alloy 4032 is used in forgings and alloy 4043 as weld filler alloy. Alloy 4032 has a low coefficient of thermal expansion and is well suited to forged engine pistons. Because of becoming black on anodizing, high silicon containing alloys are in heavy demand for structural applications.

e) 5xxx Series

Magnesium is the principal alloying element in this series of alloys. These alloys show moderate to high strength and an excellent resistance to corrosion in seawater. They have shown an excellent corrosion resistance as condenser tube material. Higher strengths are associated with higher Mg content. Alloying with more than 3% Mg makes them susceptible to stress corrosion cracking above 100°C. Alloy 5754 is used frequently in automotive body frames and panels. Alloy 5052 is used in matrix material for Al–SiC composites for aerospace and marine applications.

f) 6xxx Series

These are Al–Mg–Si alloys. They show an excellent formability, machinability, weldability and corrosion resistance. Al 6061, 6013 and 6063 are widely used. Al 6063 is well known for its excellent extrudibility and applications in automotive frames. Alloy 6061 and 6013 show excellent resistance to marine corrosion in loop tests

conducted to evaluate heat exchanger tube materials for desalination plants. Al 6061 has been widely used as a matrix material for Al–SiC composites.

g) 7xxx Series

Zinc (1–8%) is the major alloying element. These are high strength heat treatable alloys mainly used in airframe structures. Among the Al–Zn–Mg–Cu added alloys, they provide the highest strength. These alloys should be coated before exposure to atmosphere because of their low resistance to atmospheric corrosion. They also exhibit reduced resistance to stress corrosion cracking.

h) 8xxx Series

Aluminum is alloyed with Ni, Li, Fe and Sc, to provide particular functional characteristics. For instance, addition of lithium gives exceptional high strength and a good corrosion resistance for aerospace applications demanding high strength-to-weight ratio and high stiffness. Alloying with Sc increases the recrystallization temperature and reduces hot cracking.

9.19.8 CORROSION RESISTANCE OF AL 1100 SERIES (PURE AL)

This series represents commercially pure aluminum ranging from 99 to 99.5% purity. The ability of aluminum to resist corrosion depends on the inherent ability to form protective oxide film, such as Bayerite [α-Al(OH)$_3$], and Boehmite (γ-AlOOH) on its surface. As shown by Pourbaix diagrams (Fig. 9.15), the passive region (pH $= 4 - 8$) where the films of oxide forms lies between the acid range (pH $= 2-4$) and alkaline range (pH $= 8-16$). Aluminum dissolves both in acids and alkaline solutions due to instability of aluminum oxide and hydroxide films. The rate of corrosion is low between pH 5.5 and 8.5. Al is resistant to Ca(OH)$_2$, for example concrete, $NH_4OH > 10\%$, salt solutions and most dry gases. It shows a good resistance to urban and

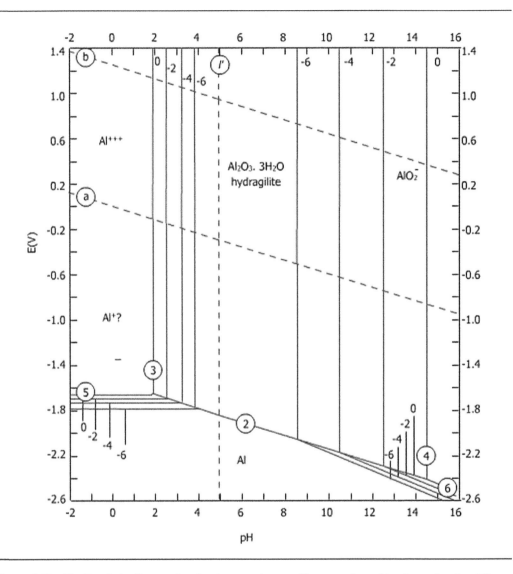

Figure 9.15   The Pourbaix diagram for aluminum and water illustrates the stable phases for the different potentials and pHs. (Reproduced by kind permission of NACE Int., Houston, USA)

marine atmospheres unless contaminated with copper. It is, however, subject to attack by chloride ions. Aluminum combines with chloride to form $AlCl_3$.

$$Al^{3+} + 3Cl^- \rightarrow AlCl_3 + 3e^-$$

As the concentration of chloride ions is increased, the passive film undergoes active dissolution due to adsorption of chloride ions, and corrosion is enhanced. Above 70%, boehmite (γ-AlOOH) replace the bayerite (α-AlOOH)$_3$. Both layers afford corrosion protection to aluminum. The corrosion rate of aluminum increases with decreasing purity levels. Aluminum is not resistant to corrosion in HCl, HF, $HClO_4$ and H_3PO_4, in $NH_4OH < 10\%$, and water-containing salts of heavy metal ions, such as Hg, Sn, Cu, Ag, Pb, CO, or Ni. It is also not resistant to moist gases, like SO_2, SO_3, NH_3 and Cl_2.

Following is a summary of electrochemical corrosion:

$$Al^{3+} + 3OH^- \rightarrow Al(OH)_3 \quad \text{(anodic reaction)}$$

$$\frac{1}{2}O_2 + H_2O + 2e \rightarrow 2OH^- \quad \text{(cathodic reaction)}$$

In case of no oxygen: $2H^+ + 2e \rightarrow H_2 \uparrow$

Upon dissolution of aluminum cation into aqueous solution, it diffuses to another area to precipitate aluminum hydrate (gelatinuous):

$$Al^{3+} + 3OH^- \rightarrow Al(OH)_3 \quad \text{(Precipitates)}$$

$$Al^{3+} + 3H_2O \rightarrow 3H^+ + Al(OH)_3 \quad \text{(Precipitates)}$$

9.19.9 CORROSION OF ALUMINUM AND ITS ALLOYS

1)   Uniform Corrosion

This attack occurs when pH is either very high or very low. The protective aluminum oxide layer dissolves rapidly under the above conditions. Prevention from uniform corrosion requires control of pH by inhibitor additions, or use of cathodic protection, or replacement with a more corrosion-resistant alloy.

2)   Galvanic Corrosion

The potential range of aluminum alloy lies between -0.6 and 0.8 V *vs* the saturated calomel electrode. If aluminum is connected to copper (corrosion potential $= -0.1$ V), a significant difference of potential (-0.5 V) would exist and aluminum, being anodic to copper, would corrode. The details of galvanic corrosion are given in Chapter 4. Alloys of 2000 series contain copper $> 1\%$ for precipitation hardening. The copper precipitates as (Al_2Cu), acts as a cathodic site to the alloy, and initiates galvanic action which decreases the corrosion resistance of the aluminum alloy. Care should be taken when using these alloys as fasteners. Appropriate designs, and good insulation practices are used to minimize galvanic corrosion.

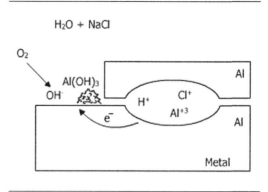

Figure 9.16 Crevice corrosion of aluminum can occur in a saltwater environment

3)   Crevice Corrosion

Crevices are gaps formed on joining two structural members. The formation of a crevice is followed by formation of a differential aeration cell. The oxygen inside the crevice becomes depleted after a length of time and the crevice becomes anodic, $Al^{3+} + 3H_2O \rightarrow 3H^+ + Al(OH)_3$. The precipitation of $Al(OH)_3$ leads to decreased pH and acidification (H^+). Crevice corrosion of aluminum is shown in Fig. 9.16. If aluminum is coated, a cathodic reduction-reaction would not occur around the crevice mouth and the cathodic reduction reaction would then take place slowly inside the crevice, $2H^+ + 2e \rightarrow H_2 \uparrow$. Crevices must be eliminated by sealing or welding.

4)   Pitting Corrosion

The mechanism of pitting corrosion of aluminum alloy is not very different from steels. The anodic reaction in the pit is $Al \rightarrow Al^{3+} + 3e$, and the cathodic reaction adjacent to the pit is $O_2 + 2H_2O + 4e \rightarrow 4OH^-$. Copper can also deposit as $Cu^{2+} + 2e \rightarrow Cu$. Al^{3+} ions within the pit attract chloride ions (Cl^-) and may form $AlOHCl^-$ and H^+: ($Al^{3+} + 3H_2O + Cl^- \rightarrow H^+ + AlOHCl^+$). The pit cavity continues to acidify and hydrogen evolves. The hydrogen bubbles also pump out $AlCl_3$ which is formed by interaction between Al^{3+} and Cl^- ions. The $AlCl_3$

formed reacts outside the pit with water and precipitates Al(OH)$_3$. This precipitation looks like white chimneys which cover the pits [40]. Such chimneys have been observed also in other aluminum alloys, such as Al 6013. Among the aluminum alloys, alloys of 5xxx series show a low resistance to pitting. Alloys containing copper increase the tendency to pit. Alloy Al 5052 has been successfully used in seawater with a minimum risk of pitting. Aluminum alloys of 2xxx series, such as 2024 are not resistant to pitting and the same is true for alloys of 7xxx series, such as alloy 7050. Alloys 6061 and 6013 show a fairly good resistance to pitting in marine environment because of smaller copper contents. High strength alloys containing copper, such as Al 2024, used in aerospace applications are protected against pitting by coating. In the 1970s, extensive work was done to examine the suitability of aluminum alloys in desalination service. A variety of heat exchanger tubes were evaluated and the resistance to pitting was a major consideration. Table 9.42 shows the results of tubes from heat reject exchangers after six months test.

The results showed excellent resistance of the above alloys to pitting. In heat recovery units, alloys 3003, 5052 and 6013 showed excellent resistance to pitting after 38 months exposure. The corrosion rate of aluminum alloys tend to significantly decrease after initial exposure due to build up of protective oxide layer. For instance, the corrosion rate of Al 1100 aluminum in hot deaerated seawater decreases linearly with an increase in pH, which suggests the importance of pH control. Pitting studies on alloys 1100, 3003, 5052, 5054 and 6061 at 80°C in oxygen saturated water showed

that Al 5052 offers the highest resistance to pitting and it was recommended for use in desalination plants [42]. The effect of oxygen and pH on pitting corrosion is shown in Fig. 9.17. Areas of very good, good and poor pitting resistance are shown.

The pitting values for different aluminum alloys for six months in Pacific Ocean are shown in Table 9.43. To summarize, aluminum alloys 3003, 5052 and 6063 can be successfully used in a marine environment by careful control of water chemistry, oxygen and pH levels.

5) Intergranular Corrosion

The mechanism of intergranular corrosion of steels is described in Chapter 4. The acid environment in the pit leads to the attack on grain boundaries. Two mechanisms have been proposed: In Al–Mg alloys (5xxx), phases such as Mg$_2$Cl$_3$, may precipitate at the grain boundaries. This active phase is preferentially attacked and dissolves leaving the grains separated from each other. In aluminum alloys, such as the 2xxx series, a more noble precipitate, such as CuAl$_2$ may be formed leaving the areas adjacent to the grain boundaries more reactive promoting their dissolution in the acid pit. Al–Mg alloys (5xxx) containing less than 3% Mg, are resistant to intergranular corrosion. Aluminum alloys of the series 6xxx show susceptibility to intergranular corrosion because of the formation of Mg$_2$Si precipitate if excessive silicon is present. Alloys of 7xxx series are prone to intergranular attack because of the formation of a Cu–Al phase. By eliminating the

Table 9.42 Pitting resistance of tube materials [41]

Alloy	Temperature°C	Max. pit depth (μm)	Pits per sq. inch	Remarks
3003	110	0	0	Excellent
5052	110	0.008	1	Shallow pits
6063	110	0.001		One minor pit
5050	110	0	0	Excellent

Source: Vernik, E. Jr, and George, P.F. (1973). *Materials Performance,* **12,** 26.

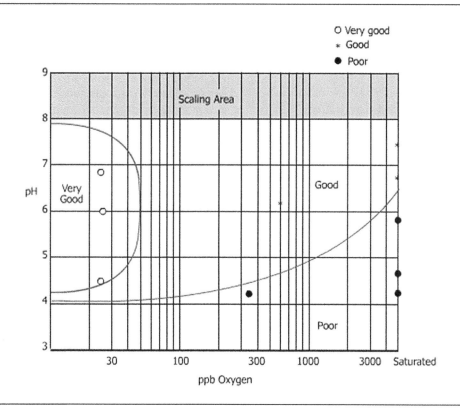

Figure 9.17 Effect of pH and oxygen content on seawater pitting of aluminum alloy tubes based on 30-day tests in environmental side unit tests. (From Vernik, E.D. and George, P.F. (1973). *Mat. Pref.* **12**, (5). Reproduced by kind permission of NACE, Int, Texas, USA)

Table 9.43 Pit depths for different aluminum alloys for six months in Pacific ocean [42]

Alloy	Temp. (°F)	Max. pit depth (inch)	Avg. pit depth (inch)	Pits per sq. inch	Remarks
3003	110	0	0	0	Excellent
5052	110	0. 008	0.002	1	Shallow pitted areas plus scattered pits
6063	110	0. 001	–	–	One minor pit, excellent
5050	110	0	0	0	Excellent

sites of pitting or crevice corrosion and minimizing the potential difference between the grain boundaries and matrix by appropriate heat treatment, intergranular attack can be minimized or eliminated.

6) Exfoliation Corrosion

This is a special version of intergranular corrosion where the grains have been subjected to deformation process, e.g. rolling. Exfoliation appears like leaves in a book. Alloys of 7xxx (such as Al7075-T6) and 2xxx series (such as Al 2024) are often found with exfoliation corrosion.

7) Flow Induced Corrosion

The resistance of aluminum alloys to flow induced corrosion depends on the stability of the protective oxide films on the surface. Dissolution of these films leads to accelerated corrosion. The protective films of bayerite and boehmite could be eroded by shear forces resulting from flow beyond a critical velocity. Aluminum alloys of series 5xxx are not adversely affected by velocities up to 3 m/s in the absence of abrasives in water. The removal of a film adjacent to a film surface sets up local corrosion cell which accelerates the corrosion process. Alloys of 5xxx series (such as 5454) show a good resistance to corrosion at velocities up to $3\,\mathrm{ms^{-1}}$ at temperatures up to 140°C. The corrosion rate increases with increased velocities in the presence of abrasive particles, which need to be controlled. The water chemistry, water velocity and pH needs to be controlled to minimize the effect of flow on localized corrosion. Maintaining pH below 9 would not allow aluminum to dissolve as AlO_2^-. The preventive measures include the minimizing of turbulent flow or changing water chemistry.

8) Stress Corrosion Cracking

Stress corrosion cracking in aluminum alloys is limited to Al–Cu–Mg alloys (2xxx), Al–Mg alloys containing more than 3% Mg and Al–Zn–Mg alloys (7xxx) series. The following is a ranking of stress corrosion cracking of major commercial aluminum alloys (Table 9.44).

In stress corrosion cracking, the crack follows the grain boundary, termed intergranular cracking. Transgranular cracking (cracking across the grain boundaries) is rare. Over the years, high strength alloys of the 7xxx series (Al 7002.T6, 7106.T63 and 7039.T64) are known to be sensitive to SCC in seawater. In alloys of the 5xxx series, the formation of Mg_2Al_3 in alloys containing more than 3% Mg, promotes

stress corrosion cracking. Alloys of the 6xxx series show a strong resistance to stress corrosion cracking as exemplified by Al 6061 which contains a stoichiometric balance of Mg and Si to form Mg_2Si. The risk of stress corrosion cracking in certain alloys of this category can be avoided by following rigidly the recommended solution heat treatment temperature and followed by slow quenching.

To avoid the risk of stress corrosion cracking, lower strength alloys may be substituted for higher strength alloys. The aerospace industry started with higher strength alloys in 1950s, such as 7079-T651, which showed a poor resistance to stress corrosion cracking. These were changed to alloys with lower stress levels. The stress levels were lowered by using appropriate tempers. For instance, alloy 7075 in temper T6 was replaced with alloy 7150 in temper T77 to minimize the risk of stress corrosion cracking. The development of aluminum–lithium alloys and other aerospace alloys represents an attempt to successfully use high strength-to-weight ratio alloys with a minimum risk of stress corrosion cracking. Readers must acquaint themselves with detailed temper systems for a better understanding of stress corrosion cracking in aluminum alloys.

9) Corrosion Fatigue

As mentioned in Chapter 4, fatigue failure is defined as the failure of metal under the combined action of cyclic stress and corrosive environment. The fatigue life of a material is defined as the total number of stress cycles of defined frequency and amplitude to cause fracture (N_f). Data from tests is displayed in the form of S–N curves (S is the stress and N is number of cycles). A fractured surface is identified by fatigue striations, described in Chapter 4. It is important to find out the fatigue life of materials in order to predict their service behavior in a defined environment. Cracks in alloys 7075 and 2024 are known to develop very rapidly. Alloys of 5xxx series (such as Al 5087-H34, Al 5086-H36) and 6xxx series (such as Al 6061) show higher fatigue strength than higher strength alloys, such as Al 7075 and Al 2024. The selection of aluminum alloys for service

Table 9.44 Ranking of stress corrosion cracking of major commercial aluminum alloys [43]

Alloy designation	Type	Tensile strength (MPa)	SCC ranking(*)
1xxx	Al	69–172	(a)
2xxx	Al–Cu–Mg–Si (3–6% Cu)	379–517	(b)
3xxx	Al–Mn–Mg	138–276	(a)
5xxx	Al–Mg (1–2.5% Mg)	138–290	(a)
5xxx	Al–Mg–Mn (3–6% Mg)	290–379	(a)
6xxx	Al–Mg–Si	152–379	(a)
7xxx	Al–Zn–Mg	379–503	(b)
8xxx	Al–Zn–Mg–Cu	517–621	(b)

*(a) No known cases of SCC,
(b) SCC occurred in service but can be avoided

must be made with due regard to their fatigue characteristics.

10) Anodizing and Al Cladding

These two corrosion prevention methods are exclusive to aluminum alloys. Anodizing is used to produce a layer of oxide by applying a specified voltage to aluminum which is made an anode in an electrolyte of sulfuric acid, oxalic acid or chromic acid. Anodizing provides longer life and aesthetic values to equipment or articles. In cladding, one of the alloys acts as a sacrificial material for the other alloy which is to be protected. High strength aluminum alloys containing copper are sensitive to corrosion, whereas Al 1100 is highly resistant to corrosion. Thin aluminum sheets (Al 1100) are hot rolled on the top and bottom of Al 2024-T3 sheet to produce the alclad version of Al 2024-T3. The pure aluminum (Al 1100) acts as a sacrificial coating to Al 2024 and corrodes in preference to Al 2024. Because of the limitation of alcladding to sheet metals, new electroplating processes have been developed to produce pure aluminum coatings on a variety of products, such as screws, fasteners, forgings, etc.

Corrosion prevention measures described in detail in Chapter 4, such as cathodic protection and inhibition treatment, are applied to aluminum. Cathodic protection is to be applied with care, as over-protection results in developing an alkaline environment, which can severely damage aluminum.

9.20 Nickel and its Alloys

Nickel is well-known as an essential alloying element in stainless steels, Ni–Cu alloys, Ni–Fe alloys, Ni–Cr–Fe alloys, super alloys, as nickel–chromium alloys and special corrosion-resistant and high temperature alloys. Nickel ferromagnetic with a density of 8.9 g/cm^3. It is ductile and malleable like steel. Nickel alloys are well-known for their high temperature strength and good resistance to corrosion.

9.20.1 Classification of Nickel Alloys

The nickel alloys can be classified in the following groups on the basis of their chemical compositions:

(1) Nickel

- Pure nickel (99.56%)
- Commercially pure nickel (wrought) 99.6–99.7%.

(2) Nickel and copper

- Low nickel alloys (2–13% Ni)
- Cupronickel (10–30% Ni)
- Non-magnetic alloys (~60% Ni)
- High nickel alloys (over 50% Ni)

(3) Nickel and iron

- Wrought alloys steels (0.5–9% Ni)
- Cast alloy steels (0.5–0.9% Ni)
- Alloy cast iron (1–6, 14–36% Ni)

(4) Iron–nickel and chromium alloys

- Stainless steels (2–25% Ni)
- Maraging steels (18% Ni)

(5) Nickel–chromium–molybdenum and iron–nickel base precipitation hardened alloys.

9.20.2 CHARACTERISTICS OF NICKEL ALLOYS

The composition of selected nickel alloys are shown in Table 9.45.

(1) Nickel–Copper Alloys

These alloys are well-known for their excellent corrosion resistance to seawater. They have been used as propellers, pump shafts, impellers and condenser tube materials. The best known is Monel (Alloy 400). It is resistant to brine and immune to stress corrosion cracking and pitting in chloride and caustic alkaline solutions. It is also resistant to HF and fluorine containing media.

Monel alloy R-405 has specified amounts of sulfur for improved machining characteristics. Monel K 500 has the dual advantage of improved mechanical strength and excellent corrosion resistance. It can retain strength up to 650°C and ductility up to 134°C.

(2) Nickel–Chrome–Iron Alloys

These alloys contain a high percentage of nickel and excellent capability to withstand high temperature oxidizing environment. Alloys, such as Inconel 600, 690, 718 and X750, belong to this category. Alloy Inconel 600 (Ni 76, Cr 15.5, Fe 8) is the basic alloy in this class with excellent corrosion resistance at elevated temperatures (~1092°C). It can, however, be subjected to pitting or crevice corrosion. Other alloys in this family include alloy 690 (29% Cr) which shows excellent resistance to SCC in chloride media and low corrosion rates at high temperatures. It is used in furnaces for petrochemical processing and in coal gasification units.

Alloy Inconel X750 contains additions of aluminum, niobium and titanium which form an intermetallic compound, $Ni_3(Al, Ti)$ to make it age hardenable and provide high strength. It is extremely resistant to SCC in chloride environment. It is used in gas turbines, vacuum envelopes, extrusion dies and springs.

(3) Nickel–Iron–Chromium Alloys

These alloys represent another version of Ni–Cr–Fe alloys and contain 30–44 % of nickel. Alloy 800 of this series has been extensively used in heat exchangers in the petrochemical industry, because of its excellent resistance to stress corrosion cracking in chloride environments and cracking in polythionic acid. It offers an excellent resistance to creep and rupture. They are used for high environments where resistance to oxidation and corrosion is required. Incoloy 825 has proved highly successful in applications in H_2SO_4, HCl, phosphoric acid and clean and polluted seawater.

(4) Nickel–Chromium–Molybdenum Alloys

This family of alloys are mainly used in the chemical processing industry and contain 45–60 % Ni. Hastelloy has been successfully used in high temperature applications (up to 1204°C). The Hastelloy C series have served the chemical industry for a long time. The modified version of Hastelloy C is Hastelloy C-276 in which silicon and C content are substantially reduced (0.005% C, 0.04% Si). It is used successfully in the petrochemical industry. Alloys 625 and 617

Table 9.45 Nominal chemical composition (wt%)

Material	Ni	Cu	Fe	Cr	Mo	Al	Ti	Nb	Mn	Si	C
Nickel											
Nickel 200	99.6	–	–	–	–	–	–	–	0.23	0.03	0.07
Nickel 201	99.7	–	–	–	–	–	–	–	0.23	0.03	0.01
Nickel–Copper											
Monel alloy 400	65.4	32	1.00	–	–	–	–	–	1.0	0.10	0.12
Monel alloy 404	54.6	45.3	0.03	–	–	–	–	–	0.01	0.04	0.07
Monel alloy R-405	65.3	31.6	1.25	–	–	0.1	–	–	1.0	0.17	0.15
Monel alloy K–500	65.0	30	0.64	–	–	2.94	0.48	–	0.70	0.12	0.17
Nickel–Chromium–Iron											
Inconel alloy 600	76	0.25	8.0	15.5	–	–	–	–	0.5	0.25	0.08
Inconel alloy 601	60.5	0.50	14.1	23.0	–	1.35	–	–	0.5	0.25	0.05
Inconel alloy 690	60	–	9.0	30	–	–	–	–	–	–	0.01
Nickel–Iron–Chromium											
Incoloy alloy 800	31	0.38	46	20	–	0.38	0.38	–	0.75	0.50	0.05
Incoloy alloy 800H	31	0.38	46	20	–	0.38	0.38	–	0.75	0.50	0.07
Incoloy alloy 825	42	1.75	30	22.5	3	0.10	0.90	–	0.50	0.25	0.01
Incoloy alloy 925	43.2	1.8	28	21	3	0.35	2.10	–	0.60	0.22	0.03
Pyromet 860	44	–	Bal	13	6	1.0	3.0	–	0.25	0.10	0.05
Nickel–Chromium–Molybdenum											
Hastelloy alloy X	Bal	–	19	22	9	–	–	–	–	–	0.10
Hastelloy alloy G	Bal	2	19.5	22	6.5	–	–	2.1	<1	1.5	<0.05
Hastelloy alloy C–276	Bal	–	5.5	15.5	16	–	–	–	<0.08	<1	<0.01
Hastelloy alloy C	Bal	–	<3	16	15.5	–	<0.7	–	<0.08	<1	<0.01
Inconel alloy 617	54	–	–	22	9	1	–	–	–	–	0.07
Udimet 600	Bal	–	<4	17	4	4.2	2.9	–	–	–	0.04

are high temperature strength alloys and exhibit a high resistance to corrosion. Alloy 625 is used extensively in seawater applications. It is highly resistant to pitting and stress corrosion cracking. Other alloys, like Udimet 500, 520, 600 and 700 retain high temperature strength up to 982°C.

9.20.3 EFFECT OF ALLOYING ELEMENTS

Table 9.46 illustrates the effect of alloying elements on the corrosion resistance of nickel.

9.20.4 LOCALIZED CORROSION SUSCEPTIBILITY OF NICKEL ALLOYS

1) Stress Corrosion Cracking

Although nickel alloys on the whole offer a better resistance to stress corrosion cracking over the steels, their application in high temperature chloride or alkaline environment and hydrogen sulfide environment may put them to the risk of stress corrosion cracking. Cases of SCC have been observed in high temperature pressurized water in steam generating turbines. Incoloy 800

Table 9.46   Effect of alloying elements on the corrosion resistance of nickel

Alloy element	Contribution to corrosion resistance
Copper	Improves resistance to non-oxidizing acids, sulfuric acid (non aerated) and HF. Addition of 2–3% Ni offers improved resistance to HCl, H_2SO_4, and H_3PO_4.
Chromium	Improves resistance to oxidizing acids (HCl, H_2SO_4 and H_3PO_4) and high temperature oxidation.
Mo	Improves resistance to pitting and crevice corrosion. High Mo content (28%) show improved resistance to HCl, H_3PO_4, H_2SO_4 and HF.
Iron	Improves resistance to de-carburization. It has no role in the improvement of corrosion resistance.
Tungsten	Alloys with 3–4% W in combination with 13–16% Mo, offer excellent resistance to corrosion. Tungsten provides a high resistance to non-oxidizing acids.
Silicon	Improves resistance to hot concentration H_2SO_4 when added in larger amounts (9–11%). It is generally added in smaller quantities.
Cobalt	Increases the resistance to carburization, like iron.
Niobium and tantalum	Reduce hot cracking during welding.
Aluminum and titanium	The combination produces aluminum scale which resists oxidation and carburization.
Carbon and carbides	The formation of carbides weakens resistance to corrosion. Ni_3C may decompose to graphite and thus weaken the grain boundaries.

is a good choice to be used in such environments. A major class of nickel-base alloys are susceptible to SCC in alkaline environment (for example, NaOH at 350°C). Alloys 400, 600 and 800 may be subject to SCC in an alkaline environment. Alloys 800, 718 and 600 have shown failures in pressurized water reactors. Increasing chromium concentration to 30% (for example alloy 690) increases the resistance to SCC. The following factors promote SCC:

(a) Temperatures above 205°C.
(b) Low pH<4.
(c) Presence of H_2S and high level of stress. Despite the risk, nickel alloys offer minimum risks against SCC.

2)   Intergranular Corrosion

Nickel–iron–chromium (for example, alloy 800) and nickel–chromium–iron (alloy 600) may become sensitized by precipitation of chromium carbides and subject to intergranular corrosion

in a highly oxidizing environment. The carbide range in chemistry is from $Cr_{23}C_6$ in simple nickel alloys to $Cr_{21}(Mo, N)C_2C_6$ in alloys containing Mo and W. Alloys 600 and 800 may become susceptible to intergranular attack. Control of elements, such as phosphorus, carbon, nitrogen and niobium, and minimizing their segregation, reduces susceptibility to intergranular attack and minimizes the risk of SCC.

3)   Hydrogen Embrittlement

Like stainless steels, some nickel alloys may fail by hydrogen embrittlement. Inconel alloy X750 is reported to be susceptible to hydrogen embrittlement.

9.20.5   RESISTANCE TO AQUEOUS ENVIRONMENT

Nickel and its alloys are known to offer an outstanding resistance to corrosion in distilled water, fresh water, high purity water and in steam hot water systems. For example, Monel alloy 400 is extensively used in valves, pumps, propeller shafts, boiler feed water heaters and heat exchangers. Alloys 600 and 690 are used in nuclear steam generators to prevent SCC. Pitting of nickel–copper alloys may be caused by soft water.

The corrosion resistance of nickel alloys has been extensively explored in seawater and saltwater (brackish water). Although stainless steel 316 is known to resist pitting in seawater, stainless steels are, in general, susceptible to pitting in the tidal zones of seawater. The nickel alloys, more expensive than steels, have been extensively used in seawater service. Inconel alloy 625 offers an excellent resistance to corrosion in seawater. It also offers an excellent resistance to SCC. Nickel alloys are best used for pump shafts, bodies and impellers while other materials, like 90-10 Cu–Ni and austenitic steels are used for other parts, such as heat exchangers and valves. Table 9.47 shows the classification of selected nickel alloys in seawater service.

It is a practice to use chromium bearing alloys for oxidizing conditions, and nickel and its alloys for reducing conditions. Thus Monel is not used with HNO_3, because it creates an oxidizing environments. Alloy 690 containing 30% Cr is highly resistant to oxidizing environments, whereas alloy 625 containing 9% Mo offers a high resistance in reducing acids.

9.20.6 NICKEL IN SEAWATER

The average corrosion rate of copper in seawater is around 20 μm/year. Nickel is resistant to flowing seawater but susceptible to pitting in quiet seawater. The corrosion rates are higher in tropical zones compared to temperate zones. Ni–Cu, Ni–Cr–Fe, and Ni–Cr–Mo alloys have been used in marine service. Nickel–copper alloys, such as Monels, have been used in wear rings, impellers and shafts of seawater pumps. Alloys 600 (Inconel) and 400 (Monel) may become susceptible in a highly corrosive marine environment. High nickel stainless steels (904 L), 25 Ni–1.5 Cu, 20 Cr–4.5 Mo, have shown very high resistance to marine corrosion. The corrosion rates in Incoloy is below 20 μm/year at higher velocities (9–15 m/s) [45]. It shows a satisfactory rate of corrosion in hot seawater.

Nickel–chromium–iron alloys offer the best resistance to seawater corrosion and stress corrosion cracking. They are virtually free from pitting except in very severe conditions. Ni–Mo alloys are not suitable for use in marine environments. The silicon containing alloys show a low corrosion rate of less than 20 μm/year, but they are susceptible to pitting [45]. With the growth of desalination plants, attention of the designers is focused on the application of these alloys. The nickel alloys represent an excellent combination of high strength, high oxidation and high corrosion resistance and represent an improvement over the application range of steels.

Table 9.47 Classification of nickel alloys in seawater [44]

Alloy	Category	Remarks
Inconel 625	A	9% Mo
Hastelloy C 276	A	16% Mo
Hastelloy G	B	1204°C
Duranickel 301	B	Contains Al and Ti
Hastelloy F	B	Generally found satisfactory
Monel K-500	C	May develop pitting
Monel 400	C	May develop pitting

9.21 TITANIUM AND ITS ALLOYS

Pure titanium is a metal with a density of 4.51 g/cm³ (~60% of the density of iron) and capable of substantial strength by alloying and deformation process. Despite being very reactive, it has a strong tendency to form an oxide layer which passivates it and provides an excellent resistance to corrosion in oxidizing solutions and seawater application.

9.21.1 PHYSICAL METALLURGY OF TITANIUM ALLOYS

Titanium undergoes an allotropic transformation from close-packed hexagonal structure (CPH) α-phase to a body-centered cubic structure β-phase. On the basis of microstructural changes titanium alloys can be classified in four categories:

(a) α alloys   Example, Ti–0.3Mo–0.8 Ni
(b) β alloys   Example, Ti–10V–2Fe–3Al
(c) Near α Example, Ti–6Al–2Sn–4Zn–
     alloys   2Mo
(d) $\alpha+\beta$ alloys   Example, Ti–6Al–4V

9.21.2 EFFECT OF ALLOYING ELEMENTS

(1) Aluminum is added to control α-phase and increase the tensile and creep strength.
(2) Tin. It stabilizes the α-phase.
(3) Zirconium. Added in amounts of 5–6% to increase the strength.
(4) Molybdenum. It increases hardenability and stabilizes the β-phase.
(5) Silicon. It mainly added to improve the creep resistance.
(6) Niobium. It stabilizes the β-phase and increases resistance to high temperature oxidation.

9.21.3 MECHANICAL AND PHYSICAL PROPERTIES

Selected physical properties of titanium are given in Table 9.48.

The composition and mechanical properties of selected titanium alloys are shown in Table 9.49.

The mechanical properties of titanium alloys are controlled by microstructural changes brought about by alloying additions which form

Table 9.48 Selected mechanical and physical properties

Density	4.51 g/cm^3
Melting point	1668°C
Boiling point	3260°C
Thermal conductivity	9.0 Btu/h-ft^2-°F
Hardness	70–74 HRB
Modulus of elasticity	102.7 GPa
Electrical conductivity	3% IACS (copper)

intermetallic components and secondary phases. Alpha (α) and beta (β) are the primary phases. The modulus of elasticity varies with the alloy type ranging from 93 GPa to 120.5 GPa. The mechanical properties are predominantly influenced by the grain size, shapes and grain boundary geometry. The yield strength is increased by smaller grain size.

Each grade of titanium exhibits characteristic elevated temperature properties. The ceiling temperature of tensile strength of commercially pure titanium is 350°C, whereas the ceiling temperature of high strength aluminum alloys is in the range of 400–540°C. Proof strength curves for a number of commercial alloys show a distinct advantage of using titanium alloys in the high temperature ranges. The tensile strength of Ti 230 (Ti + 2.5 Cu) is twice the tensile strength of the hardest titanium 160 (Ti + 0.33% O_2) alloy in the temperature range 200–300°C. This alloy could be used in chemicals at elevated temperatures where unalloyed titanium is inadequate. Commercial titanium alloys offer a good resistance to elevated temperature creep.

Figure 9.18 shows the effect of temperatures on the tensile strength of titanium and its alloys [46].

9.21.4 ELECTROCHEMICAL FACTORS

The excellent resistance of Ti and its alloys to corrosion is attributed to the formation of highly adherent and protective oxide film, such as TiO_2, Ti_2O_3 and TiO. The electrochemical parameters

Table 9.49 Composition and mechanical properties of selected titanium alloys

Alloy designation	Tensile strength MPa	0.2% yield strength MPa	N	C	H	Fe	O	Al	Sn	Zr	Mo	Others
(1) Ti (Unalloyed) ASTM Grade 1	240	170	0.03	0.08	0.015	0.20	0.18	–	–	–	–	–
α and near-α alloy												
(2) Ti–0.3 Mo–0.8 Ni	480	380	0.03	0.10	0.015	0.30	0.25	–	–	–	0.3	0.80
(3) Ti–8 Al–1 Mo–1 V	900	830	0.05	0.08	0.015	0.30	0.12	8	–	–	1.0	1.0
(4) Ti–5.8 Al–4 Sn–3.5 Zr–0.7 Nb–0.5 Mo–0.35 Si	1030	910	0.03	0.08	0.006	0.05	0.15	5.8	4	3.5	0.5	0.7 Nb 0.35 Si
α–β alloys												
(5) Ti–6 Al–4 V	900	830	0.05	0.10	0.0125	0.3	0.20	6	–	–	–	4 V
(6) Ti–6 Al–2 Sn–4 Zr–6 Mo	1170	1100	0.04	0.04	0.0125	0.15	0.15	6	2	4	6	–
β–alloys												
(7) Ti–10 V–2 Fe–3 Al	1170	1100	0.05	0.05	0.05	0.015	2.5	0.16	3	–	–	10 V
(8) Ti–8 Mo–8 V–2 Fe–3 Al	1170	1100	0.03	0.05	0.05	2.5	0.017	0.17	3	–	8	8.0 V

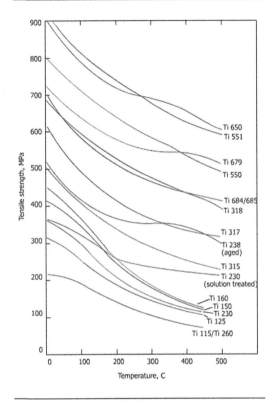

Figure 9.18 Tensile strength of titanium and its alloys (typical). (From I.M.I. (Titanium Information Bulletin), Witton Birmingham, UK)

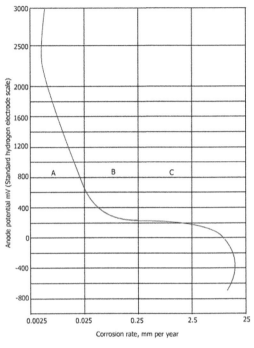

Figure 9.19 Corrosion rate *vs* anode potential for titanium in 40°C (wt) sulfuric acid at 60°C. (From I.M.I. (Titanium Information Bulletin), Witton Birmingham, UK)

primarily control the corrosion resistance of titanium and its alloys as summarized below.

(a) Suppression of anodic dissolution of titanium by addition of Ni, W or Mo.
(b) The strong tendency of titanium to passivate in an oxidizing environment makes it an unique material, shifting of the potential of the alloy to more positive value. In oxidizing conditions, titanium exhibits a strong resistance to corrosion because of the rapid formation of the protective film. In the reducing acids, such as HCl and H_2SO_4, the potential falls to negative values (active direction), because of the dissolution of the oxide films. Therefore, by raising the potential in the positive direction, the film is reformed. It can be achieved by coupling titanium with more noble (cathodic) metals, such as platinum, palladium and rhodium, or by anodic polarization. Figure 9.19 shows the anodic polarization of titanium in 40% H_2SO_4 at 60°C.

(c) Addition of 0.15% palladium to titanium decreases the corrosion rate by a factor of 500, which is significant.
(d) The passivation tendency of titanium can be improved by addition of Zr, Ta or molybdenum.

9.21.5 CORROSION BEHAVIOR OF TITANIUM IN SPECIFIC ENVIRONMENT

Service experience stretching over three decades in field studies and labs around the world has proved that titanium offers an excellent resistance to corrosion in polluted and clean seawater up to 130°C. It also shows an excellent resistance to corrosion in distilled water and tap water. It resists temperatures up to 400°C in an autoclave. It is believed that no other metallic material is more resistant to seawater corrosion than titanium.

The corrosion resistance of Ti and its alloys in to HCl, H_2SO_4, HNO_3 and H_3PO_4 is given below.

(a) **Hydrochloric acid**

Titanium can be safely used in 3% HCl at 60°C and 0.5% HCl at 100°C. It is only moderately resistant due to the non-oxidizing nature of HCl. Corrosion of titanium in HCl is given in Fig. 9.20.

(b) **Sulfuric acid**

The corrosion rate of titanium in H_2SO_4 follows a linear relationship up to 20% concentration (0.5 mm/year), followed by a peak at 40% (1.8 mmy^{-1}), a sharp drop at 60% (0.6 mm/year) and a significant rise at 80% (15 mm/year)(Fig. 9.11b). Presence of 0.25% $CuSO_4$ reduces the corrosion rates of titanium in 30% H_2SO_4 from higher rates to lower rates, such as 0.06 mm/year. Addition of 0.5% CrO_3 virtually blocks the corrosion in 30% H_2SO_4.

(c) **Phosphoric acid**

The corrosion behavior is similar to that in HCl. The corrosion rates vary from being low (0.003 mm/year in 5% H_3PO_4 at 35°C) to high corrosion rates (17.4 mm/year in 30% H_3PO_4 at 100°C). Titanium can be safely used at room temperature up to 30% concentration at 35°C (Fig. 9.21).

(d) **Nitric acid**

Due to the oxidizing nature of HNO_3, Ti exhibits low corrosion rates in HNO_3. As is to be expected, titanium is virtually unattacked

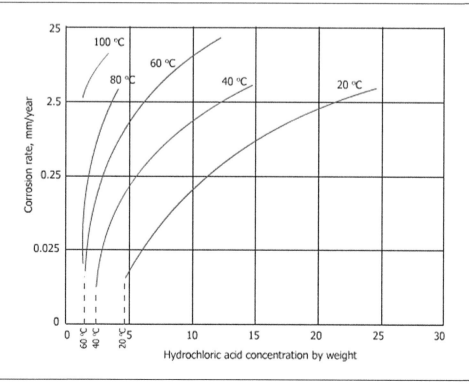

Figure 9.20 Corrosion of titanium in hydrochloric acid solutions (natural aeration). (From I.M.I. (Titanium Information Bulletin), Witton Birmingham, UK)

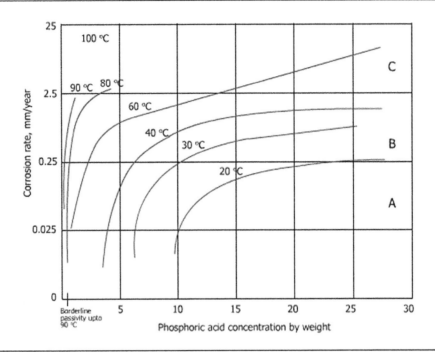

Figure 9.21 Corrosion of titanium in phosphoric acid solutions (natural aeration). (From I.M.I. (Titanium Information Bulletin), Witton Birmingham, UK)

in solutions of HNO_3 at temperatures up to the boiling point of HNO_3. Higher corrosion rates up to 10 mmy^{-1} may be observed in 20–70% HNO_3 in the temperature range 190–250°C. Addition of silica inhibits the corrosion of titanium in HNO_3 at elevated temperatures. Corrosion rates of titanium in HNO_3 are illustrated in Fig. 9.22.

Titanium is strongly resistant to attack by a mixture of concentrated H_2SO_4/HNO_3. The tendency of passivation is increased by adding oxidizing ions to the electrolyte which raise the potential to the passive region.

(e) **Chloride solution**

Ti exhibits an excellent resistance to corrosion in chloride solutions over a wide range of concentrations and temperatures. The corrosion resistance in $BaCl_2$, $CuCl_2$, $FeCl_3$, $MgCl_2$, $NiCl_3$, $SnCl_4$ is excellent. However, the corrosion resistance in $ZnCl_2$ and $CaCl_2$ and higher concentrations of $AlCl_3$ is poor.

(f) **Gaseous environment**

Exposure of titanium to higher oxygen contents are beneficial because of the protective oxide layer formation, however, it would depend on oxygen content and oxygen pressure. At a higher temperature (500°C) oxygen may be absorbed leading to oxygen embrittlement of titanium. Titanium is resistant to wet chlorine and unattacked by ammonia. It is also free from atmospheric attack by SO_2 and hydrogen sulfide.

9.21.6 LOCALIZED CORROSION OF TITANIUM

a) Galvanic Corrosion

A galvanic series of metals and alloys in flowing seawater is shown in Table 9.50. It is observed that titanium would be a cathodic member of the couple, and hence it would not undergo galvanic

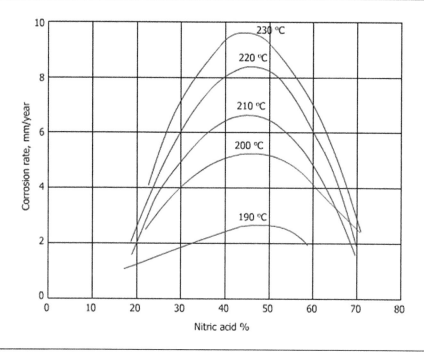

Figure 9.22 Corrosion rate of titanium in nitric acid: various concentrations and temperatures. (From I.M.I (Titanium Information Bulletin), Witton Birmingham, UK)

corrosion. It can be observed that Ti would be cathodic to naval brass, carbon steel, zinc and copper, and would not be subjected to galvanic corrosion. The extent of galvanic corrosion of the active metal would depend upon the relative anode/cathode area, the difference in the potential and the resistivity of metal functions.

b) Erosion

Titanium shows excellent resistance to erosion in flowing seawater up to 6 m/s. It is also not affected by abrasive particles present in the seawater. It also offers a high resistance to impingement corrosion.

c) Crevice Corrosion

The formation of differential aeration cells is the leading cause of crevice corrosion as described in Chapter 4. Usually titanium is resistant to

crevice corrosion and only in certain instances, it has suffered crevice corrosion, such as with wet chlorine. It was attributed to slow dehydration of the wet chlorine in crevices where there is a large metal area/gas volume ratio. When the moisture content fell below a certain value for passivation of titanium to take place, corrosion is initiated. Ti can also corrode under a gasket which keeps oxygen out such that the passive oxide film cannot be maintained. By and large, titanium is not susceptible to crevice corrosion.

d) Stress Corrosion

Unalloyed titanium does not suffer stress corrosion cracking if the oxygen content is less than 0.3%. However, certain aluminum alloys are susceptible to SCC in specified environments (Table 9.51).

Table 9.50 Galvanic series based on potential measurement in flowing seawater at 25°C

Metal	Steady-state potential, negative to a saturated calomel half-cell (volts)
18/8 stainless steel (passive)	0.08
Hastelloy C	0.08
Monel	0.08
Titanium	0.10
Silver	0.13
Inconel	0.17
Nickel	0.20
70/30 cupronickel	0.25
80/20 cupronickel	0.27
90/10 cupronickel	0.28
Admiralty brass	0.29
Gunmetal	0.31
85/15 gilding metal	0.33
Copper	0.36
63/37 brass	0.36
Naval brass	0.40
18/8 stainless steel (active)	0.53
Carbon steel	0.61
Cast iron	0.61
Aluminum	0.79
Zinc	1.03

The following are the major factors contributing to stress corrosion cracking susceptibility:

(1) **Aluminum contents.** The susceptibility to SCC increases with aluminum content ≥5 wt% (for example, Ti–10 V–2 Fe–3 Al and other β-alloys). The α-alloys have a greater tendency to form Ti_3Al intermetallic phase responsible for SCC with higher aluminum content. This is also true for α–β alloys.
(2) **Oxygen levels.** Oxygen levels above 0.25 wt% promote SCC in seawater.
(3) **Grain size.** Increased grain size increases susceptibility to SCC.
(4) **Addition of alloying elements to minimize SCC.** Tin and Al additions are not useful in

Table 9.51 Environment in which titanium and its alloys are susceptible to stress corrosion cracking

Alloy	Temperature	Contains
Ti–8 Al–1 Mo–1 V	RT	Distilled water
Ti–6 Al–4 V	RT	Red fuming HNO_3
Most alloys, except unalloy grades 1, 2, 7, 11 and 12	230–240°C	Hot chloride salts

α-alloys. Zirconium is beneficial. SCC tendency in α–β and β-alloys diminished by Mo, V, Nb and Ta additions.

Stress corrosion cracking often takes place in intergranular mode. The fracture mode in α-alloys is intergranular and transgranular in α–β and β-alloys.

9.21.7 APPLICATION OF TITANIUM AND ITS ALLOYS

Titanium is one of the most explored materials in industry. Because of its combination of good mechanical properties and an excellent resistance to corrosion in oxidizing environments, it has found wide acceptance in the petroleum, petrochemical, chemical and oil industries. Following is a brief summary of important applications:

(1) **Power generation.** The use of titanium in condenser tubes for turbine condensers is on the increase since the 1970s worldwide. Titanium tubes are used in desalination plants in the Gulf region.
(2) **Oil refineries.** Titanium tubing has been extensively specified where seawater and brackish water cooling is employed in refinery steam and overhead condensers. Seawater cooling of petroleum fractions at temperatures 110–120°C is not known to cause corrosion or erosion problems of

titanium seamless tubes and has excellent corrosion resistance.

(3) **Offshore production applications**. Titanium has been prescribed as a material for heat exchanger systems in oil production platforms. Titanium has been used in low-pressure crude coolers and auxiliary tubular coolers on production platforms.

(4) **Chemical industry**. Thick wall titanium tubes have been used in high-temperature, high-pressure, tubular heat exchangers for processes, such as manufacture of urea and for cooling sodium and potassium chloride brines up to 25% concentration. Titanium has an impressive success history in the chemical industry. Titanium has also been used as construction material in process plants and columns.

(5) **Desalination plants**. Thin walled (0.7 mm) titanium tubings have been used with success in multistage flash distillation despite tough competition with cupronickels because of the cost advantage offered by the later. The titanium tubes withstand brine temperatures up to 120°C and are free from impingement attack. Titanium tubes are standard materials in the Gulf countries for brine heaters, heat recovery and heat reject sections. Cost factors have limited their wide applications in desalination plants.

(6) **Marine applications**. Titanium has been successfully used in plate heat exchanger on ships. It has also served well in seawater cooled boiler feed water condenser where temperatures up to 130°C are encountered.

(7) **Biomedical applications**. It is extensively employed as a standard material for orthopaedic surgery, for example in hip implants.

(8) **Cryogenic applications**. Titanium alloys, like Ti–5 Al–2.5 Sn and Ti–6 Al–4 V, have shown promising application potential in cryogenic applications.

(9) **Sports industry**. Titanium tennis rackets and titanium sports bicycles are popular items because of their light weight and longer life in harsh environments.

(10) **Gas turbines**. Although on a limited scale, titanium has also been used in gas turbines, in the cool air compressor blade section.

At higher temperatures, Ti absorbs too much gas to be a high-temperature alloy.

To conclude, titanium, despite its diversified uses, suffers from economic competition because of its cost.

QUESTIONS

A. MULTIPLE CHOICE QUESTIONS

In the following questions mark one correct answer:

1. Which one of the following is not an important parameter in aggressive environment affecting the performance of the materials?

   a) Physical characteristics
   b) Strength and mechanical characteristics
   c) Corrosion and wear characteristics
   d) Fabrication parameter
   e) Recycling possibility

2. Cupronickel is an excellent material for application in

   a) potable water
   b) seawater
   c) construction industry
   d) oil drilling

3. Selection of materials is normally based on the performance according to

   a) functional requirements
   b) geometry
   c) properties
   d) low cost

4. Which one of the following material has a higher specific strength?

   a) AISI 8640
   b) Al-40% carbon
   c) Al-1060-O
   d) Titanium 6 Al–4 V

5. Which would be a preferable material to resists pitting in seawater?

   a) Titanium
   b) SS 403
   c) Al 6013
   d) Cast iron

6. Which one of the following indices is more suitable to tank corrosivity?

   a) Langelier Index
   b) Stability Index
   c) Aggressive Index
   d) Raynor Index

7. The iso-corrosion curves represent

   a) constant concentration links
   b) constant temperature lines
   c) curves showing constant corrosion rates
   d) curves where strong resistance to localized corrosion is expected

8. Which of the following are not very important in determining the properties of cast iron?

   a) Chemical composition
   b) Rate of cooling
   c) Type of graphite
   d) Physical properties

9. Which of the following does not contribute to the decreased corrosion rate of cast iron?

   a) Chromium
   b) Nickel
   c) Copper
   d) Magnesium

10. Which would you prefer to use in seawater?

    a) AISI 316
    b) AISI 430
    c) Cast iron
    d) Malleable iron

11. Bronzes are excellent materials for application in

    a) turbines
    b) propellers
    c) condenser tubes
    d) sewage pipes

12. Which of the following category of brasses would show the best corrosion resistance in seawater?

    a) Red brass (90 Cu–10 Zn)
    b) Alpha brass (70 Cu–28 Zn)
    c) Naval brass (62 Cu–36.7 Zn, 1.01 Sn, 0.23 Pb)
    d) Admiralty brass (72 Cu–26.8 Zn, 1.1 Sn, 0.03 Pb, 0.014P)

13. Alloy Al 2024 is a good material for application in

    a) high temperature applications
    b) kitchen equipment
    c) fasteners
    d) naval aircrafts

14. The major attribute of resistance of Ti alloys to seawater corrosion in oxidizing acid is

    a) tendency to resistant stress corrosion cracking
    b) tendency to resist intergranular corrosion
    c) tendency to passivate
    d) tendency to minimize galvanic corrosion

15. Titanium is attacked at an excessive

    a) non-oxidizing acids
    b) oxidizing acids

B. How and Why Questions

1. Consider equation

$$V = \left(\frac{E}{\rho} \right)^{1/2}$$

   For a fixed value of V. This equation plots E vs ρ as a straight line. What benefits does it allow?
2. What may be the general strain constraints for light, stiff and strong beams?
3. How may the recycling capability of materials affect their selection? State very briefly.

4. Why does a film of $Al(OH)_3$ promote passivation, and a film of $AlCl_3$ pitting in aluminum exposed to seawater?

5. Despite excellent corrosion characteristics of titanium, why is its use in heat exchanger tubes in salt water systems restricted?

6. What is the importance of constraint in material selection? Give three examples.

7. What are guidelines? What do they signify?

8. What is the effect of dissolved oxygen and pH on the corrosion behavior of low alloy steels?

9. What is the main reason for promotion of oxidation by vanadium?

10. Why are aluminum alloys not suitable for applications in acidic and alkaline environment?

C. Conceptual Questions

1. State briefly the difference between function, objective and constraint.

2. Derive a material index for light-stiff tie.

3. State the difference between high-strength, high-alloyed steels and ultra high-strength steels. Give examples for each.

4. State the advantages and disadvantages of ductile and cast iron pipes in an aggressive soil of low resistivity.

References

[1] Peddar, T. (2001). Steel vision for the future. *Materials World*, **9**, (12), 25.

[2] HPLL Technology (2001). Putting pressure on for improved steel forming. *Materials World*, **9**, (12), 26.

[3] Freeman, R. (2002) *CEO Freeman Technology*. Welland, Malvern Work, UK.

[4] Cebon, D. and Ashby, M.F. (1992). Computer aided materials selection for mechanical design. *Metals and Materials*, **25**, January, 9.

[5] Sulzuk, D. and Tajima, Y. (1983). The use of improved corrosion resistant aluminum alloy 6013 in the Navy P-7A aircraft. Report No. 900959/ALCOA.

[6] Snober, M.I. (1996). Impact of harsh environment on aircraft maintenance and operation, and the associated life cycle costs. *Proc. Symp. Maintenance Systems and Application of materials in the Saudi Arabian Environment*. KFUPM, Dhahran, Saudi Arabia, 2–3 November.

[7] Ahmad, Z., Paulette, P.T. and Abdul Aleem, B.J. (2000). Mechanism of localized corrosion of aluminum-silicon carbide composites in chloride containing environment. *J. of Mat. Sc.*, **35**, 2573.

[8] Christian, B.F., Siedman, D.N. and Dunard, D.C. (2003). Mechanical properties of Al (Sc, Zn) alloys at ambient and elevated temperatures. *Acta Materialia*, 4803–4814.

[9] Ahmad, Z., Hamid, A. and Abdul Aleem, B.J. (2001). The corrosion behaviour of scandium alloys Al 5052 in neutral sodium chloride solution. *Corr. Sc.*, **43**, 1227.

[10] Luster, J.W., Heath, G.R. and Kammer, P.A. (1995). Corrosion resistance of amorphous thermally sprayed coatings: Report. Research and Technology Center, Eutectic and Castolin Group, CH-1001, St. Sulpice, Lausanne, Switzerland.

[11] Singh, J. (1994). Laser beam and photon-assisted processes, and materials and their microstructure, *J. Mat. Sc.*, **28**, 5232.

[12] Rickerby, D.S. and Twancey, H.M. (1996). High coating technology for use in aero-gas turbine. *Proc. Symp. Maintenance Systems and Application of Materials in Saudi Arabian Environment*, KFUPM, Dhahran, Saudi Arabia, 2–3 November.

[13] AEPS (2001). Recycling and recognizing innovative steel packaging, *Materials World*, **4**, December.

[14] Horikawa, K. (1968). *J. Iron and Steel Institute*, **54**, 610, December.

[15] Kowaka, M. and Yamanaka, K. (1980). *J. Jpn. Inst. Metals*, **44**, 44 (in Japanese).

[16] Cebon, D., Ashby, M.F. and Lee, L. (2003). CES4, Cambridge Engineering Selection, Version 4, Design, Cambridge.

[17] Hagen, M. (2000). Corrosion of steels. In: Schutze, M. ed. *Corrosion and Environmental Degradation*. Vol. II. Weinheim, Germany: Wiley VCH, p. 1068.

[18] ASTM G48. Standard Method for Pitting and Crevice Corrosion Resistance of Stainless Steels and Related Alloys by use of Ferric Chloride Solution. Philadelphia: ASTM, USA.

[19] Avesta. *Avesta Information*. No. 9, 51, Avesta, Sweden.

[20]* Langelier, W.F. (1936). The analytical control of anti-corrosion water treatment. *JAWWA*, **28**,(10), 1500, New York, USA.

[21]* Ryznar, J.W. (1944). A new index for determining amount of calcium carbonate scale formed by water, *JAWWA*, **36**, New York, USA.

[22]* Raddick, T.M. (1944). The mechanism of corrosion of water pipes, water works and sewerage, Vol. 5, p. 133.

[23] Fontana, M.C. (1968). Corrosion by Sulfuric Acid. Corrosion Course Module 14, Metals Engineering Institute, ASM.

[24] Uhlig, H.H., Tradis, D. and Stern, M. (1955). *J. Electrochem. Soc.*, **102**, 59.

[25]* Whitman, W., Russel, R. and Altieri, V. (1924). *Ind. Eng. Chem.*, **16**, 665.

[26] OSW(1968). R and D Report, No. 3, 94.

[27] Yoshida, H. and Hagane, T. (1977). *J. Iron Steel.*, Inst. Of Japan, **63**, 1040, 1977.

[28] Ikeda, A.A., Ueda, M. and Mukai, S. (1984). *Advances in CO₂ Corrosion*. Texas: NACE, Houston, USA, p. 39. *Corrosion/83*, Paper No. 45, 1983.

[29] Masamura, K., Hashizumi, S., Nunomura, K., Sakai, J. and Matsus, I. (1984). *Advances in CO₂ Corrosion*, Texas: NACE, Houston, USA, p. 143.

[30] Baumel, A. (1986). *Werkst. u. Korr.*, **17**, 300.

[31] Kirby, G.N. (1978). Corrosion performance of carbon steel. *Chem. Eng.*, 72, March 10.

[32] Wittmoser, A. (1960). Giesserei Tech. Wiss, Beihefte, **29**, 1612, June.

[33] Roll, F. (1961). *Werkstoff Korr.*, **13**, 209–215.

[34] Collins, H.H., Fuller, A.G. and Harrison, J.T. (1968). *Proc. Conf. Corros. Prot. of Pipes and Pipelines.* London, June.

[35] Southwell, C.R. (1970). *Mater. Prot.*, **9**, 14–23, January.

[36] Kusaka, K. and Tsurumi, K. (1969). *J. Jpn. Inst. Met.*, **33**, 380 (in Japanese).

[37] INCO, Corrosion resistance of nickel containing alloys in hydroflouric acid, hydrogen fluorides, and flourines, Report 4156, UK.

[38] Katz, N. (1976). Dachema corrosion datasheet. Dachema, Frankfurt, Germany.

[39] White, J.H., Yanis, A.L. and Schide, H. (1966). *Corrosion Sc.*, **6**, 453.

[40] Burleigh, T.D., Ludwiczak, E. and Petri, R. (1995). ERRATA, *Corrosion*, **51**, (3), 248.

[41] Vernik, E. Jr., and George, P.F. (1973). *Material Performance*, **12**, 26.

[42] Legault, R.A. and Bettin, N.J. (1971). *Mat. Perf.*, **9**, March.

[43] Jacko, R.J. and Duquette, D.J. (1977). *Metallurgical Transactions*, **8A**, (11), 82.

[44] Schumacher, ed. (1979). *Seawater Corrosion Handbook*, Extracted from Table 32, p. 34.

[45] Niederberger, R.B., Ferrara, R.J. and Plummer, F.A. (1970). *Mater. Prot. Perf.*, **9**, (8), 18–22.

[46] Titanium Information Bulletin (1975). IMI, Birmingham, UK.

*(*Very old, but still very useful references.*)*

SUGGESTED READING

[47] Kutz, M. ed. (2002). *Handbook of Materials Selection*, New York: John Wiley & Sons Inc, USA.

[48] Davis, J.R. ed. (2001). *Alloying – Understanding the Basics*, Materials Park, Ohio: ASM International.

[49] Davis, M. and Scott, P.J.B. (2003). *Guide to use of Materials in Water*, Houston: NACE Inc.

KEYWORDS

Age hardening (precipitation hardening) Increase of hardness as a result of precipitation of a hard phase from a supersaturated solution.

Annealing Heating the steel from the austenitic range followed by slow cooling.

Anode A place where oxidation occurs forming electrons.

Anodizing Coating the surface with an oxide layer electrochemically. The specimen to be anodized is made the cathode connected to the positive terminal of a power source.

Austenite Face centered cubic iron (FCC).

Austenitic stainless steel A steel containing 11% chromium and containing mainly a γ-phase.

Brass An alloy of copper containing up to 40 wt% zinc.

Bronze An alloy of copper and tin, unless otherwise specified.

Cast iron Ferrous alloys containing more than 2 wt% carbon.

Cathode A place where the process of electrochemical reduction takes place.

C.C.T It is the critical crevice temperature for onset of crevice corrosion.

Constraint It is one of the parameter for design in which certain conditions are prescribed which must be met, such as load, pressure, temperature or a corrosive environment. These are non-negotiable conditions.

Corrosion Deterioration of a material in an environment.

Crevice corrosion Corrosion caused by formation of crevices in a structure. Corrosion in crevices is induced by differential aeration cells.

Critical stress intensity factor (K_{IC}). The stress at the root of the crack which is sufficient to cause crack propagation.

C.P.T. The critical pitting temperature, above which pitting occurs.

Engineering strain Elongation divided by the original length of the sample.

Engineering stress Load divided by the original cross-sectional area.

Geometric function It is one of the basic parameters of design which prescribes geometric parameters, such as specimen dimensions.

Guideline This is a family of lines in material selection charts. The guidelines of constants E/ρ, $E^{1/2}/\rho$ and E^{a}/ρ allow selection of materials for particular properties.

High strength low steel alloys (HSLA) Steels containing a total of less than 5 wt% alloy addition and exhibiting high strength.

Iso-corrosion lines Curve showing constant corrosion rates in a particular environment.

Langelier Index It is indicated by LSI and represents the scale forming tendency of water.

Lines They give the slope of the family of parallel lines belonging to a particular material index (M).

Malleable iron It is a type of cast iron with some ductility. It is obtained by heating white cast iron to change its iron carbide to clusters (rossettes) graphite.

Material index (M). The performance of material is defined by $P = f (F, G, M)$, where F is the functional requirement, G is geometrical parameter and M is material properties. The performance for all F and G is maximized by maximizing material properties which is called Material Index (merit index or material factor).

Passivity A material, such as steel, becomes passive if a protective layer of hematite or magnetite is built, which protects it from corrosion. For steel, this only occurs in certain environments.

pHs The pH at which seawater is saturated with calcium carbonate ($CaCO_3$).

Pitting factor The depth of deepest pit divided by the average penetration.

Pitting resistance equivalent (PRE) A good resistance to pitting is indicated in ferritic steels. For ferritic steels with >3% Mo, PRE = % Cr + 3.3 Mo + 16x (wt% N). Some authors use a factor of 3.3 and the factor 16 can vary from 12 to 30.

PRE (Pitting resistance equivalent of steel) a scale to determine the resistance of steel to pitting. See above.

Riddick Corrosion Index (RCI) This indicates the corrosivity of water based on numbers. For instance, RCI = 51−75 represents corrosive water and a number exceeding 100 represents highly corrosive water.

Scaling High temperature corrosion resulting in formation of thick corrosion product layers, such as deposition of insoluble materials on metal surface, usually inside water boilers or heat exchangers.

Stress intensity factor (K_I) Stress intensity at the root of a crack.

CHAPTER 10

ATMOSPHERIC CORROSION

10.1 INTRODUCTION

Atmospheric corrosion is the degradation of materials caused by air and the pollutants contained in the air. It can be precisely defined as an electrochemical process which depends upon the presence of electrolyte which may be rain, dew, humidity or melting snow. The usual electrolyte is water which is a universal solvent. Atmospheric corrosion takes place under humid conditions, where the atmospheric relative humidity exceeds the equilibrium relative humidity over any saturated solution which is present on the metal surface (usually NaCl solution, which is ubiquitous).

$$K = \sum_{t}^{n} t_n V_k(n)$$

where

K = accumulated corrosion products
t_n = time constant
V_k = average corrosion rate during the period of wetness.

The corrosion rate is controlled by time of wetness, temperature and the electrolyte composition. Atmospheric corrosion is most predominant of all the other forms of corrosion. The importance of atmospheric corrosion is exemplified by the fact that the cost of protection against atmospheric corrosion is about 50% of the total cost of all other corrosion measures. No other form of corrosion affects the materials and equipment harder than atmospheric corrosion. Its devastating range extends from small articles, such as bolts, nuts and fasteners to industrial plants and equipment and megastructures, such as skyscrapers, towers and bridges. Atmospheric corrosion has not only its own share of engineering catastrophes, such as collapse of concrete structure and bridges, which occur from time to time, but also affects the public utilities and military combat readiness. It is, therefore, important for engineers to understand the relation between the material and the environment to incorporate control measures in the design.

10.2 CLASSIFICATION OF ATMOSPHERIC CORROSION

Atmospheric corrosion can be classified in the following categories:

(a) Dry corrosion.
(b) Damp corrosion.
(c) Wet corrosion.

10.2.1 DRY CORROSION

In the absence of significant water vapor, many common metals develop films of oxide. In the presence of traces of gaseous pollutants, copper, silver and other non-ferrous metals undergo film formation which is known as *tarnishing*. The tarnishing of silver in air is well-known. Tarnishing by hydrogen sulfide may be retarded by moisture if present in very small amounts.

10.2.2 DAMP CORROSION

Damp corrosion would occur only when the relative humidity reaches 70% which is considered generally as the critical value for the onset of corrosion. The precise level of critical humidity varies with the type of contaminants, such as dust and salt particles, and the composition of metals. For instance, in the presence of marine salts corrosion is stimulated at lower values of relative humidity. The difference between the damp and wet environment is very narrow and it is more representative of a climatic condition rather than the magnitude of corrosion.

Damp environments promote the corrosion of most metals. Water saturated with dissolved gases, such as CO_2, H_2S and SO_2, cause severe corrosion of iron and steels, copper, nickel, silver and other non-metallic materials and alloys. For instance, silver loses its luster and develops a tarnished film of sulfide on coming in contact with H_2S, and copper develops tints and becomes black. In agricultural areas abundance of ammonia, particularly during the rainy seasons, subjects copper fittings to seasonal cracking and causes serious damage to water distribution systems.

10.2.3 WET CORROSION

This is the most frequently observed form of atmospheric corrosion, where the water layers or pockets are formed on the metal surface, and the metal surface remains constantly in contact with water. The rate of corrosion would depend on the solubility of the corrosion product. Higher solubility means a higher rate of corrosion, because the dissolved ions increase the electrolytic conductivity. In case of alternate dry and wet conditions, the dry corrosion product film may absorb moisture from the air which increases the rate of corrosion of the metal by bringing the moisture in contact with the metal surface. Patina formation on copper, such as brochantite, and corrosion of iron and steel structures are common examples of corrosion caused by wet atmosphere.

10.3 WEATHER AND ATMOSPHERE

Weather may be defined as the total of the atmospheric conditions at a given place and time. Atmospheres may be divided into four categories:

(a) Rural
(b) Urban
(c) Industrial
(d) Marine.

10.3.1 CLIMATIC ZONES

Broadly speaking, there are three major climatic zones – frigid, temperate and tropical-humid. The characteristics of each zone with examples are shown in Table 10.1.

10.3.2 AGGRESSIVITY OF THE ATMOSPHERE

The aggressivity of atmospheres in different climatic zones is shown in Table 10.2.

10.3.3 MAJOR POLLUTANTS

The major atmospheric pollutants are SO_2, H_2S, NO_x (oxides of nitrogen), CO_2, dust particles and marine salts. The sources of the contaminants are shown in Table 10.3.

Common air pollutants are SO_2 and NaCl. The concentration of major pollutants is shown in Table 10.4.

The corrosion rate of selected materials in different atmospheres is shown in Table 10.5.

Table 10.1 Climatic zones

Category	Characteristics	Example
Frigid	Extremely cold	Alaska
Temperate	Moderate (persistently high)	Australia
Tropical-humid	Humid	South East Asia

Table 10.2 Aggressivity of atmosphere

Climatic zones	Category	Corrosion aggressivity in atmosphere			
		Rural	Urban	Industrial	Marine
Arid or frigid	Outdoor	2	3	4	3
	Indoor	1	2	3	2
Temperate	Outdoor	3	4	5	4
	Indoor	2	3	4	3
Tropical-humid	Outdoor	4	5	5	5
	Indoor	3	4	4	4

Scale of aggressivity: 1 Very slow   4 Severe
                          2 Slight   5 Very strong
                          3 Medium

Table 10.3 Sources of contaminants in atmosphere

Molecule	Major source of pollutant	Natural source
SO_2	Combustion of coal and oil	Volcanoes
H_2S	Chemical processes, sewage treatment	Biological processes, decaying of vegetable malt
CO	Automotive emission and combustion process	Forest fires
NO_x	Combustion	Lightning
NH_3	Sewage treatment	Biological decay
Hydrocarbon	Chemical process and automotive exhaust	Biological process
CO_2	Combustion	Biological decay, release from ocean
HCl	Chemical industries	

Table 10.4 Major air pollutants in atmosphere

Types of atmosphere	Air pollutants	
	SO_2 (in g/m² day)	NaCl (in mg/m² day)
Rural	<0.01	<0.3
Urban	0.01−0.1	<0.3
Industrial	>0.01	0.3−2000
Marine	>0.01	>2000

Table 10.5 Corrosion rate of selected materials in different atmospheres

Metals	Rate of attack (in μm/year) in atmosphere			
	Rural	Urban	Industrial	Marine
Pb	0.7−1.4	1.3−2	1.8−3.7	1.8
Cd	−	2−15	15−30	−
Cu	1.9	1.5−2.9	3.2−4	3.8
Ni	1.1	2.4	4−5.8	2.8
Zn	1−3.4	1−6	3.8−19	2.4−60
Steel	4−60	30−70	40−160	64−230

It is to be observed that maximum damage to metals occur in industrial and marine environments.

10.4 MAJOR FACTORS AFFECTING ATMOSPHERIC CORROSION

The following factors largely affect the corrosion of metals in an atmosphere:

10.4.1 TIME OF WETNESS

It is the time during which the metal surface remains covered with water before inducing corrosion. At a sufficient thickness of water layer, an emf is generated which exceeds a critical value to induce corrosion.

The time of wetness of a metal surface is dependent upon:

- Temperature: a high temperature leads to decreased adsorption.
- Porosity voids: the larger the number of voids, the greater the volume of water adsorption.
- Degree of oxidation: lesser adsorption with higher oxidation.
- Grain boundaries: More adsorption with a larger number of grain boundaries.
- Nature of surface: more adsorption on a rough surface.

More often, the time during which the relative humidity exceeds 80% and the temperature stays above 32°C, is taken as a limit for time of wetness. Typical wetness values in temperate zones are 100–2700 h/year. In South East Asia (tropic-humid zone), the values are much higher (3000–5000 h/year).

The amount of water adsorbed depends upon hydrophobic and hydrophilic properties.

10.4.2 WATER ADSORPTION

Water may be adsorbed in the molecular or dissociated form on a metallic surface. The oxygen

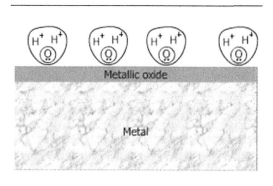

Figure 10.1 Water adsorption on a metal surface

atoms bond to the substrate and donate an electron pair (Fig. 10.1), there is a net transfer of charge from the water molecule to the substrate. Water can also bond in a dissociated form by metal-oxygen, or metal-hydrogen (OH) bond. The film of hydroxyl groups is fairly protective. The aqueous phase formed on the metal surface acts as a solvent for gaseous constituents of the atmosphere. The dissolution of corrosive gaseous species in the adsorbed layer provides sites for promotion of corrosion. When the thickness of the adsorbed layer reaches three monolayers, the properties of the adsorbed layer approach those of the bulk water and the relative humidity approaches close to 'critical humidity.' Above the critical humidity the rate of corrosion increases significantly and below, it is virtually insignificant. The critical relative humidity level is pushed to lower values in the presence of SO_2 and other atmospheric contaminants.

10.5 EFFECT OF ATMOSPHERIC FACTORS ON ATMOSPHERIC CORROSION

10.5.1 RELATIVE HUMIDITY

This is the primary driving force for atmospheric corrosion to occur. Atmospheric corrosion does not occur in dry air. The relative humidity is expressed as

$$RH = \frac{\text{Amount of water vapor in air}}{\text{Amount of water vapor required to saturate the air}}$$

The corrosion rates of metals increases sharply beyond a threshold level of relative humidity called 'critical relative humidity.' The level of the critical humidity varies with the nature of the metal and the type of contaminant. If the atmosphere is clear and uncontaminated, corrosion is negligible at a relative humidity as high as 99%. However, in the presence of contaminants, corrosion begins to increases around 80% *RH*. The critical humidity requirement in the presence of contaminants, such as KCl and NaCl is considerably reduced. For example, steel can corrode even at 35% *RH*, in the marine environment. At a relative humidity of 55%, a surface film of 15 molecules thick is formed on mild steel, which increases to a 90 molecular layer as the relative humidity increases to 100%, causing an acceleration of the corrosion process.

10.5.2 PH

Decreased pH on a metallic substrate caused by dissolution of certain contaminants like SO_4^{2-} may lead to an acidification process and acceleration of atmospheric corrosion.

10.5.3 DEW

Dew is more corrosive than rain water because of a higher concentration of atmospheric contaminants, hygroscopic salts and a lower pH value.

10.5.4 FOG

Due to the low pH of fog water (1.8–3.5) in highly contaminated regions, like the cities of Lahore and Karachi in Pakistan, fog is a corrosion promoter.

10.5.5 DUST AND SOOT

Dust has an abrasive effect on metallic surface in combination with wind velocity in desert and on shore regions, like the sea coastal environment of the Gulf countries. Impurities from emissions, such as CO_2 and CO are adsorbed in dust particles and create micro-corrosion cells. Dust particles being hygroscopic promote serious corrosion by adsorption of salts promoting corrosion (Fig. 10.2). A list of air pollutants are given in Table 10.6.

Soot particles are efficient sites for hydrogen evolution because of low hydrogen over-potential of carbon. Corrosion is localized in areas with high moisture content. Soot particles may be combined with chromium, nickel or manganese. Soot may also have toxic molecules containing cancer causing agents. Charcoal particles induce

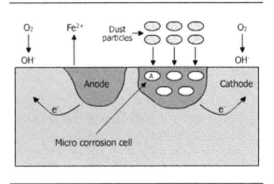

Figure 10.2 Formation of micro-corrosion cells by adsorption of dust particles

Table 10.6 Major natural and man-made sources of gaseous and particulate pollutants (worldwide)

Contaminant	Quantity evolved (10^6 metric tons)	
	Major pollutant source	Important natural source
SO_2	212	20
H_2O (org sulfide)	3	894
CO	700	2100
NO_x	75	180
N_2O	3	340
CH_4	160	1050

severe corrosion in the presence of small amounts of SO_2 because charcoal has excellent adsorption properties.

10.5.6 WIND VELOCITY

Wind velocity affects the concentration profile of salt particles in a particular area. It is responsible for the transport of pollutants which may be deposited on a metallic substrate. It may also dislodge a protective layer formed on a metal surface and promote corrosion. Wind velocity, wind direction and structural geometry are major contributors in areas subjected to strong winds.

10.5.7 CORROSION PRODUCTS

The corrosion products formed on a metallic surface (for example, Fe_2O_3, Fe_3O_4) may induce or retard atmospheric corrosion depending on the thickness, homogeneity and degree of bonding and solubility of the film. Breakdown of the protective films by corrosion-inducing ions, such as chloride ions, destroys passivity and accelerates the rate of atmospheric corrosion. The adsorption of chloride is shown in Fig. 10.3.

10.5.8 POLLUTANTS

Outdoor pollutants, like SO_2, H_2S, CO_2, HNO_3, NH_3, HCl and HCOOH, are corrosion

Table 10.7 Outdoor ranges of selected pollutants

Constituents	Range (ppm)
H_2O_2	10–30
SO_2	1–65
H_2S	0.7–24
NO_2	9–78
NH_3	7–16
HCl	0.18–30
HCOOH	4–20

promoters. They act synergistically to accelerate corrosion with humidity, temperature, wind and weather cycles. Table 10.7 shows approximate outdoor range of selected pollutants.

10.5.9 DISTANCE FROM THE SOURCE

Distance from the contamination source is an important factor in atmospheric corrosion. For instance, the fall out of salts may vary from a high level of 0.34–0.45 kg/m^2/year in the coastal areas and islands (e.g. Pacific Islands), to a low level of 5.62×10^{-4} kg/m^2/year in inland areas. Proximity to sea accelerates the corrosion rate due to abundance of sea salts. At a distance inland of 10 km or more, the corrosion rate is the same as observed further inland.

10.6 POLLUTANTS CHEMISTRY AND THEIR IMPACT ON ATMOSPHERIC CORROSION

10.6.1 CHLORINE CONTAINING COMPOUNDS

These include Cl_2 and HCl. For instance, the atmospheric corrosion rate of steel is accelerated at a critical humidity level of 70%. The value

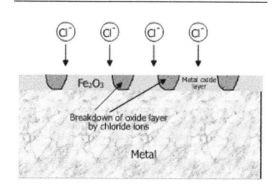

Figure 10.3 Breakdown of oxide layer by adsorption of chloride ions

Table 10.8 Selected hygroscopic salts

Salts in solution	Saturation relative humidity at 20°C (%)
Na_2SO_4	98
$(NH_4)_2SO_4$	81
NaCl	76
$CaCl_2 \cdot 6H_2O$	32
$ZnCl_2$	80

of critical relative humidity varies with the metals and the pollutants. A hygroscopic salt absorbing water from the atmosphere can produce an electrolyte which decreases the saturation relative humidity figure. Such salts are called '*hygroscopics*,' and are abundantly present in the environment. Table 10.8 shows some hygroscopic salts in equilibrium with saturated solution. Saturated solutions have an equilibrium relative humidity. If the humidity exceeds this critical value, the saturated salt solution takes up water until equilibrium is achieved and conversely the salt solution releases water if the humidity level falls below the critical value. The number of monolayers of adsorbed water formed on the substrate increases with decrease in the saturation relative humidity figure and pushes the relative humidity to critical levels for onset of corrosion. For instance, on increasing the relative humidity from 2 to 80%, the number of monolayers may increase one to five.

10.6.2 RAIN

Rain plays a dual role, increasing the rate of atmospheric corrosion or decreasing it under certain conditions.

(1) **Increase**

(a) Rain promotes thicker layers of electrolyte on the substrate. In the presence of atmospheric pollutants, like SO_2 in the air, it may wash corrosion promoters, like H^+ and SO_4^{2-} from the atmosphere and wet deposit them in the presence of electrolyte layer, thus accelerating corrosion.

(b) Degree of contamination. In a less contaminated atmosphere, rain has a pronounced effect and promotes corrosion.

(c) Orientation and design. Orientation of the structure partially exposed to rain and containing water traps would contribute to increased corrosion.

(d) Nature of substrate. The action of rain would be subject to the nature of the substrate, porous, smooth, rough, etc. A smooth surface would not hold an aqueous layer for a great length of time.

(2) **Decrease**

(a) Period of dry deposition. If the period of dry deposition is followed by rain, then this would decrease the degree of corrosion by washing out the deposits.

(b) Structure and design. Structures facing the atmosphere would receive maximum rain and be subjected to maximum washing out of deposits. Similarly, good designs, devoid of water traps would also deaccelerate atmospheric corrosion.

(c) Droplets or crystals of chloride are formed in the atmosphere in a marine environment. Their concentration decreases with increasing distance from the sea. Burning of coal is a major source of emission of HCl. Chlorine dissociates to form a chlorine radical which reacts with organic compounds to form hydrochloric acid

$$Cl_2 + h\lambda \xrightarrow[\text{radiation}]{\text{solar}} Cl : Cl \ (\lambda \leq 430 \text{ nm})$$

$$RH + Cl' \rightarrow R + HCl$$

where,

$\qquad h\lambda$ = solar radiation
$\qquad \lambda$ = wave length
$\qquad$ RH is organic compound
$\qquad$ and Cl' is the chlorine radical.

The adsorption of chloride and destruction of a protective layer is shown in Fig. 10.4. In the

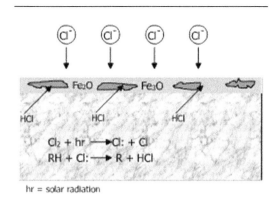

Figure 10.4 Adsorption of Cl_2 on a rust surface

presence of chloride, iron and steel corrode at significantly lower humidity levels.

10.6.3 OXIDES OF NITROGEN

Nitrogen and oxygen are present in the air and virtually no reaction occurs. These gases, however, react significantly at combustion temperature

$$NO_2 + O_2 \rightarrow 2NO_2$$
$$2NO + O_2 \rightarrow 2NO_2$$

Only 10% of the NO_2 emitted is present as NO_2. There is a small amount of nitrogen dioxide (NO_2) present in the atmosphere, some of which originates from the bacterial action on nitrogen compound in soil. NO_2 is a good absorber of light and undergoes the following reaction:

$$NO_2 + h\lambda \rightarrow NO + O^{\cdot}$$
$$O^{\cdot} + O_2 \rightarrow O_3 \text{ (Ozone)}$$

Ozone plays a predominant part in atmospheric chemistry, in particular, the ozone built-up in an environment. The effect of oxides of nitrogen on atmospheric corrosion are not well understood.

10.6.4 OXIDES OF CARBON

Sources of CO are mainly oxidation of methane in marshes (>6%), oceans (3.9%) and from other natural resources (9.1%). There are three distinct situations which produce CO.

$$2C + O_2 \rightarrow 2CO \tag{1}$$

$$2CO + O_2 \rightarrow 2CO_2 \tag{2}$$

The first reaction is ten times faster than the second. The available oxygen is not enough for combustion.

$$CO_2 + C \rightarrow 2CO \tag{3}$$

Carbon dioxide reacts with carbon-containing metals at high temperature and acts as a reducing agent. At high temperature CO_2 may dissociate to form CO and O_2.

$$2CO_2 \rightarrow 2CO + O_2$$

Carbon dioxide occurs in the atmosphere in concentrations of 0.03–0.05% by volume, which varies with time of day, and seasons of the year. Carbon dioxide is soluble in water (1.45 g CO_2 per liter) which makes it corrosive in the dissolved form

$$CO_2 + H_2O \rightarrow H_2CO_3$$

H^+ and CO_3 ions are released which stimulate corrosion.

$$CO_2(g) \rightarrow CO_2(aq) \xrightarrow{H_2O} 2H^+ + CO_3^{2-}$$

The corrosion rate is increased with an increase in CO_2 pressure.

10.6.5 HYDROGEN SULFIDE

Hydrogen sulfide is generally a minor gaseous constituent produced by petrification of organic sulfur compounds, the action of sulfate-reducing bacteria, fossil fuels, combustion and petroleum refining. H_2S in a dry atmosphere is only weakly corrosive. It dissociates into H^+ and HS^-, the later being a corrosive species.

$$H_2S(g) \rightarrow H_2S(aq) \rightarrow H^+ + HS^-$$

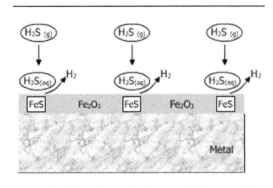

Figure 10.5 Adsorption of H_2S on a rust surface

The total global emissions of H_2S are estimated to be 70–80 Gg-S-yr^{-1} (S stands for sulfur). Corrosion induced by H_2S is a major risk in oil field equipment. The adsorption of H_2S on a metallic surface is shown in Fig. 10.5.

10.6.6 OXIDES OF SULFUR

During the combustion of sulfur-containing materials, both SO_2 and SO_3 are formed. The amount of SO_3 is relatively independent of oxygen and varies between 1 and 10% of sulfur consumed. A significant proportion of SO_2 is absorbed into aerosol particles where it is oxidized to SO_4.

$$SO_2(g) \rightarrow SO_2(aq) \Rightarrow SO_4^{2-} \text{ (Acidification)}$$

SO_2 reacts with OH^- ions on the surfaces where it is absorbed and leads to the formation of H_2SO_4.

$$SO_2(g) + OH^- \rightarrow HSO_3^- \Rightarrow H_2SO_4$$

The SO_4 ion is released again when H_2SO_4 dissolves in water. SO_2 is the main contributor to atmospheric corrosion. Little rusting appears to take place in the absence of SO_2. One of the best examples of how metallic objects are protected in the absence of SO_2 is the good state of preservation of the famous Iron Pillar at Qutub Minar, in Delhi, India, which is 1600-years old.

10.7 MECHANISM OF ATMOSPHERIC CORROSION (ELECTROCHEMICAL CYCLE)

The adsorption of water provides the electrolyte required for the formation of a corrosion cell on a metallic substrate. The rusting of steel or iron is accelerated by water and corrosion proceeds instantly after wetting. Over a period of time the water layers increase in thickness and with adsorption of salts from the atmosphere, they assume the full properties of conductive electrolyte. The primary cathodic reaction which occurs is

$$0.5O_2 + H_2O + 2e \rightarrow 2OH^- \qquad (10.1)$$

As a consequence of this reaction, a local increase in pH occurs causing the precipitation of corrosion products at some distance away from the anode. The redox potential are in the order of 0.3–0.5 V_{SHE}. If pH is below 4, the oxyhydroxide may be reduced to Fe^{2+}. Such a situation may arise in rains or fogs. Increased concentration of salts in an aqueous layer may also reduce pH. As a result of formation of corrosion products, the oxygen supply becomes restricted and the cathodic reaction slows down. The reduction of oxygen becomes the rate controlling step. The charge balance is maintained by the generation of Fe^{2+} by the anodic reaction. The generation of Fe^{2+} may occur in matter of one hour. As soon as Fe^{2+} is generated, a variety of oxides and hydroxides, such as Fe_2O_3, Fe_3O_4, $Fe(OH)_2$, $Fe(OH)_3$ and α-FeOOH, may be formed. The formation of $Fe(OH)_2$ and $Fe(OH)_3$ and release of Fe^{2+} ions is shown in Fig. 10.6. Natural rust contains a wide variety of oxides and hydroxides of iron. The stable species of FeOOH exist in several different forms (α, β and σ). The rust generally contains one inner layer of amorphous FeOOH and a crystalline Fe_3O_4 outer layer. However, this is beyond the discussion of the atmospheric corrosion mechanism. One important point to note is that the intermediate corrosion products also contain both Fe^{2+} and Fe^{3+} species. The oxidation of Fe^{2+} to Fe^{3+} may be generated by hydrogen peroxide (H_2O_2), hydroxyl

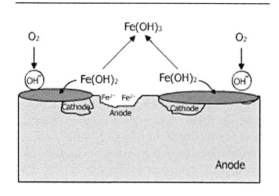

Figure 10.6 Formation of rust layers

radical (HOE), hydroperoxyl radical (HO$_2$) or super oxide radical ion (O$_2^-$). Oxidation takes place by H$_2$O$_2$ species at night as the other oxidizing species are only formed during the sunlight. The following are the reactions:

$$Fe^{2+} + HO_2^- + H_2O \rightarrow Fe^{3+} + H_2O_2 + OH^- \tag{10.2}$$

$$Fe^{2+} + O_2^- + 2H_2O \rightarrow Fe^{3+} + H_2O_2 + 2OH^- \tag{10.3}$$

where E is radical.
And also

$$Fe^{2+} + H_2O_2 \rightarrow Fe^{3+} + OH^\cdot + OH^- \tag{10.4}$$

Another cathodic process takes on the oxide covered surface:

$$Fe^{3+} + e \rightarrow Fe^{2+} \tag{10.5}$$

This reaction, however, takes place only in the absence of oxygen as the oxygen supply is markedly inhibited in the presence of corrosion products on the metallic surface. The cathodic reaction reaction (10.5) has the slowest rate, hence, it is the rate controlling cathodic reaction. It is followed by reduction of the rust formed to magnetite.

$$8FeOOH + Fe^{2+}(aq) + 2e \rightarrow 3Fe_3O_4 + 4H_2O \tag{10.6}$$

It normally takes place under wet conditions. Upon oxyhydrolysis, Fe$_3$O$_4$ reverts to rust (FeOOH) in the presence of oxygen and water.

$$3Fe_3O_4 + 0.75O_2 + 4.5H_2O \rightarrow 9FeOOH \tag{10.7}$$

The freshly formed rust is not protective. The rust layer remains attached to the metallic surface.

10.7.1 THE ROLE OF SULFUR DIOXIDE

Sulfur dioxide may be oxidized in different ways as shown in Fig. 10.7a. The absorption of SO$_2$ in an aqueous layer leads to its acidification due to the formation of SO$_4^{2-}$ ion as shown in the Fig. 10.7b. A low pH results at the site where SO$_4^{2-}$ ions are formed which prevents the precipitation of protective hydroxide. This creates conditions favorable for corrosion to proceed. A reservoir of soluble sulfate nests surrounded by a hydroxide layer are formed at the anodic site on the rusted surface. The sulfate nest

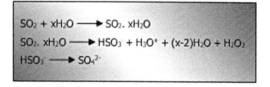

Figure 10.7a Oxidation of sulfur compounds

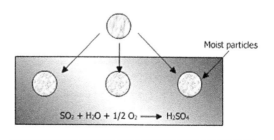

Figure 10.7b Oxidation of SO$_2$ in presence of moist particles and formation of H$_2$SO$_4$

grows by a diffusion mechanism. The sulfate nest acts as a storehouse for the supply of SO_4^{2-}. As SO_4^{2-} ions activity decreases, the corrosion rate becomes independent of the SO_4^- ion and reaches a value very close to the rate observed in the absence of SO_2.

In sulfate-containing solution, the following reactions occur:

$$Fe + H_2O \rightarrow Fe(OH)_{ads} + H^+ + e \quad (10.8)$$

$$Fe(OH)_{ads} + H_2O \rightarrow Fe(OH)_{2ads} + H^+ + e \quad (10.9)$$

$$Fe(OH)_{2ads} + SO_4^{2+} \rightarrow FeSO_4 + 2OH^- \quad (10.10)$$

$$FeSO_4 \rightarrow Fe^{2+} + SO_4^{2-} \quad (10.11)$$

The magnetite produced by cathodic reduction of FeOOH (equation (10.6)) is reverted to fresh rust (FeOOH) by oxyhydrolysis as shown in equation (10.7). Consequently, there is a net gain of one molecule provided by release of one atom of iron in the anodic area (equations (10.6) and (10.7)). Corrosion would only stop when SO_4^{2-} is removed or unless fresh FeSO$_4$ is not formed, such as in dry weather.

The formation of SO_4^{2-} in the rust leads to the formation of corrosion cells at the rust FeSO$_4$ (sulfate nests) interface. Corrosion would continue to take place as long as the supply of SO_4^{2-} is abundant [1]. A simplified diagram showing the contribution of an electrochemical cycle to atmospheric corrosion is shown in Fig. 10.8. The formation of FeSO$_4$ nests is illustrated in Fig. 10.9. To maintain corrosion, the corrosion cell requires an electronically conducting path which is provided by ferrous sulfate. Corrosion would slow down if the resistance of either of the paths is increased.

10.7.2 THE ACID REGENERATION THEORY

According to this theory [2,3], the sulfuric acid formed by oxidation of SO_2 absorbed in the rust

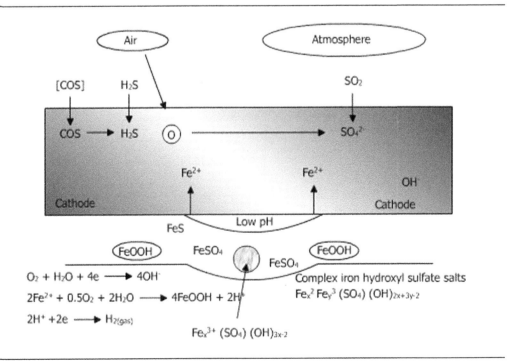

Figure 10.8 Corrosion of iron and steel electrochemical cycle

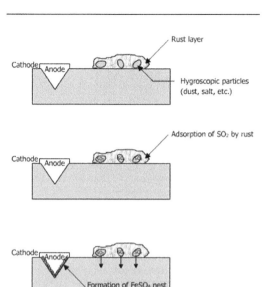

Figure 10.9 $FeSO_4$ nest formation

layer attacks steel in the presence of oxygen to form ferrous sulfate ($FeSO_4$).

$$4H_2SO_4 + 4Fe + 2O_2 \rightarrow 4FeSO_4 + 4H_2O$$

$$(10.12)$$

The sulfuric acid is reformed by oxidative hydrolysis

$$2FeSO_4 + \frac{1}{2}O_2 + 3H_2O \rightarrow 2FeOOH + 2H_2SO_4$$

$$(10.13)$$

The production of $FeSO_4$ provides the electrolyte necessary to carry the electrical current and leads to the dissolution of the protective Fe_2O_3 or Fe_3O_4 film on iron and steel surface. One SO_2 ion can cause dissolution of 100 atoms of iron.

The cycle repeats again and again and is very slow compared to the electrochemical cycle. The cycle occurs during the initial stages of atmospheric corrosion. The acidification process leads to more and more corrosion of iron and generation of $FeSO_4$. Once ferrous sulfate and rust is formed, conditions become favorable for the electrochemical cycle described above. The acid generation process is very slow requiring days

for completion depending on the environmental conditions, whereas the electrochemical process may repeat several times in one hour. The acid regeneration process is illustrated in Fig. 10.10. For more details of the mechanism, the reader may refer to reference [4].

10.8 POTENTIAL–pH DIAGRAM (FE–H₂O)

The potential–pH diagram for the Fe–H$_2$O system is shown in Fig. 10.11. The redox potentials in the ocean and fresh water systems are in the range of 0.3–0.5 V_{SHE}. More positive potentials are observed for acidic fogs. Soluble species of Fe^{2+} are formed at pH value around 4 and below. Such pH values are found in concentrated electrolytes. FeS is a stable corrosion product in the pH range of 3–6. The fog and dew regimes are shown in Fig. 10.11. FeOOH ($Fe_2O_3 \cdot H_2O$) is the predominant stable species except under conditions of acidification caused by sulfate. The pH potential diagram can, however, be used only under equilibrium conditions.

10.9 ATMOSPHERIC CORROSION OF COPPER AND ITS ALLOYS

Copper and copper alloys are well-known for their excellent resistance to corrosion in corroding environments. Copper is one of the earliest metal known to mankind and it was the most exploited material because of its ease of extraction from the ore. Copper alloy artifacts have been found in excellent conditions after having been buried for thousands of years. Copper roofing in rural areas have been found to corrode at a rate less than 0.4 mm after 200 years. Several statues of bronze and brass stood for hundreds of years before being destroyed by environmental pollutants. Copper artifacts and copper roofs, hundreds of years old, in historical buildings and castles are still in good state of preservation.

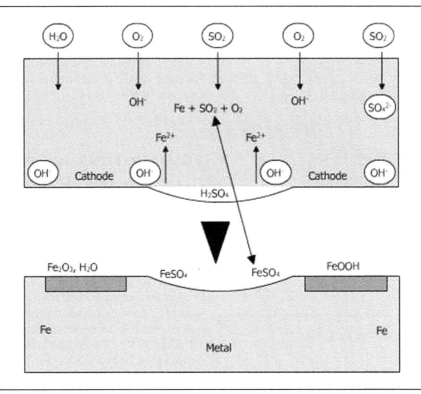

Figure 10.10 Acid regeneration cycle

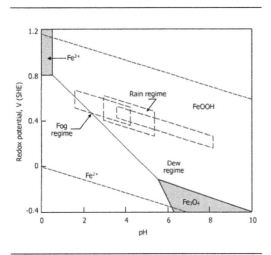

Figure 10.11 Potential–pH diagram for the Fe–H_2O system at 25°C, for a concentration of iron ionic species of 0.1 M. The approximate regimes for fog and rain are indicated

10.9.1 ENVIRONMENTAL EXPOSURE

On exposure to the atmosphere, copper and copper alloys are known to form a thin layer of greenish blue or brownish green corrosion product, called patina. It has initially a dark color which turns eventually green with a greater exposure period. The composition of patina is very complex. The time scale of patina formation varies with environment.

The following are the major copper alloys of interest in corrosion engineering:

(a) **Brasses.** These are basically copper–zinc alloys containing up to 43% zinc. The brasses have high ductility and they can be cold worked. Cartridge brass (70/30 brass) is the most commonly used brass and is used where high ductility is required.

(b) **Bronzes**. Copper–tin alloys are known as tin bronze. Alpha brasses contains up to 9% tin at room temperature and they can be cold worked.

For details of copper alloys, refer to Chapter 9 on Materials Selection.

The primary components of patina are copper oxide and salts of copper, such as:

Basic sulfate: $Cu(OH)_{1.5}(SO_4)_{0.25}$, $Cu(OH)_{1.33}$ $(SO_4)_{0.33}$
Basic copper chlorides: $Cu(OH)_{1.5}Cl_{0.5}$
Basic copper carbonate: $CuOH(CO_3)_{0.5}$, $Cu(OH)_{1.33}(CO_3)_{0.33}$, $Cu(OH)_{0.67}(CO_3)_{0.67}$

Basic sulfate occurs mainly in the urban atmosphere and basic chloride in a marine environment.

The basic ingredients of corrosion products are oxides Cu_2O and CuO which are highly protective. They react with the environmental pollutants to form salts.

$$Cu_2O + 0.5SO_2 + 1.5H_2O + 0.75O_2$$

$$\rightarrow 2Cu(OH)_{1.5}(SO_4)_{0.25}$$

The most common ingredient of patinas in urban areas is the basic sulfate salt $Cu_4(SO_4)(OH)_6$ called brochantite. A list of constituents found on the corroded copper surface are given in Table 10.9.

Table 10.9 Constituents found on corroded surface of copper

Substance	Formula
Copper	Cu
Cuprite	Cu_2O
Copper hydroxide	$Cu(OH)_2$
Chalcolite	Cu_2S
Chalcanthite	$CuSO_4 \cdot 5H_2O$
Antlerite	$Cu_3SO_4(OH)_4$
Atacamite	$Cu_2Cl(OH)_2$
Malachite	$Cu_2(CO_3)(OH)_2$
Azurite	$Cu_3(CO_3)_2(OH)_2$

10.9.2 REQUIREMENTS FOR PATINA FORMATION

The formation of brochantite in patina requires four ingredients, copper ions, an aqueous layer on copper substrate, a source of sulfur and an oxidizer which is summarized below:

(1) Copper ions (Cu^{++}): These are produced by oxidation of copper by SO_2 in the atmosphere.
(2) Aqueous layer: Several monolayers of water may be adsorbed on copper at moderate to high humidity even in the absence of precipitation. Water monolayers ranging from 5 nm to 10 nm thickness may be formed on clean copper exposed to 60–90% RH.
(3) Sulfur: Oxidized sulfur species are present in copper minerals.
(4) Oxidizer: An atmospheric gas or a product of precipitates.

10.9.3 ROLE OF SULFUR IN ATMOSPHERIC CORROSION OF COPPER

Sulfur as (SO_2) is the major source of oxidized and reduced sulfur species. SO_2 is absorbed into aerosol particles where it is oxidized to SO_4^{2-}. H_2S and carbonyl sulfide (COS) are the examples of reduced sulfur species.

Sulfur dioxide stimulates oxidation. It is the major driving force for corrosion in metropolitan areas. Most of the sulfur acquired by surface is not in the form of gas but as dry deposition. In an urban atmosphere, SO_4^{2-} is abundantly found in aerosol particles. Large particles containing ammonia are also found. H_2S, SO_2 and COS in all these participate directly in the corrosion process. The sulfur compound, COS, hydrolyzes to form H_2S and it may form Cu_2S if the quantity of COS is abundant; on the other hand, SO_2 may hydrolyze to form a bisulfate ion.

$$SO_2(g) + OH^- \rightarrow HSO_3^-$$

which is converted to SO_4^{2-} ion by either ozone or transition metal ions which may be present on the films on copper[5].

$$HSO_3^- \xrightarrow[\text{Oxidation}]{O_3, H_2O_2} SO_4^{2-}$$

Once SO_4^{2-} is formed and enters the aqueous layers, $Cu_4(SO_4)(OH)_6$ (brochantite) is formed, which is the major component of patina. On sea coasts, patinas also contain Cl^- containing species. The metal loss of copper in a marine atmosphere is around 600–700 $\mu g/cm^2$ per year. Other products formed on patinas may be posnjakite $[Cu_4(SO_4)(OH)_6H_2O]$ and malachite $[Cu_2(CO_3)(OH)_2]$.

The chemistry of patina formation in copper is highly complex. With the limits in the existing knowledge and the complexity of the process it is not easy to completely understand the patina formation mechanism.

10.10 Cracking of Copper Alloys

Ammonia and ammonium compounds are the major species associated with stress corrosion cracking (SCC) of copper alloys. Ammonia is present in certain atmospheres, such as agricultural areas where fertilizers are used. Ammonia is also present in chemicals used to treat boiler water. Water layers on metals dissolve significant quantities of ammonia from the air. In the presence of continual applied and residual stresses, ideal conditions are created for the onset of seasonal cracking. Applied and residual stresses both lead to failure and in many instances failures have been noted at stresses lower than the yield stress. Applied stresses result from fabricating techniques whereas residual stresses arise from thermomechanical processes during heating or cooling. The formation of cupric ammonium complex and the copper alloy surface is essential for the onset of stress corrosion cracking.

$$Cu(NH_3)_4^{2+} + e \rightarrow Cu(NH_3)_2^+ + 2NH_3$$

The complex provides a cathodic reduction reaction. Adsorbed layers on the copper alloy surface dissolves significant quantities of ammonia even at very low concentrations and stimulates the corrosion process by formation of complex $Cu(NH_3)_2^+$ ions which are adsorbed on the surface. The cupric ammonia complex is the major contributor to seasonal cracking. The defect structure along the grain boundaries of copper alloys in a stressed condition may favor adsorption resulting in weakening of metal bond strength and lead to cracking under repeated cycles of stress.

10.10.1 SEASONAL CRACKING OF ALPHA BRASS

Alpha brass (70/30) when subjected to a stress in the presence of oxygen and ammonia, cracks occur along the grain boundaries either in an intergranular form or in a transgranular form. Either type of cracks are known as seasonal cracking. Traces of nitrogen dioxide may also cause cracking, yellow brackets in humidifying chambers may fail. The term seasonal cracking has a historical significance as the phenomenon was frequently observed in India during the monsoon rainy season, and it was termed as seasonal cracking. Naturally, there is always an excess of ammonia during rainy seasons in rural areas.

Mechanisms proposed are conflicting. Four steps are basically involved in the process:

(1) Segregation on the grain boundaries and formation of complexes with ammonia.
(2) Creation of anodic and cathodic areas induced by electrochemical processes.
(3) Plastic deformation and by accelerating crack growth by applied stress.
(4) Propagation of the crack resulting in failure.

Although the susceptibility to SCC is decreased by annealing the brasses, they are still sensitive to seasonal cracking in ammonia contaminated environments, particularly the agricultural land. Literature is abundant on the subject and the readers may refer to some of the references cited at the end of the chapter.

The SCC may be minimized by annealing brasses and preventing them from exposure to ammonia containing environment. This may

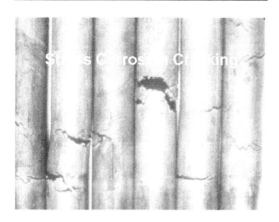

Figure 10.12 Seasonal cracking of brass

not be, however, practical in several instances. The seasonal cracking of brass is shown in Fig. 10.12.

10.11 ALUMINUM AND ITS ALLOYS

Aluminum and its alloys have been extensively used for structural applications with success. Aluminum resists corrosion from the atmosphere if there is an absence of narrow crevices. Many statues erected, over a hundreds of years ago, have not deteriorated badly which is in contrast with aluminum cables used in seawater. The corrosion resistance of aluminum is due to its tendency to form a compact oxide layer over the surface. The oxide formed offers a high resistance to corrosion. The normal surface film present in air is about 1 nm thick. The film thickness increases at the elevated temperature. The film growth is more rapid in water than in oxygen.

10.11.1 CORROSION BEHAVIOR OF ALUMINUM IN ATMOSPHERE

From the extensive weight-loss studies conducted around the world on aluminum corrosion

rates of 0.03 to 4 μm/year have been reported. The corrosivity of a location would depend upon the distance the airborne salt travels, the direction and velocity of the wind, the frequency of the prevailing wind and topography of the coast and the expanse of seawater over which the wind has come. The salt content drops rapidly with increasing distance from the sea.

Loss of tensile strength can occur due to atmospheric corrosion. In rural area's atmosphere, the corrosion rate averages 0.03 μm/year (0.001 mils/year). In industrial locations, the corrosion rates average 0.8–0.28 μm/year (0.03–0.11 mils/year). In certain polluted environments, a higher corrosion rate of 13 μm year (0.52 mils/year) was obtained. Table 10.10 shows the corrosion rates of aluminum and selected aluminum alloys in different environments.

Casting aluminum alloys containing Si and Mg show a good resistance to corrosion in all atmospheres, whereas alloys containing copper show a poor resistance in industrial and marine environments. Aluminum alloys of Series 5xxx (Al–Mg), and 6xxx (Al–Mg–Si) show a good resistance to atmospheric corrosion. The composition of selected Al alloys is shown in Table 10.11. Alloys of Al 6061, 6013, 5052 and 5054 can be used in the atmosphere without any serious risk of corrosion. Alloys of series 2xxx and 7000, such as Al 2024 and Al 7075 are susceptible to intergranular exfoliation and stress corrosion cracking in marine environments. The ranking of aluminum alloys is shown in Table 10.12.

Table 10.10 Corrosion rates of aluminum and its alloys in different environments

Environment	Material	Corrosion resistance
Rural	Al sheets	Excellent
	Al–Mn	
Marine	(compound)	Good. Poor
	Al–Mn, Al–Cu,	resistance
	Al–Zn–Mg	
Industrial	Sheets of	Becomes black.
	common	After few years,
	purity	loss of strength

Table 10.11 Nominal compositions of selected aluminum alloys

Alloy type	Si	Fe	Cu	Mn	Mg	Cr	Zn	Ti	Balance
Al 6061	0.6	–	0.27	1.0	1.0	0.3	0.25	0.1	Al
Al 6013	0.6	0.5	1.0	0.8	0.8	0.1	0.25	0.1	Al
Al 5052	0.25	0.40	1.0	0.8	0.8	0.1	0.25	0.1	Al
Al 7050	0.12	0.15	2.0–2.6	0.10	1.9–2.6	0.04	5.7–6.7	0.06	Al
Al 2024	0.5	0.5	3.8–4.9	0.3–0.9	1.2–1.8	0.10	0.25	0.15	Al

Table 10.12 Ranking of aluminum alloys in different atmospheres

Alloy	Corrosion resistance
1100	Excellent
2xxx	Poor resistance in marine environment
5xxx	Very good resistance in all environments
6xxx	Very good resistance in all environments
7xxx	Poor resistance in marine environment

10.12 POTENTIAL–PH DIAGRAM (AL–H$_2$O)

A potential–pH diagram is shown in Fig. 10.13. It shows the equilibrium phases for the aluminum water system at different pH levels and potentials. Line a in the diagram is a hydrogen line below which water is not stable and decomposes into hydrogen (H^+) and (OH^-), causing alkalization. Line b is the oxygen line above which water decomposes into oxygen and H^+ causing acidification. The region between a and b is the region of water stability. Under conditions of low pH values (1–4), aluminum dissolves as Al^{3+} and at higher pH values (9–15) aluminum dissolves as AlO_2^-. In the neutral pH range a protective passive film of $Al(OH)_3$ is formed which protects aluminum.

$$Al^{+3} + 3H_2O \rightarrow 3H^+ + Al(OH)_3$$

The passive region in the potential–pH diagram is clearly shown in the location between the two corrosion regions. In the diagram, the concentration of ionic species ranging from saturated solution (10^0) to very dilute solution (10^{-6}), are shown. Roughly 1 ppm of ionic species is in contact with the solid phase, in very dilute solutions.

Using the Pourbaix diagram it is possible to explain the corrosion tendencies of aluminum. For instance, in the absence of chloride ion aluminum would be passive in fresh water in the passivity range between pH 4 and 8. The Pourbaix diagram shows that an $Al_2O_3 \cdot 2H_2O$ film causes passivity of aluminum. The diagram, however, provides no information on the rate of corrosion of aluminum and does not guarantee that corrosion would not occur in the neutral pH range as flaws on metals, crevices and pits can cause corrosion by developing a different pH within these regions.

Nevertheless the potential–pH diagrams are very useful to predict the corrosion tendencies in atmosphere and environments. The protective oxides films formed prevent the outdoor atmosphere corrosion. Films, such as boehmite (γ-AlOOH), bayerite [α-Al(OH)$_3$] and hydragillite [γ-Al(OH)$_3$] contain varying amounts of water. These films are amorphous. The crystalline films found on corroded aluminum surfaces are shown in Table 10.13.

A variety of basic salts may be formed by deposition of SO_4^{2-} or Cl^- on the protective film formed on aluminum and stimulate corrosion. In an atmosphere containing SO_2, the corrosion rate would depend on how a low pH is produced. The protective film would dissolve in the region of low pH. The migration of Cl^- species from the environment may alter the electrode potential of the protective film and lead to low values. This may happen during pitting, whereas the H_2 or oxygen reduction occurs at the pit mouth or at the cathodic areas. The passivity of the film is destroyed by alkalization which stimulates the cathodic reduction reaction leading to

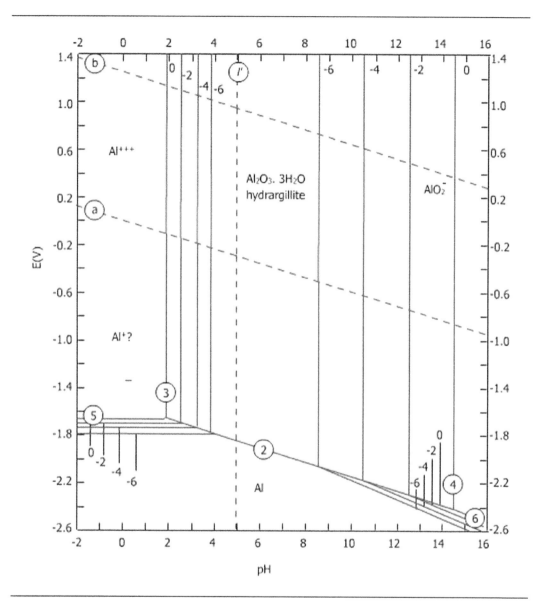

Figure 10.13 The Pourbaix diagram for aluminum and water illustrates the stable phase for the different potentials and pHs. The numbers −2, −4, etc. mean ion concentration of 10^{-2}, 10^{-4}, etc. in gram ions per liter (or kg ions per cubic meter)

the dissolution of the protective layer on the onset of corrosion.

The above example illustrates how the pH potential charts can be used to predict atmospheric corrosion. However, the effect of localized corrosion cannot be predicted from these charts which puts a limitation on the application of these charts.

10.12.1 FACTORS CONTRIBUTING TO THE CORROSION OF ALUMINUM

(a) **Absorbed water layer**. In a moderately humid atmosphere, ten monolayers of water are present on the aluminum substrate.

Table 10.13 Films formed on corroded aluminum surface

Substance	Crystal system	Formula
Aluminum oxide	Cubic	Al_2O_3
Boehmite	Orthorhombic	γ-AlOOH
Gibbsite	Monoclinic	γ-Al(OH)$_3$
Bayerite	Monoclinic	α-Al(OH)$_3$
Aluminite	Monoclinic	$Al_x(SO_4)_y(G_2O)_2$
Aluminum chloride	Monoclinic	$Al_2(SO_4)(OH)_4 \cdot$ $7(H_2O)\,AlCl_3$

The aqueous layers provide the conductive medium for mobilization of aluminum cations and adsorbed anions.

(b) **Chloride levels**. A high level of chloride in a marine–industrial environment, such as 30–40 mg/m²/day, leads to a high density of pitting of aluminum and its alloys. Incorporation of chlorides into the aqueous layer on an aluminum substrate from deposition of sea salts, aerosol particles or from an organic gas containing chlorine may lead to the formation of $AlCl_3$ or $Al(OH)_2Cl$, which are soluble in weak acidic solution and do not offer resistance to aluminum dissolution.

(c) **Sulfides and sulfates**. In aqueous layers, SO_2 dissolves and ionizes to produce HSO_3^- ion.

$$SO_2(g) \rightarrow SO_2(aq) + H_2O \rightarrow H^+ + HSO_3^-$$

Once SO_4^{2-} is released, it is readily available to form poorly soluble basic aluminum hydroxy sulfate $Al_x(SO_4)_y(OH)_2$. Aluminum reacts by anodic dissolution $(Al \rightarrow Al^{3+} + 3e)$ when the aqueous layer on the substrate becomes acidic. Sulfur compounds, such as COS and H_2S do not contribute significantly to aluminum corrosion like in case of iron and steel. The rate of dissolution of aluminum by SO_4^{2-} is higher compared to Cl^-, if its concentration is higher, which is generally the case in outdoor atmospheres.

(d) **Nature of oxide layer**. The thinning down of the protective oxide film by aggressive ions, such as in chloride or sulfate solutions leads to its breakdown and onset of localized corrosion under acidic conditions. Boehmite [γ-AlOOH or Al(OH)$_3 \cdot H_2O$] and Bayerite [Al(OH)$_3 \cdot 3H_2O$] are the major protective films found on an aluminum surface. Bayerite ($Al_2O_3 \cdot 3H_2O$) is very stable at a wide range of pH as shown by pH potential diagrams and is responsible for the outstanding corrosion resistance of aluminum.

A simplified picture of the atmospheric corrosion of aluminum is shown in Figs. 10.14a and b.

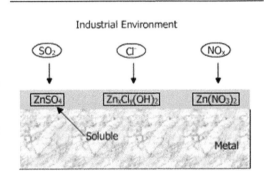

Figure 10.14a Corrosion of zinc in industrial atmosphere

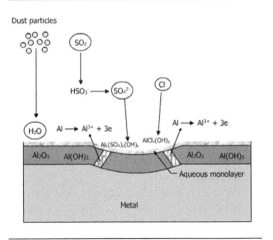

Figure 10.14b A simplified representation of mechanics of atmospheric corrosion of aluminum

10.13 ATMOSPHERIC CORROSION OF ZINC

Zinc is one of the most important tools used by engineers to control the corrosion of steel in the atmosphere. Zinc coated steels are used in general building, constructions in automobile bodies, in hardware, containers, tubes, pipe fittings, wires, ropes and hundreds of other applications. Zinc is little susceptible to corrosion even at high humidity. At 99% relative humidity the atmospheric corrosion of zinc amounts to $0.007\,g/m^2/day$. But a white hygroscopic corrosion product called wet-storage stain can form on new zinc wet surfaces unless these have been pre-stored for several weeks in low humidity air.

10.13.1 FACTORS INFLUENCING THE CORROSION OF ZINC

(a) **The duration and frequency of moisture content.** Corrosion due to condensed water is heavy if the formation of stable protective layers is impaired by reduced take up of CO_2 from the atmosphere which is the case with an insufficient access of air, a prolonged covering with condensed water and a contamination of the atmosphere. Zinc owes its corrosion resistance to the formation of an insoluble basic carbonate film.
(b) **The rate of drying of surface.** The rate of drying is an important factor, since a moisture film with higher oxygen concentration promotes corrosion. In sheltered area, the drying time is low and the corrosion rate of zinc is accelerated.
(c) **The magnitude of industrial pollution.** Fig. 10.15 shows corrosion of zinc by atmospheric pollutants.

In highly industrial environment, a $610\,g/m^2$ zinc coating would rust after 4 years and may be 80% rusted in 10 years, whereas, a similar coating would display rusting in an urban atmosphere between 30–40 years and 15–25 years in marine environment. Soot has an indirect effect because

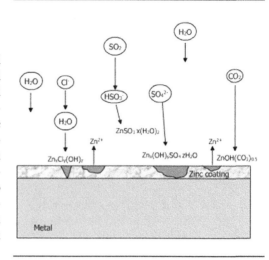

Figure 10.15 Corrosion of zinc by atmospheric pollutants

it takes up SO_2 and SO_3 from the fumes of sulfurous fuels on account of its large surface and has an abrasive effect. The corrosion rate of zinc in different environments is shown in Table 10.14. The dust contents in air affect the atmospheric corrosion of zinc. Typical values of dust contents in different environments are shown in Table 10.15. The atmospheric influences should be seen in combination with one another.

Table 10.14 Corrosion rate of zinc in different environments

Type of environment	Corrosion rate
Rural	$0.2–3\,\mu m/year$
Urban	$2–16\,\mu m/year$
Industrial	$2–16\,\mu m/year$
Marine	$0.5–8\,\mu m/year$

Table 10.15 Average dust contents in air

Type of air	Dust contents
Rural air	$0.06–0.1\,mg/m^3$
Urban air	$0.3–2.5\,mg/m^3$
Industrial air	$1–100\,mg/m^3$
Technical fuels	$100–1500\,mg/m^3$

10.13.2 INFLUENCE OF ATMOSPHERIC VARIABLE ON THE CORROSION OF ZINC

The protection offered by zinc depends upon the type of film formed on its surface which, in turn, depends on the type of atmosphere – rural, urban, industrial or marine atmosphere. Zinc oxide is the initial film which is formed in the presence of moisture or water. It is transformed into zinc hydroxide in an outdoor environment.

$$2Zn + H_2O + 0.5O_2 \rightarrow 2Zn(OH)_2$$

Zinc hydroxide is the naturally occurring species. If the pH of the moisture is sufficiently high, the zinc hydroxide formed reacts with the pollutants present to form the corresponding basic zinc salts [6].

$$Zn(OH)_2 + 0.5CO_2 + H^+$$
$$\rightarrow ZnOH(CO_3)_{0.5} + H_2O$$
$$Zn(OH)_2 + 0.25SO_2 + 0.25O_2$$
$$\rightarrow Zn(OH)_{1.5}(SO_4)_{0.25} + 0.5H_2O$$
$$Zn(OH)_2 + 0.6Cl + 0.6H^+$$
$$\rightarrow Zn(OH)_{1.4}Cl_{0.6} + 0.6H_2O$$

In the carbonate type layer, $ZnCO_3$ is most abundant. Several compounds have been reported for the basic carbonate films. The most common compounds are: $ZnCO_3$ and $Zn_5(CO_3)_2(OH)_6$. The stability domain of carbonates depends on the H_2CO_3 content and atmospheric concentration of dissolved CO_2.

Hydrozincate $[Zn_5(CO_3)_2(OH_6)]$ compound is abundantly found in protective rust layers of zinc. It is formed by a combination of $Zn(OH)_2$ and zinc carbonate.

$$3Zn(OH)_2 + 2Zn(CO_3)_3 \rightarrow Zn_5(CO_3)_2(OH)_6 \downarrow$$

The zinc hydroxide and the basic zinc salts form patina which protects the zinc surface from further attacks. Sulfur dioxide is the most harmful pollutant in the atmosphere. The presence of 0.1% SO_2 in a polluted atmosphere causes a marked increase in the rate of corrosion as the basic salts, such as $ZnOH(CO_3)_{0.5}$ and $Zn(OH)_{1.5}$ $(SO_4)_{0.25}$, formed earlier may be dissolved.

The increase in the corrosion rate is attributed to the formation of $ZnSO_4$.

$$Zn + SO_2 + O_2 \rightarrow ZnSO_4$$

The formation of basic zinc sulfate leads to increased corrosion (Fig. 10.13). $ZnSO_4$ dissolves the protective hydroxide layers and protective salts [6]. The formation of $ZnSO_4$ is shown below:

$$Zn(OH)_2 + SO_2 + 0.5O_2$$
$$\rightarrow ZnSO_4 + H_2O$$
$$ZnOH(CO_3)_{0.5} + SO_2 + O_2 + 2H^+$$
$$\rightarrow ZnSO_4 + 1.5H_2O + 0.5CO_2$$

Zinc sulfate is soluble in water and it may be washed by rain producing an increase in the corrosion rate.

In an environment containing SO_2, galvanized steel corrodes 10–20 times more slowly than unprotected steel. Zinc offers a good resistance to corrosion in urban, industrial or marine environments.

10.13.3 CORROSION RATE OF ZINC IN DIFFERENT ATMOSPHERE

(a) **Urban atmosphere.** The average corrosion rate of zinc corresponds to a material loss of 2–7 μm/year. Corrosion is more in winter months because of large SO_2 content and relative humidity of air.

(b) **Industrial atmosphere.** The average corrosion rate is about 3–20 μm/year. Fuel gases produced by combustion of coal, mineral oil and other fuels contain CO, CO_2, water vapor and organic acids besides SO_2. Chlorine, hydrogen chloride, ammonia, nitrogen gases, unburned hydrocarbons are found near chemical industries. Hence the corrosion of galvanized steel is accelerated in these environments. Major corrosion products formed in industrial atmospheres are: $ZnOH(CO_3)_{0.5}$, $Zn(OH)_{1.5}$ $(SO_4)_{0.25}$ and $ZnSO_4$. Figure 10.15 shows the corrosion of zinc by atmospheric pollution.

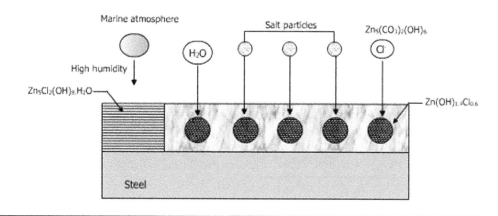

Figure 10.16 Factors contributing to corrosion of zinc coated steel in marine atmosphere

(c) **Marine atmosphere**. In marine environments, mixed salts of $Zn_5(OH)_8Cl_2$ are deposited on the zinc surface and not protective. In a highly polluted atmosphere $[Zn_5Cl_2(OH)_8·H_2O]$ called 'Simonkollite' is formed. The corrosion of zinc coated steel in a marine atmosphere is shown in Fig. 10.16.

High humidity and high salt contents increases the rate of corrosion. The average rate of corrosion of zinc is 1–7 μm/year in coastal areas. Corrosion is also intensified by erosion from sand particles. In tropical climates, zinc corrodes only a little unless high humidity causes wet storage stain on a new zinc surface (mentioned earlier). The average corrosion rate of galvanized steel is 2 μm in humid tropical atmosphere to about 6 μm in a tropical sea atmosphere.

Assuming a zinc coating thickness of 80–100 μm, the corrosion protection would amount to:

10–12 years in industrial atmosphere
20–22 years in marine climates
50 years in rural areas.

These figures are based on world-wide basis. A zinc coating of 700–800 g/m^2 is expected to have a service life of more than 30 years if the structure is located at a distance of 200 m from the coast. At a distance of 50 m, the life is reduced to 12–15 years. A simple illustration of corrosion protection offered by e-galvanized steel is shown in Fig. 10.17.

10.13.4 SUMMARY OF ENVIRONMENTAL FACTORS CONTRIBUTING TO THE CORROSION OF ZINC

The rate of corrosion of zinc is primarily influenced by the time of wetness, and the presence of pollutants, such as SO_x, Cl^- and CO_2 in the air. Corrosion products, such as $Zn(OH)_2$, $Zn_5(CO_3)_2(OH)_6$ and $ZnSO_4$, are predominantly observed. Zinc hydroxide and basic zinc salts react to form a patina which protects the surface from corrosion attack. Atmospheric pollutants, such as SO_2, lead to an increase in acidification of the electrolyte and dissolution of the protective films.

QUESTIONS

A. MULTIPLE CHOICE QUESTIONS

Circle the correct answer for the following statements:

1. The following is an example of dry corrosion.

   a) Rusting of bronze in air
   b) Water saturated with CO_2

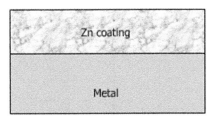

Carbon steel is coated by Zinc

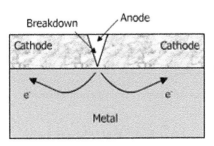

Break in coating by H_2O, CO_2, SO_2.

$Zn \longrightarrow Zn^{2+} + 2e$

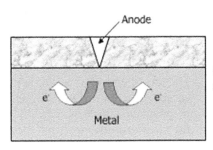

Release of electrons from zinc to steel leads to the protection of steel

Figure 10.17 Corrosion protection of E-galvanized steel

c) Corrosion of iron in an atmosphere of SO_2

d) Corrosion of iron in a salt solution

2. Which one of the following two major contaminants are present in an industrial atmosphere?

a) SO_2
b) NO/NO_2
c) Dust
d) Ammonium sulfate

3. In atmospheric corrosion caused by SO_2, corrosion is accelerated mainly by the formation of

a) $FeSO_4$
b) H_2SO_3

c) H_2SO_4
d) FeOOH

4. Pitting is an autocatalytic process, which occurs on a steel surface in a marine environment. Once oxygen is depleted corrosion continues due to the formation of

a) HCl
b) α-FeOOH
c) Fe_2O_3
d) $FeCl_3$

5. Dust particles accelerate corrosion because they

a) act as nucleation sites for the attack of corrosion
b) absorb nitrogen from the air

c) absorb moisture from the air

d) react chemically with ammonium sulfate

6. Chloride contaminant influences the atmospheric corrosion of steel principally by

   a) inducing pitting in steel
   b) destroying the passive layer on a steel surface
   c) decreasing the critical relative humidity of steel for adsorption of moisture
   d) increasing the concentration of chloride in the wet layer

7. The major source of CO_2 is mainly

   a) oceans
   b) incomplete combustion of fossil fuels
   c) oxidation of methane
   d) man-made

8. At a distance away from its emission source, NO may form NO_2. In addition, it may also be converted to HNO_3. In which form is nitrogen oxide lost to the atmosphere?

   a) As NO_2
   b) As HNO_3

9. A significant amount of SO_2 is absorbed in aerosol particle. Then it is oxidized to mostly

   a) SO_4^{2-}
   b) $H_2SO_3^-$
   c) H_2SO_4

10. Which of the following reactions would lead to a low pH and creation of acidity?

   a) $Fe(OH)_2 ads + SO_4^{2-} \rightarrow FeSO_4 + 2OH^-$
   b) $HSO_3^- + H_2O_2 \rightarrow HSO_4^- + H_2O$
   c) $2Fe^{2+} + 3H_2O + \frac{1}{2}O_2 \rightarrow 2FeOOOH + 4H$

11. Patina on a copper surface can form

   a) within six months
   b) in 8–10 years
   c) in 20 years

12. Aluminum is strongly resistant to corrosion in the pH range of

   a) 14–16
   b) 2–3
   c) 4–8

13. The level of critical humidity varies with

   a) temperature
   b) capillary condensation
   c) water vapor
   d) nature of contaminants

14. Which of the following methods cannot be used to combat atmospheric corrosion of steel?

   a) Cathodic protection
   b) Changing location of a site
   c) Installation of dust precipitators
   d) Use of weathered steels

15. The destruction of valuable monuments made of bronze is mainly caused by

   a) sulfur dioxide contaminant
   b) ammonia
   c) chloride concentration
   d) soot particles

B. HOW AND WHY QUESTIONS

1. State why:

   a) charcoal particles in the presence of 80% RH cause severe atmospheric corrosion of iron
   b) dew is highly corrosive
   c) atmospheric corrosion occurs even at 60% in the presence of certain contaminants in the atmosphere
   d) settled dust promotes corrosion

2. Explain why:

   a) the oxides of nitrogen are produced in the atmosphere
   b) SO_2 is converted eventually to H_2SO_4

c) $FeSO_4$ nests are formed

d) $FeSO_4$ promotes the corrosion of steel

3. Why does

a) steel show a maximum corrosion in the presence of NaCl particles at humidity higher than 70%?

b) an aqueous layer on steel become a highly acidified by SO_2?

c) dust particles prompt corrosion?

d) soot particles contribute significantly to atmospheric corrosion?

4. What is the role played by temperature in the atmospheric corrosion of steel? What is the effect of wind velocity, thermal capacity of the metal and its insulation properties on temperature?

5. State why the following methods are effective in minimizing atmospheric corrosion:

a) protective coatings

b) use of weathered steels

c) volatile inhibitors

d) reduction of humidity

6. What corrosion products are formed on zinc in the following environments?

a) Urban

b) Rural

c) Marine

d) Industrial

7. Zinc hydroxide forms a protective layer, however, it can be dissolved in acidic solution and weakly acidic solution characteristic of dew and rain. Is the hydroxide layer largely formed in the surface of aqueous layer or in the solid phase? How is the nucleation of hydroxide mixed salts initiated?

8. What are the major factors which contribute to corrosion by each of the following?

a) Fog

b) Dew

c) Rain

9. What are the major attributors which make relative humidity the biggest contributor to atmospheric corrosion? What is the relationship of humidity to the following

a) soot and charcoal particles

b) temperature

c) rain water

in terms of atmospheric corrosion?

10. Sulfur dioxide is a major pollutant in industrial environments. How does the adsorption of SO_2 in aqueous layers lead to acidification? What is the major reaction on moist particles?

11. What are the major sources of chlorine compounds and salt particles? How do they initiate corrosion on a steel substrate? What happens when HCl(g) comes in contact with atmospheric droplets?

12. What is the seasonal cracking of brass? How it is caused and prevented?

C. Conceptual Questions

1. State four basic conditions which prompt atmospheric corrosion.

2. Zinc hydroxide and basic zinc salts form the zinc patina which protects the surface from attack, however, the stability of zinc hydroxide and basic zinc salts depends on the pH. State the effect and specific role of pH on the protective role of the above products.

3. Explain briefly the atmospheric corrosion of steel based on an electrochemical cycle. How is $Fe(OH)_2$ converted to SO_4^{2-} and what is its effect on atmospheric corrosion?

4. Describe very briefly the corrosion mechanism of zinc in an atmosphere containing SO_2 and chlorides.

References

[1]   Evans, U.R. and Taylor, C.A.J. (1971). *Corrosion Science*, **12**, 227.

[2]   Barton, K. (1970). *Protection against atmospheric corrosion, Theories and Methods*, Vol. 3. London: John Wiley.

[3]   Evans, U.R. (1966). *The Corrosion and Oxidation of Metals*, First Supplementary Volume. London: Edward Arnold.

[4]   Graedel, T.E. and Frankenthel, R.P. (1991). *J. Of Electrochem. Soc.*, **137**, 2385–2394.

[5] Graedel, T.E., Nassay, K.N. and Franey, J.P. (1987). *Corrosion Science*, **27**, 721–740.

[6] Schweitzer, P.A. (1999). *Atmospheric Degradation and Corrosion Control.* New York: Marcel Dekker, USA.

Suggested Literature

[7] Leygraf, C. (1995). Atmospheric Corrosion. In: Marcus, P. and Oudar, J. eds. *Corrosion Mechanism in Theory and Practice.* New York: 421–449.

[8] Leygraf, C. and Graedel, T.E. (2000). *Atmospheric Corrosion*, London: John Wiley.

[9] Evans, U.R. (1961). *An Introduction to Metallic Corrosion*, 3rd ed. Edward Arnold, 981, London.

[10] Ericsson, R. (1978). The influence of sodium chloride on the atmospheric corrosion of steels. *Werkstoffe und Korrosion*, Germany: Frankfurt, **29**, 400.

[11] Martin, S. (2002). The atmospheric corrosion of iron and steel. *Proc. Electrochemical Society*, **13**, 89–103, New York, USA.

[12] Juan, J., Santana Rodriguez, F., Javier Sanatana Hevhandez, and Juan Eaonzalez, (2003), *Corrosion Science*, Oxford: Elsevier, USA, **45**, 799.

[13] Almeida, E., Morcillo, M. and Rosales, B. (2002). Atmospheric corrosion of zinc in rural and urban atmosphere, *Br. Corros. J.*, Oxford: Elsevier, UK, **35**, (41), 284–288.

[14] Knotkova, D., Kreislova, K. and Boschek, P. (1995). Trends of corrosivity based on corrosion rates and pollution data. UN-ECE International Cooperative program on effect of materials historical and cultural monuments, report 19, May.

Keywords

Acid regeneration cycle The sulfuric acid is formed by oxidation of SO_2 absorbed in rust layers and attack the steel substrate. This cycle is repeated again and again.

Adsorption The incorporation of gas in dissolved materials into liquids or solids.

Aqueous layers Monolayers of water formed on a substrate.

Bayerite This is a protective layer of $Al(OH)_3$ formed on aluminum in outdoor environment.

Boehmite This is a protective layer of γ-AlOOH ($Al_2O_3 \cdot 3H_2O$) formed on aluminum substrate. The boehmite is covered by bayerite and written as $Al_2O_3 \cdot 3H_2O$. It is a stable oxide film above 70°C.

Dew Moisture condensing on the surface of cool bodies at night. It forms when the temperature of the metal falls below the dew point of the atmosphere. Vapor begins to condense at the dew point.

Dry deposits This is the deposition of atmospheric pollutants directly on a substrate in the absence of wetting agents, like fog, dew, snow, etc. The term refers to the transfer process.

Electrochemical cycle This cycle takes place several times in an hour. The anodic reaction is balanced by the cathodic reduction of hydrated ferric oxide (rust) which forms magnetite under wet conditions. During the dry period, the magnetite is converted to freshly formed rust (FeOOH) by an electrochemical mechanism. This rust is not protective and corrosion continues. This cycle based on electrochemical reactions is called the electrochemical cycle.

Fog Formation of droplets of water with high acidity containing high concentration of sulfate and nitrate.

Galvanized steel A steel with a thin surface layer of zinc.

Green rust A greenish layer formed on iron surface. It contains iron oxyhydroxides. They may also contain carbonates and chlorides.

Industrial environment An environment with abundance of industries.

Marine environment An environment in the area of influence of the sea.

Oxyhydrolysis The oxidation of Fe^{3+}, for example, by water and oxygen to FeOOH is called oxyhydrolysis. Both oxygen and water are involved in the oxidation process.

Passive layer A thin, adherent protective layer of corrosion product formed on a metal surface.

Patina Formation of corrosion products, such as green colored brochantite on copper.

Pollutants Ingredients, such as NO_x, CO_x, SO_2 and sand particles which contaminate the air.

Rural environment Open land with little or no industries around. An agricultural land.

Rust Corrosion product consisting mainly of hydrated oxides of iron.

Seasonal cracking This is the cracking of brass and copper alloys in ammoniacal environment. This particularly happens in rainy seasons in the tropics.

Time of wetness The time taken by a metallic surface to be covered by a film of water.

Urban environment An environment characteristic of a city.

Wet deposits Deposition of pollutants on a substrate in the presence of rain, fog, dew or snow.

Zinc hydroxide [$Zn(OH)_2$] This is one of the major corrosion products of zinc. Zinc is oxidized to zinc hydroxide in a moist environment. It protects zinc from corrosion as long as the pH remains high (alkaline).

CHAPTER 11

BOILER CORROSION

11.1 DEFINITION

A boiler is a vessel in which water is continuously vaporized into steam by the application of heat. The steam that the boiler produces can be used for power generating, water heating and manufacturing processes.

11.2 INTRODUCTION

Efficient and trouble-free operation of boilers is very important when industrial plants generate their own power or when steam is required for manufacturing operations. It is, therefore, very important for engineers to understand the factors which lead to the fouling and corrosion of various components of boilers, and also to understand the methods which lead to the smooth and risk-free operation of boilers.

11.3 REQUIREMENTS OF AN EFFICIENT BOILER

To meet the minimum requirement of an efficient boiler, the boiler should:

(a) generate maximum amount of steam at required pressure and temperature with minimum fuel consumption,
(b) be light in weight and occupy minimum space,
(c) confirm to safety regulations,

(d) involve low initial installation capital and maintenance cost and
(e) exhibit easy accessibility for repair and replacements.

11.4 CLASSIFICATION OF BOILERS

Boilers are classified according to use, pressure, materials of construction, tube content, firing mode and heat source.

(a) Classification Based on Use

They are either stationary or a mobile-type. Following are their uses:

- Building heating
- Plant process steam
- Power generation
- Waste heat removal

(b) Classification Based on Pressure

Following are the major types under the above category:

- **Low-pressure boilers:** This category includes all steam boilers that do not exceed 150 psi, and all hot water boilers that do not exceed 160 psi.
- **Power boilers:** All boilers exceeding the limits of low-pressure boilers, are called power boilers.

- **Miniature boilers**: All boilers not exceeding 5 ft³ gross volume and a pressure of 100 psi are included in this category.

(c) Classification Based on Size

- Industrial units with surface area greater than 294 ft².
- Residential units with surface area from 16 to 294 ft².

Boilers may also be classified on the basis of heat source they use, such as electrical energy, nuclear energy, etc. and on the basis of mode of firing to generate steam.

11.4.1 MINIMUM REQUIREMENT FOR EFFICIENT BOILERS

To meet the minimum requirements, the boiler should:

- generate maximum amount of steam at a required pressure and temperature with minimum fuel consumption,
- be light in weight and occupy minimum space,
- confirm to the safety regulations,
- involve low initial installation capital and maintenance cost and

- exhibit easy accessibility for repair and replacement.

11.5 MAJOR DESIGN

Generally two types of boilers are designed:

(1) Fire-tube boilers
(2) Water-tube boilers

In fire-tube boilers, the flame and hot gases are within the tubes packed in bundles within a water-drum and water is circulated on the outside of the tubes (Fig. 11.1). In water-tube boilers, water is circulated within the tubes and hot gases and flames flow around the outside of the tubes (Fig. 11.2). Important details for the two basic types of boilers are given below.

11.5.1 FIRE-TUBE BOILERS

They normally produce 5–150 000 lb/h of steam at 150 lb/in². See Fig. 11.1. Above the combustion chamber, sealed by a roof, is the boiler steam drum which will be described later. As the water changes to steam, it rises in the boiler drum and exits through a steam header. Heat is transferred through the surface area of the tubes. In the above design, the boiler has a horizontal cylindrical shell for water space and boiler tubes pass along the length of this shell and extend through both ends

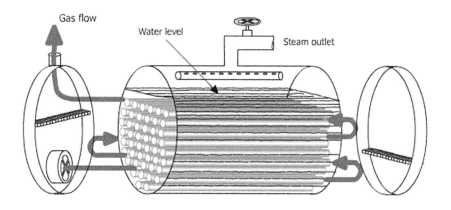

Figure 11.1   A four-pass fire-tube boiler

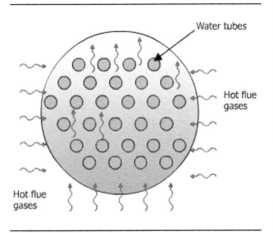

Figure 11.2a Water-tube boiler

of the plates (Fig. 11.1). The furnace is beneath the shell.

Scotch-type boilers were later developed. In these boilers, the furnace is located inside the shell. The tubes are located above and on either side of the furnace. The fire-tube boilers are available as a complete package ready for installation. The fire-tube boiler can produce steam under a variety of temperatures and pressures. They are used in heating systems, industrial processes, steam and potable steam generating units. They have a large water storage capacity. However, it takes some time to arrive at the desired operating temperature.

11.5.2 WATER-TUBE BOILERS

They are composed of drums and tubes through which the water circulates and steam is generated as shown in Fig. 11.2a. The water is converted into steam inside the tubes, while the hot gases pass around outside. Steam bubbles are formed on the heated surface of the tubes. The steam–water mixture which is formed has a lesser density than the cooler water on the unheated side of the tubes. The mixture, therefore, rises until it reaches the steam drum, as shown in Fig. 11.2b. In the

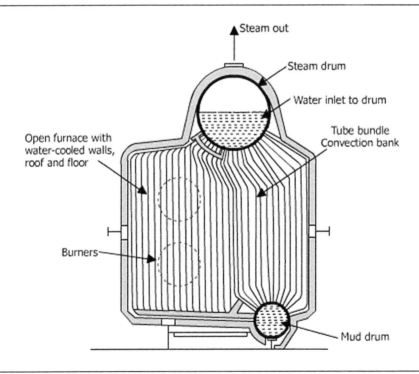

Figure 11.2b Small oil- and gas-fired package boiler. (From Kitto, Jr. J.B. and Albrecht, M.J. (1991). In: Kakag, S. ed. *Boilers, Evaporators and Condensers*. John Wiley. Published by kind permission of John Wiley, New York, USA)

steam drum, saturated steam is separated from the steam-water two-phase mixture. They can range in size from 1 to 30 m. The suspended solids are collected in the mud drum.

The steam drum is maintained half-full of water. The steam drum contains half water and half steam which are in equilibrium (both are at b.p. for a particular pressure). Water passes down the bottom of tubes forming the combustion chamber. The water which enters is at the boiling point but it requires latent heat to change to steam. The latent heat is supplied by heat of radiation from the flames. A steam–water mixture is thus returned to the drums. The steam is separated from the water. The water-tube boiler may be of horizontal straight tube design or bent tube design.

(a) Horizontal Straight Tube Design (HSTA)

They offer the advantage of good inspection, good visibility and cover a wide range of capacities and pressures, however, they show poor steam–water separation. They have been replaced by bent tube types in recent years.

(b) Bent Tube Boilers

Bent tube boilers offer many advantages over straight tube boilers. Some of these advantages include:

(1) fabrication economies,
(2) accessibility for cleaning and
(3) ability to operate at higher steaming rate.

Bent boilers are classified according to the number of drums and the arrangement of tubing. They are classified as **A**, **D** and **O** (Fig. 11.3). In Type **A**, one steam drum and two mud drums are arranged in a pattern, with the large upper drum at the apex, and two lower drums for steam generation at the bottom. Bottom blowdown is essential. Combustion takes place in the center of the boiler. In Type **D**, steam drum and mud drum are directly located above each other and off to one side of a furnace in the shape of a 'D.'

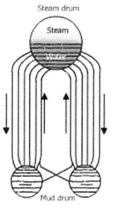
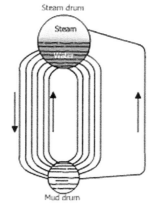
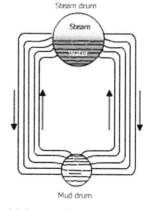

(a) A-type, with large upper drum for effective water-steam separation and lower drums (headers) for circulation

(b) D-type, with steaming tubes discharging near water line, providing more effective feed water and blowdown piping locations

(c) O-type, similar to A-type in circulation pattern with risers entering center of steam drum

Figure 11.3   Schematic of drum and tube arrangements for types A, D and O bent boiler designs. (From The NALCO Water Handbook, 2nd ed. McGraw Hill, 1987. Published by kind permission McGraw Hill, New York)

Series of tubes run vertically between the steam drum and mud drum, the rest of tubes run from steam and mud drum to the furnace wall. In Type **O**, the steam drum is located above the mud drum in the center of the boiler. The connecting tubes are in **O** shape. The burner is located in the center.

Improvements have been made to gain more efficiency and more capacity. Efforts have been made to increase the rate of heat absorption by increased circulation of water in the tubes. However, such designs have added to some deposition and corrosion problems.

11.6 CIRCULATION

The circulation in steam water circuits is of two types: (a) natural or (b) forced circulation (Fig. 11.4). Natural circulation in boilers is induced by the difference in the density between the cooled water and heated water. The tubes which transport the hotter water are called '*riser tubes*,' and those which transport the cooler water are called '*downcomer tubes*.' The difference of mean density between the cooler water in the downcomer tubes and heated water in the riser tubes (furnace evaporate tubes) generates a driving force sufficient to produce natural circulation in the steam water circuit. Too high heat may cause a decrease of the flow because the rate of flow in riser tubes would exceed the flow rate in downcomer tubes. Overheating may result in failure of tubes. This phenomenon is called '*tube starvation*.' With increasing temperature, the number of tubes acting as risers increase and the downcomers reduce correspondingly which may create two problems.

(1) The number of downcomers would be reduced restricting the supply of water to the risers.
(2) Stagnation may be caused because of some tubes acting both as risers and downcomers, causing stagnation (dead spot). This situation is to be avoided.

In the case of forced circulation boilers, a pump provides water circulation (Fig. 11.4). This system is important when boilers operate over a wide range of capacity.

11.7 WATER TUBE CIRCUIT

It consists of an upper drum (steam drum) and a lower drum (mud drum) connected by tubes. The steam generating tubes are located in the hottest area of the boiler and they are called '*risers*.' The combustion gases first pass here. The steam–water mixture is carried to the steam drum where steam is separated from water. Water flows from the steam drum, through the cooler tubes to the lower drum (mud drum). These tubes are named as '*downcomers*' (Fig. 11.5). The mud drum separates suspended solids and sludge from the water.

The steam water circuitry of a high-pressure recirculating (or drum) boiler consists typically of the following:

(1) Steam drum
(2) Superheater
(3) Economizer
(4) Reheater
(5) Steam temperature controls (attemperator)
(6) Furnace

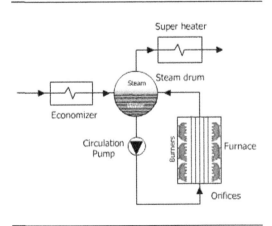

Figure 11.4 Forced circulation by pumps for boiler

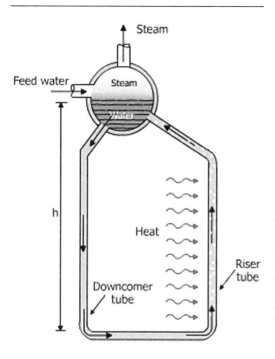

Figure 11.5   Riser and downcomer tubes in a natural circulation boiler

Following is the brief description of the above components.

11.7.1 STEAM DRUM

The main purpose of the steam drum is to separate steam from water. The following are the steam drum functions:

(1) Mixing the feedwater with the saturated liquid after steam separated.
(2) Mixing chemicals added to control corrosion.
(3) Purifying the steam to remove impurities and residual moisture prior to transfer to superheater.
(4) Removing a portion of boiling water to control boiler water chemistry.

For more efficient steam–water separation in high-pressure boilers, mechanical devices have been introduced in steam drums. One such device is a cyclone separator (Fig. 11.6). A separator consists of a single or double row of cyclones installed longitudinally on the sides of the steam drums. As soon as the steam–water mixture from the risers enters, it is transported (washed by a series of piston baffles) into cyclones. By centrifugal force, water is thrown onto the sides of the cyclones. On separation, water is returned to the steam drum below the water level while steam passes to a steam header after passing through various separating devices.

The steam drum is the entry point for boiler feedwater and internal chemical treatment. It is also the withdrawal point for continuous blowdown, purging from the solute of suspended solids to maintain the minimum level of dissolved and suspended solids. If blowdown is not done deposits of solids would form continuously on the boiler tubes and/or contaminate the system. The solid contents of steam should, in principle, be kept to zero.

11.7.2 SUPERHEATERS AND REHEATERS

Both superheaters and reheaters increase the temperature of saturated or near-saturated steam to provide the desired process condition. Simply speaking, they are single-phase heat exchangers with steam flowing inside the tubes and flue gas outside the tube. In the recirculating drum boilers, the outlet pressure of superheater is 180 MPa, while that of the reheater is only 4 MPa.

The saturated steam (at the boiling point) is separated from water in the steam drum area and is passed through the superheater tubes. The superheater heats steam above the saturation temperature for a particular boiler pressure. Superheater tubes have steam on one side and hot combustion gases on the other. Surface temperatures are higher here than the boiler tubes. The passage of heat from the hot flue gases through the tube wall increases the temperature to about 565°C. The construction material of tubes depends on the final steam temperature achieved. Carbon steel is used up to 400°C and chrome–moly steel up to 660°C. Superheater designs are based on the mode of heat transfer, convection, radiation or a combination

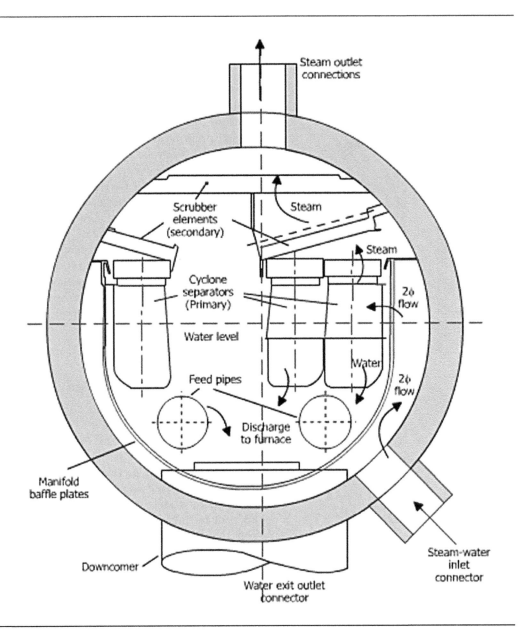

Figure 11.6 Typical steam drum internals showing scrubbers, cyclone separators and other components. (From Kitto, Jr. J.B. and Albrecht, M.J. (1991). In: Kakag, S. ed. *Boilers, Evaporators and Condensors*. JohnWiley. By kind permission of John Wiley, New York, USA)

of the two and the superheating circulation. Counterflow, parallel flow and combined parallel and counterflow can occur. A moderate temperature is developed by combined flow a maximum by counterflow. Sample superheater circuits showing different flow patterns is shown in Fig. 11.7.

11.7.3 TURBINES

These are a part of the steam generating system. The function of the turbine is to drive generators which produce electric power. The maximum energy in the steam is consumed to generate power. The exhaust steam is condensed and

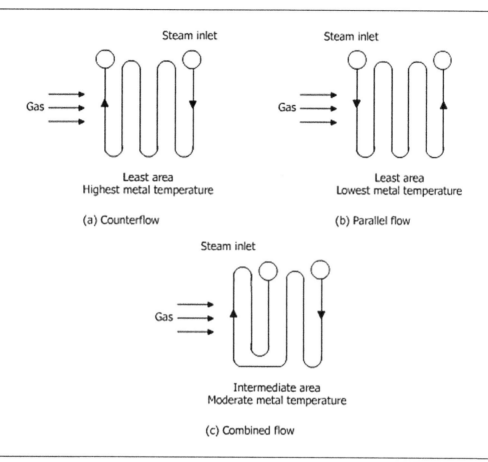

Figure 11.7   Sample superheater circuits showing different flow patterns. (From Kitto, Jr. J.B. and Albrecht, M.J. (1991). In: Kakag, S. ed. *Boilers, Evaporators and Condensors.* John Wiley. By kind permission of John Wiley, New York)

returned to the boiler as feedwater. Serious problems may be caused by turbine deposits which cause distortion of the blades causing them to become unbalanced. The steam from the outlet of the high-pressure turbine is passed back to the reheat section of the boiler. An increase of temperature up to 500°C again takes place. The steam is then passed through intermediate pressure turbine where the pressure of steam is reduced to 1 atmosphere and the temperature to slightly above 100°C. It is then passed through the low-pressure turbine where the pressure is reduced to 0.9 atmosphere and 40°C. As the vapor leaves the low-pressure turbine, it may contain droplets of water. It is passed over the outside of the condenser tubes (horizontal). Cold water is passed through the condenser tubes.

11.7.4 ECONOMIZERS

These are heat exchangers used for recovering waste heat from the combustion of products after the superheaters and reheaters, but before the air heaters. The water temperature after the feedwater heat is increased and temperature below saturation temperature and feedwater temperature is minimized. The waterside of economizer is an extension of boiler feedwater line where the heat input increases the temperature (Fig. 11.8). Tube bundles are placed in the flue gas and the water flow is counter to the flue gas flow. It is preferable for the water to flow up through the economizer tubes and for the flue gas to flow down across the tube.

Typical parts of a Water Tube Boiler include:

Economizer	Risers
Steam drum	Water walls
Downcomers	Screen tubes
Mud drum	Arches
Headers	Floor tubes
Superheater	Boiler bank
Air heater	

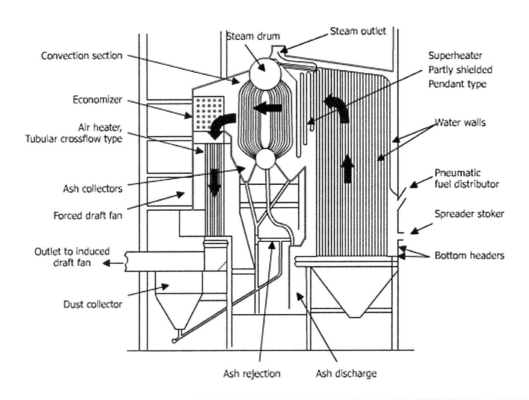

Figure 11.8 Boiler design showing economisers, air heater, water superheater, water walls and other components

11.7.5 AIR HEATERS

The air heater is located after the economizer. The heat in the flue gas from the economizer is received by incoming air thereby reducing the flue gas temperature and increasing the boiler efficiency. Combustion conditions are also improved. The location of preheater, superheater and air heater is shown in Fig. 11.8.

11.7.6 CONDENSERS

The objective of condenser is to condense the steam before it enters the reboiler. The surface condenser consists of bundles of tubes in which cooling water flows. Steam flows outside of the condenser tubes. Condensate flows out at the base of the condenser to a hot well where it is ready to re-enter the water–steam cycle. The cooling water does not cause a reduction in temperature of the vapor, the latent heat is removed thus causing condensation of water to occur. Non-condensable gases are removed by ejectors or vacuum pumps from a special air removal section. It is necessary because steam cannot be pumped and water must be returned to the boiler under pressure. The condenser pump pulls the condensate (condensed steam) and increases its pressure to 100–150 psi. The condensate passes through low-pressure feedwater heaters where its temperature is increased by heating with steam.

11.7.7 FEEDWATER HEATERS

The feedwater heaters are constructed of tubes through which feedwater is passed. The water is heated by steam from the turbine. The steam being at a higher temperature condenses on the outside of the tubes and gives up its latent heat to the feedwater. The temperature of the feedwater is thus increased. A large amount of energy as latent heat is lost in condensation. Non-condensable gases are removed by vacuum pumps from a special air removal section of a condenser. The coolant is either used on a once-through basis or recirculated through a cooling tower.

Deaerated feedwater heaters remove the undesirable dissolved gases as well as increase the feedwater temperature. Two types, such as tray and spray, are commonly used. In the tray type, a large amount of steam is condensed within the heater raising the temperature of the water to the saturation level in the heating section. There is a counterflow of steam which scrubs the dissolved gases from the feedwater heater. Spray type deaerators work on the same principles, but differ in design. Water is sprayed through steam at 20 psi, first from spring loaded nozzles in the top section of the spray unit. The two types of units are shown in Figs 11.9 and 11.10. The feedwater is passed to the economizer from the feedwater heater.

11.7.8 STEAM–WATER CIRCUIT

(a) Water Circuit

Feedwater enters the economizer through the bottom header of the economizer and passes upwards opposite to the direction of flue gases before it is collected in the outer header of the economizer. The feed is injected into the steam drums where it is mixed with the water discharged by cyclone steam–water separators. Feedwater pipes are distributed all along the drum length to provide uniform mixing of feedwater with the downcomer (unheated) outlet tubes from the steam drum. Water then flows to the risers where it is heated and converted to steam. Steam–water system showing natural circulation is shown in Fig. 11.11. Because of the difference of density between water in the downcomer and the steam–water mixture in the water walls (water tubes forming the furnace wall), natural circulation is created which causes the steam–water mixture to flow upwards in the steam drum and the water to flow to the downcomers. In the steam drum, steam is separated from the steam–water mixture produced by water walls. The saturated dry steam flows to the superheater through a number of drum outlets connected, and the remaining steam-free water is returned to the downcomer. Any residual moisture left after steam–water separation by cyclone separator is removed by secondary cyclone separators.

(b) Steam Circuit

The steam circuit serves two functions:

(a) cools the convective pass enclosure and
(b) generates the steam conditions required by superheater.

In superheaters, the saturated steam is heated and flows to the high-pressure turbines. Part of the exhaust steam is returned to reheater for reheating and goes to the inlet of reheat turbines.

(c) Airflow System

Cold air is passed to the inlet of an air heater and flows across the tubes. The air is heated by

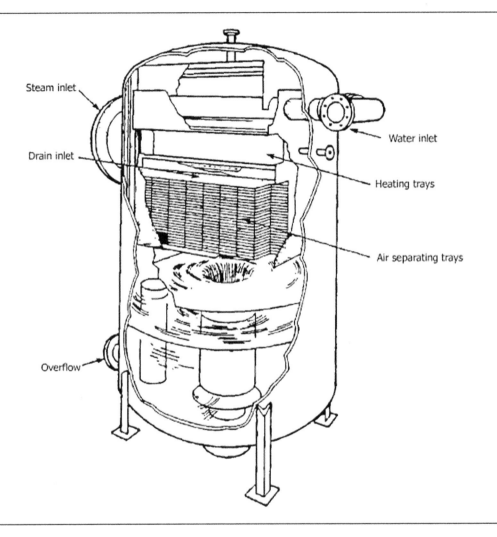

Figure 11.9 Tray-type deaerating heater. (From Principles of Industrial Water Treatment, Drew Chemicals, Boonton, NJ 07005. By kind permission of Drew Chemicals, Boonton, NJ)

flue gas to the desired temperature at the outlet of air heaters.

11.7.9 FURNACE WALL ENCLOSURE

This is composed of closed-spaced water-cooled tubes. The tube panels are connected by inlet and outlet headers. A casing provides a rigid construction to stand high pressures of the gas in the furnace. Because of high heat fluxes in the furnace, adequate cooling is provided. The flue gas temperatures are in the range of 1650–1900°C.

11.8 WATER CONDITIONING

The term '*water conditioning*' refers to the treatment of water by chemicals, change of pH, mechanical measures, etc. to prevent corrosion in boilers. Three types of water are encountered after water treatment:

(a) Raw water supply
(b) Condensate returned from steam turbines
(c) Boiler water.

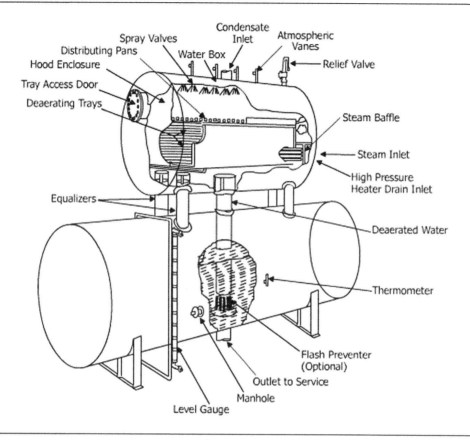

Figure 11.10 Spray-type deaerating heater

Water conditioning results in

- Freedom from deposits
- Absence of internal corrosion
- Prevention of carryover of boiler water solids into steam.

- Sludge. Settling down of solids in boiler water.
- Blowdown. Bleeding down of a portion of boiler water from the drums.

11.8.1 SOME DEFINITIONS INVOLVED IN BOILER WATER CHEMISTRY

- Condensate. Steam condensed and returned.
- Makeup water. Replacement water added to the steam.
- Feedwater. Condensate returned with makeup water.
- Carryover. Exceeding solids beyond a certain limit.

11.8.2 CONTAMINATIONS IN RAW WATER

(1) Rainwater brings into solution the atmospheric gases of oxygen, nitrogen and carbon dioxide. As it passes through the solids, it picks up minerals.
(2) Suspended solids, such as mud, salt, clay and oxides, which are insoluble in water.
(3) Dissolved solids. Dissolved minerals, such as silica, iron, calcium, magnesium and sodium chloride.

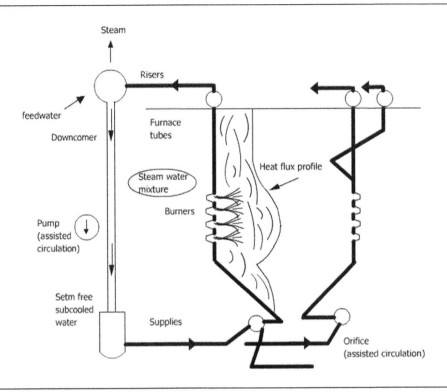

Figure 11.11 Steam–water circuit showing natural circulation

11.8.3 WATER TREATMENT

The following steps are involved in the water treatment. Coagulation and filtration of suspended materials are the oldest techniques employed for water treatment.

(a) Coagulation

This requires rapid mixing or addition of chemicals to the water to form a precipitate called '*floc*' which forms small particles. All flocs are brought together by coagulation using gentle agitation to form larger particles. The process is speeded by coagulation in which finely divided particles collect together to form larger particles which settle down rapidly. Alum and iron sulfate are well-known coagulating agents. Coagulation

is followed by chlorination for destruction of organic matter.

(b) Filtration

Finely divided suspended particles not removed by coagulation are now removed by filtration. Charcoal filters are used in filtration to remove organic matter and chlorine.

(c) Demineralization

After the coagulation and filtration process, the hardness forming constituents, such as calcium and magnesium salts present in the water, must be removed. The removal of hardness constituents is achieved by the ion exchange process described below.

The hardness constituents (Ca^{++} and Mg^{++}) are removed by treating the water with ion exchange resin. It is a process by which undesired ions of a given charge are absorbed from solution within an ion-permeable absorbent, being replaced in the solution by desirable ions of similar change from the absorbent. The hardness constituents are removed by sodium zeolite resin. This process utilizes materials that have the property of exchanging the hardness constituents of water. The hardness constituents (Ca^{++} and Mg^{++}) are replaced by sodium ions from the zeolite.

$$Ca^{2+} + Na_2Z \rightarrow CaZ + 2Na^+$$

and $$Mg^{2+} + Na_2Z \rightarrow MgZ + 2Na^+$$

Sodium ions are displaced from the resin and Ca^{2+} and Mg^{2+} ions are absorbed by the resin from water. The scale forming Ca^{2+} and Mg^{2+} ions are changed into sodium salts.

$$Na_2Z + Ca(HCO_3)_2 \rightarrow 2NaHCO_3 + CaZ$$

$$Na_2Z + MgSO_4 \rightarrow MgZ + Na_2SO_4$$

The active sodium zeolite can be regenerated by treatment with a strong salt solution (brine):

$$2NaCl + CaZ \rightarrow Na_2Z + CuCl_2$$

$$2NaCl + MgZ \rightarrow Na_2Z + MgCl_2$$

By treatment with NaCl, the base exchange resin is regenerated. After the process, the water contains bicarbonate, sulfate and chloride. Only Ca^{++} and Mg^{++} are exchanged for Na^+ ions. No reduction in dissolved solids and no reduction in bicarbonate alkalinity results.

(d) Removal of Bicarbonate Alkalinity by Soda-lime Process

To reduce total dissolved solids and bicarbonate alkalinity, the soda-lime process is used. Hydrated lime is used to react with bicarbonate. Bicarbonate precipitates when water is heated and the tubes in heat exchangers may become blocked. The reaction is represented by:

$$Ca(OH)_2 + Ca(HCO_3)_2 \rightarrow 2CaCO_3 \downarrow + 2H_2O$$

The calcium carbonate is filtered from the solution. When magnesium bicarbonate is present, twice as much lime is needed to remove the hardness. The reaction is represented by

$$2Ca(OH)_2 + Mg(HCO_3)_2$$
$$\rightarrow Mg(OH)_2 + 2CaCO_3 + 2H_2O$$

Although Mg^{++} is removed and equal amount of Ca^{++} takes its place, this requires another mole of $Ca(OH)_2$ for removal.

Water softened by the above process would contain more sodium bicarbonate in boiler water, and produce caustic soda which is highly corrosive in nature.

Another modification of the process 'hot-lime zeolite softening process' (Fig. 11.12) is used to increase silica reduction and carried out at 98°C. Calcium carbonate is filtered from the solution. To reduce silica, Mg is precipitated as $Mg(OH)_2$ which acts as an absorbent for silica. The residual hardness is removed by sodium zeolite softener. The increased temperature improves the potential of the softener to exchange sodium for hardness ions.

The raw water obtained softened by both processes contain more sodium bicarbonate than acceptable for boiler feedwater. The alkalinity content should not exceed more than 20% of total solids. The alkalinity is reduced by a split stream softening process.

(e) Demineralization

In the demineralization process, both metal ions and salt anions are removed. The metal ions are removed with the hydrogen zeolite system and the anions are removed by a resin saturated with hydroxide ions. It is desirable at drum pressures over 1000 psi to remove silica in boiler water as silica may deposit on the turbine surface. It involves two ion exchange resin, a cation

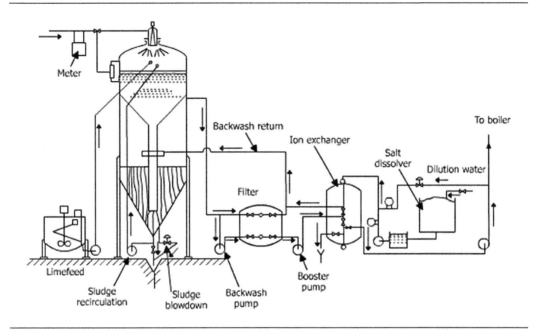

Figure 11.12 Flow sheet of a typical hot lime – hot ion exchange process. (From Source: Babcock and Wilcox Company: Tech. Information Bulletin: Water Treatment for Industrial Boilers)

exchange, and an anion exchange. The water is passed through a cation exchange resin:

$$\left.\begin{array}{c} CaSO_4 \\ Ca(HCO_3)_2 \\ NaCl \\ SiO_2 \end{array}\right] \rightarrow \text{cation H} \rightarrow \begin{array}{c} H_2SO_4 \\ H_2CO_3 \\ HCl \\ SiO_2 \end{array}$$

When all the hydrogen has exchanged for metal ions, it is regenerated with HCl or H_2SO_4, as described earlier.

The anions are removed by anion exchange resin, either weak or strong base. All the acid is removed by exchange of anion part of the acid (negatives) by hydroxide.

$$H^+ + Cl^- + Resin - OH^- \rightarrow HOH$$
$$+ \; Resin \; Cl^-$$

Hydrochloric Acid + Resin → Water + Resin chloride

(f) Mixed Bed Exchange

Water is finally passed through a mixed bed comprising of cation and strong anion resin bed.

The water after passage through billions of cations and anions comes out as silica- and carbon dioxide-free water. The two types of resin can be in separate tanks or in one tank mixed together. At this stage nearly all salts are removed.

11.8.4 BOILER FEEDWATER TREATMENT

Boiler efficiency is directly related to the quality of the feedwater. The feedwater system refers to deaerator, feedwater pumps and the piping to the boiler. Feedwater is defined as

$$\text{Feedwater } (FW) = \text{Makeup Water } (MW)$$
$$+ \text{ Return Condensate } (RC)$$

$$FW = MW + RC$$

Dissolved bicarbonates of calcium and magnesium breakdown under heat to give off carbon dioxide and form insoluble carbonate. The carbonate may precipitate directly on the boiler metal or form sludge in the boiler water that may deposit on the boiler surface. Calcium sulfate becomes less soluble and deposits with silica directly on the boiler surface. Silica forms extremely hard scale. It is, therefore, important to treat the feedwater to minimize corrosion problems in the boiler

The pre-boiler cycle consisting of feedwater heaters, feed pumps and feed lines are liable to corrode by the condensate return. The main contributors to corrosion are carbon dioxide and oxygen. Corrosion must, therefore, be minimized. This is done by feedwater control, which reduces the ingress of harmful impurities and gases to the boiler water circuit and also to the steam. The best way to prevent corrosion is by controlling the pH, which should be maintained in the range of 8.5–9.2.

The oxygen from the feedwater needs to be completely removed. Even at the levels of 7–10 ppb, oxygen can cause considerable corrosion damage, hence it must be completely removed. A complete removal of oxygen requires the use of scavengers. The most commonly known oxygen scavengers are sodium sulfite and hydrazine. Sodium sulfite is the most commonly used oxygen scavenger which reacts directly with oxygen as shown below.

$$2Na_2SO_3 + O_2 \rightarrow 2Na_2SO_4$$

It needs about 8 parts by weight of Na_2SO_3 to remove each part by weight of oxygen. The reaction is catalyzed by cobalt salts and takes place immediately between pH 5 and 8 at elevated temperature. Common sodium sulfite should contain 0.25% of cobalt sulfate. The only disadvantage is that some quantities of sulfates are introduced in the system which may cause scaling problems in the system. Hydrazine (N_2H_4) is the other commonly used scavenger, but is carcinogenic and its use should be avoided.

It is supplied as a concentrated aqueous solution and reacts directly with dissolved oxygen. One part of N_2H_4 scavenges one part of oxygen.

$$N_2H_4 + O_2 \rightarrow 2H_2O + N_2$$

Nitrogen drives off oxygen out of solution. The reaction can be direct and indirect. The direct reaction shown above proceeds rather slowly below 148.8°C (300°F), whereas indirect reaction proceeds rapidly at temperatures as low as 70°C (150°F).

The following are the reactions involved in indirect reaction:

(1) $N_2H_4 + 6Fe_2O_3 \rightarrow 4Fe_3O_4 + N_2 + 2H_2O$

(2) $4Fe_3O_4 + O_2 \rightarrow 6Fe_2O_3$

The above reactions proceed on ferrous metal. If too much chemicals are fed, sulfurous gases and ammonia may be produced by decomposition of Na_2SO_3 and N_2H_4, respectively.

$$Na_2SO_3 + N_2O + Heat \rightarrow 2NaOH + SO_2 \uparrow$$

$$4Na_2SO_3 + 2H_2O \rightarrow 3Na_2SO_4 + 2NaOH$$
$$+ H_2S$$

$$2N_2H_4 \rightarrow H_2 + N_2 + 2NH_3$$

Both of the gases produced are highly corrosive. The amounts of scavengers must, therefore, be controlled. The decomposition products of hydrogen are volatile and they do not add to the total dissolved solids.

Relative advantages and disadvantages of using Na_2SO_3 and N_2H_4 are given below:

(1) Na_2SO_3 is easy and safe to handle, whereas hydrazine causes skin irritation and must be used with care.
(2) Na_2SO_3 reacts more rapidly than hydrazine.
(3) Sodium sulfite forms sodium sulfate on reacting with oxygen and contributes solids to boiler water, whereas hydrazine does not contribute solids.
(4) Hydrazine reduces ferric oxide to a protective magnetite, whereas sulfite does not.

11.8.5 CORROSION PROBLEMS IN THE CONDENSATE SYSTEMS

The steam/condensate systems are an integral part of the water boiler system. The condensate system includes anywhere that the steam condenses to form liquid water. As the vapor leaves the low-pressure turbine it contains droplets of water. It is passed outside of homogenated tubes in the condenser through which certain water is passed. Cooling water causes condensation of the steam to water by removing the latent heat. It is necessary since a gas cannot be pumped and it is necessary to pump water back to the boiler. Every droplet in contact with the metal surface is a source of potential corrosion. As long as the steam is dry it is non-corrosive, however, soon upon condensation it poses a serious corrosion threat. It is important to maintain the heat transfer surface free from scaling or corrosion. The principal cause of condensate piping corrosion is the presence of excessive amounts of carbon dioxide and oxygen, or both.

The basic construction material of condensate systems are iron and copper. These materials must not, therefore, be subjected to corrosion. The majority of corrosion products that are deposited in the boiler originate in the condensing system. High levels of iron oxide are formed in the condensate system. The following are the major corrosion reactions:

(1) Carbonic acid corrosion:

$$Fe + 2H_2CO_3 \rightarrow Fe^{2+} + H_2 \uparrow + 2HCO_3^-$$

(2) Oxygen corrosion:

$$4Fe + 6H_2O + 3O_2 \rightarrow 4Fe(OH)_3$$

(3) Combined carbonic acid and oxygen corrosion:

$$2Fe(HCO_3)_2 + \frac{1}{2}O_2 \rightarrow Fe_2O_3 + 4CO_2 \uparrow$$
$$+ 2H_2O$$

Carbon dioxide reacts to form weakly ionized carbonic acid:

$$CO_2 \text{ (dissolved)} + H_2O \rightarrow H_2CO_3$$
$$\rightarrow H^+ + HCO_3^-$$

The pH of the water is reduced and its corrosivity is increased. In pure condensate at 65°C, 1 ppm dissolved CO_2 will decrease pH from 6.5 to 5.5. Ammonia may be sometimes present, forming

NH₄OH with water. However, copper and copper alloys in the condensate system are sensitive to corrosion; ammonia dissolves the protective Cu_2O layer on copper alloy and can cause stress corrosion cracking. In the presence of oxygen, ammonia is very destructive.

Carbon dioxide and oxygen are major corroders and both can cause serious corrosion damage of the condensate system. The important sources of CO_2 are the carbonate and bicarbonate alkalinity in makeup water. These salts undergo thermal decomposition and liberate CO_2 in the steam. The thermal decomposition of 1 ppm of bicarbonate alkalinity (as $CaCO_3$) would liberate 0.7 ppm pure CO_2.

$$Na_2CO_3 + H_2O \rightarrow 2NaOH + CO_2 \uparrow$$

$$2\left[HCO_3^-\right] + \text{Heat} \rightarrow CO_3^- + CO_2 + H_2O$$

$$CO_3^{--} + \text{Heat} + H_2O \rightarrow CO_2 \uparrow + 2OH^-$$

The carbon dioxide dissolves in water forming carbonic acid HCO_3^-, which is extremely corrosive to steel.

$$Fe + 2H^+ + CO_3^{--} \rightarrow FeCO_3 + 2H^+$$

Very little CO_2 is needed to reduce the pH of water and its corrosivity. The carbonic acid shows up as deep channeling or grooving and thinning of pipe nipples. It may also induce uniform thinning. Below a pH = 6.2, the attack is a function of pH. Above pH = 6.2, the attack is insignificant.

11.8.6 O₂ IN CONDENSATE CORROSION

(a) Oxygen

Dissolved oxygen is significantly more corrosive than carbon dioxide. Dissolved oxygen in pure water at 65°C is found to be six times more corrosive than a molar equivalent concentration of carbon dioxide. The combined corrosion caused by O_2 and CO_2 is twice that caused individually by oxygen and CO_2. Oxygen is known to produce rust and pitting in steel equipment. Oxygen picks up electrons from iron at the metal surface and produces negatively charged OH^- ions.

$$\frac{1}{2}O_2 + H_2O + 2e \rightarrow 2OH^-$$

The overall reaction of oxygen is expressed as

$$2Fe + O_2 + 2H_2O \rightarrow 2Fe(OH_2)$$

(Ferrous hydroxide)

Oxygen attack takes the form of deep pits and does not produce hydrogen gas.

(b) Carbon Dioxide

CO_2 dissolving in pure H_2O reacts to form weakly ionized carbonic acid. Very little CO_2 is needed to reduce the pH of water and increase corrosivity. The contamination of condensate by CO_2 is shown by the reactions given below:

$$2H_2CO_3 + Fe \rightarrow Fe(HCO_3)_2 + H_2 \uparrow$$

$$2H_2CO_3 + 2Fe \rightarrow 2FeCO_3 + 2H_2$$

The formation of carbonates and bicarbonates reduces the amount of CO_2, however this happens at the expense of slow dissolution of the pipe metal. The change in pH is, therefore, only transient and it can be misunderstood. When O_2 is present, the above products are again decomposed, such as:

$$Fe(HCO_3)_2 \rightarrow Fe(OH)_2 + 2CO_2$$

The following are important corrosion products reactions:

$$4Fe^{2+} + 8OH^- \rightarrow 4Fe(OH)_2$$

$$(Fe \rightarrow Fe^{2+} + 2e)$$

$$4Fe(OH)_2 + 2O_2 + 2H_2 \rightarrow 4Fe(OH)_3$$

Complete corrosion reaction between iron and dissolved oxygen in water is shown below:

$$6H_2O + 6Fe + 3H_2 \rightarrow 6Fe(OH)_3$$

The corrosion rate will be very high if the ferrous ion is oxidized to ferric oxide very rapidly. If the rate of oxidation of Fe^{2+} to Fe^{3+} is very high so that $Fe(OH)_3$ is formed on the surface and becomes protective, the rate of corrosion decreases. The attack by oxygen afterward appears in the form of pitting. The pits of varying dimensions accompanied by oxide deposits can be observed.

When both CO_2 and O_2 are dissolved in the condensate the rate of reaction between carbonic acid and iron is increased. At higher flow rates, the attack by oxygen is increased. As stated earlier, the rate of corrosion of iron is increased in the presence of both carbon dioxide and oxygen. The following are the important reactions

$$CO_2 \text{ (dissolved)} + H_2O \rightarrow H_2CO_3$$

$$\rightarrow 2H^+ + CO_3^-$$

$$2H_2CO_3 + Fe \rightarrow Fe(HCO_3)_2 + H_2 \uparrow$$

$$2HCO_3 + 2Fe \rightarrow 2FeCO_3 \downarrow + 2H \uparrow$$

Iron carbonate formation reduces the rate of corrosion. The pH of the condensate is increased by reduction in the amount CO_2 in solution. However, the pipe material is dissolved by the formation of iron carbonate ($FeCO_3$). As the condensate is replenished, the pipes continue to corrode. When oxygen is present, ferrous carbonate and bicarbonate decomposes, forming ferric oxide and hydroxide deposits, and releasing more and more carbon dioxide which lowers the pH.

When $Fe(HCO_3)_2$ is decomposed, CO_2 is released

$$Fe(HCO_3)_2 \rightarrow Fe(OH)_2 + 2CO_2 \uparrow$$

and $Fe(OH)_2$ is converted to $Fe(OH)_3$:

$$2H_2O + 4Fe(OH)_2 + O_2 \rightarrow 4Fe(OH)_3$$

$FeCO_3$ reacts with water and oxygen to form $Fe(OH)_3$, as shown below:

$$2FeCO_3 + 5H_2O + \frac{1}{2}O_2$$

$$\rightarrow 2Fe(OH)_3 + 2H_2CO_3$$

$$2FeO + 3H_2O + \frac{1}{2}O_2 \rightarrow 2Fe(OH)_3$$

On heating water containing carbonic acid, CO_2 is released:

$$H_2CO_3 \rightarrow CO_2 \uparrow + H_2O$$

$Fe(HCO_3)_2$ decomposes to give CO_2

$$2Fe(HCO_3)_2 + \frac{1}{2}O_2 \rightarrow Fe_2O_3 + 4CO_2 \uparrow$$
$$+ 2H_2O$$

Ferric oxide may again react with oxygen to give ferrous hydroxide

$$Fe_2O_3 + 3H_2O \rightarrow 2Fe(OH)_2$$

Magnetite can also be formed from $Fe(HCO_3)_2$

$$2Fe(HCO_3)_2 + \frac{1}{2}O_2 \rightarrow Fe_3O_4 + 6CO_2 + 3H_2O$$

Ferrous carbonate can also be formed from $Fe(HCO_3)_2$

$$Fe(HCO_3)_2 \rightarrow FeCO_3 + H_2O + CO_2$$

In the presence of oxygen, $Fe(OH)_2$ is oxidized to $Fe(OH)_3$ which forms deposits on the metal surface and blocks the pipe. Various deposits of iron are formed due to the reactions shown above. The formation of differential aeration cells by the deposits can also lead to pitting of the condensate system. All the above reactions lead to the destruction of condensate piping systems.

11.8.7 PREVENTION OF CONDENSATE CORROSION

The following methods can be adopted to prevent corrosion in the condensate system:

(1) Elimination of (or minimizing) oxygen and carbon dioxide contamination by mechanical and chemical methods. As shown, oxygen is minimized by mechanical deaeration and also by scavenging using sodium sulfite on hydrogen. Similarly, carbonate and bicarbonate are eliminated by soda-lime softening, split stream softening and demineralization. Alkalinity is also decreased by the above method. Chemical inhibitors, such as filming and neutralizing inhibitors (described below) are also used to prevent condensate corrosion.

(2) Minimizing the leakage of air out of the system.

11.8.8 NEUTRALIZATION OF CARBON DIOXIDE

The neutralization of carbon dioxide is essential to prevent corrosion of condensate systems and to prevent its passage to steam. One important condition is that the chemicals used in neutralization must be volatile so that they can be removed with the steam. The chemicals used for the purpose are organic compounds, like amines ($R-NH_2$). Examples are cyclohexylamine and morpholine, benzlyamine and aminomethyl propanol (AMP), and diethylaminoethanol.

(Morpholine)   (Cyclohexylamine)

These compounds neutralize the acids by hydrolyzing the condensed water to form a base.

$$R-NH_2 + H_2O \rightarrow R - N^+H_3 + OH^-$$

(hydrolysis of amine)

The more (OH^-) is produced, the more acid is neutralized. These amines volatilize with steam and condense at the same temperature as water.

In some systems, morpholine is a good choice. It has a higher distribution ratio and neutralizes carbon dioxide better in the system. The ratio of morpholine in the gas to the amount in the liquid is called the '*distribution ratio*.' Similarly, compounds with high chemical basicity (the amount of OH^- produced is measured as basicity) neutralize more acid than compounds with lower basicity. It is a measure of how strongly an acid is neutralized. An amine, such as morpholine, prefers the water phase, hence, it would be present in the initial condensate where the amount of steam is large. On the other hand, cyclohexylamine stays with steam to enter the condensate when the temperature falls and the percentage of steam becomes more than the condensate. The two amines can be used together for an effective program.

These few factors must be carefully weighed before selecting compounds for neutralization of carbonic acid. A pH of 8.5–9.2 must be maintained with amine.

11.8.9 FILMING AMINES

If carbon dioxide and oxygen concentrations are very high and they cannot be removed economically with routine chemicals, another choice is available, which is filming amines. These are long straight chain organic molecules of 10–18 carbon atoms with one amine group at the end. These organic compounds prevent corrosion by forming a film on a metal surface which acts as a barrier between the metal surface and environment. One example of this type is oxtadecylamine. These amines do not neutralize carbon dioxide, but form a non-wettable film by means of absorption on metal surface in contact with the condensate. The amine group is polar because of its being ionic. Because of its polar characteristic, the molecule is absorbed physically on the metal surface. By coating the metal surface with a thin film of the compound ($C_{18}H_{38}NH_2$) only one molecule thick, the oil-like property does not allow oxygen to come into contact with the surface. The use of filming amines has

been successful in preventing corrosion in low-pressure boiler system (3 psi). The pH range is between 6.5 and 7.5.

(a) Advantages

(1) They are economical and less costly to administer.
(2) They protect effectively against carbonic acid and oxygen corrosion.
(3) They are most widely used chemicals.
(4) They form highly protective films which resist corrosion by restricting the ingress of carbon dioxide and also oxygen ingress.

(b) Disadvantages

(1) The film can be destroyed by oily matter.
(2) The films are stripped by high-velocity water.
(3) The film is degraded by solid contents.

As hydrazine has been listed as a carcinogen, its use has been restricted. Some new compounds have been marketed since 1980, like carbohydrazide, hydroquinone and diethylhydroxylamine.

11.8.10 INTERNAL TREATMENT OF BOILER WATER

The prevention of deposit formation is important in boilers as the deposits (scales, sludges, corrosion products, etc.) act as insulators, causing a loss of heat transfer. The thicker the deposit, the greater the driving force required to penetrate the barrier and higher the temperature of the tubes. There are several methods to control deposits and some important methods are summarized below. The majority of the corrosion products that deposit in the boiler originate in the condensate system, for instance:

(a) $Fe + 2H_2CO_3 \rightarrow Fe^{2+} + H_2 \uparrow +2HCO_3^-$
$$(H_2CO_3 \text{ corrosion})$$

(b) $4Fe + 6H_2O + 3O_2 \rightarrow 4Fe(OH)_3$
$$(\text{Oxygen corrosion})$$

(c) $2Fe(HCO_3)_2 + \frac{1}{2}O_2$

$\rightarrow Fe_2O_3 + 4CO_2 \uparrow + 2H_2O$

The products are swept in the boiler and deposit on the internal surface of the tubes. The mechanical methods used to solve the problem are:

(1) Condensate polishing
(2) Condensate dumping
(3) Electromagnetic separation.

(a) Condensate Polishing

The purification of condensate can be achieved by various condensate polishing systems. The resin beds in the purification systems remove both dissolved and suspended solids. These ion exchange beds are regenerated periodically.

The polishers operate as filters to remove the oxide of metals and undissolved solids from the condensate. As iron oxide is particulate matter, it can be removed by filtration. The filter is a mixed cation and anion exchange resin. The filter removes the insoluble particles and ion exchange removes the dissolved contaminants against the formation of calcium scale while at the same time avoiding the formation of harmful free caustic in water.

(b) Condensate Dumping

When the iron levels are exceeded, the returned condensate is dumped. If the consumption rate of condensate is very high, the makeup water capacity may be exceeded. This process is rather uneconomical.

(c) Electromagnetic Separation

Cylinders are fitted with iron balls and strong magnetic fields are set up in the space between the iron balls on application of an electrical current. Electromagnetic filter is shown in Fig. 11.13. Ferromagnetic impurities are attracted to the magnetic poles and retained. For cleaning, the matrix is demagnetized and the filter is washed out. A removal efficiency of 90% has been reported.

11.8.11 METHODS USED IN INDUSTRIAL BOILERS

The following three methods are generally employed in industrial boilers:

(1) Conventional treatment method
(2) Coordinated phosphate method
(3) Chelant treatment

(1) Conventional Treatment Method

The phosphate-hydroxide or conventional treatment method is commonly used. It involves the addition of phosphate and caustic to boiler. The caustic is added to maintain a pH in the range 10.5–11.2. This practice is used in high-pressure boilers. In this process, calcium and magnesium salts are precipitated in a sludge form. The following reaction leads to the formation of sludge.

(1) $10Ca^{++} + 6PO_4^{--} + 2OH^-$

$\rightarrow 3Ca_3(PO_4^-)_2 \cdot Ca(OH)_2$
(Calcium hydroxyaptite)

(2) $2Mg^{++} + 2OH^- + 2SiO_2 + H_2O$

$\rightarrow MgSiO_3 \cdot Mg(OH)_2 + H_2O$
(Serpentine)

The products formed as above, are not very adherent and they are easily removed by blowdown.

(2) Coordinated Phosphate Method

The following are the reactions of sodium phosphate in water

$$Na_3PO_4 \rightleftarrows 3Na^+ + PO_4^{-3}$$

(Trisodium phosphate)

$$Na_3PO_4 + H_2O \rightleftarrows Na_2HPO_4 + NaOH^-$$

$$\text{or} \quad Na_3PO_4 + H_2O \rightleftarrows 3Na^+ + OH^-$$

$$+ HPO_4^{-2}$$

$$Na_2HPO_4 \rightleftarrows 2Na^+ + HPO_4^{-2}$$

$$\uparrow$$

$$H^+ + PO_4^{-3}$$

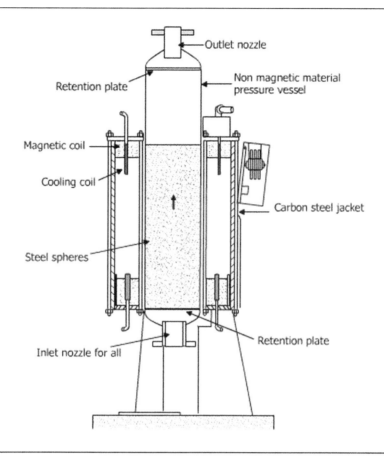

Figure 11.13 Electro-magnetic filter. (From Source: Babcock and Wilcox Company: Tech. Information Bulletin: Water Treatment for Industrial Boilers)

$$2Na_2HPO_4 + 2H_2O \rightleftarrows 2Na_2H_2PO_4 + 2NaOH$$

$$Na_2HPO_4 + 2H_2O \rightleftarrows 2Na_2H_2PO_4 + 2NaOH$$

$$NaH_2PO_4 \rightleftarrows Na^{++} + H_2PO_4^-$$

(Monosodium phosphate) $\downarrow\uparrow$

$$H^+ + HPO_4^{2-}$$

In this method, no free caustic is maintained in the boiler water. Figure 11.14 shows resulting pH versus phosphate concentration, the resulting pH on addition of trisodium phosphate. Trisodium phosphate hydrolyzes to produce hydroxide ions. Phosphate acts as buffer to minimize caustic cracking.

The type of species present depends upon pH. The control is based on the equilibrium maintained between sodium and phosphate to eliminate the formation of the free caustics by maintaining an optimum pH ($HPO_4^{2-} + OH^- \rightleftarrows PO_4^{3-} + H_2O$). A ratio of 3.0 Na to 1.0 PO$_4$ is maintained by phosphate addition. The control is called '*coordinated phosphate pH control.*' There were some problems with the system. In the present system, a ratio of 2.6 Na to 1.0 PO$_4$ is maintained. By the above program, any possibility of excess caustic built-up is eliminated.

(3) Addition of Chelants

Chelants have been used successfully to control the formation of deposits. They form soluble

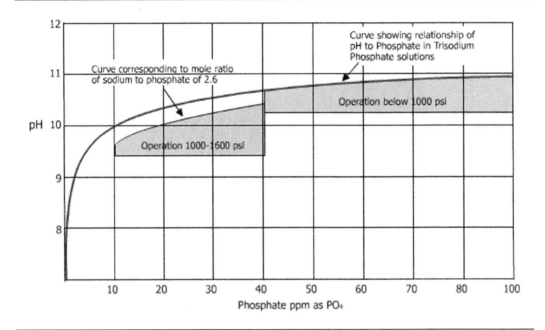

Figure 11.14 Boiler water limits (coordinated phosphate). (From Source: Babcock and Wilcox Company: Tech. Information Bulletin: Water Treatment for Industrial Boilers)

complexes with metal cations. Chelant is a compound which associates with more than one pair of electrons in a metal ion, forming a ring structure. Chelants of calcium and magnesium are soluble. Some common chelanting agents used are ethylenediamine tetraacetic acid (EDTA), nitro triacetic acid (NTA), etc. The chemical equation for chelation is similar to that for ion exchange. For instance, by adding the sodium salt EDTA to water

$$Ca^{2+} + NaEDTA \rightarrow CaNa_2EDTA + 2Na^+$$

the formation of $CaCO_3$ is eliminated, as shown in the above equation. Extreme care is needed to control this program. The following are the limitations in application of chelants:

(1) The feedwater hardness should not exceed 2.00 ppm.
(2) The ferric ion concentration in feedwater and condensate must be low.
(3) Oxygen should not enter the system, even in the shutdown period.

11.8.12 POLYMERS TREATMENT

Use is being made of polymers alone in controlling deposits in boilers up to $1500\,lb/in^2$. The treatment solubilizes Ca^{++}, Mg^{++}, Al^{+++} and maintains silica in solution. It removes scales from boilers. If polymers are to be used, oxygen must be strictly controlled. The effectiveness of a polymer is determined by its molecular weight and concentration. For example, polyacrylic acid (molecular weight = 20 000) addition reduces scale formation of only 52% compared to polymaleic acid (molecular weight = 5000) which reduces scale formation by 97%. Some typical polymers used are polycarboxylic acid, polymethacrylic acid, styrene and maleic acid.

11.9 MAJOR CORROSION PROBLEMS IN BOILERS

11.9.1 OXYGEN CORROSION

The whole boiler system is generally susceptible to corrosion caused by oxygen, mainly in the form of pitting. Superheater tubes in particular are subject to pitting corrosion due to the presence of condensed moisture and atmospheric oxygen. Saturated steam leaving the steam drum is directed to superheaters. In areas where moisture is collected deep pits are formed. The following is a summary of the corrosion mechanism leading to pitting in boiler. The subject of pitting has been discussed extensively in the chapter on forms of corrosion.

(1) In the anodic reaction, Fe is oxidized to Fe^{2+}

$$Fe \rightarrow Fe^{2+} + 2e^-$$

Iron is dissolved as a result of this anodic reaction.
(2) The cathodic reaction is the reduction of oxygen:

$$\frac{1}{2}O_2 + H_2O + 2e^- \rightarrow 2OH^-$$

The anodic and cathodic areas exist as a result of heterogeneities on the surface.
(3) As a result of hydrolysis, the pH is lowered and hydrogen ions are generated:

$$Fe^{2+} + H_2O \rightarrow FeOH^+ + H^+$$

(4) The hydrogen ions form molecular hydrogen gas in the pit

$$2H^+ + 2e^- \rightarrow H_2$$

An oxide cap is formed over the corrosion pits by the reactions given below:

$$2FeOH^+ + \frac{1}{2}O_2 + 2H^+ \rightarrow 2FeOH^+ + H_2O$$

$$FeOH^{++} + H_2O \rightarrow Fe(OH)_2 + H^+ \text{ (Alkaline)}$$

Acidic conditions are created and the pH is lowered inside the pit due to H^+ ions. Note that precipitation of Fe_3O_4 inside the pit also lowers the pH of the solution.

$$2FeOH^{++} + Fe^{++} + 2H_2O$$

$$\rightarrow Fe_3O_4 + 6H^+ \quad \text{(Magnetite)}$$

At the surface of the cap, cathodic reduction of oxygen takes place

$$\frac{1}{2}O_2 + H_2O + 2e^- \rightarrow 2OH^-$$

the hydroxyl ions react with $Fe(OH)_2$ formed under alkaline conditions and Fe_2O_3 is formed:

$$2Fe(OH)_2^+ + 2OH^-$$

$$\rightarrow Fe_2O_3 + 3H_2O \quad \text{(Hematite)}$$

The cap separates the pit from the alkaline boiler water. Inside the pit the solution is acidic. Iron corrodes as Fe^{2+} and electrons are released from the anode to the cathode. The difference in potential which exists between pit cap (acidic inside) and metal surface (alkaline) provides the driving force for the pitting reaction which is self-stimulating and continues until the metal is perforated. Although the oxygen within the pit is exhausted, the reaction once started continues once the acid is generated.

The following are the methods to control pitting in boilers:

(1) Eliminate oxygen by deaeration either mechanically or chemically. Scavenging must be done properly.
(2) Increase the pH of the boiler water.
(3) Provide a passive film of Fe_3O_4 on the metal surface or a chemical film. The corrosion of water wall and economizers also takes place depending on the protective nature of Fe_3O_4 film ($3Fe + 4H_2O \rightarrow Fe_3O_4 + 4H_2 \uparrow$). The chemical films are formed by application of film forming inhibitors, like amines and morpholine.

11.9.2 CAUSTIC CORROSION

This is the corrosion which results from the presence of sodium hydroxide in excess of the requirement. Caustic is added in boilers to maintain a thin protective film of magnetite (Fe_3O_4) on the surface of the steel tubes. The pH is controlled by sodium hydroxide. The problem arises only when the concentration of NaOH increases by 25%. The sodium hydroxide can

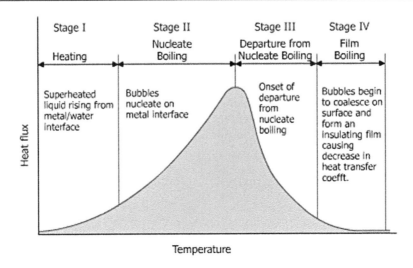

Figure 11.15   A typical boiling curve showing the different stages of boiling

become concentrated by one of the following mechanisms:

(a) Departure from Nucleate Boiling

Under turbulent flow condition in the tubes, three types of flow zones are observed – laminar sublayer, buffer zone and turbulent zone. The laminar sublayer exists at the metal–water interface and provides the main resistance to heat transfer from the tube wall to the bulk of circulating water. As the temperature gradient across the laminar sublayer increases, the concentration effect of solids on the surface also increases. The solid contents increase and the boiling point is raised. As the temperature difference between the tube wall and the bulk liquid increases, steam bubbles are formed in large numbers and they remain in the vapor state when they are transported to the bulk liquid. This is termed as *nucleate boiling*. The term nucleate boiling refers to a condition in which discrete bubbles of steam are formed at points on a metal surface. As these bubbles form, minute concentrations of boiler water solids develop at the interface of bubble and water. The formation of bubbles causes agitation and the mode of heat transfer is changed from conduction to convection. With increased agitation, the heat transfer coefficient increases.

This continues until a maximum is reached. At this point the bubbles are no longer removed from the surface. Solids begin to deposit on the metal surface and the heat transfer coefficient is decreased. The bubble now *coalesces* on the surface. This condition is called '*film boiling.*' The point at which film boiling and nucleation occur only partially, is called '*Departure from Nucleate Boiling*' (DNB). Different stages of boiling are shown in Fig. 11.15.

At the point of DNB, the rate of bubble formation exceeds the rate at which they are rinsed from the surface by water. As a consequence of this, dissolved solids, such as caustic and other types, concentrate on the surface and damage the highly protective magnetite (Fe_3O_4) on the surface.

(b) Evaporation of a Waterline

Another reason for concentration of caustic solution is the evaporation of a waterline. Where a waterline exists, corrosive species may concentrate by evaporation. Parallel trenches may form in horizontal tubes and coalesce into a single longitudinal gouge on the top of the tube. Cracking is easily identified by gouging. Gouging produces hemispherical depressions due to the reaction of high concentration of sodium hydroxide with the metal. Large quantities of magnetite crystal are present in the attacked area. Gouging eventually leads to tube failures.

11.9.3 MECHANISM OF CAUSTIC CORROSION

The undesired high concentration of caustic destroys the magnetite film and the base metal is attacked by the caustic. The reasons for the concentration of caustic in high-pressure boilers are already given above. The following reactions take place between the magnetite film and caustic:

$$Fe_3O_4 + 4NaOH \rightarrow 2NaFeO_2 + Na_2FeO_2$$
$$+ 2H_2O \qquad (11.1)$$
$$Fe + 2NaOH \rightarrow Na_2FeO_2 + H_2 \uparrow \quad (11.2)$$

The protective layer of Fe_3O_4 is destroyed as shown in equation (11.1). The base metal after it is stripped of Fe_3O_4 is attacked by NaOH. Sodium ferrite is formed and hydrogen is released. The bubbles carry the sodium ferrite formed (soluble). As sodium ferrite is removed, the surface reacts with more water and the process continues, until the strip on the tube surface is significantly thinned down. The sodium ferrite is converted to magnetite by hydrolysis when it comes into contact again with dilute (normal) boiler water. However, this time Fe_3O_4 is formed in some other region. The tube loses ductility and fails finally by thinning.

11.9.4 GENERAL METHODS OF PREVENTION OF CAUSTIC CORROSION

(1) Minimize the amount of free sodium hydroxide.
(2) Prevent condenser in-leakage as even a small in-leakage of a few ppm may be sufficient to cause localized corrosion.
(3) Prevent localized concentration of corrosive chemicals by preventing waterlines in tubes,

excessive departure from nucleate boiling and water side depositions.
(4) Prevent leakage of regenerator chemicals from the makeup water demineralizer.
(5) Prevent DNB by minimizing the hot spots. Such spots may be caused by over-firing or under-firing.

11.9.5 CAUSTIC EMBRITTLEMENT

This denotes a type of stress corrosion cracking encountered in steel exposed to high concentration of hydroxide at a temperature of 200–250°C. Caustic cracking of steel is sometimes called as caustic embrittlement (Fig. 11.16). It is a special type of stress corrosion cracking that occurs in boilers. The tube cracks under the influence of stress and high concentration of hydroxide which causes it to corrode. There are two conditions which must be satisfied:

(1) High concentration of sodium hydroxide (over 50 000 ppm).
(2) The boiler metal must be under high stress such as the areas where the boiler tubes are rolled into the drum. The welded joints become highly susceptible to caustic embrittlement. The attack proceeds on the grain boundaries, hence, it is an inter-crystalline phenomena. Post-weld heat treatment or stress relieving heat treatment can preclude cracking problems.

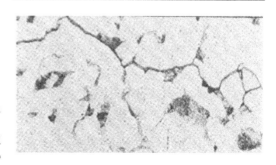

Figure 11.16 Caustic cracking of a boiler steel tube, showing the microstructure (500X)

To prevent caustic embrittlement, avoid concentration of chemicals, avoid departure from nucleate boiling, keep the surface free from deposits and minimize in leakage of alkaline salts. Sodium nitrite has been successfully used to inhibit caustic embrittlement.

11.9.6 HYDROGEN DAMAGE

This type of damage generally occurs in boilers operating above 1000 psi. When the internal surface of the boiler tube corrodes hydrogen is generated. It is usually caused by operating the boiler with a low pH water and deposition of contaminants within the deposits on the internal tube wall.

As suggested in the mechanism of caustic corrosion, caustic in a concentrated form dissolves the magnetic oxide layer on the tube wall.

$$4NaOH + Fe_3O_4 \rightarrow 2NaFeO_2 + Na_2FeO_2$$

$$+ 2H_2O$$

The protective coating is thus destroyed. The bare surface after the stripping of Fe_3O_4 reacts with water to produce hydrogen.

$$3Fe + 4H_2O \rightarrow Fe_3O_4 + 8H \uparrow$$

Hydrogen may also be produced by the reaction between iron and caustic.

$$Fe + 2NaOH \rightarrow Na_2FeO_2 + 2H \uparrow$$

Hydrogen in the molecular form is not capable of diffusion in steel, however, in the atomic form it diffuses into steel. Surface heterogeneities, voids, inclusion and/or grain boundaries are the preferred sites for accumulation of atomic hydrogen. The atomic hydrogen combines to form molecular hydrogen and builds up a high-pressure internally and lowers the fracture stress.

$$H + H \rightarrow 2H \rightarrow H_2 \uparrow$$

Some of the atomic hydrogen reacts with chromium carbide precipitated at the grain boundaries.

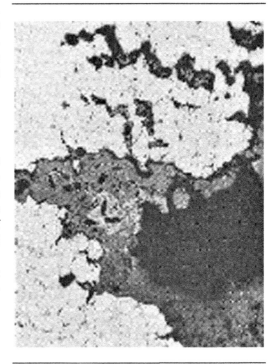

Figure 11.17 Intergranular oxidation. (From Source: NALCO Bulletin No. 61, Boiler Metal Failure Analysis, NALCO Chemical Co., Oak Brook, IL, USA)

$$Fe_3C + 4H \rightarrow CH_4 + 3Fe$$

Both the gases H_2 (molecular form) and methane (CH_4) accumulate preferentially at the grain boundaries. Intergranular oxidation is shown in Fig. 11.17. The increased pressure leads to discontinuous microcracking at grain boundaries. The grain boundaries are the preferred sites for attack as their surface energy is lower than the grain matrix and impurity segregation also takes place at the grain boundaries.

As the microcracks accumulate, the tube strength decreases. The tube bursts when the internal pressure caused by the accumulation of gas exceeds the yield strength of the metal which is still intact. The crack is longitudinal and it resembles the crack caused by creep rupture. The methods applied to control caustic corrosion are also valid for controlling hydrogen damage.

11.9.7 STRESS CORROSION CRACKING (SCC)

The phenomenon is defined as the cracking of material under the combined effects of stress and corrosion. Caustic cracking of steel is a case of stress corrosion cracking, sometimes called caustic embrittlement. The cracks are branched and propagate intergranularly and transgranularly. The stress involved may be either applied or residual. The localized corrosion on the surface acts as a stress raiser. The cracking can be inter-granular or transgranular. The austenite steels are susceptible to SCC in chloride environment. The cracking of steel may occur in caustic service by precipitation of caustic in high concentration, as discussed earlier. Failures due to SCC in feed-water heater are not very common. The material used for feedwater heaters are 90 Cu–10 Ni, car-bon steel, austenitic steels and monels. Carbon steels and 90 Cu–10 Ni are virtually immune to SCC. Much depends on the deviation from water chemistry. Chloride may concentrate on the shell side in superheating zone tubes of feedwater heaters, if there is a small leak. Traces of ammonia may cause the SCC of copper alloys in the presence of oxygen. Turbine discs may also be subjected to SCC. The evaporation of moisture causes the concentration of impurities. The impurities gen-erally deposit where the metal temperature is above the temperature of the steam. Stresses are increased by deposition of solid impurities. The problem is minimized by proper selection of materials for turbine blades. High concentra-tion of chlorides, caustics and acids must not be allowed to be built up. Generally low-carbon steel is used for low-pressure shells and cylinders and chrome–moly or austenitic steel for high pres-sure. The problem can be minimized by lowering of stresses, selection of proper material and rigid control of water chemistry.

11.9.8 CORROSION FATIGUE

This is a process under which a metal fractures prematurely under conditions of corrosion, and simultaneous cyclic loading at lower stress loads that would be required otherwise in the absence of an aggressive environment. The cracking

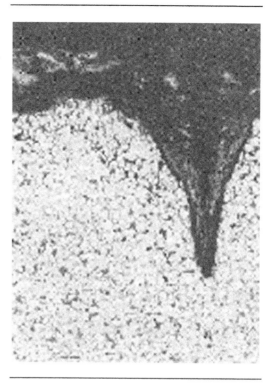

Figure 11.18 Corrosion fatigue crack. (From Source: NALCO Bulletin No. 70, NALCO Chemical Co.)

begins at surface defect, pits and irregularities and propagate transgranularly. Failure occurs by crack propagation. The cracks are unbranched and propagate perpendicular to the surface of the metal (Fig. 11.18). In the case of boilers, the con-version of metal (Fe) to Fe_3O_4 (discussed under pitting) can lead to corrosion fatigue, if sufficient cyclic stress is present. The oxide layer is brittle relative to the tube wall and it is preferentially fractured. The microscopic crack passes through the oxide layer to the metal surface. The oxide at the root of the crack forms a notch. In the next cycle, oxide fractures along the notch and the notch becomes deeper. Finally, as the pro-cess is repeated again and again a crack (needle or wedge shaped) propagates until the tube wall is penetrated or rupture of the tube occurs.

Corrosion fatigue failure is generally not a common failure in boilers. The failure is observed when the boilers are not used continuously or subjected to frequent 'startups' and shutdown. Proper control of 'startup' and 'closedown' pro-cedures can eliminate corrosion fatigue.

11.9.9 ACID ATTACK

This is an attack caused by the exposure of internal surfaces of boilers to low pH water. All areas of the boilers may be affected by acid attack Two conditions are responsible for the acid attack:

(a) The concentration of acid salts.
(b) Deviation from recommended boiler water chemistry.

The first condition is observed when seawater is used as cooling water. The second condition is observed when porous deposits or crevices occur. In the presence of deposits or a crevice, acid salts are concentrated. These salts hydrolyze and generate low pH conditions:

$$MgCl_2 + 2H_2O + Heat \rightarrow Mg(OH)_2 \downarrow + 2HCl$$

Acid attack may also occur during the acid cleaning of boiler. Caustic gouging of low carbon steel is shown in Fig. 11.19.

Mechanism

The acid attack may occur by one or more of the following processes:

(a) The protective film of magnetite is dissolved by exposure to low pH and the bare metal

surface is, therefore, attacked. The gouged areas are generally found to be covered with oxide of iron which are soluble.

(b) When the metal is subjected to low pH condition, atomic hydrogen is released which causes hydrogen damage by formation of molecular hydrogen or methane in internal defects (voids or grain boundaries).

(c) Acid salts such as $MgCl_2$ are hydrolyzed

$$MgCl_2 + 2H_2O \rightarrow Mg(OH)_2 \downarrow + 2HCl$$

$Mg(OH)_2$ precipitate until the pH falls to 4.0. The pH must, therefore, be controlled to avoid attack by HCl.

$$Fe + 2HCl \rightarrow FeCl_2 + H_2 \uparrow$$

The formation of $MgCl_2$ can lead to deposits of HCl underneath the solid deposits resulting in an extremely high rate of corrosion. Inhibitors may be used to prevent the above process.

External acid attack occurs by chemical combination of sulfur with oxygen forming SO_2 and SO_3. The SO_3 in contact with water forms H_2SO_4 (vapor) which condenses on the secondary areas at temperature between 121 and 177°C. The attack is very serious and damaging. Soot may also deposit in some areas. When vapor condenses on them, rapid corrosion is caused because of a high concentration of acid in the soot.

The control of water chemistry, selection of materials and adoption of designs which eliminate blind pockets are the means. Monitoring operations, such as control of strength of acids, deposit analysis and regulation of temperature are important control measures.

11.9.10 CHELANT CORROSION

As discussed earlier, a chelant is a compound which is capable of associating with more than one pair of electrons forming a ring structured complex of high stability. Whereas the usefulness of chelant treatment for removal of scales cannot be underestimated, high chelant concentrations are as harmful as high concentration of caustic. The chelant can be concentrated by

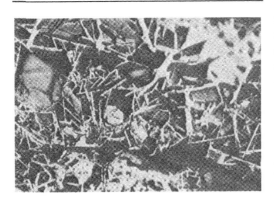

Figure 11.19 Caustic gouging of low-carbon steel. (From Wyncott, D. (1979). (NALCO), Corrosion in Boiler and Condensate Systems, *Plant Eng.*, October 18)

evaporation where departure from nucleate boiling occurs. Concentrated chelants may attack magnetite according to the following reaction:

$$Fe_3O_4 + Fe + 12H^+ + 4\,chelant$$

$$\rightarrow 4Fe(H)\,chelant + 4H_2O$$

Because of film boiling and steam blanketing, the principal source of oxygen is in-leakage to condensate systems. As steam condensation, it creates a vacuum which drains air into the system, steam traps, condensate drains and stress of in-leakage are sources of oxygen. Chelant corrosion is encountered generally in steam separating equipment. The tube surface remains smooth after chelant attack, but if the velocity is high, it may form horseshoe-shaped depressions. In the presence of oxygen, islands of unattacked metal surrounded by attacked metal may be observed by microscope.

Oxygen, even in trace amounts, must be eliminated. High chelant residuals must not be allowed under any circumstances to build in boilers. All hot spots must be removed and the quality of feedwater must be carefully controlled. Copper-base alloys must not be exposed to chelants. The protecting layer of magnetite Fe_3O_4 is removed and corrosion initiated.

11.9.11 FUEL-ASH CORROSION

Fuel-ash corrosion is caused by V_2O_5 on burning of fuel oil. Vanadium pentoxide melts at a relatively lower temperature, and forms a molten layer on the metal and causes accelerated corrosion. If the fuel oil contains sodium sulfur and vanadium, sodium sulfur and vanadium oxide are formed. But V_2O_5 is most corrosive. This phenomenon, which mainly occurs in superheaters and reheaters in the temperature range (593–816°C), is also called 'vanadium attack.' The vanadium content of heavy oil is oxidized to V_2O_5 on burning, the melting point is fairly low so that it promotes corrosion.

$$3V_2O_4 + \frac{3}{2}O_2 \rightarrow 3V_2O_5$$

$$3V_2O_4 + \frac{1}{2}O_2 \rightarrow 2V_2O_5$$

In the presence of sodium and sulfur, the melting point of vanadium compounds are reduced. The catalytic oxidation of the metal surface by V_2O_5 causes a decrease in the wall thickness and the load carrying area which introduces high stresses. These stresses in combination with high temperatures cause the failure of the superheater tubes. Modern boilers are now designed to operate at a high-temperature range (538–551°C).

11.9.12 COAL-ASH CORROSION

During the burning of coals, the mineral constituents in coal are subjected to high temperature (566–732°C), liberating surface oxide and alkali compounds. The alkali compound and sulfur compounds condense on the fly ash to form complex compounds, such as $K_3Te(SO_4)_3$ and $Na_3Fe(SO_4)_3$ which form slags at the tube wall and cause loss of wall thickness. Coal-ash corrosion is associated with coals containing more than 3.5% sulfur and 0.25% chloride. Important measures taken to minimize corrosion include over-designing of tubes, use of thermally sprayed metallic coating, and lowering of steam temperature.

QUESTIONS

A. MULTIPLE CHOICE QUESTIONS

Circle the correct answer for the following statements:

1. The primary objective in designing a boiler is to

   a) produce pure steam
   b) produce a design for reliable operation
   c) produce efficiency in heat adsorption
   d) produce economically and technically compatible unit

2. Fire-tube boilers are efficient steam generators

   a) below 150 lb/in^2 (10 bar)
   b) below 50 lb/in^2

c) above 300 lb/in^2
d) below 150 000 and 150 lb/in^2

3. Usually the radiant section (entire fire box) is surrounded by water wall tubes through which water circulates. Connection between the various tubes are obtained by headers. All this is done to

a) improve the steam quality
b) achieve fuel economy
c) increase steam generation
d) increase boiler efficiency

4. The feedwater is made of condensate return and boiler mixing water. The temperature of this water is raised to the boiler water temperature by recovering heat from the heat flue gases. This heating occurs in

a) air heaters
b) economizers
c) feedwater heaters
d) convection superheaters

5. Which one of the following is *not* the advantages of horizontal tube boiler?

a) greater economy in inspection
b) ease of replacement
c) low headroom
d) visibility of tubes

6. There are several forms of bent tube boilers, A, O and D types. In a particular boiler, the steam drum is located directly above the mud drum and both are located in the center of the boiler. This boiler belongs to

a) A class
b) O class
c) D class
d) C class

7. The major impurities in the feedwater are introduced by (Mark two)

a) makeup water
b) raw water
c) Return condensate

8. Which one of the following is the most troublesome type of turbine deposit?

a) sodium salt
b) biological deposit
c) silica
d) surface impurities

9. The pH of a condensate is generally low because of the presence of

a) ammonia
b) carbon dioxide
c) oxygen

10. The principle source of corrosion of copper tubes in the condensate system is

a) oxygen
b) carbon dioxide
c) ammonia

11. Waste gases escaping up the stack cause greatest loss of heat in a steam generating system. These waste gases are recovered by

a) reheaters
b) superheaters
c) economizers

B. How and Why Questions

1. What is a boiler? State the most important objectives of boiler.
2. How is steam formed in the boiler?
3. Why the rate of evaporation does not exceed the rate of water flow in the downcomer tubes?
4. What is the function of economizers and where are they located?
5. What is the effect of high pH and low pH substances on the protective magnetite layer of steel tubes?
6. What is condensate polishing? Why is the condensate contaminated?
7. What is the major cause of caustic cracking? Give one method to prevent it.
8. How can DNB (departure from nucleate boiling) be prevented?

9. How is atomic hydrogen liberated and what damage does it do to the steel tubes?
10. What is the effect of sodium hydroxide on steel tubes once the magnetite layer is removed?

C. Conceptual Questions

1. State three advantages of a bent tube boiler over a horizontal tube boiler.
2. State two advantages and two disadvantages of fire-tube and water-tube boilers.
3. State two methods of prevention of condensate corrosion.
4. State two critical factors which would minimize the susceptibility of steel tubes to hydrogen damage.

Suggested Reading

[1] Alfanso, A. (1994). Preventing corrosion in steam boilers, *Modern Power Systems*, **14**, (11), 02607840.
[2] Dooley, B. and McNaughton, W.P. (1997). Don't led those boiler tubes fail again, Part 1. *Power Engineering*, **101**, (6).
[3] Dooley, B. and McNaughton, W.P. (1997). Don't led those boiler tubes fail again, Part 2. *Power Engineering*, **101**, (8).
[4] Kuehen, S.E. (1995). Battling boiler corrosion. *Power Engineering*, **99**, (7), 00325961.
[5] More, J. (2001). Inlet water temperatures and boiler corrosion. *Air Conditioning Heating and Refrigeration*, **26**, 22, March.
[6] Kemmer, F.N. (1988). *NALCO Water Handbook*, 2nd ed. New York: McGraw Hill.
[7] Port, R.D. (1991). *The NALCO Guide to Boiler Failure Analysis*. New York: McGraw Hill.
[8] Kakac, S. (1991). *Boilers, Evaporators, and Condensers.* New York: John Wiley & Sons.
[9] Drew Chemicals, *Principles of Industrial water Treatment*, 3rd ed. Drew Chemical Corporation, Boonton, 1979.
[10] Newman, T.R. *Failure Analysis.* Industrial Research Development, NALCO Chemical Company, Illinois, Dec 1979.

Keywords

Air heater A device in which air is heated by flue gases flowing in the tubes to the desired temperature at the outlet.

Bent tube boiler It is a multi-drum boiler with the steam drum at top and mud drum at the bottom.

Blowdown Control of concentration of chemicals in boiler water by bleeding off a portion of water from the boiler.

Boiler Heat exchange equipment which convert water into steam.

Caustic corrosion Corrosion resulting from the concentration of caustic soda and dissolution of the protective magnetite layer.

Condensate Steam that is returned to the boiler and condenses is called a condensate.

Cyclone separators Devices in the steam drum which separate the water from steam.

Demineralization Removal of minerals from water. The term is generally restricted to ion exchange process.

Departure from nucleate boiling The rate of bubble formation exceeds the rate of bubble collapse. As a consequence, sodium hydroxide would concentrate and lead to the dissolution of magnetite film.

Downcomer Cold water flows in the tubes after separation of steam in the steam drum. The water flows in the downward direction.

Economizer A counter-flow heat exchanger for recovering additional energy from the combustion products increasing the temperature of feedwater heater and minimizing the temperature difference between saturation temperature and feedwater temperature.

Feedwater It constitutes the makeup water to replace the water lost by evaporation and the condensate returned.

Fire-tube boiler A type of boiler in which flue gases are inside the tubes in contact with water outside the tubes.

Forced circulation The use of an external device, such as a pump, to force the circulation of water.

Heat exchanger A device in which heat is transferred between two fluid streams at different temperatures in a space separated by a solid wall.

High-pressure boiler A boiler operating in the pressure range of 10–14 MPa.

Hot lime softening The softening process operates at 90°C. The increased temperature increases the potential of the softener to exchange sodium for hardness ions and results in a lower hardness than that which can be achieved.

Hydrogen cracking The damage to the tubes caused by hydrogen in the atomic form (uncombined form of hydrogen).

Low-pressure boiler A boiler operating at a pressure lower than 10 MPa (1450 psi).

Makeup water Water added in the boiler to makeup for loss caused by evaporation.

Mud drum A device for separating suspended solids and sludge from the water.

Natural circulation The circulation of water based on the density difference between the cold water (heavier) and the steam–water mixture (lighter).

Nucleate boiling A condition in which bubbles of steam nucleate at points on a metal surface and minute concentrations of solids develop at the interface of the bubbles and water.

Oil-ash corrosion A liquid phase corrosion phenomenon where metal temperatures are in the range of 593–816°C.

Reheater It works on the same principles as the superheater. The only difference being operating pressure (inlet pressure = 4 MPa) at ambient temperature.

Risers As a result of heating, the water rises in the tubes called riser tubes.

Scavenging A process for removal of oxygen by use of chemicals such as sodium sulfate and hydrogen.

Superheater Increases the temperature of the saturated or near-saturated steams. It is a heat exchanger in which steam flows inside the tubes and flue gas flows outside the tubes (outlet pressure = 18 MPa).

Water-tube boiler A boiler type in which water flow circulates inside the tubes and the tubes are heated by the flue gases.

Water walls The water tubes in the furnace constitute the water-cooled walls.

CHAPTER 12

CONCRETE CORROSION

12.1 INTRODUCTION

Concrete is the most widely used constructional material in the world. In spite of the impressive performance of concrete in several structures, the deterioration of concrete has assumed alarming proportion in harsh climatic conditions, such as the sea-coastal areas in Gulf countries. Concrete can be defined as the artificial stone produced when cement (usually Portland cement) is mixed with fine aggregate (sand), a coarse aggregate (gravel or crushed stones) and water.

12.1.1 APPLICATIONS

The intended application of concrete dictates the proportioning of major concrete ingredients, cement type, as well as particle size gradations of the fine coarse aggregates. Concrete generally provides to embedded steel a high degree of protection against corrosion by providing a highly alkaline environment which passivates the steel against corrosion. Corrosion of reinforcing steel can, however, proceed in some environments, such as a marine environment, due to the presence of chloride.

12.2 TECHNICAL TERMS

Definitions and a brief explanation of important terms related to concrete are given below to familiarize engineers with the applications of concrete.

12.2.1 AGGREGATES

Aggregates used in normal construction comprise granular materials, like limestone, dolomite, gravel, silica, etc. Table 12.1 shows the effect of properties and characteristics of aggregates which affect the characteristics of concrete. The salts in the aggregates, such as sulfates, chlorides and carbonates of calcium, magnesium and sodium, contribute to the corrosion of concrete.

The aggregate size used includes both fine and coarse fractions. The coarse fraction refers to particles greater than 4.75 mm and fine fraction to particles smaller than 4.75 mm but larger than 75 μm. The aggregate makes up from 65–75% of volume of concrete.

12.2.2 ADMIXTURES

These are inorganic or organic substances that are added to the fresh concrete before or during mixing to modify some properties, such as their setting and hardening characteristics. The most common mixtures are the air-entraining type. They form a network of voids (0.001–0.003 in diam) in the cement phase of concrete. Self-retarding admixtures increase the setting time for concrete, whereas self-accelerating mixtures decrease the setting time. Polymers are used for bonding and plasticizers for setting of concrete. Over-dosing with admixture may create problems.

Table 12.1   Properties of concrete which are influenced by properties and characteristics of the aggregates

Concrete property	Relevant aggregate property
Durability:	
■ Resistance to freezing and thawing	Soundness Porosity Pore structure Permeability Degree of saturation Tensile strength Texture and structure Presence of clay
■ Resistance to wetting and drying	Pore structure Modulus of elasticity
■ Resistance to heating and cooling	Coefficient of thermal expansion
■ Abrasion resistance	Hardness
■ Alkali-aggregate reaction	Presence of particular siliceous constituents
Strength	Strength Surface texture Cleanness Particle shape Maximum size
Shrinkage	Modulus of elasticity Particle shape Grading Cleanness Maximum size Presence of clay
Coefficient of thermal expansion	Coefficient of thermal expansion Modulus of elasticity
Thermal conductivity	Thermal conductivity
Specific heat	Specific heat
Unit weight	Specific gravity Particle shape Grading Maximum size
Modulus of elasticity	Modulus of elasticity Poisson's ratio
Slipperiness	Tendency to polish
Economy	Particle shape Grading Maximum size Amount of processing required Availability

12.2.3 MIXING WATER

The use of suitable mixing water is essential to make good concrete. Any unmineralized water can be used. Water to be excluded includes water containing total dissolved solids higher than 3.5%, sulfates in excess of 0.5%, high acidic water with a low pH (>4.0), industrial water from tanneries, cellulose factories and highly polluted water. Rain water and swamp water may be used for mixing with caution depending on the total dissolved solids.

12.2.4 WATER–CEMENT RATIO

A low water–cement ratio is necessary for the production of high strength cement. The typical design water–cement ratio is in the range of 0.4–0.5. Cement pastes of high and low density water–cement ratio are shown in Fig. 12.1. The compressive strength and density of concrete is strongly influenced by the water–cement ratio.

12.2.5 CLASSIFICATION OF CONCRETE ON THE BASIS OF COMPRESSIVE STRENGTH

They are classified in three different categories:

- Low strength   – Less than 20 MPa
- Medium strength – Between 20 and 40 MPa
- High strength   – Greater than 40 MPa.

Definition of important constituents of concrete are given in 'Selection and Use of Aggregates for Concrete,' ACI Manual of Concrete Practice, Part I, American Concrete Institute (1970).

12.3 PRODUCTION, TYPES AND CURING OF CEMENT

The production of cement involves the heating of an intimately mixed source of CaO (limestone)

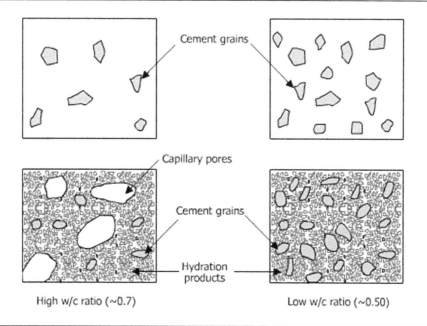

High w/c ratio (~0.7)   Low w/c ratio (~0.50)

Figure 12.1   Schematic of evolution of structure of cement pastes of low and high water–cement ratio showing hydration products, cement grains and pores

and SiO_2 (clay), in kiln, at temperatures around 1400°C (Fig. 12.2). Five main types of cements used are shown below:

English description	American (ASTM) Description
Ordinary Portland cement	Type I
Rapid hardening Portland cement	Type III
Low heat Portland cement	Type IV
Modified low-heat Portland cement	Type II
Sulfate resisting cement	Type V

Type I cement is most commonly used.

Concrete is produced by mixing the ingredients in mixers such as rotating drum or horizontal mixers. Consolidation of freshly mixed concrete is assisted by types of vibrators available to obtain a maximum density. To obtain maximum properties from concrete, it must be properly cured. Following are the basic requirements for curing:

(1) The concrete must contain an adequate water content to complete the process of hydration. Moisture losses must be prevented during the setting time (e.g. a week).
(2) The temperature of the concrete must be maintained above the freezing point.
(3) The concrete must be protected against external movements and disturbances to allow the formation of a uniform cohesive mass.
(4) Sufficient time must be allowed to ensure the completion of the hydration reaction and achieve maximum hardening of the concrete.
(5) Prevention of moisture loss by use of polyethylene sheets.

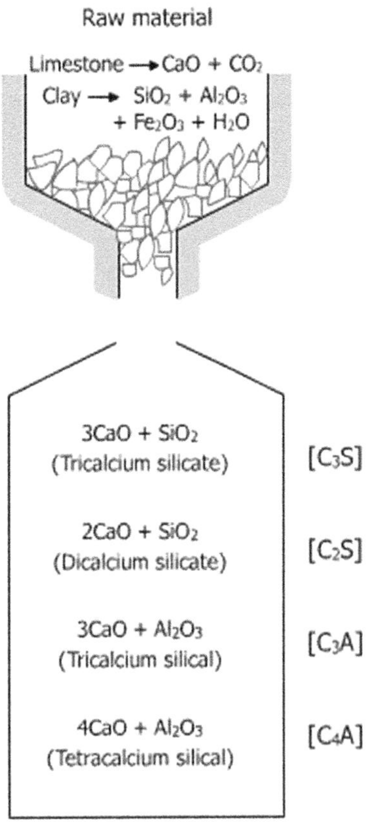

Raw material

Limestone → $CaO + CO_2$
Clay → $SiO_2 + Al_2O_3$
  $+ Fe_2O_3 + H_2O$

$3CaO + SiO_2$ (Tricalcium silicate)	[C₃S]
$2CaO + SiO_2$ (Dicalcium silicate)	[C₂S]
$3CaO + Al_2O_3$ (Tricalcium silical)	[C₃A]
$4CaO + Al_2O_3$ (Tetracalcium silical)	[C₄A]

Figure 12.2 Schematic diagram showing the production process of Portland cement

12.4 ENGINEERING PROPERTIES OF CONCRETE

Concrete offers the advantage of tailoring its properties to desired ranges. A summary of important properties of concrete is given below.

12.4.1 DENSITY

The density of concrete is controlled by the water–cement ratio, air contents and specific gravity of the aggregates used. Generally concrete using light-weight aggregates exhibit densities in the range of 0.0465 g/cm^3 (400 kg/m^3) to 0.8 g/cm^3 (800 kg/m^3). Higher densities are exhibited by concrete containing limestone in the range of 2.24 g/cm^3 (2240 kg/m^3) to 2.4 g/cm^3 (2400 kg/m^3).

12.4.2 POROSITY

The porosity include voids created by entrained air, aggregate and gel pores (-18 Å), and capillary pores in the hydrated cement paste. The magnitude of porosity is directly affected by water–cement ratio, the amount of entrained air and the nature of aggregates.

12.4.3 COMPRESSIVE STRENGTH

Concrete has a high compressive strength. Very high strength concrete (80–90 MPa) can be produced routinely. The classification of concrete on the basis of compressive strength has been shown earlier. The water–cement ratio has an important bearing on the compressive strength of concrete.

12.4.4 TENSILE STRENGTH

This is one area of weakness of concrete. Concrete is strong in comprehensive strength but is very weak in tensile strength. This inherent weakness of concrete led to the idea of using steel as a reinforcement to increase its capability to carry large tensile loads. The tensile strength of typical concrete does not exceed 2–4 MPa.

12.4.5 DRYING SHRINKAGE

If concrete is exposed continuously to water, it exhibits insignificant expansion over a period of several years. Concrete exposed to dry and wet conditions would, however, be subjected to loss of water causing it to contract and leading to its cracking due to development of internal stresses. The shrinkage caused by drying of water in concrete is called '*shrinkage*.'

12.5 CHEMISTRY OF PORTLAND CEMENT

Cement is generally termed as Portland cement. The production of cement involves heating a mixture of CaO and SiO_2 (such as, limestone and clay) in a kiln ($-1350°C$). Cement is the most

Table 12.2 Typical properties of the five major types of Portland cement

Constituents	Type of Portland cement				
	1	2	3	4	5
C_3S	51	46	58	26	39
C_2S	25	32	16	54	43
C_3A	9	4	8	2	2
C_4AF	8	12	8	12	8
Others	7	6	10	6	8

Key:
C_3S = tricalcium silicate C_2S = dicalcium silicate
C_3A = tricalcium aluminate
C_4AF = tricalcium aluminoferrite

reactive ingredient in Portland cement concrete, the others, such as aggregates, are relatively inert. The raw materials in the kiln and other major constituents of Portland cement are shown in Fig. 12.2. The typical properties of different constituents in the five major types of Portland cement are shown in Table 12.2.

The major constituents of Portland cement shown above are not pure compounds. The four major constituents make up about 93% by weight. All five types of cement contain the four major constituents (C_3S, C_2S, C_3A and C_4AF) in different proportions.

12.6 CONCRETE CHEMISTRY, HYDRATION REACTIONS

The chemistry of concrete primarily involves a study of hydration products of concrete formed on addition of water. A summary of hydration reactions is given below:

1. $2(3CaO \cdot SiO_2) + 6H_2O$
   (Tricalcium silicate) (Water)
   $= 3CaO_2 \cdot SiO_2 \cdot 3H_2O + 3Ca(OH)_2$
   (Calcium silicate hydrate) (Calcium hydrate)
2. $(2CaO \cdot SiO_2) + 4H_2O$
   (Dicalcium silicate) (Water)
   $= 3CaO \cdot 2SiO_2 \cdot 3H_2O + 3Ca(OH)_2$
   (Calcium silicate hydrate) (Calcium hydroxide)

3. $4CaO \cdot Al_2O_3 \cdot Fe_2O_3 \ + 10H_2O$
(Tetracalcium aluminoferrite)   (Water)
$+ \ 2Ca(OH)_2$
(Calcium hydroxide)
$= 6CaO \cdot Al_2O_3 \cdot Fe_2O_3 \cdot 12H_2O$
(Calcium aluminoferrite hydrate)

4. $3CaO \cdot Al_2O_3 \ + 12H_2O + Ca(OH)_2$
(Tricalcium aluminate)   (Water)   (Calcium hydroxyl)
$= 3CaO \cdot Al_2O_3 \cdot Ca(OH)_2 \cdot 12H_2O$
(Tetracalcium aluminate hydroxyl)

5. $3CaO \cdot Al_2O_3 \ + 10H_2O + CaSO_4 \cdot 2H_2O$
(Tricalcium aluminate)   (Water)   (Gypsum)
$\rightarrow 3CaO \cdot Al_2O_3 \cdot CaSO_4 \cdot 12H_2O$
(Calcium monosulfoaluminate)

The most important reactions occur with C_2S ($2CaO \cdot SiO_2$) and C_3S ($3CaO \cdot SiO_2$). The rate of hydration reaction is affected by temperature. Increased temperature leads to an increased rate of hydration, but it leads to a decreased tensile strength of concrete. The resulting hydration paste comprises heterogeneous product. The paste can be described as a crystalline gel which is made of cement gel, unhydrated cement, capillary pores containing water vapor, water and air. The gel pores are very small in size and they are filled with water. Chemically hardened cement is a complex hydrate formed of the four basic components shown above. The pores and hydration products are shown in Fig. 12.1.

12.7 DURABILITY AND DETERIORATION OF CEMENT

Concrete is a building material of the twentieth century and has been extensively used in temperature and harsh climatic conditions with success. Concrete is a very durable material, and steel embedded in concrete functions as a very effective reinforcement. Concrete provides an alkaline environment to steel and minimizes the risk of corrosion of steel. Two conditions are essential for durability of concrete:

(a) The concrete should be of high quality and should not undergo environmental degradation.

(b) The reinforcing steel must not be allowed to corrode, by maintaining the alkaline environment.

However, under certain environmental conditions, concrete may undergo degradation and reinforcing steel may also corrode. Details of degradation of concrete and corrosion of reinforcing steel are described below.

12.7.1 DEGRADATION OF CONCRETE

Conditions which exist at the time of production of concrete and its exposure to an external environment during its working life time, are two major factors which affect the durability of concrete.

(a) Conditions Prevailing at the Time of Production

Good production practices demand adhering to lower temperature ($5°C$) and upper temperature ($32°C$) limits. Deviation from the above limits may lead to decreased durability of concrete as shown below:

1. At higher temperatures in hot weather, additional water demand for mixing water leads to an increased water–cement ratio, thus adversely affecting the strength of the concrete.
2. A more rapid rate of hydration reaction may lead to an accelerated setting and a lower strength.
3. Rapid evaporation of mixing water leads to an increase in shrinkage of concrete.
4. In cold weather, the mixing water may freeze with an increase in the total volume of concrete, and delaying the setting and hardening of concrete.

(b) Environmental Factors Affecting the Durability of Hardened Concrete

1. **Freeze–thaw cycling**. The temperature of concrete is lowered, and water in the capillary

pores of the cement freezes causing an expansion of the concrete which further continues after thaw cycle. Thus, the freeze–thaw cycle has a conjoint effect on the expansion of hardened concrete. It may be recalled that cement paste has two types of pores: gel pores which are very small and filled with water and capillary pores, much larger than the gel pores and filled with air. The gel water diffuses to capillaries and ice is formed in the capillaries. The formation of ice leads to increase of volume and a dilating pressure is developed. When the pressure exceeds the tensile strength of concrete, damage to concrete occurs. The damage to freeze–thaw cycle may be minimized by using:

- dried aggregate,
- use of admixtures which provide uniformly distributed air void spaced not more than 0.005 in apart,
- use of mix with low water–cement ratio.

2. **Drying shrinkage**. This is loss of free water from concrete due to thermal cycling causing contraction and is called drying shrinkage. Internal stresses are developed causing cracking of concrete. The magnitude of drying shrinkage increased with a decrease in the relative humidity of the medium surrounding the concrete. Shrinkage can be minimized by keeping water content low and the aggregate contents as high as possible. Normal values for drying shrinkage are in the range of 0.03–0.05%.

3. **Carbonation shrinkage**. Carbon dioxide attacks concrete and produces:

$$Ca(OH)_2 + H_2O + CO_2$$
$$\rightleftarrows CaCO_3 + 2H_2O$$

Calcium carbonate formation is accompanied by a decrease in the water content and increase in the weight of concrete, thus causing the cracking of the surface. Carbonation shrinkage decreases the cover to reinforcement and increases the risk of ingress of corrosive species. Most important however, is the fall in pH as the concrete becomes less

alkaline; this can move the steel from the passive region to the corrosion region on the Pourbaix diagram.

4. **Humidity**. The humidity surrounding the concrete affects carbonation. The carbonation rate is maximum at 50% relative humidity. In carbonation, carbon dioxide molecules dissolve in water in the pores and capillaries and react with calcium hydroxide ($3CaO\cdot2SiO_2\cdot2H_2O$) and calcium ions, thus decreasing the alkalinity of pore solution to very low pH levels, such as 8.0 or less, increasing the risk of corrosion damage to reinforcing steel, as mentioned above.

5. **Sulfate attack**. Concrete in contact with ground water in clays may be subjected to attack by the sulfate salts of calcium, magnesium and sodium. Sulfate salts are also present in seawater and brackish water. Following are the major reactions:

(a) $Ca(OH)_2 + Na_2SO_4\cdot10H_2O$
$\rightarrow CaSO_4\cdot2H_2O + 2NaOH + 8H_2O$
(Gypsum)

(b) $2(3CaO\cdot Al_2O_3\cdot12H_2O)$
$+3(Na_2SO_4\cdot10H_2O)$
$\rightarrow 3CaO\cdot Al_2O_3\cdot3CaSO_4\cdot31H_2O$
(Calcium sulfoaluminate)
$+ 2Al(OH)_3 + 6NaOH + 17H_2O$

The two products from the reaction, gypsum and calcium sulfoalumate, have a larger volume than the compounds they replace, hence, the sulfate attack leads to expansion in the volume and disruption of the concrete. The sensitivity of concrete to sulfate attack can be minimized by using content with low C_3A content, and also using a low water–cement ratio. Addition of pulverized ash also helps to resist sulfate attack.

6. **Seawater**. Seawater contains sulfate like the ground water in contact with clay, however the attack is not as severe, although similar reactions take place. If an aggregate with high salt content and high porosity is used, salt would be crystallized at the sites of evaporation of water above the water level in concrete. The attack would only occur if the concrete is not impermeable. To maintain a low value of permeability, at least 20–25% of the void in the aggregates must

be filled by binder at mixing. Sufficient cover provides adequate resistance. Specifications must be followed; 50 mm of cover over steel reinforcement is sometimes specified. Many reinforced concrete structures have failed due to inadequate concrete thickness over the steel (seen as brown lines on the surface). Gypsum and calcium sulfoaluminate, which are formed as shown earlier, are more soluble in sodium chloride, hence, these are leached out by seawater. However, their absorption leads to an increase in the chloride contents which are responsible for the corrosion of reinforcing steel. The sulfate problem can be minimized by selecting sulfate resisting class of cement and selecting an impermeable concrete. The water–cement ratio should not exceed 0.45. Concrete containing C_3A content below about 8% does not encourage the reaction to occur.

12.8 Corrosion of Reinforcement (Rebar Corrosion)

Concrete, in general, provides protection to steel reinforcement because of the following two reasons:

(a) Concrete provides a highly alkaline environment to steel reinforcement which passivates the steel surface and, hence, prevents it from corrosion.

(b) Concrete prevents the ingress of corrosion species, like oxygen, chloride ions, carbon dioxide and water, in low water–cement ratio concrete.

Despite the three major factors which contribute to corrosion prevention of steel in concrete, cases of failure of reinforced concrete structures have been reported in alarming proportion from certain regions bestowed with high humidity, high temperature, persistently wet conditions and presence of hygroscopic species, such as dust and salt particles in air. To keep the reinforcing steel in a passive state, it is essential to maintain a high quality of concrete and minimize the factors which lead to its deterioration, such as quality of mixing water and aggregates, porosity, permeability, freeze–thaw cycles and sulfates. These factors coupled with inappropriate construction characteristics in harsh environmental conditions lead to the onset of corrosion of the steel reinforcement. The above factors induce corrosion mainly by destruction of the alkaline environment by fall of pH (pH of 14.0 to lower values). Once the alkaline environment is destroyed, the protective oxide layer on the steel surface is destroyed and corrosion initiates. The formation of protective oxide layers and their breakdown is shown in Figs 12.3 and 12.4. The protective oxide layers that developed are either Fe_2O_3 or Fe_3O_4, both are stable in concrete. The most stable layer in concrete is Fe_2O_3 written as γ-FeOOH. It is formed by the interaction of O_2 with $Fe(OH)_2$.

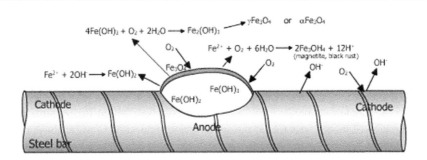

Figure 12.3 Formation of ferrous hydroxide, ferric hydroxide and magnetite layers on the steel surface

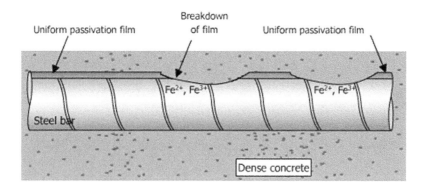

Figure 12.4   Breakdown of the protective passive film on a steel bar in concrete

$$4Fe(OH)_2 + O_2 \rightarrow Fe_2O_3 \cdot H_2O + 2H_2O$$
(Hydrated ferric oxide)

The growth of a rust layer induces mechanical stresses in the concrete and leads to its deterioration. Once corrosion initiates, controlling becomes extremely difficult unless detected in early stage.

12.9 FACTORS AFFECTING CORROSION

Before examining the mechanism of corrosion of reinforcement, it is important to understand the contribution of important factors which induce corrosion.

12.9.1 LOSS OF PASSIVATION (DEPASSIVATION)

Two major factors destroy the passivation of steels in concrete: (a) reduction of pH level by ingress of atmospheric carbon dioxide, and (b) penetration of chloride from the environment.

(a) Ingress of Carbon Dioxide (Carbonation)

The carbon dioxide molecules that penetrate in the concrete react with the solid calcium hydroxide gel, and with alkali and calcium ions in the pore solution of the concrete causing a drastic decrease in the alkalinity due to a fall in pH from 14.0 to about 8.0. At this pH, corrosion is initiated by destruction of the protective oxide layer (Fe_2O_3 or Fe_3O_4). Following is the mechanism of carbonation:

(1) $CO_2 + H_2O \rightarrow H_2CO_3$ (carbonic acid)
(2) $H_2CO_3 = H^+ + HCO_3^-$ (decrease in pH)
(3) $H_2CO_3 + Ca(OH)_2 \rightarrow CaCO_3 \cdot 2H_2O$

As soon as $Ca(OH)_2$ is removed, the pH drops from 14.0 to 8.0. The mechanism is illustrated in Fig. 12.5.

There are other secondary factors, such as depth of reinforcement cover, moisture conditions, humidity and temperature, which also influence carbonation.

Reinforcement cover. The time taken for a carbonation front to advance depends on the depth of the concrete cover. If the cover is thick enough, carbon would not reach at depths to depassivate steel. Phenolphthalein solution can be used to check depth of carbonation, by fracturing (not drilling) at a test location. A pink color indicates satisfactory concrete.

Moisture contents. The rate of diffusion of CO_2 in aqueous pore solution is very slow because of its small solubility, however, the rate of diffusion of CO_2 in air-dry concrete is greater compared to water-saturated concrete.

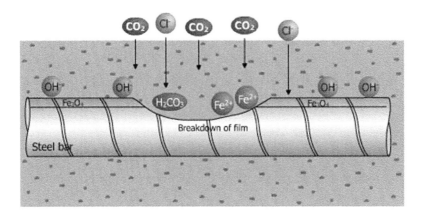

Figure 12.5 Effect of carbonation on the corrosion of steel in reinforced concrete

Humidity. Carbonation proceeds faster in humid conditions compared to dry conditions. Corrosion rates are generally lower below 75% relative humidity.

Temperature. The carbonation process proceeds faster in the temperature range of 40–60°C.

Solubility of CO_2. The solubility of CO_2 in water is affected by pressure, temperature and composition of heat. The higher the solubility, the greater would be the carbonation.

Water–cement ratio. Generally, concrete of low water–cement ratio carbonates is less than cement of higher water–cement ratio.

(b) Chloride Ion Effect

Chloride may be present in concrete from different sources. The following are some of the sources:

(a) aggregates containing chlorides salines,
(b) mixing water,
(c) sea-coastal environment,
(d) road de-icing salts, etc.

The chloride ions attack the passive layer and catalyze the corrosion process. Chloride ions breakdown the passive layer and cause pitting of the steel reinforcement.

There are two processes which proceed simultaneously, a repair process of the disrupted

protective oxide film by OH^- ions, and a breakdown process by chloride ions. The chloride ions achieve a critical concentration, dependent on the pH and pH concentration before it breakdown the passive Fe_2O_3 or Fe_3O_4 layer. The quantitative relationship between the Cl^-/OH^- ratio or between pH and Cl^- is not clearly defined. In general, the chloride threshold is taken to be 0.15% of the soluble chloride by weight of cement. The following are the major effects of chloride on the corrosion of reinforcing steel:

(a) Chloride is adsorbed in the protective oxide layer.
(b) The oxidized iron reacts to form soluble intermediate soluble iron complex at the anode.

$$Fe^2 \rightarrow Fe^{2+} + 2e^-$$
$$Fe^{2+} + 4Cl^- \rightarrow (FeCl_4)^{-2} + 2e^-$$

(c) The complex reacts with moisture to form $Fe(OH)_2$.

$$(FeCl_4)^{-2} + 2H_2O$$
$$\rightarrow Fe(OH)_2 + 2H^+ + 4Cl^-$$

The pH is lowered and the concentration of chloride is increased. Once the chlorides are released on the metal surface, the process becomes self-generating and no further chloride

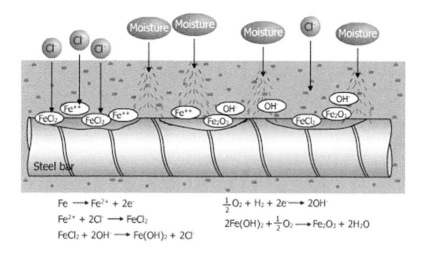

Figure 12.6 formulae:

$$Fe \longrightarrow Fe^{2+} + 2e^-$$
$$Fe^{2+} + 2Cl^- \longrightarrow FeCl_2$$
$$FeCl_2 + 2OH^- \longrightarrow Fe(OH)_2 + 2Cl^-$$

$$\tfrac{1}{2}O_2 + H_2 + 2e^- \longrightarrow 2OH^-$$
$$2Fe(OH)_2 + \tfrac{1}{2}O_2 \longrightarrow Fe_2O_3 + 2H_2O$$

Figure 12.6 Corrosion caused by ingress of chloride ions in concrete

ions are required. The repeated cycle involving the reaction of Fe^{2+} with Cl^- ions, formation of $Fe(OH)_2$ and release of H^+ and Cl^- ions, continue until the protective layer of Fe_2O_3 or Fe_3O_4 is completely destroyed. One of the major forms of corrosion brought about by chloride ion is pitting of reinforcing steel. Pitting is an autocatalytic process which continues until a hole is created in the reinforcing concrete. The corrosion of reinforcement caused by chloride ion ingress is illustrated in Fig. 12.6.

12.9.2 EFFECT OF OXYGEN

Oxygen is essential to the onset of corrosion on steel. Ingress of oxygen to reinforcing steel provide it with a basic component essential for onset of corrosion. The void space of concrete contains water, water vapor and air. Air may diffuse a considerable distance into the mass of concrete. The water in contact with air is saturated with oxygen and the total amount of oxygen would depend on the total dissolved solids and solubility of oxygen. The solubility of oxygen is a function of temperature, pressure and dissolved salt contents. The solubility of oxygen decreases beyond $80°C$ with a decrease in the corrosion rate. Areas of concrete with low water saturation would acquire a higher oxygen concentration. Differential aeration cells

may also be formed on the reinforcing steel and initiate corrosion (Fig. 12.7(a)). Cathodic reduction of oxygen would occur on the sites covered by porous concrete due to relatively more access of oxygen. In other areas, anodic reaction would occur due to little availability of oxygen. Oxygen levels are responsible for the formation of differential aeration cells which promote corrosion on the steel surface.

12.9.3 CEMENT TYPES

Chloride ions may react with hydrated tricalcium aluminate hydrate to form chloroaluminate $(C_3A \cdot CaCl_2 \cdot 10\,H_2O)$ which may contain 75–90% chloride. The amount of chloride would depend on the total chloride content, C_3A content, and degree of hydration. On exceeding the threshold chloride level, the protective passive layer is destroyed. In general, the chloride ion concentration decreases with hydration age and increased C_3A concentration.

12.9.4 POTENTIAL DIFFERENCE

As a rule, the greater the difference of potential between two sites at a metal surface, the

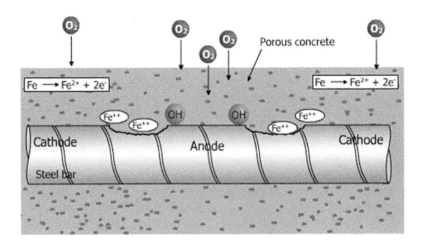

Figure 12.7 a Formation of corrosion cell due to porosity of concrete. The porosity affect the level of oxygen in the concrete

greater would be the driving force of corrosion. The formation of corrosion cells by different reasons affect the magnitude of steel reinforcement. Reinforcement steel is a heterogeneous material alloyed with carbon and other metallic elements. The alloy heterogeneity creates several corrosion cells with different electrochemical potential levels. Corrosion cells on the reinforcement can also be created by residual stresses in the steel bars or the presence of scratches or by differences of concentration in the electrolyte, such as chloride in the iron, due to its heterogeneity. Corrosion cells can be microscopic or macroscopic. In the microscopic cells or macroscopic cells, anodes and cathodes are located very close to each other on the same steel bar. Corrosion cell formation due to differences in composition of metals, aeration, stresses and electrolyte concentration, are common examples of microcell formation. The formation of different corrosion cells by difference in stress levels and segregation of impurities is shown in Fig. 12.7 (a, b and c). In macrocell formation the anodes and cathodes are located in different steel bars, for instance, in repairs of concrete by patching, the site at which the steel bars in contact with contaminants would

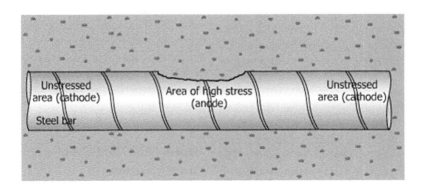

Figure 12.7 b Formation of corrosion cell by difference in stress levels of the steel bar

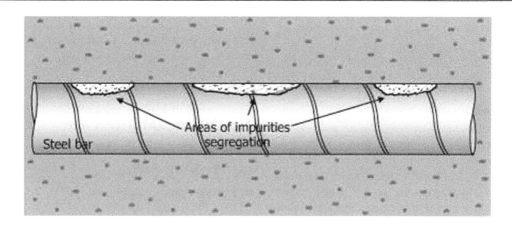

Figure 12.7c   Formation of corrosion cell due to segregation of impurities

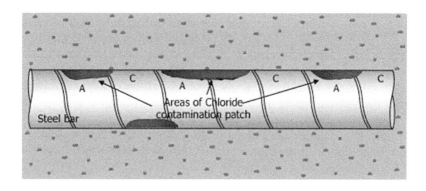

Figure 12.8   Formation of macro corrosion cells by patch work repair

form the cathode and the steel on contact with a chloride contamination patch constitutes the anode (Fig. 12.8).

12.9.5 EFFECT OF PERMEABILITY OF CONCRETE

Concrete hardens as a result of chemical reaction between Portland cement and water. The hardened Portland cement is called '*cement paste.*' As the volume of the hydration products is greater than the volume of concrete grain, the accumulation of hydration products create a space filling effect. The paste contains two different forms of pores – capillary pores and gel pores.

(a) **Gel pores**. These are fine pores within the particle of hydrated products filled with water. Their size is about 18 Å.

(b) **Capillary pores**. They are generally a thousand times larger than the gel pores and may be filled with water, water vapor and air.

The gel pores occupy most of the space in the hydrated paste. The pores create the water which dissolves ions from the cement. The aqueous solution in the pore acts as carrier of corrosive species and it is in contact with the reinforcing steel surface. In cement with high water–cement ratio, the pore volume is very high compared to pore volume in low water–cement ratio, hence low water–cement ratio cement is more impermeable than high water–cement ratio cement.

12.10 CONSTRUCTION PRACTICE

Steel embedded in concrete may be subjected to corrosion if its surroundings support the contact of steel with the species which induce corrosion, such as water traps, ingress of water, oxygen, chloride and water movement. The construction practice must not allow the freedom of movement to elements which create an environment for reinforcement corrosion.

12.10.1 CORROSION MECHANISM – COVER TO CONCRETE AND DIFFUSION

The thickness of the cover to concrete is an important factor because it affects the time required to de-passivate steel surface and the rate of reaction. For instance, corrosion can be minimized by slowing down the supply of oxygen and atmospheric CO_2 to the steel which is achieved by placing a concrete cover over the steel surface of appropriate thickness. The concrete cover acts as barrier to the environment. The cover controls the diffusion of carbon dioxide and chloride to the steel mainly through the pores in the cement paste to the steel surface. The diffusion may also occur through the aggregate. The diffusion path is very slow for very dense and low water–cement ratio concrete. The protective zone is very small in the high water–cement concrete compared to low water–cement concrete as shown in Figs 12.9 and 12.10. The thickness of the concrete cover varies with the environment and is generally 70–75 mm for seawater. It is to be remembered that diffusion is slower in the aqueous phase than the gaseous phase. Hence, diffusion would be more difficult in wet concrete compared to dry concrete. Wet concrete thus offers control over diffusion. As a separate factor, diffusion would be difficult in low water–cement ratio (denser) concrete compared to high water–cement ratio

because of the discontinuity of the pores in the former. The effect of water–cement ratio on diffusion of species to concrete is shown in Figs 12.11 and 12.12.

12.11 SUMMARY OF PRINCIPLES INVOLVED IN CORROSION REINFORCEMENT

The corrosion mechanism of reinforcement in concrete is a very complex phenomena because of several factors which contribute to corrosion. These factors have been described in Section 12.9.

Basically, corrosion of reinforcement is due to the breakdown of its passive layer which is composed of an insoluble thin oxide film. This film isolates the reinforcement from the aggressive environment. The films which grow on the reinforcement are either Fe^{2+} (ferrous) or Fe^{3+} (ferric), both being stable in the absence of carbonate. The Fe^{3+} (ferric) film is generally more stable than the ferrous film (Fe^{2+}). These films are formed by the following reaction:

$$Fe \rightarrow Fe^{2+} + 2e^- \text{ (Anodic dissolve)}$$

$$Fe^{2+} + 2OH^- \rightarrow Fe(OH)_2 \text{ (Ferrous oxide)}$$

$$2Fe(OH)_2 + O_2 \rightarrow 2FeOOH + H_2 \text{ (Ferric oxide)}$$

Details of the above reactions are given in Chapter 2. The films do not remain protective any more if the pH falls due to carbonation. They do not remain adherent to the reinforcement any longer and loose their passivity causing the reinforcement surface to be exposed to corrosive species. The ingress of CO_2 is a major factor which breaks down the passivity and dissolves the protective oxide layer on the reinforcement surface.

Another major factor is chloride ion ingress to reinforcement. The ferric oxide (Fe_2O_3) or γ-FeOOH film is resistant to chloride penetration, however ferrous oxide ($FeO \cdot nH_2O$) is sensitive to chloride ion penetration and reacts with it to

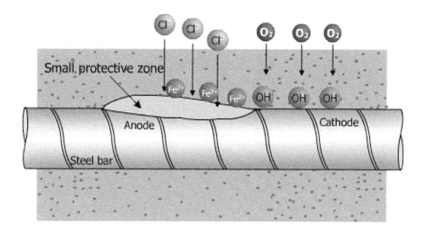

Figure 12.9   High diffusion of oxygen and chlorine in high water–cement ratio concrete

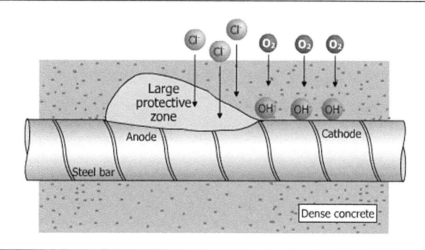

Figure 12.10   Large protective zone in low water–cement ratio concrete

form a soluble complex, leading to the breakdown of passivity and dissolution of the ferrous oxide film. Ferric oxide is red–brown color and comprises most of the rust.

The process of pitting once initiated become autocatalytic. The electron released in oxidation, $Fe + 3Cl^- \rightarrow FeCl_3 + 3e$, flows through steel to the cathode, $FeCl_3 + 3OH^- \rightarrow Fe(OH)_3 + 3Cl^-$, which results in an increase in the concentration of chloride ion and reduction of pH.

The above process would only proceed if the chloride concentration is higher than the hydroxyl ion (OH^-) concentration. If OH^- ion concentration is higher, the ferrous oxide film would convert to ferric oxide (γ-$FeOOH$) and passivity would be preserved. A higher oxygen concentration would also promote the conversion of ferrous hydroxide to ferric oxide and the passivation of the reinforcement. Carbon dioxide and chloride ions are basically the two ingredients

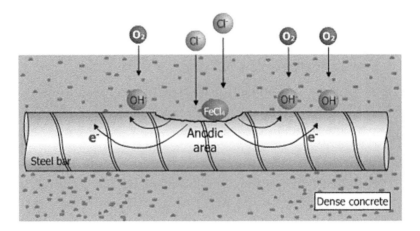

Figure 12.11   Slow diffusion of oxygen and chloride ions (low water–cement ratio)

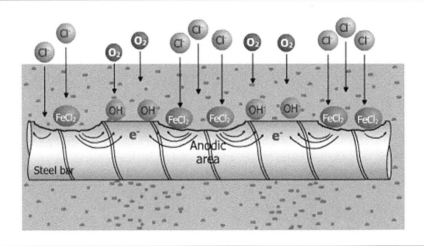

Figure 12.12   Rapid diffusion of oxygen and chloride ions (high water–cement ratio)

responsible for corrosion of reinforcement. The mechanism of corrosion of reinforcement by these two components is shown in Fig. 12.13. Other factors, such as the moisture content, temperature of oxygen, potential difference, concrete cover, act by either reducing or increasing the ease of penetration of the corrosion inducing species in concrete. The key factor is maintaining of alkalinity, as under alkaline conditions corrosion cannot take place because of the passive state of the reinforcement in this condition.

12.12 METHODS OF CORROSION CONTROL OF REINFORCING STEEL

Despite the impressive progress made in controlling the corrosion of steel in concrete in the last thirty years, reinforcement corrosion in concrete structures is still a challenging problem. Important methods to mitigate reinforcement corrosion are described below.

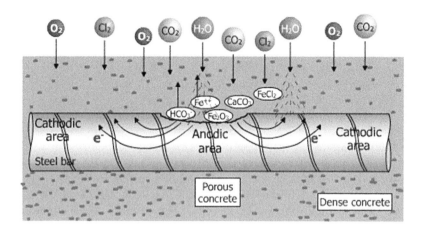

Figure 12.13   Corrosion of reinforcing steel caused by chloride ions and carbon dioxide

12.12.1 CATHODIC PROTECTION

This is one of the oldest techniques of corrosion protection of steel in a corrosive environment, such as water or soils. Cathodic protection has been applied very successfully to the protection of pipelines, storage tanks and underground structures in petroleum and oil industries. Cathodic protection has been widely applied for the protection of concrete structures. In principle, a small negative voltage is applied to the steel in concrete which makes all the steel surface the cathode and eliminates the anode areas on the steel surface. This technique has been discussed in detail in Chapter 5. The cathodic reaction at the surface leads to increased alkalinity which passivates steel.

12.12.1.1 Cathodic Protection Technique

The two systems, impressed current and galvanic anode, are used in protection of steel in soil and water and also used for the protection of steel embedded in concrete. However, the set up of a cathodic protection system for steel in concrete differs from the set up for protection of steel in soil, in the following respect:

(1) The embedded steel bars are connected to each other in the form of a cage to maintain electrical conductivity. In reinforced concrete foundations, bonding can be achieved by tack welding. The reinforcing bars are connected in series.

(2) Cathodic protection of steel is considered effective when the coated steel potential is shifted into the range of $-85\,mV_{SCE}$ to $-105\,mV_{SCE}$. However, this criterion is not always applicable to steel in concrete. The uncorroded steel in concrete has a potential in the from -0 to $-0.3\,V$ vs CSE which is about 0.30 to 0.50 more noble than coated steel in soil. Sufficient cathodic protection may be achieved by polarizing the steel bars to $-0.50\,V$ vs CSE. In case the chloride contents is higher, rebars are polarized to a more negative potential ($-0.70\,V$) vs Cu–CuSO$_4$.

The quality of concrete and the corrosive species dictate the protective potential required to stop corrosion of steel bars in concrete.

12.12.1.2 Impressed Current and Sacrificial Anode System

(a) Impressed Current System

An external positive voltage is applied to an inert anode (silicon, cast iron, titanium) which is placed in concrete. The positive terminal of a transformer–rectifier is connected to the inert anode which may be in the shape of an inert wire mesh placed over the concrete surface and sprayed with a conductive paint. The negative terminal of the transformer–rectifier is connected to the steel bar cage, comprising a network of steel bars (Fig. 12.14). For potential measurement, a copper–copper sulfate reference electrode is embedded in concrete. A voltmeter is attached to the reference electrode for reading of its potential. A simple description of the system is shown in Fig. 12.15, only one reinforcement bar is shown in the figure instead of a cage. One important requirement of cathodic protection is good electrical continuity and there must be no short circuit between the steel and the impressed current anode surface which acts as a conductor of electricity. The anode corrodes very little, unlike the sacrificial galvanic anodes, such as magnesium, which supply electrons upon corroding. It is observed that because of the cathodic reaction on the steel surface, OH^- ions are generated, which neutralize the effects of carbonation. The negative chloride ions migrate away from the steel to the anode during cathodic

protection and the passive layer is thus not damaged. Cathodic protection thus removes Cl^- ions from close to the steel (de-chloridizing). It is only necessary for Cl^- ions to move a few atomic distances away to completely stop their bad corrosive effects.

(b) Galvanic Anode System

This sacrificial anode system does not require the use of an external power source as in the impressed current system. A schematic description is shown in Fig. 12.16. Use is made of zinc and magnesium anodes which corrode and supply electrons to steel bars embedded in concrete. The current flow circuitry is same as in the impressed current system. The anode life is, however, shorter than the life of inert anode. As the current generated by the corroding anode is a function of environment, such as temperature and moisture, it is difficult to adjust and control the current. However, a major advantage is that the risk of over-protection which is inherent in impressed current system is minimized. i.e. operator dependence is removed, advantageously. The galvanic

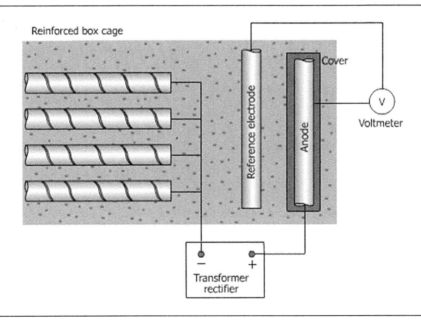

Figure 12.14   Cathodic protection by impressed current method

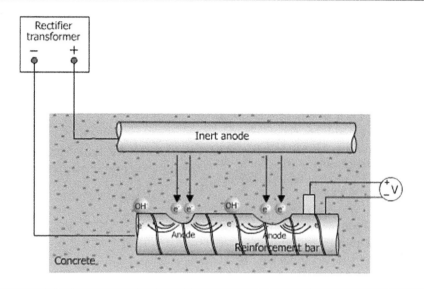

Figure 12.15 A schematic diagram of impressed current cathodic protection system. The positive voltmeter contact is brought to a post above the concrete

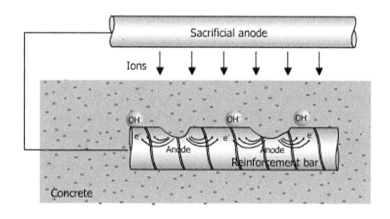

Figure 12.16 A schematic diagram of sacrificial anode cathodic protection system

system is also more economical. This type of protection is, however, limited to seawater or in buried sections of reinforcement where its resistivity is lowered by high moisture content.

Cathodic protection systems for concrete structures involve high installation and maintenance costs. Cathodic protection systems (impressed current), however, require very small power for operation as shown by the small current density ($0.2–2.0\,mA/m^2$) required for the protection of reinforced steel. The power consumption is in the range of 1–3 watts per $1000\,m^2$ of concrete for new constructions and 3–15 watts for others.

12.12.2 CRITERIA FOR CATHODIC PROTECTION OF STEEL IN CONCRETE

There are conflicting opinions about the criteria for cathodic protection. A shift in the potential to $-850\,mV_{SCE}$ may not always be the best indicator of complete cathodic protection because hydrogen may enter the steel bar in certain areas because of the heterogeneous environment of embedded steel bars. Despite certain limitations, it is a commonly used criteria.

A $-300\,mV_{SCE}$ shift in the negative direction from its original potential is widely accepted criteria for cathodic protection. Another criterion is a $100\,mV$ shift of the instant potential on switching off the current, which in many cases indicates complete cathodic protection. The potential instantly measured in the '*off*' position is called the '*depolarization potential.*' The measurement may be adversely affected by high chloride contents. The checking of effectiveness of a cathodic protection system by depolarization measurement after switching off the protective current is the most important indication. The $100\,mV$ shift test is made by measuring the pipe potential immediately after switching off the current and subtracting this from the pipe potential 4 h later; the result should be at least $100\,mV$ but some opinion recommends $150\,mV$.

Summing up, experience is the best guide in selecting the criteria for cathodic protection as it is strongly affected by quality of concrete, water saturation level, degree of contamination and water–cement ratio.

12.12.3 POTENTIAL MEASUREMENTS AND POTENTIAL MAPPING

Potential measurement is, in practice, a simple procedure as shown in Fig. 12.15. The reference electrode (Cu–CuSO$_4$) must be in good electrical contact with the concrete unless the concrete is buried underground or in the sea (e.g. concrete-covered pipeline). A reference electrode containing a cell tip spring allows a good contact. A high input impedance voltmeter capable of reading 1 mV resolution is required. Because the corrosion currents flowing in the concrete are very small, a high impedance voltmeter of about 10 megohms or more must be used. In highly dried concrete, an even higher impedance voltmeter (1000 megohms) may be used. It is necessary to drill a very small hole in the reinforcement to provide a good electrical contact. The connection is made through a self-tapping screw shown in Fig. 12.15. The corrosion status of rebar can be approximated as shown below:

Range of potential	Probability of corrosion
<-0.35	$>95\%$
>-0.20	$<5\%$
-0.20 to 0.35	-50%

The results of potential measurements can be extended to potential mapping on the surface of concrete by Jominy points of same potential. To do the mapping, grids are drawn on the concrete and readings are taken at the intersection of grid lines. Ordinary chalk can be used to draw the grid lines. It is preferable to draw grid lines while the surface is dry. The potential mapping helps to identify the areas which are being subjected to corrosion. Figure 12.17 shows isopotential mapping under wet conditions. The use of electrochemical potential to determine the areas of corrosion in reinforcing steels is described in ASTM Standard ASTM C876-91.

It is, however, hard to obtain accurate values because of the large potential drop (IR) caused by moisture and salt contents of the concrete, which affect resistivity. At the anodic sites where current is discharged, the measured potential is reduced by potential and it is increased at the cathodic sites. To minimize large potential drops, the concrete may be wetted with fresh water or preferably salt solution to decrease the potential drop. It is essential to minimize the potential (IR drop) to obtain accurate values.

Two methods for measuring potential are commonly used: a single electrode potential and a double electrode potential measurement. These methods have been described in detail in Chapter 5. The principles are the same as applied to the measurement of potential in reinforcing steels. The double electrode method, also called '*leap frog*' method, is preferable in measurement

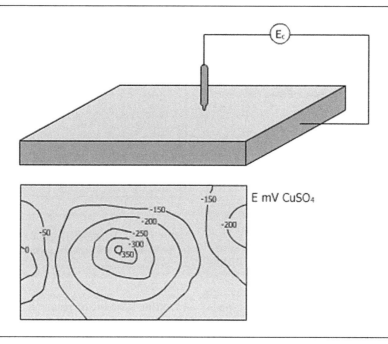

Figure 12.17 Surface electrode potential mapping pinpoints the danger areas, whatever the conditions

of potential of the reinforcement, as it is more accurate, saves time and costs.

12.13 CORROSION PROTECTION OF REINFORCEMENT

12.13.1 APPLICATION OF COATINGS

This consists of applying a thin coating on the concrete surface which acts as a barrier to the environment and does not allow penetration of corrosion species. Epoxy coatings offer excellent adhesion and corrosion resistance in harsh environments. The most commonly used coatings include acrylic, chlorinated rubber and polyurethane coating. The suitability of commonly used coatings for different environments is given in Table 12.3.

Sacrificial zinc coatings have also been applied for protection of reinforcing steel bars, even

Table 12.3 Suitability of commonly used coatings for different environments

Service conditions	Coating type
Humid conditions	Epoxy, acrylic
Temperature fluctuations	Epoxy, polyurethane
Chloride contamination	Chlorinated rubber, epoxy and polyurethane
Carbonation	Polyurethane, acrylic

though zinc is attacked by high alkalinity (see Pourbaix diagram).

12.13.2 INHIBITOR TREATMENT

Cathodic, anodic or passivating inhibitors are commonly used to prevent corrosion of steel reinforcement. In the class of anodic inhibitors, calcium nitrite, sodium nitrite, sodium benzoate and sodium chromate are commonly used. Cathodic inhibitors mainly consist of amines, phosphates, zincates, aniline and its chloroalkyl nitrosubstituted form and aminoethanol groups. The anodic inhibitors provide a larger degree of protection than cathodic inhibtors because of their direct influence on the passivation of steel surface. Appropriate dosage of nitrite can control chloride induced corrosion.

12.13.3 CONTROL OF EXTERNAL FACTORS

The quality of concrete and corrosion resistance of reinforcement are complimentary to each other. The protection of concrete against corrosive agents is as essential as the mitigation of reinforcement in concrete. As stated earlier, the water–cement ratio, concrete cover thickness, small volume of pores, degree of saturation of water, temperature and curing time are important factors which must be controlled to maintain passivation of a steel surface and minimize the ingress of chloride ions and carbon dioxide which damage the protective film. The curing period affects the durability of concrete. Curing is generally done at 80% RH (relative humidity) and ensures that most of the water is available for hydration. The curing process is adversely affected by high temperature and high winds which create dry conditions. Good curing improves the permeability of concrete 5–10 fold. Inadequate curing has a harmful influence on the ingress of chloride ions. While curing, the effect of admixture, such as fly ash and slags, must also be considered, as deficient curing makes the concrete with these ingredients more permeable to ingress of chloride.

12.13.4 USE OF SEALERS AND COATINGS

The application of coatings and sealers provides resistance to concrete attack by minimizing the ingress of corrosive species into concrete and increasing the durability of the concrete. They belong to several types depending on applications. The sealers may be applied as lining for the pores which make concrete water repellent. Silicone resins have been successfully used. They deposit

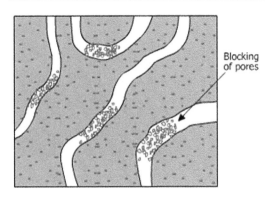

Figure 12.18 Schematic diagram of blocking of pores by formation of reaction products

a continuous film of resin on the pore surface after evaporation from the solvent on the pore surface and make the concrete water repellent. They are applied either by brush or spraying gun. On evaporation of the solvent from the paint, a continuous film is formed on the concrete surface. The film formed on the surface is maintained in an alkaline environment. Examples are acrylic, epoxy resin, polyester resin and polyurethane resin.

They may also be applied as pore lining to make the concrete surface water repellent. Silicone compound and silane are frequently used as pore liners. A schematic figure illustrating the lining of pores by sealers is shown in Fig. 12.18. The sealers can also be used to build up reaction product within the pores which block up the pores. Silicofluorides are most commonly used. They react with $Ca(OH)_2$ in the concrete to form silicofluoride compounds which block up the pores.

12.14 CONCRETE REPAIRS AND REHABILITATION

Before repairs are undertaken, it is important to determine the cause of steel reinforcement, the magnitude of the damage and the effect of damage on the structural integrity of the structure. Two major concrete failures, spalling and cracking, are caused by reinforcement corrosion, and they require patch repair by a technique called *patch repair*. These failures are the result of carbonation and chloride induced corrosion discussed earlier in this chapter. It is important to identify areas requiring repairs. Following are the characteristics of damaged areas requiring patch repair:

(1) The reduction in the cover to steel reinforcement thickness. A thickness of less than 10 mm signifies corrosion.
(2) Areas where the half cell potential exceeds -200 mV $Cu-CuSO_4$ (i.e. is more positive than -200 mV).
(3) Areas where carbonation has ingressed to within 5 mm of the reinforcement. The depth of carbonation can be determined by testing the concrete with phenolphthalein, which turns pink in an alkaline environment and remains colorless in a carbonated environment. The depth of carbonation can also be determined by: Depth of carbonates = (age in years)$^{0.5}$ at 50% RH and (age in years)$^{0.3}$ in about 90% RH. The test surface must be a fracture surface, not drilled (as grinding invalidates the phenolphthalein test).

12.14.1 PATCH REPAIR

The damaged concrete is removed until the reinforcement bars are clearly visible. It is removed beyond 50 mm of the corroded portion. Concrete must be removed to a minimum of 20 mm under the corroded bars. Rust should be completely removed from the corroded steel by grit-blasting and coating it after sandblasting with an anti-corrosive primer, if it is to be left overnight in a humid environment. Fresh repair concrete is applied to replace the damaged concrete. The repair concrete protects the steel by providing an alkaline environment which passivates the reinforcing steel. Many times external membranes are applied over the repair concrete. The aggregate in repairing concrete must be of the same size as the parent aggregates to enhance the protection and minimize the ingress of carbon dioxide and

chloride ions. Additional means, such as sealers and coatings, may be incorporated in the patch repair system. The properties of repair concrete or mortar must be very close to the properties of the damaged concrete. Material properties, such as elastic modulus, coefficient of thermal expansion, shrinkage, creep and bonding, must be carefully considered to make the replacement concrete compatible and allow minimum stresses to be introduced. Only those materials to repair concrete must be considered which provide sufficient passivity to steel.

Areas less than 10 mm thick can be hand repaired by applying proprietary cementitious mortar with or without polymer modifier. Polymeric mortars can provide physical protection to concrete but do not provide an alkaline environment. The substrate must be kept sufficiently wet to minimize the loss of water during repairing. Proprietary bonding coats may be added to provide strong bonding to the substrate. They are formulated from epoxy polymer or polyester polymer and they can be used without a primer.

12.14.2 USE OF ELECTROCHEMICAL METHODS

Electrochemical methods, such as cathodic protection and chloride extraction, can be used as a part of a repair strategy. Cathodic protection techniques, described above, provide alkalinity. The impressed current transports alkalies to the reinforcing bar and allows alkalinity to be retained. The chloride extraction system removes chloride ions (Cl^-) from the concrete electrochemically, and does not allow the breaking of passive layers. In yet another process, called re-alkalization, alkaline metal ions penetrate concrete from an external source of a suitable electrode to re-alkaline the concrete and regenerate the hydroxyl ions.

12.14.3 PATCH ACCELERATED CORROSION

The patch repair concrete material must be completely chloride-free and similar in characteristics

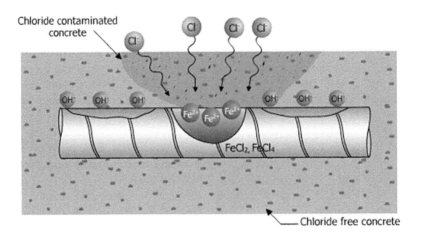

Figure 12.19 Chloride induced corrosion of reinforced steel

to the concrete to be replaced, otherwise the difference of potential between the chloride contaminated patch and chloride-free patch may induce accelerated corrosion. A potential difference exists between a chloride contaminated concrete and chloride-free patch. Figure 12.19 shows that the positive Fe^{2+} ions as a result of anodic reaction combine with the negative chloride ions from the concrete and form $FeCl_2$ at the anode (Fig. 12.20). Upon interaction with water (hydrolysis), the process become self-generating, and it starts dissolution. Hydrolysis increases the dissolution rate and the migration of chloride ions. The ferrous hydroxide which is formed is not very protective. The process becomes self-perpetuating and once initiated, the process is repeated until

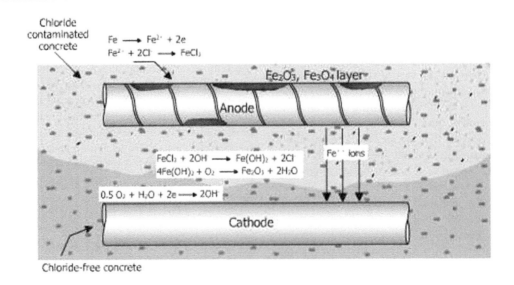

Figure 12.20 Corrosion reactions caused by improper patch repair

the passive layer on concrete is disrupted. The repair patch must be completely chloride-free to avoid the acceleration of corrosion by formation of galvanic cells. The chloride contaminated patch acts as an anode relative to the chloride-free patch which becomes the cathode. The reactions involved with the above process are shown in Fig. 12.20.

12.14.4 PRINCIPLES AND PRACTICES OF ELECTROCHEMICAL CHLORIDE REMOVAL

In chloride contaminated concrete there is an abundance of chloride, hydroxyl and calcium ions. Under the influence of an applied electric field, anions (like Cl^-) migrate towards the anode and the cations to the cathode. An inert anode and a temporary electrolyte are placed temporarily in the concrete. Catalyzed titanium is used generally as an inert anode. Steel has also been used as an anode, but it is consumed after a period of time. The electrolytes used are generally water, calcium hydroxide and lithium borate. Calcium hydroxide is a popular electrolyte as it is alkaline and

stops chlorine gas evolution. The electrolyte can be applied either in tanks clamped to concrete surface or in absorbent material, like cellulose fibers on the concrete surface. Titanium anodes are sandwiched between the layers of felt wetted with electrolyte and used in deck spans. The temporary anodes and electrolyte are removed after the process is complete. In electrochemical chloride extraction, the positive of the power supply is connected to the anode and negative to the cathode, i.e. the reinforcing bar. As soon as the power supply is switched on, chloride ions are transferred from the contaminated concrete to the anode. The electrolyte which was temporarily placed would contain the chloride ion emanating from the contaminated concrete. The electrochemical treatment system is shown in Fig. 12.21. One of the dangers of this technique is the possible evolution of chloride at the anode:

$$2Cl^- \rightarrow Cl_2 + 2e$$
$$2Cl \rightarrow Cl_2 + 2e$$

If the pH is, however, kept >7.0, the oxygen evolution reaction becomes predominant

$$2H_2O \rightarrow O_2 + 4H^+ + 4e$$

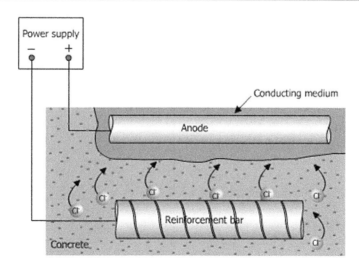

Figure 12.21   A schematic diagram of electrochemical extraction procedure. Cl^- ions are repelled from the steel bar

and chlorine evolution is minimized. The small amount which may be generated is immediately hydrolyzed in water to form hypochlorous acid and hypochlorite.

$$Cl_2 + H_2O \rightarrow HClO + Cl^- + H^+$$

$$HClO \rightarrow ClO^- + H^+$$

If a steel anode is used, a significant amount of current is consumed by its anodic dissolution, but chlorine gas is not likely to be generated. The electrolyte should be maintained at a basic pH to prevent generation of chlorine. The following are the disadvantages:

(1) the bond strength between rebars and concrete may be reduced if high current densities are evolved.
(2) Hydrogen evolution has an adverse effect on pre-stressed steel and they may be subjected to hydrogen embrittlement because of the entry of hydrogen into reinforcing bars.

There are, however, more advantages of this technique as summarized below:

(a) The corrosion rate of reinforcing bar is significantly reduced.
(b) the integrity of the passive film on the steel surface is maintained.
(c) Alkalinity is increased at the reinforcing bar/concrete surface.
(d) The service life of the concrete structure is increased.

12.14.5 RE-ALKALIZATION

Re-alkalization is another form similar to the electrochemical chloride extraction process, and is used to recreate the alkaline conditions which are lost due to carbonation. As stated in an earlier section, the ingress of carbon dioxide leads to the loss of alkalinity on the reinforcement because of the fall of pH from 14 to 9 due to the combination of carbon dioxide with calcium hydroxide [$Ca(OH)_2$] which means loss of alkalinity from the concrete.

$$Ca(OH)_2 + CO_2 + H_2O = CaCO_3 + 2H_2O$$

When the carbonated front reaches sufficient depth, the steel reinforcement corrodes due to loss of passivity. The process of re-alkalization is, therefore, employed to recreate the alkaline conditions on the reinforcement by attracting Ca^{++} ions and Na^+ ions (see below) which keeps it in a passive state. The vicinity of reinforcement is re-alkalized and a new passive layer is thus regenerated. The re-alkalization process is a consequence of the electrochemical chloride removal process. The anodes used are the same as the one used for chloride removal (titanium or steel). Sodium carbonate is used as an electrolyte. The function of Na_2CO_3 is to introduce alkali metal ions from an external source, however it would not be necessary to use Na_2CO_3 if alkali reactive aggregates are present in the concrete in which case water is used as an electrolyte. The introduction of alkaline metal ions prevents the occurrence of re-carbonation. At the anode oxidation takes place according to

$$2OH^- \rightarrow \frac{1}{2}O_2 + H_2O + 2e^-$$

At the reinforcement, hydroxyl ions are formed due to the concentration of Ca^{++} and Na^+ ions there. A schematic figure illustrating the process is shown in Fig. 12.22.

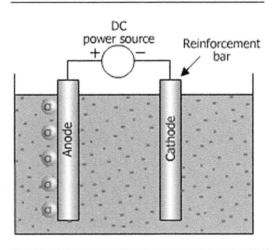

Figure 12.22 Diagram showing the re-alkalization process

Advantages

Re-alkalization is more economical and simpler than electrochemical chloride removal process. In the electrochemical chloride removal process, the evolution of chlorine gas and its removal prevents ingress of hydrogen into reinforcement making the process more complex compared to re-alkalization process. The re-alkalization technique is gradually gaining popularity in Europe and the Middle East.

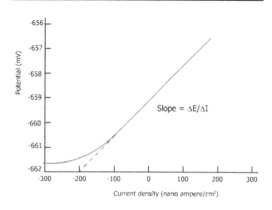

Figure 12.23   A typical polarization resistance curve

12.15 ELECTROCHEMICAL METHODS OF EVALUATION OF REINFORCEMENT CORROSION

12.15.1 POLARIZATION RESISTANCE TECHNIQUE

This is a well established technique applied to determine the corrosion rate of steel in concrete. A full theoretical description of this technique is given in Chapter 3. The specimen is polarized to within ± 25 mV of the corrosion potential (E_{corr}) as the dependence of current with potential in the vicinity of corrosion potential is linear. The open circuit potential (E_{corr}) of the reinforcement is measured against a reference electrode. Starting from this potential, the steel reinforcement is polarized to ± 25 mV from E_{corr} at a sweep rate of 5–10 mV/min and a plot of ΔE vs Δi is constructed. The polarization resistance relates the slope of the polarization curve in the vicinity of corrosion potential to the corrosion current. This technique is based on the Stern Geary equation described in Chapter 3, and given below:

$$I_{corr} = \frac{\beta_a \beta_c}{2.3(\beta_a + \beta_c)} \cdot \frac{\Delta i}{\Delta E}$$

where β_a and β_c are anodic and cathodic Tafel slopes (mV/decade). For most electrolyte systems, β_a varies from 60 to 120 mV/decade, and β_c is greater than 60 mV/decade, hence by assuming

these values, estimates of corrosion rates can be made within a factor of two.

$$\Delta i = \text{applied current (mA)}$$

$$\Delta E = \text{potential shift (mV)}$$

The polarization resistance (R_p) is given by

$$R_p(K\Omega) = \frac{\Delta E}{\Delta i} = \frac{\beta}{I_{corr}}$$

$$\text{where, } \beta = \frac{\beta_a \beta_c}{2.3(\beta_a + \beta_c)}$$

A typical polarization resistance curve is shown in Fig. 12.23. The polarization resistance measurement arrangement are shown in Figs 12.24a and b.

12.15.2 GALVANOSTATIC PULSE TECHNIQUE

The galvanostatic pulse technique is one of the most popular techniques to estimate the risk of corrosion of reinforcing steel. A simple arrangement of applying anodic pulses is shown in Fig. 12.25. A small anodic current pulse for nearly eight seconds is imposed galvanostatically on to the reinforcement and the resulting change in potential is measured as a function

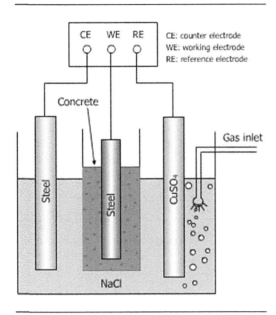

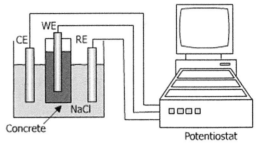

Figure 12.24 b Electrochemical polarization measurement

Figure 12.24 a Schematic of a typical polarization cell used in laboratory for polarization resistance measurement

of time. Potentials are measured with respect to a Cu–CuSO$_4$ electrode. The applied anodic current passes through a counter-electrode to the reinforcing steel in concrete. The voltage between reinforcing steel and reference electrode is connected to the AD converter of a computer. The pulse response is registered with 10 m·s/point, each curve comprising 1000 points. The pulse current is regulated in such a way to limit the polarization to 25 mV. Plots of potential (mV$_{SCE}$) measured *vs* Cu–CuSO$_4$ electrode and time (m·s) are recorded. A typical plot is shown in Fig. 12.26. From the plot, the corrosion of reinforcing bars can be estimated. The difference between a passive steel and corroding steel can be observed from Fig. 12.27. The greater the ohmic resistance (R_{ohm}), the less would be the corrosion of reinforcing steel and easier for reinforcing steel to be polarized. This is a linear relationship

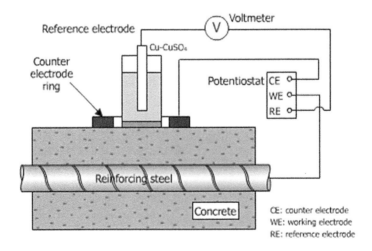

Figure 12.25 Configuration of anodic pulse technique. (From Meitz, J. and Isecke, B. Bundsanstalt fun Materials Prufung, Berlin)

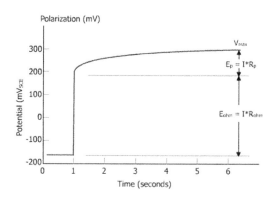

Figure 12.26 Typical potential time curve as response to a galvanostatic pulse

between ohmic resistance ($R\Omega$) and polarization resistance (R_p). This technique allows the determination of ohmic resistance and polarization resistance of reinforcement in concrete within seconds.

12.15.3 IMPEDANCE SPECTROSCOPY

Impedance spectroscopy has proved a very effective technique in analyzing corrosion of reinforcement in concrete. In DC theory resistance is defined by Ohm's Law:

$$E = IR$$

where

I = current (A)
E = potential (V)

Using Ohm's law one can apply a DC potential to a circuit and measure the resulting current, from which the resistance can be calculated. This is illustrated by DC measurements, like polarization resistance and potentiodynamic polarization discussed in Chapter 3. In the AC theory:

$$E = IZ$$

where Z is the impedance, the AC equivalent of resistance.

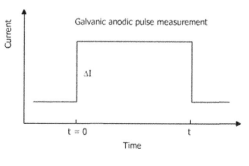

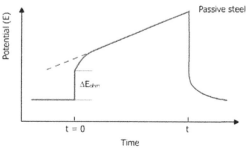

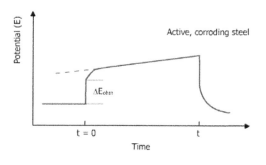

Figure 12.27 Behavior of current and potential during anodic polarization

Both terms, resistance and impedance, denote an opposition to flow of electrons or current. The opposition to flow of DC current is provided by resistors whereas in AC circuits, this effect can be produced by capacitors and inductors. The current sine wave is represented by

$$I(t) = A\sin(\omega t + \theta)$$

where

A = maximum amplitude
ω = frequency in radians per second = $2\pi f$
t = time
θ = phase shift.

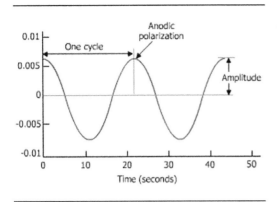

Figure 12.28 Cyclic nature of AC voltage

The cyclic nature of AC voltage is shown in Fig. 12.28. Two complete cycles are illustrated in the figure. The voltage frequency (ω) is 2 cycles/40 s or 0.05 Hz (0.05 cycles per second). In impedance spectroscopy, the polarizing voltage ranges from 5.20 mV from the open circuit potential (OCP). The frequencies are measured from 100 KHz to several millihertz.

AC currents and voltages are vector quantities. Impedance can be expressed as a complete number where the resistance is the real component, and combined capacitance and inductance is the imaging component. The vector nature of the voltage of current is shown in Fig. 12.29. In this figure, the total impedance is shown by the solid arrow and components by dashed vector.

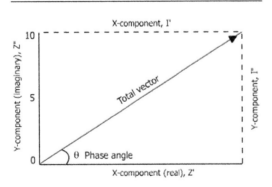

Figure 12.29 Vector nature of voltage and current

The mathematical expressions for expressing quantity in this system is to multiply I'' by $\sqrt{-1}$ or j. Hence,

$$I_{total} = I' + I''j$$

Similarly, for ac voltage

$$E' = E''j$$

the impedance vector as a quotient of the voltage and current vector is

$$Z_T = \frac{E' + E''j}{I' + I''j}$$

The resulting vector for impedance is

$$Z_{total} = Z' + Z''j$$

where

Z' = real impedance
Z'' = imaginary impedance

The absolute magnitude of impedance is

$$|Z| = \sqrt{(Z')^2 + (Z'')^2}$$

and

$$\tan\theta = \frac{Z''}{Z'}$$

When the opposition to the current is capacitive resistance, the current leads the applied voltage in phase angle. When the opposition to current flow comes for inductive reactance in the current lags behind the voltage in phase angle. The phase angle (θ) is the difference between points on x axis where current and voltage curve amplitudes are zero (Fig. 12.30). The mathematical expression for expressing quantities in this system is to multiply I'' by $\sqrt{-1}$ or 'j.'

Consider the impedance contributed by various circuits [Figs 12.31(a–g)]:

(1) $Z = R + 0j$ where $j = \sqrt{-1}$ (Fig. 12.31a)
(2) $Z_L = j\omega L$ where $j = \sqrt{-1}$ (Fig. 12.31b)
(3) $Z_C = \frac{1}{j\omega C}$ (Fig. 12.31c)

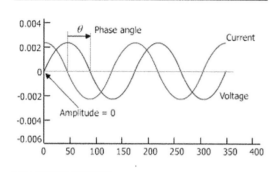

Figure 12.30 Voltage–current phase angle

(4) For simple RC circuits in series (Fig. 12.31d):

$$Z_{net} = R + \frac{1}{j\omega C}$$

(5) For simple RC circuits in parallel (Fig. 12.31e):

$$\frac{1}{Z_{net}} = \frac{1}{R} + j\omega C$$

$$\frac{1}{Z_{total}} = \frac{1}{R} + j\omega C$$

$$= \frac{1 + j\omega RC}{R}$$

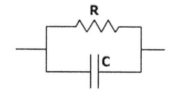

(a) Resistor   (b) Inductor   (c) Capacitor

$Z_R = R$ $\qquad$ $Z_C = 1/j\omega C$

(d) A simple R-C circuit in series   (e) A simple R-C circuit in parallel

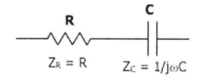

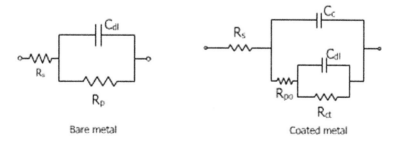

Bare metal $\qquad\qquad\qquad$ Coated metal

(f) Equivalent circuit for a bare metal and (g) a metal with porous insulating coating.

Figure 12.31 Sample elements and equivalent circuits

$$Z_{total} = \frac{R}{1+j\omega RC}$$

$$= \frac{R(1-j\omega RC)}{1+(\omega RC)},$$

since $(a+b)(a-b) = a^2 - b^2$
and $j^2 = -1$,

$$= \frac{R}{1+(\omega RC)} - j\left(\frac{\omega R^2 C}{1+(\omega RC)^2}\right)$$

$$Z_{total} = Z' - jZ'' \quad \text{For Fig. 12.31e.}$$

where

Z' = real impedance
Z'' = imaginary impedance.

(6) Equivalent circuit for a bare metal corrosion (Fig. 12.31f). The legends in the figure are:

R_s = solution/electrolyte resistance
R_p = polarization resistance
C_d = double layer capacitance.

$$Z_{total} = R_s + Z' - jZ''$$

(7) Corrosion of a metal with porous insulating coating (reinforcement in corrosion). The equivalent circuit is shown in Fig. 12.31g, where

C_c = coating capacitor
R_{po} = pore resistance
R_{ct} = charge transfer resistance. Polarization is changed to charge transfer.

This circuit is more representative of coated reinforcing bars in concrete.

12.15.4 BODE AND NYQUIST PLOTS

The corrosion resistance of metals and alloys can be estimated within a fair degree of accuracy the Nyquist or Bode plots. A Nyquist plot shows real *vs* imaginary impedance component. A typical Nyquist plot is shown in Fig. 12.32. At high frequency, $Z \rightarrow R_s$, and the impedance equals the

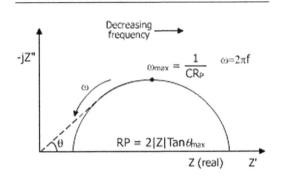

Figure 12.32   A typical Nyquist plot

solution resistance R_s, whereas at low frequency the impedance approaches $R_s = R_p$. Knowing R_s, R_p (polarization resistance) can be determined:

$$\omega_{max} = \frac{1}{C_{dl}R_p}$$

Knowing R_p, C_{dl} (capacitance of double layer) can be estimated. The higher the value of $R_s + R_p$, the greater is the resistance to corrosion. The difference between R_s and $R_s + R_p$ shows the magnitude of resistance of corrosion.

A typical Bode plot is shown in Fig. 12.33. The plot shows the total impedance and phase angle θ as a function of frequency. The plot yields

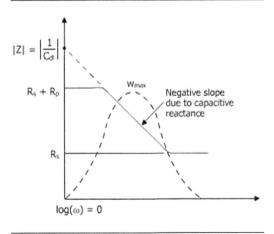

Figure 12.33   A typical Bode plot

value of R_s and $R_p + R_s$. The value of double layer capacitance is obtained by extrapolation of the linear position of the curve to log z-axis at $\omega = 1$ (log $\omega = 0$). The θ *vs* log ω plot yields a peak at a frequency corresponding to $\omega_{\phi\max}$. At this frequency ($\omega_{\phi\max}$) the phase shift is maximum. The double layer capacitance can also be calculated from frequency $\omega_{\phi\max}$ by the relationship

$$\omega_{\phi\max} = \frac{1}{C_{dl} R_p} \text{ (Walter, 1986)}$$

where $f_{\phi\max}$ (i.e. $2\pi f_{\phi\max}$).

12.15.5 PLOTS OF A METAL WITH POROUS INSULATING COATING

The equivalent circuit of a metal with porous insulating coating is given in Fig. 12.31g, where

C_c = coating capacitance
R_{po} = pore resistance
R_{ct} = charge transfer resistance

Nyquist and Bode plots for the above circuits are given in Figs 12.34 and 12.35, where τ is the time at which the exponential factor is $e^{-1} = 0.37$, the time it takes to decrease to 37% of its value. τ represents how slow or how fast is a reaction is. The Bode plot for the system is shown in Fig. 12.35. On extrapolation to the log z-axis, the value of C_c (coating capacitance) and C_{dl} (double layer capacitance) are obtained. For the known values of R_s, R_{ct} and R_c, the pore

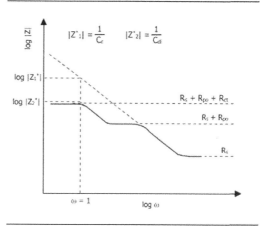

Figure 12.35 Bode plot for a metal with porous insulating coating

resistance R_{po} is obtained which indicates the resistance of the insulating coating on any metallic surface.

12.15.6 IMPEDANCE MEASUREMENT SYSTEM

Different impedance measuring systems are available from suppliers. A frequency analyzer is used in conjunction with a potentiostat to measure voltage and current signals. A DC potential controlled by potentiostat/galvanostat is mixed with sinusoidal simulation from R_{FA} applied to the cell. Two output signals are produced, one is the voltage signal developed between the cell reference electrode (ΔRE) and the other is the voltage developed across a measuring resistor by the potential which is perpetuating the current. The frequency response analyzer measures the components of the two signals and calculates their ratio, which pertains to the cell impedance and at a particular frequency which is displayed. In some systems, a lock in amplifier is used to generate excitation signals. Even very small signals can be detected by a lock in amplifier. However, they only provide a limited frequency range compared to frequency response analysis.

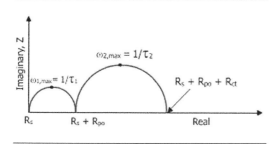

Figure 12.34 Nyquist plot for a metal with porous insulating coating

12.15.7 SCANNING REFERENCE ELECTRODE TECHNIQUE (SRET)

The anodic and cathodic reactions occur at separate sites on a corroding surface. The potential difference between these sites is small. This technique measures the micro-galvanic potentials existing on the surface of the material by using a specifically designed microelectrode probe which scans over the surface under examination. The arrangement for measurements is shown in Fig. 12.36 for cylindrical specimens. Measurement is made by means of a differential probe and a differential amplifier which gives a two-dimensional picture of the region of interest. The probe is moved vertically in defined steps of 0.5 μm and the micro-galvanic potentials over the circumference of the specimen are recorded and produce a two-dimensional mapping of the area under the probe. The line scan of potential around the circumference are also obtained. Color contours of corrosion activity can be displayed as real time. Potentiodynamic polarization and galvanostatic measurements can be conducted simultaneously with SRET experiment with a PC controlled potentiostat coupled to the SRET. Other SRET equipment is available for scanning flat surfaces.

12.15.8 POTENTIAL MAPPING

Mention has been made of potential mapping under Section 12.12.3. It is used widely in identifying the areas of corrosion risk on reinforcing steel.

12.15.9 MEASUREMENT OF ELECTRICAL RESISTIVITY

The electrical conductivity and resistivity of concrete is indirectly related to the degree of resistance offered by concrete to the passage of corrosive species to the steel reinforcement. The concrete resistivity is very high ($>4\,K\Omega$) and it can be best measured by the EIS technique (Electrochemical Impedance Spectroscopy) described earlier. ASTM Test C1202-94 can be used to determine concrete resistivity (Fig. 12.37).

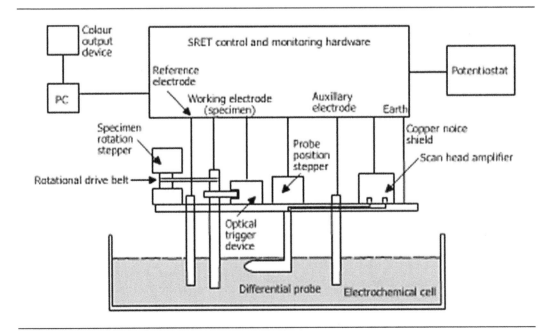

Figure 12.36 Scanning reference electrode apparatus for cylindrical specimens

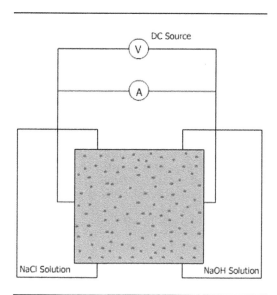
DC Source

NaCl Solution

NaOH Solution

Figure 12.37 Measurement of electrical resistivity

A concrete disk of diameter 50 mm × 100 mm is placed in separate reservoir of sodium chloride and sodium hydroxide. A DC potential is applied and the current is measured.

In addition to the major techniques used for evaluation of corrosion of the reinforcing steel described above, non-destructive techniques such as computed tomography, impact echo and magnetic field disturbance, electrochemical noise and other techniques have also been used for inspection and detection of corrosion of reinforcing steels. Reference on these techniques along with the references on major techniques are cited at the end of this chapter.

QUESTIONS

A. MULTIPLE CHOICE QUESTIONS

Circle the correct answer for the following statements:

1. The maximum aggregate size in concrete does not exceed

   a) 0.5″
   b) 1.0″
   c) 1.5″
   d) 3.0″

2. A great emphasis is placed on maintaining the low water–cement ratio necessary for the production of durable concrete. The design water–cement ratio of concrete is

   a) 0.4–0.5
   b) 0.1–0.2
   c) 0.4–0.6
   d) 1–2

3. On addition of water, Portland cement reacts immediately to form new compound, hydration products. These compounds function by

   a) decrease in the viscosity of the paste
   b) the reactants are mixed together
   c) the rate of reaction is accelerated
   d) the compound that collectively make up cement disappear, and are replaced by their hydration products

4. The cement gel resulting from hydration product has a porosity of nearly

   a) 2%
   b) 3%
   c) 5%
   d) 25%

5. The aggregates may be contaminated with salts, such as Na_2SO_4 and react with calcium aluminate hydrate to produce ettringite and also $CaSO_4·2H_2O$. These reactions products damage concrete by

   a) increase in the density of concrete
   b) decrease in the tensile strength of concrete
   c) increase in the volume and disruption of concrete
   d) increased setting rate of concrete

6. During the carbonation process calcium hydroxide is replaced by $CaCO_3$. This results

in the fall in pH into which one of the following ranges?

a) 13 to 11
b) 13 to 7
c) 13 to 6
d) 13 to 9

7. In chloride removal by electrochemical treatment, which anode is used?

a) Graphite
b) Zinc
c) Catalyzed titanium

8. The electrolyte preferably used in the electrochemical removal system is more often

a) sodium hydroxide
b) sodium borate
c) calcium hydroxide
d) potable water

9. When a direct current is passed through concrete, negatively charged ions, such as chloride, migrate toward the anode, and positive ions, such as K^+, Na^+ migrate towards the cathode. This results in

a) increase of chloride concentration is the temporary electrolyte
b) decrease of molar concentration of sodium and potassium
c) alkali-aggregate reactions
d) generation of products with lesser volume

10. Electrochemical chloride removal is only applicable if all individual bars are

a) connected
b) separated
c) coated
d) wet

B. How and Why Questions

1. Why are admixtures added to fresh concrete? Give the name of the most common class of admixture.

2. How can the damaging effect of cold weather concreting be minimized?
3. Why is potential mapping done on a concrete and how it is related to the determination of the areas of risk in concrete?
4. Why is there a greater risk of carbonation in airy–dry concrete compared to water saturated concrete?
5. Why is re-alkalization generally preferred over chloride removal?
6. Why is a galvanostatic anode pulse technique preferred over a linear polarization technique?
7. What is the role of admixture on concrete properties? State the difference between water reducing, self-retarding and set accelerating admixture.
8. Why is the attack by magnesium sulfate on concrete more severe than attack by calcium or sodium sulfate? Explain with reference to calcium aluminate hydrate.
9. What is the basic objective of applying cathodic protection to steel reinforcement in concrete with particular reference to alkalinity? How is this objective achieved?
10. Which of the reference electrodes, silver/silver chloride or $Cu–CuSO_4$, is more suitable for potential measurements? Explain why a high impedance voltmeter (1000 megaohm) is generally employed by measurement of potential.

Suggested Reading

General

[1] Mehta, P.K. and Monteiro, R.J.M. (1993). *Concrete Structure, Properties, and Material.* Int., London: Prentice Hall, UK.
[2] Lankard, D.R. (1976). Cement and concrete technology for the corrosion engineer. *Corrosion 76*, Int. Corrosion Forum, Paper # 16, Texas: Houston, March 22–26, USA.
[3] ACI Committee 222, (1985). Corrosion of metals in concrete. ACIR-85, American Concrete Institute, Detroit.
[4] Newman, J. and Chou, B.S. eds. (2003). *Advanced Concrete Technology, Testing, and Quality,* Oxford: Elsevier, UK.
[5] Tuutti, K. (1982). Corrosion of steel in concrete. *Swedish Cement and Concrete Research,* Stockholm: Institute.
[6] Bentur, A., Diamond, S. and Berke, N.S. (1997). *Steel Corrosion in Concrete,* London: E & FN Spoon.
[7] Page, C.L., Bamforth, P.B. and Figg, J.W. (1996). *Corrosion of Reinforcement in Concrete,* Royal Society of Chemist Construction, UK: Cambridge.

[8] Broomfield, J.B. (1997). *Corrosion of Steel in Concrete: Understanding, Investigation, and Repair*, London: E & FN Spoon, UK.

REFERENCES ON TECHNIQUES TO INVESTIGATE THE CORROSION OF REINFORCEMENT IN CONCRETE

[9] ASTM C876 (2001). Standard Test Method for Half Cell. Potentials of Uncoated Reinforcing in Concrete, 452–457.

[10] ASTM G59 (1994). Standard Reference Method for Conducting Potentiodynamic Polarization Resistance Measurement, ASTM Annual Book of Standards, Vol. 02, 03.

[11] ASTM G5 (1994). Standard Reference Test Method for Making Potentiostatic and Potentiodynamic Anode Polarization Measurement, ASTM Annual Book of Standards, Vol. 02, 03.

[12] Broomfield, J.P., Longford, P.E. and Ewins, A.J. (1992). The use of potential wheel to survey reinforced concrete structures. Berke, N.S. Chaker, V. and Whiting, D. eds. *Corrosion Rates of Steel in Concrete*. Philadelphia: ASTM STP 1065, 157–173.

[13] Arubi, H. (1984). Potential mapping of reinforced concrete structures. The Danish Corrosion Center Report, January.

[14] Elsener, B. and Bohni, H. (1992). Potential mapping and corrosion of steel in concrete, *Corrosion Rates of Steel in Concrete*, Berke, N.S. Chaker, V. and Whiting, D. eds. Philadelphia: ASTM STP 1065, 143–156.

[15] Stern, M. and Geary, A.L. (1957). Electrochemical polarization, *J. Electrochem Soc.*, **104**, 50–63.

[16] Newton, C.J. and Sykes, J.M. (1988). A galvonostatic pulse technique for investigation of steel corrosion in concrete, *Corrosion Science*, **28**, 1051–1074.

[17] Elsener, B., Wojtas, H. and Bohni, H. (1994). Galvanostatic pulse measurement - Rapid on-site corrosion monitoring, *Proc. Int. Conf.*, Univ. of Sheffield, **I**, 236–246, 24–28 July.

[18] Meitz, J. and Isecke, P.B. (1997). Electrochemical potential monitoring on reinforced concrete structure using anodic pulse technique, Bungey, H. ed. *Non-destructive Testing in Civil Engineering*. The British Institute of NDT, **2**, 567–577.

[19] SHRP-S-324 (1992). Method for measuring the corrosion rate of reinforcing steel. Washington, D.C: National Research Council.

[20] Lemoine, L., Wenger, F. and Gallard, J. (1992). Study of the corrosion of concrete reinforcement by electrochemical impedance measurement. Berke, N.S., Chaker, V. and Whiting, D. eds. *Corrosion Rates of Steel in Concrete*. Philadelphia: ASTM STP 1065, 118–133.

[21] Jafar, M.I., Dawson, J.L. and Johnson, D.G. *Electrochemical impedance and harmonic analysis measurements of steel in concrete in electrochemical impedance*, Analysis and Interpretation. ASTM Standard 1188, 384–403.

[22] Walter, G.W. (1986). *Corrosion Science*, 26, 39 and 681.

[23] Akid, R. (1990). Application of scanning reference electrode technique to study steel corrosion in cement based coatings, *Conference held at the University of Sheffield*, **I**, 287–289.

[24] Cheng, C. and Sansalone, M. (1993). Effect of impact-echo signals caused by steel reinforcing bars and voids around bars, *ACT Materials Journal*, 421–434, Sep–Oct.

[25] Robert, J.L. and Brachel-Rolland, M. (1992). Survey of structures by using acoustic emission monitoring, *IABSZ Symposium*, Washington, **39**, 33–38.

REFERENCES ON CHLORIDE REMOVAL, RE-ALKALIZATION AND REHABILITATION OF CONCRETE

[26] Bennett, J. and Thomas, J.S. (1993). *Evaluation of NORCURE processes for electrochemical chloride removal from steel reinforced concrete bridge components*. Washington, D.C: National Academy of Sciences, No. HRP-C-620.

[27] Manning, D.G. and Bianca, F. (1990). *Electrochemical removal of chloride ions from reinforced concrete*. Ministry of Transportation, Canada: Ontario, No. MAT-90–14.

[28] Anderson, G. (1991). Chloride extraction and re-alkalization of concrete, *Hong Kong Contractor*, 19–25 July.

[29] Grantham, M. (2003). Concrete repairs, Newman J. and Chou B.S. eds. *Advanced Concrete Technology*. Elsevier.

[30] Pullar, P. and Strecker, C.E. (2002). *Design and Practice Guide: Concrete Reinforcement Corrosion*. London: Thomas Terford Publishing.

KEYWORDS

Additives Inert or reactive inorganic materials added to concrete to achieve specific properties.

Admixture Inorganic or organic materials added in small quantities during mixing to modify concrete properties.

Aggregate contamination Contamination in concrete caused by aggregate, such as contamination with chlorides.

Alkalinity environment As the pH of a water solution exceeds 9.6 to 8.81, a measurable concentration of OH^- ions begins to appear. Alkalinity is a measure of OH^- ion concentration. A strongly alkaline environment has a pH greater than 11.0.

Anodes An electrode or a site on a metal surface where oxidation reaction occurs.

Anode pulse technique Application of a small anodic current by means of a counter-electrode and a galvanostat to measure the changes in potential over a period of time.

Carbonation The reduction of the alkalinity of the pore fluids in concrete by ingress of atmospheric carbon dioxide and its interaction with calcium hydroxide causing the lowering of pH from 11 or higher to 8.

Cathode A cathodic area is a site on a metal surface where the reduction reaction occurs. A cathodic area consumes electrons released by the anode.

Cathodic protection A means of protecting a metal from corrosion by causing a direct current to flow from its electrolytic environment into the metal to be protected. The reinforcement is cathodically protected (made negative) by causing a direct current to flow from an external electrode into it through the concrete (electrolyte).

Cement contents and types It is the proportion of cement in concrete. There are five major types of Portland cement. The main difference is the proportioning of anhydrous mineral constituents and the fineness of ground products.

Curing This refers to the process of hydration of cement in the presence of moisture, particularly at the surface, over a period of time.

Diffusion This is a term used for processes whereby particles (usually ions or atoms) spontaneously spread out over a period of time. Random motion of particles in gases and liquids is responsible for diffusion.

Drying shrinkage Contraction of concrete on removal of its free water. This contraction can produce cracks if it is not restrained.

Electrochemical chloride removal Electrochemical extraction to reduce chloride extraction near reinforcement. It is achieved by using an external anode powered by a DC source. The negatively charged chloride ions migrate toward this anode and are repelled by the steel reinforcement (negative).

Electrochemical impedance spectroscopy By application of small AC voltages in the system of interest, the response of the electrochemical interface of interest to a controlled electrochemical current (degree of polarization, temperature and concentration of species) is determined. This information expressed in terms of impedance is utilized in analyzing the corrosion resistance of the system.

Electrochemical potential mapping This is a map showing contours of equal electrode potential to local chloride contaminated areas (anodic areas) over predetermined distances on a grid.

Electrochemical re-alkalization This is an electrochemical process which introduces hydroxyl ions and alkaline metal ions in the concrete cover from an external source of electrolyte, such as sodium carbonate. This process reverses re-carbonation. The steel reinforcement is made the negative electrode which attracts cations such as Ca^{++} and Na^+ if present.

Freeze–thaw cycle As the temperature of the hardened concrete is lowered, the water in the capillary pores in the cement phase freezes. On thawing, not all expansion is lost and on freezing further expansion takes place so that freeze–thaw cycle has a cumulative effect.

Hardened concrete The strength achieved by concrete upon hardening.

Passivity The formation of a thin layer of γ-FeOOH (iron oxide) on the reinforcement which acts as a barrier between the reinforcement and the environment and makes the reinforcement surface inert to corrosion. This phenomena is called 'passivity.' See Pourbaix diagram, high pH zone.

Patch repair A repair procedure in which small areas of chloride contaminated concrete are replaced by fresh chloride-free concrete.

Pitting A form of localized corrosion which leads to destruction of the passivity by chloride ions and creation of pits on the surface.

Resistivity This is the electrical resistance offered by 1 cm cube of concrete. It is the inverse of conductivity. Also called specific resistance or resistivity.

Spalling This is the separation of a surface layer from the body of a concrete without being completely detached. Can be caused in defective or carbonated concrete if the steel corrodes, due to expansion caused by corrosion products (rust).

INDEX

Printed and bound by CPI Group (UK) Ltd, Croydon, CR0 4YY

03/10/2024

01040331-0014